AF836134

This item must be returned or renewed by the last date shown below. The loan period may be shortened if it is reserved by another reader. A fine will be due if it is not returned on time.

DATE OF RETURN

-5 FEB 1999

4 FEB 2000

05 DEC 2005

30 JAN 2006

08 JAN 2007

WITHDRAWN

UNIVERSITY COLLEGE LONDON
Gower Street London WC1E 6BT

LD4D

RLO323.7.8.97

Bioconversion of Waste Materials to Industrial Products

VISIT OUR FINE CHEMISTRY SITE ON THE WEB

http://www.finechemistry.com

e-mail orders: direct.orders@itps.co.uk

Bioconversion of Waste Materials to Industrial Products

Second edition

Edited by

A.M. MARTIN
Department of Biochemistry
Memorial University of Newfoundland
St John's
Canada

BLACKIE ACADEMIC & PROFESSIONAL
An Imprint of Chapman & Hall
London · Weinheim · New York · Tokyo · Melbourne · Madras

Published by Blackie Academic & Professional,
an imprint of Thomson Science, 2–6 Boundary Row, London SE1 8HN

Thomson Science, 2–6 Boundary Row, London SE1 8HN, UK

Thomson Science, 115 Fifth Avenue, New York NY 10003, USA

Thomson Science, Suite 750, 400 Market Street, Philadelphia, PA 19106, USA

Thomson Science, Pappelallee 3, 69469 Weinheim, Germany

First edition 1991
Second edition 1998

© 1998 Thomson Science

Thomson Science is a division of International Thomson Publishing I(T)P®

Typeset in 10/12pt Times by Cambrian Typesetters, Frimley, Surrey

Printed in Great Britain by St Edmundsbury Press Ltd, Bury St Edmunds, Suffolk

ISBN 0 7514 0423 3

All rights reserved. No part of this publication may be reproduced, stored in a retrieval system or transmitted in any form or by any means, electronic, mechanical, photocopying, recording or otherwise, without the prior written permission of the publishers.

The publisher makes no representation, express or implied, with regard to the accuracy of the information contained in this book and cannot accept any legal responsibility or liability for any errors or omissions that may be made.

A catalogue record for this book is available from the British Library

Library of Congress Catalog Card Number: 97–76802

UNIVERSITY COLLEGE LONDON
LIBRARY

∞ Printed on permanent acid-free text paper, manufactured in accordance with ANSI/NISO Z39.48-1992 (Permanence of Paper).

Contents

List of contributors **xiii**

Preface **xvii**

Preface to the first edition **xix**

Part One: The Principles of Bioconversion of Waste Materials

1 The enzymic treatment of waste materials **3**
PETER GACESA and JOHN HUBBLE

1. Introduction 3
1.2 Factors influencing enzyme use 3
1.2.1 Sources of enzymes 3
1.2.2 Enzyme stability 7
1.3 Application of enzymes 12
1.3.1 Hydrolases 12
1.3.2 Nonhydrolytic enzymes 16
1.4 Enzymes with modified activities 19
1.4.1 Applications of molecular techniques 19
1.4.2 Nonaqueous/low water systems 20
1.5 Conclusions 24
References 25

2 Processes with immobilized enzymes and cells **29**
SEVERIAN DUMITRIU and ESTEBAN CHORNET

2.1 Current status of immobilized enzyme technology 29
2.1.1 Advantages and disadvantages of enzyme and cell immobilization 29
2.1.2 Immobilization of microorganisms or enzymes? 31
2.2 Immobilization procedures 31
2.2.1 Carriers 31
2.2.2 Methods of immobilization 32
2.3 Reactors for immobilized biomaterial systems 54
2.4 Waste conversion in the dairy industry 57
2.4.1 Bioconversion of whey 57
2.4.2 Milk processing 59
2.5 Bioconversion of cellulosic wastes 60
2.5.1 Conversion of cellulose to ethanol 60
2.6 Hemicellulose conversion 61
2.6.1 Conversion of xylose 62
2.7 Bioconversion of starch wastes 62
2.7.1 Simultaneous saccharification and fermentation of starch 63
2.7.2 Recovery of waste glucose solutions 69
2.7.3 Recovery of waste from beet sugar industry 71

2.8 Immobilized enzymes in organic solvents 73
2.8.1 Bioconversion of lipids 76
2.9 Waste treatment 78
2.9.1 Methane bioconversion of wastes 78
2.9.2 Immobilized cells and waste water treatment 83
2.10 Immobilized microorganisms in waste gas purification 89
References 91

3 Solid substrate fermentation: a biotechnological approach to bioconversion of wastes 103

O. PAREDES-LÓPEZ, S.H. GUZMÁN-MALDONADO and A. ALPUCHE-SOLÍS

3.1 Introduction 103
3.2 Critical factors for microbial growth on solid substrates 105
3.2.1 Water activity and moisture 105
3.2.2 Temperature 107
3.2.3 pH 109
3.2.4 Aeration and oxygen transfer 110
3.2.5 Mixing 112
3.3 Microbial growth patterns and control of fermentation 113
3.3.1 Microbial types and inoculum 113
3.3.2 Microbial growth patterns and growth rate 114
3.3.3 Control by physical and nutritional factors 117
3.4 Genetic engineering for biodegradation of lignocellulosic wastes 118
3.4.1 Lignin biodegradation 119
3.4.2 Cellulose bioconversion 123
3.4.3 Practical applications of a lignin biodegradation system 123
3.5 Reactors for solid substrate fermentation 124
3.5.1 Tray fermenter 125
3.5.2 Rotating drum fermenter 126
3.5.3 Packed-column fermenter 126
3.5.4 Auger tube fermenter 127
3.5.5 Helical screw fermenter 127
3.5.6 Fluidized biomass fermenter 127
3.5.7 Miscellaneous types 128
3.6 Fermentation processes and compositional changes 130
3.6.1 SSF processes 130
3.6.2 Some currently practiced SSF processes 132
3.7 Advantages, disadvantages and future prospects of SSF 146
3.7.1 Advantages and disadvantages 146
3.7.2 Future prospects 147
Acknowledgements 148
References 148

4 Composting processes 154

S.P. MATHUR

4.1 Introduction 154
4.2 Definition and principles of composting 156
4.2.1 Definition 156
4.2.2 Principles 156
4.2.3 Compost feedstocks 157
4.2.4 Requirements of optimal composting 160
4.3 Chemistry and biology of the composting process 176
4.4 The technology of composting 178

4.4.1 Open systems 179
4.4.2 In-vessel (or reactor confined) systems 184
4.5 Criteria of compost maturity 184
4.5.1 C/N ratio 186
4.5.2 Absence of plant inhibitors 186
4.5.3 Absence of human pathogens 186
4.5.4 Other criteria 187
4.6 Uses of composts 187
4.7 Summary 188
References 189

Part Two: Bioconversion Applications

5 Bioprocessing of agro-residues to value added products 197
V.S. BISARIA

5.1 Introduction 197
5.2 Characteristics of lignocellulosic materials and their pretreatment 201
5.2.1 Lignocellulosic materials 201
5.2.2 Physical and chemical constraints in enzymatic hydrolysis of cellulose 204
5.2.3 Pretreatment of lignocellulosic residues 204
5.3 Properties, production and applications of cellulolytic enzymes 210
5.3.1 Properties of cellulases 210
5.3.2 Production of cellulases 213
5.3.3 Properties of hemicellulases 218
5.3.4 Production of xylanases 219
5.3.5 Application of cellulases and xylanases 220
5.4 Bulk chemicals from cellulose and hemicellulose 222
5.4.1 Glucose and xylose 222
5.4.2 Ethanol 228
5.4.3 Acetone–butanol 235
5.4.4 2,3-Butanediol 236
5.5 Future prospects 237
Acknowledgement 238
References 238

6 Use of photosynthetic bacteria for the production of SCP and chemicals from organic wastes 247
KEN SASAKI, TOHRU TANAKA and SHIRO NAGAI

6.1 Introduction 247
6.1.1 General characteristics of photosynthetic bacteria 247
6.1.2 Application of photosynthetic bacteria for SCP and chemical production from organic wastes 248
6.2 SCP production from waste 250
6.2.1 Pineapple waste 250
6.2.2 Soybean waste 253
6.2.3 Cassava solid waste 254
6.2.4 Mandarin orange peel 256
6.2.5 Swine and cow dung waste 260
6.2.6 Cell yields and composition of PSB 263
6.3 Vitamin production 266
6.3.1 Vitamin B_{12} 266
6.3.2 Ubiquinone 269

6.4 5-Aminolevulinic acid production 270
6.4.1 ALA production from swine waste 272
6.4.2 ALA production from sewage sludge 275
6.4.3 ALA production by aerobic fermentation 276
6.4.4 Applications of ALA 278
6.5 Problems and future prospects 288
6.5.1 Problems 288
6.5.2 Future prospects 289
References 290

7 Utilization of starch industry wastes 293
SUDIP K. RAKSHIT

7.1 Introduction 293
7.2 Nature of cereal and tuber starches 293
7.3 Starch-based industrial products 294
7.3.1 Hydrolytic products and sweeteners 295
7.3.2 Food applications 296
7.3.3 Paper industry applications 298
7.3.4 Fermentative products from starch 299
7.4 Extraction procedures and starch industry waste streams 301
7.4.1 General extraction procedure 301
7.4.2 By-product and effluent streams 303
7.5 Utilization and treatment of starch industry wastes 304
7.5.1 Production of single cell proteins 304
7.5.2 Protein extraction from potato processing 308
7.5.3 Energy recovery from liquid streams 309
7.5.4 Miscellaneous 311
7.6 Conclusion 312
References 312

8 Bioconversion of food processing wastes 316
G.TH. KROYER

8.1 Introduction 316
8.2 Characteristics of food processing wastes 317
8.3 Biotechnological processes in food processing waste treatment 318
8.4 Production of biomass from food processing wastes 319
8.5 Meat and fish processing wastes 322
8.6 Fruit and vegetable processing wastes 324
8.7 Dairy industry wastes 329
8.8 Wastes from the fermentation industry 332
8.9 Conclusion and future outlook 333
References 335

9 Bioconversion of cheese whey to organic acids 342
R.D. TYAGI and D. KLUEPFEL

9.1 Introduction 342
9.2 Production of whey 342
9.3 Pollution control 343
9.4 Current disposal methods of whey 344
9.5 Global utilization of whey 346
9.6 Management strategies 346
9.7 Lactic acid 347
9.7.1 Microorganisms involved in lactic acid fermentation 348

9.7.2 Batch process 349
9.7.3 Continuous process 353
9.7.4 Product inhibition in lactic acid fermentation 357
9.7.5 Immobilized cell process 362
9.8 Acetic acid and propionic acid 367
9.9 Conclusions 371
Acknowledgement 372
References 372

10 Lignocellulosic wastes: biological conversion 376
P.S. CHAHAL and D.S. CHAHAL

10.1 Introduction 376
10.2 Composition and structure of lignocelluloses 377
10.2.1 Cellulose 379
10.2.2 Hemicelluloses 384
10.2.3 Lignin 385
10.2.4 Protein 388
10.2.5 Extraneous materials 388
10.3 Pretreatment of lignocelluloses 388
10.4 Biological conversions 388
10.4.1 Liquid-state fermentation 389
10.4.2 Solid-state fermentation 398
10.5 Utilization of the lignin component of lignocelluloses 409
10.5.1 Ligninase/ligninolytic enzymes 409
10.5.2 Production of ligninases 412
10.6 Problems in bioconversion and future trends 415
References 416

11 Bioconversion of waste water from the pulp and paper industry 423
K. EL HAJI, V. SACHDEVA and R.D. TYAGI

11.1 Introduction 423
11.2 Source of effluent from the pulp and paper industry 424
11.2.1 Pulping process 425
11.2.2 Bleaching process 426
11.3 Characteristics of waste water from pulp and paper mills 427
11.3.1 Biodegradable part 427
11.3.2 Wood compounds 428
11.3.3 Parts with difficulty in or absence of biodegradability 429
11.3.4 Toxic substances 430
11.4 Treatment technologies 430
11.4.1 Internal treatment 430
11.4.2 External treatment 432
11.5 Biotechnological applications in the pulp and paper industry 434
11.5.1 Pulp manufacture 434
11.5.2 Bleaching of pulp 435
11.6 Evaluation of the potential for effluent use from the pulp and paper industry in bioconversion 436
11.7 Suitability of spent sulfite liquor for the bioconversion of by-products 437
11.8 Effluent treatment by conversion to by-products 438
11.8.1 Bioconversion of cellulose and lignocellulose materials present in pulp and paper waste waters 439
11.8.2 Production of ethyl alcohol from cellulosic by-products 441
11.9 Major difficulties in bioconversion 443

11.10 Conclusions 444
Acknowledgements 445
References 445

12 **Fisheries waste biomass: bioconversion alternatives** **449**
A.M. MARTIN

12.1 Introduction 449
12.1.1 Antecedents of the recovery of fisheries wastes and by-products 450
12.2 Hydrolytic processes for the recovery of fish protein 452
12.2.1 Enzymatic methods 456
12.2.2 Methods employing microorganisms 457
12.3 Biological methods for the recovery of chitin and chitosan 459
12.4 Biological water treatment of fisheries wastes 463
12.5 Composting of fisheries offal 464
12.6 Other products from fisheries waste biomass 465
12.6.1 Fermentation substrates 465
12.6.2 Enzymes from fish biomass 466
12.6.3 Media for the cultivation of edible mushrooms 467
12.7 Conclusions 469
12.7.1 Present developments 469
12.7.2 Future trends 471
References 471

13 **Production of *Bacillus thuringiensis* biopesticides using waste materials** **480**
MARIA DE LOURDES TIRADO MONTIEL, RAJESHWAR D. TYAGI and JOSÉ R. VALÉRO

13.1 Introduction 480
13.2 Characteristics of *Bacillus thuringiensis* 481
13.2.1 Taxonomy 481
13.2.2 Metabolism 482
13.3 Genetic characteristics 483
13.3.1 Localization and organization of crystal producing genes 483
13.4 Toxicity (crystal–spore complex) 484
13.4.1 Characteristics 484
13.4.2 Synthesis 484
13.4.3 Specificity 485
13.4.4 Mode of action 486
13.5 Effect of medium composition and operation conditions on the production of crystal–spore complex 487
13.5.1 Temperature and pH 487
13.5.2 Process options for Bt production 488
13.5.3 Aeration 490
13.5.4 Mineral elements 491
13.5.5 Nitrogen and amino acids 492
13.5.6 Carbon source 493
13.6 Alternative raw materials for Bt biopesticide production 495
13.6.1 Production of Bt subsp. *thuringiensis* on alternate protein-rich raw materials 495
13.6.2 Production of Bt subspecies *entomocidus*, *kurstaki*, *aizawai*, *finitimus* and *galleriae* from various raw materials 496
13.6.3 Production of Bt subsp. *israelenis* (Bti) using different raw materials 500
13.7 Toxicity determinations 504
13.7.1 Bioassays 505
13.7.2 Tests *in vitro* 506

13.8 Applications of Bt biopesticides 507
13.8.1 Utilization of Bt for control of lepidopteran pests 507
13.8.2 Utilization of Bt for control of dipteran pests 508
13.8.3 Utilization of Bt for control of coleopteran pests 508
13.9 Conclusions 509
Acknowledgements 509
References 510

14 Biorecovery of metals from mining wastes 517
DAVID S. HOLMES

14.1 Historical perspective 517
14.2 Significance of biomining 518
14.3 Copper dump bioleaching 519
14.3.1 Economics of dump bioleaching 521
14.3.2 Microbiology 522
14.3.3 Problems 526
14.3.4 Technical solutions 526
14.4 Heap bioleaching 527
14.5 Concentrate bioleaching 527
14.6 *In situ* bioleaching 528
14.7 Uranium bioleaching 528
14.8 Bio-oxidation of gold ore 529
14.8.1 Principles 529
14.8.2 Opportunities 532
14.9 Development of new strains of microorganisms 533
14.9.1 Introduction 533
14.9.2 Fauna and flora of a bioleaching operation 533
14.9.3 Isolation of new strains from the environment 537
14.9.4 Selection and adaptation of naturally occurring strains 537
14.9.5 Classical genetic mutation 539
14.9.6 Genetic engineering 539
14.10 Conclusions 540
14.11 Summary 542
Acknowledgements 542
References 542

Index 547

Contributors

A. Alpuche-Solís Depto. de Biotecnología y Bioquímica, Unidad Irapuato, Centro de Investigación y de Estudios Avanzados del IPN, Apdo. Postal 629, 36500 Irapuato, Gto., Mexico

V.S. Bisaria Department of Biochemical Engineering and Biotechnology, Indian Institute of Technology – Delhi, Hauz Khas, New Delhi – 110 016, India

D.S. Chahal DC Enterprises, Inc., 3979 Acadia, Laval, Quebec, Canada, H7T 1G3

P.S. Chahal Biotechnology Research Institute, National Research Council of Canada, 6100 Royalmount Avenue, Montreal, Quebec, Canada, H4P 2R2

E. Chornet Department of Chemical Engineering, University of Sherbrooke, Sherbrooke, Québec, Canada, J1K 2R1

S. Dumitriu Department of Chemical Engineering, University of Sherbrooke, Sherbrooke, Québec, Canada, J1K 2R1

K. El Haji Institut National de la Recherche Scientifique, Université du Québec, INRS-Eau, 2700 rue Einstein, CP 7500, Sainte-Foy, Québec, Canada, G1V 4C7

P. Gacesa Faculty of Science and Engineering, Manchester Metropolitan University, Manchester M1 5GD, UK

S.H. Guzmán-Maldonado Depto. de Biotecnología y Bioquímica, Unidad Irapuato, Centro de Investigación y de Estudios Avanzados del IPN, Apdo. Postal 629, 36500 Irapuato, Gto., Mexico

D.S. Holmes Department of Biological Sciences, University of Santiago, Avda. Bernardo O'Higgins, Santiago, Chile

J. Hubble School of Chemical Engineering, University of Bath, Claverton Down, Bath, BA2 7AY, UK

D. Kluepfel Institut Armand-Frappier, 531 boul. Des Prairies, CP 100, Succ. L-Q-R, Ville de Laval, Québec, Canada, H7N 4Z3

G. Th. Kroyer Institute of Food Chemistry and Technology, Technical University Vienna, Getreidemarkt 9, A-1060 Vienna, Austria

A.M. Martin Department of Biochemistry, Memorial University of Newfoundland, St John's, Newfoundland, Canada, A1B 3X9

S.P. Mathur Compost & Peat Specialist, Inc., 169 Castlefrank Road, Kanata, Ontario, Canada, K2L 1T3

S. Nagai Yaegaki Research Institute, Mukudani, Hayashidacho, Himeji 679–42, Japan

O. Paredes-López Depto. de Biotecnología y Bioquímica, Unidad Irapuato, Centro de Investigación y de Estudios Avanzados del IPN, Apdo. Postal 629, 36500 Irapuato, Gto., Mexico

S.K. Rakshit Bioprocess Technology Program, Asian Institute of Technology, PO Box 4, Khlong Luang, Pathum Thani 12120, Thailand

V. Sachdeva Institut National de la Recherche Scientifique, Université du Québec, INRS-Eau, 2700 rue Einstein, CP 7500, Sainte-Foy, Québec, Canada, G1V 4C7

K. Sasaki Material Science and Engineering, Graduate School of Hiroshima, Denki Institute of Technology, Nakano, Akiku, Hiroshima 739–03, Japan

T. Tanaka Cosmo Research Institute, Gongendo, Satte, Saitama 340–01, Japan

M. de L. Tirado Montiel Comisión Nacional del Agua, 15 Poniente 1317, Puebla, Pue., Mexico, CP 72000

R.D. Tyagi Institut National de la Recherche Scientifique, Université du Québec, INRS-Eau, 2700 rue Einstein, CP 7500, Sainte-Foy, Québec, Canada, G1V 4C7

J.R. Valéro Laurentian Forestry Center, 1055 rue du PEPS, PO Box 3800, Sainte-Foy, Québec, Canada, G1V 4C7

Preface

The general objectives of the first edition of this book, published in 1991, still remain valid. The existence of pollution-associated problems created by wastes, the scarcity of places to dispose of the wastes and the need to save valuable resources which are part of the refuse of modern society are universally acknowledged. Recycling, which could contribute to solving some of the most serious problems affecting human economic performance at present and in the future, is gaining appreciation as a viable commercial activity. Since the publication of the first edition of this book, increased recognition has been given to bioconversion of wastes as one of the most appropriate methods to return to the environment resources previously extracted from it.

This book is designed as a study of the biotechnological methods for the recovery of wastes. As its name indicates, it emphasizes the recycling objective of the bioconversion, i.e. the production of industrial products from wastes. The chapters deal with the scientific and technological bases of the bioconversion processes involved, the problems and advantages associated with each, the products arising from the operations, and trends and future possibilities.

Although relatively few years have passed from the publication of the first edition, accelerated advances in the areas to which this book is devoted have resulted in a significant overhaul of the book's content. In addition to updating the information presented, a new edition provides the opportunity to review the work done previously, and to try to add to it. As a consequence, most of the chapters in the first edition have been thoroughly revised, some of them becoming completely new chapters. Also, some subjects not included in the first edition have been added.

The contents of the book have been organized into two parts. The chapters in Part One are oriented more towards the principles of the bioconversion operations, while the chapters in Part Two consider the characteristics of the main groups of waste materials, and specific technologies for their bioprocessing and the production of valuable products.

As was indicated in the first edition, although a single book cannot cover all of the areas of such a large and expanding field, this book will provide useful and up-to-date information to the academic, industrial and scientific communities. The inclusion of technological examples should illustrate, to those working on the solution of waste disposal problems as administrators, consultants or in government, the advantages and potential of bioconversion methods.

Antonio M. Martin
St John's, Newfoundland
Canada

May 1997

Preface to the first edition

When focusing on the latest stages of human development, many factors have been long identified as representative of both mankind's successes and failures. Industrialization, rising standards of living and the exploitation of new materials and energy sources, while characteristics of progress, are also a source of new problems such as overpopulation, increased urbanization, the energy crisis and pollution, to mention only a few. More recently, the problem of wastes from processing operations and their disposal has gained full-fledged public recognition.

In the past, problems associated with wastes were not given special treatment by society, and generally they were recognized as specific problems of the institutions which generated them: cities, industries, and agriculture. Indeed, before the advent of the modern chemical and processing industry of the present century, most of the wastes were recycled, and before the population explosion of the last decades, it appeared that there was enough space on eath to simply dump wastes and allow nature to dispose of those which were biodegradable.

This situation is no longer sustainable. There is increasing recognition of the pollution-associated problems created by wastes, the scarcity of places to dispose of them, and the need to save the valuable resources which are part of the refuse of our present 'throwaway' society. The recycling of resources is becoming a valid and viable economic activity and is increasingly mentioned as a solution to some of the most pressing problems which will affect mankind's future economic performance.

Bioconversion of wastes has been the natural way to return to the environment the resources previously extracted from it. It is expected that the development of biotechnology will facilitate the acceleration of this natural recycling process, which is being made necessary by our present and future levels of population densities and their increasing demands.

This book is designed as a study of the biotechnological methods for the recovery of wastes, emphasizing the recycling objective of the bioconversion, i.e. the production of industrial products from wastes. In conducting this study, it is this book's objective that its various chapters deal with the scientific and technological bases of the bioconversion processes involved, the problems and advantages associated with each, the products arising from the operations, and future trends and possibilities. If relevant to their content, individual chapters also deal with processing methods required to concentrate and purify the complex mixture of waste materials, applied before the biological step ('upstream' operations), and present an overview of the economic basis for the bioconversion process discussed.

This book is not intended to be a specialized study of the biodegradation processes involved, which will be presented in a second volume currently being prepared. Although the bioconversion processes have been traditionally applied to products of biological origin, such as agricultural, fisheries, forestry and food processing wastes, the present volume also deals with areas where novel bioconversion processes are also applicable, such as some mineral- and hydrocarbon-based industrial operations.

The contents of the book have been organized in two sections. Chapters in Section 1 are oriented more toward the principles or fundamentals of the bioconversion operations, while Chapters in Section 2 consider the characteristics of the main groups of waste materials and the specific technologies for bioprocessing and recycling them.

This book cannot cover all of the areas of a presently increasing, expanding field of research. However, it is expected that this book will provide useful and updated information to the academic, industrial and scientific communities, including ecologists and environmentalists. By including technological examples which will allow and encourage the use of bioconversion methods for the solution of waste disposal problems, this book could act as a guide to administrators, consultants, and governments.

Antonio M. Martin

Part 1
The Principles of Bioconversion of Waste Materials

1 The enzymic treatment of waste materials

PETER GACESA AND JOHN HUBBLE

1.1 Introduction

The use of enzymes in biotechnological processes is part of a long and established tradition. In some cases this has involved the specific extraction of an enzyme, e.g. calf chymosin for cheese making, whereas in other applications the activity of endogenous enzymes has been utilized *in situ*, e.g. the malting of barley. The utilization of enzymes for waste processing is a relatively recent development and has grown out of the increasing demands, both economic and environmental, for acceptable methods for the disposal or recycling of these materials. In most countries the disposal of wastes is strictly regulated and is likely to incur some form of taxation.

1.2 Factors influencing enzyme use

1.2.1 Sources of enzymes

There are some 2500 different enzyme-catalysed reactions listed in the International Union Handbook of Enzyme Nomenclature (Webb, 1984). However, this is a significant underestimate of the total number of enzymes that have been discovered because many of the reactions may be catalysed by a multiplicity of proteins with different properties. Of this number only 20–30 enzymes are produced on the industrial scale, i.e. kilogram to tonne quantities per annum.

Most enzymes are produced from approximately 20 microbial sources, either bacterial or fungal. This has resulted largely because many of the enzyme manufacturers are significant producers for the food industry where legislation proscribes the range of acceptable microorganisms (for a review, see Denner, 1983). Therefore, the enzyme producers have tended to use these microorganisms in preference to others with the result that, in many cases, more than one enzyme is extracted from each organism (Table 1.1).

Enzymes are also obtained from several traditional nonmicrobial sources. A group of structually and catalytically similar sulphydryl proteinases are obtained from a variety of tropical plants including papaya (papain), pineapple (bromelain) and fig (ficin). These are rich sources of

Table 1.1 The major microbial sources of bulk enzymes

Organism	Enzyme	Applications
Aspergillus sp.	Amyloglucosidase	Sugar industry
	Cellulase	Cellulose processing
	Hemicellulase	Food processing
	Lactase	Whey processing
	Lipase/esterase	Fat processing
	Pectinase	Pectin degradation
Bacillus sp.	α-Amylase	Saccharification
	β-Glucanase	Glucan degradation
	Glucose isomerase	Sugar industry
	Alkaline proteinase	Detergents
	Neutral proteinase	Protein hydrolysis
	Subtilisin	Detergents
Klebsiella aerogenes	Pullulanase	Debranching of starch
Kluyveromyces lactis	Lactase	Whey processing
Mucor miehei	Microbial rennet	Dairy products
Saccharomyces spp.	Invertase	Confectionery industry
Streptomyces sp.	Glucose isomerase	Sugar industry
Trichoderma sp.	Cellulase	Cellulose processing

proteolytic enzymes which may be obtained readily using unskilled labour. For example, ficin comprises approximately 90% of the protein in the latex tapped from *Ficus glabrata* (Gacesa and Hubble, 1987). As the enzymes are relatively concentrated in these sources, they are rarely purified to any degree.

A number of enzymes have been obtained from animal-derived sources. For example, the proteolytic enzyme chymosin, which is extracted from the fourth stomach of unweaned calves, has been used for cheese making for as long as records go back. Canine excrement has proved to be an invaluable (if somewhat unpleasant!) source of pancreatic proteinases for the bating of hides. Although the use of chymosin continues and may be enhanced by the cloning and expression of the recombinant enzyme (Garg and Johri, 1994), pancreas-derived enzymes are being replaced by similar products obtained from microorganisms. This is largely because the availability of offal is restricted as the production of enzymes is merely a sideline of meat processing and manufacture.

It has been estimated (Godfrey and Reichelt, 1983) that approximately 80% of enzymes are produced by fermentation. Most of these enzymes are destined for use in detergents and for the processing of starch and dairy-related products (Frost and Moss, 1987) (Figure 1.1a). The major categories of industrial enzymes are also shown (Figure 1.1b). There is no simple analysis of the quantities of enzymes used for the recycling of waste *per se*. However, it could be argued that, as a minimum, the proportion attribution to starch processing would fall into this category. Therefore, it is probably safe to estimate that at least 30% of total enzyme production is concerned with the recycling of waste into usable products.

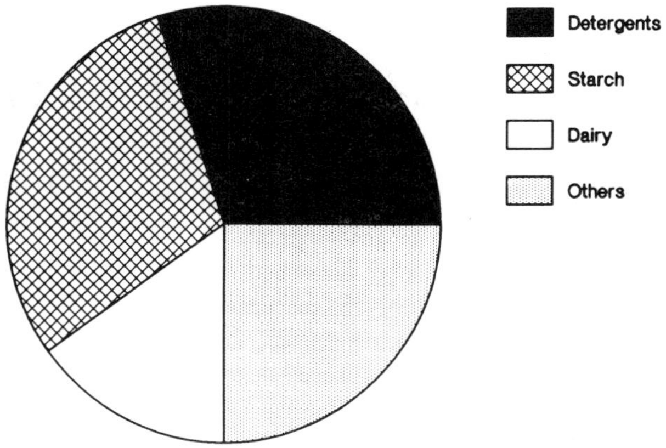

(a)

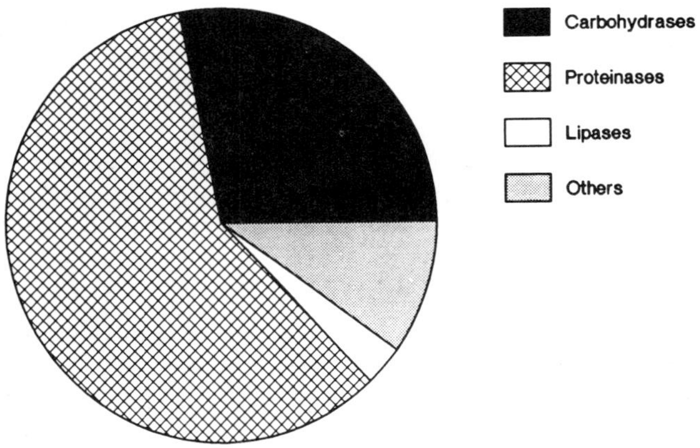

(b)

Figure 1.1 The major applications (a) and the major types of enzymes (b) used in industrial processes.

The ability of an organism to produce extracellular enzyme is an important consideration as it simplifies subsequent processing (Figure 1.2). Consequently, only one enzyme is obtained on the large scale from a Gram-negative bacterium as the outer membrane of these organisms is an additional barrier to export of protein from the cell. The pullulanase (an essential enzyme in starch processing, see later) of the Gram-negative bacterium *Klebsiella aerogenes/pneumoniae* is a complex lipoprotein which is excreted from the cell but accumulates in the periplasm of this organism, where it associates with various membrane structures (Sauvonnet *et al.*, 1995). However, the advent of more industrially applicable heat-stable

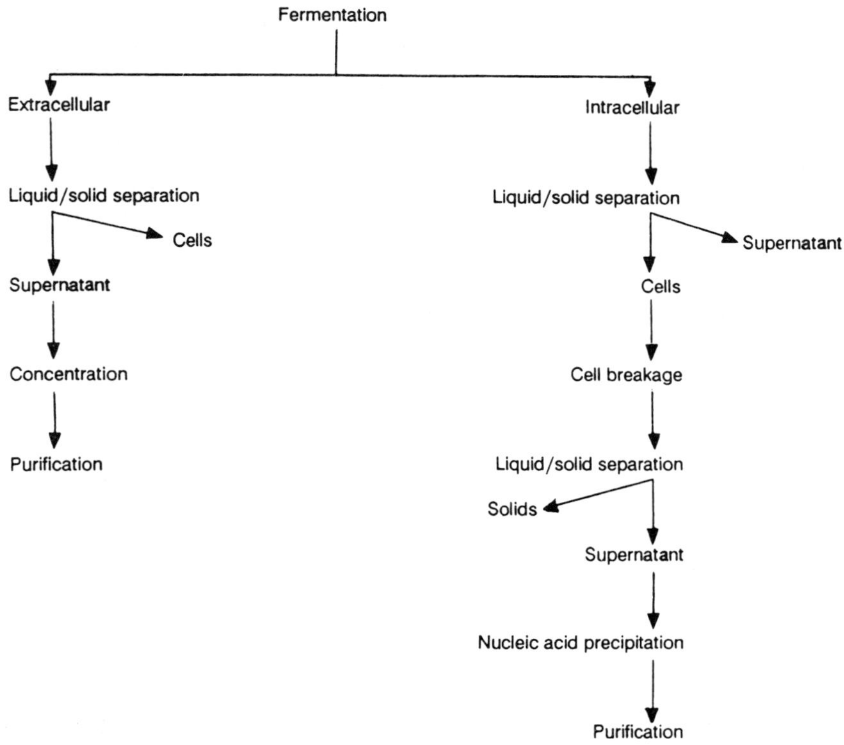

Figure 1.2 Comparison of the processing steps involved in the extraction of intra- and extracellular enzymes. (Adapted from Hacking, A.J., *Economic Aspects of Biotechnology*; published by Cambridge University Press, 1986).

forms of the enzyme will overcome these problems (Antranikian *et al.*, 1995; Rudiger *et al.*, 1995).

Most enzymes that are used for industrial processes require only a minimum of purification and in some cases it is a crude disrupted-cell extract that is used as the catalyst. Numerous combinations of methods have been used to purify enzymes on a laboratory scale (Scopes, 1982; Brummer and Gunzer, 1987) and, although the same principles are applicable to large-scale manufacture, due note has to be paid to the scale of the operation and the problems that this introduces (Atkinson *et al.*, 1987; Bucke, 1988).

The quantities of enzymes and the conditions under which they can be produced may be altered radically by mutagenesis and recombinant DNA techniques. The biosynthesis of most proteins is controlled by induction, e.g. induction of cellulase by cellulose (Merivuori *et al.*, 1984), and/or by catabolite repression (Demain, 1983), thus limiting the quantities of enzyme that may be produced on non-defined media. These problems may

Table 1.2 Some examples of industrially important enzymes which have been cloned

Enzyme	Source strain	Host strain	Reference
α-Amylase	*Bacillus licheniformis*	*B. subtilis*	Piggott *et al.* (1984)
α-Amylase	*B. amyloliquefaciens*	*Escherichia coli*	Palva (1982)
Pullulanase	*Klebsiella aerogenes*	*E. coli*	Takizawa and Murooka (1985)
Cellulases	*Cellulomonas fimi*	*E. coli*	Gilkes *et al.* (1984a,b)
Pectate lyases	*Erwinia crysanthemi*	*E. coli*	Keen *et al.* (1984)
Lactase	*B. stereathermophilus*	*B. subtilis*	Hirata *et al.* (1985)
Neutral proteinase	*B. stereathermophilus*	*B. subtilis*	Fujii *et al.* (1983)
Alkaline proteinase	*B. stereathermophilus*	*B. subtilis*	Vasantha *et al.* (1984)
Subtilisin	*B. amyloliquefaciens*	*B. subtilis*	Wells *et al.* (1983)
Chymosin	Bovine	*Saccharomyces cerevisiae*	Mellor *et al.* (1983)
Xylose isomerase	*E. coli*	*Schizosaccharomyces pombe*	Ueng *et al.* (1985)
Lignin peroxidase	*Phanerochaete crysosporium*	*E. coli*	Tien and Kirk (1983)

be overcome by using random mutagenesis procedures (Dale, 1988) to obtain constitutive mutants and overproducers.

Recombinant DNA techniques have been widely applied to the cloning and overexpression of industrially significant enzymes (Table 1.2). In most cloning experiments *Escherichia coli* has been utilized as the host organism because it is the best genetically characterized microorganism, but *Bacillus subtilis* is a more suitable bacterium for industrial processes. Overproduction of enzyme can be achieved by insertion of the DNA fragment into a high copy number plasmid, i.e. a gene-dosage effect, preferably downstream of a high-efficiency promoter sequence. Also, DNA encoding an N-terminal signal peptide is needed if the protein is to be exported from the cell. There is now a vast range of vectors and hosts available for the overexpression and secretion of recombinant enzymes (Gacesa and Ramji, 1994). Other bacteria, particularly *Bacillus* spp., are better suited for the secretion of proteins into the medium, as the absence of an outer membrane precludes the possibility of periplasmic accumulation of the cloned gene product.

1.2.2 Enzyme stability

The economics of any process which utilizes enzymes will be considerably influenced by the cost and stability of the enzyme preparation used. Stability factors would usually be expected to have a more significant effect on continuous rather than batch operation, although in situations like waste processing where the feed stream may contain many contaminants, enzyme stability may be severely compromised even in the context of a batch reaction.

The limited stability of enzyme preparations results from their critical dependence on a specific molecular conformation. Given that the conformation of a protein results from the collective effects of a number of weak, usually non-covalent, interactions, the free energy difference between an enzyme in an active and an inactive state can be as little as 40 kJ mol^{-1}. Enzyme inactivation is usually attributed to thermal effects but the temperature dependence of the rate constant for deactivation will be strongly influenced by changes in the local chemical environment. As protein conformation depends on a range of weak interactions (Daniel, 1996) whose individual contributions will vary for each case, it is clear that prediction of the resultant changes in enzyme activity is unlikely to be feasible except over a limited range of conditions.

(a) Quantification of stability. Although not universally applicable, first-order deactivation kinetics have been widely used to predict the effects of enzyme inactivation (Gacesa and Hubble, 1987). As this simple analytical approach is a gross simplification of the range of factors affecting a real process, e.g. it ignores changes related to throughput rather than time, it must be used with caution.

In practice it would be usual to monitor reactor performance throughout the life of the catalyst and to adjust the feed flow rate to maintain fractional conversion. For economic reasons it is usually only practical to operate for three half-life decay periods before changes in feed rate become unacceptable. Within this constraint the use of reactors in parallel allows the overall process throughput to be maintained at a constant rate.

(b) Stabilization of enzymes. For many applications, particularly where continuous operation is required, it is advantageous to immobilize the enzyme on an insoluble support. Ideally this improves the handling properties of the enzyme, improves the ease of reuse and often confers some stability advantages. The basic approaches for enzyme immobilization have been established for many years and are widely reported in the literature (e.g. Zaborsky, 1973; Messing, 1985). However, the apparent kinetic properties of the enzyme often change as a consequence of enzyme immobilization. This can result from one of a number of factors including conformational changes, microenvironmental effects and mass transfer limitations (Bailey and Ollis, 1986; Gacesa and Hubble, 1987).

The immobilization process may also affect enzyme stability (Manson and Combes, 1988) but, although widely quoted as improving stability, this cannot be assumed as a potential benefit in all cases. Both the thermal stability and the ability to reverse conformational changes can be dependent on the number of bonds between enzyme and support (Koch-Schmidt and Mosbach, 1977). There is a compromise to be made between the benefits of locking the enzyme into an active conformation and the disadvantages of the molecule losing its ability to 'renature' once a conformational change has been induced. Hence the effects of immobilization on protein conformation are complex and in some ways analogous to the effects of intramolecular crosslinking (see section 1.2.2(b)). Given the heterogeneous nature of the bonds formed between enzyme and support, the benefits on stability must again be assessed on a case by case basis.

In the context of waste treatment, there are probably two significant stability benefits which may be gained from immobilization. The first applies specifically where proteinases are being used to modify the structure of a waste protein. A major problem is autolysis of the catalyst preparation. Immobilization of the enzyme will tend to eliminate this effect by preventing contact between individual proteinase molecules. The second is the use of encapsulation of the enzyme within a semi-permeable membrane to protect against potential denaturants in the process feed.

In addition to those factors described above which affect the intrinsic stability of an enzyme preparation, there are a number of other effects which can be introduced as a consequence of immobilization and lead to deviations from first-order deactivation kinetics. Examples include the formation of subpopulations of immobilized enzymes with different stabilities, and also mass transfer limitations which can mask enzyme decay. In this case, while mass transfer is limiting, the reactor may appear to give a constant performance over an extended period. However, once a critical enzyme level is reached, the effects of inactivation become apparent. Therefore, mass transfer effects can lead to erroneously optimistic assessments of the benefits of immobilization on enzyme stability (Cheetham, 1983).

(c) Stabilization of soluble enzymes. Although there has been much interest in the development and use of immobilized enzyme preparations (Gestrelius and Mosbach 1987), many enzymes are still used in a soluble form. For this reason there has been significant research on the stabilization of enzymes using additives and by direct chemical modification (Fagain, 1995). Clearly, the cost and benefits of any stabilization protocol must be carefully balanced. In the case of additives, e.g. chelating agents and microbial growth inhibitors, the costs of subsequent removal must be added to the raw materials cost.

A more direct approach to stabilization is to modify the native structure of the enzyme using chemical agents (Fontana, 1991), for example, bifunctional agents have been used successfully to stabilize enzyme molecules by the introduction of intramolecular crosslinks (Martinek and Torchilin, 1988). The potential advantages to be gained from internal crosslinking for both monomeric and multimeric proteins are shown in Figure 1.3. Increases in stability have been demonstrated experimentally

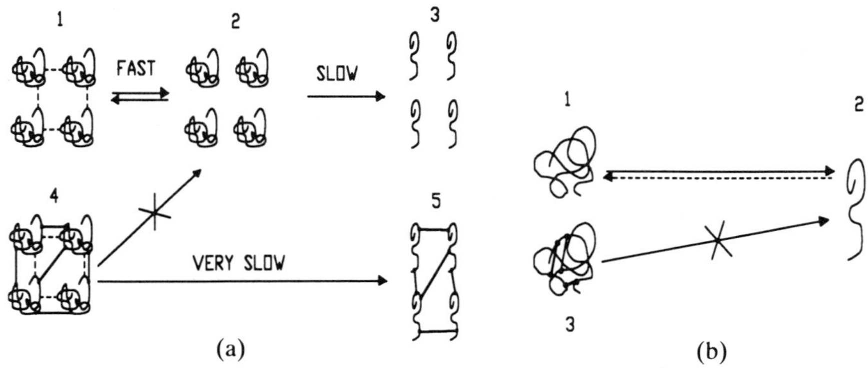

Figure 1.3 General scheme of enzyme stabilization effected by intramolecular crosslinking. (a) 1 = Native oligomeric enzyme; 2 = reversibly dissociated subunits; 3 = irreversibly denatured subunits; 4 = crosslinked enzyme; 5 = irreversibly denatured crosslinked enzyme. (b) 1 = Native monomeric enzyme; 2 = denatured enzyme; 3 = crosslinked enzyme. (Reproduced from Martinek, K. and Torchelin, V.P. *Methods in Enzymology*, **137**, 615, 1988).

for both chymotrypsin (Torchilin *et al.*, 1978) and glyceraldehyde 3-phosphate dehydrogenase (Torchilin *et al.*, 1983). An alternative approach to chemical modification is to use site-directed mutagenesis to introduce additional disulphide bridges into an enzyme molecule. This approach has been demonstrated using the enzyme lysozyme (Perry and Wetzel, 1984).

(d) Enzymes from extremophiles. As an alternative to the stabilization of enzymes, increasing attention is now being paid to the use of naturally stable enzymes from extremophilic organisms which are adapted for life under conditions of high temperature or salinity (Jaenicke, 1991). In the case of temperature it is not uncommon to find microorganisms inhabiting natural ecosystems at temperatures approaching 100°C (Daniel, 1996). Clearly the survival of such organisms depends on the stability of their enzymes and hence they are seen as potentially valuable sources of commercial enzymes (Coolbear *et al.*, 1992).

With respect to the enzyme categories discussed in this chapter there are a number of interesting examples which can be obtained from thermophilic organisms. One of the most stable enzymes known is an amylase obtained from *Pyrococcus furiosus* (Koch *et al.*, 1990) which has a significant half-life at 130°C. Other thermophilic carbohydrases include xylanase (Simpson *et al.*, 1991), amyloglucosidase (Oren, 1983), cellulase and hemicellulase (Durand *et al.*, 1984; Patchett *et al.*, 1989; Hreggvidsson *et al.*, 1996). Similarly a number of thermostable lipases (Ammar and McDaniel, 1984; Kambourova *et al.*, 1996; Schmidtdannert *et al.*, 1996) and proteinases (Cowan and Daniel, 1982; Fujii *et al.*, 1983; Wilson *et al.*, 1994; Bryan *et al.*, 1986) have been discovered which are now starting to find application in the area of waste treatment (see section 1.3.1(b)).

Although offering enhanced thermal stability, in general, enzymes from thermophilic organisms show similar performance at their 'native' temperature to enzymes obtained from mesophilic organisms. Used at elevated temperatures the thermophilic enzyme will show a normal decay profile; if the temperature is reduced the stability increases but the activity falls. In consequence, the total production expected from a thermophilic enzyme may not be greater than that which might be obtained from a mesophilic enzyme. This has been interpreted in terms of a required flexibility for catalytic activity such that the two enzymes would show similar flexibility, stability and activity at their optimum temperature of use, which might be 20–30°C apart (Daniel, 1996).

Studies of the factors responsible for the stability of thermophilic organisms have shown that in many cases the stabilizing effect of an additional hydrophobic interaction would be sufficient to explain the observed increase in stability. However, work on using an understanding of the molecular basis of stability to engineer new thermostable enzymes has shown that, in practice, the interactions involved are significantly more complex (Russell and Taylor, 1995).

The advantages offered by thermophilic enzymes in waste processing are most likely to be seen where the waste is generated at elevated temperatures. In such cases the use of thermostable enzymes allow direct treatment prior to cooling and could potentially reduce problems of microbial contamination. A peripheral advantage is also offered by potential cost savings in the production of thermostable enzymes from genetically modified mesophilic organisms. In this case the thermostable enzyme can be rapidly purified by using thermal denaturation to remove the other proteinaceous cellular components (Patchett *et al.*, 1989).

In addition to thermophilic organisms there are other species which are halophilic (adapted to life in a highly saline environment), alkaliphilic (adapted to environments above pH 9) and psychrophilic (adapted to life at low temperature). Organisms which have adapted to life in extremely saline environments, e.g. the Dead Sea, survive as a result of adapting their internal composition to match their environment rather than attempting to maintain an osmotic gradient across their cell wall. In consequence, their enzymes must show activity under conditions which would be highly denaturing for normal proteins (Frolow *et al.*, 1996). The potential use of these organisms and their enzymes in biotechnological applications has been considered by Ventosa and Nieto (1995) and they have obvious potential for use with waste streams where salt concentrations would preclude the use of normal enzymes. Alkaline proteinases from alkaliphilic organisms have already found use in the formulation of detergents and have significant potential in a number of areas including waste processing (Horikoshi, 1996). Psychrophilic organisms, adapted to life at low temperatures, show higher catalytic activities than their mesophilic counterparts making them ideally suited for processing waste streams where the ambient temperature is low (Margesin and Schinner, 1994).

This recent work on the structure–function relationships which determine protein thermostability raises significant questions regarding the benefits which might be expected from stabilization studies. While improvements might be envisaged in specific cases, the flexibility considerations referred to above suggest that an expectation of widespread benefits might be unrealistic. Perhaps of more significance is the development of enzymes from a range of organisms adapted to extreme environments which allow their native properties to be closely matched with the conditions of use.

1.3 Application of enzymes

1.3.1 Hydrolases

(a) Carbohydrates. Polysaccharides are probably the most abundant waste materials available for recycling into useful products. Cellulose and starch are major by-products of a variety of commercial processes and both are amenable to enzyme treatment. Similarly, lactose, the major waste product from the dairy industry, will also be a major target for enzyme treatment.

Starch utilization. The processing of waste starch into a variety of products has been one of the most useful applications of enzyme technology (Figure 1.4). Plentiful supplies of inexpensive amylases are available and there has been no serious attempt to produce an immobilized enzyme. Traditionally, α-amylase has been obtained from *Aspergillus oryzae* but the enzyme obtained from various thermophilic *Bacillus* spp. have the advantage of temperature stability. Although there is considerable sequence homology between the *Bacillus* α-amylases (Mercier and Colonna, 1988) some are much more thermostable than others. The main cause of thermal denaturation in these enzymes is the deamination of asparagine and glutamine residues (Tomazic and Klibanov, 1988). The advent of *Bacillus licheniformis* α-amylase, which operates at temperatures between 95° and 105°C, has allowed the heat disruption of the starch grain and the enzymic hydrolysis to occur simultaneously, thus simplifying the whole process (Reichelt, 1983). There have been numerous reports describing the cloning and expression of α-amylase genes (e.g. Kobayashi *et al.*, 1994; Iefuji *et al.*, 1996).

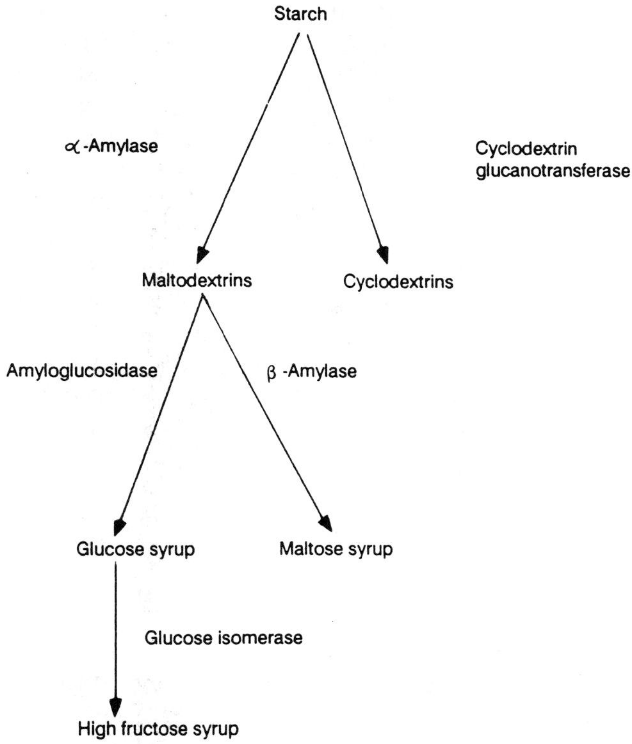

Figure 1.4 The use of enzymes for the modification of starch waste into useful products (after Mercier, C. and Colonna, P., *Proceedings of the 8th International Biotechnology Symposium*, Vol. II, p. 1042; published by Société Française de Microbiology, Paris, 1988).

The dextrins produced by α-amylase action are further processed by a variety of enzymes including β-amylases for the production of maltose syrups and amyloglucosidase for the production of glucose syrups. Pullulanase, usually derived from *Klebsiella pneumoniae*, is required to hydrolyse the α1–6 branch points in the dextrin structure. Amyloglucosidases are produced by a variety of organisms, although the commercial product is obtained principally from *Aspergillus* or *Rhizopus* (Frost and Moss, 1987). Janse and Pretorius (1995) have developed recombinant yeast (*Saccharomyces cerevisiae*) containing genes for α-amylase, pullulanase and amyloglucosidase which enables a one-step hydrolysis of starch. Glucose syrups produced using amyloglucosidase are a useful starting material for a wide variety of processes, including fermentation to products such as alcohol and organic acids (Lehtonen, 1988), or for isomerization into high fructose syrups by the use of glucose (xylose) isomerases (see section 1.3.2(b)).

Cellulose utilization. Cellulose is an abundant waste material for the production of glucose and hence other products. It has been estimated that 1.8×10^{12} tons of cellulose are biosynthesized annually (Alfani and Cantarella, 1987). Although cellulose is a simple polysaccharide (poly β1–4-glucose), enzymic degradation is hampered by its crystalline structure and complex mixture of other components which occurs in plant cell-wall material. Typical agricultural wastes, such as straw, contain 38–42% cellulose, 31–34% hemicellulose, 7–11% lignin and 10–13% of other components (Ladisch *et al.*, 1983). Therefore, cellulose-containing material is normally pretreated to disrupt the fibres using a variety of chemical or physical methods prior to enzyme degradation.

Native cellulose is degraded by a complex mixture of enzymes, including endoglucanases, cellobiohydrolases and β-glucosidases, which act in concert (Eriksson and Pettersson, 1975). There is now a much better understanding of the multienzyme complex termed the 'cellulosome'. The potential of constructing hybrid cellulosomes should facilitate the use of cellulosic wastes (Bayer *et al.*, 1994). *Trichoderma reesei* has been one of the most useful organisms for the production of cellulases and overproducing variants have been obtained by mutagenesis (Montenecourt, 1983). A major disadvantage of *T. reesei* is the low specific activity of the enzymes obtained from this organism (Beguin *et al.*, 1988). Consequently other cellulolytic organisms, e.g. *Cellulomonas*, have been investigated and many of the genes for the cellulose-degrading enzymes have been cloned (Gilkes *et al.*, 1984a,b; Han *et al.*, 1995).

Attempts have been made to use cellulases to process materials contained in mixed and defined wastes. For example, municipal solid waste has been separated into its various components and the cellulose-containing fraction has been saccharified with bacterial and fungal cellulases (Clanet *et al.*, 1988). Similarly, sawdust (Kumakura and Kaetsu, 1988) and wheat straw (Gonzalez *et al.*, 1988) have been processed using cellulases. It is these types of applications that will result in the large-scale use of cellulases for waste processing.

Lactose utilization. The UK dairy industry produces some 16 Mt year^{-1} of combined waste (Wheatley, 1987) of which a significant proportion contains lactose. The lactose in dairy waste is a considerable problem and is the major concern in recycling the material into new products. It has been estimated that up to a half of the whey produced during cheese making is dumped directly into the drains (Marwaha and Kennedy, 1984), which causes severe pollution problems because of the high biological oxygen which averages 35 000–40 000 ppm (Maiorella and Castillo, 1984).

Attempts to use β-galactosidases to hydrolyse lactose into its constituent sugars, glucose and galactose have met with some success, and a plant in Italy processes 10 000 l of milk per day for human consumption (Harrison,

Table 1.3 Costs of hydrolysis of the lactose component of whey

Method	Cost($)/te dry solids[a]
Batch acid	40–60
Acid resin	90–110[b]
Batch enzyme	50–100
Immobilized enzyme	20–100

[a]These are estimates of total hydrolysis costs (enzyme + reactor costs + labour) but exclude the costs of concentration, transport and other charges.
[b]Includes the cost of pretreatment.
Adapted from Hacking (1986).

1987). Although this and other uses of β-galactosidase have been developed primarily to increase the digestibility of milk rather than to process waste material derived from the dairy industry, these processes are applicable to the recycling of wastes, e.g. in the use of lactose-containing whey as an animal-feed supplement.

The use of enzymes for the degradation of lactose has been compared to other methods such as acid hydrolysis (Burgess and Shaw, 1983) and also to microbial fermentation (Mariorella and Castillo, 1984). The syrup derived from the enzymic hydrolysis of lactose in whey has been used commercially for the production of bread (Matveeva *et al.*, 1989). At present enzyme degradation of lactose cannot be considered as the major route for whey processing. A number of appropriate enzymes have been obtained from a wide range of organisms (Frost and Moss, 1987), although many of these are of limited commercial value, often because of susceptibility to product inhibition (van Griethuysen-Dilber *et al.*, 1988). However, the economics of enzyme hydrolysis of lactose in whey can be favourable (Table 1.3) and there is no doubt that this will be an expanding area of waste recycling.

(b) Proteinases. Proteinases have been widely used to modify the structure, and hence texture and flavour of proteins in a range of food industry applications (Alder-Nissen, 1986). The ability to change structure and hence modify the digestibility of protein materials offers the potential for upgrading protein wastes for use as animal or human feed supplements. A number of raw materials have been targets for this approach including, horn and hoof (Kida *et al.*, 1995), blood (Quaglia and Massacci, 1982), turkey waste (Fonkwe and Singh, 1996) and feathers (Dalev, 1994). Although the reports cited appear to be preliminary studies, the results described for the work with hide trimmings shows a significant reduction of protein molecular weight attributable to a two-stage alkali/enzymic

treatment with the product material showing 86% digestibility *in vitro*. Given the scale of the animal processing industry, this offers obvious process opportunities.

In addition to improving the nutritional properties of waste proteins, enzymes have been used to upgrade waste materials for other applications, including the use of proteolytic enzymes to process waste silk (Gulrajani and Gupta, 1995), to facilitate the extraction of high-grade fat from fleshing machine waste (Taylor *et al.*, 1989) and the use of alkaline proteinase for leather treatment (Hameed *et al.*, 1996).

Although it does not lead to a biological product, there have been a number of reports on the use of proteinases for the recovery of silver from photographic emulsions (Phillips, 1977; Krusteva et al., 1987). Fujiwara *et al.* (1989) report on the commissioning of a pilot plant capable of processing 900–1000 kg of X-ray film per day. While the product is valuable, it is difficult to assess the viability of this enzymic treatment compared with a simple acid digestion, although the authors claim a reduction in processing odours.

A somewhat different slant on the relationship between proteinases and products from waste is given by the work of Morimura *et al.* (1994) who use waste water arising from the manufacture of sochu as the growth medium for an *Aspergillus* fermentation designed to produce proteinase. In this case the proteinase is produced as a valuable by-product of a fermentation which also serves to treat the distillery waste.

1.3.2 Nonhydrolytic enzymes

(a) Oxidoreductases. The oxidoreductases are not a group of enzymes normally associated with the processing of waste materials into industrially useful compounds. However, there are one or two key applications emerging. One of particular interest is the possibility of using an NADH-dependent nitrate ester reductase for the enzymic disposal of explosives (White and Snape, 1993). Explosive nitrate esters such as nitroglycerine and pentaerythritol tetranitrate have been manufactured in large quantities and now present a major waste disposal problem. It has been estimated that remediation of these types of materials from sea dumps or newly discarded munitions has a market potential of $10 billion (Jannsen, 1993, 1994).

The degradation of lignin, which is essentially an oxidative process, represents the most significant potential application of an oxidoreductase to the recycling of waste materials. Lignin is a complex heteropolymer comprised of a variety of phenolic compounds linked by ether and carbon–carbon bonds. Furthermore, the lignins obtained from softwoods, hardwoods and grasses differ significantly in structure (Sarkanen and Ludwig, 1971; Adler, 1977). The potential of lignin as a raw material has been recognized (Haemmerli *et al.*, 1988), although the difficulties inherent in selectively degrading the polymer present many problems.

It has been recognized for some time that white rot fungi, in particular, are able to degrade lignin (for a review, see Kirk and Farrell, 1987); however, it is only relatively recently that a lignin-degrading enzyme has been isolated and characterized (Tien and Kirk, 1983). A gene for lignin peroxidase has subsequently been cloned and sequenced (Tien and Tu, 1987) and overexpression in a suitable host would make industrial use of the enzyme a more feasible proposition.

The lignin peroxidase enzyme catalyses a variety of reactions, including the oxidation of phenolics, hydroxylation of alkene bonds, cleavage of ether bonds and aromatic ring opening (Palmer *et al.*, 1987). The enzyme, which exists in multiple isoenzyme forms, contains a single iron protoporphorin IX prosthetic group and catalyses a one-electron oxidation of the substrate to yield radical cations (Schoemaker *et al.*, 1985). It has been argued that lignin peroxidase produces the initial radical cation and that subsequent degradation of the substrate occurs by free-radical attack. However, some workers (Alfani and Cantarella, 1987) believe that free-radical attack might be less important than the combined action of several different extracellular peroxidase-type enzymes. The lignin peroxidase is relatively non-specific for substrate which would be an advantage in commercial applications as the structures of lignins are so heterogeneous.

A perhaps unexpected finding was the discovery that other peroxidases, e.g. horseradish peroxidase, were able to degrade lignin in nonaqueous systems (Van Brunt, 1986). At the time when lignin peroxidases were difficult to purify, the use of other peroxidase enzymes was a particularly attractive idea. However, now that the lignin peroxidase has been cloned, adequate supplies of the enzyme are no longer an insurmountable problem.

The potential of using ligninase for the degradation of waste lignin is vast and the generation of low molecular weight compounds would provide precursors for chemical processing, possibly substituting for raw materials derived from the petrochemical industry (Buswell and Odier, 1987). As lignocelluloses are a major waste-disposal problem, e.g. from paper mills, a process that would enable the effluent water to be treated and also produce new materials would be economically attractive.

Although the degradation of lignin represents a major potential use for peroxidase-type enzymes, other applications have been proposed. For example, horseradish peroxidase might be used for the removal of carcinogenic aromatic amines from water (Klibanov and Morris, 1981). A comprehensive study of the reaction products from the oxidation of phenolic pollutants using various enzymes illustrated clearly that care must be taken to ensure that the toxic pollutant is not inadvertantly converted into an even more toxic product (Aitken *et al.*, 1994). It is essential that the composition of the polluting stream is characterized and that appropriate reaction conditions and enzyme, e.g. chloroperoxidase, horseradish peroxidase, lignin peroxidase, manganese peroxidase or mushroom polyphenol oxidase are used. For example, these experiments indicated that compounds such as 2-nitrophenol, 4-nitrophenol, 2-chlorophenol, *o*-cresol and phenol could be made more toxic by enzymic treatment whereas others, e.g. 4-chlorophenol, were detoxified (Aitken *et al.*, 1994). The potential for the use of peroxidases and related enzymes in viable treatment processes is great.

(b) Isomerase enzymes. The enzymic treatment of by-products from the corn industry currently represents the largest scale application of enzyme technology in the waste treatment area. The success of the process is critically dependent on the ability to meet very stringent glucose/fructose ratios in the final product (Daniels, 1986) and it was the successful solution of the problems associated with the development of an enzyme-catalysed glucose isomerization that was crucial to the formation of a large-scale industry. The history of the development of the isomerization process has been documented by Jensen and Rugh (1988) and provides an interesting insight into the factors affecting the growth of an industry based on the production of a valuable product from a waste material.

It was shortcomings in the chemical isomerization step that first provided the incentive to investigate an enzymic alternative. Unlike the hydrolytic enzymes that had been used previously on a large scale, the isomerase enzyme was intracellular, showed significantly lower stability and required close control of reaction conditions. Not withstanding these problems, the advantages of an enzyme-catalysed isomerization were such that development of the batch process was almost inevitable. However, the batch process was still labour intensive and difficult to operate, although it did solve some of the problems of by-product formation associated with the chemical process (Coker and Venkatasubramanian, 1985). These shortcomings led to the introduction of a continuous process, hence in this case it was market 'pull' rather than technology 'push' which provided the impetus for development.

There are a number of advantages in the use of continuous plug flow reactors compared with the batch process (Table 1.4). The pertinent factors are the reduction in residence time with concomitant reduction in reactor size and hence plant installation costs, and the significant increase (>50%) in enzyme productivity. Given the relatively small 'value added' in this process, enzyme costs have been critical in determining economic viability (Daniels, 1988).

Recent developments have focused on the identification and development of highly stable isomerase enzymes. Given that the starch liquefaction and saccharification stages are typically operated at a temperature of 20–30°C

Table 1.4 Glucose isomerase: comparison between batch and plug flow processes

Parameter	Batch process (1974–1976)	Plug flow processes	
		1976	1984
Syrup conc.	40–45% w/w	40–45% w/w	43–47% w/w
pH	6.5–7.0	8.4 (inlet)	7.5–7.8 (inlet)
Temperature (°C)	65	65	60
Co^{2+}	3.5×10^{-4}M	–	0
Mg^{2+}	$20 \times (Ca^{2+})$	$20 \times (Ca^{2+})$	$20 \times (Ca^{2+})$
Half-life (enzyme) (h)	300	500	1200–1500
Productivity (kg syrup/kg enz.)	900	1500	3000–4000

Source: Jensen and Rugh (1987).

higher than the isomerization stage there are considerable benefits which might be envisaged from the use of a more thermostable preparation (Brown *et al.*, 1993; Deshmukh and Shankar, 1996). The industrial applications of glucose isomerase have recently been reviewed by Bhosale *et al.* (1996) who, in addition to discussing the production of fructose syrups, also identify a potential role in the conversion of hemicellulose to ethanol.

(c) Lyases. The only lyase of commercial significance is pectin trans-eliminase which is one of a series of enzymes involved in pectin degradation. This and other enzymes have been used extensively for the extraction of fruit juice (Fogarty and Kelly, 1983) but not for the recycling of waste materials. However, low-grade agricultural waste materials have been used in solid substrate fermentations for the production of these enzymes (Miura and Endo, 1962).

1.4 Enzymes with modified activities

1.4.1 Applications of molecular techniques

The application of enzymes to waste processing has often been hampered by a number of factors, including availability, stability and substrate specificity. However, the advent of molecular biological and other techniques provides opportunities for overcoming these problems. Many examples have already been cited in this chapter of enzymes that have been cloned and overexpressed in appropriate host organisms (Table 1.2), and there have been many reviews on the topic (e.g. Nygren *et al.*, 1994).

The techniques of molecular biology have been further extended by the advent of site-directed mutagenesis. It is now possible to alter the properties of an enzyme in a predetermined manner (Knowles, 1987), although not all of the modifications produce the expected results! In the field of waste treatment it is often the stability of an enzyme that is of major concern. Several attempts have been made to increase the stability of proteins by the introduction of additional sulphydryl bridges (Perry and Wetzel, 1984; Katz and Kossiakov, 1986; Wells and Powers, 1986) or by the replacement of particular amino acids in the primary structure. For example, subtilisin is unstable in the presence of oxidizing agents but this problem has been overcome by the replacement of a single methionine residue whilst still retaining activity at 80% of the original k_{cat} (Estell *et al.*, 1985). It is clear that site-directed mutagenesis and related techniques will be of paramount importance for the development of novel enzymes; however, a fuller description of these methods is outside the scope of this review.

1.4.2 Nonaqueous/low water systems

This is an area of research which has attracted a great deal of interest (Lilly *et al.*, 1987; Bell *et al.*, 1995; Almarsson and Klibanov, 1996), with particular importance for the treatment and conversion of wastes containing hydrophobic compounds. The problems encountered in the treatment of non-polar materials stems from the fact that most traditional enzyme technology applications have been based on aqueous systems. Until relatively recently the view was that most enzymes were inherently unstable in organic solvents and therefore could only be used in an aqueous environment. However, it is now appreciated that polar and non-polar solvents have completely different effects on enzyme stability (Dordick, 1991). Polar solvents (e.g. ethanol and acetone) compete for water molecules, which are essential for maintaining enzyme structure, resulting in inactivation. Conversely, non-polar solvents can support highly active enzyme conversions providing that a critical level of 'essential' water is present in the system (Halling, 1987a).

Although it has been possible to demonstrate the use of enzymes suspended in organic solvents, the possibility also exists of various intermediate approaches for the modification of non-polar molecules. These are mainly based on the use of surface-active agents to create micelles in a two-phase system, i.e. where droplets of one phase are dispersed throughout the second phase. Some experimental techniques for the production of enzyme-containing micelles have been described (Martinek *et al.*, 1986) but in essence these simplify to two basic systems:

- normal micelles – aqueous bulk phase – dispersed organic phase;
- reverse micelles – organic bulk phase – dispersed aqueous phase.

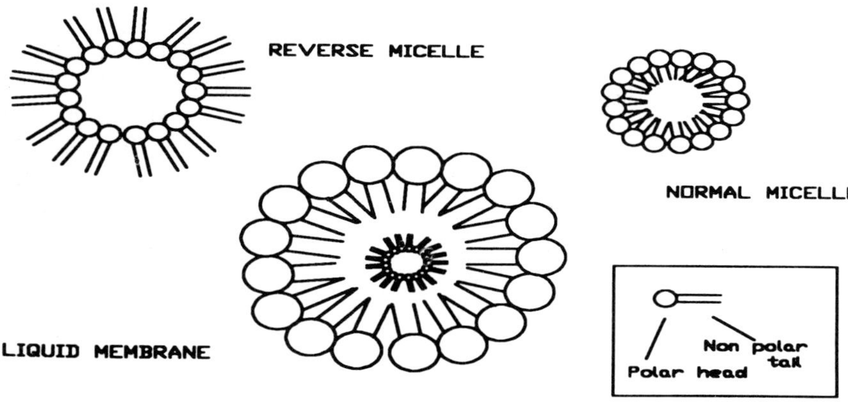

Figure 1.5 Possible micelle conformations in two-phase aqueous/organic systems.

A refinement of the use of micelles is the liquid membrane, where there is a double micelle arrangement (Figure 1.5), which can allow additional selectivity to be achieved.

The advantages of micelles is that the enzyme remains in a localized aqueous environment but there is a large interfacial surface area which facilitates mass transfer of substrate and product. An additional advantage is that the choice of organic solvent can be optimized to control the partitioning of reactants and products between the two phases. This has been demonstrated in experimental systems where inhibitory products have been removed by partition from the aqueous phase (Schwartz and McCoy, 1977). In the design of a two-phase reactor based on micelles, two aspects are critical:

- optimization and prediction of mass-transfer effects (Lilly *et al.*, 1988);
- separation and recovery of the two phases after completion of the reaction (Lilly *et al.*, 1987).

It was the difficulties encountered in the separation of the two phases which provided the incentive for development of the two-phase membrane bioreactor (Hoq *et al.*, 1985a). In this case the two phases are physically separated by a microfiltration or ultrafiltration membrane. The reactor geometry (Figure 1.6) has been used for both the synthesis and hydrolysis of triglycerides using lipases (Hoq *et al.*, 1985a,b; Basheer *et al.*, 1995; Giorno *et al.*, 1995) and it would appear to have significant potential applications for conversions based on lipid- or fatty-acid-containing waste streams.

The five possible approaches to the use of enzymes in nonaqueous/low water streams are summarized schematically in Figure 1.7. Of these, the use of a homogeneous organic phase containing enzyme with the quantity

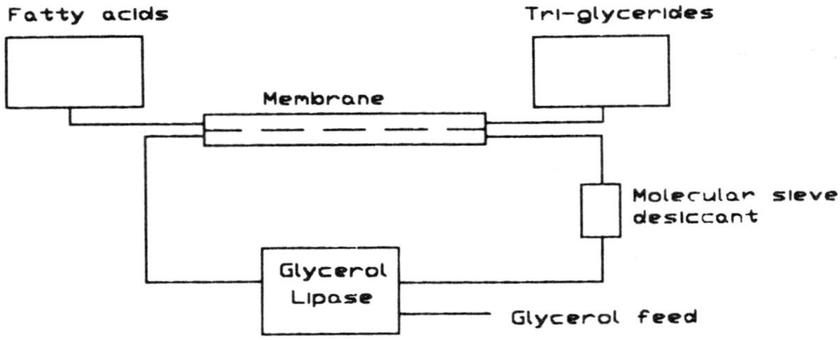

Figure 1.6 Two-phase membrane reactor for triglyceride synthesis (adapted from Hoq. 1985b).

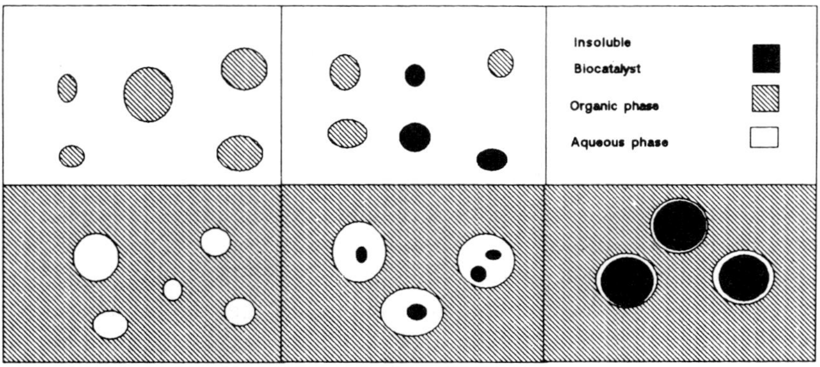

Figure 1.7 Diagrammatic representation of reaction systems involving an organic liquid phase (Lilly *et al.*, 1987).

of water restricted only to that essential for enzyme activity represents the most radical approach. The exclusion of nonessential water (>90%) can lead to significant changes in enzyme specificity in addition to changes in the equilibrium position of the reaction. One of the most interesting possibilities lies in the use of hydrolytic enzymes to drive synthetic rather than degradative reactions. By excluding water from the reaction mixture, the equilibrium of the hydrolytic reaction is shifted in favour of polymer formation rather than degradation. In this way it is possible to synthesize lipids and peptides from simple precursors (Halling, 1987b; Napier *et al.*, 1996).

The influence of organic solvents on the enzyme molecule stems from the significance of water molecules in maintaining and stabilizing the noncovalent interactions which determine the three-dimensional conformation of the protein molecule. By using a reaction medium based on an organic solvent, it is possible to control the amount of water present in the system closely and hence vary the hydration state of the enzyme preparation. This allows stability, specificity and catalytic efficiency to be modified systematically to match process requirements (Klibanov, 1989).

Data obtained for chymotrypsin show the effect of solvent hydrophobicity on catalyst activity (Zaks and Klibanov, 1988). In this case there is a 10 000-fold increase in the activity observed in octane compared with pyridine. This difference is attributed to the polar solvent (pyridine) stripping 'essential' water from the enzyme and reducing catalytic effectiveness. A similar solvent effect has been reported by Desampaio *et al.* (1996) who investigated the activity of subtilisin in three different solvents. In addition to partition effects these authors also identified inhibition as a factor influencing solvent suitability. The control of 'essential' water can also have remarkable effects on enzyme stability. For example, chymotrypsin has a half-life of several hours at 100°C in octane compared with only several minutes at 60°C in an aqueous solvent. This effect is attributed to a decrease in conformational mobility resulting from the restricted availability of water, which in turn leads to greater structural stability (Zaks and Klibanov, 1985; Volkin *et al.*, 1991). Similar effects have been reported for lipases (Zaks and Klibanov, 1984).

An interesting finding from work with enzymes in organic solvents is that the enzyme retains a 'memory' of its chemical environment prior to the intermediate drying phase before transfer into the new solvent. Hence, the activity of the enzyme will vary with the pH of the solution from which it was lyophilized (Zaks and Klibanov, 1985) and therefore the appropriate activity can be 'preprogrammed' (Gupta, 1992). Another opportunity offered by the use of enzymes in organic solvents is the possibility of changing specificity (Cowan and Plant, 1992), particularly where hydrophobic interactions between enzyme and substrate are important. Further refinements are possible by the formation of enzyme/polyethylene glycol conjugates which can improve solubility in organic solvents and in some cases lead to significant stabilization (Matsushima *et al.*, 1996).

In the area of waste processing the most interesting opportunities for the use of enzymes in nonaqueous solvents concern the treatment of phenolic-containing streams and the treatment of lipids. Traditionally, phenols would be microbially degraded to prevent pollution; however, tyrosinase can be used directly to oxidize phenols in organic solvents (Kazandjian and Klibanov, 1986; Robb, 1995). In aqueous solution the oxidized products rapidly polymerize, whereas in chloroform the reaction produces quinones which can be readily recovered. There are a wide range of lipid bioconversions which might potentially be applied to waste streams, including both synthetic and degradative reactions. These have been reviewed by Balcao *et al.* (1996).

In his review of the use of enzymes in anhydrous solvents, Klibanov

(1989) considered the potential of 'solvent engineering', where solvent properties are adjusted to optimize the appropriate reaction. Further advances in this field have been such that it is now possible to make predictions about the effects of solvent choice on enzyme performance as an aid in design (Halling, 1994). In conjunction with modified enzymes developed using protein engineering techniques, non-aqueous solvent systems offer the potential for a wide range of novel waste recovery processes. Specific examples which have appeared in the literature include the development of a continuous reactor for the enzymatic glycerolysis of butteroil (Garcia *et al.*, 1995) and the use of enzymes for the extraction of edible oil from oil seeds (Rosenthal *et al.*, 1996). The attractive feature of these applications is that the use of appropriate enzymes removes the necessity for exogenous solvents, hence reducing potential environmental impact.

1.5 Conclusions

The last five years have seen very significant advances in our understanding of the factors which determine enzyme activity, specificity and stability. The work which underpins these advances has been conducted on a number of fronts, including studies aimed at fundamental aspects of stability, molecular properties of enzymes from extremophiles and work to investigate the performance of enzymes in nonaqueous solvent systems. In addition to an improvement in fundamental understanding, this work has greatly increased the range of catalysts and conditions which might be envisaged for any prospective enzyme-based reaction. This is particularly important in the area of waste treatment, where to be cost effective the catalyst must be capable of carrying out the desired conversion in the presence of a range of potentially denaturing compounds, under conditions which in the past would have been considered to be incompatible with enzyme activity. The benefits of these advances are already starting to be seen in certain areas, e.g. the use of lipases in organic solvents, and a continued expansion can be expected.

While these scientific and technological developments are undoubtedly important, developments in environmental management are likely to be equally significant in waste-treatment applications. The last few years have seen a major change in industrial emphasis from waste treatment to waste reduction or minimization (Cheremisinoff and Ferranti, 1992). The driving pressure to eliminate wastes at source has already had significant impacts on traditional waste treatment industries, e.g. incineration, and will undoubtedly influence the development of enzyme-based processes.

While pollution prevention and clean technology programmes have and will continue to reduce the volume of waste streams to be treated, it is equally clear that waste cannot be completely eliminated. The tenets of waste minimization require that, where waste cannot be avoided, its effects should be mitigated by recycling or by conversion into useful and/or valuable by-products. It is likely that these trends will lead to smaller volumes of more concentrated and possibly more toxic wastes such that niche processes will be needed for their treatment which are capable of maximizing the recovery or generation of valuable products. Given their inherent specificity and our increasing ability to engineer their stability, enzymes are likely to play an increasingly important role in this area.

References

Adler, E. (1997) *Wood Sci. Technol.*, **11**, 169.
Aiba, S., Kitai, K. and Imanaka, T. (1983) *Appl. Environ. Microbiol.*, **46**, 1059.
Aitken, M.D., Massey, I.J., Chen, T.P. and Heck, P.E. (1994) *Water Res.*, **28**, 1879.
Alder-Nissen, J. (1986) *Enzymic Hydrolysis of Food Proteins*. Elsevier Applied Science, London.
Alfani, F. and Cantarella, M. (1987) In *Biotechnology of Waste Treatment and Exploitation*, (eds J.M. Sidwick and R.S. Holdom), Ellis Horwood, Chichester, p. 256.
Almarsson, O. and Klibanov, A.M. (1996) *Biotechnol. Bioeng.*, **49**, 87.
Antranikian, G., Rudiger, A., Carrganella, F., Klingeberg, M. and Sunna, A. (1995) *J. Macromolec. Sci.*, **A32**, 661.
Atkinson, T., Scawen, M.D. and Hammond, P.M. (1987) In *Biotechnology*, Vol. 7a (ed. J.F. Kennedy), V.C.H., Weinheim, p. 279.
Bailey, J.E. and Ollis, D.F. (1986) *Biochemical Engineering Fundamentals*, 2nd edn, McGraw-Hill, New York.
Balcao, V.M., Paiva, A.L. and Malcata, F.X. (1996) *Enzyme Microbial Technol.*, **18**, 392.
Basheer, S., Mogi, K. and Nakajima, M. (1995) *Process Biochem.*, **30**, 531.
Bayer, E.A., Morag, E. and Lamed, R. (1994) *Trends Biotechnol.*, **12**, 379.
Beguin, P., Grepinet, O., Millet, J. and Aubert, J.P. (1988) In *Proceedings of the 8th International Biotechology Symposium*, Vol. II (eds G. Durand, L. Bobichon and J. Florent), Société Francaise de Microbiologie, Paris, p. 1015.
Bell, G., Halling, P.J., Moore, B.D., Partridge, J. and Rees, D.G. (1995) *Trends Biotechnol.*, **13**, 468.
Bhosale, S.H., Rao, M.B. and Deshpande, V.V. (1996) *Microbiol. Rev.*, **60**, 280.
Brown, S.H., Sjoholm, C. and Kelly, R.M. (1993) *Biotechnol. Bioeng.*, **41**, 878.
Brummer, W. and Gunzer, G. (1987) In *Biotechnology*, Vol. 71 (ed. J.F. Kennedy), V.C.H., Weinheim, p. 123.
Bryan, P.N., Rollance, M.L., Pantoliano, M.W. *et al.* (1986) *Proteins*, **1**, 326.
Bucke, C. (1988) In *Principles of Biotechnology*, 2nd edn (ed. A. Wiseman), Surrey University Press, London, p. 143.
Burgess, K. and Shaw, M. (1983) In *Industrial Enzymology* (eds T. Godfrey and J. Reichelt), Macmillan, Byfleet, p. 260.
Buswell, J.A. and Odier, E. (1987) *CRC Crit. Rev. Biotechnol.*, **6**, 1.
Cheetham, P.S.J. (1983) In *Principles of Biotechnology* (ed. A. Wiseman), Surrey University Press, Glasgow, p. 172.
Cheremisinoff, P. and Ferranti, L. (1992) *Waste Reduction and Pollution Prevention*, Butterworth Heinemann, Oxford.
Clanet, M., Durand, H. and Tiraby, G. (1988) *Biotechnol. Bioeng.*, **32**, 930.
Coker, L.E. and Venkatasubramanian, H.J. (1985) *Comprehens. Biotechnol.*, **3**, 778.
Coolbear, T., Daniel, R.M. and Morgan, H.W. (1992) *Adv. Biochem. Eng. Biotech.*, **45**, 57.
Cowan, D.A. and Daniel, R.M. (1982) *Biochem. Biophys. Acta*, **705**, 293.

Cowan, D.A. and Plant, A.R. (1992) *ACS Symposium Series*, Vol. 498, p. 86.
Dale, J.W. (1988) In *Principles of Biotechnology*, 2nd edn (ed. A. Wiseman), Surrey University Press, London, p. 44.
Dalev, P.G. (1994) *Bioresource Technol.*, **48**, 265.
Daniel, R.M. (1996) *Enzyme Microbial Technol.*, **19**, 74.
Daniels, M.J. (1986) In *Process Engineering Aspects of Immobilised Cell Systems* (eds C. Webb, G.M. Black and B. Atkinson), Institution of Chemical Engineers, Rugby, p. 218.
Daniels, M.J. (1988) *Meth. Enzymol.*, **136**, 371.
Demain, A.L. (1983) In *Overproduction of Microbial Products* (eds V. Krumphanzl, B. Sikyta and Z. Vanek), FEMS Symposium No. 13, Academic Press, London, p.3.
Denner, W.H.B. (1983) In *Industrial Enzymology* (eds T. Godfrey and J. Reichelt), Macmillan, Byfleet, p. 111.
Desampaio, T.C., Melo, R.B., Moura, T.F., Michel, S. and Barreiros, S. (1996) *Biotechnol. Bioeng.*, **50**, 257.
Deshmukh, S.S. and Shankar, V. (1996) *Biotechnol. Appl. Biochem.*, **24**, 65.
Dordick, J.S. (1991) *Curr. Opin. Biotechnol.*, **2**, 401.
Durand, H., Soucaille, P. and Tiraby, G. (1984) *Enzyme Microbiol. Technol.*, **6**, 175.
Eriksson, K.E. and Pettersson, B. (1975) *Eur. J. Biochem.*, **51**, 193.
Estell, D.A., Graycar, T.P. and Wells, J.A. (1985) *J. Biol. Chem.*, **260**, 6518.
Fagain, C. (1985) *Biochim. Biophys. Acta*, **1252**, 1.
Fogarty, W.M. and Kelly, C.T. (1983) In *Microbial Enzymes and Biotechnology* (ed. W.M. Fogarty), Applied Science Publishers, London, p. 131.
Fonkwe, L.G. and Singh, R.K. (1996) *Process Biochem.*, **31**, 605.
Fontana, A. (1991) *Curr. Opin. Biotechnol.*, **2**, 551.
Frolow, F., Harel, M., Sussman, J.L., Mevarech, M. and Menachem, S. (1996) *Nature: Struct. Biol.*, **3**(5), 452.
Frost, G.M. and Moss, D.A. (1987) In *Biotechnology*, Vol. 7a (ed. J.F. Kennedy), VCH Weinheim, p. 65.
Fujii, M., Takagi, M., Imanaka, T. and Aiba, S. (1983) *J. Bacteriol.*, **154**, 831.
Fujiwara, N., Tsumiya, T., Katada, T., Hosobuchi, T. and Yamamoto, K. (1989) *Process Biochem.*, **24**, 155.
Gacesa, P. and Hubble, J. (1987) *Enzyme Technology*. Open University Press, Milton Keynes.
Gacesa, P. and Ramji, D.P. (1994) *Vectors – Essential Data*. John Wiley & Sons, Chichester.
Garcia, H.S., Yang, B. and Parkin, K.L. (1995) *Food Res. Int.*, **28**, 605.
Garg, S.K. and Johri, B.N. (1994) *Food Rev. Int.*, **10**, 313.
Gestrelius, S. and Mosbach, K. (1987) *Meth. Enzymol.*, **136**, 353.
Gilkes, N.R., Kilburn, D.G., Langsford, M.L. *et al.* (1984a) *J. Gen. Microbiol.*, **130**, 1377.
Gilkes, N.R., Kilburn, D.G., Miller, R.C. and Warren, R.A.J. (1984b) *Biotechnology*, **2**, 259.
Giorno, L., Molinari, R., Drioli, E., Bianchi, D. and Cesti, P. (1995) *J. Chem. Technol. Biotechnol.*, **64**, 345.
Godfrey, T. and Reichelt, J.R. (1983) In *Industrial Enzymology* (eds T. Godfrey and J. Reichelt), Macmillan, Byfleet, p. 1.
Gonzalez, G., Caminal, G., deMas, C. and Lopez-Santin, J. (1988) *Process Biochem.*, **24**, 62.
Gulrajani, M.L. and Gupta, S.V. (1995) *Indian J. Fibre Textile Res.*, **20**, 192.
Gupta, M.N. (1992) *Eur. J. Biochem.*, **203**, 25.
Hacking, A.J. (1986) *Economic Aspects of Biotechnology*. Cambridge University Press, Cambridge.
Haemmerli, S.D., Fiechter, A. and Leisola, M.S.A. (1988) In *Proceedings of the 8th International Biotechnology Symposium*, eds G. Gurand, L. Bobichon and J. Florent), Société Francaise de Microbiologie, Paris, p. 1030.
Halling, P.J. (1987a) In *Bioreactors and Biotransformations* (eds G.W. Moody and P. Baker), Elsevier Applied Science, London, p. 109.
Halling, P.J. (1987b) In *Dechema Biotechnology Conferences*, Vol. 1. Dechema, Frankfurt, p. 35.
Halling, P.J. (1994) *Enzyme Microbial Technol.*, **16**, 178.
Hameed, A., Natt, M.A. and Evans, C.S. (1996) *World J. Microbiol. Biotechnol.*, **12**, 289.
Han, S.J., Yoo, Y.J. and Kang, H.S. (1995) *J. Biol. Chem.*, **270**, 26012.

Harrison, L.A. (1987) In *Biotechnology of Waste Treatment and Exploitation* (eds J.M. Sidwick and R.S. Holdom), Ellis Horwood, Chichester, p. 81.
Hirata, H., Negoro, S. and Okada, H. (1985) *Appl. Environ. Microbiol.*, **49**, 1547.
Hoq, M.M., Koke, M., Yamane, T. and Shimizu, S. (1985a) *Agric. Biol. Chem.*, **49**, 3171.
Hoq, M.M., Tagami, H., Yamane, T. and Shimizu, S. (1985b) *Agric. Biol. Chem.*, **49**, 335.
Horikoshi, K. (1996) *FEMS Microbiol. Rev.*, **18**, 259.
Hreggvidsson, G.O., Kaiste, E., Holst, O. *et al.* (1996) *Appl. Environ. Microbiol.*, **62**, 3047.
Iefuji, H., Chino, M., Kato, M. and Iimura, Y. (1996) *Biochem. J.*, **318**, 989.
Jaenicke, R. (1991) *Eur. J. Biochem.*, **202**, 715.
Janse, B.J.H. and Pretorius, I.S. (1995) *Appl. Microbiol. Biotechnol.*, **42**, 878.
Janssen, J. (1993) *Jane's Defence Weekly*, November.
Janssen, J. (1994) *Jane's Defence Weekly*, September.
Jensen, V.J. and Rugh, S. (1987) *Meth. Enzymol.*, **136**, 358.
Kambourova, M., Emanuilova, E. and Dimitrov, P. (1996) *Folia Microbiol.*, **41**, 146.
Katz, B.A. and Kossiakov, A. (1986) *J. Biol. Chem.*, **261**, 15480.
Kazandjian, R.Z. and Klibanov, A.M. (1986) *J. Am. Chem. Soc.*, **107**, 5448.
Keen, N.T., Dahlbeck, D., Straskawicz, B. and Belser, W. (1984) *J. Bacteriol.*, **159**, 825.
Kida, K., Morimura, S., Noda, J. *et al.* (1995) *J. Ferment. Bioeng.*, **80**, 478.
Kirk, T.A. and Farrell, R.L. (1987) *Ann. Rev. Microbiol.*, **41**, 465.
Klibanov, A.M. (1989) *Trends Biochem. Sci.*, **14**, 141.
Klibanov, A.M. and Morris, E.D. (1981) *Enzyme Microbiol. Technol.*, **3**, 119.
Knowles, J.R. (1987) *Science*, **236**, 1252.
Kobayashi, T., Kanai, H., Aono, R., Horikoshi, K. and Kudo, T. (1994) *J. Bacteriol.*, **176**, 5131.
Koch, R., Zoblowski, P., Spreinant, A. and Antranikian, G. (1990) *FEMS Microbiol Lett.*, **71**, 21.
Koch-Schmidt, A.C. and Mosbach, K. (1977) *Biochemistry*, **16**, 2101.
Krusteva, M., Peev, G. and Iotova, L. (1987) *Acta Biotechnol.*, **7**, 93.
Kumakura, M. and Kaetsu, I. (1988) *Process Biochem.*, **23**, 51.
Ladisch, M.R., Lin, K.W., Voloch, M. and Tsao, G.T. (1983) *Enzyme Microbiol. Technol.*, **5**, 82.
Lehtonen, P.O. (1988) In *Proceedings of the 8th International Biotechnology Symposium*, Vol. II (eds G. Durand, L. Bobichon and J. Florent), Société Francaise de Microbiologie, Paris, p. 1060.
Lilly, M.D., Brazier, A.J., Hocknull, M.D., Williams, A.C. and Woodley, J.M. (1987) In *Biocatalysts in Organic Media* (eds C. Laane, J. Tramper and M.D. Lilly), Elsevier, Amsterdam, p. 3.
Lilly, M.D., Harbron, S. and Narendranathan, T.J. (1988) *Meth. Enzymol.*, **136**, 615.
Maiorella, B.L. and Castillo, F.J. (1984) *Process Biochem.*, **19**, 157.
Manson, P. and Combes, D. (1988) *Meth. Enzymol.*, **137**, 584.
Margesin, R. and Schinner, F. (1994) *J. Biotechnol.*, **33**, 1.
Martinek, K. and Torchilin, V.P. (1988) *Meth. Enzymol.*, **137**, 615.
Martinek, K., Levashov, A.V., Klyachko, N., Khmelnitski, Y.L. and Berezin, I.V. (1986) *Eur. J. Biochem.*, **155**, 453.
Marwaha, S.S. and Kennedy, J.F. (1984) *Process Biochem.*, **19**, 79.
Mateeva, I.V., Rogovskikh, I.V. and Puchkova, L.I. (1989) *Khleboprodukty*, 32.
Matsushima, A., Kodera, Y., Hiroto, M., Nishimura, H. and Inada, Y. (1996) *J. Molec. Catal. B-Enzymatic*, **2**, 1.
Mellor, J. Dobson, J.H., Roberts, N.A. *et al.* (1983) *Gene*, **24**, 1.
Mercier, C. and Colonna, P. (1988) In *Proceedings of the 8th International Biotechnology Symposium*, Vol. II (eds G. Durand, L. Bobichon and J. Florent), Société Francaise de Microbiologie, Paris, p. 1042.
Merivuori, H., Siegler, K.M., Sands, J.A. and Montenecourt, B.S. (1984) *Biochem. Soc. Trans.*, **13**, 411.
Messing, R.A. (1985) In *Comprehensive Biotechnology*, Vol. 2 (ed. M. Moo-Young), Pergamon, Oxford, 191.
Miura, Y. and Endo, A. (1962) Process for the preparation of pectolytic enzymes. US Patent 3058890.
Montenecourt, B.S. (1983) *Trends Biotechnol.*, **1**, 156.

Morimura, S., Kida, K. and Sonada, Y. (1994) *J. Fermentation Bioeng.*, **77**, 183.
Napier, P.E., Lacerda, H.M., Rossel, C.M. *et al.* (1996) *Biotechnol. Progres*, **12**, 47.
Nygren, P.A., Stahl, S. and Uhler, M. (1994) *Trends Biotechnol.*, **12**, 184.
Oren, A. (1983) *Curr. Miocrobiol.*, **8**, 225.
Palmer, J.M., Harvey, P.Y. and Schoemaker, H.E. (1987) *Phil. Trans. Roy. Soc. A*, **321**, 494.
Patchet, M.L., Neal, T.L., Schofield, L.R. *et al.* (1989) *Enzyme Microb. Technol.*, **11**, 13.
Palva, I. (1982) *Gene*, **19**, 81.
Perry, L.J. and Wetzel, R. (1984) *Science*, **226**, 555.
Phillips, P.A. (1977) Stripping photographic materials, US Patent 4150977.
Piggott, R.P., Rossiter, A., Ortlepp, S.A., Pembroke, J.T. and Ollington, J.F. (1984) *Biochem. Biophys. Res. Commun.*, **122**, 175.
Quaglia, G.B. and Massacci, A. (1982) *J. Sci. Food Agric.*, **33**, 634.
Reichelt, J.R. (1983) In *Industrial Enzymology* (eds T. Godfrey and J. Reichelt), Macmillan, Byfleet, p. 375.
Robb, D.A. (1995) *ACS Symposium Series*, Vol. 600, p. 159.
Rosenthal, A., Pyle, D.L. and Niranjan, K. (1996) *Enzyme Microbial Technol.*, **19**, 402.
Rudiger, A., Jorgensen, P.L. and Antranikian, G. (1995) *Appl. Environ. Microbiol.*, **61**, 567.
Russell, R.J.M. and Taylor, G.L. (1995) *Curr. Opin. Biotechnol.*, **6**, 370.
Sarkanen, K.V. and Ludwig, C.H. (1971) *Lignins. Occurrence, Formation, Structure and Reactions*, Wiley-Interscience, New York.
Sauvonnet, N., Poquet, I. and Pugsley, A.P. (1995) *J. Bacteriol.*, **177**, 5238.
Schmidtdannert, C., Rua, M.L., Atomi, H. and Schmid, R.D. (1996) *Biochim. Biophys. Acta*, **1301**, 105.
Schoemaker, H.E., Harvey, P.J., Bowen, R.M. and Palmer, J.M. (1985) *FEBS Lett.*, **183**, 7.
Schwartz, R.D. and McCoy, C.J. (1977) *Appl. Environ. Microbiol.*, **34**, 47.
Scopes, R.K. (1982) *Protein, Purification: Principles and Practice*, Springer-Verlag, New York.
Simpson, H.D., Haufler, U.R. and Daniel, R.M. (1991) *Biochem. J.*, **277**, 413.
Takizawa, N. and Murooka, Y. (1985) *Appl. Environ. Microbiol.*, **49**, 294.
Taylor, M.M., Diefendorf, E.J., Foglia, T.A., Bailey, D.G. and Feairheller, S.H. (1989) *J. Am. Leather Chem. Soc.*, **84**, 71.
Tien, M. and Kirk, T.K. (1983) *Science*, **221**, 661.
Tien, M. and Tu, C.P.D. (1987) *Nature*, **326**, 520.
Tomazic, S.J. and Klibanov, A.M. (1988) *J. Biol. Chem.*, **263**, 3092.
Torchilin, V.P., Maksimenko, A.V., Klibanov, A.M. Berezin, I.V. and Martinek, K. (1978) *Biochim. Biophys. Acta*, **522**, 277.
Torchilin, V.P., Trubetskoy, V.S. and Martinek, K. (1983) *J. Molec. Catal.*, **19**, 291.
Ueng, P.P., Volpp, K.J., Tucker, J.V., Gong, C.S. and Chen, L.F. (1985) *Biotechnol. Lett.*, **7**, 153.
Van Brunt, J. (1986) *Biotechnology*, **4**, 611.
Van Griethuysen-Dilber, E., Flaschel, E. and Renken, A. (1988) *Process Biochem.*, **23**, 55.
Vasantha, N., Thompson, L.D., Rhodes, C. *et al.* (1984) *J. Bacteriol.*, **159**, 811.
Ventosa, A. and Nieto, J.J. (1995) *World J. Microbiol. Biotechnol.*, **11**, 85.
Volkin, D.B., Staubli, A., Langer, B. and Klibanov, A.M. (1991) *Biotechnol. Bioeng.*, **37**, 843.
Webb, E.C. (1984) *Enzyme Nomenclature, 1984*, Academic Press, London.
Wells, J.A. and Powers, D.B. (1986) *J. Biol. Chem.*, **261**, 6564.
Wells, J.A., Ferrari, E., Henner, D.I., Estell, D.A. and Chen, E.Y. (1983) *Nucleic Acid Res.*, **11**, 7911.
Wheatley, A.D. (1987) In *Biotechnology of Waste Treatment and Exploitation* (eds J.M. Sidwick and R.S. Holdom) Ellis Horwood, Chichester, p. 173.
White, G.F. and Snape, J.R. (1993) *J. Gen. Microbiol.*, **139**, 1947.
Wilson, S.A., Peek, K. and Daniel, R.M. (1994) *Biotechnol. Bioeng.*, **43**, 225.
Zaborsky, O.R. (1973) *Immobilised Enzymes*, CRC Press, Cleveland, OH.
Zaks, A. and Klibanov, A.M. (1984) *Science*, **224**, 1249.
Zaks, A. and Klibanov, A.M. (1985) *Proc. Natl. Acad. Sci. USA*, **82**, 3192.
Zaks, A. and Klibanov, A.M. (1988) *J. Biol. Chem.*, **263**, 3194.

2 Processes with immobilized enzymes and cells

SEVERIAN DUMITRIU AND ESTEBAN CHORNET

2.1 Current status of immobilized enzyme technology

The immobilization of enzymes is a technique extensively studied since the late 1960s (Silman and Katchalski, 1966). The knowledge base accumulated on enzyme and cells immobilization studies has grown to very large proportions (Klibanov, 1983; Ariga *et al.*, 1993; Crumbliss *et al.*, 1993; Champagne *et al.*, 1994). This wealth of information is one of the primary reasons for the present advances in enzyme engineering. The introduction of immobilized enzyme systems into commercial use, which was slower than predicted, has been the result of numerous factors, such as the long time required for approval of new processes for use in food applications, the need to control microbial contamination in biological reactor systems and some enzyme characteristics that limit the economic success of the immobilization process. The engineering of enzymes with better characteristics will overcome some of the problems encountered that have prevented commercial processes from developing.

Biotechnological applications of immobilized biocatalysts include several fields of general interest, in particular, clinical and analytical chemistry, medicine, food and pharmaceutical technology, organic synthesis, and industrial production of chemical compounds. In the literature there are available many reviews devoted to those subjects (Carr and Bowers, 1980; Klibanov, 1983; Bowers, 1986; Monsan and Combes, 1988; Tramper, 1990; Dumitriu, 1991; Ariga *et al.*, 1993; Mustranta *et al.*, 1993; Reddy and Shankar, 1993; Northon and Vuillemard, 1994; Fernandez-Lafuente *et al.*, 1995; Toshifumi *et al.*, 1995). Enzymatic reaction mechanisms (Doubradi *et al.*, 1980; Amarant and Bohak, 1981; Martinek and Moshaev, 1985; Bender, 1987; Evnin and Craik, 1988; Lortie *et al.*, 1993; Ganapathi *et al.*, 1995) and allosteric regulation mechanisms (Bickerstaff, 1982; Chiancone and Gattoni, 1987) have also been elucidated through the immobilization of the investigated enzymes.

2.1.1 Advantages and disadvantages of enzyme and cell immobilization

The use of freely suspended enzymes has the disadvantages that:

1. Most enzymes are labile under normal operating conditions and therefore have only a limited lifetime.

2. Since enzymes are water soluble, they are difficult to separate from their substrates and products. Therefore their reuse is difficult, which increases the cost of the process considerably.

The development of immobilized enzyme processes has the following advantages:

1. The immobilization of an enzyme into water-insoluble particles has been shown to increase its stability considerably.
2. The ability to separate the enzyme easily from the products and substrate allows its reuse in a continuous process.

The disadvantage of using immobilized enzymes is that the heterogeneous nature of such catalysts can impose diffusion limitations which can affect their overall activity.

The use of immobilized cells in place of normally grown cells has similar advantages to that of immobilized enzymes:

1. Batch fermentation can be replaced by continuous reactors, which increase fermenter productivity by allowing the use of high flow rates in continuous operations while avoiding wash out.
2. Immobilized cells allow the use of considerably higher cell density.
3. Many metabolites or enzymes are active only in resting or stationary phase cells; in an immobilized system, cells can be maintained in this state.
4. Biological catalysts can be reused.
5. Interfacial inactivation can be prevented.
6. Immobilized cells provide protection against a turbulent environment.
7. Immobilized cells allow the use of improved methods for process control and methods for product recovery (continuous extraction).

Among the techniques for immobilizing living cells, physical entrapment in porous granular matrices is favored by numerous authors (Scott, 1987; Philips and Poon, 1988; Tanaka and Nakajima, 1990).

The disadvantages of using immobilized cells are similar to those for the enzymes and are mainly related to system diffusion limitations.

The advantages and disadvantages of immobilized enzymes compared with immobilized cells depend upon the system involved, but can be summarized as follows:

1. The use of immobilized cells removes the need to extract and purify the enzyme.
2. Often, the whole cell system is less sensitive to changes in operating conditions such as pH.
3. Immobilized cells allow a high loading of the support. With isolated enzymes, high loading may reduce activity owing to protein–protein interaction.

4. If the reaction system involved requires many enzymes and cofactor renewals, the immobilized cell system is the best solution.

2.1.2 *Immobilization of microorganisms or enzymes?*

Many extracellular microbial enzymes are produced in quantities large enough to be used in industrial processes. However, the cost of isolating and purifying intracellular enzymes for commercial processes can affect the profitability sufficiently for their use to become prohibitive. Nonetheless, there are advantages to the isolated enzymes that must be balanced against cost based on the nature of the conversion process. Isolated enzymes offer greater purity and the possibility of modification which, in turn, may lead to higher conversions and yields and less contamination. Immobilized microorganisms containing the catalyst of interest combine the advantages inherent with the use of immobilized enzymes with those of microbial fermentation processes.

Immobilizing the whole cell should be considered when the extracted enzyme is unstable, since retaining the enzyme within its natural surroundings preserves its stability. Additional considerations for cell immobilization are that:

1. no interfering side reactions should be present within the cells (although often these can be deliberately inactivated);
2. the cells retain greater enzyme activity than is possible with immobilized enzymes;
3. the reactor volume per unit of product is much smaller for the continuous, immobilized cells method than for the conventional fermentation process.

A conceivable shortcoming could be encountered by the resistance of the cell wall and the cell membrane to the substrate and the product's transport.

2.2 Immobilization procedures

2.2.1 *Carriers*

In choosing the support material, one should realize that the characteristics of the final immobilized enzyme preparations strongly depend on this material. The ideal support material for the immobilization of enzymes should have the following characteristics:

1. large specific surface and sufficient permeability;
2. relatively high diffusion coefficient for substrate, after immobilization;

Table 2.1 Matrices for enzyme immobilization

Organic	Inorganic
Cellulose	Nonporous glass
Starch	Controlled pore glass
Dextran	Bentonite
Chitin and chitosan	Kieselgur
Carrageenan	Metals
Xanthan	Porous titanium
Agar and agarose	Porous silica
Silk	Silane derivative of inorganic carriers
Collagen	
Gelatin	
Synthetic polymers	
Polyacrylates and polymethacrylates	
Poly(maleic anhydride)	
Polyamides	
Cellulose derivatives	

3. presence of functional groups to which the enzyme can be attached under mild conditions;
4. hydrophilicity;
5. insolubility;
6. chemical, mechanical and thermal stability;
7. rigid particles of a proper shape;
8. resistance to microbial breakdown;
9. regenerability in case of expansive support material;
10. general recognition as a safe material.

Carriers can be classified according to their chemical composition as organic and inorganic supports (Table 2.1).

2.2.2 Methods of immobilization

In general, six different types of immobilization methods are known (Figure 2.1):

1. binding to carriers by adsorptive interactions;
2. entrapment in gels, beads or films;
3. crosslinking or co-crosslinking with bifunctional reagents;
4. encapsulation in microcapsules or membranes;
5. binding to carriers by covalent bonds.

(a) Immobilization by noncovalent procedures

Adsorption and ion exchange. Adsorption is based on the physical adsorption or ionic binding, or both, of the enzyme to the surface of the

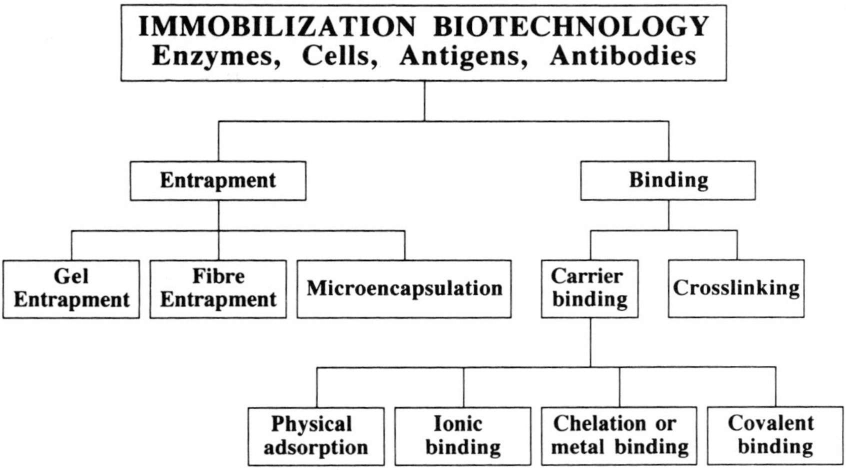

Figure 2.1 Immobilization methods.

support. Immobilization by physical adsorption or ionic binding, both low-cost procedures, is simple and effective, and, unlike covalent procedures, usually brings about little change in the overall conformation of the enzyme or of the active site.

Entrapment in a matrix. The entrapping method is based on confining enzymes or cells in the lattice of a polymer matrix or enclosing them in semipermeable membranes. For this method, the following matrices are employed: agar, alginate, chitosan, xanthan, carrageenan, gelatin, collagen, polyacrylamide, polyurethanes and a few additional polymeric matrices (Chibata *et al.*, 1986).

The entrapment techniques currently used may be grouped into the following three basic categories:

- Gel formation by ionic crosslinking of a ionic polymer (xanthan, chitosan, alfinate, carrageenan, etc.).
- Gel formation by cooling a heated polymer solution (agar, agarose, etc.).
- Gel formation by a chemical crosslinking.

Calcium alginate has been chosen as an important and widely used entrapping agent because of its good characteristics (high carrier activity, availability in quantity, low cost of immobilization, ease of scale-up of operation, mechanical strength for length of operation).

Immobilization in polyelectrolyte complex. Ionotropic gelation of polymers is applied to water-soluble polyelectrolytes. When mixed with the appropriate, usually multivalent, counterions, solidification by polysalt formation occurs (Tsuchida and Abe, 1986; Kim and Rha, 1989; Mireles *et al.*, 1992).

The polymer–counterion systems which have been successfully applied to whole cell immobilization are summarized in Table 2.2. The large range of possible utilizations of these systems in the treatment of residual waters (phenol degradation) as well as in the productions of ethanol should be noted (Table 2.3).

The complex formation process between polycations and polyanions has been used for the recovery of proteins (Sheih and Glatz, 1994; Taravel and Domard, 1994; Xia and Dubin, 1994) and other bioproducts (Chavasit *et al.*, 1988). During the 1990s, increasing attention has been given to polyelectrolyte coagulation (Niederauer and Glatz, 1994; Wang *et al.*, 1996) to aid the separation of colloidal and dispersed particles from food processing wastes (Green and Kramer, 1979; Kargi and Shuler, 1980; Dubin *et al.*, 1994).

Several types of encapsulation, including the formation of polyelectrolyte complexes and thermoplastic permselective membranes, have been developed (Vorlop and Klein, 1987; McKnight *et al.*, 1988; Natthew *et al.*, 1993; Burgess, 1994; Zielinski and Aebischer, 1994; Crescenzi *et al.*, 1995) (Table 2.4).

The complex coacervation concept is attractive because of its simplicity, gentle nature and the availability of a large number of biocompatible ionic polymers and polysaccharides capable of interacting to form membranes.

Polyelectrolyte complex formation between chitosan and polyanions, such as alginates (Daly and Knorr, 1988), heparin (Kikuchi and Noda, 1976), carboxymethyl cellulose (CMC) (Fukuda, 1980), xanthan (Dumitriu *et al.*, 1994; Chu *et al.*, 1995, 1996; Ikeda *et al.*, 1995; Dumitriu and Chornet, 1996a,b), acidic glycosaminoglycans (Hirano *et al.*, 1978) and dextran sulfate (Kikuchi and Fukuda, 1974), has been reported previously.

We have immobilized protease (EC 3.4.2.1.19) in a hydrogel obtained by complexation of xanthan with chitosan. The protease immobilization yield can be as high as 98%, this being a function of the concentration of protease dissolved in the xanthan solution (Figure 2.2). The activity of the immobilized protease depends on the concentration of protease used within the xanthan solution (Figure 2.3). The decrease of activity observed for the range of protease concentrations between 0.96 and 1.88% may be explained as follows:

1. The hydrogel has a limited binding capacity in the presence of large quantities of enzyme which results in decreased immobilization (Figure 2.2) and, consequently, a decreased amount of activity.
2. The protease in the hydrogel exists in two forms: one complexed with xanthan–chitosan while the other is free. It is possible that the protease

Table 2.2 Ionic polymers and related counterions in the preparation of ionotropic gels for the entrapment of whole cells

Polyions	Counterions
Alginate–COO^-	Ca^{2+}, Al^{3+}, Zn^{2+}, Co^{2+}, Ba^{2+}, Fe^{2+}
Pectin	Ca^{2+}, Al^{3+}, Zn^{2+}, Co^{2+}, Mg^{2+}
Carboxymethylcellulose	Ca^{2+}, Al^{3+}, Ti^{4+}
Guar-guar	Ca^{2+}, Al^{3+}
Carrageenan–SO_3^{2-}	K^+, Ca^{2+}
Furcellaran	K^+, Ca^{2+}
Cellulose sulfate	K^+
Chitosan–NH_3^+	Polyphosphates Xanthan Alginic acid Hyaluronic acid Polypeptide

Table 2.3 Application of ionotropic gels for the entrapment of whole cells

Matrix	Cells	Reaction	Reference
Calcium alginate	*Saccharomyces cerevisiae*	Glucose/ethanol	Kierstan and Bucke (1977), Galazzo and Bailey (1990), Vives *et al.* (1993), Roca *et al.* (1996)
	Kluyveromyces marxianus	Glucose/ethanol	Nolan *et al.* (1994), Riordan *et al.* (1996)
	Trichosporon pullulans	Cellobiose/glucose	Adami *et al.* (1988)
	Pichia etchellsii	Cellobiose/glucose	Jain and Ghose (1984)
	Lactobacillus helveticus	Lactic acid	Roy *et al.* (1987), Boyaval and Goulet (1988)
	Rhizopus orizae	L(+)-Lactic acid	Hang *et al.* (1989)
	Aspergillus awamori and *Streptococcus lactis*	Starch/L-lactic acid	Kurosawa *et al.* (1988)
	Scenedesmus bicellularis	Ammonium and orthophosphate removal	Kaya and Picard (1995)
Chitosan/alginate	*S. cerevisiae*	Glucose/ethanol	Vorlop and Klein (1983)
Aluminium alginate	*C. tropicalis*	Phenol degradation	Hackel *et al.* (1975)
Ca,Al carboxy-methylcellulose	*C. tropicalis*	Phenol degradation	Klein *et al.* (1979)

Table 2.4 Ionic polymers used in microencapsulation techniques

Polysaccharides
Chondroitin sulfate A
Heparin
Alginate
Hyaluronate
Chitosan
Dextran sulfate
Polygalacturonate
Pectin
Synthetic and artificial polymers
Polyacrylate
Poly-L-lysine
Polyethyleneimine
Carboxymethylcellulose
DEAE-dextran
Poly(1-hydroxy-1-sulfonate-2-propene)
Polyphosphate
Poly(styrene sulfonic acid)
Poly(ethylene sulfonic acid)
Poly(vinyl sulphate)

may participate in autoproteolysis resulting in a greater mobility. This process of autoproteolysis has been observed for all proteases immobilized by inclusion in these types of gel (Kasche, 1989).

The thermal stability of the immobilized protease is quite different from the free form of the enzyme (Figure 2.4). Identical behavior is observed for the immobilized protease in a polyacrolein microsphere (Hayashi and Ikada, 1990a).

Protease and xylanase were both immobilized by inclusion in a xanthan–chitosan hydrogel. The co-immobilization was carried out in order to obtain an enzymatic system capable of hydrolysing protein and xylan present in the waste waters of the food industry. Protease activity increases with longer incubation times without any observable inhibition by urea (Figure 2.5). Proteolytic activity is a function of the concentration of the amount of both protease and xylanase that are co-immobilized. When the concentration of xylanase is set at 1%, an increase of the protease activity is observed as its concentration increases to 1%, after which the activity decreases (Figure 2.5).

The co-immobilization of protease and xylanase causes a synergistic effect; the proteolytic activity is increased owing to the presence of the xylanase (Figure 2.6). The synergy observed on the protease activity by the xylanase causes an increase of up to 85% for a ratio of protease/xylanase = 1/1 (g/g). This effect may be explained by the possibility that xanthan–chitosan–xylanase–protease interaction in the hydrogel stabilizes the

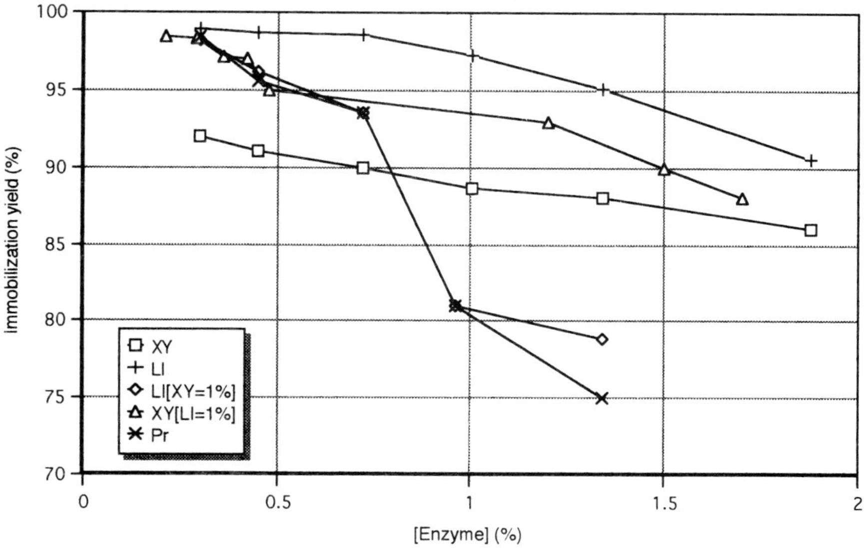

Figure 2.2 Variation of the immobilization yield as a function of enzyme concentration in the xanthan solution. XY = Xylanase; Li = lipase; Pr = protease.

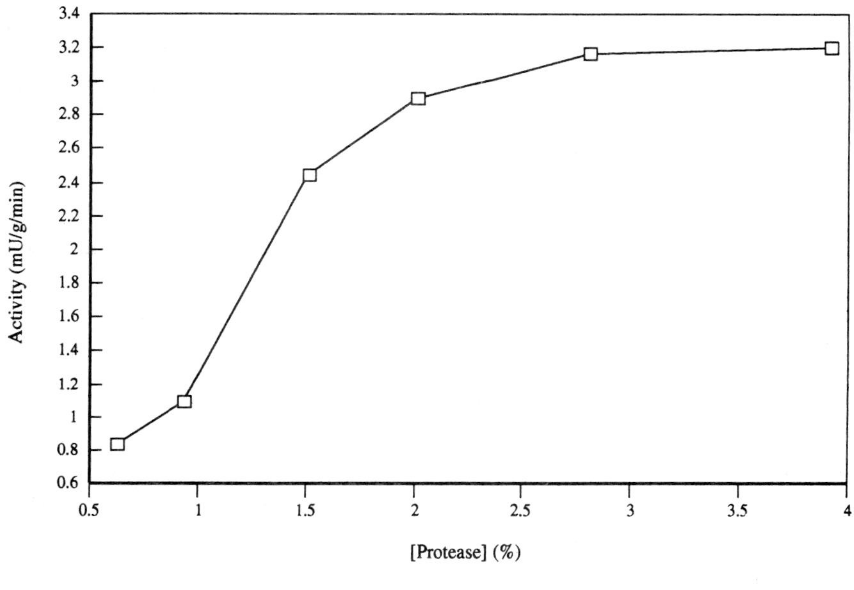

Figure 2.3 Variation of the activity of the immobilized protease as a function of the protease concentration in the xanthan solution.

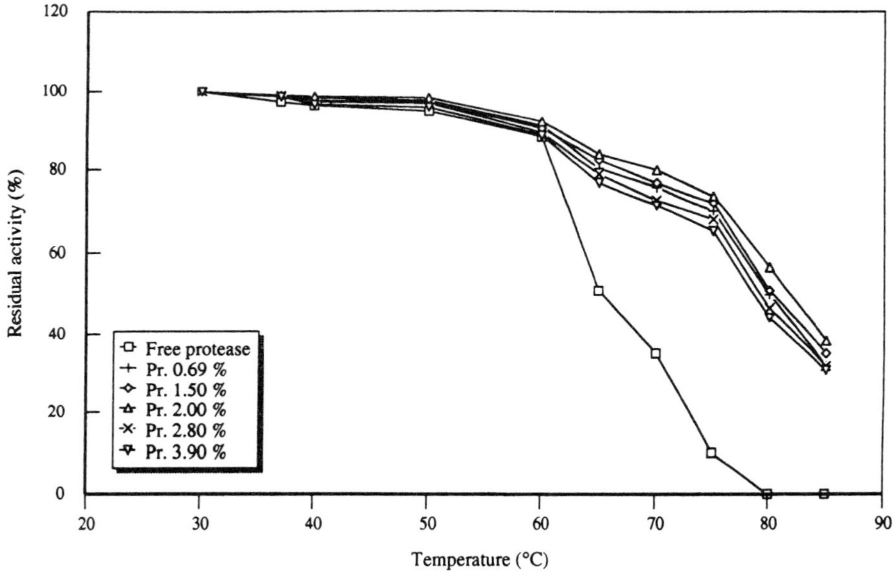

Figure 2.4 Effect of heat treatment, at a given temperature at pH 5.6 for 1 h, on the residual activity (RA) of hemoglobin hydrolysis at pH 7.2 at 37 °C. Pr. = protease.

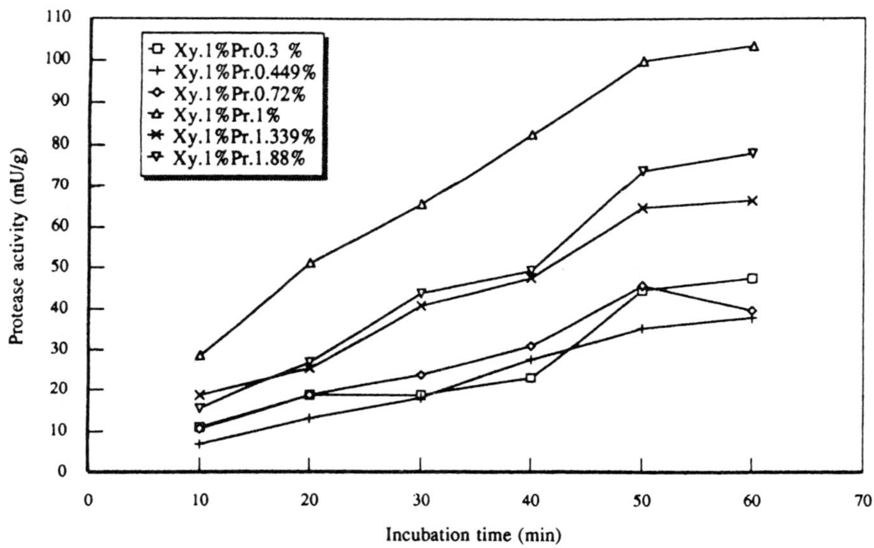

Figure 2.5 Variation of the protease (Pr.) activity co-immobilized with xylanase (Xy.) as a function of incubation time. [Xylanase] = 1%; temperature = 37 °C.

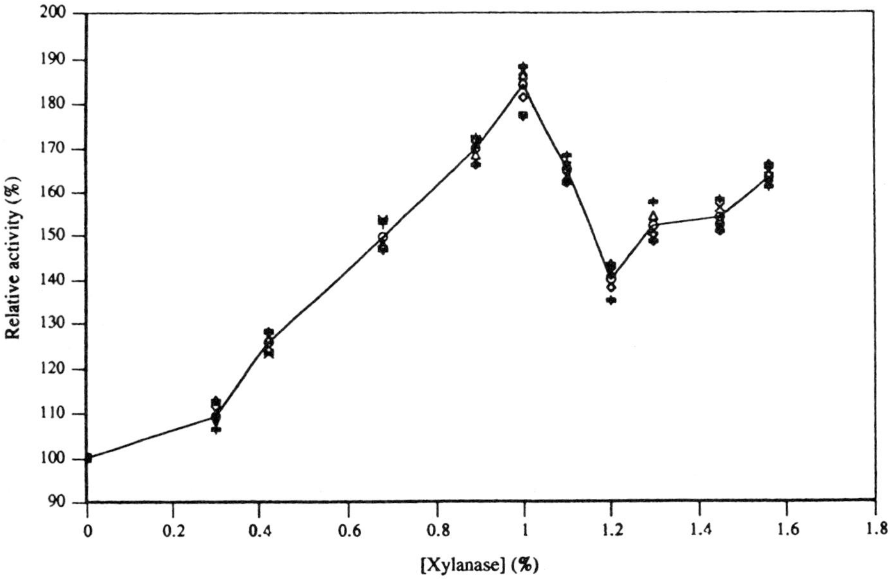

Figure 2.6 Relative protease activity as a function of varying concentrations of co-immobilized xylanase. [Protease] = 1%; incubation time = 50 min.

protease structure. A similar effect has been observed in the co-immobilization of protease with peroxidase (Grzywnowicz *et al.*, 1983). At the same time chitosan may itself activate immobilized enzymes as has been demonstrated for protease.

The xylanase activity in this protease–xylanase system is negatively influenced by the presence of protease (Figure 2.7). For every tested xylanase concentration (in the presence or absence of protease) there is an increase in the xylanase activity as compared to the free enzyme, but the protease present in the system reduces xylanase activity without, however, going lower that the activity of the free xylanase. This is observed especially for the large concentrations of xylanase used when the xanthan–chitosan–protease–xylanase matrix is saturated with xylanase. In this case an important quantity of xylanase remains in the hydrogel which allows the protease, also free in the system, to function. The proteolytic–xylanasic activity in series is dependent on xylanase incubation time (Figures 2.8 and 2.9). There is an important decrease in proteolytic activity as incubation times for the xylanase reaction increase, especially over 20 min.

We have studied the properties of immobilized xylanase in polyionic hydrogels. The amount of activity of the immobilized xylanase is proportional to the enzyme concentration in the xanthan solution (Table 2.5). A maximal temperature between 85 and 90 °C is found for

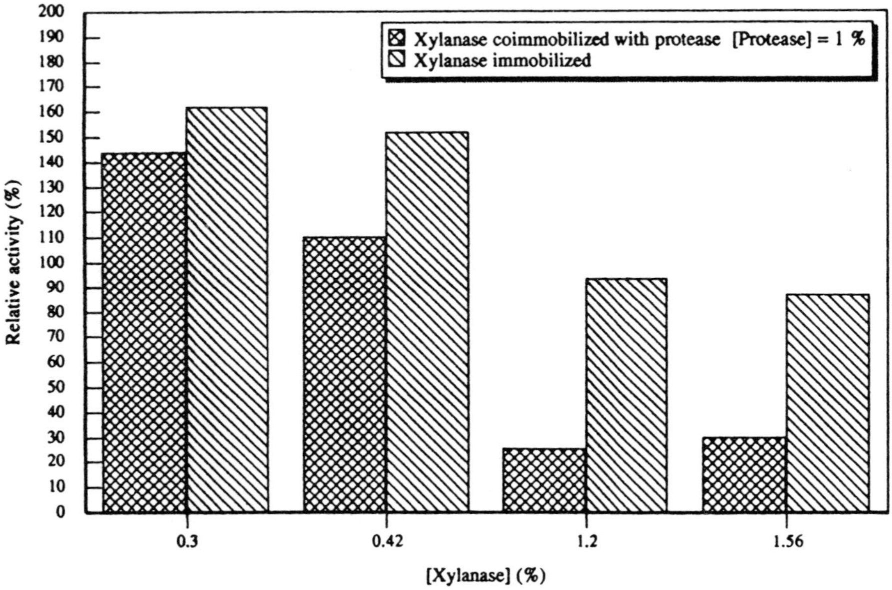

Figure 2.7 Relative xylanase activity as a function of varying concentrations of xylanase in the system. [Protease] = 1%; substrate = Remazol Brilliant Blue (RBB)–xylan; incubation time = 60 min; temperature = 30 °C.

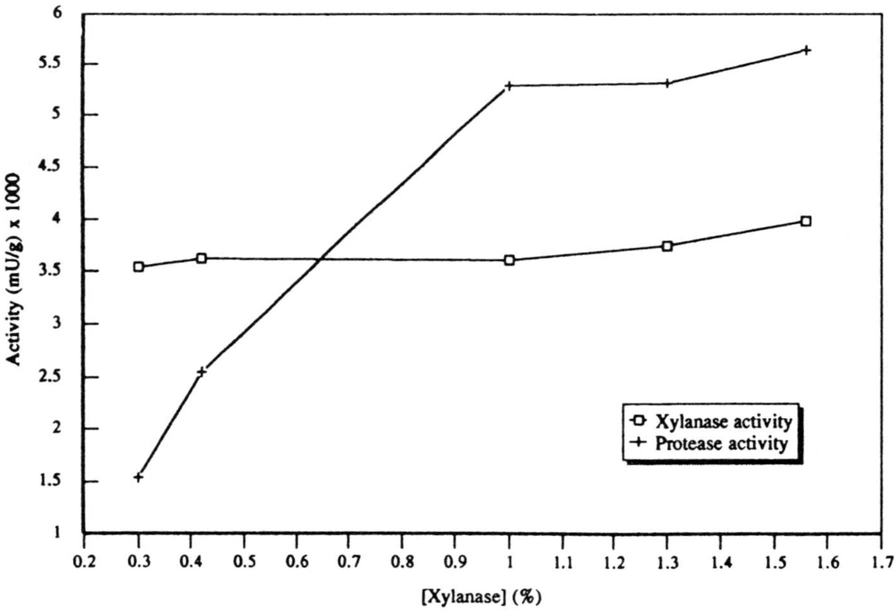

Figure 2.8 Serial variation of protease and xylanase activities as a function of varying concentrations of xylanase in the hydrogel. [Protease] = 1%; incubation time for xylanase reactions = 60 min; temperature = 30 °C; substrate = RBB–xylan; incubation time for protease reactions = 10 min; temperature = 37 °C; substrate = hemoglobin.

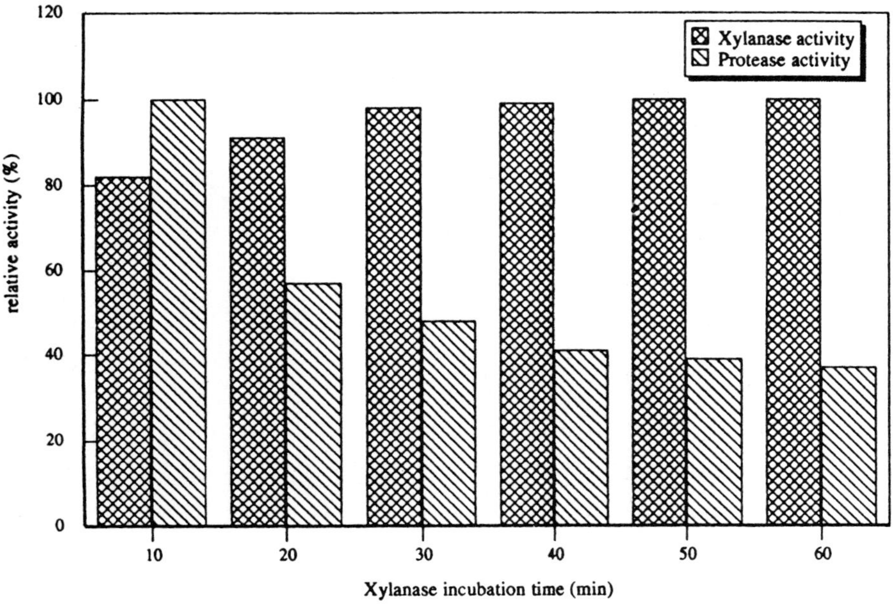

Figure 2.9 Variations of relative protease and xylanase activities as a function of xylanase reaction incubation times. [Protease] = 1%; [xylanase] = 1%. Incubation times for protease reactions = 50 min at 37 °C; substrate = hemoglobin.

Table 2.5 Activity of the immobilized xylanase as a function of its concentration in the xanthan solution

[Xylanase] in xanthan solution (%)	Observed activity (mU g^{-1})	Expected activity (mU g^{-1})	Increase in activity[a] (%)
0.21	8 889	5 250	+69
0.29	11 737	7 250	+61
0.36	16 000	9 000	+77
0.42	15 927	10 500	+51

[a]Calculated as a function of initial activity of xylanase (2.5×10^6mU g^{-1}). Reaction time in the chitosan solution = 10 min.

immobilized xylanase (Figure 2.10) while for free xylanase this value is between 40 and 50 °C. The values for the Michaelis–Menten (K_M) constants depend on the xylanase concentration in the hydrogels (Figure 2.11) and are greater in comparison with the free enzyme.

Images obtained by electron microscopy of the xanthan–chitosan matrices, with or without xylanase, show a fibrillar structure in which

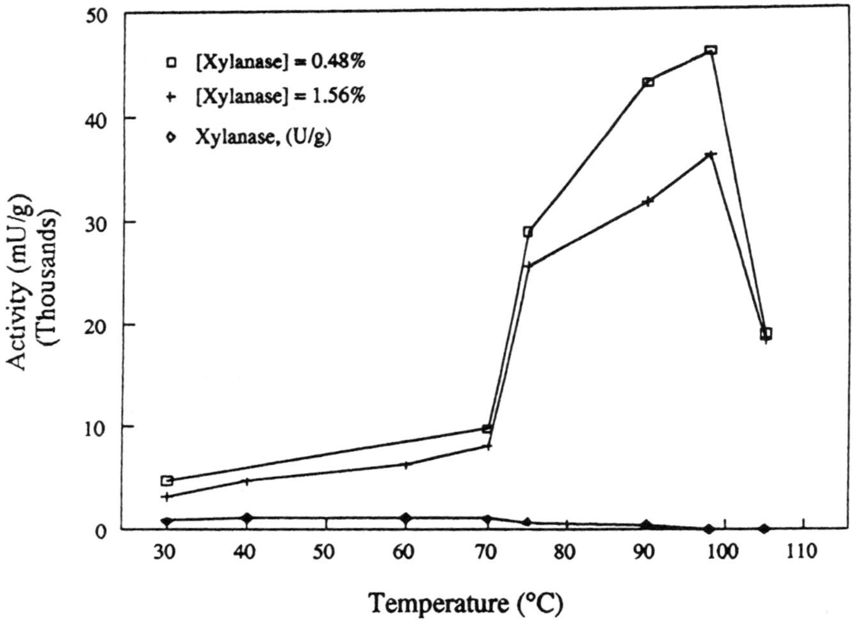

Figure 2.10 Variation of activity of immobilized or free xylanase as a function of temperature.

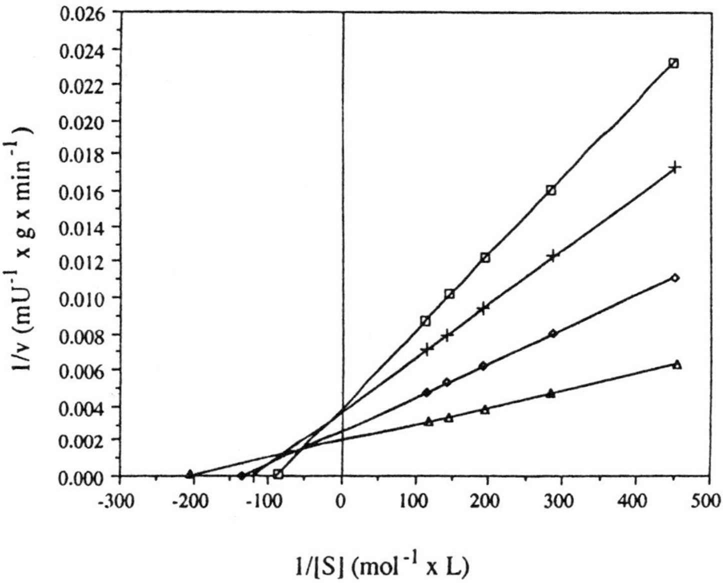

Figure 2.11 Lineweaver–Burk representation of the immobilized xylanase in the xanthan–chitosan matrix. □, [Xylanase] = 0.69%, K_m = 88.52 M; +, [xylanase] = 0.87%, K_m = 118.82 M; ◊, [xylanase] = 1.00%, K_m = 133.45 M; △, [xylanase] = 3.26%, K_m = 207.31 M.

globular formations are lodged when xylanase is present (Figure 2.12). These globular formations are likely to be formed by the triple complexation between xanthan, chitosan and xylanase. In a cross-section of the beads lamellae of fibrils in a layer approximately 4 μm thick followed by an unorganized fibrous structure have been observed. Because of the mass transfer resistance of hydrophobic substrates into the pores of the matrix the immobilization of lipase represents an interesting challenge in enzyme technology. So far, the reported activities of the immobilized lipases are considerably lower than those reported for other enzymes.

The immobilization of lipase and lipase–xylanase in xanthan–chitosan hydrogels were studied in order to determine the immobilization capacity of this type of gel and to attempt to obtain a biocatalyst capable of hydrolysing lipids and hemicellulose present in food and industrial waste waters. The immobilized lipase is quite stable in this type of hydrogel. The loss of activity is approximately 17% after 100 min of washing for those samples containing a high concentration of lipase (1.8%) and 1.8% for those samples containing 0.35% lipase. The activity of the immobilized lipase in an emulsion of olive oil depends on reaction time, temperature and pH (Figures 2.13–2.15). According to these figures, the optimal conditions for lipase activity in aqueous medium are incubation time = 10 min, incubation temperature = 37 °C, pH = 7.5.

The co-immobilization of lipase with xylanase changes the properties of both enzymes. For lipase we observe different kinetics with inhibition occurring after 14 min of incubation (Figure 2.16). Moreover, as the concentration of co-immobilized xylanase increases, this causes a synergy of the lipase activity. In an organic medium the activity of the co-immobilized lipase decreases (Figure 2.17). Xylanase activity, in a system co-immobilized with lipase, is increased (Figure 2.18).

A polyelectrolyte complex prepared from xanthan and chitosan was applied to the immobilization of Corynebacterium glutamicum (Chu *et al.*, 1996) having fumarase activity. The fumarase activity of immobilized cells was about five times that of intact cells. It was suggested that the interaction between the chitosan and the *C. glutamicum* cells contributed to the enhancement of fumarase activity.

Lloyd-George and Chang (1993, 1995) have studied the use of whole-cell tyrosine phenol–lyase activity of *Erwinia herbicola* immobilized for the conversion of ammonia and pyruvate along with phenol or catechol, respectively, into L-tyrosine or dihydroxyphenyl-L-alanine (L-dopa). Whole cells of *E. herbicola* were encapsulated within alginate–polylysine–alginate microcapsules. The apparent kinetic parameters for the production of tyrosine by whole-cell tyrosine phenol–lyase (EC 4.1.99.2) were found for both free and microencapsulated cells.

When the monovalent ion of sodium from alginate is replaced by divalent or trivalent cations, ionic crosslinking among the carboxylic acid

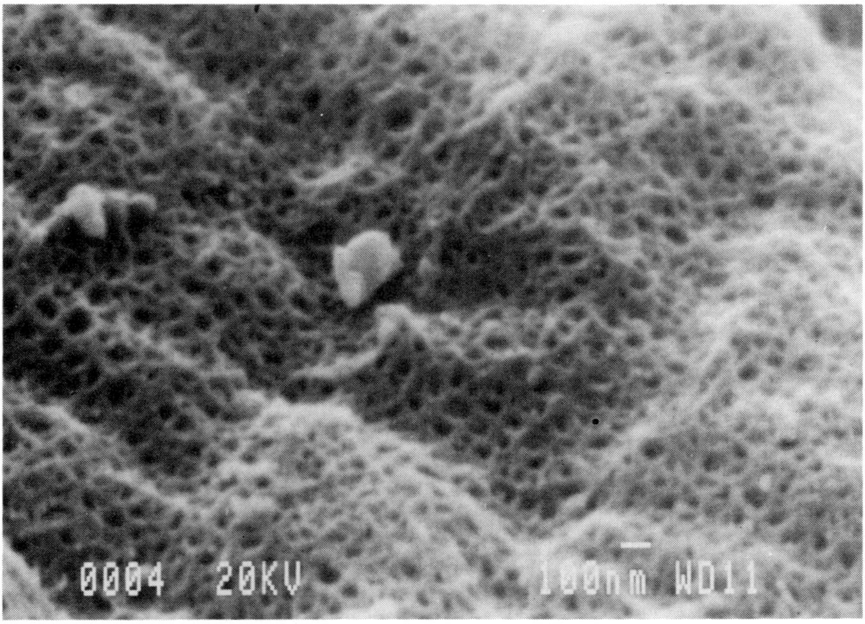

a

b

Figure 2.12 Scanning electron microscopy of hydrogels with or without immobilized xylanase. (a) External surface of xanthan–chitosan beads; (b) external surface of the hydrogel xanthan–chitosan beads containing immobilized xylanase, [xylanase] = 1.56%;

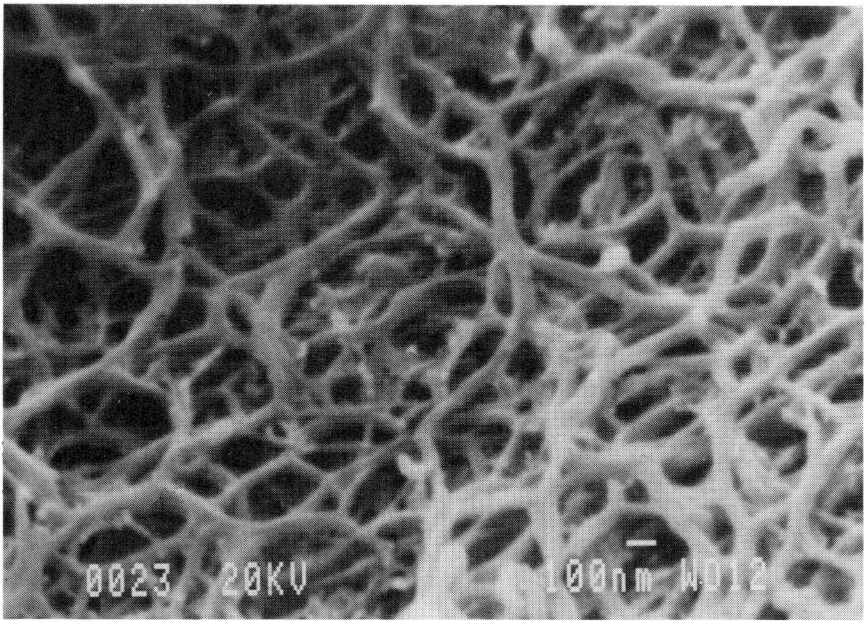

c

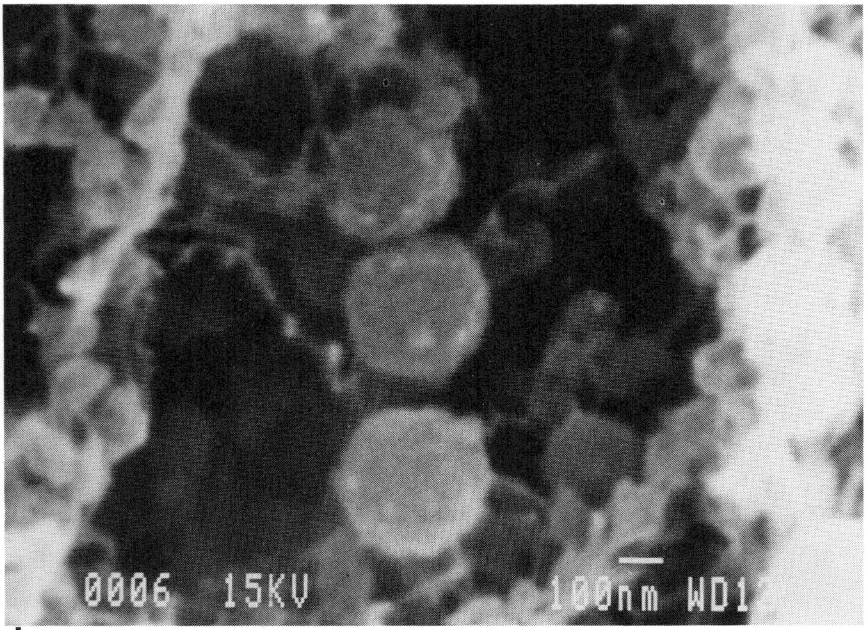

d

Figure 2.12 (*continued*) (c) internal structure of the hydrogel xanthan–chitosan beads; (d) internal structure of the hydrogel xanthan–chitosan beads containing immobilized xylanase, [xylanase] = 1.56%.

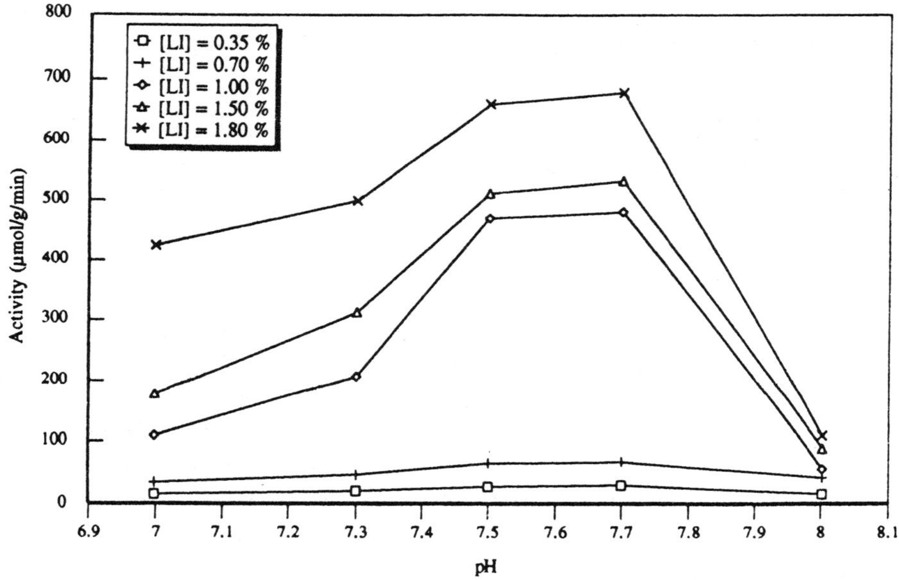

Figure 2.13 Reaction rate of immobilized lipase (LI) in the polyionic hydrogel as a function of the pH of the olive oil emulsion. Incubation time = 10 min; temperature = 37 °C.

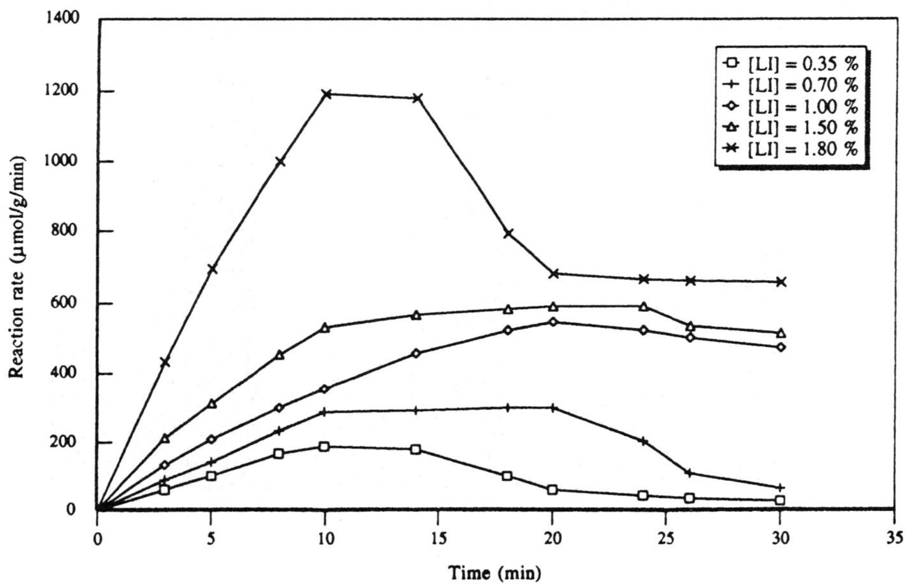

Figure 2.14 Reaction rate for immobilized lipase (LI) in the polyionic hydrogen as a function of incubation time. Incubation temperature = 37 °C, pH = 7.5.

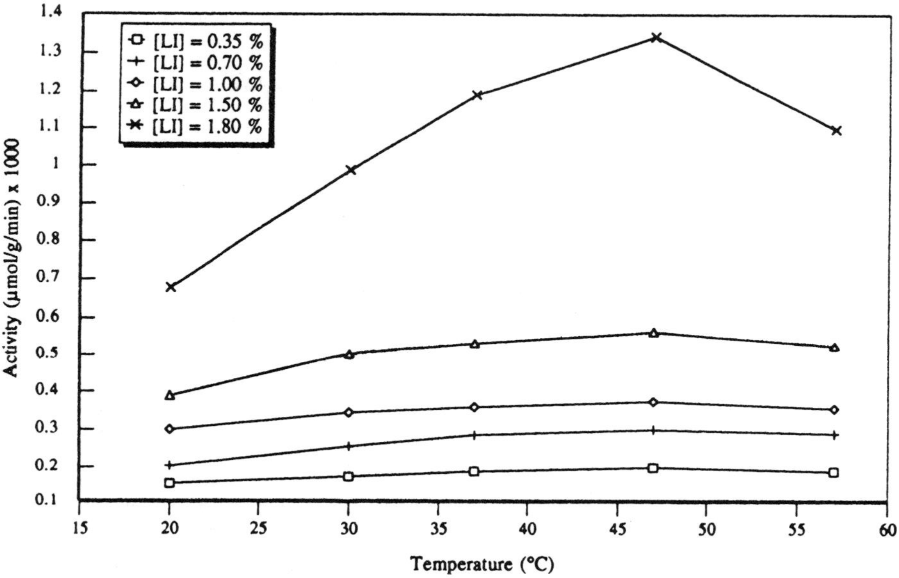

Figure 2.15 Reaction rate for immobilized lipase (LI) in the polyionic hydrogel as a function of incubation temperature. Incubation time = 10 min, pH = 7.5.

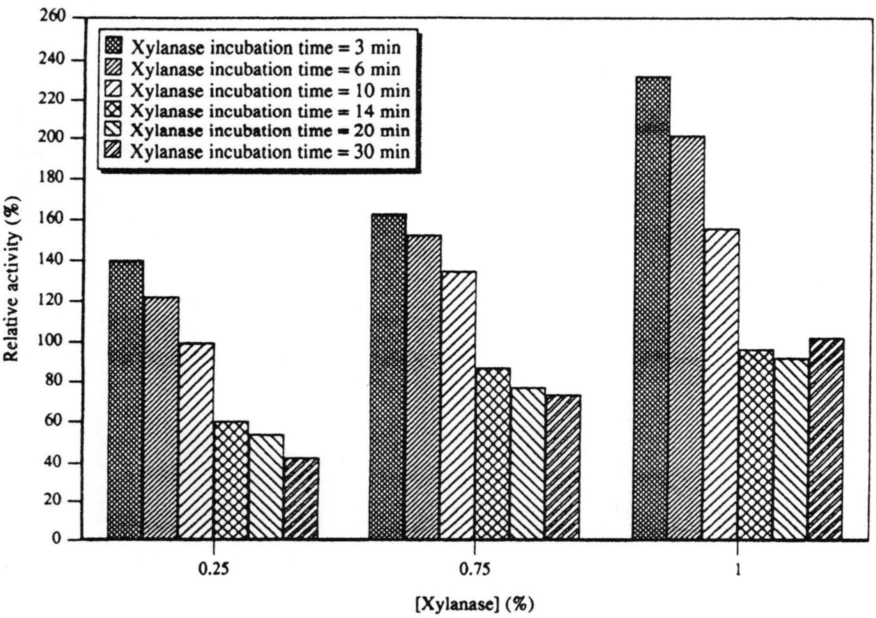

Figure 2.16 The relative activity of lipase as a function of the concentration of co-immobilized xylanase and incubation time. [Lipase] = 1%; incubation temperature = 37 °C; substrate = olive oil emulsion.

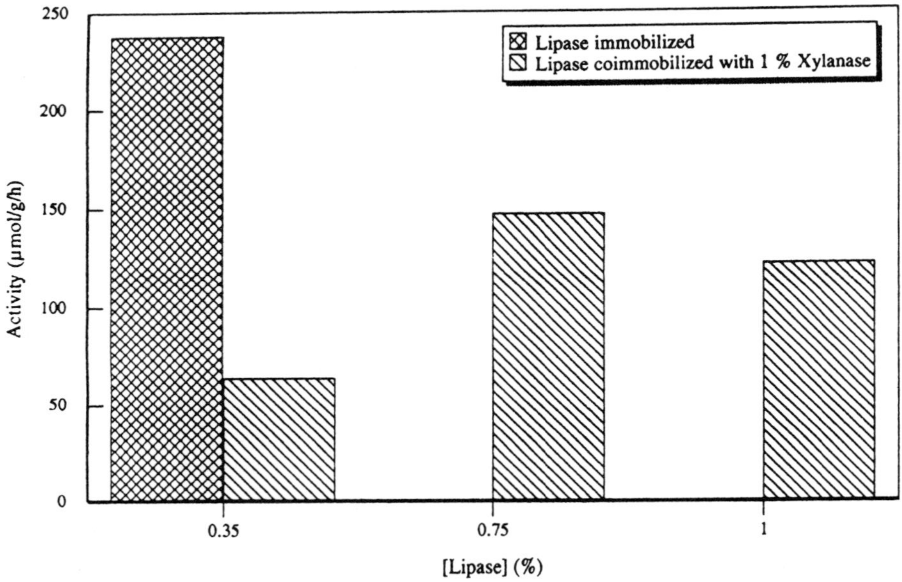

Figure 2.17 Lipase activity in the lipase–xylanase system in iso-octane medium. [Xylanase] = 1%; [olive oil] = 20.27%.

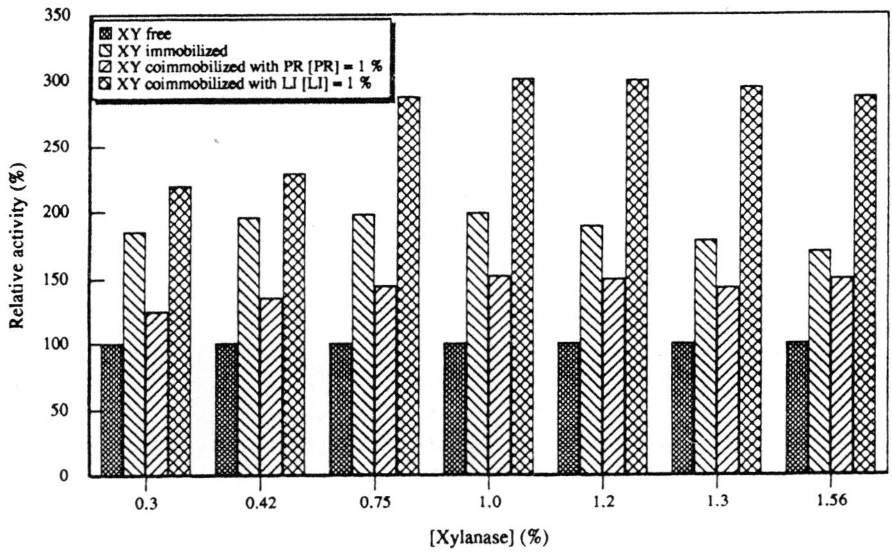

Figure 2.18 Activity of the xylanase co-immobilized at different concentrations with lipase (LI) in a xanthan–chitosan matrix. Substrate = RBB–xylan; incubation temperature = 30 °C; incubation time = 30 min. PR = protease.

groups occurs and the polysaccharide molecules form a polymeric network (Smidsord and Haug, 1972; Martinsen *et al.*, 1989). Entrapment of cells in beads of calcium alginate is one of the most widely used techniques (Brodelius *et al.*, 1979; Brodelius and Nilsson, 1980; Bucke, 1987; Guiseley, 1989; Ogbonna *et al.*, 1989). Chibata *et al.* (Tanaka *et al.*, 1977; Tosa *et al.*, 1979; Chibata *et al.*, 1987) have used carrageenan as the matrix for cell immobilization.

It has been reported that urease (Ristau *et al.*, 1985) and invertase (Mansfeld *et al.*, 1991) can be encapsulated in polyelectrolyte complex membranes from cellulose sulfate and poly(dimethyldiallylammonium chloride). Cytochrome C, liver microsomes and hemoglobin have also been successfully immobilized by this method (Pommerening *et al.*, 1983). Invertase and *Yarrowia lipolytica* cells were also co-immobilized in polyelectrolyte complex (Mansfeld *et al.*, 1995). The co-immobilized enzyme/cell system was used to produce citric acid.

***Encapsulation or confinement in a membrane structure*.** Gels of, for example, calcium alginate, polyionic complexes and starch are formed in the presence of the enzyme. The enzyme is captured within the polymer matrix; substrates and reaction products move relatively freely. Entrapment is probably the most popular method of immobilization of microbial cells, although other methods have been successfully utilized in recent years (Sharma *et al.*, 1986; Kaul *et al.*, 1987; Hunik *et al.*, 1994; Danzing *et al.*, 1995). Materials like alginic acid and carrageenan have also been employed. They form beads when the cell-containing solution of these materials is extruded into salt solutions such as calcium chloride.

Cell entrapment in a matrix may be achieved by applying the following methods (Klein *et al.*, 1984):

- gelation of polymers;
- precipitation of polymers;
- ionotropic gelation of polymers;
- polycondensation;
- polymerization.

Gelation is a temperature-controlled phase transition in polymer–solvent systems, where a homogeneous polymer solution is tranferred to a homogeneous gel without change in composition. Immobilization through polymer precipitation employs concentrated solutions of polymers containing cells that are coagulated by addition of nonsolvent. Solutions of cellulose triacetate cellulose (Dinelli, 1972; Hackel *et al.*, 1975; Linko *et al.*, 1980) are commonly employed.

***Microencapsulation*.** The techniques used to obtain microencapsulated enzymes can be classified as follows:

1. phase-separation method;
2. interfacial polymerization method;
3. liquid-drying method;
4. liquid–surfactant–membrane method;
5. immobilization in liposomes.

Formation of composite granules. The process of cell association may be stimulated by the use of a skeleton creating zones protected from attrition by the fluid and favoring accumulation of microorganisms. For instance, mesh particles made of stainless steel wire allow accumulation of mycelium into beads with a diameter of about 6 mm and a porosity of the order of 80%, which can be used in a stirred reactor. This method is applicable to bacteria.

Another example is provided by immobilization of cells occurring in a continuous process of anaerobic digestion and methane production (Binot, 1983). After introduction of glass beads in the fermenter, the formation of a coating on their surface takes place progressively following the main steps already observed in natural environments, for instance, in marine fouling:

- adsorption of polymeric substances forming a continuous film;
- retention of certain microbial species;
- increase in the number of species and the biomass held on the surface.

The particles tend to agglomerate and are maintained together by the growing biomass. The granules formed may reach a size of 1–3 mm and contain a few thousand glass particles. Within the granule there exists a certain spatial organization with a network of channels which gives a spongy appearance to the granule. The granules are maintained in a fluidized mode and follow a cyclic process of destruction and reconstruction. As compared with a homogeneous stirred digester, this method of immobilization has allowed:

- obtaining a ratio of 10–100 (instead of 1) between the biomass residence time and the fluid residence time;
- increasing by a factor of about 10 the biomass concentration in the reactor;
- increasing by a factor of about 10 the rate of methane production per reactor volume unit.

Cell flocculation is ruled by physical processes similar to those mentioned for cell adhesion. It is very well known in brewing, and provides a direct connection between continuous fermentation and immobilization of a cell used as biocatalyst. Flocculation is the only mode of immobilization that is well adapted to production requiring a high rate of multiplication.

Flocculation is strongly dependent on the physicochemical properties of the cell surface and on the composition of the medium. It may be promoted by adsorption of various substances (ions, polyelectrolytes, polymers) or by the presence of particles of opposite surface charge (heteroflocculation), which are brought as such (Kayem and Rouxhet, 1983) or are produced in contact with the cells (precipitation of metallic hydroxides).

Techniques of microbial and animal cell immobilization. Immobilization of microorganisms on the surface of solids was an ancient practice used for the manufacture of vinegar and in trickling filters for the treatment of drinking water. Furthermore, the immobilization of living cells can, in specific circumstances, provide a useful means for carrying out biochemical syntheses (Kerel *et al.*, 1985). A wide range of techniques has been developed in which cells can be immobilized (Barbotin *et al.*, 1992):

- binding to an organic or inorganic support matrix;
- entrapment in porous polymers or microcapsules;
- flocculation (self-aggregation of the biomass).

(a) Covalent attachment. Nonessential amino-acid residues of the enzymes are attached to chemically activated supports, such as natural polymers (cellulose, starch, dextran, agarose, carrageenan, chitin and chitosan, collagen gelatin, albumin, silk) or synthetic polymers (polystyrene, polyacrylates and polymethacrylates, polyacrylamide, polyamide, etc.). This is by far the most intensively studied technique of immobilization (Figure 2.19).

To design glutaraldehyde-based methods for the immobilization of restriction endonucleases to inorganic supports, Olzewski and Wasserman (1986) made a systematic investigation on the effect of glutaraldehyde on the activity of several enzymes. Because glutaraldehyde reaction primarily involves ε-amino groups of Lys, the sensitivity to glutaraldehyde points towards the probable involvement of Lys in the catalytic activity of some enzymes and/or conformational changes as a result of crosslinking. Nasri and Thomas (1987) have demonstrated that the glutaraldehyde-mediated inactivation is in fact due to the modification of catalytically active Lys residues. The involvement of Lys in substrate binding has also been demonstrated (Gite *et al.*, 1992a,b; Gite and Shankar, 1993). The retention of activity in the presence of the substrate, however, was attributed to the protection of catalytically active Lys residues (Reddy, 1989).

Chitin and chitosan have been extensively used as carriers for immobilized enzymes using glutaraldehyde as a crosslinking reagent (Muzzarelli, 1980; Mitsutomi *et al.*, 1985). A variety of enzymes – lactase, acid phosphatase, chymotrypsin (Stanley *et al.*, 1975), glucose isomerase (Stanley *et al.*, 1976), papain (Finley *et al.*, 1977), glucoamylase (Stanley *et al.*, 1978), glucose oxidase (Liu *et al.*, 1978) – have been immobilized on chitin by this procedure.

Azide Method

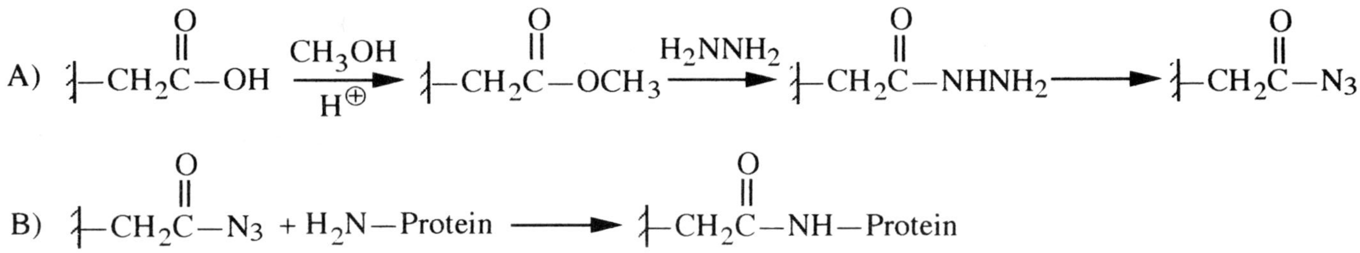

Carbodiimide Method

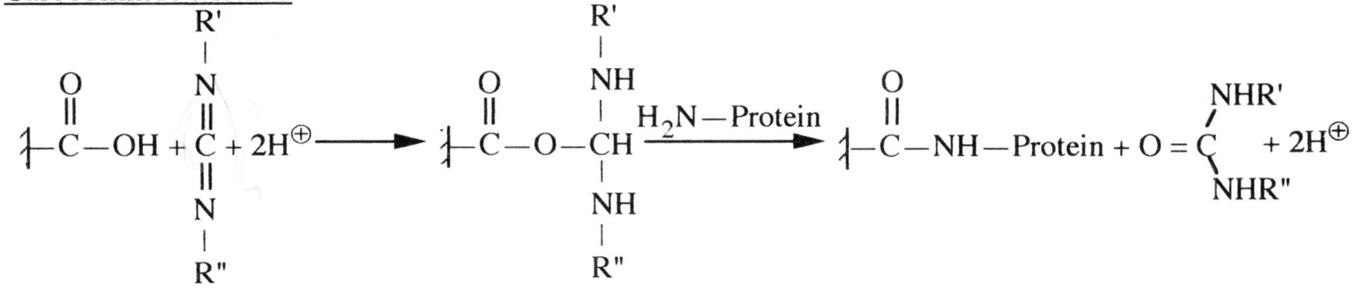

Isothiocyanate Method

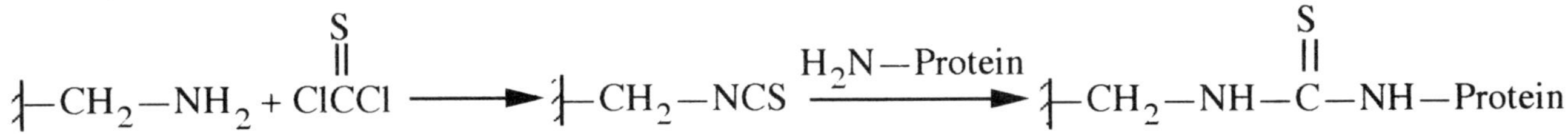

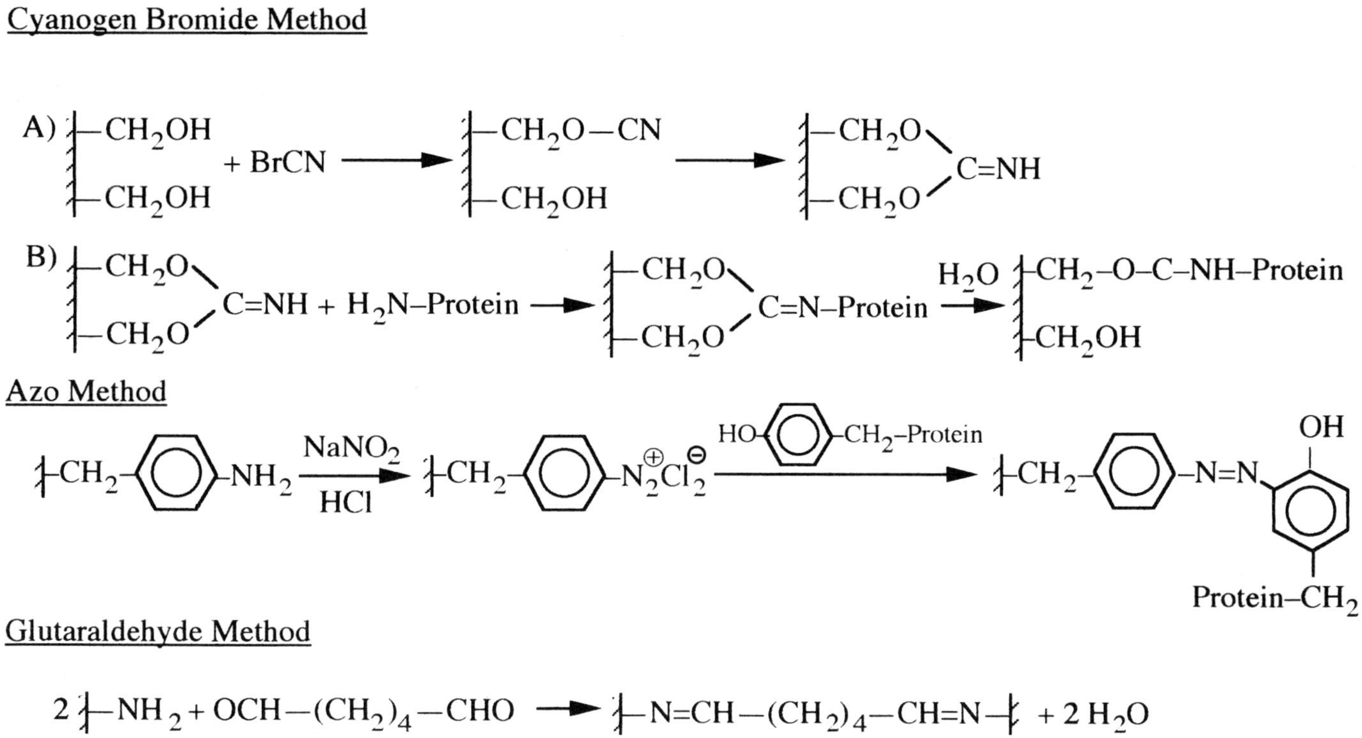

Figure 2.19 Methods most commonly used for immobilization of enzymes by covalent bonding.

(b) Crosslinking between enzyme molecules. The immobilization of enzymes or cells occurs by the formation of intermolecular crosslinkages between the enzyme molecules or the cells by means of bifunctional or multifunctional reagents. The crosslinking reagents that have been employed are glutaraldehyde, bis-isocyanate derivative, bis-diazobenzidine and so on. Recently, an original strategy to obtain intramolecular crosslinking has been proposed (Ferandez-Lafuente *et al.*, 1995). This strategy consists of three consecutive steps to form intramolecular crosslinks:

1. Enzymes are partially modified with the glutaraldehye.
2. The excess of reagent is removed.
3. The modified enzyme is incubated to allow a crosslinking reaction without the competition of additional single-point modifications.

In this way it has been possible to immobilize penicillin G acylase using glutaraldehyde as a crosslinking reagent under very controlled conditions (Ferandez-Lafuente *et al.*, 1995). Dubay *et al.* (1989) stabilize nuclease by crosslinking with various bifunctional agents, viz. glutaraldehyde, dimethyl adipamide, dimethyl suberimide and dimethyl 3,3′-dithiobispropionimidate.

2.3 Reactors for immobilized biomaterial systems

Eight types of immobilized enzyme reactors, presented schematically in Figure 2.20, are currently being employed. Details of each type are given below.

In **batch reactors**, the initial substrate concentration needs to be high. The final product concentration is also high. Because the entire batch of enzymes is initially exposed to high substrate concentrations and to high product concentrations (at the end of the process), this type of reactor is not suitable for reactions with substrate or product inhibition.

The **continuously stirred tank reactor (CSTR)** is little affected by pH shifts. Owing to the desired high product concentration, it should not be used in the case of product inhibition. Material balances are as follows:

$$rV_1 = Q_0S_0X \tag{2.1}$$

$$V_1 = Q_0S_0X(K'_m + S)/r_{max}S \tag{2.2}$$

where V_1 = reactor volume, Q_0 = input flow the substrate, S_0 = substrate concentration in input flow, S = substrate concentration in the reactor at time t; X = conversion degree of the substrate $(S_0 - S)/S_0$, K'_m = Michaelis apparent constant, and r = overall rate of the enzymatic reaction per unit reactor volume = $r_{max}S/(K'_m + S)$. Substituting for S in equation

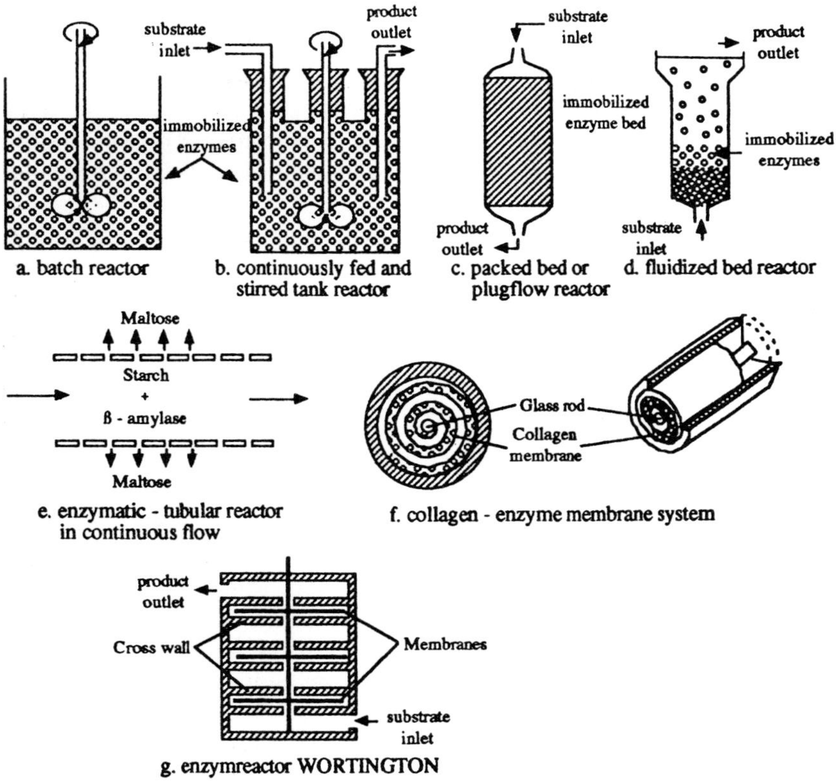

Figure 2.20 Immobilized enzyme reactors. (a) Batch reactor; (b) continuously fed and stirred tank reactor; (c) packed-bed or plug flow reactor; (d) fluidized bed reactor; (e) enzymatic–tubular reactor in continuous flow; (f) collagen–enzyme membrane system; (g) enzymereactor Wortington.

2.2, the volume required for a given conversion and flow rate is given by:

$$V_1 = Q_0X\,[K'_m/V_{max}(1 - X) + S_0/V_{max}] \tag{2.3}$$

In the **plug flow packed-bed reactor**, the substrate is converted into product during downflow (or upflow) passage over a column packed with immobilized enzymes. When there is no substrate inhibition and no pH shift, it is very difficult to control the pH in the column without using buffer. In an industrial situation this not generally feasible but this type of reactor has advantages over the first two types mentioned. This type of reactor has been largely used for ethanol production (Godia *et al.*, 1987). Packed bed reactors with immobilized cells have also been used on an industrial scale (Chibata *et al.*, 1979).

The **fluidized bed reactor** is theoretically the most attractive type, in terms of easy mass transfer. The substrate is passed through the column in an upflow mode with a velocity that fluidizes the immobilized enzyme particles. In contrast with the packed bed reactor, the fluidized bed reactor facilitates solid–fluid mixing and gas removal, and minimizes pressure drop. Fluidized bed reactor operation with continuous product removal has been used to reduce product inhibition (Davison, 1989; Bajpai *et al.*, 1990).

In comparing these types of reactors, one further point should be borne in mind: when the immobilized enzyme reactor is of the stirred type, the immobilized enzyme particles should be elastic to avoid abrasion. With a packed bed reactor, the immobilized enzyme particles should be relatively rigid, otherwise pressure drop problems will occur. If one considers that there is (1) no substrate inhibition, (2) no, or hardly any, pH shift and (3) an equilibrium reaction in the enzymatic conversion, such as in the case of glucose to fructose, plug flow reactors ought to be used.

Membrane reactors contain membrane systems representing simple physical barriers retaining the macromolecules in the reaction compartment. The enzyme may either be in a native soluble form or fixed on solid particles. For substrates with high molecular weights, such as starch, Glosset *et al.* (1974) proposed a **tubular membrane reactor** (Figure 2.20e). The main shortcoming of this type of reactor is the phenomenon known as 'polarization', i.e. an accumulation of products with high molecular weight, when entering the reactor, and the blockage of the pores.

Several reactors are being equipped with **enzymatically active membranes**; the enzyme is adsorbed on the surface or entrapped in the support's skeleton. The biocatalytic modulus of Wang and Vieth (1973), is made of collagen and enzyme film, formed by electrode siting, and finally fixed on a cellulose acetate film (Figure 2.20f). The membrane is wrapped on an axis, the various layers being separated by small glass rods.

In **enzymreactor Wortington** (Figure 2.20g) (Durand and Monsan, 1974), a series of membrane-bearing disks are fixed on an axis, to be introduced in the reaction medium. This is a multifloored system, with each compartment offering the possibility of being equipped with membranes containing various enzymes.

In **fixed-film reactors**, bacteria are grown on an inert support medium within the reactor. The contaminated water passes over the attached bacteria and forms a thin water film into which the contaminants and oxygen diffuse. A relatively new biological design is a combination of suspended-growth and fixed-film reactor designs, generally referred to as submerged fixed-film reactors. In these units, the plastic medium is submerged in the water in the reactor tank. The bacteria are grown on the plastic membrane as in a fixed-film system. The medium is aerated from below and, as it is released, the air pushes water in front of it as it rises, creating an air-lift pumping action. The main advantages of this design are ease of operation and high-quality performance. The main disadvantages of the submerged fixed-film design are the high cost of oxygen transfer and lack of scalability.

Some factors that influence the choice of reactor type are:

1. the nature of the biomaterial support;
2. the nature of the substrate to be processed;
3. the reaction kinetics;
4. the cost of reactor construction and operation;
5. the method of regeneration or replacement of the immobilized biomaterial;
6. the ease of reactor scale-up;
7. the operational stability and requirements;
8. the activity per unit volume or productivity required.

2.4 Waste conversion in the dairy industry

2.4.1 Bioconversion of whey

Champagne *et al.* (1989), proposed the use of cells immobilized by *Propionibacterium shermanii* to produce propionic acid from whey in a two-step process. Whey lactose was partially converted to lactic acid by *Lactobacillus helveticus* followed by the *P. shermanii* fermentation.

Recycle batch fermentation using immobilized *Propionibacterium acidipropionici* has been studied (Yang *et al.*, 1995) for propionate production from whey permeate, D-lactose whey permeate and acid whey. Cells were immobilized in a spirally wound fibrous sheet recently developed by Yang *et al.* (Lewis and Yang, 1982; Yang *et al.*, 1992). High fermentation rates and conversions were obtained with these whey media without nutrient supplements. The highest propionic acid concentration obtained with the recycle batch reactor was 65 g/L, which is much higher than the normal maximum concentration of 35–45 g l^{-1} reported in the literature (Playne, 1985; Woskow and Glatz, 1991; Babuchowski *et al.*, 1993).

(a) Methods of lactase immobilization and properties of the immobilized lactase. An immobilization technique for lactase (from *A. niger*) has been suggested using chitosan flakes activated with glutaraldehyde (Leuba and Widmer, 1977), but their low density and their tendency to agglomerate has prevented its use in fluidized beds reactors. A method has been developed to render the support more rigid and to increase its density (Van Griethuysen *et al.*, 1986). The activity of the immobilized enzyme depended strongly on the pH during the contact period between the activated carrier and the enzyme (Figure 2.21) (Van Griethuysen *et al.*,

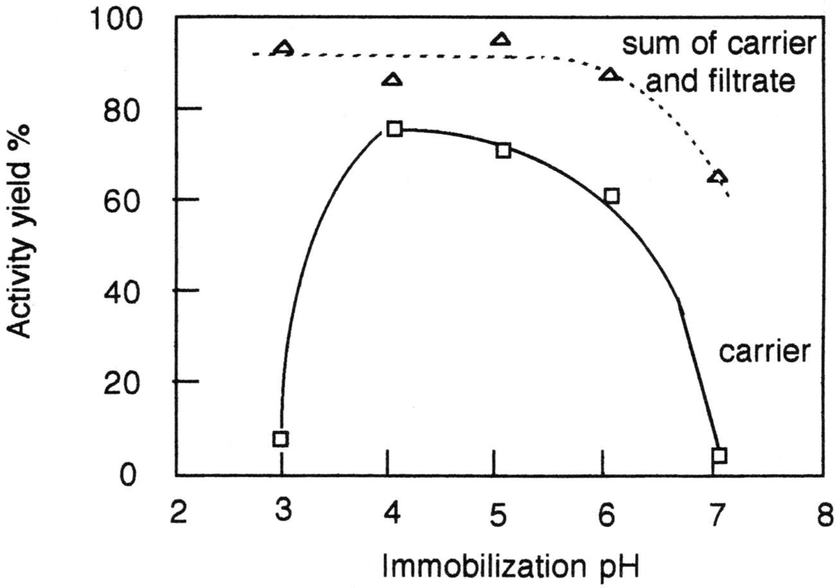

Figure 2.21 pH profile for the immobilization of *A. niger* lactase (immobilization in a stirred tank for 2 h, mean particle size of silica gel, d_p = 0.100–0.125 mm). (Reproduced with permission from Van Griethuysen *et al.*, in *Chitin in Nature and Technology*, edited by R. Muzzarelli *et al.*; published by Plenum Press, New York, 1986.)

1986). The pretreatment of the catalyst with glutaraldehyde was introduced after having observed a rapid decrease in the activity of the immobilized lactase during continuous operation with whey as substrate (Figure 2.22) (Van Griethuysen *et al.*, 1986).

In a batch fermentation process for the removal of lactose from raw buffalo milk, immobilized *Kluyveromyces fragilis* in alginate gels was used (Rao *et al.*, 1988). Decleire *et al.* (1985) studied the continuous hydrolysis of whey by *K. bulgaricus* immobilized in calcium alginate gels. Hydrolysis rates of 80–90% were obtained. Cells of *K. lactis* and *K. fragilis* were permeabilized and immobilized in a glass wool matrix (Champluvier *et al.*, 1989), polystyrene (Champluvier *et al.*, 1988) and gelatin fibers (Castillo and Cases, 1990) respectively. The steady-state enzymatic activity was found to be 50–60% of the initial activity.

(b) Milk proteins hydrolysis. The continuous production of small peptides from buttermilk protein proteolysis has been studied using whole free and immobilized *Serratia marcescens* cells as an enzyme source (Veillemard *et al.*, 1988). Production of small peptides (200–1000 Da) by *S. marcescens* cells immobilized in calcium alginate hydrogels has been discussed with results comparable to specialized dietary products obtained using purified enzyme preparations.

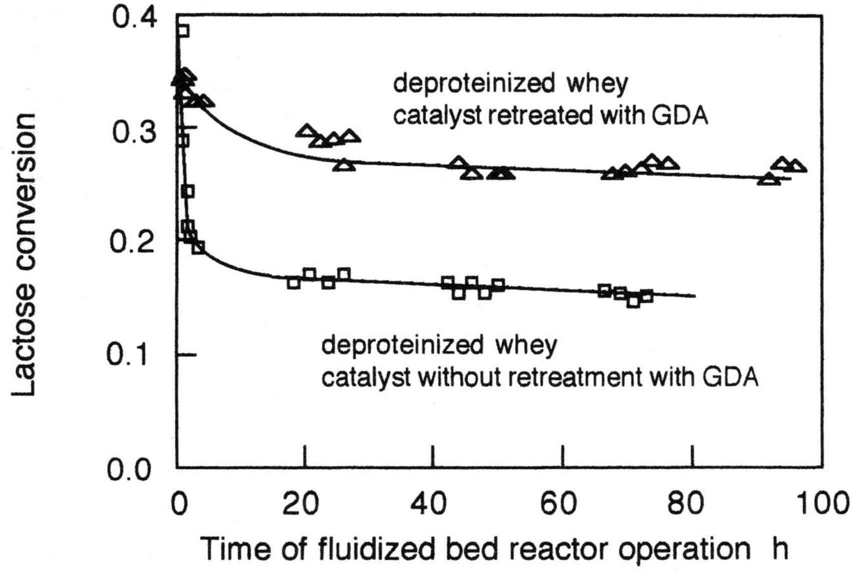

Figure 2.22 Operational stability of immobilized *A. niger lactase* (substrate: deproteinized whey), GDA = glutaraldehyde. (Reproduced with permission with Van Griethuysen, E. *et al.*, in *Chitin in Nature and Technology*, edited by R. Muzzarelli *et al.*; published by Plenum Press, New York, 1986.)

2.4.2 Milk processing

The immobilization of lactic acid bacteria has been used for pre-acidification and continuous inoculation of milk (yogurt, cream, fresh cheese, cheddar cheese) (Champagne and Côté, 1987; Prévost, 1987; Prévost and Diviès, 1987, 1992; Rossi *et al.*, 1990), starter production (Kearney *et al.*, 1990; Morin *et al.*, 1992) and metabolites such as lactic acid (Northon, 1992; Westrin and Axelsson, 1991); and diacetyl (Takahashi *et al.*, 1990).

Cachon *et al.* (1995) report the study of the lactose–citrate co-metabolism by *Lactococcus lactis* spp *lactis* bv. *diacetylactis* immobilized in calcium alginate gel beads. A dynamic diffusion–reaction–growth model was proposed for the study of lactic fermentation, the bioconversion of citric acid and cell release in a pH-stat continuous stirred tank reactor with immobilized cells. Results obtained with the model show that the activity of the gel and cell release are concentrated in the superficial zones of the beads. Effects of external and internal diffusion limitations, and the microenvironment on the activity of immobilized cells, cell release and the activity of free cells on the fermentation have been studied.

Advantages of immobilized cell technology in the dairy industry include continuous utilization, high cell densities, protection of cells from oxygen found in growth medium, ease of recuperation, protection from bacteriophages, less contamination problems, better survival to freeze-drying (Steenson *et al.*, 1987; Audet *et al.*, 1988, 1989, 1990; Kearney *et al.*, 1990; Champagne *et al.*, 1992, 1993).

2.5 Bioconversion of cellulosic wastes

2.5.1 Conversion of cellulose to ethanol

Utilization of cellulosic wastes for the production of food and chemicals involves depolymerization by chemical, physical or biological methods to sugars and subsequent fermentation to desired products. Acid hydrolysis is the oldest approach for the depolymerization of cellulose.

In addition to producing ethanol from starch and sugar crops, it can also be made from lignocellulosic biomass. Examples of existing sources of lignocellulosic biomass include agricultural and forestry residues, a major fraction of municipal solid waste, waste paper, various industrial waste streams, herbaceous (grasses) and wood crops. Residual lignocellulosic biomass is available and of low cost. However, the cost of conversion to ethanol has always been high owing to the recalcitrant nature (i.e. difficult hydrolysis) of lignocellulosic materials.

The saccharification of wood residues (sawdust) has been studied using *Aspergillus niger* immobilized onto glass beads (Wilkins and Ramesh, 1991). Saccharification results indicate that pretreatment (thermal treatment of the substrate, milling and treatment with 1.5% sodium hydroxyde) is essential to reach high yields of glucose. The glucose yield is roughly 1.5–2.0 times faster when pretreatment is carried out prior to enzymatic saccharification. Simultaneous saccharification–fermentation (SSF) studies also have been conducted. The optimum pH and temperature for SSF are 4.0–5.0 and 45 °C, respectively.

The immobilization of the yeast *Saccharomyces cerevisiae* has generally been achieved by adsorption (surface immobilization) onto solid supports (Ghose and Bandyopadhyay, 1980; Daugulis *et al.*, 1981) by entrapment in gel matrices such as calcium alginate or carrageenen (Wada *et al.*, 1979; McGhee *et al.*, 1982) or by crosslinking (Compere and Griffith, 1981).

Immobilized cell bioreactors may be subject to rate limitations arising from mass transfer resistances, i.e. diffusion of substrate from the fermentation medium to the cells and diffusion of products (ethanol and carbon dioxide) in the reverse direction.

Ethanol production from lignocellulosic raw materials has been reported by Saida *et al.* (1985). The process consists of several steps: pretreatment, saccharification and fermentation as shown in Figure 2.23.

A conceptual flow diagram for continuous fermentation is shown in

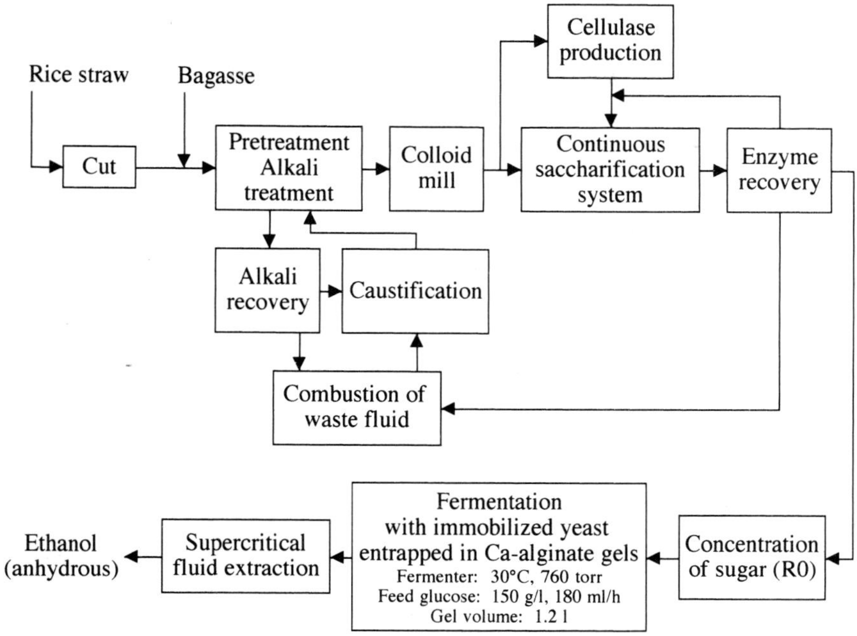

Figure 2.23 Block flow diagram to produce ethanol from biomass. (Reproduced with permission from Saida, T. *et al.*, in *Bioenergy 84*, Vol. III, *Biomass Conversion*, edited by H. Egneus and A. Ellegard; published by Elsevier Applied Science, London, 1985).

Figure 2.24 (Saida *et al.*, 1985). The experimental results obtained by a bench scale plant operated continuously for 300 h are shown in Figure 2.25. The volume efficiency of this fermenter is 20–30 g l^{-1} h^{-1} compared to conventional suspended-cell batch fermenters of 1.3–1.5 g l^{-1} h^{-1}.

2.6 Hemicellulose conversion

Considerable interest has been shown for the use of hemicellulose-rich agricultural crops for fuel ethanol production by free and immobilized cells of *S. cerevisiae*, *Kluyveromyces marxianus* and *Zymomonas mobilis* (Margaritis and Bajpai, 1982, 1983; Rosario and Pamatong, 1985; Bajpai and Margaritis, 1987a,b; Kana *et al.*, 1989a,b).

The production of ethanol from carob pod extract by free and immobilized *S. cerevisiae* cells in calcium alginate beads in batch and fed-batch culture has been investigated (Roukas, 1994). In fed-batch cultures, both free and immobilized S. cerevisiae cells gave the same maximum concentration (62 g l^{-1}) of final ethanol at an initial sugar concentration of 300 g l^{-1} and feeding rate (F) = 167 ml h^{-1} (Figure 2.26) (Roukas, 1994).

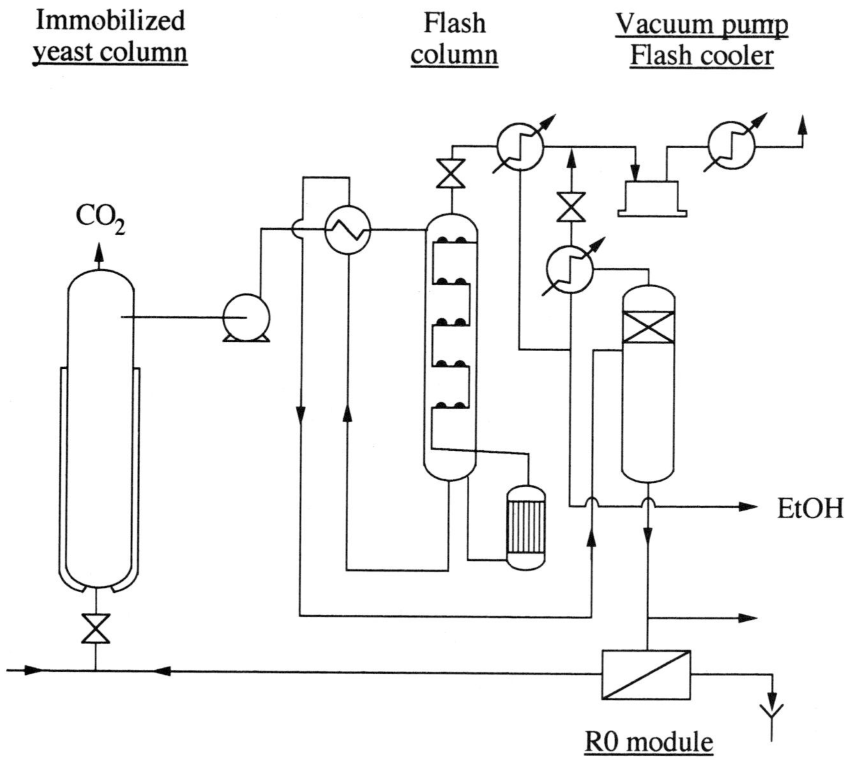

Figure 2.24 Flash fermentation system. (Reproduced with permission from Saida, T. *et al.*, in *Bioenergy 84*, Vol. III, *Biomass Conversion*, edited by H. Egneus and A. Ellegard; published by Elsevier Applied Science, London, 1985.)

In fed-batch cultures, immobilized *S. cerevisiae* cells gave a higher overall ethanol concentration compared with free cells (Figure 2.27) (Roukas, 1994). The immobilized *S. cerevisiae* cells in calcium alginate beads retained their ability to produce ethanol for 10 days.

2.6.1 Conversion of xylose

Kautola and Linko (1989) studied the production of fumaric acid from xylose by immobilized *Rhizopus arrhizus* cells. The highest mass yield was 23.7% with a product concentration of 15.3 g l^{-1}. Immobilized cells produced 3.5 times more fumaric acid than free cells.

2.7 Bioconversion of starch wastes

Starchy materials are abundant in nature and have great potential as alternative energy sources. Residual solutions of starch can be recovered in

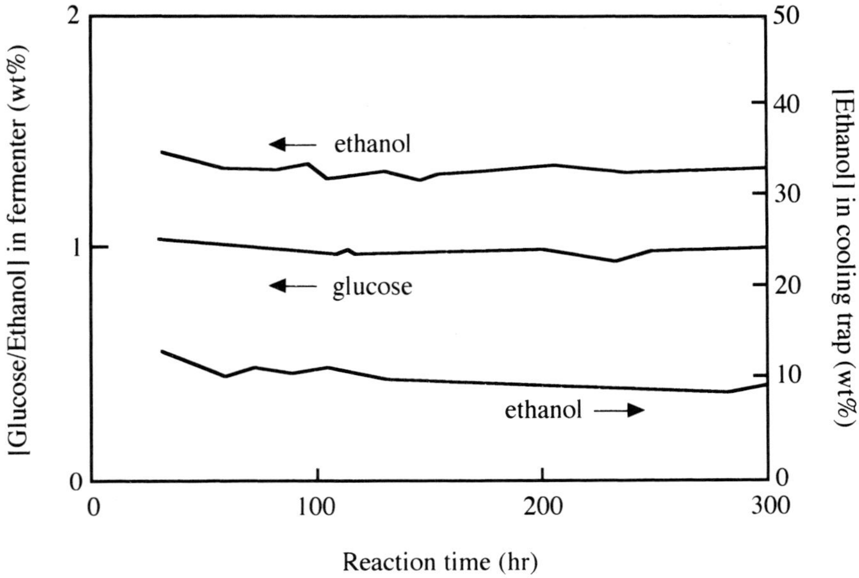

Figure 2.25 Continuous operation of flash fermentation with immobilized yeast entrapped in calcium alginate gels. Fermenter: 30 °C; flash column: 30 °C (L), 25 °C (V), 30 torr; feed glucose: 150 g l^{-1}, 180 ml h^{-1}; gel volume: 1.2 l. (Reproduced with permission from Saida, T. *et al.*, in *Bioenergy 84*, Vol. III, *Biomass Conversion*, edited by H. Egneus and A. Ellegard; published by Elsevier Applied Science, London, 1985.)

processes of glucose production and used for ethanol production. Residual solutions of glucose may also be utilized. With this aim in view, a series of immobilized enzymes and cell systems have been studied.

The development of a process for simultaneous hydrolysis and fermentation of untreated raw starch would both reduce the energy input and increase the efficiency of the utilization of starch as a substrate in industrial fermentations. There are many reports on the efficient utilization of microorganisms for the production via fermentation of various polysaccharides. Of all the methods described, entrapment of enzymes or microorganisms in polymers has been studied most extensively because this method is simple and the immobilized cells remain active, making it possible to use them for long-term processes.

2.7.1 Simultaneous saccharification and fermentation of starch

The direct conversion to ethanol of residual solutions of starch was first discussed by Lee and Long (1974). They employed a continuous fermenter with hollow-fiber ultrafiltration (UF) for processing a recirculating system containing soluble amyloglucosidase (AMG) and *Zymomonas mobilis*. The process temperature was limited by the use of *Z. mobilis* at 35 °C.

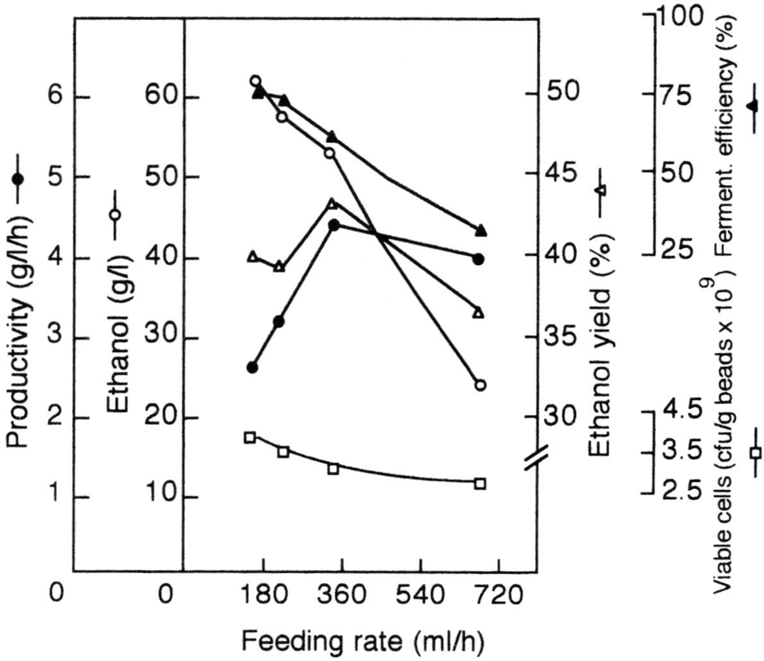

Figure 2.26 Effect of feeding rate on kinetic parameters of carob pod extract fermentation by immobilized cells of *S. cerevisiae* in fed-batch culture. (Reproduced with permission from Roukas, T. Ethanol production from nonsterilized carob pod extract by free and immobilized *Saccharomyces cerevisiae* cells using fed-batch culture. *Biotechnol. Bioeng.*, **43**, 189–201; published by John Wiley & Sons, Inc., 1994.)

However, AMG functions best at approximately 60 °C, therefore an enzyme dose much higher than normal was required to prevent overall rate limitation by the hydrolysis reaction.

Charles and Phillips (1985) studied and developed an installation of starch saccharification/fermentation (S/F) that can be easily adapted for starch wastes. Fermentation occurs at 37 °C, while saccharification occurs at 60 °C. Two variations of the process were discussed: one used soluble glucoseamylase (GA), the other used immobilized (IME) GA. In each case, the fermentation broth was circulated constantly, through a hollow-fiber, microporous, cross-flow filter until the hydrolysis was complete.

For continuous production of ethanol using thinned corn starch as substrate, two bioreactors were coupled:

1. A reactor with GA covalently immobilized on a polyacrylamide bead support containing carboxylic groups activated by a water-soluble carbodiimide.
2. A reactor with *S. cerevisiae* cells entrapped in calcium alginate gel beads (Gombin *et al.*, 1994).

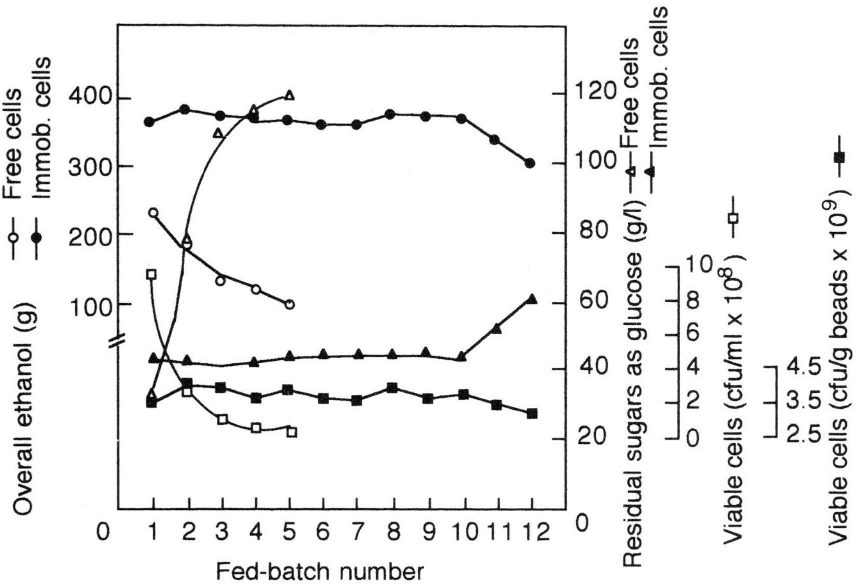

Figure 2.27 Effect of fed-batch numbers on fermentation kinetics using free and immobilized cells of *S. cerevisiae* during ethanol production from nonsterilized carob pod extract in repeated fed-batch cultures. (Reproduced with permission from Roukas, T. Ethanol production from nonsterilized carob pod extract by free and immobilized *Saccharomyces cerevisiae* cells using fed-batch culture. *Biotechnol. Bioeng.*, **43**, 189–201; published by John Wiley & Sons, Inc., 1994.)

The production of ethanol from starch by a co-immobilized mixed culture system of aerobic and anaerobic microorganisms in calcium alginate gels beads was investigated (Tanaka *et al.*, 1986). The mold *Aspergillus awamori* was used as an aerobic amylolytic microorganism and an anaerobic bacterium, *Z. mobilis*, as an ethanol producer. Starch hydrolysis and glucose accumulation by immobilized *A. awamori* in calcium alginate gel beads has been investigated. A simplified scheme for the correlation of the metabolism of both microorganisms in the system is shown in Figure 2.28 (Tanaka *et al.*, 1986).

The ethanol productivity in the flask culture system was particularly affected by shear stress (dependent on the shaking speed) which controlled the mycelial growth on the surface of the gel beads (Table 2.6) (Tanaka *et al.*, 1986). From the above experiments, it was concluded that the co-immobilized mixed culture system was an excellent system for the conversion of starch to ethanol. The co-immobilized mixed culture system has two excellent advantages (Tanaka *et al.*, 1986):

1. Aerobic and anaerobic microrganisms spontaneously exhibit 'habitat segregation' on the surface and in the center of the gel beads, respectively.

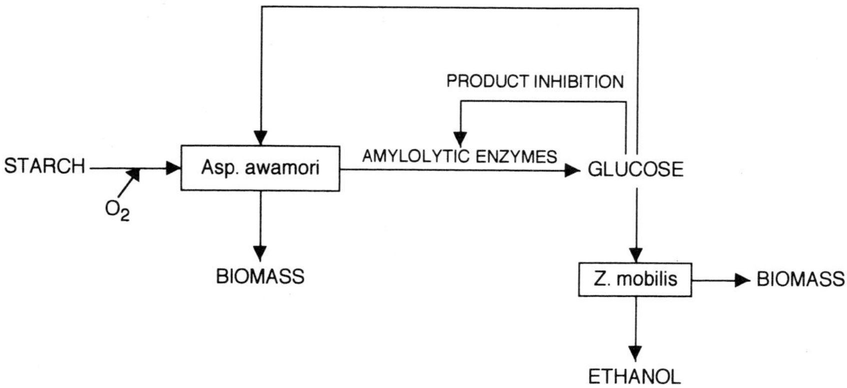

Figure 2.28 Scheme for the correlation of the metabolism of *Aspergillus awamori* and *Zymomonas mobilis* in the co-immobilized mixed culture system. (Reproduced with permission from Tanaka, H. *et al.* Ethanol production from starch by a co-immobilized mixed culture system of *Aspergillus awamori* and *Zymomonas mobilis*. *Biotechnol. Bioeng.*, **28**, 1761–68; published by John Wiley & Sons, Inc., 1986.)

Table 2.6 Ethanol productivities from glucose or starch by various culture systems

		Ethanol productivity[a] under various shaking conditions (g l^{-1})		
Carbon source	Culture system (microorganisms)	Static	100 rpm	220 rpm
Glucose	Submerged (*Z. mobilis*)	11.0	8.0	4.2
	Immobilized (*Z. mobilis*)	8.2	7.0	7.0
Starch	Submerged mixed (*Z. mobilis*, *A. awamori*)	No data	4.2	4.2
	Immobilized mixed (*Z. mobilis*, *A. awamori*)	No data	4.2	4.5
	Co-immobilized mixed (*Z. mobilis*, *A. awamori*)	No data	6.6	6.6

[a]Ethanol concentration was measured at the end of the culture.
(Reproduced with permission from Tanaka, H. *et al.* Ethanol production from starch by co-immobilized mixed culture system of *Aspergillus awamori* and *Zymomonas mobilis*. *Biotechnol. Bioeng.*, **28**, 1761–68; published by John Wiley & Sons, Inc., 1986.)

2. The anaerobic condition in the central part of the gel beads can be greatly accelerated by covering it with a thick layer of aerobic microbial cells on the surface of the gel beads.

The development of a co-immobilized mixed culture system of aerobic (*A. awamori*) and facultative anaerobic microorganisms (*S. cerevisiae*) in calcium alginate gel beads and the ethanol production from starch has been investigated (Kurosawa *et al.*, 1989). Neither microorganism exhibited 'habitat segregation' in the gel beads and leaked yeast cells grew aerobically without ethanol production in the broth. Ethanol productivity was low under these conditions. A more desirable co-immobilized mixed culture system of *Asp. awamaori* and *S. cerevisiae* was established by adding Vantocil IB (a biocidal compound) to the production medium. The system with Vantocil IB produced ethanol at 4.5 and 12.3 g l^{-1} from 16 and 40 g l^{-1} starch, respectively (Figure 2.29) (Kurosawa *et al.*, 1989). A continuous culture using this system (with Vantocil IB) was also carried out, and a stable steady state could be maintained for six days without leakage of yeast cells and contamination (Kurosawa *et al.*, 1989). The suitability of co-immobilized *Aspergillus awamori* (A) and *Zymomonas mobilis* (Z) (A–Z system), and of *Rhizopus japonicus* (R) and *Z. mobilis* (R–Z system) in calcium alginate for direct ethanol production from raw starch has been studied by Lee *et al.* (1993).

The production of ethanol or lactic acid from starch by a co-immobilized mixed culture system of aerobic and anaerobic microorganisms has been developed by several investigators (Tanaka *et al.*, 1986; Kurosawa *et al.*,

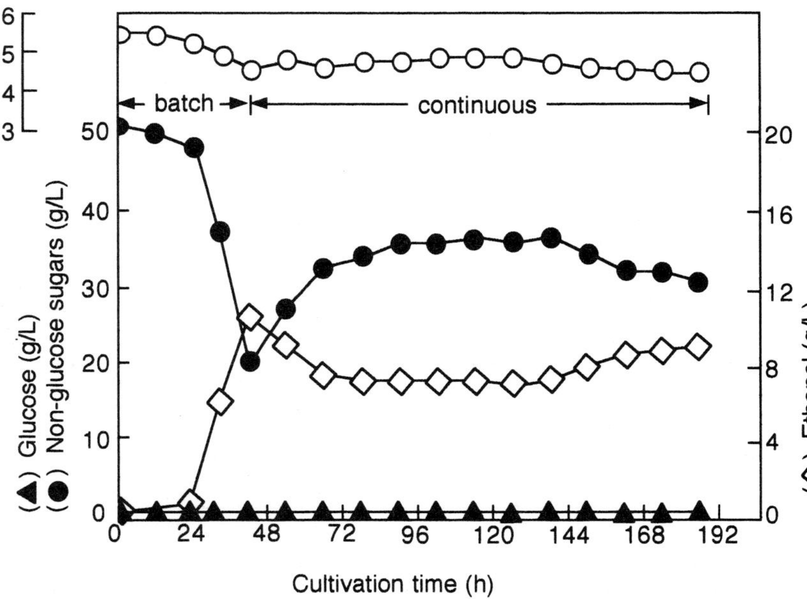

Figure 2.29 Continuous ethanol production from 5% starch by an *Aspergillus awamori–Saccharomyces cerevisiae* system with 0.2% Vantocil. IB. (Reproduced with permission from Kurosawa, H. *et al.* Ethanol production from starch by a coimmobilized mixed culture system of *Aspergillus awamori* and *Saccharomyces cerevisiae*. *Biotechnol. Bioeng.*, **33**, 716–23; published by John Wiley & Sons, Inc., 1989.)

1988; Kurosawa and Tanaka, 1990). The co-immobilized system mentioned above has the advantage of solving the problem of the noncompatibility of an aerobic and anaerobic mixed culture system (Tanaka *et al.*, 1986; Kurosawa *et al.*, 1988; Kurosawa and Tanaka, 1990). When three strains of microorganisms were co-immobilized (A–R–Z system), there was a higher rate of raw starch hydrolysis but the ethanol production was only 50% of that obtained with the A–Z system (Figure 2.30) (Lee *et al.*, 1993).

Kurosawa *et al.* (1988) have investigated the production of lactic acid from starch by a co-immobilized mixed culture system of *Aspe. awamori*

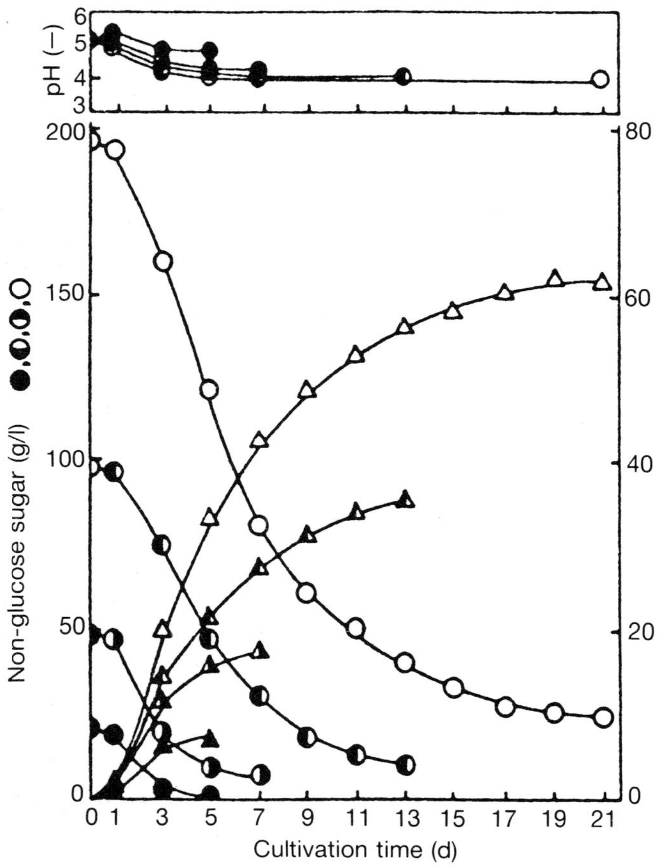

Figure 2.30 Effect of initial raw starch concentration on rate of hydrolysis and ethanol production by the A–R–Z 18 system. Symbols: 2% (●, ▲); 5% (◑, ▲); 10% (◐, ▲); 20% (○, △). (Reproduced with permission from Lee, S.-W. *et al.* Co-immobilization of three strains of microorganisms and its application in ethanol production from raw starch under sterile conditions. *J. Ferment. Bioeng.*, **75**, 36–48; published by The Society for Fermentation and Bioengineering, Japan, 1993.)

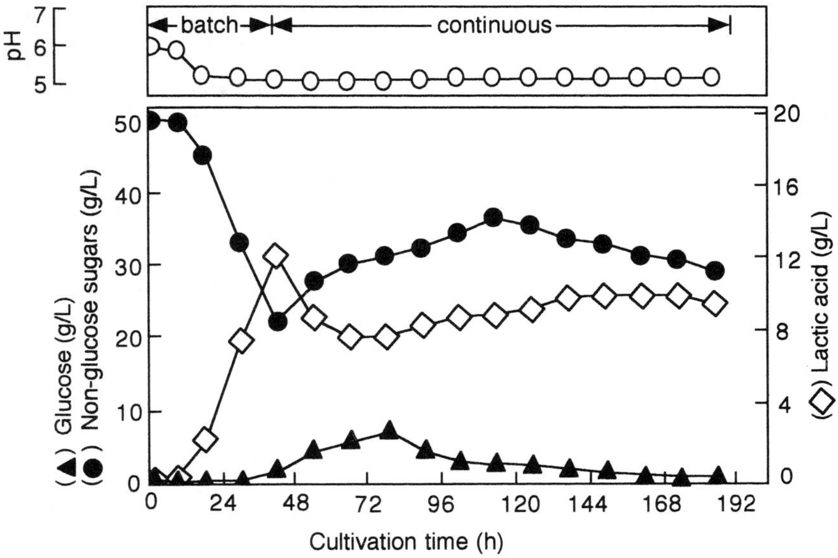

Figure 2.31 Lactic acid production from 5% starch by a co-immobilized mixed culture system of *Aspergillus awamori* and *Streptococcus lactic* in continuous culture. (Reproduced with permission from Kurosawa, H. *et al.* L-Lactic acid production from starch by coimmobilized mixed culture system of *Aspergillus awamori* and *Streptococcus lactic*. *Biotechnol. Bioeng.*, **31**, 183–7; published by John Wiley & Sons, Inc., 1988.)

and *Streptococcus lactic*. The final concentration of lactic acid produced from 50 g l^{-1} starch was 25 g l^{-1} (Figure 2.31).

2.7.2 Recovery of waste glucose solutions

Several bioreactor configurations (fixed/fluidized bed, gas-lift, pulsed, pulsing–packed-bed membrane fermenters, reciprocating bioreactors) were employed to carry out the ethanolic fermentation of glucose using *Saccharomyces cerevisiae* immobilized in calcium alginate (Godia *et al.*, 1987; Chamy *et al.*, 1990; Tyagi *et al.*, 1992; Gilson and Thomas, 1993; Roca *et al.*, 1996). The use of a new pulsing device (Elastic Membrane Pulsator) coupled to a packed-bed bioreactor has proved useful for increasing the efficiency of several laboratory-scale fermentation processes (Roca *et al.*, 1996). A more stable operation was obtained for larger periods in the semipilot scale bioreactor when a pulsation is applied to the feed stream. The nonpulsed operation leads frequently to a bed fragmentation and bed-slug formation. The overall productivity reached a value of up to 36 g l^{-1} h^{-1} in the pulsed bioreactor which implies an improvement of 18% with respect to the nonpulsed system.

Saccharomyces cerevisiae ATCC 39859 was immobilized onto small cubes of wood in order to produce very enriched fructose syrup from synthetic glucose–fructose mixtures through the selective fermentation of glucose (Guenette and Duvnjak, 1996). The ethanol production rate and cell growth rate were found to be inhibited linearly by both substrate and product concentrations. A maximum ethanol productivity of 21.9 g l^{-1} h^{-1} was attained from a feed containing 10% wt glucose and 10% wt fructose. At lower ethanol productivity levels the fructose:glucose ratio increased.

The thermotolerant ethanol-producing yeast strain *Kluyeromyces marxianus* IMB3 has been shown to produce ethanol from a variety of carbohydrate substrates, including molasses, sucrose, lactose and cellobiose (Banat *et al.*, 1992; Fleming *et al.*, 1993; Barron *et al.*, 1994; Brady *et al.*,

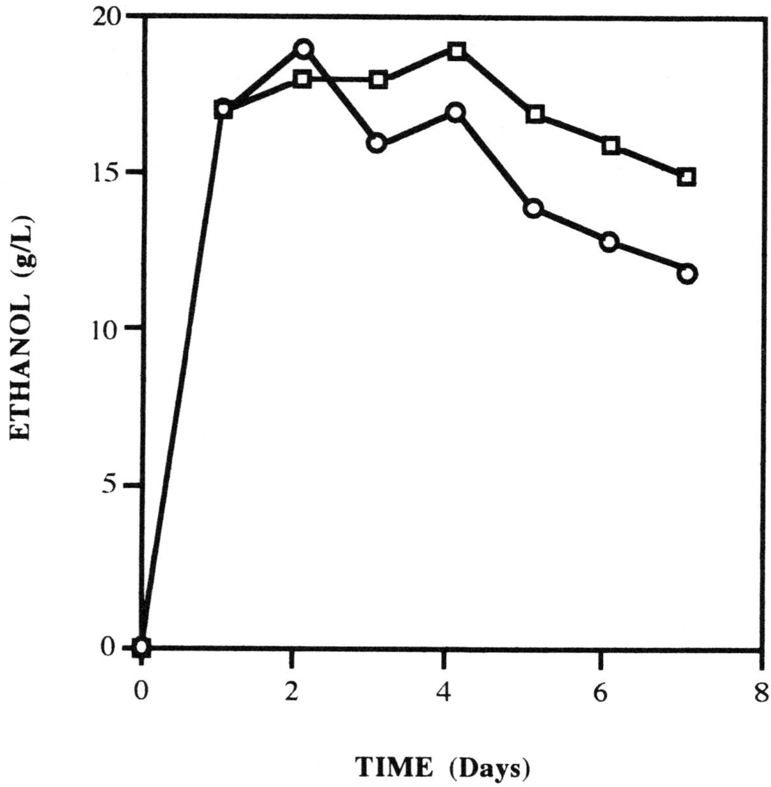

Figure 2.32 Ethanol production by free (○–○) and immobilized (□–□) *Kluyeromyces marxianus* IMB3 during growth on media containing 4% (w/v) sucrose. Samples were harvested at the indicated times, clarified by centrifugation and subsequently analysed for ethanol concentration. Each point in the curves represents the average value obtained from four experiments. (Reprinted from *Bioresource Technol.*, Vol. 55, Riordan, C. *et al.* Production of ethanol from sucrose at 45 °C by alginate-immobilized preparations of thermotolerant yeast strain *Kluyeromyces maxianus* IMB3, pp. 171–3. Copyright 1996, with kind permission from Elsevier Science Ltd, The Boulevard, Langford Lane, Kidlington OX5 1GB, UK.)

1994). Recently it has been demonstrated that immobilization of this yeast in calcium alginate increased the productivity of ethanol (Nolan *et al.*, 1994; Riordan *et al.*, 1996) (Figure 2.32).

The ethanolic bacteria *Zymomonas* have been shown to produce ethanol (Rogers *et al.*, 1979; Lawford *et al.*, 1983; Cross *et al.*, 1988; Beavan *et al.*, 1989; Cross and Clausen, 1993) using simple sugar substrates. The performance of *Zymomonas mobilis*, immobilized (using glutaraldehyde) on gelatin-coated spherical glass beads was evaluated for feed glucose concentrations of 150 and 200 g l^{-1} (Cross *et al.*, 1993).

Aspergillus niger and *Gluconobacter oxydans* was immobilized in alginate gel beads, ceramic and polyurethane foam (Tramper *et al.*, 1983; Sakurai *et al.*, 1989; Shiraishi *et al.*, 1989a,b; Vassilev *et al.*, 1993) for the production of gluconic acid by oxidation of glucose. Immobilized *G. oxydans* converted glucose to gluconic acid with a yield of 90% and a productivity of 10 g l^{-1} h^{-1}, while *A. niger* produced a productivity of 60 g l^{-1} h^{-1} in batch fermentations.

In a continuous fermentation of glucose to butanol and isopropanol by immobilized *Clostridium beyerinkii* cells in calcium alginate beads, the products were removed continuously by pervaporation (Groot *et al.*, 1984). The glucose conversion and the total amount of alcohols produced are increased by 65–70%.

2.7.3 Recovery of waste from beet sugar industry

α-Galactosidase is useful in the beet sugar industry (Kobayashi and Suzuki, 1972; Reynolds, 1974) and soymilk processing (Smiley *et al.*, 1976; Cruz *et al.*, 1981). Six polymers, shown in Table 2.7, were tested as carriers for the immobilized α-galactosidase (Mitsutomi *et al.*, 1985).

Table 2.7 Immobilization of α-galactosidase[a] on several compounds

Compounds	Activity yield (%)
Chitin	27.5
Colloidal chitin	72.1
Chitosan	23.4
Colloidal chitosan	45.0
DEAE–Sephadex A-50	39.2
DEAE–cellulose DE-23	42.3

[a]α-Galactosidase (9.98 units) was mixed with 0.2 g of each compound treated with glutaraldehyde and the mixture was shaken for 1 h at 25 °C and at pH 5.0. The activity yield was expressed as the percentage of immobilized enzyme activity to added enzyme activity. (Reproduced with permission from Mitsutomi, M. *et al.* Immobilization of thermostable α-galactosidase from *Pycnoporus cinnabarinus* on chitin and some properties of the immobilized enzyme. *J. Ferment. Technol.*, **63**, 325–29; published by The Society for Fermentation and Bioengineering, Japan, 1985.)

A vertical upflow, continuous bioreactor consisting of *Saccharomyces cerevisiae* (NRRL Y-132), which is surface immobilized by adsorption onto presterilized beech wood chips, has been developed and tested for producing ethanol from glucose and molasses (456 kg reducing sugar per m^3) (Lamptey, 1983; Campbell *et al.*, 1985). Steady-state operation has been achieved for up to 90 days, during which period a product having an ethanol concentration of 120 g l^{-1} was obtained with an ethanol productivity of 20 g l^{-1} h^{-1} for a residence time of only about 5 h (Campbell *et al.*, 1985). Table 2.8 compares the results of the bioreactor performance with immobilized cells (packed-bed bioreactor) with those of previous investigations on the basis of performance number (PN, kg m^{-3} liquid s^{-1}) and performance economy (PE, kg m^{-3} liquid s^{-1}) in addition to ethanol productivity (EP, kg m^{-3} liquid s^{-1}).

$$\begin{aligned} \text{PN} &= (\text{Ethanol productivity}) \times (\text{Product factor}) \times (\text{Substrate factor}) \\ &= (\text{EP}) \times (P/120) \times (P \times 0.49 S_0) \end{aligned} \tag{2.4}$$

where P = ethanol concentration (kg m^{-3}) and S_0 = substrate concentration (kg m^{-3}).

Incorporating a product recovery factor, equation 2.4 can be extended to define the overall fermenter-recovery system (PE):

$$\begin{aligned} \text{PE} = {} & (\text{Ethanol productivity}) \times (\text{Product factor}) \times (\text{Substrate factor}) \\ & \times (\text{Product recovery factor}) = (\text{EP}) \times (P/120) \times (P \times 0.49 S_0) \\ & \times (P/80) \end{aligned} \tag{2.5}$$

Table 2.8 Bioreactor performance comparison

Reactor type	S_0 (kg m^{-3})	P (kg m^{-3})	EP (kg m^{-3} s^{-1} × 10^3)	PN (kg m^{-3} s^{-1} × 10^3)	PE (kg m^{-3} s^{-1} × 10^3)	References
Bioreactor with *S. cerevisiae* immobilized by adsorption	270	118	6.0	5.2	11.7	Campbell *et al.* (1985)
Adsorption	294	83	6.4	2.5	3.97	Gencer (1981)
Adsorption	120	48	14.7	4.8	4.32	Daugulis *et al.* (1981)
Entrapment	126	56	5.0	2.1	2.22	Williams and Mannecke (1981)
Entrapment	127	12	15.0	0.3	0.07	Williams and Mannecke (1981)
Continuous-flow stirred tank bioreactor	160	60.5	8.9	3.4	3.93	Ghose and Tyagi (1979)

P = ethanol concentration; S_0 = substrate concentration; EP = ethanol productivity; PN = performance number; PE = performance economy.

(Reproduced with permission from Campbell, W.R. *et al.* (1985) Ethanol production in a surface-immobilized yeast, packed-bed bioreactor, in *Bioenergy 84*, Vol. III, Biomass Conversion, edited by H. Egneus and A. Ellegard, pp. 220–8; published by Elsevier Applied Science, London, 1985.)

Batch and continuous glycerol fermentation of sugar beet molasses have been studied using a vertical packed-bed reactor with *Saccharomyces cerevisiae* yeast cells immobilized in sintered glass rings (Bisping and Rehm, 1986; Benito *et al.*, 1994). A maximum glycerol concentration of 30 g dm^{-3} and a productivity of 36.6 g dm^{-3} d^{-1} at 28 °C, pH 6.9 with 25 g dm^{-3} sodium sulphite, for a dilution rate of 1.22 day^{-1}, was achieved (Benito *et al.*, 1994).

2.8 Immobilized enzymes in organic solvents

A promising, current trend in the field of biocatalysis is the replacement of part of the aqueous reaction medium by water-immiscible organic solvents (Lilly and Woodley, 1985; Brink and Tramper, 1986).

The production of propene oxide by *Mycobacterium* cells entrapped in calcium alginate (Brink and Tramper, 1986) has been used as a model to study the application of organic solvents in biotechnological processes. The nature of the organic solvent has significant effects. High activity retention of the immobilized cells relate to low polarities and high molecular weights of the solvents.

Co-immobilization of biocatalyst (*Arthrobacter simplex*) and substrate (hydrocortisone) in alginate has been suggested as a method to increase the conversion rate by transformation to prednisolone (Kaul *et al.*, 1987).

The efficiency of enzymes in organic solvents is determined by the catalytic parameters, which are changed with respect to those in aqueous solutions. The interplay of these processes was studied on soluble and immobilized glucoamylase and invertase using mixtures of buffer and several water-miscible solvents such as methanol, glycerol, dimethylformamide and dioxane (Ulbrich-Hofmann and Selisko, 1992). Three types of solvents can be differentiated (Ulbrich-Hofmann and Selisko, 1992): glycerol which does not induce denaturation; dimethylformamide, which induces a strong denaturation; aliphatic alcohols, which show moderate denaturation. The use of immobilized enzymes is attractive for the application in organic solvents because it circumvents the problem of protein precipitation.

The increasing use of predominantly organic reaction systems has motivated research towards solvent selection, a very important factor, for example, in the successful application of immobilized lipases.

Lipases immobilized with Celite or synthetic prepolymers such as urethane prepolymer and photocrosslinkable resin prepolymer have been applied for the asymmetric syntheses of complex molecules and kinetic resolution of many water-insoluble substrates (Akita, 1996). The urethane prepolymer gel-trapped lipase was used for the enantioselective esterification of D,L-menthol with various kinds of organic acids (Koshiro *et al.*, 1985).

Newly developed phospholipid–lipase aggregates containing phospholipid analogs with ether linkages have also been found to function effectively as immobilized biocatalysts and can enhance the enzymatic activity in comparison with the native lipase in water-saturated organic media (Akita *et al.*, 1989, 1993; Akita, 1996).

The hydrolysis of olive oil by immobilized lipase in xanthan–chitosan hydrogels was found to be high in nonpolar solvents such as iso-octane, heptane, cyclohexane, whereas moderate hydrolysis activity was found using benzene. However, in semipolar solvents, such as chloroform and methylene chloride, the activity was very low or equal to zero. These results suggest that the immobilized lipase activity in organic solvents decreases as the polarity of the solvent increases. We have selected iso-octane as the most suitable solvent in fat splitting in biphasic organic–aqueous systems on the basis of the lipase activity and the maintenance of the lipase activity.

The lipase activity in iso-octane has been studied as a function of the concentration of olive oil, the incubation time and the percentage of water in the system. The optimum time and temperature of incubation are different than those derived for the aqueous emulsion of olive oil. Optimum time is 24 h (Figure 2.33) and temperature is 34 °C (Figure 2.34). The optimal values for the reaction in organic medium are:

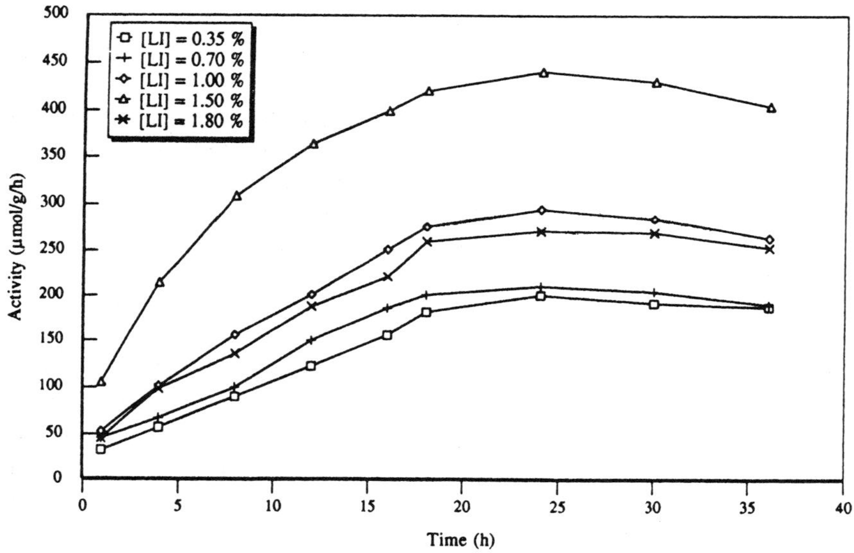

Figure 2.33 Variation of the reaction rate for immobilized lipase (LI) in the polyionic hydrogel as a function of time. [Olive oil] = 34.44%; temperature = 34 °C; solvent = iso-octane.

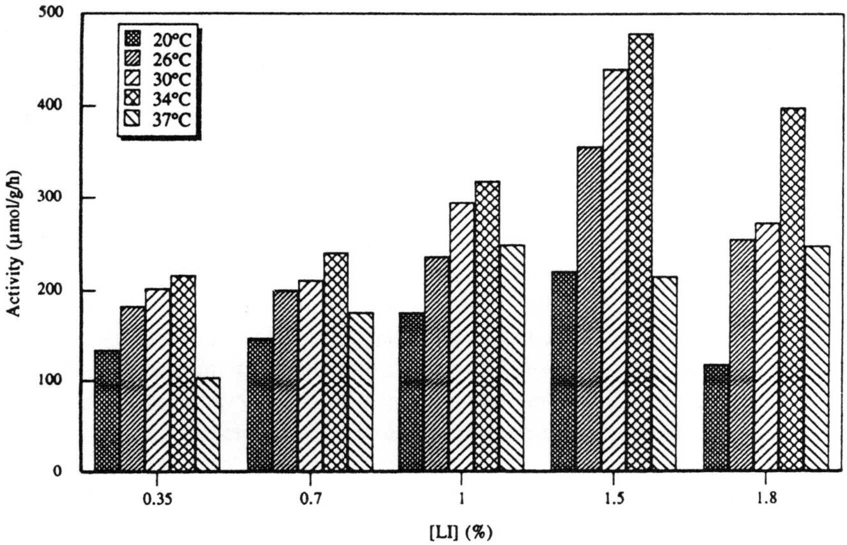

Figure 2.34 Lipase (LI) activity in iso-octane as a function of lipase concentration in the polyionic hydrogel and of temperature. [Olive oil] = 34.44%; solvent = iso-octane; incubation time = 24 h.

concentration of lipase in the hydrogel = 1.5%; concentration of olive oil = 34.44%.

Synthesis of triglycerides from free fatty acids and glycerol catalysed by immobilized lipases has been reported (Ergan *et al.*, 1990; Lortie *et al.*, 1993). *Mucor miehei* lipase immobilized on anion-exchange resins has been studied. Conversion of more than 85% of the fatty acid into the form of the triglyceride was observed. The 1,3-specific lipase can catalyse the synthesis of triolein because the ester enzymatically formed with the primary alcohol isomerizes, through acyl migration, to an ester on the secondary hydroxyl (Lortie *et al.*, 1993). The freed primary hydroxyl may undergo further enzymatic conversion.

The esterification of a long-chain fatty acid was conducted using a nylon-immobilized lipase from *Candida cylindracea* in a nearly anhydrous, nonpolar organic medium, hexane (Carta *et al.*, 1991, 1992; Zaidi *et al.*, 1995). Butyl laurate was produced from lauric acid and *n*-butanol at a maximum initial reaction rate of 37 mmol h^{-1} g^{-1} immobilized enzyme when the substrates were present in equimolar amounts at an initial concentration of 0.5 mmol l^{-1}. The rate of reaction increased with temperature, reaching a maximum between 35 and 45 °C (Zaidi *et al.*, 1995).

Procedures have been developed for the preparation of a number of biologically active oligosaccharide glycosides, by means of glycosidase catalysed transglycosidations using α- or β-glycosides as acceptor (Nilsson, 1986). The reactions were carried out with both soluble and immobilized glycosidases in aqueous solvents containing 0–45% *N,N*-dimethylformamide.

2.8.1 *Bioconversion of lipids*

The importance of free fatty acids for the characteristic flavors of various dairy foods has become increasingly apparent to food technologists in the past two decades. Current commercial practice for producing flavor compounds from milk fat uses an enzyme system that involves a final heat-processing step for controlling the desirable flavor profile of the product.

A lipase from *Aspergillus niger*, immobilized by adsorption on polypropylene flat-sheet membranes, was used to effect the continuous hydrolysis of the glycerides of melted butterfat (Malcata and Hill, 1991). The half-life of the immobilized lipase is of the order of 20 days at 35 °C. Malcata *et al.* (1992a,b) used a lipase from *A. niger* immobilized on

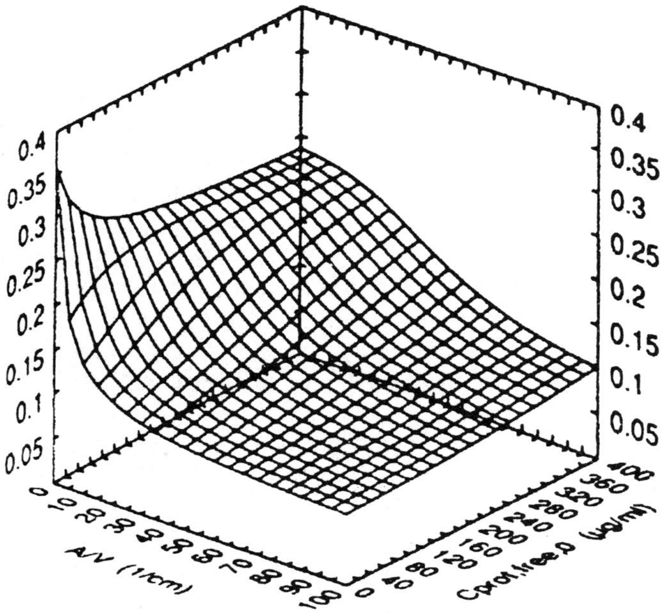

Figure 2.35 Specific activity ($a_{\text{lip,ads}}/C_{\text{prot.ads}}$) of the lipase adsorbed on the hollow fibers as a function of the protein concentration of the initial supernatant solution ($C_{\text{prot.ads,0}}$) and the ratio of area of membrane to volume of supernatant solution (A/V). Temperature = 20 °C. (Reproduced with permission from Malcata, X.F. *et al.* Hydrolysis of butteroil by immobilized lipase using a hollow-fiber reactor: part I. Lipase adsorption studies. *Biotechnol. Bioeng.*, **39**, 647–57; published by John Wiley & Sons, Inc., 1992.)

microporous polypropylene hollow fibers to effect the hydrolysis of the glycerides of melted butterfat at 40 °C and pH 7.0. Lipase is selectively adsorbed relative to the other proteins in the crude preparation (Figure 2.35). Statistical analysis of the data indicates that a 'ping-pong' mechanism controlled by the rate of deacylation of the lipase accurately describes the lipase-catalysed hydrolysis of butteroil.

Lipoprotein lipase was covalently immobilized onto the surface of porous chitosan beads (Itoyama *et al.*, 1994) and the surface of polyacrolein microspheres (Hayashi and Ikada, 1990b). The relative activity of the lipoprotein lipase immobilized without spacer decreased gradually with the decreasing surface concentration of the immobilized lipoprotein lipase. On the other hand, lipoprotein lipase immobilized with oligoglycine spacer gave an almost constant activity.

Many workers have reported lipase immobilization using macroporous weak anion exchange resins (Sugihara *et al.*, 1988; Kosugi *et al.*, 1990; Otero *et al.*, 1990; Kosugi and Suzuki, 1992). Lipase from *Pseudomonas fluorescences* was immobilized on anion exchange resin using glutaraldehyde (Kosugi and Suzuki, 1992). The immobilized lipase was not substantially inhibited by oleic acid or sodium oleate. A fluidized bed reactor (Figure 2.36) was found to be superior to a fixed bed reactor since, in the latter, the separation rate of the two layers was slow and the flow rate of the reactor consequently had to be reduced (Kosugi *et al.*, 1990).

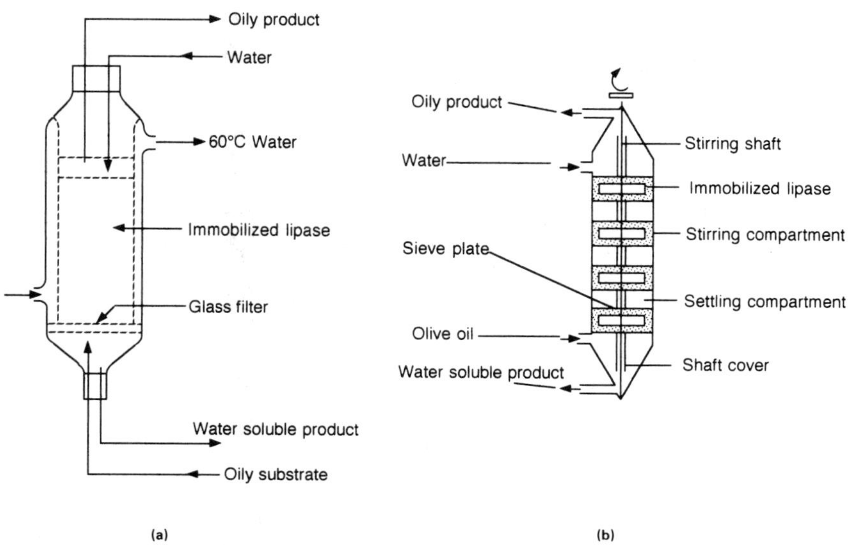

Figure 2.36 Fixed bed reactor (a) and fluidized bed reactor (b). (Reproduced with permission from Kosugi, Y. *et al.* Continuous hydrolysis of oil by immobilized lipase in a contracurrent reactor. *Biotechnol. Bioeng.*, **36**, 617–22; published by John Wiley & Sons, Inc., 1990.)

Kimura *et al.* (1983) immobilized lipase from *Candida cylindracea* on different supports and found that hydrophobic matrices exhibit the highest activity in the hydrolysis of olive oil.

Chromobacterium viscosum lipase which was adsorbed on liposome and solubilized in microemulsion droplets of glycerol containing a small amount of water could catalyse the continuous glycerolysis of olive oil (Chang *et al.*, 1991). The glycerolysis was dependent on several important operational parameters. Optimum initial substrate (olive oil), flow rate of oil and water content in the glycerol phase for the conversion were 2.0% (v/v), 2.5 ml h^{-1} and 8.0% (w/v), respectively.

2.9 Waste treatment

2.9.1 Methane bioconversion of wastes

It is generally known that organic residues subjected to condition of anaerobic digestion are being converted into methane. Anaerobic digestion involves four stages, namely, hydrolysis, acidogenesis, acetogenesis and methanogenesis. Methane, as such, is obtained mainly through the following reactions (Figure 2.37).

(a) Fermentation procedures. According to the way in which fermenters are supplied, there are **discontinuous processes**, characteristic of wastes having solid layers, and **continuous processes**, employing liquid, pumpable discharges. Depending on the type of cell cultures, the anaerobic fermentation may be classified into:

1. cultures of microorganisms in free suspension;
2. cultures of immobilized microorganisms.

Cultures of immobilized microorganisms permit the maintenance of a high concentration of active biomass into the reactor, consequently accepting a strong pollution charge, with no dilution or recycling being necessary.

The technique employing immobilized cells involves:

1. Digestion on specially arranged supports in reactors. An ascendant or descendant flow is employed, by passing the discharges through plastic systems or through mineral beds containing immobilized cells.
2. Digestion on granular supports. Microorganisms are fixed on macroporous granular supports kept in the fluidizing layer.

The main reactors employing immobilized cells for methane production are presented in Figure 2.38. Reactors with a fixed mud bed (Figure 2.38a) contain a layer of granulated mud permeated by the upflow of ascending liquid. The biomass may attain 120 g l^{-1} at the base of fermenter. Such

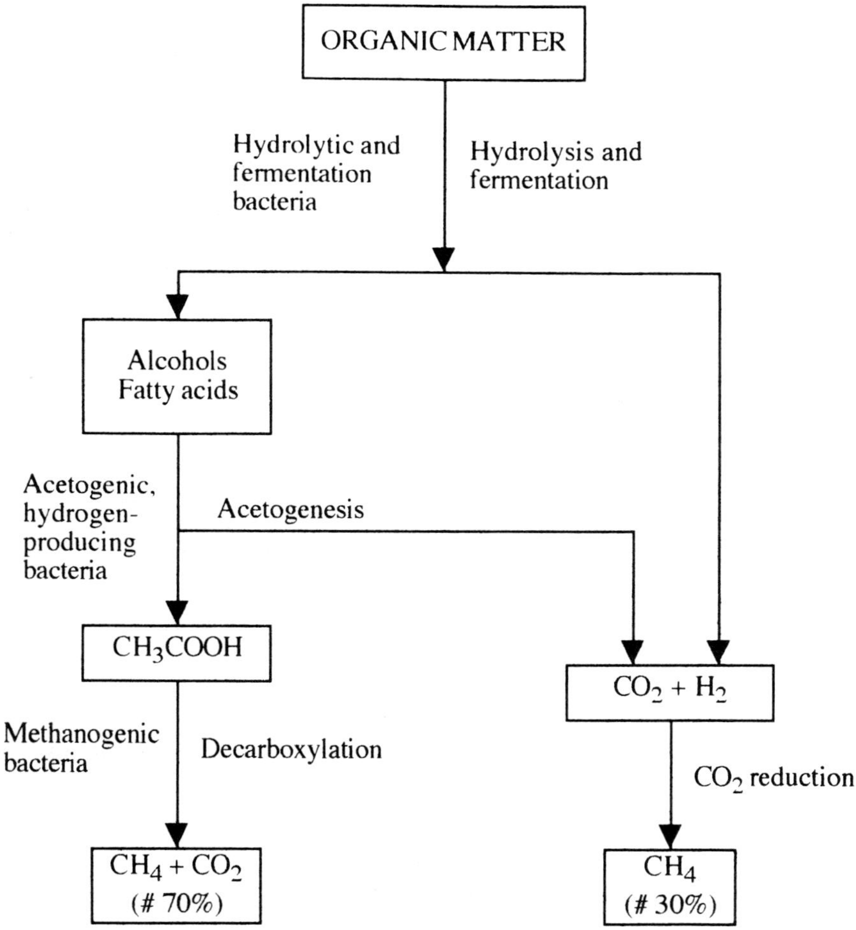

Figure 2.37 Simplified scheme of methane biosynthesis.

reactors may be applied to residual waters resulting from the agricultural and food industries. The chemical oxygen demand (COD) is 2–10 g l^{-1}, the stationing time is 5–6 h and the volumic charge is 10–45 kg COD m^{-3} day^{-1}. In the case of reactors of the anaerobic filter type, in the fixed layer (Figure 2.38b), the biomass is fixed on a mineral (sand, gravel) or synthetic (polyvinyl chloride and other polymer) support, among which the liquids circulate in upflow or downflow. These reactors may be used for residual waters having small quantities of solid materials as there is a risk of clogging. The COD is 2–7 g l^{-1}, the stationing time is a few hours and the volumic charge is 4–8 kg COD m^{-3} day^{-1}. This anaerobic filter saw its first full-scale application in the treatment of wheat starch waste water. For

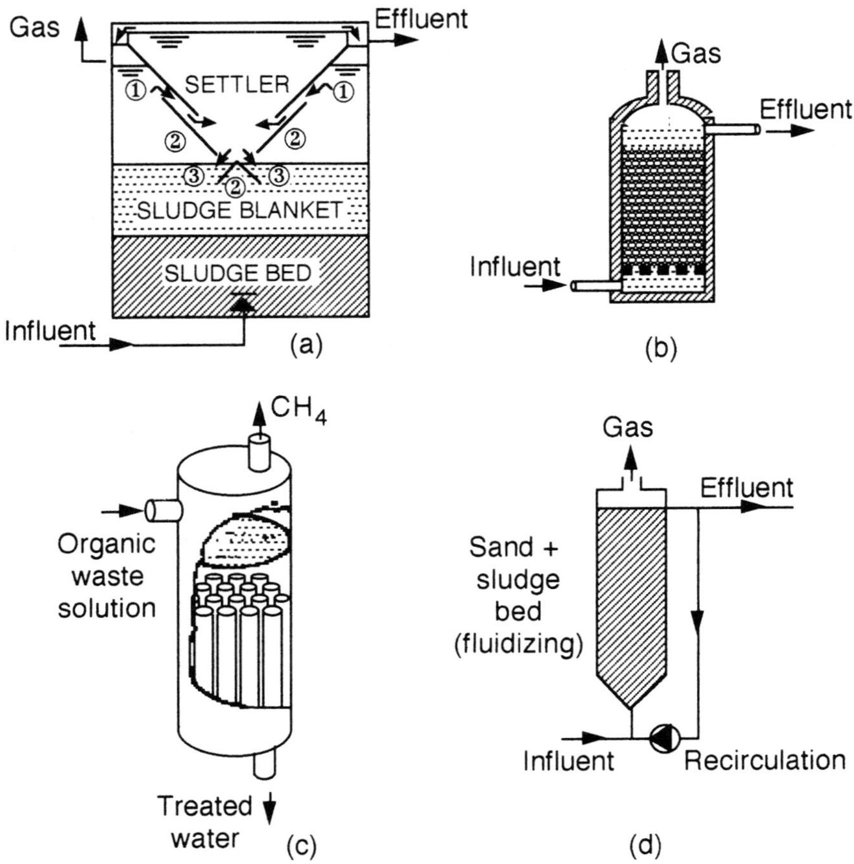

Figure 2.38 Common systems of methane fermentation. (a) Scheme of an upflow anaerobic sludge blanket reactor: 1 = sludge–liquid mixture inlet; 2 = gas screens; 3 = settled sludge return opening. (b) Anaerobic filter containing inert support material onto which the microorganisms are attached. (c) Thin-film reactor in which the methanogenic bacteria adhere to the inside surface of the tubes. (d) Fluidized bed reactor in which the waste water is passed upwards through a bed of particles carrying the bacteria.

carbohydrate-based waste waters this process readily achieves 80–90% removal of COD at space loadings of 2–4 kg COD m^{-3} day^{-1}. The most impressive performance is for purification of waste waters containing acetic acid, methanol and formic acid, which are direct substrates for the methane bacteria and which have very low biomass yield coefficients. Here the anaerobic filter's outstanding ability to retain biomass permits COD removal of 90% at a space loading of 10–20 COD m^{-3} day^{-1} with virtually no surplus sludge production.

Another recent reactor is the thin-film reactor (Figure 2.38c). In this system the acetogenic and methanogenic bacteria adhere to the inside surface of tubes made from red drain-tile clay. Each reactor contains several tubes which have an inside diameter of 5–10 cm and a length of about 140 cm (Figure 2.38c). By admitting waste water from above, rather than below as is done with anaerobic filters, problems of channeling and plugging are avoided. The film formed on clay was about 1–3 mm thick and had an activity of about 0.8–1.2 COD g^{-1} day^{-1}. This fixed biofilm concept was extended by developing the fluidized bed reactor (Figure 2.38d). This contains the biomass fixed on fine particles, forming aggregates that are kept suspended through the upward flowing liquid.

The process has been successfully employed in the treatment of municipal waste water, and industrial effluents for nitrogen control as well as the aerobic biochemical oxygen demand (BOD) treatment. Pilot-scale testing has shown that anaerobic fluidized bed reactors are very effective for the treatment of various wastes including dairy, chemical, food processing, soft drink bottling and heat treatment liquors.

(b) Waste sources and methods of bioconversion. Among biological degradation processes, methanogenesis appears ideally suited for an environmental biotechnology: biomethanation. The sludge remaining after the process is stabilized, i.e. it no longer undergoes fast spontaneous biological degradation.

Anaerobic decomposition is generally considered to progress in two stages, an acid production stage and a methane production stage. The first step in the acid production stage is the hydrolysis and liquefaction of the polymeric molecules: proteins to amino acids, polysaccharides to oligo-saccharides and monosaccharides. The low molecular weight products obtained in the acid production stage are further degraded to methane and carbon dioxide by a highly specialized group of bacteria: *Methanobacterium*, *Methanobacillus*, *Methanococcus* and *Methanosarcina*. Each species within the four genera is quite restrictive with regard to the organic acid or alcohol it can use as a carbon source. Pyruvic and acetic acid occupy key intermediate positions from which further biochemical reactions originate. In the absence of oxygen, some heterotrophic bacteria utilize the sulphur and nitrogen of the sulfate and nitrite anions as electron acceptors.

Agricultural and food wastes. Disposal of processing wastes is a major problem for the food industry. Studies have been undertaken (Keenan and Kormi, 1977; Hills and Dykstra, 1980; Price and Cheremisinoff, 1981) to evaluate the practicality of processing waste disposal using anaerobic digestion with the concurrent production of methane. Through their methane fermentation, up to 12% of the energy consumed in the sugar beet and wine industries is being recovered. Methane-producing processes for BOD reduction and sludge stabilization will play a significant but small role in energy production, but mainly they will alleviate or prevent environmental pollution. Research has demonstrated that, for the treatment of sugar refinery wastes, anaerobic treatment is more energy-saving than aerobic treatment. Promising results have been obtained from anaerobic digestion studies of brewery by-products. One of the advantages is the generation of a saleable microbial biomass, for the bacteria generated are nearly 70% protein (Price and Cheremisinoff, 1981).

Fixed-film biological processes for anaerobic treatment of agricultural waste water have been developed. The active biomass can be immobilized on large-size particles with mean diameters of a few centimeters and more, or oven on continuous matrices. These processes include trickling filters, rotating biological contactors, activated biofilters, biological towers, submerged filters (contact aerators) and biological fluidized beds (Figure 2.39) (Bishop and Kinner, 1986). The organisms responsible for the

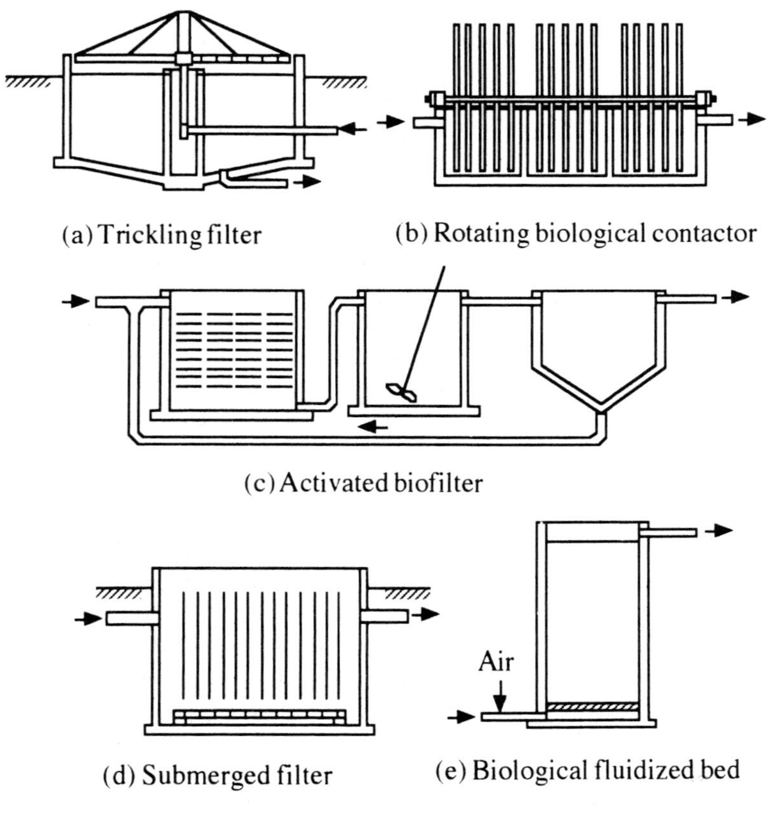

Figure 2.39 Process schematics for fixed-film biological processes. (Reproduced with permission from Bishop, P.L. and Kinner, N.E. Aerobic fixed-film processes, in *Biotechnology*, Vol. 8, edited by H.J. Rehm and G. Reed, pp. 113–76; published by VCH Weinheim, Germany, 1986.)

treatment are present primarily in the form of a biofilm attached to a solid support surface. The biofilm contains bacteria, protozoa and fungi. As waste water passes over the surface of the biofilm, soluble organics diffuse into the film and are adsorbed onto the microorganisms and metabolized. Colloidal macromolecules adsorb onto the biofilm surface and can be solubilized by extracellular enzymes.

The treatment efficiency on a milk-based synthetic organic waste [2-butanol, 16.7 ml, acetic acide, 9.1 ml, NH_3 (conc. soln), 9.4 ml, NaOH for pH = 7.0, milk powder, 11.1 g, water up to 1000 ml] has been examined on a fixed-bed anaerobic reactor with immobilized cells in an organic support (straw) (Guitonas *et al.*, 1994). The biogas produced has a high methane content (70% CH_4), while the biogas production at 15 °C was negligible. It is possible to treat waste waters with low or high concentrations of waste in different operating temperatures (at 35 °C for an applied influent concentration C_i = 4.0 kg COD m^{-3}, the treatment efficiency is 82%; at 25 °C, the treatment efficiency is 85%).

2.9.2 *Immobilized cells and waste water treatment*

Nitrogen salt emissions have led to environmental problems like acidification and increased oxygen demand in surface waters. Also relationships between these emissions and algal blooms in marine environments have been established (Wyatt, 1979; Underdal *et al.*, 1988; Ibrekk *et al.*, 1991). This has been the reason for the reduction of nitrogenous compounds in waste streams.

Immobilized microalgae have been used for biotechnological processes because of the high stability and yields for many biologic conversions (Brouers, 1986). The use of photosynthetic microalgae, which are able to consume the inorganic nitrogen compounds for growth, has been investigated for contaminant bioelimination (Painter, 1977; Jeanfils and Thomas, 1986).

The biologic viability and nitrite uptake by *Chlamydomonas reinhardtii* cells entrapped in calcium alginate beads were studied using discontinuous and continuous flow air-lift reactors (Vilchez and Vega, 1994, 1995). Immobilized cells showed 60–70% of their initial photosynthetic activity, which could be maintained during 3 weeks in the reactor. Immobilized *C. reinhardtii* cells may be used in air-lift reactors, preferentially with continuous flow systems, for water purification purposes.

A dynamic model was developed to study the growth and substrate consumption of immobilized *Nitrobacter agilis* and *Nitrosomonas europaea* cells in an air-lift loop reactor with oxygen as the limiting substrate (Wijffels *et al.*, 1993). It was shown that, in this system, high nitrification capacities could be reached. It was also shown to have promising possibilities to combine nitrification and denitrification in a single reactor.

Activation energies of suspended and immobilized *Nitrobacter agilis* and *Nitrosomonas europaea* in κ-carrageenan were determined (Wijffels *et al.*, 1995). By determination of the effective diffusion coefficient, apparent affinity constants, maximum respiration rates and actual overall respiration rates, it was shown that immobilized cells were relatively insensitive to temperature changes as the result of both diffusion limitation and increased affinity to substrate.

Chevalier and De la Noue (1985) immobilized *Scenedemus obliquus* in κ-carrageenan for nutrient removal from waste waters and they suggested that immobilized microalga may provide a method to prevent waterbody eutrophication. De la Noue and Pruix (1988) and De la Noue and Brasseres (1989) have found on passing secondary-treated sewage waste water through chitosan-immobilized *Phormidium laminosum* and *Phormidium bohneri* a removal of 98% of nitrogen and 60% of phosphorus at a hydraulic retention time of 24 h.

In order to determine whether chitosan enables an adequate viability of microalgal cells, the performance of green microalgae (*Scenedesmus bicellularis*) immobilized on chitosan screens to remove nitrogen and phosphorus ions from simulated urban waste water was monitored (Kaya and Picard, 1996). Sodium phosphate was the chelating agent used for crosslinking and did not affect the diffusion of inorganic nutrients. After the second uptake, the elimination of $NH_4^+ - N$ and $PO_4^{3-} - P$ was 100% within 3 h.

Travieso *et al.* (1996) have studied the immobilization capacities of three different eukaryotic algae (*Chlorella vulgaris*, *Chlorella kessleri* and *Scenedesmus quadricauda*) using internal immobilization with sodium alginate and κ-carrageenan and external immobilization using polystyrene and polyurethane foams. The use of sodium alginate pellets of *C. vulgaris* and *C. kessleri* gave good results with raw sewage. The ammonia nitrogen removal was in the range of 66–82% for the column with immobilized *C. vulgaris*, while the range for *C. kessleri* was 26–64%. Natural light increased the nutrient removal compared with the artificial light.

Immobilization of *Acinetobacter* cells can provide a method which could possibly be used to explain the mechanism of biological phosphorus removal from activated sludge plants (Stephenson, 1987; Toerien *et al.*, 1990) and an efficient phosphate removal process from waste water (Gersberg and Allen, 1984). Various immobilization matrices [κ-carrageenan, ι-carrageenan, polyacrylamide (Tanaka *et al.*, 1977) and alginate (Poncelet *et al.*,. 1992)] have been used.

Pure cultures of *Acinetobacter johnsonii* and *Acinetobacter calcoaceticus* cells were therefore immobilized within alginate beads (Myima and Cloete, 1995) to assess their behavior and, particularly, their survival, leakage, growth and phosphate uptake abilities. The effect of immobilization of the viability depended both on the alginate concentration, the strain and the

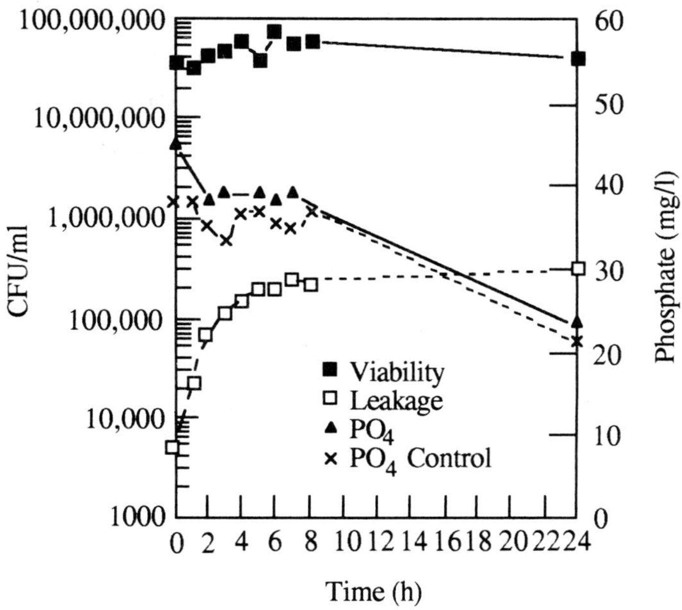

Figure 2.40 Viability, leakage and phosphate uptake relationship of immobilized *A. johnsonii* cells within 3% alginate beads suspended in activated sludge mixed liquor. (Reprinted from *Water Res.*, Vol. 29, Myima, N.Y.O. and Cloete, T.E. Growth and phosphate uptake of immobilized *Acinetobacter* cells suspended in activated sludge mixed liquor, pp. 2461–6, copyright 1995, with kind permission from Elsevier Science Ltd, The Boulevard, Langford Lane, Kidlington OX5 1GB, UK.)

cell density (Figure 2.40) (Myima and Cloete, 1995). The largest quantities of phosphate were removed within the first hour (Figure 2.40).

The potential of the co-immobilized system for denitrifying bacteria and methanogenic bacteria has been recognized in several water purification and fermentation processes (Hashimoto and Furukawa, 1987). Kokufuka *et al.* (1988) studied the process of simultaneous nitrification and denitrification by *Nitrosomonas europaea* and *Paracoccous dentrificans* cells co-immobilized in polyelectrolyte complexes (Beunink and Rehm, 1988, 1990).

A co-immobilized mixed-culture system of denitrifying bacteria and methanogenic bacteria in polyvinyl alcohol (PVA) gel beads, has been developed (Lin and Chen, 1995). Denitrification and methanogenesis with methanol employed as a carbon substrate using the sludges collected from a sewage treatment plant in Taipei City were investigated under anoxic conditions (Lin and Chen, 1995) (Figure 2.41). The simultaneous production of nitrogen and methane gases, in continuous culture, has elucidated that denitrification and methanogenesis could proceed within one reactor (Figure 2.42). Microscopic observation revealed that a

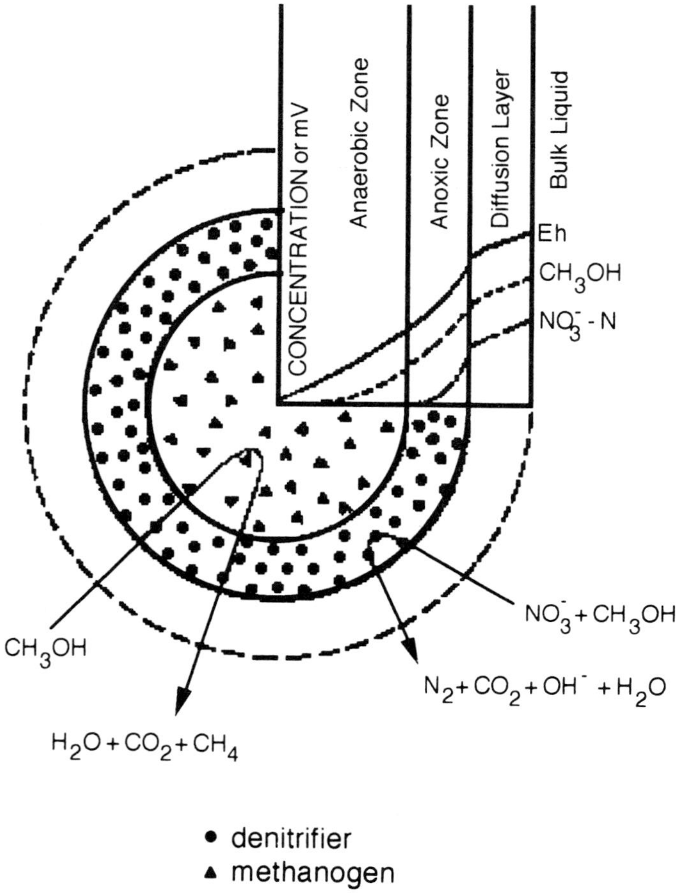

Figure 2.41 Schematic metabolic process for simultaneous denitrification and methanogenesis of the co-immobilized mixed culture system in a polyvinyl alcohol gel bead. (Reprinted from *Water Res.*, Vol. 29, Lin, Y.F. and Chen, K.C. Denitrification and methanogenesis in a co-immobilized mixed culture system, pp. 35–43, copyright 1995, with kind permission from Elsevier Science Ltd, The Boulevard, Langford Lane, Kidlington OX5 1GB, UK.)

Hypomicrobium-like bacterium (a denitrifier) grows mainly on the peripheral surface while a *Methanosarcina*-like bacterium (a methanogen) grows in the inner part of the gel beads.

A new process for waste water biotreatment has been developed using a green microalgae (*Scenedesmus bicellularis*) immobilized on alginate screens, which were starved in air, to obtain a high rate of nutrient removal (Kaya and Picard, 1995). The screens were then inserted in a photochamber saturated at 100% relative humidity and subjected at a photoperiod of 16 h. After 48 h of nutrient starvation, the immobilized *S. bicellularis* were

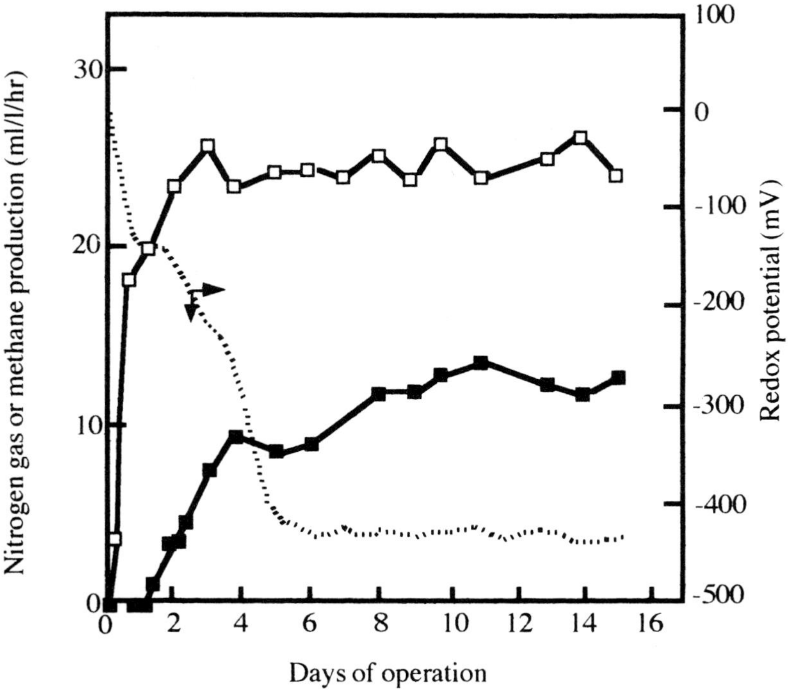

Figure 2.42 Gas production and the monitoring redox potential (E_h) of the co-immobilized mixed culture system during continuous operation. □ = Nitrogen production rate; ■ = methane production rate; .... = E_h. (Reprinted from *Wat. Res.*, Vol. 29, Lin, Y.F. and Chen, K.C. Denitrification and methanogenesis in a co-immobilized mixed culture system, pp. 35–43, copyright 1995, with kind permission from Elsevier Science Ltd, The Boulevard, Langford Lane, Kidlington OX5 1GB, UK.

used for removal of ammonium and orthophosphate from waste water (Figure 2.43).

Enzymatic detoxification has been employed for the removal of aromatic compounds from waste and drinking water because polymeric products are insoluble and therefore easily removed by filtration (Klibanov *et al.*, 1980; Klibanov and Morris, 1981; Atlow *et al.*, 1984). Phenoloxidases have been successfully immobilized on supports such as soils, clays and porous glass beads (Leonowicz *et al.*, 1988; Ruggiero *et al.*, 1989; Sarkar *et al.*, 1989). Davis and Burns (1990) used laccase, tyrosinase and peroxidase entrapped in alginate beads for color removal from phenolic industrial effluents.

Gelatine gels obtained from water in oil microemulsions in which the ternary system consists of iso-octane/sulfosuccinic acid bis-[2-ethyl hexyl] ester/water were used to immobilize 2-phenoloxidase, a laccase from *Trametes versicolor* and a tyrosinase from mushroom (Crecchio *et al.*, 1995). Laccase and tyrosinase are firmly retained inside the gels without any release of activity in the aqueous phase and without any change in their

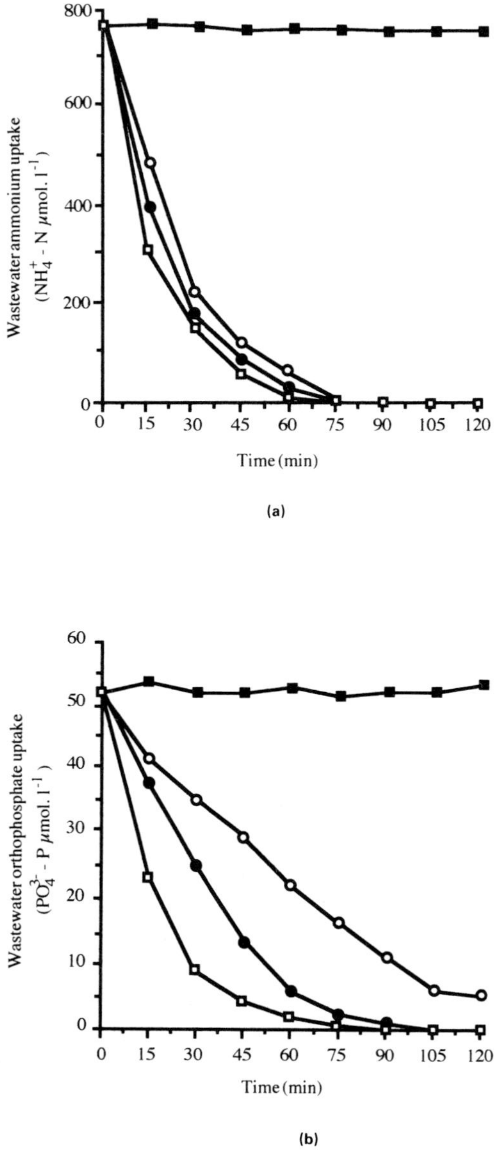

Figure 2.43 Uptake of ammonium (a) and orthophosphate (b) from artificial effluent by *Scenedesmus bicellularis* immobilized on alginate screens after nutrient starvation in air substrate at 100% relative humidity. ■ = Calcium alginate without microalge cells (control); ○ = first trial after 48 h of nutrient starvation in moist air; ● = second trial after the first uptake followed by 48 h of nutrient starvation; □ = third trial after the second uptake followed by 48 h of nutrient starvation. (Reproduced with permission from Kaya, V.M. and Picard, G. The viability of *Scenedesmus bicellularis* cells immobilized an alginate screens following nutrient starvation in air at 100% relative humidity. *Biotechnol. Bioeng.*, **46**, 459–64; published by John Wiley & Sons, Inc., 1995.)

Table 2.9 Removal of natural and xenobiotic aromatic compounds by laccase and tyrosinase immobilized in organic gels

Substrate ($100\ mg\ l^{-1}$)	Immobilized laccase (0.95 U)		Immobilized tyrosinase (60 U)	
	Incubation time (h)	Maximum removal (%)	Incubation time (h)	Maximum removal (%)
1-Naphthol	1	98	4	40
Vanilic acid	1	90	16	10
4-Chloroaniline	1	95	4	45
2,6-Dimethoxyphenol	1	95	4	50
2,4-dichlorophenol	16	35	16	42
Syringic acid	4	70	4	75
Catechol	1	70	1	92
Caffeic acid	4	75	4	90

(Reproduced with permission from Crecchio, C. *et al.* Polyphenoloxidases immobilized in organic gels: properties and applications in the detoxification of aromatic compounds. *Biotechnol. Bioeng.*, **48**, 585–91; published by John Wiley Sons, Inc., 1995.)

pH optimum. Table 2.9 shows that 1-naphthol, 2,6-dimethoxyphenol, 4-chloroaniline and vanilic acid were almost completely removed in laccase in a relatively short period (less than 1 /h).

Degradation of quinoline by *Comamonas acidovorans* in a stirred tank reactor using batch and continuous cultivation has been investigated by Miethling *et al.* (1983). *Comamonas acidovorans* is able to use quinoline as the sole source of carbon, nitrogen and energy. Quinoline degradation by *C. acidovorans* was investigated using a three-phase airlift reactor (Ulonska *et al.*, 1995). Porous glass beads were used for immobilization of the microorganisms. The biofilm efficiency was about 54% compared to suspended organisms. The immobilized cells in the reactor were able to survive for at least 25 h.

Biodegradation of dyes and dye-industry waste effluents has been reported (Bumpus and Brock, 1989; Kanekar and Sarnaik, 1991). A culture of *Pseudomonas alcaligenes* was immobilized on rock media packed to treat waste effluent from dye-industry manufacturing methyl violet, rhodamine B, nigrosine and chrysoidine (Kanekar and Sarnaik, 1995). The treatment resulted in the removal of 51% COD, 82% BOD, 74% total organic carbon (TOC) and 60% color in terms of methyl violet.

2.10 Immobilized microorganisms in waste gas purification

For the purification of waste air containing a few known chemicals, the degrading microflora may be restricted to a few species of microbes. In these cases it has become a common practice to inoculate the filter bed using pure cultures of microorganisms known to degrade the pollutants actively. This process is well established in Europe and Japan where it has been used as an air pollution-control technology successfully for controlling odors, volatile organic compounds and air-toxic compounds.

Hydrogen can be continuously produced using immobilized photosynthetic bacterium. The photosynthetic bacterium (*Rhodospirillium rubrum*) immobilized with calcium alginate has resulted in continuous hydrogen production for over 20 days at a rate of 20–30 l min^{-1}. The energy conversion efficiency from organic acids to hydrogen has been 28% (Karube, 1989). A bacterial fuel cell has been manufactured which produces hydrogen from molasses using a combination of an immobilized

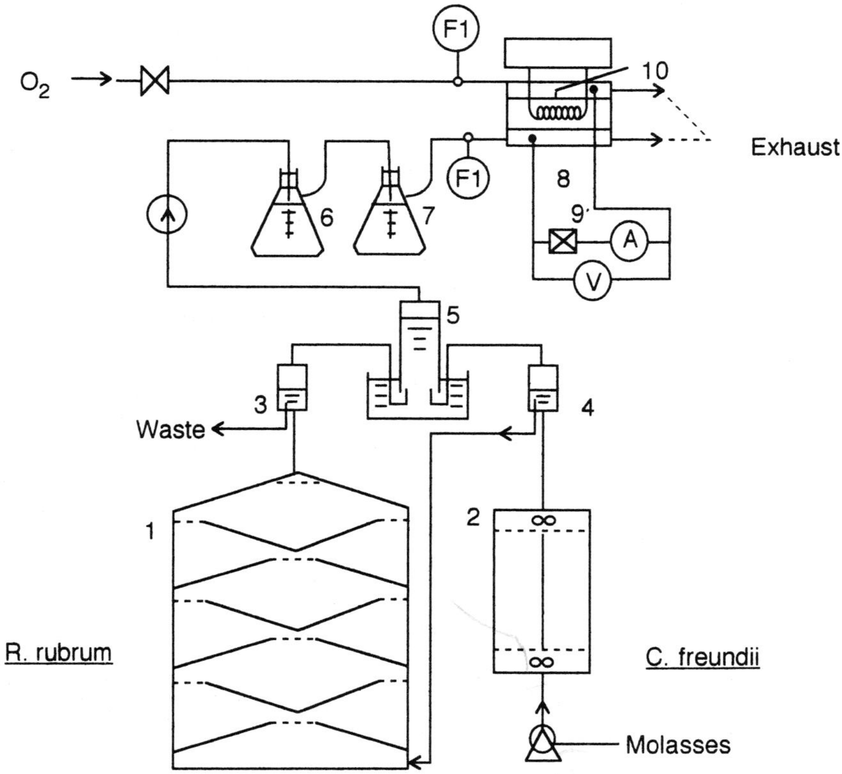

Figure 2.44 Bacterial fuel cell using immobilized hydrogen-producing bacteria and photosynthetic bacteria. 1 = Bioreactor for immobilized photosynthetic bacteria; 2 = bioreactor for immobilized hydrogen producing bacteria, 3,4,5 = H_2 reservoir; 6 = KOH solution; 7 = water; 8 = fuel cell; 9 = variable electronic load; 10 = heater; F1 = flow indicator; A = ammeter; V = voltameter. (Reproduced with permission from Karube, I. Bioelectric cells, in *Biomass Handbook*, edited by O. Kitani, C.W. Hall, pp. 809–18; published by Gordon and Breach Science Publishers, New York, 1989.)

hydrogen-producing bacterium (*Citrobactor freundii*) and an immobilized photosynthetic bacterium (*R. rubrum*) (Figure 2.44) (Karube, 1989).

References

Adami, A., Cavazzoni, V., Trezzi, M. and Craveri, R. (1988) Cellobiose hydrolysis by *Trichosporon pullulans* in calcium alginate. *Biotechnol. Bioeng.*, **32**, 391–402.

Akita, H. (1996) Recent advances in the use of immobilized lipases directed toward the asymmetric syntheses of complex molecules. *Biocat. Biotransf.*, **13**, 141–56.

Akita, H., Umezawa, I., Matsukura, H. and Oishi, T. (1989) A lipid–lipase aggregate as a new type of immobilized enzyme. *Chem. Pharm. Bull.*, **39**, 1632–33.

Akita, H., Umezawa, I., Tisnadjaja, D., Matsukura, H. and Oishi, T. (1993) Enantioselective acetylation of an α-hydroxy ester by using ether-linked lipid–lipase aggregates in organic solvents. *Chem. Pharm. Bull.*, **41**, 16–20.

Amarant, T. and Bohak, Z. (1981) Immobilization of protein as a tool for studying primary structure around their cysteinyl residues. *Appl. Biochem. Biotechnol.*, **6**, 237–45.

Ariga, O., Kato, M., Sano, T., Nakazawa, Y. and Sano, Y. (1993) Mechanical and kinetic properties of PVA hydrogel immobilizing beta-galactosidase. *J. Ferment. Bioeng.*, **76**, 203–10.

Atlow, S.C., Bonadonna-Aparo, L. and Klibanov, A.M. (1984) Dephenolization of industrial waste waters catalyzed by polyphenol oxidase. *Biotechnol. Bioeng.*, **26**, 599–603.

Audet, P., Paquin, C. and Lacroix, C. (1988) Immobilization growing lactic acid bacteria with $\varkappa$-carrageenan-locust bean gum gel. *Appl. Microbiol. Biotechnol.*, **29**, 11–19.

Audet, P., Paquin, C. and Lacroix, C. (1989) Sugar utilization and acid production by free and entrapped cells of *Streptococcus salivarius* subsp. *thermophilus*, *Lactobacillus delbrueckii* subsp. *bulgaris*, and *Lactococcus lactis* subsp. *lactis* in a whey permeate medium. *Appl. Environ. Microbiol.*, **55**, 185–91.

Audet, P., Paquin, C. and Lacroix, C. (1990) Batch fermentations with a mixed culture of lactic acid bacteria immobilized separately in $\varkappa$-carrageenan locust bean gum gel beads. *Appl. Microbiol. Biotechnol.*, **32**, 662–73.

Babuchowski, A., Hammond, E.G. and Glatz, B.A. (1993) Survey of propionibacteria for ability to produce propionic acid. *J. Food Product.*, **56**, 493–6.

Bajpai, P. and Margaritis, A. (1987a) The effect of temperature and pH on ethanol production by free and immobilized cells of *Kluyveromyces marxianus* grown on Jerusalem artichoke extract. *Biotechnol. Bioeng.*, **30**, 306–13.

Bajpai, P. and Margaritis, A. (1987b) Kinetics of ethanol production by immobilized *Kluyveromyces marxianus* cells at varying sugar concentrations of Jerusalem artichoke juice. *Appl. Microbiol. Biotechnol.*, **26**, 447–9.

Bajpai, R., Thompson, J.E. and Davison, B.H. (1990) FBR, with extractive fermentation to remove butanol. *Appl. Biochem. Biotechnol.*, **24**, 25–31.

Banat, I.M., Nigram, P. and Marchant, R. (1992) Isolation and thermotolerant yeasts growing at 52 °C and producing ethanol at 45 °C and 50 °C. *World J. Microbiol. Biotechnol.*, **8**, 259–63.

Barbotin, J.N., Saucedo, J.E.N., Bazinet, C. *et al.* (1992) Immobilization of whole cells and somatic embryons: coating process and cell-matrix interactions in *SYNSEEDS, Applications of Synthetic Seeds to Crop Improvement* (ed. K. Redenbauch), CRC Press, Boca Raton, pp. 66–103.

Barron, N., Marchant, R., McHale, L. and McHale, A.P. (1994) Growth of thermotolerant ethanol-producing strain of *Kluyeromyces marxianus* on cellulosic-containing media. *Biotech. Lett.*, **16**, 625–30.

Beavan, N., Zawadzki, B., Droniuk, R., Lawford, H. and Fein, J. (1989) Comparative performance trials with yeast and *Zymomonas* for fuel alcohol production from corn. *Appl. Biochem. Biotechnol.*, **20/21**, 319–26.

Bender, M.L. (1987) Kinetic studies of immobilized α-chymotripsin in aprotic solvents, in *Methods in Enzymology*, Vol. 135, Part B (ed. K. Mosbach), Academic Press, Orlando, pp. 537–56.

Benito, G.G., Ozores, M. and Pena, M. (1994) Continuous glycerol production in a packed-bed bioreactor with immobilized cells of *Saccharomyces cerevisiae. Bioresource Technol.*, **49**, 209–12.

Beunink, J. and Rehm, H.J. (1988) Synchronous anaerobic and aerobic degradation of DDT by an immobilized mixed culture system. *Appl. Microbiol. Biotechnol.*, **29**, 72–80.

Beunink, J. and Rehm, H.J. (1990) Coupled reductive and oxidative degradation of 4-chloro-2-nitrophenol by a co-immobilized mixed culture system. *Appl. Microbiol. Biotechnol.*, **34**, 108–15.

Bickerstaff, G.F. (1982) Applications of immobilized enzymes to fundamental studies on enzyme structure and function, in *Topics in Enzyme and Fermentation Biotechnology 9* (ed. A. Wiseman), Ellis Harwood, New York, pp. 162–75.

Binot, R.A. (1983) Biométhanisation en lit fluidisé de microorganismes immobilisés. Thèse de Doctorat en Sciences Agronomiques, Université Catholique de Louvain, Belgium.

Bishop, P.L. and Kinner, N.E. (1986) Aerobic fixed-film processes, in *Biotechnology*, Vol. 8 (eds H.J. Rehm and G. Reed), VCH, Weinheim, pp. 113–76.

Bisping, B. and Rehm, H.J. (1986) Glycerol production by cells of *Saccharomyces cerevisiae*, immobilized in sintered glass. *Microbiol. Biotechnol.*, **23**, 174–9.

Bowers, L.D. (1986) Applications of immobilized biocatalysts in chemical analysis. *Anal. Chem.*, **58**, 513–20.

Boyaval, P. and Goulet, J. (1988) Optimal conditions for production of lactic acid from cheese whey permeate by Ca-alginate-entrapped *Lactobacillus helveticus*. *Enzyme Microbiol. Technol.*, **10**, 725–34.

Brady, D., Marchant, R., McHale, L. and McHale, A.P. (1994) Production of ethanol by thermotolerant yeast *Kluyeromyces marxianus* IMB3 during growth on lactose-containing media. *Biotech. Lett.*, **16**, 737–40.

Brink, L.E.S. and Tramper, J. (1986) Design of an organic-liquid-phase/immobilized-cell reactor for the microbial epoxidation of propene, in *Biocatalysis in Organic Media* (eds C. Laane, J. Tramper and M.D. Lilly), Elsevier Science Publishers, Amsterdam, pp. 133–46.

Brodelius, P. and Nilsson, K. (1980) Entrapment of plant cells in different matrices. A comparative study. *FEBS Lett.*, **122**, 312–8.

Brodelius, P., Deus, B., Mosbach, K. and Zenk, M.H. (1979) Immobilized plant cells for the production and transformation of natural products. *FEBS Lett.*, **103**, 93–99.

Brouers, M. (1986) Hydrogen production by immobilized Scenedesmus cells and ammonia production by immobilized Anabaena filaments, in *Process Engineering Aspects of Immobilized Cell Systems*, Manchester, England, 1984, Institute of Chemical Engineers, Rugby, pp. 272–6.

Bucke, C. (1987) Cell immobilization in calcium alginate. *Meth. Enzymol.*, **135**, 175–95.

Bumpus, J.A. and Brock, B.J. (1989) Biodegradation of crystal violet by the white rot fungus Phanerochete cryssosporium. *Appl. Environ. Microbiol.*, **54**, 1141–50.

Burgess, D.J. (1994) Complex coacervation: microcapsule formation, in *Macromolecular Complexes in Chemistry and Biology* (eds p. Dubin, J. Bock, R. Davis, D.N. Schulz and C. Thies), Springer-Verlag, Hiedelberg, pp. 285–300.

Cachon, R., Molin, P. and Diviès, C. (1995) Modeling of continuous Ph-stat stirred tank reactor with *Lactococcus lactis* spp *lactis* bv. *diacetylactis* immobilized in calcium alginate gel beads. *Biotechnol. Bioeng.*, **47**, 567–74.

Campbell, W.R., Lamptey, J. and Murray, M.Y. (1985) Ethanol production in a surface-immobilized yeast, packed-bed bioreactor, in *Bioenergy 84*, Vol. III, Biomass Conversion, (eds H. Egneus and A. Ellegard), Elsevier Applied Science, London, pp. 220–28.

Carr, P.W. and Bowers, L.D. (1980) *Immobilized Enzymes in Analytical and Clinical Chemistry*. J. Wiley and Sons, New York.

Carta, G., Gainer, J.L. and Benton, A.H. (1991) Enzymatic synthesis of esters using an immobilized lipase. *Biotechnol. Bioeng.*, **37**, 1004–9.

Carta, G., Gainer, J.L. and Gibson, M.E. (1992) Synthesis of esters using a nylon-immobilized lipase in batch and continuous reactors. *Enzyme Microbiol. Technol.*, **14**, 904–10.

Castillo, E. and Casas, L.T. (1990) Reutilization of free and immobilized *Kluyveromyces fragilis* yeast cells with a controlled permeabilization treatment, in *Physiology of Immobilized Cells* (eds J.A.M. de Bont, J. Visser, B. Mattiasson and J. Tramper), Elsevier Science Publishers, Amsterdam, pp. 213–31.

Champagne, C.P. and Côté, C.B. (1987) Cream fermentation by immobilized lactic acid bacteria. *Biotechnol. Lett.*, **9**, 329–32.
Champagne, C.P., Baillargeon, C.B. and Goulet, J. (1989) Whey fermentation by immobilized cells of *Propionibacterium shermanii. J. Appl. Bacteriol.*, **66**, 175–80.
Champagne, C.P., Morin, N., Couture, R. *et al.* (1992) The potential of immobilized cell technology to produce freeze-dried, phage-protected cultures of *Lactobacillus lactis. Food Res. Int.*, **25**, 419–31.
Champagne, C.P., Girard, F. and Rodriguez, N. (1993) Production of concentrated suspensions of thermophilic lactic acid bacteria in calcium-alginate beads. *Int. Dairy J.*, **3**, 257–65.
Champagne, C.P., Lacroix, C. and Sodini-Gallot, I. (1994) Immobilized cell technologies for the dairy industry. *Cr. Rev. Biotechnol.*, **14**, 109–34.
Champluvier, B., Kamp, B. and Rouxhet, P.G. (1988) Immobilization of beta-galactosidase retained in yeast: adhesion of the cells on a support. *Appl. Microbiol. Biotechnol.*, **27**, 464–72.
Champluvier, B., Marchal, F. and Rouxhet, P.G. (1989) Immobilization of lactase in yeast cells retained in a glass wool matrix. *Enzyme Microbiol. Technol.*, **11**, 422–31.
Chamy, R., Nunes, M.J. and Lema, J.M. (1990) Optimization of the hardening treatment of *S. cerevisiae* bioparticles. *Enzyme Microbiol. Technol.*, **12**, 749–54.
Chang, P.S., Rhee, J.S. and Kim, J.J. (1991) Continuous glycerolysis of olive oil by *Chromobacterium viscosum* lipase immobilized on liposome in reversed micelles. *Biotechnol. Bioeng.*, **38**, 1159–65.
Charles, M. and Phillips, J. (1985) Combined immobilized enzyme/cell systems, in *Comprehensive Biotechnology*, Vol. 2 (eds M. Moo-Young, C.W. Robinson and J.A. Howell), Pergamon Press, New York, pp. 225–9.
Chavasit, V., Kienzle-Sterzer, C. and Torres, J.A. (1988) Formation and characterization of an insoluble polyelectrolyte complex: chitosan–polyacrylic acid. *Polym. Bull. (Berlin)*, **19**, 223–9.
Chevalier, P. and De la Noue, J. (1985) Efficiency of the immobilized hyperconcentrated algae on ammonia and orthophosphate removal from wastewaters. *Biotechnol. Lett.*, **7**, 395–400.
Chiancone, E. and Gattoni, M. (1987) Use of immobilized subunits for the purification of oligomeric and self-associating proteins, in *Methods in Enzymology*, Vol. 135, Part B, (ed. K. Mosbach), Academic Press, Orlando, pp. 484–512.
Chibata, I., Tosa, T. and Sato, T. (1979) *Microbial Technology*, 2nd edn (eds H.J. Peppler and D. Perlman), Academic Press, San Diego.
Chibata, I., Tosa, T. and Sato, T. (1986) Biocatalysis: immobilized cells and enzymes. *J. Macromol. Catalysis*, **37**, 1–24.
Chibata, I., Tosa, T., Sato, T. and Takata, I. (1987) Immobilization of cells in carrageenan. *Meth. Enzymol.*, **135**, 189–98.
Chu, C.H., Sakiyama, T. and Yano, T. (1995) pH-sensitive swelling polyelectrolyte complex gel prepared from xanthan and chitosan. *Biosci. Biotech. Biochem.*, **59**, 717–19.
Chu, C.H., Kumagai, H. and Nakamura, K. (1996) Application of polyelectrolyte complex gel composed of xanthan and chitosan to the immobilization of *Corynebacterium glutamicum. J. Appl. Polym. Sci.*, **60**, 1041–7.
Compere, A.L. and Griffith, W.L. (1981) Microorganism immobilization, U.S. patent 4,287,305 (Sept. 1).
Crecchio, C., Ruggiero, P. and Pizzigallo, M.D.R. (1995) Polyphenoloxidases immobilized in organic gels: properties and applications in the detoxification of aromatic compounds. *Biotechnol. Bioeng.*, **48**, 585–91.
Crescenzi, V., Imbriaco, D., Velasquez, C.L., Dentini, M. and Ciferri, A. (1995) Novel types of polysaccharide assemblies. *Macromol. Chem. Phys.*, **196**, 2873–80.
Cross, J.S. and Clausen, E.C. (1993) Effects of organic buffers on batch fermentations of Zymomonas mobilis in a synthetic and complex medium. *Biomass Energ.*, **4**, 277–81.
Cross, J.S., Vega, J.L., Clausen, E.C. and Gaddy, J.L. (1988) Performance of a biofilm reactor with *Zymomonas mobilis* for ethanol production. *196th ACS National Meeting*, Los Angeles, California, Sept. 25–30.
Cross, J.S., Vega, J.L. and Clausen, E.C. (1993) Performance of a cross-linked immobilized

cell reactor with *Zymomonas mobilis* using synthetic and complex media for ethanol production. *Biomass Energ.*, **4**, 283–91.
Crumbliss, A.L., Stonehuerner, J.G. and Henkens, R.W. (1993) Carrageenan hydrogel stabilized colloidal gold multi-enzyme biosensor electrode utilizing immobilized horseradish peroxidase and cholesterol oxidase/cholesterol esterase to detect cholesterol in serum and whole blood. *Biosensor Bioelectronics*, **8**, 331–9.
Cruz, R., Batistela, J.C. and Wosiacki, G. (1981) Microbial alfa-galactosidase for soymilk processing. *J. Food Sci.*, **46**, 1196–204.
Daly, M.M. and Knorr, D. (1988) Chitosan–alginate complex coacervate capsules: effects of calcium chloride, plasticizers, and polyelectrolytes on mechanical stability. *Biotechnol. Prog.*, **4**, 76–95.
Danzing, J., Tischer, W. and Wandrey, C. (1995) Continuous enzyme-catalyzed production of 6-aminopenicillanic acid and product concentration by reverse osmosis. *Chem. Eng. Technol.*, **18**, 256–62.
Daugulis, A.J., Brown, N.M., Cluett, W.R. and Dunlop, D.B. (1981) Production of ethanol by adsorbed yeast cells. *Biotechnol. Lett.*, **3**, 651–9.
Davis, S. and Burns, R.G. (1990) Decolorization of phenolic effluents by soluble and immobilized phenol oxidase. *Appl. Microbiol. Biotechnol.*, **32**, 721–6.
Davison, B.H. (1989) Dispersion and holdup in a three-phase fluidized bed. *Appl. Biochem. Biotechnol.*, **20/21**, 449–53.
Decleire, M., Van Huynh, N., Motte, J.C. and De Cat, W. (1985) Hydrolysis of whey by whole cells of *Kluyveromyces bulgaricus* immobilized in calcium alginate gels and in hen white. *Appl. Microbiol. Biotechnol.*, **22**, 438–45.
De la Noue, J. and Brasseres, A. (1989) Biotreatment of anaerobically digested swine manure with microalgae. *Biol. Wastes*, **29**, 17–31.
De la Noue, J. and Pruix, D. (1988) Biological tertiary treatment of urban wastewater with chitosan-immobilized *Phormidium*. *Appl. Microbiol. Biotechnol.*, **29**, 292–7.
Dinelli, D. (1972) Fiber-entrapped enzymes. *Process Biochem.*, **7**, 9–15.
Doubradi, V., Hjdu, J., Bot, G. and Friedrich, P. (1980) Structural changes in glycogen phosphorylase as revealed by cross-linking with bifunctional diimidates: phosphodephospho hybrid and phosphorylase A. *Biochemistry*, **19**, 2295–302.
Dubay, A.K., Bisaria, V.S., Mukhopadhyay, S.N. and Ghose, T.K. (1989) Stabilization of restriction endonuclease Bam HI by cross-linking reagents. *Biotechnol. Bioeng.*, **33**, 1311–20.
Dubin, P.L., Gao, J. and Mattison, K.W. (1994) Protein purification by selective phase separation with polyelectrolytes. *Sep. Purif. Meth.*, **23**, 1–7.
Dumitriu, S. (1991) Processes with immobilized enzymes and cells, in *Bioconversion of Waste Materials to Industrial Products* (ed. A.M. Martin), Elsevier Applied Science, London, New York, pp. 64–115.
Dumitriu, S. and Chornet, E. (1996a) Functional versatility of polyionic hydrogels, in *Chitin Enzymology*, Vol. 2 (ed. R.A.A. Muzzarelli), Atec Edizioni, Italy, pp. 543–64.
Dumitriu, S. and Chornet, E. (1996b) Polyionic hydrogels as supports for enzyme immobilization, in *Chitin Enzymology*, Vol. 2 (ed. R.A.A. Muzzarelli), Atec Edizioni, Grottammare, Italy, pp. 527–42.
Dumitriu, S., Magny, P., Montane, D., Vidal, P.F. and Chornet, E. (1994) Polyionic hydrogels obtained by complexation between xanthan and chitosan: their properties as supports for enzyme immobilization. *J. Bioactive Compat. Polym.*, **9**, 184–209.
Durand, G. and Monsan, P. (1974) *Serie Syntheses Bibliographiques*, No. 5, CDIUPA-APRIA, Massy, Paris.
Ergan, F., Trani, M. and André, G. (1990) Production of glycerides from glycerol and fatty acid by immobilized lipases in non-aqueous media. *Biotechnol. Bioeng.*, **35**, 195–200.
Evnin, L.B. and Craik, C.S. (1988) Development of an efficient method for generating and screening active trypsin and trypsin variants. *Ann. N.Y. Acad. Sci.*, **542**, 61–7.
Fernandez-Lafuente, R., Rosell, C.M., Rodriguez, V. and Guisan, J.M. (1995) Strategies for stabilization by intramolecular crosslinking with bifunctional reagents. *Enzyme Microbiol. Technol.*, **17**, 517–23.
Finley, J.W., Stanely, W.L. and Watters, G.G. (1977) Removal of chill haze from beer with papain immobilized on chitin. *Biotechnol. Bioeng.*, **19**, 1895–903.
Fleming, M., Barron, N., McHale, L., Marchant, R. and McHale, A.P. (1993) Studies on the

growth of thermotolerant yeast strain, *Kluyeromyces marxianus* IMB3 on sucrose-containing media. *Biotechnol. Lett.*, **15**, 1195–8.
Fukuda, H. (1980) Polyelectrolyte complexes of chitosan with sodium carboxymethylcellulose. *Bull. Chem. Soc. Jpn.*, **53**, 837–40.
Galazzo, J.L. and Bailey, J.E. (1990) Growing *Saccharomyces cerevisiae* in calcium-alginate beads induces cell alterations which accelerate glucose conversion to ethanol. *Biotechnol. Bioeng.*, **36**, 417–25.
Ganapathi, S., Butterfield, D.A. and Bhattacharyya, D. (1995) Flat-sheet and hollow fiber membrane bioreactors: a study of the kinetics and active site conformational changes of immobilized papain including sorption studies of reaction constituents. *J. Chem. Technol. Biotechnol.*, **64**, 157–64.
Gencer, M.A. (1981) Ethanol fermentation in a yeast immobilized tubular fermenter, Ph.D. thesis, Drexel University, USA.
Gersberg, R.M. and Allen, D.W. (1984) Phosphorus uptake by *Klebsiella pneumoniae* and *Acinetobacter calcoaceticus*, in *Proceedings of IAWPRC Conference on Enhanced Biological Phosphorus Removal from Waste Water*, Paris, pp. 146–51.
Ghose, T.K. and Tyagi, R.D. (1979) Rapid fermentation of cellulose hydrolysate. I: Batch vs. continuous systems. *Biotechnol. Bioeng.*, **21**, 1387–95.
Ghose, T.K. and Bandyopadhyay, K.K. (1980) Rapid ethanol fermentation in immobilized yeast cell reactor. *Biotechnol. Bioeng.*, **22**, 1489–96.
Gilson, C.D. and Thomas, A. (1993) A novel fluidized bed bioreactor for fermentation of glucose to ethanol using alginate immobilized yeast. *Biotechnol. Technol.*, **7**, 397–400.
Gite, S. and Shankar, V. (1993) Preparation and properties of RNase T1 immobilized on aminoethyl Bio-Gel P-2. *J. Biotechnol.*, **28**, 339–48.
Gite, S., Reddy G. and Shankar, V. (1992a) Active-site characterization of S1 nuclease. I. Affinity purification and influence of amino-group modification. *Biochem. J.*, **285**, 489–95.
Gite, S., Reddy G. and Shankar, V. (1992b) Active-site characterization of S1 nuclease. II. Involvement of histidine in catalysis. *Biochem. J.*, **288**, 571–83.
Glosset, G.P., Cobb, J.T. and Shah, Y.T. (1974) Study of performance of a tubular membrane reactor for an enzyme catalyzed reaction. *Biotechnol. Bioeng.*, **16**, 345–51.
Godia, F., Casas, C., Castellano, B. and Sola, C. (1987) Immobilized cells behaviour of carrageenan entrapped yeast during continuous ethanol fermentation. *Appl. Microbiol. Bioeng.*, **30**, 836–40.
Godia, F., Casas, C. and Sola, C. (1987) Mathematical modelisation of a packed-bed reactor performance with immobilized yeast for ethanol fermentation. *Biotechnol. Bioeng.*, **30**, 836–43.
Gombin, D., Klamar G., Toth, M. and Szajani, B. (1994) Production of ethanol from thinned starch. *Hungarian J. Ind. Chem.*, **22**, 273–7.
Green, J.H. and Kramer, A. (1979) *Food Processing Waste Management*. AVI Publishing Co., Westport, CT.
Groot, W.J., Schoutens, G.H., Van Beelen, P.N., Van den Oever, C.E. and Kossen, N.W.F. (1984) Increase of substrate conversion by pervaporation in the continuous butanol fermentation. *Biotechnol. Lett.*, **6**, 789–92.
Grzywnowicz, K., Greppin, H., Brzyska, M. and Labarzewski, J. (1983) Changes in activity of soluble and immobilized peroxidases after treatment by various proteases and some metal ions. *J. Molec. Cat.*, **80**, 117–23.
Guenette, M. and Duvnjak, Z. (1996) Wood blocks as a carrier for *Saccharomyces cerevisiae* used in the production of ethanol and fructose. *Chem. Eng. J. Biochem. Eng. J.*, **61**, 233–40.
Guiseley, K.B. (1989) Chemical and physical properties of algal polysaccharides used for cell immobilization. *Enzyme Microbiol. Technol.*, **11**, 706–11.
Guitonas, A., Paschalidis, G. and Zoubpulis, A. (1994) Treatment of strong wastewaters by fixed bed anaerobic reactors with organic support. *Water Sci. Technol.*, **29**, 257–63.
Hackel, U., Klein, J., Meguet, R. and Wagner, F. (1975) Immobilization of microbial cells in polymeric matrices. *Eur. J. Appl. Microbiol.*, **1**, 291–302.
Hang, Y.D., Hamamei, H. and Woodams, E.E. (1989) Production of L(+)lactic acid by *Rhizopus orizae* immobilized in calcium alginate gels. *Biotechnol. Lett.*, **11**, 1198–204.
Hashimoto, S. and Furukawa, K. (1987) Immobilization of activated sludge by PVA-boric acid method. *Biotechnol. Bioeng.*, **30**, 52–9.

Hayashi, T. and Ikada, Y. (1990a) Protease immobilization onto polyacrolein microspheres, *Biotechnol. Bioeng.*, **35**, 518–23.

Hayashi, T. and Ikada, Y. (1990b) Lipoprotein lipase immobilization onto polyacrolein microspheres. *Biotechnol. Bioeng.*, **36**, 593–600.

Hills, D.J. and Dykstra, K. (1980) Anaerobic digestion of cannery tomato solid wastes. *J. Environ. Eng. Div.*, ASCE 106 (EE2).

Hirano, S., Mizutani, C., Yamaguchi, R. and Miura, D. (1978) Formation of the polyelectrolyte complexes of some acidic glycosaminoglycans with partially *N*-acetylated chitosans. *Biopolymers*, **17**, 805–10.

Hunik, J.H., Bos, C.G., van den Hoogen, M.P., De Gooijer, D.C. and Tramper, J. (1994) Co-immobilization of *Nitrosomonas europaea* and *Nitrobacter agilis* cells: validation of a dynamic model for simultaneous substrate conversion and growth in κ-carrageenan gel beads. *Biotechnol. Bioeng.*, **43**, 1153–62.

Ibrekk, H.O., Molvaer, J. and Faafeng, B. (1991) Nutrient loading to Norwegian coastal waters and its contribution to the pollution of the North Sea. *Water Sci. Technol.*, **24**, 239–49.

Ikeda, S., Kumagai, H., Sakiyama, T., Chu, C.H. and Nakamura, K. (1995) Method for analyzing pH-sensitive swelling of amphoteric hydrogels – Application to a polyelectrolyte complex gel prepared from xanthan and chitosan. *Biosci. Biotechnol. Biochem.*, **59**, 1422–7.

Itoyama, K., Tokura, S. and Hayashi, T. (1994) Lipoprotein lipase immobilization onto porous chitosan beads. *Biotechnol. Prog.*, **10**, 225–9.

Jain, D. and Ghose, T.K. (1984) Cellobiose hydrolysis using *Pichia etchellsii* cells immobilized in calcium alginate. *Biotechnol. Bioeng.*, **26**, 340–8.

Jeanfils, J. and Thomas, D. (1986) Culture and nitrite uptake in immobilized *Scenedesmus obliquus*. *Appl. Microbiol. Biotechnol.*, **24**, 417–22.

Kana, K., Kanellaki, M., Papadimitriou, A., Psarianos, C. and Koutinas, A. (1989a) Immobilization of *Saccharomyces cerevisiae* on γ-alumina pellets and its ethanol production in glucose and raisin extract fermentation. *J. Ferment. Bioeng.*, **68**, 213–15.

Kana, K., Kanellaki, M., Psarianos, C. and Koutinas, A. (1989b) Ethanol producion by *Saccharomyces cerevisiae* immobilized on mineral kissiris. *J. Ferment. Bioeng.*, **68**, 144–7.

Kanekar, P. and Sarnaik, S. (1991) An activated sludge process to reduce pollutional load of a dye-industry waste. *Environ. Pollution*, **70**, 27–33.

Kanekar, P. and Sarnaik, S. (1995) Microbial process for treatment of phenol bearing dye-industry effluent in a fixed film bioreactor. *J. Environ. Sci. Health*, **A30**, 1817–26.

Kargi, F. and Shuler, M.L. (1980) An evaluation of various flocculants for the recovery of biomass grown on poultry waste, in *Agricultural Wastes* (ed. T.R. Miles), Applied Science Publishers, London.

Karube, I. (1989) Bioelectric cells, in *Biomass Handbook* (eds O. Kitani and C.W. Hall), Gordon and Breach, New York, pp. 809–18.

Kasche, V. (1989) Protease in peptide synthesis, in *Proteolytic Enzymes. A Practical Approach* (eds R.J. Beynon and J.S. Bons), IRL Press, Oxford University Press, Oxford, New York, Tokyo, pp. 140.

Kaul, R., Adlercreutz, P. and Mattiasson, B. (1987) Co-immobilization – an alternative to biocatalysis in organic media, in *Biocatalysis in Organic Media* (eds C. Laane, J. Tramper and M.D. Lilly), Studies in Organic Chemistry, Vol. 29, Elsevier, Amsterdam, pp. 107–14.

Kautola, H. and Linko, Y.Y. (1989) Fumaric acid production from xylose by immobilized *Rhizopus arrhizus* cells. *Appl. Microbiol. Biotechnol.*, **31**, 448–52.

Kaya, V.M. and Picard, G. (1995) The viability of *Scenedesmus bicellularis* cells immobilized an alginate screens following nutrient starvation in air at 100% relative humidity. *Biotechnol. Bioeng.*, **46**, 459–64.

Kaya, V.M. and Picard, G. (1996) Stability of chitosan gel as entrapment matrix of viable *Scenedesmus bicellularis* cells immobilized on screens for teriary treatment of wastewater. *Bioresource Technol.*, **56**, 147–55.

Kayem, G.J. and Rouxhet, P.G. (1983) Adsorption of colloidal hydrous alumina on yeast cells. *J. Chem. Soc. Faraday Trans. I*, **79**, 561–70.

Kearney, L., Upton, M. and McLaughlin, A. (1990) Enhanching the viability of *Lactobacillus plantarum* inoculum by immobilizing the cells in calcium alginate beads incorporating cryoprotectants. *Appl. Environ. Microbiol.*, **56**, 3112–20.

Keenan, J.D. and Kormi, I. (1977) Methane fermentation of brewery by-products, *Biotechnol. Bioeng.*, **19**, 867–78.
Kerel, S.F., Libicki, S.B. and Robertson, C.R. (1985) The immobilization of whole cells engineering principles. *Chem. Eng. Sci.*, **10**, 1321–30.
Kierstan, M. and Bucke, C. (1977) The immobilization of microbial cells, subcellular organelles, and enzymes in calcium alginate gels. *Biotechnol. Bioeng.*, **19**, 387–96.
Kikuchi, Y. and Fukuda, H. (1974) Polyelectrolyte complexes of sodium dextran sulfate with chitosan. *Makromol. Chem.*, **175**, 3593–6.
Kikuchi, Y. and Noda, A. (1976) Polyelectrolytic complexes of heparin with chitosan. *J. Appl. Polym. Sci.*, **20**, 2561–72.
Kim, S.K. and Rha, C.K. (1989) Chitosan for the encapsulation of mammalian cell culture, in *Chitin and Chitosan* (eds G.S. Braek, T. Anthonsen and P. Sandford), Elsevier Applied Science, London, pp. 617–21.
Kimura, Y., Tanaka, A., Sonomoto, K., Nihira, T. and Fukui, S. (1983) Application of immobilized lipase to hydrolysis of triacylglyceride. *Eur. J. Appl. Microbiol. Biotechnol.*, **17**, 107–12.
Klein, J., Hackel, U., Schara, P. and Eng, H. (1979) Polymer networks for entrapment of microorganisms. *Angew. Makromol. Chem.*, **76/77**, 329–41.
Klein, J., Becke, J.W., Kressdorf, B., Steinert, H.J. and Vorlop, K.D. (1984) Matrix development for specific requirements, in *3rd European Congress of Biotechnology*, Vol. 4, VCH, Weinheim, p. 383.
Klibanov, A.M. (1983) Immobilized enzymes and cells as practical catalysts. *Science*, **219**, 722–30.
Klibanov, A.M. and Morris, E.D. (1981) Horseradish peroxidase for the removal of carcinogenic aromatic amines from water. *Enzyme Microbiol. Technol.*, **3**, 119–22.
Klibanov, A.M., Alberti, B.N., Morris, E.D. and Felshim, Z.M. (1980) Enzymatic removal of toxic phenols and anilines from waste waters. *J. Appl. Biochem.*, **2**, 414–21.
Kobayashi, H. and Suzuki, H. (1972) Decomposition of raffinose by alpha-galactosidase of mold. II. Formation of mold pellet and its enzyme activity. *J. Ferment. Technol.*, **50**, 625–32.
Kokufuka, E., Shimohashi, M. and Nakamura, I. (1988) Simultaneously occurring nitrification and denitrification under oxygen gradient by polyelectrolyte complex-coimmobilized *Nitrosomonas europaea* and *Paracoccus denitrificants* cells. *Biotechnol. Bioeng.*, **31**, 382–4.
Koshiro, S., Sonomoto, K., Tanaka, A. and Fukui, S. (1985) Stereoselective esterification of dl-menthol by polyurethane-entrapped lipase in organic solvent. *J. Biotechnol.*, **2**, 47–57.
Kosugi, Y. and Suzuki, H. (1992) Functional immobilization of lipase eliminating lipolysis product inhibition. *Biotechnol. Bioeng.*, **40**, 369–74.
Kosugi, Y., Tanaka, H. and Tomizuka, N. (1990) Continuous hydrolysis of oil by immobilized lipase in a contracurrent reactor. *Biotechnol. Bioeng.*, **36**, 617–22.
Kurosawa, H. and Tanaka, H. (1990) Advances in immobilized cell culture: development of coimmobilized mixed culture system of aerobic and anaerobic micro-organisms. *Process Biochem.*, Dec., pp. 189–96.
Kurosawa, H., Ishikawa, H. and Tanaka, H. (1988) L-Lactic acid production from starch by coimmobilized mixed culture system of *Aspergillus awamori* and *Streptococcus lactic*. *Biotechnol. Bioeng.*, **31**, 183–7.
Kurosawa, H., Nomura, N. and Tanaka, H. (1989) Ethanol production from starch by a coimmobilized mixed culture system of *Aspergillus awamori* and *Saccharomyces cerevisiae*. *Biotechnol. Bioeng.*, **33**, 716–23.
Lamptey, J. (1983) Production of ethanol in an immobilized-yeast, packed-bed bioreactor, Ph.D. thesis, University of Waterloo, Ontario, Canada.
Lawford, G.R., Lavers, B.H., Good, D., Charley, R.R. and Fein, J. (1983) *Zymomonas* ethanol fermentations: biochemistry and bioengineering, in *International Symposium on Ethanol from Biomass*, Royal Society Canada, Ottawa, Canada, pp. 482–506.
Lee, G.K. and Long, M.E. (1974) US Patent 3939041.
Lee, S.-W., Ebata, T., Liu, Y.-C. and Tanaka, H. (1993) Co-immobilization of three strains of microorganisms and its application in ethanol production from raw starch under sterile conditions. *J. Ferment. Bioeng.*, **75**, 36–48.

Leonowicz, A., Sarkar, J.M. and Bollog, J.M. (1988) Improvement in stability of an immobilized fungal laccase. *Appl. Microbiol. Biotechnol.*, **29**, 129–35.
Leuba, J.L. and Widmer, F. (1977) Immobilization of the β-galactosidase from *A. niger* on chitosan. *J. Solid Phase Biochem.*, **2**, 257–69.
Lewis, V.P. and Yang, S.T. (1992) Continuous propionic acid fermentation by immobilized *Propionibacterium acidipropionici* in a novel packed-bed bioreactor. *Biotechnol. Bioeng.*, **40**, 465–71.
Lilly, M.D. and Woodley, J.M. (1985) Biocatalytic reactions involving water-insoluble organic compounds, in *Biocatalysts in Organic Synthesis* (eds J. Tramper, H.C. van der Plas and P. Linko), Elsevier Science Publishers, London, pp. 179–92.
Lin, Y.F. and Chen, K.C. (1995) Denitrification and methanogenesis in a co-immobilized mixed culture system. *Water Res.*, **29**, 35–43.
Linko, P., Poutanen, K., Weckstron, L. and Linko, Y.Y. (1980) Preparation and kinetic behavior of immobilized whole cell biocatalysts. *Biochimie*, **62**, 387–98.
Liu, W.H., Wang, S.D. and Su, Y.C. (1978) Preparation and application of immobilized glucose oxidase of *Pullularia pullulans*. *Proc. Natl. Sci. Council*, **2**, 275–82.
Lloyd-George, I. and Chang, T.M.S. (1993) Free and microencapsulated *Erwinia herbicola* for the production of tyrosine. *Artif. Cells Immob. Biotechnol.*, **21**, 323–33.
Lloyd-George, I. and Chang, T.M.S. (1995) Characterization of free and alginate-polylysine-alginate microencapsulated *Erwinia herbicola* for the conversion of ammonia, pyruvate, and phenol into L-tyrosine. *Biotechnol. Bioeng.*, **48**, 706–14.
Lortie, R., Trani, M. and Ergan, F. (1993) Kinetic study of the lipase-catalyzed synthesis of triolein. **Biotechnol. Bioeng.**, **41**, 1021–26.
Malcata, F.X. and Hill, C.G. (1991) Use of a lipase immobilized in a membrane reactor to hydrolyze the glycerides of butteroil. *Biotechnol. Bioeng.*, **38**, 853–68.
Malcata, X.F., Garcia, H.S., Hill, C.G. and Amundson, C.H. (1992a) Hydrolysis of butteroil by immobilized lipase using a hollow-fiber reactor: Part. I. Lipase adsorption studies. *Biotechnol. Bioeng.*, **39**, 647–57.
Malcata, X.F., Hill, C.G. and Amundson, C.H. (1992b) Hydrolysis of butteroil by immobilized lipase using a hollow-fiber reactor: Part. II. Uniresponse kinetic studies. *Biotechnol. Bioeng.*, **39**, 984–1001.
Mansfeld, J., Forster, M., Schellenberger, A. and Dautzenberg, H. (1991) Immobilization of invertase by encapsulation in polyelectrolyte complexes. *Enzyme Microbiol. Technol.*, **13**, 240–4.
Mansfeld, J., Forster, M., Hoffmann, T. and Schellenberger, A. (1995) Coimmobilization of *Yarrowia lipolytica* cells and invertase in polyelectrolyte complex microcapsules, *Enzyme Microbiol. Technol.*, **17**, 11–17.
Margaritis, A. and Bajpai, P. (1982) Ethanol production from Jerusalem artichoke (*Helianthus tuberosus*) using *Kluyveromyces marxianus* and *Saccharomyces rosei*. *Biotechnol. Bioeng.*, **24**, 941–53.
Margaritis, A. and Bajpai, P. (1983) Effect of sugar concentration in *Jerusalem artichoke* extract on using *Kluyveromyces marxianus* growth and ethanol production. *Appl. Environ. Microbiol.*, **45**, 723–5.
Martinek, K. and Moshaev, V.V. (1985) Immobilization of enzymes: an approach to fundamental studies in biochemistry, in *Advances in Enzymology*, Vol. 57 (ed. A. Meister), John Wiley & Sons, New York, pp. 179–99.
Martinsen, A., Skjak-Braek, G. and Smidsord, O. (1989) Alginate as immobilization material I. Correlation between chemical and physical properties of alginate beads. *Biotechnol. Bioeng.*, **33**, 79–88.
McGhee, J.E., St Julian, G. and Detroy, R.W. (1982) Ethanol production by immobilized *S. cerevisiae*, *S. uvarum* and *Z. mobilis*. *Biotechnol. Bioeng.*, **24**, 1155–63.
McKnight, C.A., Ku, A. and Goosen, M.F.A. (1988) Synthesis of chitosan–alginate microcapsule membranes. *J. Bioact. Comp. Polym.*, **3**, 334–55.
Miethling, R., Hecht, V. and Deckwer, W.D. (1983) Microbial degradation of quinoline: kinetic studies with *Comamonas acidovorans* DSM 6426. *Biotechnol. Bioeng.*, **42**, 589–95.
Mireles, C., Martino, M., Bouzas, J. and Torres, J.A. (1992) Complex formation of chitosan and naturally occurring polyanions, in *Advances in Chitin and Chitosan* (eds C.J. Brine, P.A. Sansford and J.P. Zikakis), Elsevier Applied Science, Amsterdam, pp. 506–12.
Mitsutomi, M., Uchida, Y. and Ohtakara, A. (1985) Immobilization of thermostable α-

galactosidase from *Pycnoporus cinnabarinus* on chitin and some properties of the immobilized enzyme. *J. Ferment. Technol.*, **63**, 325–9.
Monsan, P. and Combes, D. (1988) Enzyme stabilization by immobilization, in *Methods in Enzymology* (ed. K. Mosbach), Vol. 137, Academic Press, New York, p. 584.
Morin, N., Benier-Cardou, M. and Champagne, C.P. (1992) Production of *Lactococcus lactis* biomass by immobilized cell technology. *J. Ind. Microbiol.*, **9**, 131–5.
Mustranta, A., Forssell, P. and Poutanen, K. (1993) Applications of immobilized lipases to transesterification and esterification reactions in nonaqueous systems. *Enzyme Microbiol. Technol.*, **15**, 133–43.
Muzzarelli, R.A. (1980) Immobilization of enzymes on chitin and chitosan. *Enzyme Microbiol. Technol.*, **2**, 177–84.
Myima, N.Y.O. and Cloete, T.E. (1995) Growth and phosphate uptake of immobilized *Acinetobacter* cells suspended in activated sludge mixed liquor. *Water Res.*, **29**, 2461–6.
Nasri, M. and Thomas, D. (1987) Immobilization of the restriction endonucleases Pvu II and Hind III. *Appl. Biochem. Biotechnol.*, **15**, 119–26.
Natthew, H.W., Salley, S.O., Peterson, W.D. and Klein, M.D. (1993) Complex coacervate microcapsules for mammalian cell culture and artificial organ development. *Biotechnol. Prog.*, **9**, 510–19.
Niederauer, M.Q. and Glatz, C.E. (1994) Model of the polyelectrolyte precipitation of genetically eningeering enzymes processing charged polypeptide tails. *Pure Appl. Chem.*, **31**, 127–35.
Nilsson, K.G.I. (1986) A comparison of the enzyme catalyzed formation of peptides and oligosaccharides in various hydroorganic solutions using the nonequilibrium approach, in *Biocatalysis in Organic Media* (eds C. Laane, J. Tramper and M.D. Lilly), Elsevier Science Publishers, London, pp. 369–74.
Nolan, A.M., Barron, N., Brady, S. *et al.* (1994) Ethanol production at 45 °C by an alginate-immobilized strain of *Kluyeromyces marxianus* following growth on glucose-containing media. *Biotechnol. Lett.*, **16**, 849–52.
Northon, S. (1992) Etude de la production d'acide lactique par fermentation continue du perméat de lactosérum à l'aide d'une souche de *Lactobacillus helveticus* immobilisée. Ph.D. dissertation, Université Laval, Quebec, Canada.
Northon, S. and Vuillemard, J.C. (1994) Food bioconversions and metabolite production using immobilized cell technology. *Cr. Rev Biotechnol.*, **14**, 193–24.
Ogbonna, J.C., Amano, Y. and Nakamura, K. (1989) Elucidation of optimum conditions for immobilization of viable cells by using calcium alginate. *J. Ferment. Bioeng.*, **67**, 92–8.
Olzewski, J. and Wasserman, B.P. (1986) Effect of glutaraldehyde on the activity of some DNA restriction endonuclease. *Appl. Biochem. Biotechnol.*, **13**, 29–38.
Otero, C., Pastor, E., Fernandez, V.M. and Ballesteros, A. (1990) Influence of the support on the reaction course of tributyrin hydrolysis catalyzed by soluble and immobilized lipases. *Appl. Biochem. Biotechnol.*, **23**, 237–47.
Painter, H.A. (1977) Biotechnology of waste water treatment. *Prog. Water Tehcnol.*, **8**, 8–29.
Philips, C.R. and Poon, Y.C. (1988) *Immobilization of Cells*. Springer-Verlag, Berlin.
Playne, M.J. (1985) Propionic and butyric acids, in *Comprehensive Biotechnology*, Vol. 3 (ed. M. Moo-Young), Pergamon, New York, pp. 731–59.
Pommerening, K., Ristau, O., Rein, H., Dautzenberg, H. and Loth, F. (1983) Immobilisierung von Protein durch ein neues Verfahren der Mikroverkapselung. *Biomed. Biochem. Acta.*, **42**, 813–23.
Poncelet, D., Lencki, R., Beaulieu, C. *et al.* (1992) Production of alginate beads by emulsification/internal gelation-methodology. *Appl. Microbiol. Biotechnol.*, **38**, 39–45.
Prévost, H. (1987) Pré-fermentation du lai en continu á l'aide de bactéries lactiques incluses. Application à la production de yaourt et de fromages frais. Ph.D. dissertation, Université de Bourgogne, Dijon, France.
Prévost, H. and Diviès, C. (1987) Fresh fermentation cheese production with continuous pre-fermented milk by a mixed culture of mesophilic streptococci entrapped in Ca-alginate. *Biotechnol. Lett.*, **9**, 789–94.
Prévost, H. and Diviès, C. (1992) Cream fermentation by a mixed culture of *Lactococci* entrapped in two-layer calcium alginate gel beads. *Biotechnol. Lett.*, **14**, 583–8.
Price, E.C. and Cheremisinoff, P.N. (1981) *Biogas. Production and Utilization*. Ann Arbor Science Publisher, Ann Arbor MI, pp. 69.

Rao, B.Y.K., Godbole, S.S. and D'Souza, S.F. (1988) Preparation of lactose free milk by fermentation using immobilized *Saccharomyces fragilis*. *Biotechnol. Lett.*, **10**, 427–32.

Reddy, L.G. (1989) Immobilization of enzymes: purification and immobilization of S1 nuclease, Ph.D. thesis, University of Poona, Pune, India.

Reddy, L.G. and Shankar, V. (1993) Immobilized nucleases. *Cr. Rev. Biotechnol.*, **13**, 255–73.

Reynolds, J.H. (1974) Immobilization alpha-galactosidase continuous flow reactor. *Biotechnol. Bioeng.*, **16**, 135–42.

Riordan, C., Love, G., Barron, N. *et al.* (1996) Production of ethanol from sucrose at 45 °C by alginate-immobilized preparations of thermotolerant yeast strain *Kluyveromyces maxianus* IMB3. *Bioresource Technol.*, **55**, 171–3.

Ristau, O., Pommerening, K., Jung, C., Rein, H. and Scheler, W. (1985) Aktivitatseigenschaften von Urease in Mikrokapseln. *Biomed. Biochem. Acta.*, **44**, 1105–11.

Roca, E., Flores, J., Nunes, M.J. and Lema, J.M. (1996) Ethanolic fermentation by immobilized *Saccharomyces cerevisiae* in a semipilot pulsing packed-bed bioreactor. *Enzyme Microbiol. Technol.*, **19**, 132–9.

Rogers, P.L., Lee, K.J. and Tribe, D.E. (1979) Kinetics of alcohol production by *Zymomonas mobilis* at high sugar concentration. *Biotechnol. Lett.*, **2**, 165–70.

Rosario, E. and Pamatong, F. (1985) Continuous-flow fermentation of banana fruit pulp sugar into ethanol by carrageenan-immobilized yeast. *Biotechnol. Lett.*, **7**, 819–20.

Rossi, J., Gobetti, M. and Soccio, P. (1990) Milk prefermentation process in fresh cheese manufacture. I. Study of immobilized lactic acid bacteria cells. *Sienza Tecnica Lattiero-Casearia*, **41**, 401–13.

Roukas, T. (1994) Ethanol production from nonsterilized carob pod extract by free and immobilized *Saccharomyces cerevisiae* cells using fed-batch culture. *Biotechnol. Bioeng.*, **43**, 189–201.

Roy, D., Goulet, J. and LeDuy, A. (1987) Continuous production of lactic acid from whey permeate by free and calcium alginate entrapped *Lactobacillus helveticus*. *J. Dairy Sci.*, **70**, 506–11.

Ruggiero, P., Sarkar, J.-M. and Bollag, J.M. (1989) Detoxification of 2,4-dichlorophenol by a laccase immobilized on soil or clay. *Soil Sci.*, **147**, 361–70.

Saida, T., Michiki, H., Miyakawa, H. *et al.* (1985) EtOH production from cellulosics, in *Bioenergy 84*, Vol. III, Biomass Conversion, (eds H. Egneus and A. Ellegard), Elsevier Applied Science, London, pp. 212–20.

Sakurai, H., Lee, H.W., Sato, S., Mukataka, S. and Takahashi, J. (1989) Gluconic acid production at high concentration by *Aspergillus niger* immobilized on a nonwoven fabric. *J. Ferment. Bioeng.*, **67**, 404–12.

Sarkar, J.-M., Leonowicz, A. and Bollag, J.M. (1989) Immobilization of enzymes on clays and soils. *Soil Biol. Biochem.*, **21**, 223–30.

Scott, C.D. (1987) Immobilized cells. A review of recent literature. *Enzyme Microbiol. Technol.*, **9**, 66–75.

Sharma, B.P., Bayley, L.F and Messing, R.A. (1986) Immobilized biomaterials – technology and applications. *Angew. Chem. Int. Ed. Engl.*, **21**, 469–75.

Sheih, J. and Glatz, C.E. (1994) Precipitation of proteins with polyelectrolytes: role of polymer molecular weight, in *Macromolecular Complexes in Chemistry and Biology* (eds P. Dubin, J. Bock, R. Davis, D.N. Schulz and C. Thies), Springer-Verlag, Heidelberg, pp. 273–84.

Shiraishi, F., Kawakami, K., Tamura, A., Tsuruda, S. and Kusunoki, K. (1989a) Continuous production of free gluconic acid by *Gluconobacter suboxydans* IFO 3290 immobilized by adsorption on ceramic honeycomb monolith: effect of reactor configuration on further oxydation of gluconic acid to keto-gluconic acid. *Appl. Microbiol. Biotechnol.*, **31**, 445–50.

Shiraishi, F., Kawakami, K., Tamura, A., Tsuruda, S. and Kusunoki, K. (1989b) Characterization of production of free gluconic acid by *Gluconobacter suboxydans* adsorbed on ceramic honeycomb monolith. *Biotechnol. Bioeng.*, **33**, 1413–21.

Silman, I.H. and Katchalski, E. (1966) Water-insoluble derivatives of enzymes, antigens, and antibodies. *Ann. Rev. Biochem.*, **35**, 873–91.

Smidsord, O. and Haug, A. (1972) Dependence upon gel-sol state of the ion-exchange properties of alginates. *Acta Chem. Scand.*, **26**, 2063–72.

Smiley, K.L., Hensley, D.E. and Gasdorf, H.J. (1976) Alpha-galactosidase production and use in a hollow-fiber reactor. *Appl. Environ. Microbiol.*, **31**, 615–23.
Stanley, W.L., Watters, G.G., Chan, B. and Mercer, J.M. (1975) Lactase and other enzymes bound to chitin with glutaraldehyde. *Biotechnol. Bioeng.*, **17**, 315–21.
Stanley, W.L., Watters, G.G., Kelly, S.H. *et al.* (1976) Immobilization of glucose isomerase on chitin with glutaraldehyde and by simple adsorption. *Biotechnol. Bioeng.*, **18**, 439–45.
Stanley, W.L., Watters, G.G., Kelly, S.H. and Olson, A.C. (1978) Glucoamylase immobilized on chitin with glutaraldehyde. *Biotechnol. Bioeng.*, **20**, 135–42.
Steenson, L.R., Klaenhammer, T.R. and Swaisgood, H.E. (1987) Calcium alginate immobilized cultures of lactic streptococci are protected from bacteriophage. *J. Dairy Sci.*, **70**, 1121–29.
Stephenson, T. (1987) Acinetobacter: its role in biological phosphate removal, in *Advances in Water Pollution Control, Biological Phosphate Removal from Wastewater* (ed. R. Ramadori), IAWPRC, pp. 313–16.
Sugihara, A., Shimada, Y. and Tominaga, Y. (1988) Enhanced stability of a microbial lipase immobilized to a novel synthetic amine polymer. *Agric. Biol. Chem.*, **52**, 1589–90.
Takahashi, M., Ochi, H., Kaneko, T., Suzuki, H. and Tanaka, H. (1990) Diacetyl production by immobilized citrate positive *Lactococcus lactis* spp lactis 3022 in the fibrous Ca-alginate gel. *Biotechnol. Lett.*, **12**, 569–74.
Tanaka, A. and Nakajima, H. (1990) Application of immobilized growing cells. *Adv. Biochem. Eng. Biotechnol.*, **42**, 97–104.
Tanaka, I., Tosa, T. and Chibata, I. (1977) Screening of matrix suitable for immobilization of microbial cells. *J. Solid-Phase Biochem.*, **2**, 225–36.
Tanaka, H., Kurosawa, H. and Murakami, H. (1986) Ethanol production from starch by a co-immobilized mixed culture system of *Aspergillus awamori* and *Zymomonas mobilis*. *Biotechnol. Bioeng.*, **28**, 1761–68.
Taravel, M.N. and Domard, A. (1994) Competition between gel and polyanion–polycation complex formation in a collagen-chitosan binary system. *Macromol. Rep.*, **A31**, 1237–45.
Thananunkul, D., Tanaka, M., Chichester, C.O. and Lee, T.C. (1976) Degradation of raffinose and stachyose in soybean milk by alpha-galactosidase from Mortierella vinacea. Entrapment of alpha-galactosidase within polyacrylamide gel. *J. Food Sci.*, **41**, 173–80.
Toerien, D.F., Gerber, A., Lotter, L.H. and Cloete, T.E. (1990) Enhanced biological phosphorus removal in activated sludge system. *Adv. Microbial Ecol.*, **11**, 173–230.
Tosa, T., Sato, T., Mori, T. *et al.* (1979) Immobilization of enzymes and microbial cells using carrageenan as matrix. *Biotechnol. Bioeng.*, **21**, 1697–709.
Toshifumi, S., Nobuko, T., Mamoru, Y., Keiji, F. and Haruna, K. (1995) Enzyme immobilization on thermosensitive microspheres. *Colloids Surfaces B: Biointerfaces*, **4**, 267–80.
Tramper, J. (1990) Conversions by immobilized cells vs. traditional fermentations, in *Physiology of Immobilized Cells* (eds J.A.M. de Bont, J. Visser, B. Mattiasson and J. Tramper), Elsevier, Amsterdam, pp. 123–39.
Tramper, J., Luyben, K.C.A. and van den Tweel, W.J.J. (1983) Kinetic aspects of glucose oxidation by *Gluconobacter oxydans* cells immobilized in calcium alginate. *Eur. J. Appl. Microbiol. Biotechnol.*, **17**, 13–19.
Travieso, L., Benitez, F., Weiland, P. *et al.* (1996) Experiments on immobilization of microalgae for nutrient removal in wastewater treatments. *Bioresource Technol.*, **55**, 181–6.
Tsuchida, E. and Abe, K. (1986) Polyelectrolyte complexes, in *Development in Ionic Polymers* (eds A.D. Wilson and H.J. Prosser), Elsevier Applied Science, London, pp. 191–266.
Tyagi, R.D., Gupta, S.K. and Chand, S. (1992) Process engineering studies on continuous ethanol production by immobilized *S. cerevisiae*. *Process Biochem.*, **27**, 23–32.
Ulbrich-Hofmann, R. and Selisko, B. (1992) Soluble and immobilized senzymes in water-miscible organic solvents: glucoamylase and invertase. *Enzyme Microbiol. Technol.*, **15**, 33–41.
Ulonska, A., Deckwer, W.D. and Hecht, V. (1995) Degradation of quinoline by immobilized *Comamonas acidovorans* in a three-phase airlift reactor. *Biotechnol. Bioeng.*, **46**, 80–7.
Underdal, B., Skulberg, O.M., Dahl, E. and Aune, T. (1988) Disastrous bloom of

Chrysochromulina polylepis (*Prymnesiophyceae*), in *Norwegian Coastal Waters 1988 – Mortality in Marine Biota, AMBIO*, Vol. 18, pp. 265–70.
Van Griethuysen, E., Flaschel, E. and Renken, A. (1986) Immobilization of lactase on chitosan-coated silica gel particles, in *Chitin in Nature and Technology* (eds R. Muzzarelli, C. Jeuniaux and G.W. Gooday), Plenum Press, New York, pp. 422–7.
Vassilev, N.B., Vassileva, M.C. and Spassova, D.I. (1993) Production of gluconic acid by *Aspergillus niger* immobilized on polyurethane foam. *Appl. Microbiol. Biotechnol.*, **39**, 285–90.
Veillemard, J.C., Goulet, J., Amiot, J., Vijayalakshmi, M.A. and Terré, S. (1988) Continuous production of small peptides from milk proteins by extracellular proteases of free and immobilized *Serratia marcescens* cells. *Enzyme Microbiol. Technol.*, **10**, 2–8.
Vilchez, C. and Vega, J.M. (1994) Nitrite uptake by *Chlamydomonas reinhardtii* cells immobilized in calcium alginate. *Appl. Microbiol. Biotechnol.*, **41**, 137–41.
Vilchez, C. and Vega, J.M. (1995) Nitrite uptake by immobilized *Chlamydomonas reinhardtii* cells growing in airlift reactors. *Enzyme Microbiol. Technol.*, **17**, 386–90.
Vives, C., Casas, C., Godia, F. and Sola, C. (1993) Determination of the intrinsic fermentation kinetics of *Saccharomyces cerevisiae* cells immobilized in Ca-alginate beads and observations on their growth. *Appl. Microbiol. Biotechnol.*, **38**, 467–73.
Vorlop, K.D. and Klein, J. (1983) New developments in the field of cell immobilization – formation of biocatalysts by ionotropic gelation, in *Enzyme Technology* (ed. R.M. Lefferty), Springer International, New York, pp. 219–35.
Vorlop, K.D. and Klein, J. (1987) Entrapment of microbial cells in chitosan. *Meth. Enzymol.*, **135**, 259–68.
Wada, M., Kato, J. and Chibata, I. (1979) A new immobilization of microbial cells. *Eur. J. Appl. Microbiol. Biotechnol.*, **8**, 241–8.
Wang, S.S. and Vieth, W.R. (1973) Collagen–enzyme complex membranes and their performance in biocatalytic modules. *Biotechnol. Bioeng.*, **15**, 93–9.
Wang, Y., Gao, J.Y. and Dubin, P.L. (1996) Protein separation via polyelectrolyte coacervation: selectivity and efficiency. *Biotechnol. Prog.*, **12**, 356–62.
Westrin, B.A. and Axelsson, A. (1991) Diffusion in gel containing immobilized cells: a critical review. *Biotechnol. Bioeng.*, **38**, 439–46.
Wijffels, R.H., Leenen, E.J.T.M. and Tramper, J. (1993) Possibilities of nitrification with immobilized cells in waste-water treatment: model or practical system? *Water Sci. Technol.*, **27**, 233–40.
Wijffels, R.H., Englund, G., Hunik, J.H. *et al.* (1995) Effects of diffusion limitation on immobilized nitrifying microorganisms at low temperatures. *Biotechnol. Bioeng.*, **45**, 1–9.
Wilkins, E. and Ramesh, Y. (1991) Performance of immobilized enzyme on saccharification and fermentation of agricultural wastes and wood residues. *J. Environ. Sci. Health*, **A26**, 883–98.
Williams, D. and Mannecke, D.M. (1981) The production of ethanol by immobilized yeast cells. *Biotechnol. Bioeng.*, **23**, 1813–25.
Woskow, S.A. and Glatz, B.A. (1991) Propionic acid production by a propionic acid-tolerant strain of *Propionibacterium acidipropionici* in batch and semicontinuous fermentation. *Appl. Environ. Microbiol.*, **57**, 2821–8.
Wyatt, T. (1979) Global patterns of discolored water and related events in the nineteenth and twentieth centuries, in *Developments in Marine Biology*, Vol. 1 – *Toxic Dinoflagellate Blooms* (eds D.L. Taylor and H.H. Seliger), Elsevier, Amsterdam, pp. 263–8.
Xia, J. and Dubin, P.L. (1994) Protein–polyelectrolyte complexes, in *Macromolecular Complexes in Chemistry and Biology* (eds P. Dubin, J. Bock, R. Davis, D.N. Schulz and C. Thies), Springer-Verlag, Heidelberg, pp. 247–71.
Yang, S.T., Tank, I.C. and Zhu, H. (1992) A novel fermentation process for calcium magnesium acitate (CMA) production from cheese whey. *Appl. Biochem. Biotechnol.*, **34/35**, 569–83.
Yang, S.T., Huang, Y. and Hong, G. (1995) A novel recycle batch immobilized cell bioreactor for propionate production from whey lactose. *Biotechnol. Bioeng.*, **45**, 379–86.
Zaidi, A., Gainer J.L. and Carta, G. (1995) Fatty acid esterification using nylon-immobilized lipase. *Biotechnol. Bioeng.*, **48**, 601–5.
Zielinski, B.A and Aebischer, P. (1994) Chitosan as a matrix for mammalian cell encapsulation. *Biomaterials*, **15**, 1049–56.

3 Solid substrate fermentation: a biotechnological approach to bioconversion of wastes

O. PAREDES-LÓPEZ, S.H. GUZMÁN-MALDONADO AND A. ALPUCHE-SOLÍS

3.1 Introduction

Solid substrate fermentation processes have been used by man for many centuries. The term 'solid substrate fermentation' (SSF) describes the microbiological transformation on and/or within the particles of solid matrix (solid substrate) where the liquid content, bound with them, is at the level corresponding to the water activity (a_w) assuring growth and metabolism of cells, but not exceeding the maximal water-holding capacity of the solid matrix (Cannel and Moo-Young, 1980a; Durand *et al.*, 1988; Paredes-López and Harry, 1988; Pandey, 1992). In other words, in SSF the moist water-insoluble solid substrate is fermented by microorganisms in the absence or near-absence of free water, resulting in semisolid or solid fermentation systems (Hesseltine, 1977a). 'Solid state' and 'solid substrate' fermentations are terms used by different workers but they are essentially the same. On the other hand, in liquid state fermentations (LSF), the substrate is solubilized or suspended as fine particles in a large volume of water. SSF were used long before the microbiological or biochemical processes involved were understood. Various SSF processes with histories reaching far back in time are still practiced today (Moo-Young *et al.*, 1983; Steinkraus, 1983a). These traditional SSF systems deal with fermented foods (e.g. tempeh in Indonesia, miso in Japan, pozol in Mexico), mold-ripened cheeses, starter cultures for fermented brews, and silage and compost. The use of soy sauce koji, a fermented oriental food preparation used in China, Japan and south-east Asia goes back as far as 1000 BC and probably 3000 BC in China (Cannel and Moo-Young, 1980a; Pandey, 1992).

The koji process may be considered the prototype of SSF. At present, SSF processes are used on a commercial scale for the production of different types of fermented foods, particularly in Asia and in some developing countries (Aidoo *et al.*, 1982). In addition to these types of foods, such processes have been used successfully in recent years for large-scale production of protein-enriched feed from starchy wastes, single cell protein (SCP) from a variety of wastes, fungal metabolites and bioconversion of plant/animal/domestic wastes into useful products (Aidoo *et al.*, 1982; Tengerdy, 1985).

There are abundant organic wastes in most countries of the world (Figure 3.1). The net productivity of dry biomass due to photosynthesis by plants on the earth is estimated to be close to 150×10^9 t per year (Bassham, 1975; Rajarathnam and Bano, 1989), of which about 11% is represented by the primary production of materials with food/feed properties. Wastes of various kinds (e.g. inedible parts of plants, wastes generated during the harvesting season, wastes during handling and processing), suitable for consideration as sources of raw materials, may amount to more than 13×10^9 t per year. The potential for the transformation of these wastes via SSF into valuable products is remarkable.

The attraction of SSF comes from its closeness to the natural way of life for many microorganisms and from its simplicity (Tengerdy, 1985; Hesseltine, 1987). A good example of these assessments is a compost pile where multiple microbial cultures grow (e.g. bacteria, actinomycetes,

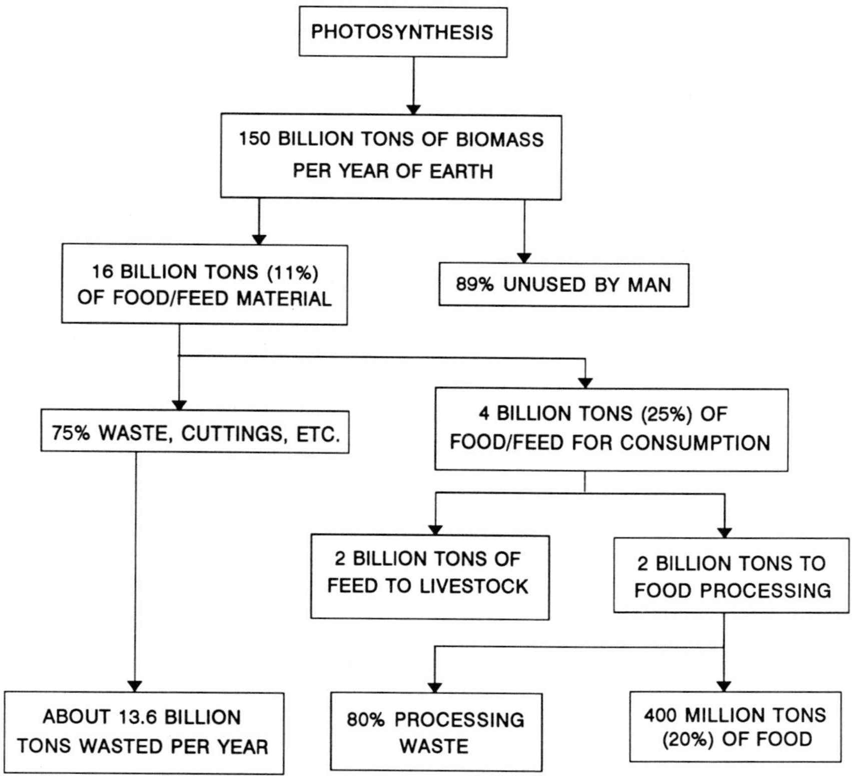

Figure 3.1 Biomass production on earth by photosynthesis and annual availability of wastes.

yeasts, fungi). Another attraction of SSF is the low volume of water present in the media per mass of substrate which can substantially reduce the space occupied by the fermenter without severely sacrificing the yield of products, and the cost of water removal at the end of fermentation is minimized. Aeration and mixing requirements may also be easily met (Hesseltine, 1977a,b; Moo-Young *et al.*, 1983). This type of fermentation requires relatively low capital investments and reduced energy inputs; it is also especially suitable for growing mixed cultures of microorganisms where symbiosis stimulates better growth and productivity (Bushell and Slater, 1981). These advantages can outweigh the disadvantages of SSF in relation to LSF, which are the lower product yields and the slowness of the fermentation. They also present heat build-up problems, bacterial contamination, problems in scale-up and restrictions for growth measurement, and are more difficult to control owing to the lack of adequate sensors and efficient solids-handling techniques, especially for continuous operations. This chapter focuses on the fundamental principles of SSF, both scientific and technological, and it also pays attention to the various opportunities to convert agroindustrial wastes into useful products employing SSF.

3.2 Critical factors for microbial growth on solid substrates

3.2.1 Water activity and moisture

In general, the type of the microorganisms that can grow in SSF systems are determined by the water activity factor a_w. Water activity is defined as the relative humidity of the gaseous atmosphere in equilibrium with the substrate; the a_w of the substrate quantitatively expresses the water requirements for microbial activity (Smith, 1985; Pandey, 1992; Sarrette *et al.*, 1992; Antier *et al.*, 1993a,b). On the other hand, water potential is directly related to the energy state of water in organic substrates; at equilibrium, conditions between liquid, solid and vapor phases, the water potential may be related to the relative humidity of the vapor phase or to the a_w of the substrate (Gervais *et al.*, 1988a). The influence of water potential, conditioned by the a_w values of the medium, on the development of microorganisms has been clearly demonstrated (Mossel, 1975). It is important to know the optimal value of a_w for individual physiological phenomena, such as microbial growth, sporulation and production of main primary and secondary metabolites (Grajek and Gervais, 1987).

The microorganisms which can grow and are capable of carrying out their metabolic activities at low a_w values are suitable for SSF processes. It has been shown that in the course of fungal growth in SSF, high water activities favor sporulation while low water activities favor spore germination or mycelial growth (Molard *et al.*, 1985; Pandey, 1992). Thus, it is very important to know the a_w of a system at a given time, which is an estimate of the proportion of free water that is available for biological and physicochemical activity. During SSF, substrate polymers are degraded into smaller molecules and CO_2 while metabolites and water are produced. This means that the composition and structure of the medium changes during fermentation together with its sorption properties (Paredes-López and Harry, 1988). Because of this, it is important that the a_w of solid substrate can be controlled by way of the relative humidity (RH) of the air. Gervais *et al.* (1986) reported a SSF process allowing the control of a_w. Moreover, a sensor was designed in order to control a_w on-line during submerged and solid substrate fermentations. The sterilizable sensor allows the measurement of the relative humidity of the atmosphere in a small chamber by means of an element separated from the medium by a membrane (Gervais, 1989).

Perhaps the most widely sought parameter with regard to the effects of a_w on microorganisms is that relating to the minimal level for growth. Theoretically, this point defines the limit below which a microorganism or group of microorganisms can no longer reproduce (Beuchat, 1981; Troller, 1987). The specificity and uniqueness of minimal levels can be illustrated in Table 3.1. Bacteria usually are capable of subsisting at higher a_w ranges than are yeasts and molds. The minimum a_w at which molds and yeasts are capable of growing is about 0.60. This corresponds to a water content of 10–20% and it can be concluded that SSF may take place at moisture levels between 10–20% and 80%. The exploitation of combinations of environmental factors together with lowered a_w levels is common in industrial microbiology. The optimum a_w values for microbial growth are usually lower than those required for production of the respective metabolites.

Numerous experiments have demonstrated the influence of a_w on the metabolism of microorganisms. Gervais *et al.* (1988b) reported the influence of a_w on a solid substrate on growth rate and sporogenesis of filamentous fungi. Generally, as the minimal a_w for growth of a microorganism is approached, alteration of other environmental factors (e.g. pH) will have a greater impact on growth rates (Troller, 1987). On the other hand, the a_w of the medium is a fundamental parameter for mass transfer of the water and solutes across the cell membrane; the control of this parameter could be used to modify the metabolic production or excretion of a microorganism (Gervais, 1990). Other studies have been conducted on the interacting effects of a_w for growth and for production of metabolites (Table 3.2) (Norholt *et al.*, 1977, 1979; Narahara *et al.*, 1982; Grajek and Gervais, 1987; Oriol *et al.*, 1988). A kinetic model which relates the rate constant of the death of the microbial cells to water activity and temperature has been proposed by Moser (1988). Besides, it has been reported that a_w can be depressed by the addition of salts, sugars and different kinds of polyols; their effects on microbial growth and physiology have been also reported (Brown, 1978; Hanh-Hagerdel, 1986).

Table 3.1 Water activity and growth of microorganisms in various substrates

Range of a_w	Microorganisms inhibited by lowest a_w of this range	Substrate/food
1.000–0.950	*Escherichia*, *Pseudomonas*, *Shigella*, *Bacillus*, *Clostridium perfringens*, some yeasts	Fresh fruits and vegetables, meat, fish, milk, breads, substrates containing up to 40% sucrose or 70% sodium chloride
0.950–0.910	*Lactobacillus*, *Salmonella*, *Clostridium botulinum*, some yeasts	Some cheeses, cured meat, coffee pulp, substrates containing 55% sucrose or 12% sodium chloride
0.910–0.870	Many yeasts (*Candida*, *Hansenula*, *Torulopsis*)	Dry cheeses, fermented sausage, substrates containing 65% sucrose or 15% sodium chloride
0.870–0.800	Most molds, most *Saccharomyces*	Most fruit juice concentrates, most cereal flours, pulses containing 15–17% moisture, syrups
0.800–0.750	Most halophilic bacteria	Glacé fruits, marmalade
0.750–0.650	Xerophilic molds (*Aspergillus chevalieri*, *A. candidus*)	Molasses, raw cane sugar
0.650–0.600	Osmophilic yeasts (*Saccharomyces rouxii*), few molds (*Aspergillus Candidus*, *Wallermia sebi*), *Saccharomyces bisporus*	Dried fruit and vegetable wastes containing 15–20% moisture, honey
0.500	No microbial proliferation	Spices containing 10% moisture, starchy substrates containing 12% moisture
0.400	No microbial proliferation	Grain hulls, egg powder containing 5% moisture
0.300	No microbial proliferation	Bread crusts containing 3–5% moisture
0.200	No microbial proliferation	Dried fruit and vegetable wastes containing 5% moisture, whole milk powder containing 2–3% moisture

3.2.2 Temperature

Temperature is yet another critical parameter that can affect SSF process. This factor is intimately connected with water activity control, partly because of the dependence of thermal diffusion on substrate moisture content, but partly because water evaporation is the greatest source of cooling that SSF may have (Tengerdy, 1985). Heat build-up causes evaporative water loss, stops vegetative growth and induces protective but nonproductive sporulation of the molds. It has been reported that about 59% of heat generated during SSF is used to evaporate water produced during the metabolic activity while the remaining 41% is left in the substrate to be dissipated (Chahal, 1983). This energy, if not dissipated immediately, as it is released, reduced productivity, causes evaporative

Table 3.2 Water activity for growth and for metabolite production in solid substrate fermentation

Microorganism	Product/metabolite	Optimum growth a_w	Production		Reference
			Minimum a_w	Optimum a_w	
Aspergillus oryzae	Koji	>0.900	0.942	0.975	Narahara *et al.* (1982)
Trichoderma viride TS	Polygalacturonase	<0.995	–	0.995	Grajek and Gervais (1987)
	D-Xylanase	<0.995	–	0.995	
	β-Glucosidase	>0.980	–	0.960–0.980	
Aspergillus niger	Microbial protein	0.980			Oriol *et al.* (1988)
Aspergillus flavus	Aflatoxin	>0.800	0.830–0.870	–	Norholt *et al.* (1977)
Penicillum cyclopium	Ochratoxin	>0.810	0.870–0.900	–	Norholt *et al.* (1979)
Exophiala jeanselmei	Styrene	>0.910	–	1.000	Cox *et al.* (1996)

water loss, stops vegetative growth or may kill the organism (Hayes, 1977). Heat transfer problems in SSF also induce temperature gradients that may cause delated microbial activity and undesirable metabolic deviations (Sargantanis *et al.*, 1993); thus, it is extremely important to maintain the temperature in the range 25–32 °C, which is the usual temperature range employed in SSF (Lonsane *et al.*, 1985).

Heat removal difficulties are due to low transfer coefficients and low thermal conductivity of the heterogeneous materials used in the process. Different methods are used (e.g. forcing large quantities of air through the fermenter, circulation of water in a jacket around the fermenter) to remove the excessive heat produced. Among conventional methods, convective mechanisms are more efficient than conductive ones, but require large-scale aeration rates which may result in undesirable dehydration of the media. Other less conventional methods make use of evaporative cooling (latent heat of water vaporization) to eliminate metabolic heat accumulation (Barstow *et al.*, 1988). Recent studies have shown the effectiveness of such evaporative methods (Ryoo *et al.*, 1991; Sargantanis *et al.*, 1993). However, swift changes in the temperature of the medium provoked strong effects on biomass morphology (Sargantanis *et al.*, 1993), which may result in associated metabolic deviations of the organisms.

On the other hand, some efforts have been made in modeling some parameters in SSF including heat transfer. Rathbun and Shuler (1983) studied the heat and mass transfer effects in static SSF. They observed that, during a tempeh fermentation, temperature gradients could be as high as 3 °C cm^{-1}. Lai *et al.* (1989) described a mathematical model for the estimation of thermal diffusivity in SSF process of sorghum brewing. Habib (1990) described a bioengineering model for fungal growth on lignocelluloses in SSF. Saucedo-Castañeda *et al.* (1990) developed a model and tested it to simulate the generation and transfer of heat in SSF. Grajek (1988) studied the cooling aspects of solid-state cultures of mesophilic and thermophilic fungi. This author discussed heat generation and heat removal in relation to the a_w; heat removal from the culture media was found to be due to the enthalpy changes and water vaporization.

Temperature is also an important factor for controlling metabolite production. Ohno *et al.* (1995a) reported the production of two different metabolites by *B. subtilis* RB14 in the solid-substrate fermentation of okara. The optimal temperatures for producing those metabolites were 25 and 37 °C in spite of their dependency on a common gene.

3.2.3 pH

Although pH is one of the critical factors, the monitoring and control of pH during fermentation is not usually attempted in SSF (Lonsane *et al.*,

1985; Paredes-López and Harry, 1989). Most fungi are able to grow in a wide pH range of 5–8 (Chahal, 1983). Some of the substrates (e.g. straws, grains) have a good buffering capacity and there may not be any need to adjust the pH during SSF. This advantage is, therefore, exploited in the initial adjustment of the pH of the solids in the range 4.5–5.0 during moistening by using water at the desired pH level. However, local changes in the pH of the agglomerates produced cannot be checked and result in low productivity in unagitated fermenters.

In relation to the effect of pH on productivity, it is well known that the synthesis of extracellular enzymes by several microorganisms is regulated by the pH value of the medium. Chahal (1985) reported that maintenance of pH around 4.5 in crucial for cellulase production with *Trichoderma reesei*. Roche *et al.* (1994) found that the optimum pH for α-L-arabinofuranosidase production by *Thermoascus aurantiacus* on sugar beet pulp was 8.4. Jaleel *et al.* (1992) studied amyloglucosidase production from *A. niger* using media prepared with tap water, tap water and trace elements, and tap water and trace elements plus HCl. They found that enzyme production was comparable when using those three different media, nevertheless the optimum pH for enzyme production when *A. niger* was grown in tap water medium was 6.3 compared with 5.7 with the other two media. Recently, Cavalitto *et al.* (1996) studied the effect of different concentrations of HCl during thermal pretreatments of the substrate (wheat bran) on pectinase production using *Aspergillus foetidus*; the pH values in culture extracts (4.0, 3.0 and 2.3) under different tested acidic conditions showed no substantial variations with time. In all cases, a small increase of around 0.2 and 0.4 units of pH was observed at different times. The highest pectinase productivity occurred at pH around 2.3. In contrast, George *et al.* (1995) found that protease secretion by *B. amyloliquefaciens* into the medium was accompanied by a sharp increase in pH, followed by a sharp decline which affected enzyme production.

3.2.4 Aeration and oxygen transfer

In a solid-state culture, aeration is needed for three important functions:

1. To maintain aerobic conditions, because the partial pressure of O_2 and CO_2 in the gas environment of a SSF are critical factors for growth and product formation. Increased aeration stimulated α-galactosidase and invertase production in SSF of wheat bran with *Aspergillus awamori*. Similarly, aflatoxin production by *A. flavus* was enhanced considerably when the aeration was increased (Silman, 1980). In contrast, ochratoxin biosynthesis by *A. ochraceus* in a rotating drum fermenter stopped when the air-flow rate was greater than 0.1 l kg^{-1} min^{-1} (Lindenfelser and Ciegler, 1975).

2. To control substrate temperature because, in comparison with submerged culture, the low moisture environment in SSF creates difficult conditions for heat transfer and temperature regulation.
3. To regulate water content/a_w of the substrate, CO_2 and volatile metabolites produced during metabolism. As described previously, the a_w markedly influences the cell growth and the metabolism in SSF. Narahara *et al.* (1982) and Grajek and Gervais (1987) reported that production of enzymes was strongly affected by the a_w of the substrate.

The aeration of moist solid medium is one of the critical factors which governs productivity since it not only supplies O_2 but also removes metabolic heat, gaseous and volatile products from the fermenting mass (Lonsane *et al.*, 1992). Moreover, in a solid-state culture, aeration is needed to maintain aerobic conditions, because the partial pressures of O_2 and CO_2 in the gas environment of a SSF are critical factors for growth and product formation.

Aeration is needed to control substrate temperature because, in comparison with submerged culture, the low moisture environment in SSF creates difficult conditions for heat transfer and temperature regulation (Pandey *et al.*, 1996). Aeration also regulates water content/a_w of the substrate, CO_2 and volatile metabolites produced during metabolism. As pointed out previously, the aeration of moist solid medium is one of the critical factors which governs productivity since it not only supplies O_2 but also removes metabolic heat, and gaseous and volatile products from the fermenting mass. Recently, the rate of aeration has been integrated with the control of temperature and moisture content by evaporative cooling water (Barstow *et al.*, 1988; Ryoo *et al.*, 1991).

Fermentation kinetics are sensitive to the variation in ambient and internal gas composition. The methodology for controlling the gaseous environment in SSF systems has been described by several authors (Bajracharya and Mudgett, 1980; Saucedo-Castañeda *et al.*, 1990; Desgrenges and Dureng, 1990). Gaseous components such as O_2 and CO_2 can be monitored with specific gas analysers or by gas chromatography (Pandey, 1992).

Ulmer *et al.* (1981) found that biomass production by *Chaetomium cellulolyticum* in SSF of manure fibers was not oxygen limited but depended on the ventilation of CO_2. Inhibitory effects of CO_2 under conditions of adequate O_2 supply were also observed in submerged fermentations (Paredes-López *et al.*, 1974). Bajracharya and Mudgett (1980) reported that oxygen enrichment enhanced enzyme productivity in a traditional solid-state fermentation. In a SSF of rice with *A. oryzae*, Bajracharya and Mudgett (1980) also found that growth and biomass accumulation was insensitive to pO_2 but was sensitive to pCO_2 in a closed gas system.

As it has been pointed out by Lonsane *et al.* (1985), although SSF systems are highly aerobic in nature, except those in composting/ensiling and some food fermentations (Paredes-López and Harry, 1988), little information is available on the mechanism of O_2 transfer. This mechanism is probably similar to that which takes place in hydrocarbon fermentation (Prokop *et al.*, 1971) and hence may occur from the gas phase directly and also from the O_2 dissolved in the water which keeps the substrate moist, though the contribution of the latter will be negligible (Lonsane *et al.*, 1985). Many operating parameters and media characteristics can affect O_2 transfer rates including the air pressure and flow rate, the porosity of the moist solids, the bed depth of the moist fermenting solids, perforations in the culture vessel, the moisture content of the medium, the reactor geometry, impeller rotation speed and geometry. The agitation of the medium can also promote O_2 transfer by dispersing air as small bubbles in the medium.

3.2.5 Mixing

Mixing is an additional control parameter used in connection with aeration. Mixing of the fermenting mass has beneficial effects like providing new surfaces to aeration, distribution of inoculum, promotion of homogeneity and of growth on individual particles of the substrate, and prevention of aggregate formation and of localized changes (Hesseltine, 1977a; Moo-Young *et al.*, 1983). The speed of mixing or rotation of the fermenter is dictated by the same factors as those governing the rate of aeration.

It must be emphasized that agitation is not employed in many aerobic SSF processes such as tray fermentations which are carried out in static reactors (Lonsane *et al.*, 1985; Viesturs *et al.*, 1987). In contrast, mixing is an essential process on periodically or continuously agitated SSF bioreactors. Different products such as aflatoxins, ochratoxins and enzymes are produced in greatly enhanced yields in agitated systems (Lindenfelser and Ciegler, 1975; Hesseltine, 1977a; Silman, 1980). Opposite to Kargi and Curme (1985) who reported decreased ethanol productivity by *Saccharomyces cerevisiae* in agitated system, Christakopoulos *et al.* (1993), Lezinou *et al.* (1994) and Mamma *et al.* (1996) observed promising yields of ethanol from shorgum by mixed microbial culture with *S. cerevisiae* plus *Fusarium oxysorum*.

Agitation has adverse effects on substrate porosity owing to the compacting of the substrate particles, disruption of fungal attachment to the solids and damage to fungal mycelia owing to shear forces in SSF systems. However, no beneficial or adverse effects of agitation in the production of gibberellic acid and bacterial α-amylase have been reported (Kumar and Lonsane, 1987a, 1988). Moreover, the agitation may promote or prevent aggregate formation of the fermenting mass depending on the nature of the solids.

3.3 Microbial growth patterns and control of fermentation

3.3.1 Microbial types and inoculum

Based upon the type of microorganism involved, SSF processes can be categorized into two main groups: natural (indigenous) SSF and pure culture SSF (individual or mixed). Ensiling and composting are the two best SSF processes which utilize natural microflora. On the other hand, SSF with pure culture is known since antiquity (i.e. the koji process with *A. oryzae*). Pure cultures are generally used in industrial SSF processes to improve the control of the substrate utilization and end-product formation. Bioconversion of agricultural residues (e.g. straw) to fungal protein using two different microorganisms is a typical example of mixed-culture SSF systems. In nature, SSF is often carried out by mixed cultures in which several microorganisms show development and symbiotic cooperation (Pandey, 1992).

Many microorganisms can grow on solid substrates but only filamentous fungi can grow to a significant extent in the absence of free water (Moo-Young *et al.* 1983; Smith, 1985; Pandey, 1992). Bacteria and yeasts will grow on solid substrates at 40–70% moisture levels, such as in composting, anaerobic and aerobic ensiling, but the growth and the propagation of single cell organisms always require free water. Thermophilic bacteria grow predominantly in the first stage of composting. Lactic acid bacteria grow anaerobically and aerobically in ensiling processes causing souring and flavor changes (Tengerdy, 1985). Yeasts usually grow on solid substrates in symbiosis with other microorganisms in composting, ensiling and some industrial SSF processes involving fruit wastes and other sugar-containing natural substances.

The cultivation of the filamentous fungi on solid substrates has been widely used for different purposes at laboratory scale, e.g. for koji and lignocellulose fermentations, for fungal spores, and for mycotoxin production (Matteau and Bone, 1980; Bhumiratna *et al.*, 1980; Lotong and Suwarnarit, 1983). Among the filamentous fungi, three classes have gained practical importance in SSF:

1. the Phycomycetes such as the genera *Mucor* and *Rhizopus*;
2. the Ascomycetes with the genera *Aspergillus* and *Penicillium*;
3. the Basidiomycetes, especially white rot fungi, among them certain edible mushrooms.

The former two classes grow naturally on fruits and grains, and the latter on decaying lignocellulose, wood, straw, and other forestry and agricultural wastes.

Table 3.3 gives a view of the different microorganisms used in recent years in various SSF processes. Malathi and Chakrabarty (1991) obtained proteases using *A. flavus*. *A. niger* has most commonly been used for enzyme production as glucoamylase (Pandey, 1990a,b; Jaleel *et al.*, 1992), cellulase and amylase (Desgrenges and Dureng, 1990) as well as for citric acid production (Grewal and Kalra, 1995). Another *Aspergillus* species has been used to study different processes and/or to obtain a variety of products; *A. oryzae* has been used for the production of alcohol, aldehydes and ketones (Ito *et al.*, 1990). Paredes-López *et al.* (1991) used chickpea to obtain an outstanding protein food using *R. oligosporus*. Several other species have been used to produce lipases (Muñoz *et al.*, 1991), lipopeptidases (Ohno *et al.*, 1995a), tempeh (Peñaloza *et al.*, 1992), α-L-arabinofuranosidase (Roche *et al.*, 1994), L-glutamic acid (Nampoothiri and Pandey, 1996) as well as to study the effect of water content and water activity in *E. jeanselmei* performance (Cox *et al.*, 1996).

The inoculum is generally used at a high ratio in most fermentation processes for the production of secondary metabolites, with the aim of producing the desired level of product in a short period. Similarly, most of the SSF processes involve the use of a high ratio of inoculum, but the purpose in this case is to prevent contamination during fermentation (Lonsane *et al.*, 1992). For many SSFs inoculum can be prepared and used in the same fashion as in liquid media. Either preformed inoculum or dry spores in a carrier may be used to inoculate the substrate (Hesseltine, 1987). However, in the case of fungi used commercially that do not sporulate in culture, such as *Agaricus*, *Lentinus*, *Pleurotus* and *Volvariella*, the inoculum has to be prepared as spawn (Rajarathnam and Bano, 1989). In this procedure, the fungus is grown on solid particles such as rye kernels and the infected kernels are dispersed evenly throughout the substrate.

3.3.2 Microbial growth patterns and growth rate

(a) Growth patterns. In nature, bacteria grow best only when in a liquid phase or at least when the nutrients are in free water. Likewise, single-celled fungi, the yeasts, grow well when the nutrients are in a soluble form. Growth of such microorganisms in SSF becomes difficult where substrate carbon is not available in soluble form (Chahal, 1983). A limited success in converting lignocellulosic wastes into animal feed by using bacteria (e.g. *Cellulomonas* sp., *Alcaligenes faecalis*) or yeasts (e.g. *Aureobasidium pullulans*, *Candida utilis*) has been reported in semisolid state fermentation (Han and Anderson, 1975). Low protein yields in the final product might

Table 3.3 Microbial types, processes and substrates in SSF

Microbial type	Process/product	Substrate	Reference
Aspergillus flavus	Protease	Wheat bran	Malathi and Chakrabarty (1991)
Aspergillus niger	Growth and kinetic studies	Cassava	Saucedo-Castañeda *et al.* (1990)
	Glucoamylase	Wheat bran	Pandey (1990a; 1990b; Jaleel *et al.* (1992)
	Kinetic studies	Wheat bran	Gowthaman *et al.* (1995)
	Citric acid	Simple sugars	Grewal and Kalra (1995)
Aspergillus oryzae	Alcohol, aldehydes, ketonas	Rice	Ito *et al.* (1990)
Bacillus coagulans	α-Amylase	Wheat bran	Babu and Satyanarayana (1995)
Bacillus licheniformis	α-Amylase	Wheat bran	Ramesh and Lonsane (1989, 1990, 1991)
Bacillus subtilis	Lipopeptidase	Okara	Ohno *et al.* (1995a)
Brevibacterium spp.	L-Glutamic acid	Sugar-cane bagasse	Nampoothiri and Pandey (1996)
Coprinus fimetarius	Growth studies	Wheat straw	Singh *et al.* (1990)
	Protein	Wheat straw	Kumar and Singh (1990)
Exophiala jeanselmei	Kinetic	Styrene	Cox *et al.* (1996)
Lentinula edodes	Extracellular enzymes	Lignocellulosics	Mishra and Leathman (1990)
Mucor meihei	Rennet	Wheat bran	Thakur *et al.* (1990)
Penicillum spp., *Mucor meihei*, *Rhizopus arrhizus* and *R. delemer*	Lipases	Wheat bran	Muñoz *et al.* (1991)
Polyporus spp.	Protein	Bagasse	Nigam (1990)
Rhizopus oligosporus	Kinetic studies	Cassava	Mitchell *et al.* (1990)
	Protein food	Chickpea	Paredes-López *et al.* (1991)
Saccharomyces cerevisiae, *Fusarium oxysparum*	Ethanol	Shorgum	Mamma *et al.* (1996)
Thermoascus aurantiacus, *Trichoderma reesei*	α-L-Arabinofuranosidase	Sugar beet pulp	Roche *et al.* (1994, 1995)
Trichoderma viride and *A. niger*	Growth and kinetics, cellulase and amylase	Sugar beet pulp	Desgrenges and Dureng (1990)
Tubercularia vulgaris	Pectic enzymes	Citrus pulp pellets	Vieira *et al.* (1991)

be attributed to the fact that in such systems the unicellular organisms (bacteria and yeasts) were unable to penetrate deep into the tissue for complete utilization of the substrate.

It has been demonstrated that bacteria and yeast are promisory microorganisms for SSF. Ramesh and Lonsane (1989, 1990) and Babu and Satyanarayana (1995) reported the production of bacterial α-amylase with *B. megaterium*, *B. licheniformis* and *B. coagulans*, respectively. Moreover, the production of lipopeptidase antibiotics from okara (Ohno *et al.*, 1995a) and L-glutamic acid from sugar cane bagasse (Nampoothiri and Pandy, 1996) by *Bacillus* and *Brevibacterium* species has also been reported. Several yeasts have been used for protein enrichment or ethanol fermentation in SSF. Smith *et al.* (1986) reported protein production by *Sporotrichum pulverulentum*. *Saccharomyces cerevisiae* has most commonly been used for ethanol production (Cochet *et al.*, 1988; Mamma *et al.*, 1996).

On the other hand, filamentous fungi grow on solid surfaces such as wood, seeds, roots and leaves of plants, much as roots grow in the soil. A high protein content in the finished product has been recorded in SSF of cellulosic materials with *Chaetomium cellulolyticum* (Abdullah *et al.*, 1985). The high protein content in the final product has been attributed to the fact that the hyphae of this fungi have the power to penetrate deep into the intercellular and intracellular spaces for better utilization of the substrate. In SSF experiments on the growth of *R. oligosporus* on 'hard-to-cook' common beans used as a substrate, it was found that the long tubular body of the fungal filament grows alongside the solid particle using the available surface liquid film as a source of moisture and nutrients, and penetrating cracks and pores in the substrate with special appendages to scavenge for further nutrients (Paredes-López, O., unpublished results). Filamentous fungi grow differently in LSF and SSF. In a stirred fermenter, the fungi multiply by mycelial fragmentation. On an agar plate, fungi grow typically by apical extension in a radical fashion toward an increasing concentration gradient of nutrients. The growth rate is linear and depends on the width of the peripheral growth zone where hyphal growth is exponential and approximately equal to the specific growth rate (0.11–0.15 h^{-1}) observed in liquid cultures (Moo-Young *et al.*, 1983). The low packing density of mycelia in the internal part of the substrate has been suggested to be partially due to geometric limitations (Laukevics *et al.*, 1985); these authors also reported that the available space per unit mass is 6–10 times smaller in SSF than in LSF. Such conditions set an upper theoretical limit on the achievable biomass concentration in SSF.

Georgiou and Shuler (1986) presented a relatively simple computer model for the growth of a mold colony on a solid surface with a defined medium utilizing glucose. Unlike submerged cultures the model needs to account for both cellular differentiation and the spatial heterogeneity in the system. Results of the simulation studies suggest that mass transfer limitations are at least partially responsible for the proliferation of differentiated structures on solid substrates as compared to liquid cultures. The major limitation of the model may well be the assumption that simple saturation kinetics with external nutrient control can be used for a dynamic system.

(b) Growth rate. One of the disadvantages of SSF is the technical difficulty for growth rate assessment in a direct way. Most of the standard procedures used in LSF for biomass measurement are of limited value or are indeed useless. The content of biomass has been estimated by measuring some of the mycelial components such as protein, DNA and RNA. Corrections need also to be made for the measured component present in the substrate. However, not much attention has been given to the fact that qualitative and quantitative changes of these components take place in actively growing cultures. Outstanding differences in protein content may be given by the standard assays at the various developmental stages of the culture (Paredes-López *et al.*, 1988). Some of the indirect methods reported for cell growth estimations are based on O_2 uptake (Sato *et al.*, 1983; Oriol *et al.*, 1988), glucose uptake (Mitchell *et al.*, 1991), glucosamine formation (Smits *et al.*, 1996), CO_2 evolution (Narahara *et al.*, 1982; Sato and Yoshizawa, 1988; Raghava-Rao *et al.*, 1993) and infrared techniques (Durand and Chereau, 1988).

For fungi, the average mass doubling time in LSF is 3 h. In SSF of beet pulp with *A. oryzae* a maximum specific growth rate (μ) of 0.10 h^{-1} was observed (Czajkowska and Ilnicka-Olejniczak, 1988). However, Oriol *et al.* (1988) reported a μ of 0.45 h^{-1} for SSF of cassava by *A. niger*. For some yeasts, a μ of 0.47 h^{-1} in LSF has been reported (Paredes-López *et al.*, 1976) and of 0.48 h^{-1} in SSF (Sato *et al.*, 1988).

3.3.3 Control by physical and nutritional factors

As mentioned above, the partial pressures of O_2 and of CO_2 in the gas environment of SSF may influence microbial growth and product biosynthesis. High pO_2 increased amylase production in SSF of rice with *A. oryzae* and increased pCO_2 deteriorated amylase production in the same fermentation (Bajracharya and Mudgett, 1980). Desgrenges and Dureng (1990) studied the effect of pCO_2 on growth and on enzyme production; they found that culturing fungal strains at different CO_2 partial pressures in the gaseous environment showed an increase of growth with pCO_2 at 10. Mitchell *et al.* (1986) developed a model solid substrate to simplify studies of the basic aspects to control SSF. The substrate consisted of starch and other nutrients embedded in a κ-carrageenan gel matrix. The model system was designed to mimic the growth of *R. oligosporus* on cassava tubers, a process which may be used for protein enrichment of this substrate for use as animal feed. A large improvement in protein content was obtained by simultaneously decreasing the particle size of the substrate and increasing the nitrogen content of the fermentation medium (Mitchell *et al.*, 1988).

On the other hand, in the production of the enzyme amyloglucosidase by *A. niger* from wheat bran, it was possible to control the SSF process by changing substrate autoclaving time and HCl content; it was found that as substrate autoclaving time increased, enzyme titles decreased. After fermentation, the amyloglucosidase yields were higher when acid and trace elements were eliminated and substrate autoclaving time was 20 min (Jaleel *et al.*, 1992). Emelyanova (1996) demonstrated that the treatment of cereal substrates, addition of nutrients to the grain and thickness of the solid substrate influenced fungal growth and product yield as well. When wheat bran was used for producing gibberellic acid by *Gibberella fujikuroi*, Kumar and Lonsane (1987b) found that particle size of the substrate greatly affected gibberellic acid production. Other physical factors (e.g. pH, temperature, mixing) may be used to control the SSF as well.

In SSF the restriction effects of nutritional factors may be very severe owing to the limited diffusion rate of the substrate and the limited access of microorganisms to the substrate (Moo-Young *et al.*, 1983). The C/N ratio is an important indicator in SSF for nutritional regulation of growth. For composting, an ideal C/N ratio is 20 (Cannel and Moo-Young, 1980b). Low biomass production yields of *A. niger* have been found with a C/N ratio of 85 and 114. This low production was obtained since ammonium sulfate was used as nitrogen source at limiting concentrations to favor citric acid accumulation; assays without nitrogen limitation (C/N = 12) produced higher biomass concentration. The C/N ratio of garbage from North America ranges from 20 to 70 because of its high paper content. In this situation another source of nitrogen (e.g. sewage sludge) may be added to speed up the fermentation process. Nitrogen starvation favors lignin degradation whereas higher nitrogen levels favor cellulose depolymerization. Fermentation media with a high C/N ratio are optimal for secondary metabolite formation (Hutter, 1982).

3.4 Genetic engineering for biodegradation of lignocellulosic wastes

The most abundant organic compounds in the biosphere are cellulose and lignin which are available as agroindustrial residues. Nowadays, we are in a world of diminishing resources, pollution problems and increasing needs, therefore optimization of lignocellulosic wastes is required. In nature the agroindustrial wastes undergo microbial colonization and transformation, so most of the organic wastes are converted or mineralized, thereby restabilizing the environment. Nevertheless, high concentrations of wastes can overwhelm the capacity of the natural degradation process and cause pollution (Kanotra and Mathur, 1995).

Some organic wastes are burned but this practice has been criticized because of the resulting air pollution and the danger of soil erosion. Another approach has been the use of chemicals to enhance the digestibility of organic wastes, but their use can be tedious and costly, and could require further treatment to eliminate side effects. Alternately, microbiological treatments can be economically and environmentally attractive for degradation of lignocellulosic wastes.

Chemical, enzymatic or microbiological conversion of lignocellulosic residues is affected mainly by lignin and cellulose crystallinity, leading to an ineffective degradation. Lignin impedes enzymatic and microbiological access to the cellulose, and cellulose crystallinity affects the attack rate of the mechanisms on cellulose. However, lignin degradation may be overcome by treatment with white-rot basidomycetes by SSF because of its lignolytic enzymes (Kaal *et al.*, 1995). The conversion of crystalline cellulose has been achieved using mutants with high cellulase yielding, such as those from *Trichoderma reesei* consisting of endo-1,4-β-glucanase, exo-1,4-β-glucanase and 4-β-glucosidase. These enzymes act sinergystically and convert the crystalline cellulose into cellobiose and glucose (Kanotra and Mathur, 1995).

Recently, recombinant microorganisms have been employed in SSF systems for different purposes (Ohno *et al.*, 1995b); one of these is the improvement of lignin biodegradation systems. The study and characterization of waste-degrading organisms can lead to the identification of the genes responsible for lignocellulolytic activity. Employing genetic engineering techniques, it may be possible to transfer these characteristics to other organisms with the potential for SSF but with the restriction of low or inefficient ligninocellulolytic activity. Several pieces of work relating to gene cloning, sequence analysis, microorganism transformation, expression and purification of recombinant lignocellulolytic enzymes have been published (Table 3.4); these investigations offer the alternative to improve the SSF systems in which lignocellulolytic waste is biodegraded.

3.4.1 Lignin biodegradation

The biodegradation of lignin by white-rot fungi is a complex reaction which appears to be achieved by the concerted action of extracellular enzymes such as lignin peroxidases (LPO) and Mn(II)-dependent peroxidases (MnP). Also, hydrogen peroxidase-producing enzymes, such as glyoxyl oxidase and laccase, are considered important in the process of lignin biodegradation. The first lignin-degrading enzyme, ligninase (lignin peroxidase) from the white rot basidomycete *Phanerochaete chrysosporium*

Table 3.4 Cloned genes and identified mutants, related with biodegradation of lignocellulosic residues

Sequence/mutant	Organism	Role in biodegradation	Reference
ADD, aryl-alcohol dehydrogenase	*Phanerochaete chrysosporium* BKM-F-1767	Lignin degradation	Reiser *et al.* (1994)
CDH, cellobiose dehydrogenase	*Phanerochaete chrysosporium* OGC101	Lignin degradation	Li *et al.* (1996)
GLG5, lignin peroxidase	*Phanerochaete chrysosporium* BLM-F-1767	Lignin degradation	Gaskell *et al.* (1991)
GLG6, lignin peroxidase	*Phanerochaete chrysosporium* BKM-F-1767	Lignin degradation	Naidu *et al.* (1990)
LAC2, laccase	*Podospora anserina*	Lignin degradation	Fernández-Larrea and Stahl (1996)
LCC1, LCC2, laccase	*Trametes villosa*	Lignin degradation	Yaver *et al.* (1996)
LPOA, lignin peroxidase	*Phanerochaete chrysosporium* BKM-F-1767	Lignin degradation	Smith *et al.* (1988)
LPOB, lignin peroxidase linked to LPOA	*Phanerochaete chrysosporium* BKM-F-1767	Lignin degradation	Huoponen *et al.* (1990)
LPO811, lignin peroxidase	*Phanerochaete chrysosporium*	Lignin degradation	Reiser *et al.* (1993)
0282, lignin peroxidase	*Phanerochaete chrysosporium*	Lignin degradation	Schalch *et al.* (1989)
POX1, POX2, laccases	*Pleurotus ostreatus*	Lignin degradation	Giardina *et al.* (1995)
V4, lignin peroxidase	*Phanerochaete chrysosporium* BKM.1667	Lignin degradation	Schalch *et al.* (1989)
VS1, VS7, VS27 mutants	*Phanerochaete chrysosporium* BKAM.F.1767	Conversion of crystalline cellulose into cellobiose and glucose	Kanotra and Mathur (1995)

was described in 1983 (Tien and Kirk, 1983). The LPOs of *P. chrysosporium* are present as isoenzymes; LPO cDNA and genomic sequences from *P. chrysosporium* have been analysed (Table 3.4). In addition, the regulation of LPO gene expression has been investigated by using Northern (RNA) blots (James *et al.*, 1992) and more recently using competitive polymerase chain reaction (PCR) (Stewart *et al.*, 1992).

Another approach to investigate the expression of lignin peroxidase genes, has been described by Reiser *et al.* (1993), using nuclease protection assays. They worked out a sensitive procedure capable of differentiating transcripts derived from different LPO genes. These authors employed the 3′-end region of a LPO811 gene to make a labeled RNA probe because the sequence of this region varies quite substantially among all LPO genes, and thus it was possible to differentiate the transcripts from different LPO genes. Furthermore, Reiser *et al.* (1993) showed that the study of LPO or MnP gene expression using different conditions, such as carbon and nitrogen limitation, can help to distinguish the best conditions for an efficient lignin biodegradation.

Kaal *et al.* (1995) demonstrated that several commercially important and commonly occurring white-rot fungi produce higher ligninolytic enzyme activities in response to a nitrogen-rich medium, in contrast to the physiological model proposed for *P. chrysosporium.* Another important factor to be considered is the manganese concentration, which, according to Kerem and Hadar (1995a), in general enhances the specific enzymes involved in ligninolysis during the secondary growth phase of *Pleurotus ostreatus* at SSF. Some studies have shown that some lignin peroxidase isoforms are more abundant in SSF than in liquid fermentation such as LiP2 from the white-rot fungus *Phebia radiata*. The fungi may utilize mechanisms of lignin depolymerization in wood that are different from those used in liquid cultures.

It is important to mention that laccases are believed to participate in lignin degradation, but their role has not been well established. However, it has been suggested that lignin degradation is performed through the oxidation of phenolics, but phenolic subunits make up only a small proportion of the lignin polymer (Bonnen *et al.*, 1994). Nevertheless, laccase produced by *Trametes versicolor* is able to oxidize nonphenolic substances, which would allow for a greater role in total lignin degradation. One approach to clone laccase genes in basidiomycetes has been the study of laccase-less mutants using protoplasts and cell fusion. Cloning genes have contributed to the understanding of the molecular basis of enzymatic catalysis and the regulatory mechanism controlling the production of different isoforms of laccases (Table 3.4) (Giardina *et al.*, 1995). Recently Yaver *et al.* (1996) cloned two laccase genes from *Trametes villosa* and expressed one of them (Lcc 1) in *A. oryzae* using the protoplasts transformation method. In this work they used the pDSY2 expression vector which has the *A. oryzae* α-amylase gene promoter, in order to produce larger quantities of the laccase LCC1.

Additional contribution to the genetic manipulation of fungi has been published by An *et al.* (1996). They constructed new cosmid vectors for the ascomycete *Magnaporthe grisea* and the basidiomycete *Ustilago maydis*, and these vectors could be extensively used for transformation, because the promoter functions in other ascomycetes and basidiomycetes.

In addition to the enzymes mentioned above, it should be considered that the constant oxidation of lignin by peroxidases will produce a highly oxidized state of the polymer, and degradation products such as aldehyde, quinone and perhaps acidic groups will increase and, as a consequence, peroxidase catalysis will stop. Furthermore, laccase oxidative reduction proceeds first through the formation of very unstable radical intermediates, which can undergo rapid association and hence polymerization. Because of that, reductive reactions are also important for lignin biodegradation to occur. Subsequent action of oxidative peroxidases and reductive intracellular enzymes is needed for complete degradation of the lignin biopolymer.

Reiser *et al.* (1993) found that veratryl alcohol does not exert its effect at the RNA level as far as the LPO811 gene is concerned, although the LPO activity was 25 times higher in the presence of the veratryl alcohol. Following up this work, Reiser *et al.* (1994) published the DNA sequence of the aryl-alcohol dehydrogenase (AAD) cDNA clone from *Phanerochaete chrysosporium* and demonstrated, by Southern blot analysis, the presence of multiple AAD gene-related sequences in other white-rot fungi including *Bjerkandera adusta* and *Fomes lignosus*. In this study they found that, under conditions of nitrogen limitations, the AAD mRNA levels are higher than in carbon-limited cultures. Furthermore, they found a correlation between the appearance of AAD-specific transcripts and LPO-specific transcripts which is in line with the proposed synergystic interaction of the two enzymes in lignin biodegradation. These studies achieved the heterologous expression of the AAD cDNA in *E. coli* and purified the recombinant enzyme; this cloned gene could be employed to improve any nonefficient reductive system of lignin biodegradation. More recently, another approach to lignin degradation has been proposed, which consists of the modification of lignin profiles of plants through genetic engineering (antisense and sence suppression of gene expression) (Boudet and Grimapettenati, 1996).

The first transgenic plants with modification of monomeric composition have been obtained. Down-regulation of *O*-methyl transferase (OMT) gene induces dramatic reduction of syringyl units. In addition to this, cinnamyl alcohol dehydrogenase (CAD) down-regulation is important. In both OMT and CAD down-regulated plants, no changes in phenolic characteristics were observed. One of the chemical modifications was an increase in cinnamaldehydes in the polymer structure, which is removed more easily during the pulping process.

3.4.2 Cellulose bioconversion

The cellulase from *T. reesei* have the most efficient activity against microocrystalline cellulose. Endo-1,4-β-glucanase, exo-1,4-β-glucanase and β-glucosidase act synergystically to bring the conversion of crystalline cellulose into cellobiose and glucose. Kanotra and Mathur (1995) proposed a method for identifying mutants of *T. reesei* with altered characteristics. The best mutants alone or combined with *Pleurotus sajor-caju* were used in SSF of paddy straw; the aim was to bring about maximum degradation of cellulose, hemicellulose and lignin of lignocellulosic residues. The combination of *T. reesei* VS27 mutant and *P. sajor-caju* has shown a synergystic effect in the biodegradation of lignocellulosic residues.

3.4.3 Practical applications of a lignin biodegradation system

The efficient bioconversion of lignocellulosic residues may have many practical purposes including production of edible mushrooms and animal fodder, degradation of organopollutants as well as biopulping and biobleaching processes (Das and Karim, 1995; Katagiri *et al.*, 1995). Currently, the production of some edible fungi by SSF such as *Agaricus bisporus*, requires a composting phase. This consists of a substrate pretreatment where raw materials are wetted and allowed to decompose using microorganisms present in the raw material followed by a pasteurization phase (Nigam and Singh, 1994). The composting phase facilitating biodegradation by *A. bisporus* is not required for wood-rotting fungi such as *P. chrysosporium* because the latter has a better lignin degradation system. However, Bonnen *et al.* (1994) demonstrated that *A. bisporus* produces MnPs which are regulated in a developmental manner similar to laccase activity and thus correlates with lignin loss from compost and with its lignin-degrading activity. However Iiyama *et al.* (1994) found that, during composting and mushroom growth, lignin monomers are chemically modified and lignins are to some extent polymerized by condensation reactions and some cleavage of inert monomer linkages may occur, but an oxidative depolymerization, such as that achieved by certain lignin-degrading fungi, does not take place.

The application of biotechnology to the cultivated mushroom *A. bisporus* has been hindered by the lack of an efficient transformation system. Recently, Vanderhee *et al.* (1996a) achieved the transformation of both a homokaryotic and heterokaryotic strain of *A. bisporus* to hygromycin B resistance; the DNA was integrated into the genome and stably maintained throughout vegetative growth. Vanderhee *et al.* (1996b) used a plasmid (pHAG3-1) carrying the hygromycin resistance gene and 3.2 kb genomic fragment from *A. bisporus*. They linearized the pHAG3-1 plasmid at three different positions and in the vegetative mycelium of a transformant with tandemly integrated pHAG3-1 plasmids at the homologous position. Exoglucanase mRNA was strongly increased without any apparent effect on growth rate or morphology. This work opens up the possibility for the introduction of some enzymes participating in the lignin degradation system from *P. chrysosporium* as a 'selective' lignin degrader, such as *Lentinus edodes* or the white-rot fungus IZU-154 (Nishida *et al.*, 1988), to *A. bisporus* so as to improve the degradation system which may result in savings in time and money.

3.5 Reactors for solid substrate fermentation

The design of SSF systems has been for the most part highly empirical. The absence of accurate and reliable experimental data for such systems is the major reason for the lack of sound knowledge in this area (Aidoo *et al.*, 1982; Moo-Young *et al.*, 1983). The information is very scarce on the instrumentation used to monitor fermentation processes (e.g. dissolved oxygen electrodes, pH measurement). Nevertheless, different types of bioreactors have been recently described and used for various purposes, incorporating several modifications for improved operation and performance. An ideal fermenter should have several characteristics:

1. In particular, the material of construction should be nontoxic and able to withstand pressure.
2. It should not be affected by chemical corrosion.
3. There should be proper arrangements for aeration/agitation, with sampling, charging and discharging ports.
4. A cooling mechanism may be required to control the generated metabolic energy.
5. Furthermore, the bioreactor systems should be capable of operating under aseptic operations (Pandey, 1991a).

For pilot-plant and large-scale fermentation of moist solid substrate, some factors require special consideration for the construction or selection of the fermentation equipment (Durand and Chereau, 1988; Durand *et al.*, 1988):

1. Inoculation, sampling and transfer techniques must be simple.
2. Homogeneity of the culture medium is important to prevent particle agglomeration and the settling of substrate during fermentation.
3. Agitation must be suitable for the substrate, not destructive for the microorganism.

4. An appropriate aeration rate should be chosen and heat build-up problems prevented.
5. The equipment must be sterilized.
6. Monitoring and control devices of fermentation parameters must be used.
7. Handling (empty, refill, clear out, maintenance) must be simple.
8. The choice of materials used for construction of the fermentation vessel.
9. Low capital costs and operation expenses.

There are a number of different types of reactors presently used for SSF. These are summarized below.

3.5.1 Tray fermenter

Traditionally, most SSFs are conducted in shallow trays where the substrate is spread in about 10 cm layers or in specially formed cakes. Heat build-up is avoided in such shallow trays (Tengerdy, 1985). The fermenter is humidified by humidifiers or by forcing in humid air. This equipment is used with success in small-scale operations at the pilot or commercial level (Lonsane *et al.*, 1985); it gives a final product uniform in nature and with high enzymatic activity. It is widely used in the traditional oriental food preparation schemes, such as in the koji process (Steinkraus, 1983b; Paredes-López *et al.*, 1987). However, the operation of this equipment is labor intensive. The need for a large area is another disadvantage of tray fermenters.

The use of trays to make soy sauce koji goes back for 1000 years in Japan and South West Asia and probably for 3000 years in China. Miso, which is a fermented food in Japan, China, Taiwan, The Philippines, Indonesia, has traditionally been prepared in shallow pans (Steinkraus, 1983b; Paredes-López, *et al.*, 1987; Pandey, 1991a).

In recent years, mathematical models for SSF in tray bioreactors have been developed. As a result, Raghava-Rao *et al.* (1993) proposed a simple mathematical model for the interaction of mass transport with biochemical reactions under isothermal conditions. Unfortunately, the application of the model to practical systems requires that some kinetic parameters have to be obtained separately. Another disadvantage is the assumption of isothermality in the tray bioreactor which may simplify the mathematical approach but may not be the situation obtained in practice. On the other hand, Rajagopalan and Modak (1995) were capable of modeling oxygen consumption, heat generation and cell growth of *A. niger* in a tray fermenter. The model incorporates the transport of oxygen in a bed of moist substrate particles, simultaneous transport and consumption of oxygen in the biofilm around the particle and heat transport in the bed.

The important feature of this model is the incorporation of local transport phenomena within the biofilm phase. Previous models of SSF do not incorporate this phenomena. The analysis of biofilm expansion around a single particle is carried out to gain some insight into cell growth.

3.5.2 Rotating drum fermenter

Drum fermenters basically consist of a drum-shaped reactor equipped with a rotating device and usually an inlet and an outlet for air. The air-inlet tube may almost reach the bottom of the drum (Pandey, 1991a) and it employs forced aeration. The drum may be provided with baffles or may be sectioned into four parts. The drum is placed on a shaft rotating on two half-bearings and is powered by a variable speed drive motor. The unit is provided with a port for charging the fermenter, water addition, inoculation and sampling (Lindenfelser and Ciegler, 1975). For the most effective agitation, baffles permit the substrate to be elevated and subsequently dropped at some point during each revolution (Hesseltine, 1977a).

Microbial growth in these type of fermenters is considered to be better and more uniform than in tray fermenters. Operation is simple and cleaning is fast, but they present some problems such as microbial contamination, medium aggregation and heat build-up. They have been used successfully for ochratoxin production by *A. ochraceus* (Lindenfelser and Ciegler, 1975). Pou *et al.* (1987) used a rotating reactor to increase the protein content of an acid-pretreated bagasse pith with *Trichosporum penicillatum*.

3.5.3 Packed-column fermenter

As the name suggests, this consists of a glass or plastic column with lids at both ends. It may be fitted with a jacket for the circulation of water to control the temperature during fermentation. Alternatively, the whole column may be placed in a temperature-controlled water bath (Pandey, 1991a). Rodríguez *et al.* (1985) carried out solid-state fermentation of dried citrus peel by *A. niger*; the fermentation system consisted of a packed-bed column reactor with bottom to top air circulation. Larroche *et al.* (1986) reported spore production of *Penicillium roqueforti* by simulated solid-state fermentation in a jacketed Pyrex column filled to a depth of about 500 cm^3 and supplied with air at its top. Gowthaman *et al.* (1995) used a packed-bed column for estimating overall oxygen-transfer coefficients.

The packed-column reactor uses as substrate sugar-rich wastes such as pineapple residues (González *et al.*, 1985), and granulated starchy materials such as cereal grain residues, cassava (Raimbault and Alazard,

1980) and sweet potato residues (Yang, 1988). These substrates, loosely packed into the columns, are inoculated with molds and fermented for various times. The final product is used for animal feed purposes. Also, Cochet *et al.* (1988) reported the use of a tubular reactor in a semisolid state fermentation to produce ethanol.

3.5.4 Auger tube fermenter

The auger tube fermenter has been designed for batch and continuous operations (Gibbons and Westby, 1988; Gibbons, 1989). It consists of:

1. a nonported steam pasteurization chamber to destroy bacterial contaminants;
2. an inoculum port;
3. an auger tube that simultaneously conveys and mixes the substrate under fermentation.

The horizontal auger measures 15.2 cm in diameter and is 470 cm long. The fermentation chamber is surrounded by a plastic heating/cooling jacket. This fermenter has been used to produce ethanol from sweet sorghum stalks and Jerusalem artichoke tubers. Starchy wastes may also be processed in this reactor.

3.5.5 Helical screw fermenter

According to Tengerdy (1985), a promising new design is the helical screw fermenter. Its form of operation avoids compacting the substrate, does not damage the mycelia by shearing forces and provides thorough mixing. Tengerdy and his group have tested a 15 l reactor coupled to a microcomputer-regulated temperature and moisture control device for batch and continuous cultivation. Tests have been run with the fungus *Chaetomium cellulolyticum* and wheat straw.

3.5.6 Fluidized biomass fermenter

The productivity of a microbial process is a strong function of the number of microorganisms or of microbial biomass per unit of volume of bioreactor and per unit of time. In general, the volumetric concentration of microbial biomass in any type of SSF bioreactor is small (Laukevics *et al.*, 1985). In a fluidized biomass reactor, the fluidizing medium is air or an inert gas (Mishra *et al.*, 1982; Brauer, 1988). The advantages observed for this reactor at laboratory level are:

1. high biomass concentration;
2. intensive mixing of all ingredients in the system;
3. optimal mass transfer;
4. continuous operation (substrate input and product removal);
5. reduction of inhibitory effects by final products;
6. ease of control.

Despite the fact that this type of reactor has a high productivity because of its high cell density compared to submerged cultures, Mishra *et al.* (1982) found a relatively long generation time of 14–17 h for growth of *Saccharomyces cerevisiae*. However, Sato *et al.* (1988) studied the production of ethanol with a thermophilic yeast in a SSF system with circulation of CO_2, and found that this procedure had, compared with the conventional SSF, a higher fermentation efficiency and a shorter fermentation time. Further studies on growth rate kinetics and scaling-up of the fluidized system are necessary.

3.5.7 Miscellaneous types

A laboratory scale reactor (2.5 kg dry-matter capacity) was designed and constructed to allow control of temperature and moisture level of the substrate without agitation (Figure 3.2). A fermenter unit is composed of two reactors. Each cylindrical vessel contains a removable basket with a perforated bottom (cylinder of 40 cm diameter and 60 cm height). Forced aeration is accomplished by means of thermostated air injected at the bottom of the reactor. By bubbling in a water bath and after heating, the thermostated air allows the regulation of temperature and moisture content of the medium during the cultivation. The two reactors have strictly the same conditions for temperature and relative himidity of the inlet air. This system has a great advantage in that experiments can be made in duplicate for studying certain parameters without varying the environmental conditions. Previous experience (Durand *et al.*, 1988) with this type of equipment has demonstrated that mechanical agitation is not necessary and that there is a uniform temperature distribution.

The development and investigation of high-efficiency SSF bioreactors is one of the fundamental problems biochemical engineers have to solve. By increasing efficiency of the reactor the downstream processes will be much simplified, and consequently energy and costs will be reduced. Based on the findings of the fundamental research, the following requirements for high-efficiency reactors may be established:

1. a high rate of microbial conversion per unit of biomass;
2. identical conditions for enzymatic/microbial reactions in all volumetric segments of the reactor (e.g. uniform distribution of microorganisms, temperature, pH, substrate concentration);

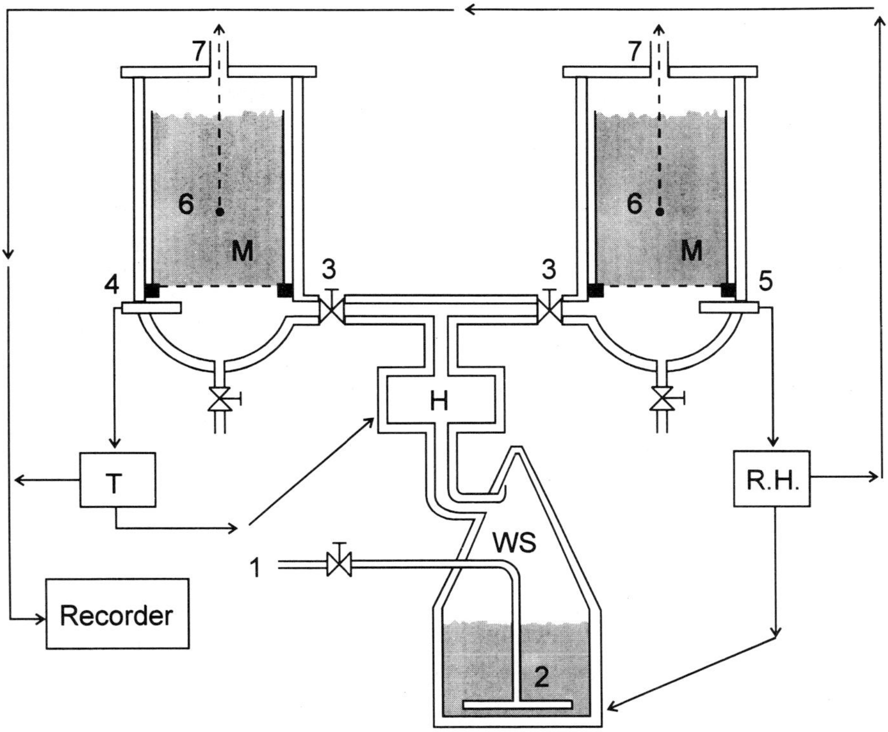

Figure 3.2 Schematic representation of a fermenter unit composed of two reactors for solid-state fermentation. H = heating device; M = individual reactors with perforated basket; R.H. = relatively humidity control; T = temperature control; WS = water system for air humidification. 1 = air inlet; 2 = air sparger; 3 = values for air-flow regulation; 4, 5 = probes for temperature and relative humidity, respectively; 6 = probe for temperature of sample under fermentation; 7 = air exhaust.

3. low energy requirements;
4. uniform distribution of energy/oxygen in the reactor;
5. small volume and site requirement of reactor;
6. simplicity of design;
7. applicability for batch and continuous operation;
8. possibility of integration into production systems.

Finally, in a few words, a modern solid substrate fermenter should be able to mix large amounts of substrate for a uniform fermentation, allow maximum automation and scale-up of the process, allow continuous operation and facilitate fermentation controls (Laukevics *et al.*, 1984; Tengerdy, 1985; Durand *et al.*, 1988; Almanza *et al.*, 1995).

3.6 Fermentation processes and compositional changes

3.6.1 SSF processes

Figure 3.3 shows the basic steps involved in a biotechnological process, in solid-state conditions, to use wastes from different origins. Some of these steps are optional (e.g. sterilization, product isolation) according to the objective of the particular SSF. SSF processes may be classified in two broad categories, which are briefly reviewed here.

(a) Typical SSF processes. In nature, solid substrates such as plant and animal wastes undergo microbial colonization and transformation by

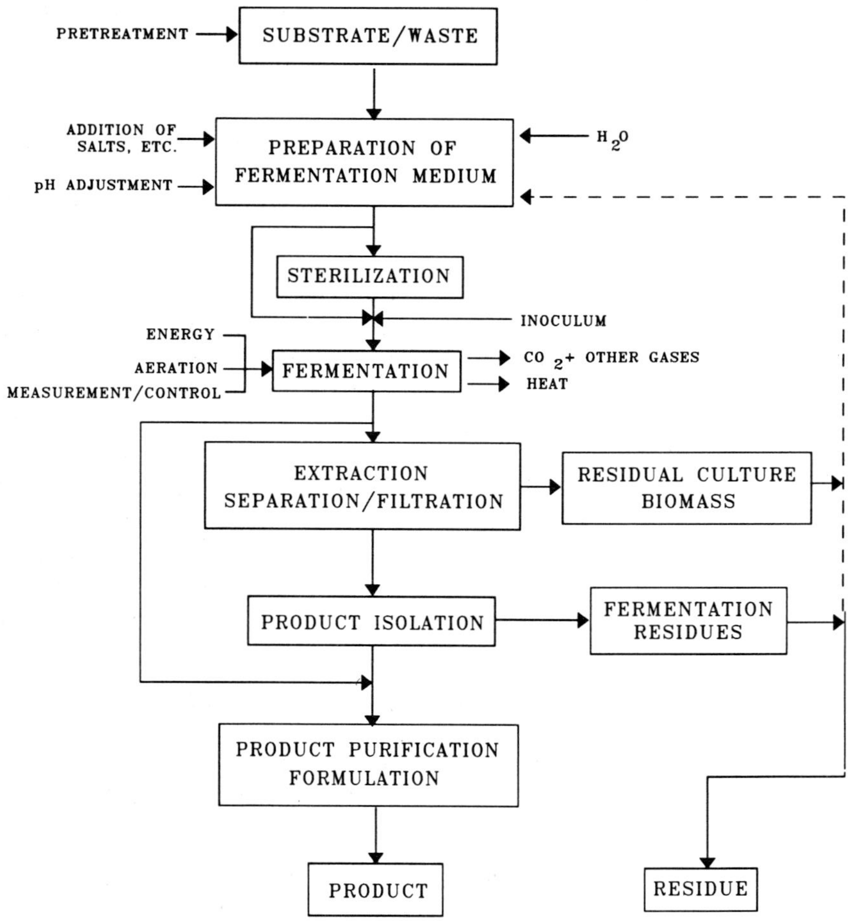

Figure 3.3 General description of the fundamental steps of a biotechnological process for the transformation of wastes from different origins.

various microbiological processes. Therefore, studies on solid-state cultures of molds have traditionally been restricted to substrates comprising mostly polymeric materials (e.g. starch, cellulose) with a given water-holding capacity in their porous matrix. Bacteria and yeast grow on the surface of the substrate while fungal mycelia penetrate into the particles of the substrate (Pandey, 1992). Pandey and Radhakrishnan (1993) and Ralph (1976) described these fermentations as those in which a solid material serves as the main source of nutrients for an organism which intimately associates itself with the surface and interstices of the material.

The physical morphology, especially porosity and particle size of the substrate, governs the surface area to the organism (Kumar and Lonsane, 1987b; Nampoothiri and Pandy, 1996). Pandey (1991b) and Moloney *et al.* (1984) found that a higher enzyme productivity and substrate, respectively, improved degradation because substrate contained finer particles. Similar results were found by Zadrazil and Puniya (1995) in the fermentation of sugar-cane bagasse into animal feed using white-rot fungi.

(b) SSF processes on inert solid supports. Use of synthetic polymers and substances like agar or gelatin provides homogeneous solid substrates which are used by some workers in SSF for kinetic studies (Thomas and Turner, 1981; Gervais *et al.*, 1988c). Georgiou and Shuler (1986) used nutrient agar as a model substrate. Gelatin and $\varkappa$-carrageenean have been used by Wei *et al.* (1981) and Mitchell *et al.* (1986, 1988) to produce a model substrate to mimic the growth of microorganisms. On the other hand, several attempts have been made to grow fungi on solid inert materials impregnated with a liquid nutrient. Ralph (1976) described these processes as those in which a nutritionally inert solid support provides some advantages to the organism in respect of access to nutrients. It has been found that production rates of amylase with *A. oryzae* cultured on vermiculite impregnated with a starch solution were higher than those in submerged cultures (Aidoo *et al.*, 1982). This type of fermentation has been also investigated by Larroche and Gros (1989). Aidoo *et al.* (1982) and Larroche and Gros (1989) reported that fermentation with inert porous particles allows:

1. the use of low-molecular-weight carbohydrates directly available to the microorganisms;
2. the handling of high concentrations of substrate;
3. direct determinations of biomass at any given time as in LSF;
4. simplified product recovery.

Recently, Prabhu and Chandrasekaran (1995) produced L-glutaminase from *Vibrio costicola* by SSF using polystyrene as an inert carrier. Interestingly, glucose at 10 g kg^{-1} enhanced the enzyme yield by 66%. The support system allowed glutaminase to be recovered with higher specific activity and lower viscosity than when a wheat bran system was used. Moreover, Zhu *et al.* (1996) used polyurethane foam for the production of nuclease P1 by *Penicillium citrinum*; enzyme productivity was ninefold higher than in SSF of wheat bran and 500-fold higher than in submerged fermentation. Christen *et al.* (1995) reported that *Phyzopus delemar* showed a good capacity to grow on amberlite (polymeric resin) with various carbon sources in SSF to produce lipase; this enzyme was produced in shorter times than in submerged culture. In fact, the lipase was adsorbed onto the support allowing the obtention of a ready-to-use immobilize enzyme.

3.6.2 Some currently practiced SSF processes

Table 3.5 describes the basic characteristics of some of the most important SSF processes, most of which may use wastes from plant and animal origin. Oriental-type foods have always been very popular in several Eastern countries. Pozol is one of the SSF products with the longest tradition in Latin America. Lactic acid fermentation has a strong potential for use with various agro-industrial wastes, and mushroom cultivation is steadily gaining popularity and economic viability. Animal feed production involves the use of an array of wastes and some of these processes are practiced on a commercial scale. Processes for ethanol (Gibbons, 1989), enzymes (Blain, 1975; Grajek, 1987; Ramesh and Lonsane, 1987) and the production of speciality biochemicals (Hang, 1988) appears to have a very sound economic future. Recent reports deal with the production of feedstuffs from vegetable wastes which have been subjected to SSF for eliminating toxic and/or antinutritional substances (Bau *et al.*, 1994; Ofuya and Obilor, 1994; Masaphy *et al.*, 1996).

(a) Food production. Lactic acid fermentation is an ancient process whereby a varied group of bacteria ferment carbohydrates producing lactic acid as the major end-product. This type of fermentation is used for the production of dairy products, sauerkraut, bread, meat and silage. In particular, SSF fermentation of legumes, cereal and starchy substrates have been associated in many regions of the world with the activity of lactic acid bacteria (Fukushima, 1985; Paredes-López and Harry, 1988). Several processes have been developed to produce fermented foods involving a fermentation by lactic acid bacteria (e.g. *Pediococcus halophilus*, *Klebsiella pneumoniae*) followed by a fungal fermentation, using as substrate wastes and by-products of plant origin (Steinkraus, 1983a; Paredes-López and Harry, 1988).

Tempeh is a generic term that is applied to foods that have been fermented by a filamentous fungus to obtain a compact cake of substrate particles knitted together by the mycelium (Steinkraus, 1983a). Tempeh is a food of Indonesian origin and is now produced at the industrial level in various countries (Harrigan, 1985). It is a food most commonly prepared from soya; however, this fermentation technique is extremely versatile and has been applied to a wide variety of plant wastes and by-products (Paredes-López and Harry, 1988; Paredes-López *et al.*, 1990; Shambuyi *et al.*, 1992). In recent years tempeh has been prepared from lupin (*Lupinus* spp.) (Agosin *et al.*, 1989) and quinoa (*Chenopodium quinoa*) (Peñaloza *et al.*, 1992), the latter being an indigenous marginal crop in Ecuador and other Andean countries. Lupin seeds have been suggested as an economically attractive alternative to soybean in the production of tempeh (Fudiyansyah *et al.*, 1995).

One of the most important steps in tempeh production is the soaking of the substrate during which, at ambient temperatures in the tropics, lactic acid bacteria ferment the solutes leached from the substrate to lactic acid (Figure 3.4). The consequent acidification prevents the growth of spoilage microorganisms that may impede the growth of the fungus used in the second fermentation step (Harry, 1988). Variations to the basic procedure were made by Fudiyansyah *et al.* (1995) by washing the lupin/soybeans in cold tap water, then soaking overnight in tap water again. After dehulling by hand, the kernels were partially cooked in boiling water, pH adjusted with acetic acid and cooled. The beans were inoculated and incubated for an appropriate time. On the other hand, Peñaloza *et al.* (1992) simplified tempeh production procedure by soaking quinoa seeds at 25–30 °C in an excess of tap water (low pH) including previous soak water (ragi water); a vigorous lactic acid fermentation resulted. *R. oligosporus* was added to the substrate for fermentation.

There are significant amounts of fresh bananas in the banana-producing regions (e.g. Central America) which are unacceptable to both internal and external markets. The use of these wastes has become important for economic and ecological reasons. The natural pH of banana purée (4.8–5.0) requires a drastic heat sterilization process followed by aseptic canning. This preservation process has a relatively small market. An alternative may be the fermentation process developed by De Porres *et al.* (1985) for the preparation of a purée by lactic acid bacteria, which may be associated more with yoghurt than with natural banana purée (Figure 3.5). Enzymatic browning of the ripe, peeled fruit is prevented by blanching in boiling water before pulping. Lactic bacteria inoculum is added to the purée, which reduces the pH to below 4.5 in a short time and produces an acceptable flavor. The fermented purée is suitable for use in banana flour speciality products. This technology may be implemented by cooperatives in rural areas. On the other hand, banana wastes have been used as a substrate for α-amylase production by *B. subtilis* under SSF (Krishna and Chandrasekaran, 1996).

Traditional SSFs of Latin America have received scarce attention as

Table 3.5 Typical solid substrate fermentation processes

Microorganisms	Oxygen need	Process	Products	References
Mixed fungal cultures and fungal monocultures	AE	Koji process	Oriental food and beverages	Steinkraus (1983b), Paredes-López *et al.* (1987), Peñaloza *et al.* (1992), Fudiyansyah *et al.* (1995)
Mixed microbial flora	AE	Dough formation	Mexican food (pozol)	Ulloa-Sosa (1974)
Saccharomyces cerevisiae/ Lactobacillus sanfrancisco	AE/AN	Bread dough formation	Bread	Aidoo *et al.* (1982)
Composting flora, spawn culture, mushroom cropping	AE	Mushroom cultivation	Edible mushrooms	Präve *et al.* (1987); Rajarathnam and Bano (1989), Jwanny *et al.* (1995), Kerem and Hadar (1995b)
Mixed lactic acid bacteria	AE/AN	Ensiling of crops and animal excreta	Preserved and enriched animal feed	Moo-Young *et al.* (1983), Iñiguez-Covarrubias *et al.* (1989, 1990a, 1990b), Joshi and Sandhu (1996)
Fungal monocultures	AE	Animal feed production	Converted lignocellulose	Ulmer *et al.* (1981), Laukevics *et al.* (1985), Hatakka *et al.* (1989), Zadrazil and Puniya (1995), Das and Karim (1995)

Mixed fungal cultures and fungal monocultures	AE	Detoxification	Animal feed and human food	Ofuya and Obilor (1994), Zvauya and Muzondo (1995), Kuo *et al.* (1995), Masaphy *et al.* (1996)
Mixed microbial cultures and fungal monocultures	AE	Animal feed from fruit and vegetable wastes	Protein-rich feedstuff	González *et al.* (1985), Rolz *et al.* (1988), Yang (1988), Durand and Chereau (1988)
Mixed microbial flora	AE/AN	Composting	Soil conditioner	Rajarathnam and Bano (1989)
Mixed microbial flora	AN	Methane production	Methane	Molnar and Bartha (1989)
Yeast	AN	Alcohol production	Alcohol	Sato and Yoshizawa (1988), Mamma *et al.* (1996)
Fungal monoculture	AE	Enzyme production	Various specific enzymes (e.g. cellulases proteases)	Tengerdy (1985), Couri and Defarias (1995), Babu and Satyanarayana (1995), Murado *et al.* (1997), Pandey *et al.* (1996), Ramadas *et al.* (1995)
Fungal monoculture	AE	Speciality biochemicals	Various biochemicals (e.g. mycotoxins organic acids, antibiotics, saccharin)	Lindenfelser and Ciegler (1975), Hang (1988), Yang and Ling (1989), Ohno *et al.* (1992), Khare *et al.* (1995), Emelyanova (1996), Valino *et al.* (1996)
Acetobacter species	AE	Vinegar production	Vinegar	Aidoo *et al.* (1982)

AE = aerobic; AN = anaerobic.

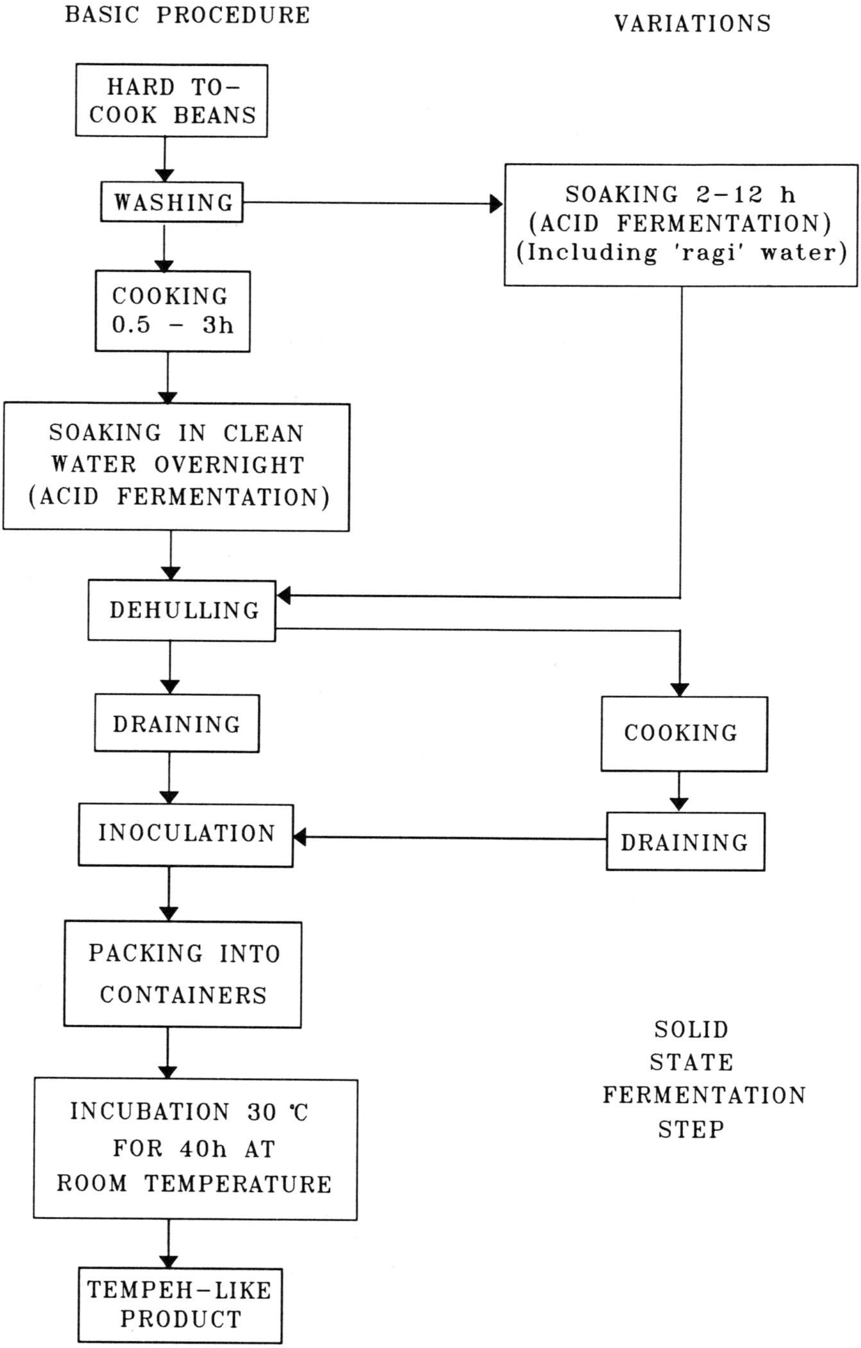

Figure 3.4 Technological procedure to produce a tempeh-like food using *Rhizopus oligosporus* and 'hard-to-cook' common beans as the substrate.

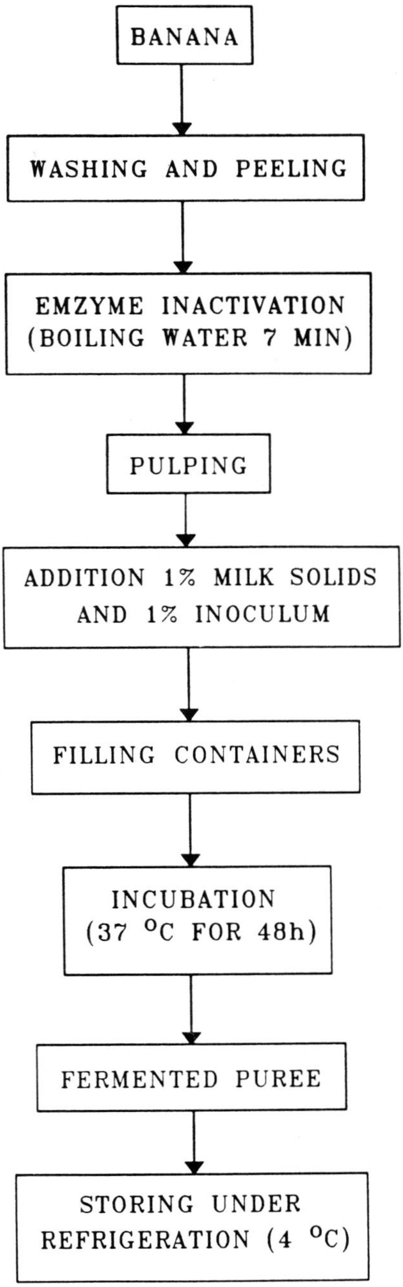

Figure 3.5 Utilization of fresh banana wastes and lactic bacteria as an inoculum, under solid substrate fermentation, to produce a fermented purée to be used in yoghurt-like products.

compared to food fermentations of Asia (Paredes-López and Harry, 1988). Lactic acid fermentation of maize in the solid state have been used in Mexico at village level for centuries (Ulloa-Sosa, 1974). The transformation of the substrate gives an end-product, termed pozol, with improved nutritional properties and with promising characteristics for industrialization (Cravioto *et al.*, 1955).

The demand for edible fungi has risen greatly in the last decade and has been satisfied by an expansion of the production of cultivated mushrooms using a variety of agricultural and lignocellulosic wastes as substrate (Dare *et al.*, 1988; Rajarathnam and Bano, 1989; Jwanny *et al.*, 1995). The mushroom species cultivated in North America, Latin America and Europe are mostly *Agaricus bisporus*, *A. bitorquis*, *A.* spp., *Pleurotus ostreatus* and *P. eryngii* (Martínez *et al.*, 1994; Kerem and Hadar, 1995b), whereas in Asia *Lentinus edoides* and *Volvariella volvacea* are cultivated (Präve *et al.*, 1987). The biotechnology of modern cultivation of mushrooms includes two basic steps: preparation of suitable compost, and growth of the mycelium and fructification (Figure 3.6). The replacement of traditional manure compost with municipal waste compost, and the availability of uniform reliable spawn cultures have increased the economic viability of mushroom growing (Zadrazil and Grabbe, 1983; Nigam and Singh, 1994). After 8–9 days the compost is packed into boxes and subjected to pasteurization. This kills animal pests and harmful microorganisms. The free ammonia present in the compost, which is toxic for fungi, is eliminated by thermophilic microorganisms. After inoculation and mycelial growth, the compost is covered with a mixture of soil, peat and chalk. Yields of 0.5–1 kg of mushrooms per kg of compost dry matter may be obtained. Besides being a delicacy, the mushrooms are also an important source of food protein for human consumption (Gray, 1970). The residual compost could be used as an upgraded form of ruminant feed (Gujral *et al.*, 1987), as a source of degradatory enzymes or as a fertilizer (Rajarathnam and Bano, 1989).

Mushroom growing combined with composting converts a considerable portion of agricultural residues partly to edible mushrooms and partly to protein-enriched animal feed; Das and Karim (1995) found that in SSF *P. ostreatus* increased the protein content of substrate when barley straw was subjected to degradation of lingnin by this mushroom.

(b) Feed production. The process of ensilage is traditionally used for animal feeding in various regions of the world and is an alternative to the drying of green crops. This process involves the controlled fermentation of green crops such as grass, maize plants and other available by-products and wastes from agricultural activities (Carrizales and Ferrer, 1984; Das and Karim, 1995). Molasses are sometimes added to promote fermentation and increase palatability. As the fermentation proceeds, the temperature rises

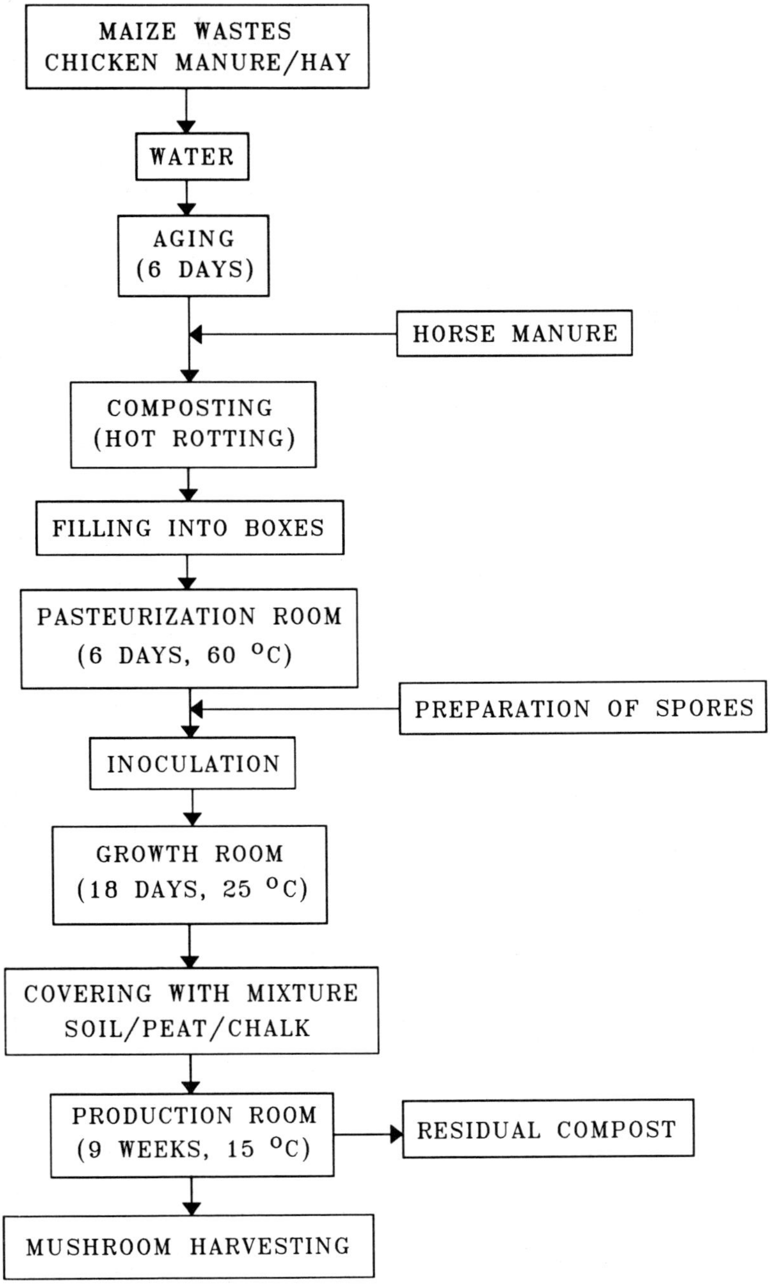

Figure 3.6 Mushroom production using wastes of plant and animal origin.

and the product becomes more acid because lactic acid-producing bacteria ferment water-soluble carbohydrates to lactic and acetic acids. The production of acids and the rapid establishment of anaerobic conditions suppress the activities of many putrefactive microorganisms (Iñiguez-Covarrubias *et al.*, 1990a,b). As the oxygen is consumed, strict aerobe microorganisms are inhibited, but facultative and strict anaerobes continue to grow, but these are replaced by *lactobacilli* and *streptococci* as pH declines. Lactic, butyric, propionic and acetic acids are synthesized giving flavours very acceptable to animals.

One of the agroindustrial wastes available in some tropical countries is the coffee pulp from wet processing plants. Approximately half of the world coffee harvest is processed by the wet method in which the coffee berry is subjected to mechanical and biological operation in order to separate the bean or seed from the exocarp (skin), mesocarp (mucilagenous pulp) and the endocarp (parchment) (Rolz *et al.*, 1988). The skin and most of the pulp is separated in the pulpers. This fraction represents about 40% of the weight of the fresh fruit and presently is underutilized, causing serious pollution problems. The use of coffee pulp as an animal feed has been mentioned as an attractive possibility; however, such utilization is limited by antiphysiological factors occurring naturally in the material (Peñaloza *et al.*, 1985). Figure 3.7 describes the basic steps for ensiling and for SSF, under controlled conditions, of coffee wastes. The most important changes in composition are given in Table 3.6. Both processes appear to promise improvement in the nutritive features of coffee wastes for monogastric animal feeding (Carrizales and Ferrer, 1984; Peñaloza *et al.*, 1985).

Ensiling is also an economical method of preserving and rendering animal excreta silages safe from potentially pathogenic microorganisms (Iñiguez-Covarrubias *et al.*, 1989, 1990a). Animal wastes represent one of the most underutilized resources. Utilization of animal excreta as feed can alleviate pollution problems, decrease feed costs and increase the supplies of available nitrogen and essential mineral resources. Although animal wastes have been used satisfactorily in feed mixtures without apparent harmful effects to animals, the practice of feeding unprocessed wastes may be a potential health hazard as excreta may contain agents harmful to human and animal health. The ensiling process suppresses the activities of undesirable microorganisms. This technology has been also reported to eliminate pathogenic bacteria such as *Salmonella* ssp., *Mycobacterium* ssp. and *E. coli* from beef cattle wastes (McCaskey and Wang, 1985), and to reduce the viability of bovine coccidia and clostridia from animal wastes (Iñiguez-Covarrubias, 1989). This process was evaluated at commercial level by Iñiguez-Covarrubias *et al.* (1990b) by ensiling different mixtures of swine waste, wheat straw and cane molasses in full-scale bunker silos with a 16.5 ton capacity. After 3 months, silos were opened and the product was

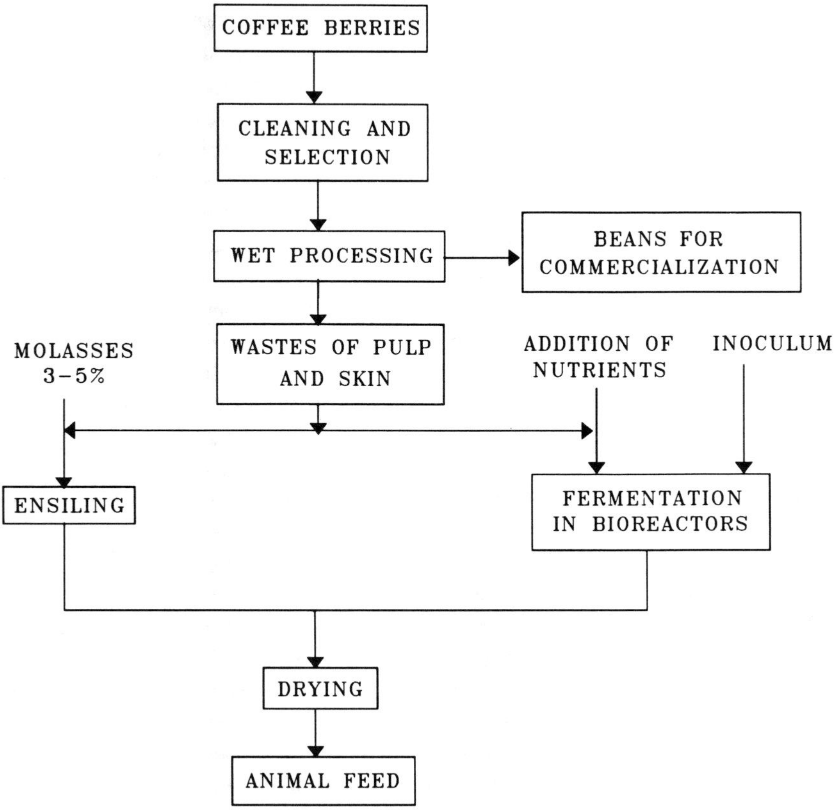

Figure 3.7 Ensiling and solid substrate fermentation of coffee pulp and skin under controlled conditions.

evaluated for physicochemical, microbiological and nutritional characteristics. Moreover, two performance trials on sheep were conducted. They found that the samples had a pleasant aroma and appearance similar to that of a good-quality haylage; typical manure aroma was not present in waste silages. *Clostridia* and aerobic bacteria tended to decrease by ensiling. The authors concluded that the product resulted in a safe and preserved feed that supported efficient sheep growth when balanced with a basal diet. Carrasco *et al.* (1996) explored the fermentation of sugar cane with bovine feces, finding a protein enrichment of the product. Another alternative for manure is the use of SSF under anaerobic conditions for methane production (Molnar and Bartha, 1989).

The use of fruit and vegetable wastes in SSF for feed production has been thoroughly investigated by several workers. Wastes from citrus (Nicolini *et al.*, 1987), banana (Baldensperger *et al.*, 1985; De Porres *et al.*,

Table 3.6 Effect of ensiling and fermentation under controlled conditions on chemical composition of coffee pulp (% dry basis)

Component	Ensiling		SSF under controlled conditions[a]	
	Fresh pulp	Ensiled product	Fresh pulp	Fermented product
Protein ($N \times 6.25$)	11.60	13.60	10.90	36.00
Crude fiber	15.20	17.20	23.70	13.10
Ether extract (fat)	5.00	2.20	2.20	2.40
Ash	6.70	8.10	5.00	12.60
Free-nitrogen extract	61.50	58.90	58.20	35.90
Caffeine	0.95	0.82	0.68	0.46
Cell wall constituents	–	–	31.40	23.50
Hemicellulose	–	–	0.98	0.05
Acid detergent fiber	–	–	30.50	23.40
Cellulose	–	–	18.60	12.00
Tannins	–	–	2.30	2.80

SSF = solid substrate fermentation.

1985), potato (Yang, 1988), sugar beet pulp (Durand and Chereau, 1988), sugar-cane bagasse (Zadrazil and Puniya, 1995; Carrasco *et al.*, 1996), cassava (Zvauya and Muzondo, 1994) and apple pomance (Joshi and Sandhu, 1996) industrial processing have been converted into feed supplements of high nutritional value. González *et al.* (1985) reported that during pineapple processing each tonne of fresh fruit produces not less than 500 kg of wastes. These wastes may be transformed by SSF with *T. viride* into a valuable feed product. SSF is also a useful technology for enriching the feed value of lignocellulosic materials (Abdullah *et al.*, 1985; Hatakka *et al.*, 1989). In relation to costs associated with SSF, Durand and Chereau (1988) estimated that, on an industrial scale, the most significant costs for the whole process are for drying.

(c) Ethanol production. Various SSF processes for ethanol production using sugar-rich substrates have been studied. For this purpose, crops such as fodder beets (Gibbons and Westby, 1988), sugar beets (Cochet *et al.*, 1988), sweet sorghum and Jerusalem artichoke (Gibbons, 1989) have been tested. These SSF processes appear to be economically competitive compared to traditional LSF procedures based on corn, sugar beets and sugar cane substrates. The use of SSF for ethanol production from sugar-rich wastes may be a promising alternative as well.

Increasing volumetric productivity and saving in distillation costs make the SSF process attractive (Tengerdy, 1985). Recently, Mamma *et al.* (1996) investigated SSF for ethanol production from the carbohydrate fraction of sweet sorghum by a mixed culture *F. oxysporum* and *S. cerevisiae* in a bioreactor. As a result, simultaneous saccharification and fermentation yielded an enhancement of the theoretical yield (51 g ethanol/100 g glucose), ranging from 20% to 32%, underlying the significance of SSF of sorghum cellulose and hemicellulose to bioethanol in a reactor. The authors concluded that the considerable increase of ethanol yields makes the process worthy of further scale-up investigation.

(d) Enzyme production. It is estimated that SSF will have truly significant economic importance in high volumetric productivity systems where recovery expenses are considerably reduced or even obviated (Aidoo *et al.*, 1982). Table 3.7 shows some of the most important enzymes produced by SSF. Different microorganisms and substrates have been used for the production of enzymes. Moreover, there is an increasing interest in using inert supports as amberlite (Christen *et al.*, 1995), polystyrene (Prabhu and Chandrasekaran, 1995) or polyurethane foams (Zhu *et al.*, 1996; Murado *et al.*, 1997) impregnated with a liquid nutrient for enzyme production. The use of inert particles allows:

1. the use of low-molecular-weight carbohydrates directly available to the microorganisms;
2. the handling of higher substrate concentrations;
3. simplified product recovery;
4. shorter production times;
5. in some cases, obtention of a ready-to-use immobilized enzyme (Christen *et al.*, 1995).

For enzyme production, SSF may be attractive owing to its low-level technology, the high product concentration and reduced cost of dewatering (Laukevics *et al.*, 1984; Macris *et al.*, 1987). Hatakka *et al.* (1989) postulated that protein enrichment of lignocellulosic wastes may be economically feasible if at the same time wood-rotting fungi are used for the production of extracellular enzymes (e.g. cellulases, lignases). In these SSF processes, lingnin could be selectively removed, the protein content of the residue increased, and extracellular enzymes extracted from the residue prior to use as animal feed. Soccol *et al.* (1994a) demonstrated the feasibility of production of α-amylase, glucoamylase and protein enrichment of cassava by *Rhyzopus* strains in SSF. They found that the results indicated that raw cassava could prove an inexpensive source of starch and constitute a strategic biological matter for production of commercially valuable microbial metabolites and feed products as well.

SSF is also a very valuable technology for the production of crude enzyme preparation for the food industry (Deschamps and Huet, 1984; Considine *et al.*, 1987; Nakadai and Nasuno, 1988). Apparently, there are no reports related to scale-up experiments in enzyme production except for the preliminary scale-up trials conducted for the production of alkaline proteases by *A. flavus* (Karanth and Lonsane, 1988).

Table 3.7 Production of enzymes by solid substrate fermentation

Enzymes	Microorganisms	Substrates	References
Amylases	*Aspergillus oryzae*	Broken rice kernels	Narahara *et al.* (1982)
		Inert support	Murado *et al.* (1997)
	Aspergillus kawachii	Rice	Sudo *et al.* (1994)
	Bacillus coagulans	Wheat bran	Babu and Satyanarayana (1995)
	Bacillus megaterium	Wheat bran	Ramesh and Lonsane (1987)
Amyloglucosidase	*Aspergillus niger*	Wheat bran	Jaleel *et al.* (1992), Pandey and Radhakrishnan (1993), Pandey *et al.* (1996)
	Rhizopus spp.	Wheat bran	Blain (1975)
α-L-Arabinofuranosidase	*Thermoascus aurantiacus*	Sugar beet pulp	Roche *et al.* (1994)
	Trichoderma reesei	Sugar beet pulp	Roche *et al.* (1995)
Cellulases	*Neurospora crassa*	Wheat straw	Macris *et al.* (1987)
	Penicillum chrysogenum	Spent residues	Sharma *et al.* (1995)
	Phanerochaete chrysosporium	Soyhull	Jha *et al.* (1995)
	Thermoascus aurantiacus	Sugar beet pulp	Grajek (1987)
	Trichoderma reesei	Wheat straw	Chahal (1983, 1985)
		Wood residues	Hatakka *et al.* (1989)
Lipases	*Rhizopus delemar*	Polymeric resin	Christen *et al.* (1995)
Phytase	*Aspergillus carbonarius*	Canola meal	Al-Asheh and Duvnjak (1995)
	Aspergillus ficuum	Canola meal	Ebune *et al.* (1995a,b)
Pectinases	*Aspergillus foetidus*	Wheat bran	Cavalitto *et al.* (1996)
	Penicillum capsulatum	Sugar beet pulp	Considine *et al.* (1987)
	Ttichoderma viride	Sugar beet pulp	Grajek and Gervais (1987)
Proteases	*Aspergillus oryzae*	Wheat bran	Nakadai and Nasuno (1988)
	Aspergillus oryzae	Broken rice kernels	Narahara *et al.* (1982)
	Bacillus amyloliquefaciens	Wheat bran	George *et al.* (1995)
D-Xylanase	*Chaetomium globosum* and *A. niger*	Agroresidues	Wiacek-Zychlinska *et al.* (1994)
	Melanocarpus albomyces	Agroresidues	Jain (1995)
	Trichoderma viride	Sugar beet pulp	Grajek and Gervais (1987)
β-Glucosidase	*Aspergillus phoenicis*	Sugar beet pulp	Deschamps and Huet (1984)
	Humicola grisea	Wheat bran	Filho (1996)
	Penicillium capsulatum	Sugar beet pulp	Considine *et al.* (1987)

Table 3.8 Biochemicals produced by solid state fermentation

Biochemical products	Substrates/wastes	Microorganisms	References
Aflatoxin	Rice, corn	*Aspergillus flauvus*	Norholt *et al.* (1977)
Citric acid	Sugar-cane bagasse, apple pomace	*Aspergillus niger*	Lakshminarayana *et al.* (1975), Hang (1998)
	Wheat bran		Shankaranand and Lonsane (1994)
	Okara	*Aspergillus terreus*	Khare *et al.* (1995)
Gibberellic acid	Wheat bran	*Gibberella fujikuroi*	Kumar and Lonsane (1987b)
Glutamic acid	Sugar-cane bagasse (inert support)	*Brevibacterium* spp.	Nampoothiri and Pandey (1996)
Koji acid	Sucrose	*Aspergillus* spp.	Aidoo *et al.* (1982)
L(+)-Lactic acid	Sugar-cane bagasse (inert support)	*Rhizopus oryzae*	Soccol *et al.* (1994b)
γ-Linolenic	Cereal substrates	*Cunninghamella japonica*	Emelyanova (1996)
Ochratoxin	Wheat	*Aspergillus ochraceus*	Lindenfelser and Ciegler (1975), Hesseltine (1977a), Norholt *et al.* (1979)
Penicillin	Sugar-cane bagasse (inert support)	*Penicillum chrysogenum*	Barrios-González *et al.* (1988)
Tetracycline	Sweet potato residue	*Streptomyces viridifaciens*	Yang and Ling (1989)
Iturin-surfactin (antibiotics)	Soybean/wheat bran	*Bacillus subtilis*	Ohno *et al.* (1992, 1995b, 1996)

(e) Speciality biochemicals. Various procedures for production of speciality biochemicals by SSF have been reported. The increased volumetric productivity and resulting cost reductions in product recovery make the SSF a viable alternative to the LSF. Some of these biochemical products are mycotoxins (Norholt *et al.*, 1977), organic acids (Lakshminarayana *et al.*, 1975; Kumar and Lonsane, 1987b; Nampoothiri and Pandey, 1996) and antibiotics (Yang and Ling, 1989; Ohno *et al.*, 1992, 1995b, 1996) (Table 3.8). Cereals and agricultural wastes (Emelyanova, 1996; Soccol, 1994a,b; Ohno *et al.*, 1995b, 1996) may be used as substrates for the production of these metabolites.

(f) Detoxification. Some agricultural products present undesirable characteristics which prevent the use of those materials for animal feed or human consumption. Cassava peel, which is highly toxic because of high levels of cyanohydrins content, is a good example (Ofuya and Obilor, 1994; Zvauya and Muzondo, 1995). The use of defatted rapeseed meals as a protein source in livestock rations and human diets is severely limited owing to the presence of many antinutritional substances and toxic compounds associated with the protein fraction (Bau *et al.*, 1994). On the other hand, although seeds of *Lathyrus sativus* (grass pea or chickling pea) seeds are tasty and rich in protein, overconsumption can cause an upper motor neurone disease because of the neurotoxin 3-*N*-oxalyl-L-2,3-diaminopropanoic acid (*β*-ODAP) present in the seed (Kuo *et al.*, 1995). It has been demonstrated that those toxic/antinutritional compounds can be drastically reduced through SSF processes (Bau *et al.*, 1994; Ofuya and Obilor, 1994; Kuo *et al.*, 1995; Zvauya and Muzondo, 1995). SSF has also been proposed as a system for bioremediation because it permits the successful degradation of the herbicide atrazine which is present in cotton and wheat straw (Masaphy *et al.*, 1996).

3.7 Advantages, disadvantages and future prospects of SSF

3.7.1 Advantages and disadvantages

There are many advantages of SSF processes for bioconversions of wastes over the conventional LSF systems at both the laboratory and the industrial scale (Aidoo *et al.*, 1982; Laukevics *et al.*, 1984; Lonsane *et al.*, 1985; Hesseltine, 1977b; Paredes-López and Harry, 1988). The most attractive advantages of SSF may be the high fermenter volumetric productivity, and the reduction in product recovery expenses, pollution problems and total cost of production. There also are some important disadvantages of SSF (Aidoo *et al.*, 1982; Paredes-López and Harry, 1988), such as:

1. the limited type of organisms which can grow at reduced water activities, namely fungi, some yeasts, and some bacteria and streptomycetes;
2. technical problems in controlling the heat generated during fermentation;
3. low product yields;
4. the slowness of fermentation;
5. the difficulty of using reliable devices to measure/control some of the fermentation parameters (e.g. moisture, temperature, pH, oxygen, product concentration).

Disadvantages such as problems in controlling the heat build-up and the use of reliable monitoring devices are being reduced or eliminated as investigations on those factors are being developed (Gowthaman *et al.*, 1995; Rajagopalan and Modak, 1995; Pandey *et al.*, 1996; Smits *et al.*, 1996).

3.7.2 Future prospects

As the availability of agricultural land declines and pollution problems increase, the bioconversion of wastes from plant and animal origin through SSF into valuable products is becoming more pressing than ever in both developed and developing countries. Food is becoming more expensive in terms of basic inputs such as water and energy. The increasing use of wastes and agroindustrial residues to produce SSF foods and feeds may optimize the indigenous resources, increase the availability of nutritious products and reduce pollution problems. A combination of anaerobic and aerobic fermentation procedures appears also to be an attractive way of increasing the ease of storage, sensory properties and nutritional value of starchy residues. Technological procedures for composting will be used increasingly for agricultural, industrial and municipal solid waste disposal. SSF is expected to have a significant economic impact in the near future on the production of metabolites of relatively high price per unit of weight (e.g. enzymes, organic acids, antibiotics).

More research is required for the development of new SSF bioreactors with improved performance in the utilization of wastes and residues on a commercial scale. Also, more attention should be given to SSF processes based on the use of solid inert material, impregnated with suitable media, to improve the control of some fermentation parameters and reduce recovery expenses of metabolites. Research into the genetic improvement of fermentative microorganisms by using different molecular biology techniques in order to attain a more efficient utilization of a variety of substrates, and conversion into more valuable products, is pending.

Acknowledgements

Research on fermentation was supported by Consejo Nacional de Ciencia y Tecnología – México and Organización de los Estados Americanos.

References

Abdullah, A.L., Tengerdy, R.P. and Murphy, V.G. (1985) *Biotechnol. Bioeng.*, **27**, 20–6.

Agosin, E., Diaz, D., Aravena, R. and Yañez, E. (1989) *J. Food Sci.*, **54**, 102–7.

Aidoo, K.E., Hendry, R. and Wood, B.J.B. (1982) *Adv. App. Microbiol.*, **28**, 201–6.

Al-Asheh, S. and Duvnjak, Z. (1995) *World J. Microbiol. Biotechnol.*, **11**, 228–31.

Almanza, S., Durand, A., Renaud, R., Maratray, J. and Diez, M. (1995) *Biotechnol. Technol.*, **9**, 395–400.

An, Z.Q., Farman, M.L. and Budde, A. (1996) *Gene*, **176**, 93–6.

Antier, P., Minjares, A., Roussos, S., Raimbault, M. and Viniegra-González, G. (1993a) *Enzyme Microbiol. Technol.*, **15**, 254–60.

Antier, P., Minjares, A., Roussos, S. and Viniegra-González, G. (1993b) *Biotechnol. Adv.*, **11**, 429–40.

Babu, K.R. and Satyanarayana, T. (1995) *Process Biochem.*, **30**, 305–9.

Bajracharya, R. and Mudgett, R.E. (1980) *Biotechnol. Bioeng.*, **22**, 2219–35.

Baldensperger, J., Le Mer, J., Hannibal, L. and Quinto, P.J. (1985) *Biotechnol. Lett.*, **7**, 743–8.

Barrios-González, J., Tomasini, A., Viniegra-González, G. and López, J. (1988) *Biotechnol. Lett.*, **10**, 793–8.

Barstow, L.M., Dale, B.E. and Tengerdy R.P. (1988) *Biotechnol. Techniques*, **2**, 237–42.

Bassham, J.A. (1975) In *Cellulose as a Chemical and Energy Resource*, Biotech. Bioeng. Symp. No. 5 (ed. C.R. Wilke), Interscience, New York, p. 9.

Bau, H.-M., Villaume, C., Lin, C.-F., Evrad, J. *et al.* (1994) *J. Sci. Food Agric.*, **65**, 315–22.

Beuchat, L.R. (1981) *Bioproc. Eng.*, **3**, 11–5.

Bhumiratna, A., Flegel, T.W., Glinsukon, T. and Somporana, W. (1980) *Appl. Environ. Microbiol.*, **39**, 430–6.

Blain, J.A. (1975) In *The Filamentous Fungi*, Vol. IV (eds J.E. Smith, D.R. Berry and B. Christiansen), Edward Arnold, London, p. 210.

Bonnen, A.M., Anton, L.H. and Orth, A.B. (1994) *Appl. Environ. Microbiol.*, **60**(3), 960–5.

Boudet, A.M. and Grimapettenati, J. (1996) *Molecular Breeding*, **2**(1), 25–39.

Brauer, H. (1988) *Bioproc. Eng.*, **3**, 11–7.

Brown, A.D. (1978) *Adv. Microb. Physiol.*, **17**, 181–242.

Bushell, M.E. and Slater, J.H. (1981) *Mixed Cultured Fermentations*, Special Publ. No. 5. Society of General Microbiology, Academic Press.

Cannel, E. and Moo-Young, M. (1980a) *Process Biochem.*, **15**(3), 2–6.

Cannel, E. and Moo-Young, M. (1980b) *Process Biochem.*, **15**(4), 24–9.

Carrasco, E., Valino, A.E.E., Rodrígues, Z. and Febles, I. (1996) *Cuban J. Agricul. Sci.*, **30**(2), 171–3.

Carrizales, V. and Ferrer, J. (1984) In *Symposium on the Importance of Lactic Acid Fermentation in the Food Industry*, Mexico City, Mexico.

Cavalitto, S., Arcas, J.A. and Hours, R.A. (1996) *Biotechnol. Lett.*, **18**, 251–6.

Chahal, D.S. (1983) In *Foundations of Biochemical Engineering* (eds H.W. Blanch, E.T. Papoutsakis and G. Stephanopoulos), American Chemical Symposium Series 207, Washington, DC, p. 421.

Chahal, D.S. (1985) *Appl. Environ. Microbiol.*, **49**, 205–10.

Christakopoulos, P., Li, L.-W., Kekos, D. and Macris, B.J. (1993) *Bioresource Technol.*, **45**, 89–92.

Christen, P., Angeles, N., Corzo, G., Farres, A. and Revah, S. (1995) *Biotechnol. Lett.*, **9**, 597–600.

Cochet, N., Nonus, M. and Lebeault, J.M. (1988) *Biotechnol. Lett.*, **10**, 491–6.

Considine, P.J., Hackett, T.J. and Coughlan, M.P. (1987) *Biotechnol. Lett.*, **9**, 131–3.
Couri, S. and Defarias, A.X. (1995) *Revista Microbiol.*, **26**(4), 314–17.
Cox, H.H.J., Magielsen, F.J., Doddema, H.J. and Harder, W. (1996) *Appl. Microbiol. Biotechnol.*, **45**(6), 851–6.
Cravioto, O.R., Cravioto, Y.O., Massieu, G. and Guzmán, J. (1955) *Ciencia* (Mexico), **15**, 27.
Czajkowska, D. and Ilnicka-Olejniczak, O. (1988) *Acta Biotechnol.*, **8**, 407–13.
Dare, P.H., Clarke, T.A. and Chu-Chou, M. (1988) *Process Biochem.*, **23**, 156–61.
Das, P. and Karim, M.N. (1995) *J. Ind. Microbiol.*, **15**, 25–31.
De Porres, E., de Arriola, M.C., García, R. and Rolz, C. (1985) *Lebensm.-Wiss. u. Technol.*, **18**, 379–82.
Deschamps, F. and Huet, M.C. (1984) *Biotechnol. Lett.*, **6**, 55–60.
Desgrenges, C. and Dureng, A. (1990) *Enz. Microbial Technol.*, **12**, 546–51.
Durand, A. and Chereau, D. (1988) *Biotechnol. Bioeng.*, **31**, 476–86.
Durand, A., de la Broise, D. and Blachère, H. (1988) *J. Biotechnol.*, **8**, 59–66.
Ebune, A., Al-Asheh, S. and Duvnjak, Z. (1995a) *Bioresource Technol.*, **53**, 7–12.
Ebune, A., Al-Asheh, S. and Duvnjak, Z. (1995b) *Bioresource Technol.*, **54**, 241–7.
Emelyanova, E.V. (1996) *Process Biochem.*, **31**, 431–4.
Fernández-Larrea, J. and Stahl, U. (1996) *Mol. Gen. Genet.*, **252**(5), 539–51.
Filho, E.X.F. (1996) *Can. J. Microbiol.*, **42**, 1–5.
Fudiyansyah, N., Petterson, D.S., Bell, R.R. and Fairbrother, A.H. (1995) *Ind. J. Food Sci. Technol.*, **30**, 297–305.
Fukushima, D. (1985) *Food Rev. Int.*, **1**, 149–54.
Gaskell, J., Dieperink, E. and Cullen, D. (1991) *Nucleic Acids Res.*, **19**, 599–603.
George, S., Raju, V., Krishnan, M.R.V., Subramanian, T.V. and Jayaraman, K. (1995) *Process Biochem.*, **30**, 457–62.
Georgiou, G. and Shuler, M. (1986) *Biotechnol. Bioeng.*, **28**, 405–16.
Gervais, P. (1989) *Biotechnol. Bioeng.*, **33**, 266–71.
Gervais, P. (1990) *Appl. Microbiol. Biotechnol.*, **33**, 72–5.
Gervais, P., Bazelin, C. and Bana, E.N.S. (1986) *Biotechnol. Lett.*, **8**, 191–6.
Gervais, P., Bensoussan, M. and Grajek, W. (1988a) *Appl. Microbiol. Biotechnol.*, **27**, 389–93.
Gervais, P., Grajek, W., Bensoussan, M. and Molin, P. (1988b) *Biotechnol. Bioeng.*, **31**, 457–63.
Gervais, P., Belin, J.M., Grajek, W. and Sarrett, M. (1988c) *J. Ferment. Technol.*, **66**(4), 403–7.
Giardina, P., Cannio, R., Martirani, L. *et al.* (1995) *Appl. Environ. Microbiol.*, **61**(6), 2408–13.
Gibbons, W.R. (1989) *J. Ferment. Bioeng.*, **67**(3), 258–63.
Gibbons, W.R. and Westby, C.A. (1988) *Biotechnol. Lett.*, **10**, 665–9.
González, E.E., Vernon, E.J. and Moctezuma, A.R. (1985) *Ind. Environ.*, **8**(4), 19–24.
Gowthaman, M.K., Raghava-Rao, K.S.M.S., Ghidyal, N.P. and Karanth, N.G. (1995) *Process Biochem.*, **30**, 9–15.
Grajek, W. (1987) *Appl. Microbiol. Biotechnol.*, **26**, 126–9.
Grajek, W. (1988) *J. Ferment. Technol.*, **66**, 675–9.
Grajek, W. and Gervais, P. (1987) *Enzyme Microbiol. Technol.*, **9**, 658–62.
Gray, W.D. (1970) *The Use of Fungi as Food and in Food Processing*. CRC Press, Cleveland, OH.
Grewal, H.S. and Kalra, K.L. (1995) *Biotechnol. Adv.*, **13**, 209–15.
Gujral, G.S., Bisaria, R., Madan, M. and Vasudevan, P. (1987) *J. Ferment. Technol.*, **65**, 101–6.
Habib, M.U. (1990) *Diss. Abstr. Int.*, **50**, 32–44.
Han, Y.W. and Anderson, A.W. (1975) *Appl. Microbiol.*, **30**, 930–4.
Hang, Y.D. (1988) *Biotechnol. Lett.*, **10**, 421–6.
Hanh-Hagerdel, B. (1986) *Enzyme Microbiol. Technol.*, **8**, 322–7.
Harrigan, W.F. (1985) *Food Flav. Ingr. Proc.*, **10**, 32–8.
Harry, G.I. (1988) M.Sc. thesis, Unidad Irapuato, CIEA-Inst. Politécnico Nal., Irapuato, Mexico.
Hatakka, A.I., Mohammadi, O.K. and Lundell, T.K. (1989) *Food Biotechnol.*, **3**, 45–50.

Hayes, W.A. (1977) *Composting*. The Mushroom Growers' Association, London.
Hesseltine, C.W. (1977a) *Process Biochem.*, **21**, 24–7.
Hesseltine, C.W. (1977b) *Process Biochem.*, **12**, 29–32.
Hesseltine, C.W. (1987) *Internat. Biodeter.*, **23**, 79–84.
Huoponen, K., Ollikka, P., Kälin, M. *et al.* (1990) *Gene*, **89**, 145–50.
Hutter, R. (1982) In *Bioactive Microbial Products: Search and Discovery* (eds J.D. Bullock, L.J. Nisbet and D.T. Winstanley), Academic Press, London, p. 112.
Iiyama, K., Stone, B.A. and Macauley, B.J. (1994) *Appl. Environ. Microbiol.*, **60**(5), 1538–46.
Iñiguez-Covarrubias, G. (1989) Ph.D. thesis, Inst. Investigaciones Biomédicas, UNAM, Mexico.
Iñiguez-Covarrubias, G., Franco-Gómez, M.J. and Ruis-Carrera, V. (1989) *Biol. Wastes*, **28**, 293–7.
Iñiguez-Covarrubias, G., de la Torre-Martínez, M., Cuarón-Ibargüegoitia, J.A., Pérez-Gavilán, P. and Magaña-Plaza, I. (1990a) *Biol. Wastes*, **34**, 227–39.
Iñiguez-Covarrubias, G., Cuarón-Ibargüengoitia, J.A., Pérez-Gavilán, P., de la Torre-Martínez, M. and Magaña-Plaza, I. (1990b) *Biol. Wastes*, **34**, 281–99.
Ito, K., Yoshida, K., Ishikawa, T. and Kabayashi, S. (1990) *J. Ferment. Bioeng.*, **70**(3), 169–72.
Jain, A. (1995) *Process Biochem.*, **30**, 705–9.
Jaleel, S.A., Srikanta, S. and Karanth, N.G. (1992) *J. Microbiol. Biotechnol.*, **7**, 1–8.
James, C.M., Felipe, M.S.S., Sims, P.F.G. and Broda, P. (1992) *Gene*, **114**(3), 217–22.
Jha, K., Khare, S.K. and Gandhi, A.P. (1995) *Bioresource Technol.*, **54**, 321–2.
Joshi, V.K. and Sandhu, D.K. (1996) *Bioresource Technol.*, **56**, 251–5.
Jwanny, E.W., Rashad, M.M. and Abdu, H.M. (1995) *Appl. Biochem. Biotechnol.*, **50**, 71–8.
Kaal, E.E.J., Field, J.A. and Joyce, T.W. (1995) *Bioresource Technol.*, **53**, 133–9.
Kanotra, S. and Mathur, R.S. (1995) *J. Environ. Sci. Health*, **A30**(6), 1339–60.
Karanth, N.G. and Lonsane, B.K. (1988) In *Solid State Fermentation in Bioconversion of Agro-Industrial Raw Materials* (ed. M. Raimbault), ORSTOM Centre, France, p. 113.
Kargi, F. and Curme, J.A. (1985) *Biotechnol. Bioeng.*, **27**, 1122–5.
Katagiri, N., Tsutsumi, Y. and Nishida, T. (1995) *Appl. Environ. Microbiol.*, **61**(2), 617–22.
Kerem, Z. and Hadar, Y. (1995a) *J. Cell. Biochem.*, **5221A SIA**, 48.
Kerem, Z. and Hadar, Y. (1995b) *Appl. Environ. Microbiol.*, **61**, 3057–62.
Khare, S.K., Jha, K. and Gandhi, A.P. (1995) *Bioresource Technol.*, **54**, 323–5.
Krishna, C. and Chandrasekaran, M. (1996) *Appl. Microbiol. Biotechnol.*, **40**(2), 106–11.
Kumar, N. and Singh, K. (1990) *Biol. Wastes*, **33**(4), 231–42.
Kumar, P.K.R. and Lonsane, B.K. (1987a) *Biotechnol. Lett.*, **9**, 179–82.
Kumar, P.K.R. and Lonsane, B.K. (1987b) *Biotechnol. Bioeng.*, **30**, 267–71.
Kumar, P.K.R. and Lonsane, B.K. (1988) *Process Biochem.*, **23**, 43–7.
Kuo, Y.-H., Bau, H.-M., Quemener, B., Khan, J.K. and Lambein, F. (1995) *J. Sci. Food Agric.*, **69**, 819.
Lai, M.V., Wang, H.H. and Chang, F.W. (1989) *Biotechnol. Bioeng.*, **34**, 1337–40.
Lakshminarayana, K., Chaudhary, K., Ethiraj, S. and Tanyo, P. (1975) *Biotechnol. Bioeng.*, **17**, 291–3.
Larroche, C. and Gros, J.B. (1989) *Process Biochem.*, **24**, 97–101.
Larroche, C., Desfarges, C. and Gross, J.B. (1986) *Biotechnol. Lett.*, **8**, 453–6.
Laukevics, J.J., Apsite, A.F., Viesturs, U.E. and Tengerdy, R.P. (1984) *Biotechnol. Bioeng.*, **26**, 1465–74.
Laukevics, J.J., Apsite, A.F., Viesturs, U.E. and Tengerdy, R.P. (1985) *Biotechnol. Bioeng.*, **27**, 1687–91.
Lezinou, V., Christakopoulos, P., Kekos, D. and Macris, B.J. (1994) *Biotechnol. Lett.*, **16**, 983–8.
Li, B., Nagalla, S.R. and Renganathan, V. (1996) *Appl. Environ. Microbiol.*, **62**(4), 1329–35.
Lindenfelser, L.A. and Ciegler, A. (1975) *Appl. Microbiol.*, **29**, 323–7.
Lonsane, B.K., Ghildyal, N.P., Budiatman, S. and Ramakrishna, S.V. (1985) *Enzyme Microbiol. Technol.*, **7**, 258–65.
Lonsane, B.K., Durand, A., Renaud, R. *et al.* (1992) In *Solid Substrate Cultivation* (eds H.W. Doelle, D.A. Mitchell and C.E. Rolz), Elsevier Science Publishers, London.
Lotong, W. and Suwarnarit, P. (1983) *Appl. Environ. Microbiol.*, **46**, 1224–6.

Macris, B.J., Kekos, D., Evangelidou, X., Galiotou-Panayotou, M. and Rodis, P. (1987) *Biotechnol. Lett.*, **9**, 661–4.
Malathi, S. and Chakrabarty, R. (1991) *Appl. Environ. Microbiol.*, **57**, 712–6.
Mamma, D., Koullas, D., Fountoukidis, G., Kekos, D., Macris, B.J. and Koukios, E. (1996) *Process Biochem.*, **34**(4), 377–81.
Martínez, A.T., Camarero, S., Guillen, F. *et al.* (1994) *FEMS Microbiol. Rev.*, **13**, 265–74.
Masaphy, S., Levanon, D. and Henis, Y. (1996) *Bioresource Technol.*, **56**, 207–14.
Matteau, P.P. and Bone, D.H. (1980) *Biotechnol. Lett.*, **2**, 127–32.
McCaskey, T.A. and Wang, Y.D. (1985) *J. Dairy Sci.*, **68**, 1401–7.
Mishra, C. and Leathman, G.F. (1990) *J. Ferment. Bioeng.*, **69**(1), 8–15.
Mishra, I.M., El-Temtamy, S.A. and Schügerl, K. (1982) *Eur. J. Appl. Microbiol. Biotechnol.*, **16**, 197–202.
Mitchell, D.A., Greenfield, P.F. and Doelle, H.W. (1986) *Biotechnol. Lett.*, **8**, 827–32.
Mitchell, D.A., Doelle, H.W. and Greenfield, P.F. (1988) *Biotechnol. Lett.*, **10**, 497–501.
Mitchell, D.A., Greenfield, P.F. and Doelle, H.W. (1990) *World J. Microbiol. Biotechnol.*, **6**, 201–8.
Mitchell, D.A., Do, D.D., Greenfield, P.F. and Doelle, H.W. (1991) *Biotechnol. Bioeng.*, **38**, 353–62.
Molard, D.R., Lesage, L. and Cahagnier, B. (1985) In *Influence of Water on Food and Stability* (eds D. Simatos and J.L. Multon), Martinus Nijhoff, Boston.
Molnar, L. and Bartha, I. (1989) *Biol. Wastes*, **28**, 15–19.
Moloney, A.P., O'Rorke, A., Considine, P.S. and Coughlan, M.P. (1984) *Biotechnol. Bioeng.*, **26**, 714–18.
Moo-Young, M., Moreira, A.R. and Tengerdy, R.P. (1983) In *The Filamentous Fungi*, Vol. IV (eds J.E. Smith, D.R. Berry and B. Christiansen), Edward Arnold, London, p. 117.
Moser, A. (1988) In *Bioprocess Technology, Kinetics and Reactors*. Springer-Verlag, Berlin, p. 198.
Mossel, D.A.A. (1975) In *Water Relations of Foods* (ed. R.B. Duckworth), Academic Press, London, p. 347.
Muñoz, G.R., Valencia, J.R., Sánchez, S. and Farres, A. (1991) *Biotechnol. Lett.*, **13**, 277–80.
Murado, M.A., González, M.P., Torrado, A. and Pastrana, L.M. (1997) *Process Biochem.*, **32**, 35–42.
Naidu, P.S., Zhang, Y.Z. and Reddy, C.A. (1990) *Biochem. Biophys. Res. Commun.*, **173**, 994–1000.
Nakadai, T. and Nasuno, S. (1988) *J. Ferment. Technol.*, **66**, 525–30.
Nampoothiri, K.M. and Pandey, A. (1996) *Biotechnol. Lett.*, **18**, 199–204.
Narahara, H., Koyama, Y., Yoshida, T. and Taguchi, H. (1982) *J. Ferment. Technol.*, **60**, 311–19.
Nicolini, L., von Hunolstein, C. and Carilli, A. (1987) *Appl. Microbiol. Biotechnol.*, **26**, 95–100.
Nigam, P. (1990) *Enzyme Microbial Technol.*, **12**, 808–11.
Nigam, P. and Singh, D. (1994) *J. Basic Microbiol.*, **34**(6), 405–23.
Nishida, T., Kashino, Y., Mimura, A. and Takahara, Y. (1988) *Mokuzai Gakkaishi*, **34**, 530–6.
Norholt, M.D., Van Egmond, H.P. and Paulsch, W.E. (1977) *J. Food Protect.*, **49**, 778–83.
Norholt, M.D., Van Egmond, H.P. and Paulsch, W.E. (1979) *J. Food Protect.*, **42**, 485–89.
Ofuya, C.O. and Obilor, S.N. (1994) *Process Biochem.*, **29**, 25–8.
Ohno, A., Ano, T. and Shoda, M. (1992) *Biotechnol. Lett.*, **14**, 817–22.
Ohno, A., Ano, T. and Shoda, M. (1995a) *J. Ferment. Bioeng.*, **80**(5), 517–19.
Ohno, A., Ano, T. and Shoda, M. (1995b) *Biotechnol. Bioeng.*, **47**, 209–14.
Ohno, A., Ano, T. and Shoda, M. (1996) *Process Biochem.*, **31**(8), 801–6.
Oriol, E., Raimbault, M., Roussos, S. and Viniegra-González, G. (1988) *Appl. Microbiol. Biotechnol.*, **27**, 498–503.
Pandey, A. (1990a) *Biol. Wastes*, **34**, 11–19.
Pandey, A. (1990b) In *International Symposium on Industrial Biotechnology*, Hyderabad, India, November, 1990, CP. 33.
Pandey, A. (1991a) *Process Biochem.*, **26**, 355–61.
Pandey, A. (1991b) *Bioresource Technol.*, **37**, 169–72.

Pandey, A. (1992) *Process Biochem.*, **27**(1), 109–17.
Pandey, A. and Radhakrishnan, S. (1993) *Process Biochem.*, **28**, 305–9.
Pandey, A., Selvakumar, P. and Ashakumary, L. (1996) *Process Biochem.*, **31**, 43–46.
Paredes-López, O. and Harry, G.I. (1988) *CRC Crit. Rev. Food Sci. Nutr.*, **27**, 159–87.
Paredes-López, O. and Harry, G.I. (1989) *J. Food Sci.*, **54**, 968–70.
Paredes-López, O., Vital-Bori, G. and Camargo-Rubio. E. (1974) *J. Ferment. Technol.*, **52**, 592–7.
Paredes-López, O., Camargo-Rubio, E. and Ornelas-Vale, A. (1976) *Appl. Environ. Microbiol.*, **31**, 487–91.
Paredes-López, O., Harry, G.I and Montes-Rivera, R. (1987) *Biotechnol. Lett.*, **9**, 333–7.
Paredes-López, O., Guevara-Lara, F., Schevenin-Pinedo, M.L. and Montes-Rivera, R. (1988) *Plant Foods Hum. Nutr.*, **39**, 137–48.
Paredes-López, O., Harry, G.I. and González-Castañeda, J. (1990) *J. Food Sci.*, **55**, 123–6.
Paredes-López, O., Castañeda-González, J. and Cárabez-Trejo, A. (1991) *J. Ferment. Bioeng.*, **71**(1), 58–62.
Peñaloza, W., Davey, C.L., Hedger, J.N. and Kell, D.B. (1992) *J. Sci. Food Agric.*, **59**, 227–35.
Peñaloza, W., Molina, M.R., Gómez, R. and Bressani, R. (1985) *Appl. Environ. Microbiol.*, **49**, 388–91.
Pou, J. Vittori, N., Fernández, M.J. and Garrido, J. (1987) *Interferón Biotecnol.*, **4**, 121–6.
Prabhu, G.N. and Chandrasekaran, M. (1995) *World J. Microbiol. Biotechnol.*, **11**, 683–4.
Präve, P., Faust, U. and Sitting, W. (1987) *Fundamentals of Biotechnology*. VCH, Weinheim.
Prokop, A., Erickson, L.E. and Paredes-López, O. (1971) *Biotechnol. Bioeng.*, **13**, 241–4.
Raghava-Rao, K.S.M.S., Gowthaman, M.K., Ghildyal, N.P. and Karanth, N.G. (1993) *Bioprocess Eng.*, **8**, 255–62.
Raimbault, M. and Alazard, D. (1980) *Eur. J. Appl. Microbial Biotechnol.*, **9**, 199–204.
Rajarathnam, S. and Bano, Z. (1989) *CRC Crit. Rev. Food Sci. Nutr.*, **28**, 33–55.
Rajagopalan, S. and Modak, J.M. (1995) *Bioprocess Eng.*, **13**, 161–9.
Ralph, B.J. (1976) *Food Technol. Austr.*, **28**(7), 247–51.
Ramadas, M.V., Holst, O. and Mattiasson, B. (1995) *Biotechnol. Technol.*, **9**(12), 901–6.
Ramesh, M.V. and Lonsane, B.K. (1987) *Biotechnol. Lett.*, **9**, 323–8.
Ramesh, M.V. and Lonsane, B.K. (1989) *Biotechnol. Lett.*, **11**, 49–52.
Ramesh, M.V. and Lonsane, B.K. (1990) *Appl. Microbiol. Biotechnol.*, **33**, 501–5.
Ramesh, M.V. and Lonsane, B.K. (1991) *Appl. Microbiol. Biotechnol.*, **35**, 591–3.
Rathbun, B.L. and Shuler, M.L. (1983) *Biotechnol. Bioeng.*, **25**, 929–38.
Reiser, J., Walther, I.S., Fraefel, C. and Fiechter, A. (1993) *Appl. Environ. Microbiol.*, **59**, 2897–903.
Reiser, J., Muheim, A., Hardegger, M., Frank, G. and Fiechter, A. (1994) *J. Biol. Chem.*, **269**(45), 28152–9.
Roche, N., Desgranges, C. and Durand, A. (1994) *J. Biotechnol.*, **38**, 43–50.
Roche, N., Berna, P., Desgranges, C. and Durand, A. (1995) *Enzyme Microbial Technol.*, **17**, 935–41.
Rodríguez, J.A., Echevarria, J., Rodríguez, F.J., Sierra, N., Daniel, A. and Martínez, O. (1985) *Biotechnol. Lett.*, **7**, 577–80.
Rolz, C., de León, R. and Arriola, M.C. (1988) *Biol. Wastes*, **26**, 97–103.
Ryoo, D., Murphy, V.G., Karm, M.M. and Tengerdy, R.P. (1991) *Biotechnol. Techniques*, **5**, 19–24.
Sargantanis, J., Karim, M.N., Murphy, V.G., Ryoo, D. and Tengerdy, R.P. (1993) *Biotechnol. Bioeng.*, **42**, 149–58.
Sarrette, M., Nout, M.J.A., Gervais, P. and Roumbouts, F.M. (1992) *Appl. Microbiol. Biotechnol.*, **37**, 420–5.
Sato, K. and Yoshizawa, K. (1988) *J. Ferment. Technol.*, **66**, 667–72.
Sato, K., Nagatani, M., Nakamura, K. and Sato, S. (1983) *J. Ferment. Technol.*, **61**, 621–9.
Sato, K., Miyazaki, S., Matsumoto, N. and Nakamura, K. (1988) *J. Ferment. Technol.*, **66**, 173–80.
Saucedo-Castañeda, G., Gutierrez-Rojas, M., Bacquet, G., Raimbault, M. and Viniegra-González, G. (1990) *Biotechnol. Bioeng.*, **35**, 802–8.
Schalch, H., Gaskell, J., Smith, T.L. and Cullen, D. (1989) *Mol. Cell Biol.*, **9**, 2743–7.

Shambuyi, M., Beuchat, L.R., Hung, Y.C. and Nakayama, T. (1992) *Int. J. Food Microbiol.*, **15**(1–2), 77–85.
Shankaranand, V.S. and Lonsane, B.K. (1994) *Process Biochem.*, **29**(5), 29–37.
Sharma, D.K., Tiwari, M. and Behera, B.K. (1995) *Appl. Biochem. Biotechnol.*, **51/52**, 495–501.
Silman, R.W. (1980) *Biotechnol. Bioeng.*, **22**, 411–20.
Singh, K., Rai, S.N., Neelkantan, S. and Han, Y.W. (1990) *Ind. J. Anim. Sci.*, **60**(8), 484–90.
Smith, J.E. (1985) In *Biotechnology Principles* (ed. J.E. Smith), American Society for Microbiology, USA, p. 39.
Smith, R.E, Osothsilp, C., Bicho, P. and Gregory, K.F. (1986) *Biotechnol. Lett.*, **8**, 31–6.
Smith, T.L., Schalch, H., Gaskell, J., Covert, S. and Cullen, D. (1988) *Nucleic Acids Res.*, **16**(3), 1219.
Smits, J.P., Rinzema, A., Tramper, J., Schlösser, E.E. and Knol, W. (1996) *Process Biochem.*, **31**(3), 669–78.
Soccol, C.R., Iloki, I., Marin, B. and Raimbault, M. (1994a) *J. Food Sci. Technol.*, **31**(4), 320–3.
Soccol, C.R., Marin, B., Raimbault, M. and Lebeault, J.-M. (1994b) *Appl. Microbiol. Biotechnol.*, **41**, 286–90.
Steinkraus, K.H. (1983a) In *Handbook of Indigenous Fermented Foods* (ed. K.H. Steinkraus), Marcel Dekker, New York, p. 131.
Steinkraus, K.H. (1983b) In *The Filamentous Fungi*, Vol. IV (eds J. Smith, D.R. Berry and B. Christiansen), Edward Arnold, London, p. 171.
Stewart, P., Kerste, P., Vanden-Mymelenberg, A., Gaskell, J. and Cullen, D. (1992) *J. Bacteriol.*, **174**, 5036–42.
Sudo, S., Ishikawa, T., Sato, K. and Oba, T. (1994) *J. Ferment. Bioeng.*, **77**(5), 483.
Tengerdy, R.P. (1985) *Trends Biotechnol.*, **3**, 96–9.
Thakur, M.S., Karanth, N.G. and Nand, K. (1990) *Appl. Microbiol. Biotechnol.*, **32**(4), 409–13.
Thomas, T.D. and Turner, K.W. (1981) *Appl. Environ. Microbiol.*, **41**, 1289–94.
Tien, M. and Kirk, T.K. (1983) *Science*, **221**, 661–3.
Troller, J.A. (1987) In *Water Activity: Theory and Applications to Food* (eds L.B. Rockland and L.R. Beuchat), Marcel Dekker, New York, p. 101.
Ulloa-Sosa, M. (1974) *Bol. Soc. Mex. Mic.*, **8**, 17–21.
Ulmer, D.C., Tengerdy, R.P. and Murphy, V.G. (1981) *Biotechnol. Bioeng. Symp.*, **11**, 449–61.
Valino, E., Elias, A., Alvarez, E. and Albelo, N. (1996) *Cuban J. Agric. Sci.*, **30**(1), 65.
Vanderhee, M.D., Graca, P.M.A., Huizing, H.J. and Mooibroek, H. (1996a) *Mol. Gen. Genet.*, **250**(3), 252–8.
Vanderhee, M.D., Mendes, O., Werten, M.W.T., Huizing, H.J. and Mooibroek, H. (1996b) *Curr. Genet.*, **30**(2), 166–73.
Vieira, M.J.F., Spadaro, A.C.C. and Said, S. (1991) *Biotechnol. Lett.*, **13**, 39–42.
Viesturs, V.E., Steinkraus, S.V., Leite, M.P., Berzines, A.J. and Tengerdy, R.P. (1987) *Biotechnol. Bioeng.*, **30**, 282–8.
Wei, C.-J., Tanner, R.D. and Woodward, J. (1981) *Biotechnol. Bioeng. Symp.*, **11**, 541–53.
Wiacek-Zychlinska, A., Czakaj, J. and Sawicka-Zukowska, R. (1994) *Bioresource Technol.*, **49**, 13–16.
Yang, S.S. (1988) *Biotechnol. Bioeng.*, **32**, 886–90.
Yang, S.S. and Ling, M.Y. (1989) *Biotechnol. Bioeng.*, **33**, 1021–8.
Yaver, D.S., Xu, F., Golightly, E.J., Brown, K.M. *et al.* (1996) *Appl. Environ. Microbiol.*, **62**, 834–41.
Zadrazil, F. and Grabbe, K. (1983) In *Biotechnology 3* (eds H.J. Rehm and G. Reed), Verlag Chemie, Weinheim, p. 145.
Zadrazil, F. and Puniya, A.K. (1995) *Bioresource Technol.*, **54**, 85–7.
Zhu, Y., Knol, W., Smits, J.P. and Bol, J. (1996) *Enzyme Microbial Technol.*, **18**(2), 108–12.
Zvauya, R. and Muzondo, M.I. (1994) *Food Sci. Technol.*, **27**, 590–1.
Zvauya, R. and Muzondo, M.I. (1995) *Int. J. Food Sci. Nutr.*, **46**(1), 13–16.

4 Composting processes

S.P. MATHUR

4.1 Introduction

Newly heightened concerns or realizations about climate change, depletion of stratospheric ozone, environmental pollution, wildlife, shrinking biodiversity, health, food safety, land degradation, pressures on nonrenewable resources and intergenerational justice, have vindicated the sustainable development ethos expressed by the Brundtland Commission in its report, 'Our Common Future'. In practical terms, the new paradigm promoted the 4R strategy of Reduce, Re-use, Recover and Recycle wastes to conserve resources and reduce pollution by emulating, adopting and maximizing beneficial processes of an ecosystem. An ecosystem is maintained in essence by the cycling of carbon and mineral nutrients so that the losses of nutrients from the system are not greater than the relevant mineral ions gained through soil weathering and the nitrogen fixed by biological agents or the electrochemical force in lightning.

The carbon is cycled through the food chain of herbivores, omnivores, carnivores and soil-inhabiting decomposer organisms, because without it photosynthesis will cease within a decade or so for lack of atmospheric CO_2. The beneficial biological activity involved in carbon cycling also helps to control pathogens, pests and parasites through antagonism and competition in a balanced ecosystem. A portion of the carbon is maintained in soil humus, an amphoteric heteropolycondensate of carbon, nitrogen, sulphur and phosphorus compounds that is relatively biostable (Paul and Clark, 1989). Humus shelters and nourishes soil organisms while it buffers and moderates physical, chemical and biological processes and attributes of soil.

When man first cultivated land and removed portions of the organic produce as food or materials, the loss of nutrients and carbon was not a serious problem as long as agriculture was confiend to the river valleys and deltas which recouped the losses through the alluvial silt deposited by flood waters, or when the slash and burn strategy was used on forest lands at low frequency (Altieri, 1987). Agriculture was sustainable, particularly on lands out of the two aforesaid areas, only where the organic wastes were recycled to maintain soil humus and fertility. The organic wastes, however, ideally were not returned to the cultivated land in the raw state as that would have been unhygenic for both plants and man, and would have overloaded the system's capacity to cycle the carbon in the wastes in the near-absence of on-site herbivores present in a balanced ecosystem.

Processing of the raw wastes in composite mixtures in some ancient cultures reduced bulk, dissipated part of the carbon, sanitized the wastes and conserved nutrients (Howard, 1943). Agriculture was thus sustained in China, where the connection between disease and faecal matters was first realized, for 5000 years, and in the Minoan civilization of Greece for a shorter period. In contrast, according to Hughes (1980), the Mayas of middle America, who had no domestic animals, exhausted the lands they cultivated, and had to move and resettle repeatedly, for lack of the cycling of organic matter and nutrients through manuring.

Awareness of the oriental practice of recycling wastes through the composting process was gained strikingly through King's (1911) account of agriculture there in the Far East, although biblical references to advisability of not applying manures without storage have long existed.

Krishna Murthy (1978) and Howard (1943) have traced the history of development of the scientific principles of composting, mostly in India, culminating in the aerobic Indore method and the partially aerobic Bangalore method. Indore and Bangalore are cities in the erstwhile Holkar and Mysore states, the Maharajahs of which employed researchers such as Howard, Wad, Jackson and Acharya, during the British colonial rule of India.

With the advent of cheap chemical fertilizers, and owing to other economic and social factors, recycling of organic matter through composting lost ground to the 'artificial manures'. Liebig is credited with initiating the chemical approach by discovering that plants absorb nutrients only as mineral ions. Liebig did, however, state that 'there is but one manure which permanently keeps up the fertility of land, and that is farmyard manure' (Krishna Murthy, 1978). The rediscovery of this type of thinking has reawakened interest in composting which had in the interim held the interest mostly of the avid home gardener and the sanitary engineer, almost irrespective of the labour in the former and cost in the latter, while food-processing wastes and manures were being disposed contaminatively or buried in landfills to pose insidious, slow but definite threat to the environment. The threats included generation of the very potent greenhouse gases CH_4 and N_2O. At the same time disposal of the bioresidues, particularly from agricultural and food industries, exacerbated land degradation caused by depletion of soil organic matter.

This chapter, therefore, deals with the composting processes, with special reference to problems faced by compost facilities, causing many closures, and the research that is needed to tackle them. The focus is on industrial-scale composting not on the backyard small-scale composting that is being promoted in some regions, such as in Canada, probably as an expediency, because it is inherently incapable of achieving sanitization (Linteau and Beauregard, 1992).

4.2 Definition and principles of composting

4.2.1 Definition

Composting is a managed bioconversion of putrescible organic materials into a humus-rich hygienic product that improves soils and nourishes plants.

Composting and the product compost are thus defined by the multiple objectives achieved together, not singly. For example, the sanitization of wastes by radiation or biostabilization by desiccation will not produce humus or conserve nutrients. The addition of soluble nutrients to organic wastes and their joint application to land will neither sanitize the wastes nor produce all the possible humus, as the leachable nutrients may not remain with the organic matter in the surface layer. Decomposition of certain wastes alone, e.g. bark or sawdust, may reduce bulk, which is important mainly in the disposal perspective, and generate sanitizing heat owing to oxidation of organic matter without producing the nutrient-rich compost beneficial to plants. Organic matter merely left to rot also does not produce compost any more than rotting grapes on the table produces champagne. Composting indeed is akin to the making of other biological products like beer, wine and cheese, and, therefore, needs to be achieved under well-defined conditions of environment and materials.

Many definitions of composting can be found in the literature, varying in their foci, priorities and specificities. For example, Zucconi and De Bertoldi (1987) proposed the definition that composting is a controlled bioxidative process that:

1. involves a heterogeneous organic substrate in the solid state;
2. evolves by passing through a thermophilic phase and a temporary release of phytotoxin;
3. leads to production of carbon dioxide, water, minerals and stabilized organic matter.

All acceptable definitions recognize humus production, biostabilization, nutrient conservation, carbon-recycling, nuisance-removal and sanitization as essential features of composting. That is, composting is an accelerated simulation of the concomitant soil processes of mineralization and humification of organic detritus.

4.2.2 Principles

In essence, composting involves oxidative dissipation of part of the carbon in the waste to CO_2 and H_2O while the plant nutrients are assimilated and mineralized by organisms. Some by-products of the carbon oxidation, bioresistant compounds or their derivatives, secondary metabolites and microbial biomass fractions, are condensed and polymerized into humus through the presence of reactive free radicals.

The heat generated during the oxidation of carbon, particularly by heat-loving (thermophilic) organisms, antagonistic compounds produced by decomposer organisms, and the competitive ability of a versatile consortium of heterotrophic organisms, combine to kill plant and animal pathogens or at least to suppress them to benign levels. The survival capacity of pathogens is limited in composts owing to the narrow requirements of disease organisms for specific substrates and temperatures for optimal activity. It is, therefore, obvious that conditions in the composting mass must foster optimal bioxidation by a suitable group of organisms for an adequate period.

4.2.3 Compost feedstocks

Any biodegradable or biological material can be composted, alone if it has the right physical and chemical properties but more commonly with other suitably matched material(s). All materials of biological origin contain, in various proportions, water-soluble organic compounds, proteins, lipids (fats and waxes), structural or storage polysaccharides (starches, cellulose, hemicelluloses; chitin, glycogen, etc.), and, in the case of plants, phenylpropanoid compounds and their polymers such as lignin. As animals and their intestinal flora do not digest all of these fully, the same are also present in excreta along with their products and microbial biomass. The protein-rich bacterial biomass may account for as much as 50% of the carbon in the faecal matter (see Mathur *et al.*, 1990b). Animal excreta particularly urine, is therefore a valuable source of nitrogen in compost mixes.

Modern methods of high-density indoor rearing of some pigs, poultry (mainly layers) and dairy cattle, with little or no litter, in highly specialized farms, generally result in collection and storage of animal excreta as slurries (Gray *et al.*, 1987; Mathur *et al.*, 1989, 1990a,b).

The C/N ratio in the urine-containing slurries is so low that, during decomposition, much of the N is lost as it is in excess of the requirement for all the microbial biomass synthesis possible from the C present (Witter and Lopez-Real, 1987). Also much of the N is in forms readily degraded by exothermic enzymatic hydrolysis. The microbial proteins, urea and uric acid thus yield foul-smelling aliphatic amines, sulphides and NH_3 even under conditions suboptimal for microbial activity (Miner and Hazen, 1969, 1977; Elliott *et al.*, 1971; Mathur, 1982). The aerobic microbial activity that transforms the malodorous compounds into NO_3^-, SO_4^{2-}, CO_2 and H_2O is impaired by lack of oxygen and by low temperature. The microbial activity that does occur anaerobically produces CH_4, amines, NH_3, CO_2, and the toxic and malodorous sulphides including H_2S. High concentrations of the CO_2 promote retention of the NH_3 as the unstable

$(HN_4)_2CO_3$ even at neutral to slightly alkaline pH. But the high viscosity needed to occlude and maintain CO_2 at the suitably high concentrations curtails the slurry's capacity to retain amines, H_2S and other sulphides is acqueous solution, thus heightening the problem of maladours. On the other hand, the NH_3 lost from manures, and the nitrogen oxides from partly aerated systems, promote eutrophication of water bodies and the acidification of soils in more than the immediate vicinity as the NH_3 is oxidized to nitrates on soil and water (Schroder, 1985; Voorburg, 1988). Even the highly viscous and malodorous slurries would lose ammonia rapidly during handling and application to soil, as the CO_2 is degassed to the atmosphere from the solution. The less odorous dilute, NH_3-poor slurries, on the other hand, are costly to transport and apply, as larger application tankers simply compact soils more.

Consequently, the available treatment systems for animal slurries probably do not eliminate all adverse environmental impacts as they cannot both conserve the N and contain the malodours at all stages (Leger *et al.*, 1991). Composting with 5–10% sphagnum moss peat as a bulking agent has been shown to be a viable and acceptable alternative (Mathur *et al.*, 1988b, 1990a,b). The cost of the peat can be met by the ammonia saved (Mathur, 1993; Mathur *et al.*, 1989; Barrington *et al.*, 1990).

Composting of slaughterhouse wastes, meat leftovers, small bones, blood, seafood-processing wastes and other animal matter can be accomplished by the use of special care and practices (Mathur, 1991, 1992; Mathur *et al.*, 1996; Dalzell *et al.*, 1987).

Organic seafood-processing waste, including 'morts' (dead fish) from aquaculture, are particularly attractive as a feedstock for composting through natural or tranquil methods of aeration (Hayes *et al.*, 1993; Mathur, 1992; Mathur *et al.*, 1996). Harvesting of roe (eggs) only renders about 95% of the catches, such as of herring, as unutilized waste. Shellfish scraps contain about 33% chitin (poly-*N*-acetyl glucosamine) and 33% protein (Martin and Patel, 1991). Fin fish wastes include whole waste fish (bycatch), offal that contains viscera and fish scrap that is the residue of filleting. The scraps (racks) contain skin, heads, tails, fins, scales and backbones. About 30% of fishes in aquaculture die, while processing of fin fish, crab, shrimp and lobster can generate 30–60%, 75–85% and 30–80% waste, respectively. According to Swanson *et al.* (1980), the waste is only slightly less in protein and nutrients than the utilized portions.

Vegetable materials left over from extraction of juices, oils, fibres and pulps are valuable as compost materials (Krishna Murthy, 1978; Chen and Hadar, 1987; Pudelski, 1987), as are the filters used for refining oils and juices (Mathur *et al.*, 1996). Residues of tea and coffee processing and residues from microbe-based beverage and pharmaceutical industries belong to the same group.

Putrescible town and home refuse are, of course, major sources of compostable materials (Gotaas, 1956; Dalzell *et al.*, 1987; de Bertoldi *et al.*, 1987), as are cullings, weeds, fallen leaves and grass clippings, commonly called leaf and yard materials. Landfilling of such wastes is inimical to the environment.

Municipal wastewater treatment generally includes aerobic microbial decomposition of dissolved and suspended organic matter, followed by anaerobic fermentation. Both processes involve growth of microorganisms, and release of some of the carbon mainly as carbon dioxide (CO_2) under aerobic conditions and CH_4 under anaerobiosis. Under aerobiosis, carbon mineralization is accompanied or closely followed by humification that occurs only in the presence of 'free' gaseous oxygen (Mathur and Farnham, 1985). Anaerobiosis suppresses humification. The lack of oxygen also prevents the ammonia produced to be oxidized to nitrate, a process that is accomplished almost solely by mesophilic aerobic bacteria (Mathur, 1991; Miller *et al.*, 1989).

The dewatered sewage sludge of treated wastewater, called biosolids, therefore, are rich in intermediate products of decomposition, mostly 'raw' humus and microbial proteins. The proteins are about 37% by weight of the solids (Poincelot, 1975), and proteins contain about 16% N.

Moist organic materials rich in protein and other N-compounds, such as the biosolids, manure slurries, dead fishes or their parts, and some slaughterhouse wastes, decompose rapidly even if decomposer microbes themselves are not active. The decomposition is initially achieved by exoenzymes, i.e. biocatalysts that can act independently of their parent cells from which they may be far removed in both time and space. They have an important natural role in soil and detritus (Mathur, 1982). These biocatalysts cause cleavage of proteins, urea, uric acid and some other N-compounds by exothermic hydrolysis, i.e. by releasing rather than using energy. The products include peptides, amino acids, ammonia, amines and sulphamines with highly evocative names, such as skatoles, putrescine and cadaverine (Hayes *et al.*, 1993; Mathur, 1994). In a soil or any habitat with an appropriate mixture of water and air, the malodorous amines and sulphamines are not produced. Instead, the amino acids are degraded into ammonia and acids. The acids are used as constituents of microbial bodies and as sources of energy. The excess ammonia is also utilized in synthesizing microbial bodies using N-poor organic (C-rich) compounds as sources of carbon for building bodies and to provide energy through oxidation (Mathur, 1991). Oxygen in the air is used to produce CO_2. The ammonia can also be microbially oxidized to the odourless plant-nutrient nitrate. Because the N-rich wastes mentioned above are too dense to allow air movement, and lack C-rich organic compounds, the main degradation products during their handling and storage are ammonia, amines, sulphamines and hydrogen sulphide (rotten-egg gas). The result is a distinctive malodour that deters storage and land application.

The maladour associated with biosolids, seafood wastes and manure slurries is sought by regulations to be subdued during land application by the requirement that the materials be incorporated into soil during or immediately after application. As soil injection is ruled out mainly by a lack of fluidity, the most effective means of incorporating biosolids would be moldbord ploughing in which the top soil layer is flipped over. For various reasons of cost, opportunity and soil conservation, this type of ploughing is not recommended any more. The next choice is harrowing with tines, or shear or chisel ploughing, as discing is less effective. None of these incorporation practices, however, can be undertaken while the soil is waterlogged, frozen or covered with a high crop, thus severely limiting opportunities for utilization or disposal of these bioresidues on agricultural land. More so, because they can not be stored under ambient conditions in a stable form that does not create nuisance, hazard or threat to the environment, mostly due to uncontrolled biological activity. Composting is recognized as a means of biostabilizing such highly putrescible organic materials.

Wastes from wood-based industries tend to be low in N and P and, therefore, are composted best in combination with some other materials or with N and P fertilizers. Wood wastes have phenolic compounds and polymers which release bioinhibitory phenols, terpenes and tannins during decomposition. When ammoniacal-N is added or released into the wood wastes, e.g. from urea or urine, the slightly alkaline pH in the presence of air causes neutralization and autooxidation of phenols to produce semiquinone free radicals and hydroxyquinones which polymerize, by themselves, or through enzymatic catalysis, into insoluble humus-like polymers, thus removing the bioinhibitory compounds from an active role (Mathur and Farnham, 1985).

4.2.4 Requirements of optimal composting

A special issue of the journal *Biomass and Bioenergy*, in its introduction, contained a quote from the futurist Arthur C. Clarke: 'Solid wastes are the only raw materials we are too stupid to use'. The editors of this issue, Hyatt and Richard (1992) made a simple, short and sharp diagnosis: 'We need to be smarter in how we manufacture and use compost'. This is driven home by the recent failure and even closure of many compost facilities in North America.

We manufacture compost by employing accelerated simulation of a soil process where decomposer organisms fulfil their pivotal role in global cycling of carbon and plant nutrients as mentioned earlier. The decomposers act optimally within certain specific boundaries. For example, the soil particles and their aggregates should form a stable structure that allows movement of both water and air while enough fluid is retained to provide a medium for free movement of both solutes and the motile organisms. The air is needed for respiration that releases energy from oxidation of organic compounds. Microbes obtain their oxygen from the dissolved form and their respiration is independent of soluble oxygen concentrations above 0.1 ppm (Chance, 1957). Concentration in water is influenced by the O_2 in the air and the temperature of the water, with less dissolved in hotter water.

The water is held in soil by capillarity and by physical and chemical absorption to extents that depend on various attributes of soil such as colloidal clay and humus contents. The gravimetric (per cent by weight) water content is, therefore, not a good measure of how much water is free to move or be used by plant roots or microbes. A more meaningful measure is related to the total capacity of the soil to hold water against a pull, e.g. gravity. One such commonly used measure is water-holding capacity (WHC) against a 1/3 bar suction. It has been known for more than 50 years (Waksman, 1938) that soil decomposer activity is at its maximum when the soil contains water equal to 2/3 of its WHC. In the field of compost science, however, the designs and control are still based on gravimetric water content, although some attention has recently been given to the true criterion, e.g. by Miller (1989) who has shown the curvilinear relationship between gravimetric content of a compost and its matric water potential (water held by physical attraction). Incidentally, Miller (1989), described the gravimetric water content in composting as 'a crude tool providing little fundamental insight'. Crude indeed, when we say that ideal moisture content for composts lies between 50% and 85%, it means 1 part of dry matter is to be mixed with 1–6 parts of water. On the other hand, peat with 50% water has no bioavailable moisture, it is air-dry.

In soil and composts, when there is more water there is less air and less of the air movement needed to replenish the oxygen used by the microbes. Amenability of a soil to air movement, i.e. air permeability, has been recognized as an important property for more than a hundred years (Smith, 1986).

Air moves in and through soil owing to the diffusion that tends to equalize individual gas concentrations between the air in the matrix and the open atmosphere, but it also moves conductively owing to the pressure difference created by barometric changes, temperature gradients, wind gusts, water movement within matrix and the extraction of water by roots. The structure of the soil or compost, i.e. the pore sizes, their distribution, tortuosity and continuity of the pore spaces, and their stability influence the ease of penetration and rate of movement of water and air through them. For a given stable medium, therefore, the rate of flow of water or air is determined mostly by the pressure gradient. However, in addition, in the case of air, flow through a single pore varies as the fourth power of the pore radius so that a major part of the moving air passes through large pores and wide cracks in soil (or compost) (Grant and Groenvelt, 1993) leaving behind a few nooks and crannies (microniches) in a oxygen-depleted state. Consequently, even aerobic soils (and aerated composts) always have some anaerobic pockets (Op den Camp, 1987; Bellamy *et al.*, 1992) but the products of anaerobic decomposition in these pockets are further oxidized within the aerobic zones in the undisturbed soil (or compost). However, when the compost is disturbed by turning or forced air, some of the products of anaerobic decomposition, generally malodorous, are released (Jiminez and Garcia, 1989; Kissell *et al.*, 1992).

The water and air conditions for optimal decomposition in soil have not been, it seems, simulated in composts with all the possible veracity. Harking back to Hyatt and Richard (1992), we could have been 'smarter'. Some allowances have had to be made for the differences between soils and composts. In the composts, the debris is concentrated so that heat, O_2 consumption and CO_2 generation are more intense in a medium that conducts heat less readily than soil. The retention of heat is beneficial to an extent in that it can optimize conditions for the thermophilic micro-organisms most active at about 55 °C (cf. Mathur, 1991). At the same time the heat helps to sanitize the wastes. The larger CO_2 and O_2 concentration differences between matrix and atmosphere create higher potential for diffusion of air into compost than in normal soil. In addition to diffusion, movement of air into a compost is driven by convection due to heat. Unfortunately, early interest in the relationships between physical properties of percentage moisture, bulk density, percentage free air space and percentage porosity (McCauley and Shell, 1956) was not sustained sufficiently so that quantitative modelling of heat and gas transfer in composts is not yet possible owing to lack of basic data (Miller *et al.*, 1989). However, a new start has been made such as by Op den Camp (1987) who reported convection mass flow in composts of 6.1–7.8 $m^3\ m^{-2}\ h^{-1}$ and by Miller *et al.* (1989) who discussed the subject with clarity. The area is promising because theoretically natural convection alone can provide sufficient aeration to a compost (Kuchenrither *et al.*, 1985; Ishaii *et al.*, 1991). The research is needed because Hyatt and Richard (1992) stated that 'time is ripe for innovative approaches in air management and filtration' and because aeration relates to most of the interconnected concerns that have been generated by failure of many compost facilities and poor quality of many products.

The enzymology and microbiology of composting parallel those of organic matter decomposition in soils and, therefore, the optimum conditions required also have some similarities. Optimization is achieved not by having only one or two conditions at their best but by manipulating the whole set of conditions to be near the most favourable set achievable under the circumstances. The requirements are discussed below.

(a) C/N Ratio. The activity of microorganisms involved in composting is directed at synthesizing microbial protoplasm which contains about 50% C, 5% N and 0.25–1% P on a dry weight basis (Alexander, 1977). Since the atomic weights of C and N are so close, the C/N ratios on atomic and percentage bases are 8 and 10, respectively, which, not coincidentally, match the ratio in mature humus (Kononova, 1966). The assimilation and maintenance of every atom of C in microbial biomass require the oxidation of about two atoms of C for energy. Consequently, Waksman (1938) established that the microbes overall utilize about 30 parts of carbon for each part of nitrogen. Since these are approximate values, and not all the C and N in a compost mixture is bioavailable, nor is all the mass converted to humus or microbial biomass, the optimal ratios for different materials have been found to vary from 26 to 35 (Gotaas, 1956). Satisfactory results are obtainable mostly with C/N ratios between 30 and 40 (University of California, 1953).

Table 4.1 presents the C/N ratios of various materials.

When the C/N ratio is too narrow, N will be lost from the system as ammonia. If the C/N ratio is too wide, the synthesis of biomass and its breakdown for new formation of protoplasm will be repeated many times before a stable state is reached, protracting the composting process. As the word compost itself implies, different materials have to be mixed

Table 4.1 Approximate nitrogen content and C/N ratios of some compostable materials (dry weight basis)

Materials	%N	C/N
Urine	15–8	0.8
Blood	10–14	3
Mixed fish scrap	6.5–10	–
Filleted fish scrap	8.2	4
Crab scrap	8.2	3.5
Lobster scrap	4.6	–
Night soil	5.5–6.5	6–10
Sewage sludge	5.0–6.0	6–8
Animal manures (without litter)	1.7–3.7	11–25
Farmyard manure (average, with litter)	2.15	14
Grass clippings (average)	2.4	19
Green vegetable wastes, weeds	2.0–3.7	11–20
Cereal straw	0.3–1.05	45–130
Combined refuse	1.05	34
Rotted sawdust	0.25	208
Raw sawdust	0.11	511
Peat (horticultural)	1.0	50
Newspaper	Negligible	Infinity
Paper sludge	Negligible	Infinity

Sources: Gotaas (1956); Mathur *et al.* (1990a,b)

(composited) or layered to achieve an optimal C/N ratio, and thus optimal composting. Obviously, for reasons similar to some of the above, a C/P ratio of 100–200 is also desirable but this is less critical than the C/N ratio as most organic materials contain sufficient P for the purpose.

Conservation of N in the waste is one of the objectives of composting (Meyboom, 1993), although this fact is sometimes played down or ignored when the focus is mistakenly on disposal or biodrying (Finstein *et al.*, 1993; Hogan *et al.*, 1989) or on producing a 'source of carbon' for soils (Bellamy *et al.*, 1992). Loss of N as ammonia contributes to odour, protracts the process and may result in a compost that robs crop plants of their soil-N supply. The ammonia in the atmosphere helps precipitate acid ions thus intensifying acid rain, augmenting eutrophication, and causes further acidification as the NH_3 is oxidized to NO_3 (Witter and Lopez-Real, 1988). If this or the nitrate leached from compost piles is bacterially reduced, it will generate some N_2O. N_2O is also known to be produced in manure storages where both aerobic and anaerobic zones exist. Similarly, N_2O may be generated in some compost piles. The overall loss of N during some composting processes renders them less valuable than the raw material as a N fertilizer (Cooper and Warman, 1993; Rynk, 1992).

Compost contains ammonium, nitrate, amino compounds and humic substances so that it supplies some N immediately and the rest over a period of time at various rates (e.g. Mathur *et al.*, 1990b). The nitrate is susceptible to leaching and denitrification under anaerobic soil or storage conditions. The ammonia, if in excess, can limit plant growth. Full benefits of the slow-release N in composts can be manifested only in long-term experiments.

(b) pH. As most biological fluids are balanced in their cationic and anionic ions near neutrality, pH 7.0, that is also the ideal for soils and a composting mass. At a pH much higher than 7.0, the aqueous medium of activity will contain dissolved base ions at a concentration higher than those in the microorganisms, causing an osmotic loss of water from their vital fluids. The excess of the OH^- ions will also cause loss of the ammonium ions as ammonia, and hydroxylation of essential biological elements like Cu and Zn, rendering them amenable to precipitation as insoluble mixed carbonates. At the other extreme, mass action of the excessive H^+ ions at low pH can cause decomplexation and desorption of essential base ions, e.g. Ca and Mg from organisms, and toxic metal ions like Al, Mn and Cu from minerals and organic matter.

The pH of manures is usually slightly above 7.0; that of plant materials and most household garbage is between 5.0 and 7.0, more towards the lower figure if the material is partly putrified as decomposition initially releases organic acids. On the other hand, if the compostable material is high in substances like proteins, urea, uric acid and chitin, enzymatic hydrolysis initially will release ammonia, thus raising the reaction to alkalinity.

Mixtures of alternating layers of acidogenic and ammonifying materials do not pose a problem provided the interlayer and overall pH is between 6.5 and 7.5, which is optimal for most decomposer organisms. Since composting is a dynamic biochemical process, marginal fluctuations in reaction will occur temporarily and spatially without affecting the overall process.

For most materials, therefore, pH adjustments are not necessary, but exceptions do occur. Wood wastes and sludges from pulp and paper mills may have a low pH of 5–6 and a high buffering capacity. As these materials also have a wide C/N ratio, necessary addition of ammonia-releasing compounds like blood, urine or urea can be made to neutralize the acidity until base cations are released by decomposition of the wood waste. In the interim, loss of ammonia can be prevented by placing a 100–150 mm thick layer of acidic peat or mature compost as an envelope over the composting mass. The other example is of seafood wastes or other food processing wastes where alkaline materials may have been used in the factory. The alkalinity can be neutralized by use of peat or acidic wood materials in layers or mixtures.

Amelioration of pH can also be achieved by using sulphur to create acidity, and lime to neutralize acidity. In so doing, one must take into account the buffering capacities of the materials involved, and pretest different rates of application and their effect over time. At the same time an excess of lime should be avoided as it may cause deficiency of some plant nutrients like Mn when applied to soil.

(c) Physical state (particle size). Various shredders, raspers, rotary drums or hammer mills are sometimes used to break down the materials into a suitable physical state.

Composting involves actions of enzymes, microorganisms and oxygen. Therefore, the smaller the size of the material acted on, the greater is the surface area and easier the contact between various reactants, except perhaps oxygen. Oxygen and the water produced and present do not move freely in narrow interstitial spaces between small particles that pack together tightly, nor does air diffuse easily into large particles to a distance greater than 2.5 cm. Consequently, the composting is optimal in a mixture of particles of various sizes in which the largest is less than 5 cm in its greatest dimension. Ideally, as in a soil, both micropores and macropores are needed between particles so that water can be retained in the former to provide a medium for the microorganisms, while excess water, the CO_2 produced and the necessary air can move easily in the latter.

Recent experience has indicated that convective air flow can nearly continuously meet oxygen requirements of a turned windrow and prevent

Table 4.2 Effect of initial mix density on oxygen levels and production of mercaptan (an indicative malodorous sulphur compound)

Initial mix ratio (yard debris: produce waste)	4:0	4:1	4:1	2:1
Preprocessing (hammer mill)	None	None	Ground	Ground
Bulk density (lb yd^{-3})	300	500	1500	1400
Oxygen concentration (%) at 4 ft depth	19.9	18.8	0.3	0
Total mercaptans on surface	0.2	0.5	25	100

odour generation if the porosity of the mix is high enough, as shown in Table 4.2 from CWC (1993).

(d) Moisture content. The main site of microbial activity in soil or composts is the thin film of water on the surface of particles. Water is essential also for dissolving and transporting the nutrients and substrates that the organisms can absorb only as solutions. Motile organisms and free enzymes also move through aqueous medium. However, for optimal aerobic activity, as in soils, the interstitial spaces, voids or pore spaces between particles need to be occupied by and accessible to both water and air.

The water absorption capacity of soil particles and materials varies widely. In soil science, therefore, it has long been known that maximum aerobic microbial activity occurs when the soil moisture content is two-thirds of its (WHC) at ⅓ bar suction (Waksman, 1938). Surprisingly, no such criterion seems to have been applied to composts. Research findings for the appropriate water content on a moist weight basis, therefore, vary from 45% to 90% for different materials (Gotaas, 1956; Poincelot, 1975; Biddlestone *et al.*, 1987), with 50–60% being found to be suitable for most materials. Thostrup (1985) observed that 70–80% moisture in solid manures of a bulk density equal to 300 kg m^{-3} gave the best results. The author of this chapter has also effectively composed peat-based composts with 70–90% moisture content and paper mill sludges with 70% water. Pieces of wood in the sludge, and peat have the necessary structural strength owing to their fibrosity. Peat and the sludge have low bulk densities of 60–300 kg m^{-3}. In contrast, any material that tends to form a tight mat may not compost well at higher than 60% moisture content.

As stated in section 4.2.2, the area of moisture content nonetheless needs more research to seek a relationship between the water retention capacity of the compost mix and the optimal moisture content for it. In any case, one should also be aware of the changes that might occur in water relationships in the composting process, the system of aeration used and the climate. In warm climates the evaporative loss of moisture may necessitate periodic irrigation of the compost mix. Some materials like paper and seaweeds may become soggy on wetting or decomposition.

Frequent or continuous mixing of the hot compost-mix may change the structure of the mass and cause excessive moisture loss. Moisture loss can be minimized by the use of vapour barrier films as covers in tranquil composting systems (Mathur, unpublished observation; Mathur *et al., 1996).*

(e) Air spaces and aeration. A substantial portion of the organic carbon is oxidized to CO_2 and H_2O. Jeris and Regan (1973) found that each gram of volatile organic matter in municipal refuse required 144 mg of O_2. Their results suggested that a minimum of 30% free air space should be maintained for a wide variety of composting mixtures. Thostrup (1985) on the other hand obtained best results with 70% free air space in solid manures. As the air within a composting mass may be highly enriched in CO_2, it is important to consider the oxygen content itself. Wiley and Spillane (1962) reported that the proportion of oxygen 38 cm inside a pile was 18.6% but fell to as low as 1–2% at 61 cm. It has been suggested that a minimum concentration of 5% oxygen in the gas phase is essential for high-temperature (>45 °C) composting. Biddlestone *et al.* (1987) have recommended 10–18% oxygen in the gas phase of the matrix, although maintaining those levels may involve energy-consuming mechanical aeration or turning to an extent that may be deleterious to structure and retention of H_2O and NH_3. Thostrup's (1985) experiment revealed that the O_2 content of exhaust air should be not less than 10%, although even 7% is not limiting. It would seem that the proper criterion for monitoring and maintaining oxidative conditions should be the redox potential in the compost mix.

It is important to maintain a proper redox condition to avoid the malodour-generating anaerobiosis where reduced sulphur and nitrogen compounds like sulphides and amines are produced. That and the lack of heat sanitization were indeed among the reasons why partial or total anaerobic composting generally fell out of favour. Composting methods (section 4.4), therefore, now vary mostly on the basis of the manner in which the aerobic state is achieved and maintained. The optimal is considered to be a supply of 0.6–1.8 m^3 air kg^{-1} dry volatile solids during the thermophilic stage of composting.

Agitation and/or forced aeration are used in many systems to attempt maintenance of aerobic conditions in hot composts, but designing of aeration requirements is based mostly on the need to prevent or correct overheating (Finstein *et al.*, 1983; Haug, 1986). In some cases high forced aeration may actually make the compost too dry to sustain thermophilic activity, giving a false impression of maturity (Hay and Kuchenrither, 1990). Questions being raised now come from various perspectives.

1. Is forced air removing and dispersing intermediate and odorous products of decomposition from microaerobic pockets before they have a

chance to be oxidized in fully aerobic microniches or zones within the compost? For example, in a stack of compost not aerated forcefully, one sees evidence that reduced malodorous sulfur compounds are oxidized to inodorous elemental sulfur which appears as a yellow layer in a zone below the surface (Miller and Macauley, 1988) where more oxygen is available owing to diffusion of air from the outside. Eventually this sulphur may become bacterially oxidized to the acidic sulphate which will help retain any escaping ammonia as a stable nonvolatile salt while the ammonia is itself oxidized to nitrate, another nonvolatile acid ion. When air is being forced through, the reduced sulfur compounds are dispersed into the atmosphere before they can be oxidized to sulphur. Also, Reddy and Patrick (1975) refer to studies showing how some alternate aerobic and anaerobic conditions can be as effective in net oxidation of organic matter as total aerobicity. Miller and Macauley (1988) refer to how the 5% H_2S in anaerobic areas disappeared in nearby zones with just 2% O_2.

2. Is forced excessive air causing chemical oxidation that generates further heat and odours of the smouldering fire type? It is known that chemical oxidations that cause humus formation, melaninization, Maillard-type browning and caramelization that happens on heating sugars do occur in composts (Miller *et al.*, 1990). The only argument being given against chemical oxidation as a factor is that the temperature never rises above 83 °C (Miller *et al.*, 1990), but that could be because there is a limitation on how much substrate is available for chemical oxidation at a given time, and due to evaporative cooling. The question needs to be examined because, as Miller *et al.* (1990) suggested 'Maillard-type reactions inhibit subsequent ammonia incorporation into the compost'. In other words, the browning by heat delays humification and thus prolongs the maturation process.

3. Is the passing of excessive air or repeated overexposure to air causing loss of ammonia thus increasing the proportion of carbon that has to be oxidized before attainment of biostability? (Acharya *et al.*, 1945). We know that within a manure lagoon or stockpile the ammonia is neutralized by the carbonic acid resulting from dissolution of the CO_2 produced by decomposition, and that the product ammonium carbonate is stable under high CO_2 concentrations within the matrix but not under ambient atmospheric CO_2 levels. That is the reason additions of phosphate or gypsum into animal litter helps to retain the ammonia in manures as nonvolatile and stable phosphate or sulphate. The Ca in sulphate or phosphate compounds may help maintain good structure as it does in soils. Conversely, composting has been used to valorize insoluble rock phosphate into soluble forms (Mathur *et al.*, 1987).

4. Does the answer to the problem lie in maintaining aerobic conditions without having the same oxygen concentration in the compost air as in the atmosphere (20.9%)? By various accounts the aerobic condition can be maintained by as little as 5% oxygen, while even 1% oxygen prevents H_2S

formation (Miller *et al.*, 1989), although, as Nakasaki *et al.* (1987) bemoaned, 'no quantitative treatment of the relationship between the rate of oxygen supply and microbial activity has been reported to date'. Using oxygen as a control criterion, rather than temperature, was rejected earlier by Haug (1986) as he felt that more air has to be pushed through anyway to prevent overheating than that needed to provide the oxygen to the microorganisms. An attempt is now being made to control the generation of heat in the first place by restricting air without jeopardizing aerobicity, preventing odour dispersal and ammonia loss as side benefits (Jacob, 1993). The oxygen concentration during the active phase is maintained between 7% and 15% by diminishing air supply when the 15% is exceeded, thus subduing excessive heat generation. It is noteworthy that the United States Department of Agriculture (USDA) system of composting was originally intended to provide 5–15% oxygen levels (Richard, 1992). According to Miller *et al.*, (1990) 'Implementing aerobic conditions and achieving lower temperature uniformly . . . could alleviate odour problems, shorten process time and conserve substrate, and be more reliable'.

5. Is it possible to have a system where the process is controlled entirely by O_2 feedback, if only to save costs? Bertoldi *et al.* (1988) conducted a comprehensive study to establish that a closed bioreactor aerated by pressure ventilation can be controlled actively to maintain O_2 between 15% and 20%, with good effects. Leton and Stentiford (1990), however, do not find oxygen content as a control point to be necessary for a static pile aerated on the basis of temperature alone according to prejudged stages of composting.

6. Is it possible that the chimney effect and diffusion relied upon in the passive (natural) aeration systems (section 4.4.1(c)) through basal perforated pipes or plena (Mathur, 1992; Hayes *et al.*, 1993; Mathur *et al.*, 1996) are inadequate when the C/N ratio of the feedstock is low (and the heat generation is low) (Ishii *et al.*, 1991), as also when the exterior surface freezes under some winter conditions. The feedstock Zhan *et al.* (1992) used had a C/N ratio of 13 to yield a product with a C/N of 10 in a pile that did not reach sanitizing hot temperatures in some pockets. Freezing of the cover layer would prevent diffusion and smother convective movement of air. Both in the author's experience and in that of Lynch and Cherry (1996), freezing of the exterior is never solid enough to prevent air movement (Mathur *et al.*, 1996).

Research is needed on the air permeability of feedstock mixtures at various stages of decomposition, and pore size distribution characterized to optimize formulation and aeration strategies.

Research is also needed on the quantitative relationship between interstitial oxygen content, overall microbial activity, N conservation and odour generation. For example, is it better to control both oxygen (7–15%?) and temperature, or temperature only at 18–21% O_2 content?

This topic is discussed from other perspectives in section 4.2.4(h) on odours and in the introduction to section 4.2.4.

(f) Temperature. The temperature of a compost mix is important in several ways. As heat is generated by oxidation of organic matter it is symptomatic of progress of the process. Microbial processes of thermophiles and chemical reactions are faster at higher temperatures but beyond 70 °C the organisms and many exoenzymes are inactivated and 'thermo-killed', causing the process to stall. High temperatures are also conducive to excessive evaporative loss of water and odour emanation (Finstein and Miller, 1985; Van der Hoek and Oosthoek, 1985). Retention of heat is necessary not only for fostering thermophilic microbial activity but also for sanitizing the waste by 'pasteurization-type' killing of pathogens. Retention and continual generation of heat are influenced by ambient temperatures, configuration and size of the compost mass, its insulating property and the chemical nature of its constituents.

The heats of combustion of the three most common organic constituents of biomass proteins, carbohydrates and lipids – range from 9 to 40 kJ g^{-1} with lipids yielding about twice as much heat per unit weight as the other two. Wiley (1957) found that 75% of the 22 000–26 500 BTU of heat released on oxidation of 1 kg volatile solids into CO_2 and H_2O was derived from the lipid fraction. Pulverized waste was experimentally determined to release in an 8–10 day period 7 kJ g^{-1} of volatile solids. It is, therefore, not surprising that within 2–5 days of initiation a composting mass attains a temperature of >40 °C which inhibits mesophilic (moderate-temperature-loving) organisms, causing organic acids, the intermediate products of their metabolism, to accumulate. Above 40 °C the mesophiles are killed, except some spores and the thermophiles start to dominate the population. These increase the temperature to beyond the 60 °C that is lethal to most of the fungi in the compost. Above 60 °C the decomposer activity is carried out mainly by actinomycetes such as *Streptomyces* whose spores generate the good-earth smell of a mature compost. The process may also be affected in the above succession as proteins in the original mix or the killed biomass are hydrolysed by enzymes to produce ammonia. At the high temperature, a combination of the ammonia or amino compounds, sugars or phenols may theoretically cause Maillard-type reactions to produce brown humus-like polymers, although this has apparently not been demonstrated.

The rate at which the heat is lost at this stage is determined by the size and shape of the compost mass. Readily available substrates such as starches, sugars, lipids and protein are the main 'fuel' at this stage until the temperature falls below 60 °C to allow fungi and actinomycetes to attack cellulose, hemicellulose and lignin. The need to maintain the temperatures at high levels has been perceived to be due to their lethality to pathogenic microorganisms and ova of annelid parasites and nuisance insects. The earlier perception that the higher the temperature the greater is the decomposition has been altered by the realization that few organisms are active at above 70 °C. Based on various criteria, optimal temperatures for decompositon of different materials have been found to range from 48 °C to 71 °C (Poincelot, 1975). Recently, heightened awareness of (1) the dependence of pathogen inactivation on both the period and temperature of exposure rather than the latter alone; (2) the role of antagonistic organisms and bioinhibitory properties of certain intermediate products of decomposition; and (3) the need for odour control, as well as detailed observations of maximal decomposition at near 55 °C, have seemingly led to a consensus that 55–60 °C is the optimal temperature range (Bollen, 1985; Finstein and Miller, 1985; Lopez-Real and Foster, 1985). Significantly, Lopez-Real and Foster (1985) found that the plant pathogens they tested survived higher temperatures *in vitro* than in the compost.

Of special importance to agriculture is the presence of seeds of weeds in farm wastes. Certain seeds survive or even need passage through animal tracts while others are gained by manures on exposure to air during storage in dumps or lagoons. Field application of raw farm wastes consequently often increases weed infestation. Aerobic composting of the wastes kills or inactivates the weed seeds (Gotaas, 1956; Dalzell *et al.*, 1987).

(g) Heap size. It is well documented that a minimum height of 1–5 m and width of 2–5 m is necessary to retain enough heat in a composting mass to promote the desirable thermophilic activity (Biddlestone *et al.*, 1987), although Mathur *et al.* (1985, 1990a) showed that a height of 1 m was sufficient when the medium of composting is peat and mixtures of peat with fibre-containing manure, both of high thermal-insulation capacities.

The length of the heap and its total size will also depend on the aeration and agitation system used. In any case, an important fact to remember is that large heaps tend to allow pockets of anaerobic conditions owing to compression by the overlying burden, and should be avoided. Eight feet height is considered to be a maximum for windrows and naturally aerated static piles (Rynk, 1992; Mathur *et al.*, 1996).

(h) Odour and fly control. Decomposition of organic material naturally may release ammonia and aliphatic acids under aerobic conditions and foul-smelling H_2S, mercaptans, diamines, methyl amines and skatoles under anaerobic conditions. Some of these products are cadaverine and putrescine which need no elaboration of their implications as they are related to the words cadaver and putrefaction.

In active composts, and in soils where organic matter is being decomposed, lack of gaseous oxygen drives the decomposer organisms to seek oxygen present in other substances to produce reduced (oxygen-deficient) compounds in the following sequence (Figure 4.1).

Redox potential (E_h) at pH 7.0					
+700 –					
–				Oxidized soil	Aerated
+500 –	O_2	→	H_2O		
–					
+300 –	Mn^{4+}	→	MN^{2+}	Moderately reduced	
–	NO_3	→	N_2		
+100 –					
–	Fe^{3+}	→	Fe^{2+}	Reduced	Anaerobic soil
−100 –					
–	SO_4^{2-}	→	H_2S		
−300 –	CO_3^-	→	CH_4	Highly reduced	
–					

Figure 4.1 Critical redox potential at which oxidized inorganic species begin to undergo reduction in submerged soils (adapted from Lindau *et al.*, 1993).

Most of the reduced or oxygen-deficient compounds of N and S have malodours.

Besides the aesthetic undesirability of the aforesaid compounds, their presence on the compost mass surface attracts nuisance insects such as fruit, vinegar and house flies. The eggs laid by these insects, and their subsequent hatching in compost bags or beds, render such composts unattractive to vendors, purchasers and users of the product, and pose some health risks owing to vectoring of diseases by some insects.

The classic control of these two problems, used since ancient times in China, was a mud plaster of 5–10 cm in thickness, often containing binding agents such as chopped rice straw (Dalzell *et al.*, 1987). The latter two materials may also help shed rainwater when the heaps are sloped. The control of odour by avoiding high heat has already been dealt with in the previous section.

Periodic or persistent malodours from compost facilities may cause irreversible loss of support from neighbours and perhaps represents suboptimal quality of the product, and of the process or its control. It is the most common cause of closure of compost facilities.

The following is a list of compounds identified or implicated in composting odours, with the more diagnostic ones indicated by asterisks.

Olfaction, or sense of smell, is the least understood sense (Summer, 1971). Unlike the senses of sight and sound, interpretation of olfactory stimulation occurs only in the brain and is therefore purely subjective. Also, the area of the brain that interprets smell has an important role in

Sulphur compounds

*Hydrogen sulphide	H_2S
*Carbonyl oxysulphide	COS
Carbon disulphide	CS_2
Dimethyl sulphide	$(CH_3)_2S$
Dimethyl disulphide	$(CH_3)_2S_2$
Dimethyl trisulphide	$(CH_3)_2S_3$
*Methanethiol (mercaptan)	CH_3SH
Ethanethiol	CH_3CH_2SH
Butanethiol	$CH3(CH_2)_2SH$

Nitrogen compounds

*Ammonia	NH_3
*Amino methane	CH_3HN_2
Dimethylamine	$(CH_3)_2NH$
Trimethylamine	$(CH_3)_3N$
*3-Methyl indole (skatole)	

Volatile fatty acids

Formic acid	$HCOOH$
Acetic acid	CH_3COOH
*Butyric acid	$CH_3(CH_2)_2COOH$
V*Propionic acid	CH_3CH_2COOH
Valeric acid	$CH_3(CH_2)_3COOH$
Isovaleric acid	

Aldehydes and ketones

*Acetaldehyde	CH_3CHO
*Acetoin	$CH_3CHOHCOCOOH$
Acetone	CH_3COCH_3
Butanone	$C_2H_5COCH_3$
*Ketene	CH_2CO
*Acrolein	CH_2CHCHO

Phenolics

Phenol	C_6H_5OH
Benzene	C_6H_6
Ethyl benzene	$C_6H_5C_2H_5$
Benzene thiozole	C_6H_4SCHN

emotions and memory, which naturally vary widely among people. Osmoreceptors, where the smell is first received, are high up in the nose. From here the sensation is conveyed to the brain where the odour is evaluated. Smell perception is therefore difficult to research in man and, because of its subjective nature, use of animal models limited in value. It is difficult to understand why we have so many anomalies and the turnaround cases. For example, hydrogen sulphide or the sewer gas that smells like rotten eggs at low and nonlethal concentrations smells pleasant at high lethal concentrations. Trimethylamine, $(CH_3)_3N$, smells fishy at low concentration but only like ammonia at high concentrations. Indole smells like the flower jasmine at low concentration, but like fecal matter at high concentrations. We are even more easily fooled by combinations of smells. For example, Walker (1991) talks about the fact that at a composting plant one may smell only ammonia while some distance from the plant the perception is mainly of the reduced sulphur compounds that smell like dead rat or fish, because by the time the air reaches there the masking odour of ammonia has been dissipated by dilution in the atmosphere or by absorption on moist surfaces. Ammonia is perceptible by the human nose at about 55 ppm, mercaptan (CH_3SH) at a few parts per 100 million, while demethyltrisulphide $(CH_3)_2S_3$, is detectable at 1 part in one hundred thousand million (10^5 ppm) (Op den Camp, 1987; Derikx *et al.*, 1990). Even more striking is the fact that human nose can smell ketene, CH_2:CO, at a concentration below the chemical detection limit. Ketene is produced by smouldering fires, along with acrolein, CH_2:CH·CHO. The human nose can detect smouldering fires quite easily, although the organ seems to be designed primarily for sniffing what goes into our mouth. The nostrils open right over and in parallel to the mouth, unlike in animals where the nostrils open outwards or sideways to be most receptive to odours of the surroundings where may lurk preys or predators (Summer, 1971). It may be a matter of survival that man is particularly sharp at smelling food that is spoiled, as that may contain biotoxins, or contaminated with materials that may carry disease, e.g. faecal matter. Faecal matter owes its smell to decomposition in the lower intestines under anaerobic conditions – the same type of anoxic environment that generates malodours in composts or soils. It is fortuitous, or natural by design, that the products of anaerobic decomposition, the reduced compounds of C, H, N and S, are mostly volatile, because that way they can escape to environments where there is oxygen and aerobic organisms that can complete the cycling of these elements by producing CO_2, H_2O, NO_3 and SO_4. As mentioned in the other sections, this may happen within an undisturbed compost pile (Miller and Macauley, 1988). However, when the piles are disturbed, or air is forced through, the odours are broadcast. As Williams and Miller (1992ab) have pointed out, biofilters can not trap odours in water and oxidize them in air, both optimally. A need for understanding odour and designing pre-emptive steps is therefore obvious, but excessive aeration alone may delay maturation, disperse odours, and generate smouldering fire smells.

It is being recognized that chemical scrubbers and biofilters are not infallible, 'do not represent the entire universe of effective options' (Hyatt and Richard, 1992), whereas wet scrubbers may never be totally effective (Mahoney and MacFarlane, 1992) partly because many of the odorous compounds are not very soluble in water (Miller and Macauley, 1988), fundamental knowledge needed to design biofilters is lacking (Haug, 1990; Kissel *et al.*, 1992). It is recognized that odours are released by turning, wet activated carbon is not effective, biofilters are susceptible to short-circuiting particularly under cold conditions and that a pre-emptive approach is needed for odour control (Kissell *et al.*, 1992). Diaz (1987) has said that 'research on process control may be more beneficial than money spent on extensive application of odour control equipment'. Miller and Macauley (1988) mention the additional problem that ammonia, hydrogen sulphide and organic acids can rapidly damage many common construction materials. Smells can also be absorbed by structural materials and released slowly over a long period (Summer, 1971).

Koe and Ng (1987) identified 20 different odorous compounds including carboxylic acids and methyl sulphides from refuse waste from a food centre by simply blowing air over it within a 200 l drum containing 30 kg of waste for 1 day only, showing that much of the smell at a municipal solid waste

(MSW) facility could be from the feedstock itself, if it is not immediately set up as a compost in a proper manner.

Research is needed to examine whether E_h, H_2S and volatile organic compounds (VOC) measurements can be used to detect potential problems, optimize the process and evaluate complaints even after the episodes.

(i) Additives (optional). The first step in compost-mix formulation is the matching of different materials for their diverse C/N ratios, water absorption capacity, moisture content and physical structure to obtain optimal conditions for the composting process (Gotaas, 1956). In many systems air is added continuously or periodically by mixing, natural ventilation or the use of fans (see section 4.4).

Liming of compost mixes, except to mitigate biologically limiting acidity, may hasten decomposition of acidogenic materials but is generally inadvisable because the high pH generated by the lime and the base elements released on decomposition are conducive to loss of ammonia (Poincelot, 1975).

Addition of sources of bioessential mineral elements, where they are lacking, may promote microbial activity. Theoretically, addition of degradative exoenzymes may also hasten decomposition. However, the natural microflora of most wastes are capable of producing the biocatalysts quite rapidly. Also some autolytic enzymes may already be present in the biomass being degraded.

Addition of sources of N and P may be desirable when the materials to be composted lack these elements markedly, such as when bark, sludge from pulp and paper mills, and certain wastes of food processing are to be composted. For example, urea or ammonium nitrate, preferably the former, can be used for composting certain wood wastes lacking in N (Pudelski, 1987), but steps need to be taken to avoid loss of ammonia as the urea is hydrolysed quite rapidly in the compost by the almost ubiquitous enzyme urease.

The addition of soluble phosphate compounds in composts was researched by Acharya (1954), originally to increase plant availability of the fertilizer-P in soils of high phosphate-fixation capacity. Subsequently, sparingly substantiated claims were made that phosphate addition suppresses denitrification (NO_3^- reduced to N_2) and promotes biological nitrogen fixation by fulfilling the microbial need for P.

Addition of insoluble phosphorus compounds is also made to solubilize unavailable forms of P from plants such as soft and hard (sedimentary and igneous) rock phosphate, and basic slag from old-type steel mills. The heat, the high concentrations of CO_2, the Ca-chelating properties of acidic intermediate products of decomposition and humus, as well as the anions CO_3^{2-}, SO_4^{2-} and NO_3^-, may help mobilize the insoluble phosphate materials. The literature on this subject was reviewed by Mathur *et al.* (1987) who reported on the composting of an igneous rock phosphate.

Another type of additive is represented by the starters and inocula touted by many commercial enterprises to activate the composting process owing to their high levels of enzymes, 'special vitamins and hormones needed by compost organisms' and decomposer organisms. Any ecosystem at equilibrium has all the niches covered by the type of organisms it can sustain in their most suitable form. Addition of any extraneous organisms is both unnecessary and an added stress on the normal population. Therefore, any beneficial effects of such additives to wastes that already contain normal saprophytes, and are open to natural seeding by soil and dust, are unlikely to be due to the microbial inocula except where the wastes are low in population due to some pretreatment (Gotaas, 1956; Poincelot, 1975). Experiments with inoculations of composts, therefore, give neutral, positive or negative results (Nakasaki *et al.*, 1987). This does not contravene the evidence of rapid and beneficial decomposition of straw alone by cellulolytic fungi and N-fixing organisms, as the latter may hasten cellulose hydrolysis by consuming the product, glucose (Biddlestone *et al.*, 1987). When the readily available sources of energy for the microorganisms are extremely low, inclusion of soluble carbohydrates in sugar-related industry by-products or materials which yield soluble carbohydrates readily, e.g. seaweeds and starchy wastes, may hasten initiation of the thermophilic phase (Krishna Murthy, 1978; Mathur *et al.*, 1986; Dalzell *et al.*, 1987).

The best additive for a compost mix is old but moist mature compost to provide a suitable starting population throughout the composting mass and the bioavailable minor elements essential for life.

4.3 Chemistry and biology of the composting process

The multitude of reactions and alterations that occur during composting are analogous to those that biotransform organic detritus in soil to humus (Paul and Clark, 1989) except for the thermophilic stage, and therefore the speed of the bioconversion. Also, the dismutation or physical breakdown of the plant residues in or on soil by the soil fauna (centipedes, millipedes, springtails, mites, nematodes, earthworms, etc.) is limited in composting to the initial and maturation stages, owing to the characteristic heat-intolerance of cold-blooded animals. This, of course, does not apply to the mild temperature composting of wastes specifically by earthworms using, hopefully, only pathogen-free materials.

At the initial stage, mesophilic microorganisms that occur naturally on soil, dust particles and waste begin the decomposition process in which sugar-utilizing fungi and acid-producing bacteria are the most active in depleting soluble sources of carbon in the compost mix. The oxidation of the sugars, amino acids and alcohols generates heat. Temperatures beyond 45 °C inactivate or kill most mesophiles, except their spores. At this stage the more readily accessible polysaccharides, such as starch, chitins and pectins, proteins and lipids, may be hydrolysed by exoenzymes amylase, chitinase and pectinase, protease and lipase to release sugars, amino sugars, uronic acids, amino acids, alcohols and fatty acids. These fuel thermophilic fungal activity until the temperature goes beyond about 60 °C when the heat-tolerant actinomycetes begin their domination.

Many bacteria and fungi produce exoenzymes (extracellular enzymes) to hydrolyse insoluble amylose and amylopectins outside the parent cells to produce soluble compounds they can absorb and use intracellularly. α-amylases cleave amylose and amylopectins into glucose directly while β-amylase produces the dimer maltose. Both maltose and dextrins are generated from amylopectins. Pectins like other hemicelluloses are copolymers of hexoses, pentoses and, in some case, uronic acid. Their complete decomposition involves several enzymes called pectinases. A polysaccharide of special interest is cellulose, a major constituent of plant materials and municipal wastes (owing to their paper content). The degradation of a cellulose by bacteria and fungi requires not one but three major enzymes as a complex. Endo-β-1,4-glucanase hydrolyses insoluble crystalline polymers of cellulose randomly to release glucose and cellobiose. The smaller chains of glucose are then the substrate for the exo-β-1,4-glucanase which cleaves off cellobiose dimers from the nonreducing end of the polymer chain. The disaccharide cellobiose is hydrolysed by cellobiase into glucose. Thermophilic bacteria and fungi utilize cellulose mostly during the maturation process and while the temperature is below about 60 °C. At higher temperatures, actinomycetes dominate. They are specifically active in decomposting hemicelluloses.

A polymer that does not seem to have been researched extensively in this context is chitin (poly-*N*-acetyl glucosamine), and its homologue poly-*N*-acetyl muramic acid found in bacterial cells walls. Chitin is the main component of exoskeletons of crustaceans and of fungal cell walls. Lack of research in this area is surprising because faecal excreta contain as much as 50% of their carbon in microbial biomass and detritus while insects are commonly present in many urban and industrial wastes. Chitin decomposition is of special importance to the composting of shellfish wastes. Extracellular chitinase enzymes do occur in soils (Mathur, 1982).

Proteins are the major N-containing polymer in living bodies. Proteolytic enzymes hydrolyse the peptide links between amino-acid monomers of the protein polymer. Microbial proteases, pronase and subtilisin, depolymerize proteins by cleavage of terminal amino acids. Decomposition of most proteins in soils and composts should be fairly rapid except when the protein is highly insoluble. Keratin, the S-rich highly crosslinked insoluble protein in hair, feathers, hooves and horns, is degraded, albeit slowly, by actinomycetes and some fungi. The microbial proteins produced in soils and composts may resist decomposition owing to their protective tanning by free phenolic compounds and humic substances, and that is the reason for the desirable slow-release nature of part of the N in composts.

Monomers of both lipids and proteins are easily utilized by soil organisms. Lipids are depolymerized by the lipolytic enzymes, lipases, except when the lipids are water-insoluble waxy materials. Aliphatic compounds of long-chain fatty acids have lately been recognized to persist in soils where they may have a role in waterproofing soil aggregates (e.g. Preston *et al.*, 1986).

Another material that resists decomposition is lignin as it is not a specific polymer of identical repeating units and thus is not degraded by specific enzymes. Also its phenylpropanoid units or their phenolic products tend to be biostatic. In the latter stages of the thermophilic range, when fungi begin to return to the core from the cooler periphery, they transform and break by attrition the layers of lignin that encrust cellulose and hemicellullose fibres. The oxidized by-products of lignin attrition are semiquinones, hydroxyquinones and their derivatives, the aminoquinones. All these free radicals condense with each other and with amino acids, peptides and other N compounds to form the recalcitrant humic substances. The phenolic compounds that contribute to humus formation may also be microbial metabolites or products of other aromatic substances in the detritus.

The amino acids and other N-compounds are converted to ammonia which, in turn, is oxidized to nitrates by a specific group of mesophilic acid-intolerant autotrophic organisms. The reader is referred to Paul and Clark (1989) for the microbiology and biochemistry of soils, and to Poincelot (1975) for those of composts.

4.4 The technology of composting

This section outlines a number of aerobic 'hot' composting methods but excludes wormicomposting and anaerobic systems. In wormicomposting (Edwards *et al.*, 1985) the main objective is to harvest the nutrients as earthworms which then have to be used as hog, fish or fowl feed to recycle the nutrients in the waste. The composts cannot be allowed to be hot enough to kill the worms. The whole multistep system requires intensive management and extremely careful environmental controls.

The anaerobic or partly anaerobic systems were designed mainly to prevent loss of ammonia from night soil or to produce methane ('gobar gas', biogas), which is a subject in itself.

Aerobic systems of solid composting can be classified into open systems and confined, in-vessel, or container systems. The former are based on the Indore method of Howard and Wad (1931), severally modified and refined.

4.4.1 Open systems

(a) The Indore method. The Indore method requires a heap of trapezoidal cross-section. The heap is about 2 m wide at the bottom and 1.4 m wide at the top, with sloping sides 1.3 m high and a minimum of 2 m in length. The heap contains, starting with carbonaceous wastes at the bottom, 20 cm of carbon-rich and 10 cm of nitrogen-rich materials in layers. A depression at the top, in hot climate areas, traps rainwater. The heap is covered with soil or hay as thermal insulation. Where drainage channels are provided, the liquid is added back to the pile. The pile is turned over at 6- and 12-week intervals until the compost is mature. The Indore method also works when the heaps are built in partially filled trenches (allowing space for mixing and turning), as long as aeration ducts and drainage channels are provided.

(b) The simple turned windrow. A compost windrow should be a minimum of 1–5 m high and 2.5 m wide, of any convenient length. The materials are mixed and placed manually or mechanically. At the farm scale, such heaps can be built by a modified solid manure spreader which can also effect mixing and shredding of the materials. If the outside of the windrow does not have to be subjected to sanitizing heat, such as when composting fallen leaves, no steps are necessary for remixing and relaying the materials (turning) or for forcing air into the mass. This type of composting can, however, take a long time to be completed, up to two summers in temperate climate areas. To promote faster decomposition and better hygiene, and to further break the materials physically, any of many different types of turning equipment can be used. For certain types of mixtures of good structural strength, vertical aeration channels (chimneys) can be created by using rods which are pushed in, jiggled and then withdrawn (Biddlestone *et al.*, 1987). This may be repeated at the first few turnings.

Advantages of the windrow systems are the low capital costs, faster product maturation compared to the nonaerated static piles, and generally good quality of the product. Disadvantages are the cost of turning (where practised), odour emanation, vermin, the high requirements for land; loss of ammonia and dispersal of allergenic spores and dust during turning, and vulnerability to adverse conditions of temperature and precipitation.

Turning by power buckets or loaders is preferable to the use of windrow turners as the latter decrease porosity through maceration, thus diminishing air permeability well before compost maturation.

(c) The passive aeration windrow or natural aeration atatic pile system of composting. As recently summarized by Mathur *et al.* (1996), the small-scale passive aeration windrow system (PAWS) and the large-scale natural aeration static pile system (NASP) are both based on the following:

1. The heat generated by aerobic decomposition during composting is enough to energize air movement through the oxidizing mass due to convection, i.e. the chimney effect, particularly so if the medium is porous enough and air can enter at the base through pipes, or a plenum (with holes directed at the interior).
2. Hot vapours condense in cooler environments such as an envelope layer of peat or mature compost that is itself not undergoing carbon odixation as is the composting mass, thus providing more oxygen for aerobic transformation of malodorous nitrogen and sulphur compounds into odourless products, particularly because ammonia-oxidizing bacteria prefer cooler temperatures (Paul and Clark, 1989; Mathur, 1991).
3. As aerobic microbes can use only water-dissolved oxygen, and its concentration reaches its effective maximum in a composting system with the air containing about 5% oxygen, higher oxygen content, such as the 20.9% in ambient air, at high temperatures, only causes undesirable exothermic thermochemical reactions and overheating. However, as little as 2% oxygen is sufficient to transform malodours to inodorous products so that no bad smell will emanate as long as the mass is aerobic in the overall and tranquil, even if some anaerobic pockets exist transiently or spottily. More than a dozen field trials have confirmed this.

The abovementioned chimney effect moves air through the mass in PAWS and NASP both from the perforated floor or pipes at the base, and through the open sides of the compost. In addition, there is diffusion of oxygen such as from the sides as gas concentrations tend to equilibrate between connected volumes of air. The more important mechanism, however, is the chimney effect when the composting mass is hot and this is even greater in winter as the difference in temperatures within and outside the oxidizing mass is greater in winter than in summer. The higher temperature gradient in winter causes more flow through the holes at the base to the top even if the sides of compost piles freeze (but not if an inappropriately wet bottom layer freezes solidly).

Operations using the NASP composting utilize feedstocks that are originally free of inert contaminants or separated at source, or are immature composts from other systems that exclude or remove inert contaminants. A NASP can be built either on:

1. a perforated floor with about 3% openings of a minimum of ½ inch diameter or width over a basal plenum that is laterally open to the atmosphere; or
2. on a pad embedded with perforated pipes.

Pads of a minimum of 20 feet width supporting piles of 6–9 feet height are needed for all-season NASP composting in Canada. The pad can be made of either gravel, or clinkers, or overscreen (>½ inch) mature compost, or chips of weathered wood or bark, or chipped urban (or C&D) wood waste. The pad is 9–12 inch thick and contains 4 inch-diameter pipes with ½ inch round holes, 4 inch apart, in two rows, at 2 and 10 o'clock positions facing upwards. The pipes are placed across the width, 14 inch apart, and kept open at both ends.

Another essential feature of NASP is a 6–9 inch-thick envelope of mature compost or peat on the exterior of the compost pile. The envelope acts as a buffer and an integrated biofilter. A basal layer of mature compost or peat is needed only when the pile contains materials like whole fish that may exude liquids.

The compost piles are built by using a front-end loader or excavator, or a conveyor belt system. Routinely, the piles are monitored for temperature and percentage oxygen only. NH_3, pH, E_h and H_2S are also measured but only when a recipe, formulation or configuration is being tried at the location for the first time.

In times of up to 10 days, sections of the pile heat up owing to the biological use of the oxygen in the air within the mix. As the oxygen content declines and temperature rises, diffusion increases and the convection or chimney effect is triggered by the heat. The movement of air through the pipes or plena and the pile is then driven and controlled by the oxidative activity of the microorganisms themselves. Overheating is avoided naturally mainly by the nature of the process itself as it deters the presence of atmospheric level oxygen in areas where it may engender excessive chemical or biological oxidation that causes overheating. The oxidation is thus controlled by the living bodies in the compost, not by forced air. Therefore, there is no need to push air through to prevent or reverse overheating. There is no excessive drying because (1) the hot material is not exposed to the atmosphere and (2) it is covered by a cooler envelope.

Consequently, no environmental controls are needed during the process itself. Ameliorative measures that can be taken to counter any problems created by mistakes or mishaps have been devised and tested. For example, if the feedstock mix is or has become too tight or too wet, vertical vents can be created by a rod to aid the airflow that would kick-start the chimney effect.

NASPs of proper size, shape and content do not generate leachates. As the bottom plane is not in direct contact with ground, it does not become waterlogged by soaking up moisture by the wick effect, thus retaining its capacity to absorb any moisture coming from the upper layer. The composting interior is also never exposed directly to rain or snow. The envelope layer that is exposed has a high moisture absorption capacity as it is made of mature compost or peat. As neither mature compost nor peat have a significant proportion of their organic matter in forms that dissolve in water, water management systems for a NASP site need to be only like those for runoffs.

There is no need for structures to deal with condensates or air emissions. The only air that is emitted is naturally through the envelope layer. As the envelope is made of biostable material and therefore not heating up on its own, it is cooler than the interior. Consequently, hot vapours coming out of the interiors condense in the envelope where more oxygen is available, owing to diffusion and to lack of competitive oxygen demand of the organic matrix. As the emission is without artificial force, it does not, in a proper envelope, create channels through which odours may escape. Any smells of nitrogen and sulphur compounds present are captured and oxidized to innocuous products in the envelope layer.

NASP is a one-step composting process. Both the active and maturation phases are completed without disturbance to the pile, until temperature and oxygen content indicate biostability. When finished, the compost can be stored with or without screening in larger piles preferably with just a few natural aeration pipes at the base to avoid development of any anaerobic pockets. The rough (>½ inch) organic material that remains after screening can be used as a base or an envelope for a new pile or can be included with a new compost mix.

The moisture content and the physical state of the feedstocks are specially important in this method as, once set up, the compost mix is **not** disturbed by agitation, dumping, vibration, toppling, forced air or turning over until the product is mature.

The NASP described above is based on Agriculture Canada's PAWS (Mathur, 1991, 1992; Mathur *et al.*, 1996) that was specifically designed to compost, without odours, strong smelling fisheries wastes and manure slurries, and to require only low capital outlay, and low input of technical skill (Mathur *et al.*, 1986; Hayes *et al.*, 1993; NESFI-ACA, 1993). NASP avoids odour problems partly by not exposing the decomposing mass directly to outside air, and by postponing physical maceration until after aerobic biomaturation (Mathur *et al.*, 1996).

During the scaling up of the process, and its testing and demonstration for wastes from farming (13 different types of manures), food processing and consumption, forestry (pulp and paper mills), fisheries (22 different species), aquaculture, and urban activities (leaf and yard waste (L&YW), construction and demolition waste, wood and biosolids), it was discovered that this method generates a mature product faster than some others. That is perhaps because the system avoids both the loss of ammonia and the overheating that causes the caramelization or premature browning that delays maturation. Conservation of the nitrogen thus hastens maturation.

The area needed for NASP composting depends primarily on the bulk density of the material and the time required for process completion. For example, a foot length of a 23.5 feet wide pile of 9 feet height contains a 2.7 t mixture of biosolids with chipped brushwood or L&YW. PAWS piles of similar size containing manures with bedding, or L&YW, compost and mature within 4 months in Ottawa summers or winters.

Many operations in Canada, USA and elsewhere are using the PAWS or NASP for compost production, or for maturation of composts generated by other methods. For example, Lynch and Cherry (1996) recently found PAWS to be effective in composting manures even in winter in Idaho, USA, when the ambient temperatures were as low as −27 °C. Both the Audobon Society and the Agricultural Composting Association in the USA have published guidebooks on PAWS (NESFI-ACA, 1993).

(d) Forced-aerated static piles. A mixture of solid wastes is stockpiled in the open air and turned occasionally for aeration. This low-technology method is considered to waste both ammonia and energy. Both of these disadvantages have been substantially mitigated by mechanical forcing of air through perforated pipes at the bottom of the static pile by (1) continuous air suction, (2) continuous air blowing, (3) alternate blowing and sucking, or (4) intermittent air blowing to keep the temperature below 60 °C.

The static pile system with an air suction component was created by the US Department of Agriculture staff at Beltsville, for the US Environmental Protection Agency, to compost a mixture of sewage sludge and wood chips. Undecomposed wood chips are screened out and reused in a new pile (Epstein *et al.*, 1976). Air is sucked through continuously, at a rate, into the blower, of about 0.2 m^3 min^{-1} 1000 kg^{-1} of the composting mass to maintain an oxygen concentration of nearly 15% in the mixture. The inevitable odour in the exhaust air has to be filtered through piles or towers of mature compost. Several operations in the USA are based on this system.

Sucking in air provides the least amount of oxygen in the centre of the pile where it is needed most, it does not aid desiccation and it allows temperatures to exceed 60 °C – the maximum tolerated by most lignolytic and cellulolytic fungi. These problems were avoided in a system formulated by a team at Rutgers University, New Jersey (Finstein *et al.*, 1983). The Rutgers system involves blowing air through the compost mass in response to a thermistor and fan, both controlled by a microcomputer to maintain temperatures below 60 °C. To overcome the problem that temperatures in

Table 4.3 Examples of confined composting systems

System	Configuration	Aeration	Agitation
Dana	Horizontal flow	Forced	Continuous
Earp-Thomas	Vertical flow	Forced	Continuous
Ewesen	Horizontal flow	Forced	Continuous
Fairfield	Mixed flow	Forced	Continuous
Paygro	Walled container	Forced	Intermittent
Peabody	Vertical flow	Forced	Intermittent
Tollemache	Walled container	Natural and forced	Intermittent

the compost pile may not be high enough everywhere to sanitize the wastes, several systems now provide for sucking in air through the pile in the primary phase and then reversing the air flow in conjunction with temperature control.

4.4.2 In-vessel (or reactor confined) systems

Several such systems (Table 4.3) have been developed since the first in-vessel method was designed by Becarri and modified by Verdier and Bordas (Gotaas, 1956).

The main advantages of these systems are rapidity of the process (10–14 days), low requirements for land, complete process control and consistency of end-product. In many cases, part of the composting process and all of the maturation occurs in stockpiles outside the enclosed systems.

In-vessel systems were reviewed by Gotaas (1956), Gray *et al.* (1973) and de Bertoldi *et al.* (1985). The main feature of these systems is that the compost materials are mixed and the mass advanced and aerated automatically and mechanically. They are costly to install and operate, and need intensive and skilful management.

4.5 Criteria of compost maturity

Effective composting transforms organic wastes into a useful product which is rich in available forms of plant nutrients, poor in readily biodegradable carbon, devoid of phytoinhibitory substances and relatively free of plant pathogens. Subjectively, a mature compost should be: dark brown or black; granular, spongy or fibrous in feel; and smell mouldy or rather like good earth, reminiscent of the aroma from a sunbaked garden patch that has first received a summer shower (owing to the release of spores of thermophilic *Streptomyces*). In objective analysis, attributes of a compost should indicate that the purposes of the process have been achieved. A lack of reheating of the compost mix indicates that the amount of readily biodegradable C sources is insufficient to support thermophilic activity.

Biomaturity relates to many other concerns and is generally regarded to need urgent attention, particularly for evaluating claims of 'fast' composting. Immature composts pose fire and health hazards, cause odour and vector nuisance, and inhibit plant growth, but maturity is hard to determine and regulate (Mathur *et al.*, 1993b). Fires in piles of immature composts are particularly hard to put out as they are fueled by the methane produced by bacteria within the pile through anaerobic decomposition. Inbar *et al.*, (1990) contain references to studies showing that immature composts support proliferation of surviving or introduced human, animal and plant pathogens. At present there seems to be a consensus that, until a better test is available, biomaturity should be determined by a battery of three tests, i.e. C/N ratio, O_2 consumption rate and seed germination.

A recent study in Canada indicated that hot water extracts of immature composts of manures and shredded paper are coloured, owing to intermediate products of decomposition and/or new water-soluble humus partly solvated by ammonia, until maturity when the ammonia is oxidized, water-soluble organics are depleted and the humus autopolymerized oxidatively into solids. A simple, inexpensive and insubvertible test has been proposed, but not tested yet on composts from feedstocks other than manures and paper (Mathur *et al.*, 1993a; Schnitzer *et al.*, 1993).

It is noteworthy that the premises of the Canadian studies were recently confirmed by Keeling *et al.* (1994) in the UK. They worked with municipal solid waste (MSW). The premises were as follows.

1. Irrespective of the components of a composting mass, and regardless of the presence or absence of transitory or permanent bioinhibiting factors, a substantial portion of the organic content occurs in a readily bioavailable, i.e. water-soluble form, as long as the compost is immature.

2. Aerobic decomposition of organic matter involves both mineralization and humification. Mineralization to CO_2 and H_2O can proceed to a limited extent even in the absence of free oxygen as some microbes can utilize fixed oxygen (e.g. in NO_3^-, SO_4^{2-}, CO_3^{2-} and PO_4^{3-}). In contrast, true humification or formation of water-insoluble humus involves free oxygen radicals. A lack of formation of water-insoluble humus, or the presence of water-soluble humus, is therefore a characteristic of incomplete decomposition, as in peatlands.

3. Although the solution phase of an immature compost may contain various aliphatic acids, phenols and ammonia, its intermediate stage will always contain water-soluble humic substances which can be measured photometrically in ultraviolet or near-visible light regions without significant interference from iron compounds. Also, the presence of any readily auto-oxidizable free phenols will indicate lack of maturation of humus.

Any water-soluble aliphatic acids, amino acids, proteins and polysaccharides present will compete with humus for the metal ions that complex and coagulate humus; any free ammonia present will help solvate humus, as ammonium salts of humus are water-soluble. Similarly, excess of Na^+ and K^+ ions in solution, due to salinity, will help peptize and solvate humus.

4. Measurement of the optical density of the water extract of a compost at a fixed wavelength, as in a glucometer, will be inexpensive and easy to perform at any ordinary composting facility, as a test for compost maturity. Such a test may supplant the many current tests discussed below.

4.5.1 C/N ratio

The C/N ratio of a mature compost should ideally be about 10 but this is hardly ever achievable owing to the presence of recalcitrant organic compounds or materials which resist decomposition due to their physical or chemical properties. For example, peat does not decompose readily when added to mineral soils because even 75% moisture content in the total mass may be held too strongly physicochemically to be bioavailable. Lignin, on the other hand, resists decomposition because it does not have identical monomers and therefore no specific enzyme deploymerizes it, and because intermediate phenolic products of lignin decomposition tend to be biostatic or bioinhibitory. The critical point about the C/N ratio is, therefore, whether decomposition of the mature compost in soil will release mineral nitrogen or cause competition with plants for the N in soil solution. C/N ratios of 14–20 in mature composts are, therefore, acceptable as long as their further decomposition is slow and does not require additional N from the soil. Gotaas (1956) refers to studies where composting a material with a C/N ratio of 78 produced a biostable product with a C/N ratio of 35, but that is an exception.

4.5.2 Absence of plant inhibitors

Perhaps the best indicator of whether a compost is biomature is the absence of bioinhibitory aliphatic acids (Devleeschauwer *et al.*, 1981) and phenolics which can be assayed chromatographically or by seed germination tests. Small seeds placed on a filter paper soaking in a water extract of the underlying compost in a Petri dish should germinate to the same percentage level as those placed on the paper soaked in pure water (e.g. Mathur *et al.*, 1986).

4.5.3 Absence of human pathogens

For microbiological properties, Zucconi and De Bertoldi (1987) proposed that a compost should be considered hygienic if a 100 g sample contains no salmonellae, no infective parasitic ova and no more than 5×10^4 faecal coliforms and 5×10^5 faecal streptococci. In certain specific types of composts, other microbiological criteria may be advisable. For example, Mathur and Johnson (1987) tested fish wastes composts for the toxins produced by pathogenic microbes by using tissue cultures of Y-1C mouse adrenal tumour cells, vero (monkey kidney cells), Chinese hamster ovary (CHO) cells and the suckling mice test. These tests ensured that neither the pathogens nor the microbial toxigens that actually cause the diseases were present at significant levels in the mature composts. Some of the toxins were present in the raw materials and immature composts.

4.5.4 Other criteria

Other criteria one may use are related to the actual rate of decomposition of the compost and the amount of soluble humus present in the compost. The former may be limited in value owing to the presence of bioinhibitory metals or organic compounds in a compost, and the latter by the presence of metals that complex the humic substances into insoluble substances. Preston *et al.* (1986), therefore, subjected the whole composts of seafood wastes to nuclear magnetic resonance (NMR) spectroscopy to find that the proteins, lipids and chitin present in raw seafood wastes were not present in mature composts, and that the C, N, and P in the composts were distributed as in humus from peat or soils. NMR spectroscopy, where feasible, may be a highly reliable means of testing compost maturity.

4.6 Uses of composts

As implied earlier, the historical use of composts has been as a soil conditioner and fertilizer, to restore and maintain soil humus and productivity. Compost application is the backbone of true organic farming and some definitions of 'organically produced foods' that fetch premium prices, particularly from consumers concerned about and vulnerable to agrichemical residues, insist on the practice. Other major uses of composting are landscaping, turf production and stabilization of steep slopes in vineyards and orchards.

The composts used in mushroom culture are not the bioconverted mature product dealt with in this chapter but mostly bioavailable carbon-rich mixtures of various materials composited for the specific purpose. These mixtures may include the biologically produced composts, in the mature or immature state.

Recent impetus to the use of composts in or as media for greenhouse production of seedlings, flowers and vegetables comes from evidence that soil-borne pathogens are naturally controlled by their antagonists in mature compost (Hoitink *et al.*, 1987). Pudelski (1987) has recently described such horticultural use of composts, particularly in greenhouses where the heat generated in maturing compost at a lower layer of the on-ground growth medium helps save the energy used for heating the glasshouses in temperate climates. A recent study in Canada has shown that up to 75% of the peat used in greenhouse media can be replaced by a good-quality compost to increase plant growth and profits, and save wetland ecosystems (Mathur and Voisin, 1996). Some studies have indeed been made to examine the feasibility of using the heat produced during composting (Thostrup *et al.*, 1987) and the CO_2 generated by the decomposition to enhance photosynthesis in greenhouses.

Hoitink and Kuter (1984) predicted that the biological agents of pest control, particularly fungi and bacteria antagonistic to plant pathogens, may be grown and marketed in a compost medium. This may become a significant industrial market for composts as more enterprises that supply organic farmers and pesticide manufacturing companies respond to the new consumer and market conditions favourable to alternate agriculture with minimal or no chemical inputs.

Composts can be used as a base for biofilters to trap and biotransform foul odours from animal houses, abattoirs and industry into harmless products. The nitrates and phosphates produced in the compost biofilters can be leached and the leachate used for algae culture, or for irrigation of field and greenhouse crops.

Organic wastes have been used in aquaculture both to feed detritus-eating fishes (e.g. tilapia, carp) directly and to increase the production of phytoplankton, the main natural fish food algae. Composts are better than the raw wastes for this purpose as the former create much lower demand for biological oxygen in the water as it is already biostable. Consequently, composts do not significantly deplete the dissolved oxygen needed by the fish. Banerjee and Srinivasan (1983) demonstrated the advantage of compost application to fish ponds in India, and Edwards *et al.* (1983) in Africa, for tilapia culture.

4.7 Summary

Composting is the managed bioconversion of waste into a hygienic soil conditioner and fertilizer. Heightened awareness of it is due to the new ethos of sustainable development that blends economic and environmental considerations with a sense of intergenerational justice. Restoration and maintenance of soil humus conserves resources and represents one of the most essential and beneficial ecological processes.

To compost, one needs to match different materials to achieve ideal conditions of C/N ratio, pH, moisture content and redox. To optimize the oxidation needed to dissipate carbon and generate the sanitizing heat, one needs to mix and relay the material, or use various strategies of moving air through the composting mass or of moving the mass through air, in or out of enclosed systems. The costly mechanization and intensive skilful management required in high-technology methods have been the main reasons for the virtual limitation of the practice of composting to home gardens and to sanitization of residues containing pathogen-laced human wastes.

A new and inexpensive system of composting, the passively aerated windrow system, harnesses the heat produced during decomposition to move air naturally through the mass via specific types of pipes placed suitably at or near the base of the windrow (elongated triangular or trapezoidal piles). The pipes are open ended and have perforations only on the top. The need to invert the outside areas of the piles into the centre of the mass, to sanitize them with the heat there, is obviated by the use of hygienic materials such as mature compost as the base and the envelope of the compost. PAWS has been successfully used for seafood wastes and peat, and for manure slurries bulked with peat, and has now been scaled up and extended to all types of wastes as the natural aerated static pile on a perforated floor over an open plenum.

The criteria for a mature compost reflect achievement of the objectives of composting – the production of a biostable, hygienic, humus-rich product that benefits plants.

A multitude of chemical and biological processes combine to transform wastes into humus, microbial biomass, waxy compounds, and mineral nutrients such as NH_4^+, NO_3^-, $P_2O_5^-$ and SO_4^{2-}.

Composts can be used:

- to condition soils, fertilize plants, and stabilize slopes;
- to grow mushrooms, nursery plants, ornamental or vegetable crops in greenhouse;
- as biofilters of odours from certain industries;
- as a feed for detrivorous fishes in aquaculture and a promoter of plankton growth;
- as a medium for mycorrhizae, and organisms used for their antagonism to plant pathogens.

References

Acharya, C.N. (1954) *Indian Farming*, **4**, 9.

Acharya, C.N., Sabnis, C.V. and Menezes, F.G.T. (1945) *Ind. J. Agric. Sci.*, **15**, 214–21.

Alexander, M. (1977) *Introduction to Soil Microbiology*, John Wiley, New York.

Altieri, M.A. (1987) *Agroecology: The Scientific Basis of Alternative Agriculture*. IT Publications, London, p. 227.

Banerjee, R.K. and Srinivasan, K.V. (1983) *Agric. Wastes*, **7**, 209.

Barrington, S.F., Scheupp, P., Cap, R. and Blanchette, J. (1990) *Proceedings of the Sixth International Symposium on Agricultural and Food Processing Wastes*, Chicago, IL, American Society for Agricultural Engineering Publication #05–90, pp. 434–441.

Bellamy, K., Smith, D.K., Meadd, E., Varangu, L. and Buggelin, R.G. (1992) In *Proceedings of the Second Annual Meeting, Composting Council of Canada*, Ottawa, pp. 101–128.

Biddlestone, A.J., Gray, K.R. and Day, C.A. (1987) In *Environmental Biotechnology* (eds C.F. Forster and D.A.J. Wase), Ellis Horwood, Chichester, p. 136.

Bollen, G.J. (1985) In *Composting of Agricultural Wastes* (ed. J.K.R. Gasser), Elsevier Applied Science, London, p. 282.

Chance, B. (1957) *Fed. Am. Soc. Exp. Biol. Fed. Proc.*, **16**, 671–80.

Chen, Y. and Hadar, Y. (1987) In *Compost: Production, Quality and Use* (eds M. DeBertoldi, M.P. Ferranti, P. L'Hermite and F. Zucconi), Elsevier Applied Science, London, p. 71.

Cooper, J.M. and Warman, P.R. (1993) *Proceedings of the Third Annual Meeting, Composting Council of Canada*, Ottawa, pp. 247–64.

CWC (The Clean Washington Centre) (1993) *Best Management Practices Manual for Compost Produce Waste and Wax Coated Cardboard Using a Low Technology Approach*, Report No. B8.

Dalzell, H.W., Biddlestone, A.J., Gray, K.R. and Thurairajan, K. (1987) *FAO Bull.*, **56**, 177.

de Bertoldi, M., Vallini, G. and Pera, A. (1985) In *Composting of Agricultural and Other Wastes* (ed. J.K.R. Gasser), Elsevier Applied Science, London, p. 27.

de Bertoldi, M., Ferranti, M.P., L'Hermite, P. and Zucconi, F. (eds) (1987) *Compost: Production, Quality and Use*. Elsevier Applied Science, London, p. 853.

de Bertoldi, M., Rutili, A., Citterio, B. and Civilini, M. (1988) *Waste Man. Res.*, **6**, 239–59.

Derikx, P.J.L., Op den Camp., H.J.M., vand der Drfit, C., van Griensven, L.J.L.D. and Vogels, G.D. (1990) *Appl. Environ. Microbiol.*, **56**, 176–80.

Devleeschauwer, D., Verdonk, O. and Van Assche, P. (1981) *Biocycle*, **22**, 44–46.

Diaz, L.F. (1987) *Biocycle*, **28**, 54–8.

Edwards, C.A., Burrows, I., Fletcher, K.E. and Jones, B.A. (1985) In *Composting of Agriculture and Other Wastes* (ed. J.K.R. Gasser), Elsevier Applied Science, London, p. 229.

Edwards, P., Polprasert, C., Pacharaprakiti, C., Rajput, V.S. and Suthirawut, S. (1983) *Agriculture*, **32**, 409.

Elliott, L.F., Schuman, G.E. and Viets, F.G., Jr (1971) *Soil Sci. Soc. Amer. Proc.*, **35**, 752.

Epstein, E., Willson, G.B., Burge, W.D., Mullen, D.C. and Nekri, N.K. (1976) *J. Water Pollut. Control Fed.*, **48**, 688.

Finstein, M.S. and Miller, M.C. (1985) In *Composting of Agricultural and Other Wastes* (ed. J.K.R. Gasser), Elsevier Applied Science, London, p. 243.

Finstein, M.S., Miller, F.C., Strom, P.F., MacGregor, S.T. and Psarianos, K.M. (1983) *Biotechnology*, **1**, 347–53.

Gotaas, H.B. (1956) *Composting Sanitary Disposal and Reclamation of Organic Wastes.* World Health Organization, Monograph Series, No. 31, WHO, Geneva.

Grant, C.D. and Groenvelt, P.H. (1993) In *Canadian Society of Soil Science* (ed. M.R. Carter), Lewis Publishers, Ottawa.

Gray, K.R., Biddlestone, A.J. and Clark, R. (1973) *Process Biochem.*, **8** (10), 11.

Haug, R.T. (1986) *Biocycle*, October, 53–7.

Haug, R.T. (1990) *Biocycle*, October, 60–7.

Hay, J.C. and Kuchenrither, R.D. (1990) *J. Environ. Eng.*, **116**, 746–63.

Hayes, L.A., Richards, R. and Mathur, S.P. (1993) *Proceedings of the Third Annual Meeting of the Composting Council of Canada*, Ottawa, p. 135–54.

Hogan, J.A., Miller, F.C. and Finstein, M.S. (1989) *Appl. Environ. Microbiol.*, **55**, 1082–92.

Hoitink, H.A.J. and Kuter, G.E. (1984) *Biocycle*, May/June, 40.

Hoitink, H.A.J., Chen, W., Trillas-Gay, M.I. and Chung, Y.R. (1987) In *Compost: Production, Quality, and Use* (eds M. De Bertoldi, M.P. Ferranti, P. L'Hermit and F. Zucconi), Elsevier Applied Science, London, p. 414.

Howard, A. (1943) *An Agricultural Testament*, Oxford University Press, London.

Howard, A. and Wad, Y.D. (1931) *The Waste Products of Agriculture: Their Utilization as Humus*, Oxford University Press, London.
Hyatt, G.W. and Richard, T.L. (1992) *Biomass Bioenergy*, **3**, 121–5.
Inbar, Y., Chen, Y., Hadar, Y. and Hoitink, H.A.J. (1990) *Biocycle*, December, 64–9.
Ishii, H., Tanaka, K., Aoki, M., Murakami, T. and Yamada, M. (1991) *Water Sci. Technol.*, **23**, 1979–89.
Jacob, M.T. (1993) *Proceedings of the Third Annual Meeting of the Composting Council of Canada*, Ottawa, pp. 275–94.
Jeris, J.S. and Regan, R.W. (1973) *Compost Sci.*, **14**, 10.
Jiminez, E.I. and Garcia, V.P. (1989) *Biol. Wastes*, **27**, 115–42.
Keeling, A.A., Mullett, J.A. and Paton, I.K. (1994) *Soil Biol. Biochem.*, **26**, 773–6.
King, F.H. (1911) *Farmers of Forty Centuries or Permanent Agriculture in China, Korea and Japan*, Harcourt Brace and Co., New York, p. 441.
Kissel, J.C., Henry, C.L. and Harrison, R.B. (1992) *Biomass Bioenergy*, **3**, 181–94.
Koe, L.K. and Ng, W. (1987) *Water Air Soil Pollution*, **33**, 199–204.
Kononova, M.M. (1966) *Soil Organic Matter*, Pergamon Press, Oxford.
Krishna Murthy, R. (1978) *A Manual on Compost and Other Organic Manures*, Today and Tomorrow Printers and Publishers, New Delhi.
Kuchenrither, R.D., Martin, W.J., Smith, D.G. and Williams, D.W. (1985) *J. Water Pollution Cont. Fed.*, **57**, 217–19.
Leger, D.A. Patnki, N.K. and Ho, S.K. (1991) *Proceedings of a National Workshop on Land Application of Animal Manure*, Canadian, Agriculture Research Council, Ottawa.
Leton, T.G. and Stentiford, E.I. (1990) *Waste Man. Res.*, **8**, 299–306.
Lindau, C.W., Patrick, W.H., Jr and DeLaune, R.D. (1993) In *Agricultural Ecosystem Effects on Trace Gases and Global Climate Change*. American Society of Agronomy, Special Publication, Vol. 55, pp. 157–66.
Linteau, J.-P. and Beauregard, B. (1992) *Proceedings of the Second Annual Meeting*, Composting Council of Canada, Ottawa, pp. 193–230.
Lopez-Real, J. and Foster, M. (1985) In *Composting of Agricultural and Other Wastes* (ed. J.K.R. Gasser), Elsevier Applied Science, London, p. 291.
Lynch, N.J. and Cherry, R.S. (1996) *Compost Sci. Util.*, **4**, 44–52.
Mahoney, E.M. and MacFarlane, S. (1992) *Proceedings of the Second Annual Meeting*, Composting Council of Canada, Ottawa, pp. 129–42.
Martin, A.M. and Patel, T.R. (1991) In *Bioconversion of Waste Materials to Industrial Products* (ed. A.M. Martin) 1st edn, Elsevier Applied Science, London, pp. 417–40.
Mathur, S.P. (1982) *Transactions of the 12th Congress of the International Society for Soil Science*, Vol. 1, p. 125.
Mathur, S.P. (1991) In *Bioconversion of Waste Materials to Industrial Products* (ed. A.M. Martin), 1st edn, Elsevier, London, New York, pp. 147–86.
Mathur, S.P. (1992) *Second Annual Meeting of The Composting Council of Canada*, Ottawa, Environment Canada, pp. 387–97.
Mathur, S.P. (1993) *Proceedings 1993, Newfoundland Peat Opportunities – An International Conference*, Corner Brook, Newfoundland, Newfoundland and Labrador Peat Association, paper #17, p. 20.
Mathur, S.P. (1994) *Proceedings of the 4th Annual Meeting of the Composting Council of Canada*, Toronto, Environment Canada, pp. 259–79.
Mathur, S.P. (1995) *Composting Technologies and Practices. A Guide for Decision Makers*. The Composting Council of Canada, Ottawa.
Mathur, S.P. and Farnham, R.S. (1985) In *Humic Substances in Soil Sediment and Water* (eds G.R. Aiken, D.M. McNights, R.L. Warshaw and P. MacCarthy), John Wiley, New York, p. 53.
Mathur, S.P. and Johnson, W.M. (1987) *Biol. Agric. Hort.*, **4**, 235.
Mathur, S.P. and Voisin, B. (1996) *The Use of Compost as a Greenhouse Media*, Ontario Ministry of Environment and Energy, PIBS 3458E, Queens Printer for Ontario, p. 37.
Mathur, S.P., Daigle, J.-Y., Levesque, M. and Dinel. H. (1986) *Biol. Agric. Hort.*, **4**, 27.
Mathur, S.P., Proulx, J.G., Levesque, M. and Sanderson, R.B. (1987) In *Agrogeology in Africa*. Commonwealth Science Council, Technical Publication Series no. 226, p. 129.
Mathur, S.P. Daigle, J.-Y., Brooks, J.L., Levesque, M. and Arsenault, J. (1988a). *Biocycle*, **29**, 44.

Mathur, S.P., Patni, N.K. and Levesque, M.P. (1988b) *Canadian Society of Agricultural Engineering Annual Meeting*, Paper no. 88–123.

Mathur, S.P., Proulx, J.G. and Daigle, J.-Y. (1989) *Proceedings of the 1989 International Peat Society Sumposium, Québec City* (eds R.P. Overend and J.K. Jeglum), International Peat Society, Helsinki.

Mathur, S.P., Patni, N.K. and Levesque, M.P. (1990a) *Biol. Waste*, **34**, 323.

Mathur, S.P., Schnitzer, M. and Schuppli, P. (1990b) *Biol. Agric. Hort.*, **7**, 153–63.

Mathur, S.P., Dinel, H., Owen, G., Schnitzer, M. and Dugan, J. (1993a) *Biol. Agric. Hort.*, **10**, 87–108.

Mathur, S.P., Owen, G., Dinel, H. and Schnitzer, M. (1993b) *Biol. Agric. Hort.*, **10**, 65–85.

Mathur, S.P., Voisin, B. and Riddell, A. (1996) *Proceedings of the Annual Meeting of the Composting Council of Canada*, Toronto, 21 pp (in press).

McCauley, R. and Shell, B.J. (1956) *Proceedings of the 11th Industrial Waste Conference*, Purdue University, USA, p. 4536–453.

Meyboom, P. (1993) *Proceedings of the Third Annual Meeting of the Composting Council of Canada*, Ottawa, pp. 1–18.

Miller, F.C. (1989) *Microb. Ecol.*, **18**, 59–71.

Miller, F.C. and Macauley, B.J. (1988) *Aust. J. Exp. Agric.*, **28**, 553–60.

Miller, F.C., Harper, E.R. and Macauley, B.J. (1989) *Aust. J. Exp. Agric.*, **29**, 741–50.

Miller, F.C., Harper, E.R., MaCauley, B.J. and Gulliver, A. (1990) *Austral. J. Exp. Aric.*, **30**, 287–96.

Miner, J.R. and Hazen, T.E. (1969) *Trans. Amer. Soc. Agric. Eng.*, **12**, 772.

Miner, J.R. and Hazen, T.E. (1977) In *Soils for Management of Organic Wastes and Waste Waters* (eds L.F. Elliott and F.J. Stevenson), American Society of Agronomy, Madison, WI, p. 379.

Nakasaki, K., Nakano, Y., Akiyama, T., Shoda, M. and Kubota, H. (1987) *Ferment. Technol.*, **65**, 43–8.

NESFI-ACA (New England Small Farms Institute and the Agricultural Composting Association of America), Belchertown, MA (1993), p. 25.

Op den Camp, H.J.M. (1987) *De Champignoncultuur*, **31**, 513–19. [Cited by Miller *et al.* (1989).]

Paul, E.A. and Clark, F.E. (1989) *Soil Microbiology and Biochemistry*, Academic Press, San Diego.

Poincelot, S.P. (1975) *Biochemistry and Methodology of Composting.* Connecticut Agricultural Experimental Station Bulletin, 754.

Preston, C.M., Ripmeester, J.A., Mathur, S.P. and Levesque, M. (1986) *Can. J. Spectrosc.*, **31**, 63.

Pudelski, T. (1987) In *Compost: Production, Quality and Use* (eds M. De Bertoldi, M.P. Ferranti, P. L'Hermite and F. Zucconi), Elsevier Applied Science, London, p. 20.

Reddy, K.R. and Patrick, W.H. Jr (1975) *Soil Biol. Biochem.*, **7**, 87–94.

Richard, J.L. (1992) *Biomass Bioenergy*, **3**, 163–80.

Rynk, R. (ed.) (1992) *On-farm Composting Handbook. Cooperative Extension, Northeast Regional Agricultural Engineering Service*, NRAES #54, Ithaca, NY, p. 186.

Schnitzer, M., Dinel, H., Mathur, S.P., Schulten, H.R. and Owen, G. (1993) *Biol. Agric. Hort.*, **10**, 109–23.

Schroder, H. (1985) *Agric. Ecosyst. Environ.*, **14**, 279.

Smith, J.E. (1986) M.Sc. thesis, University of Guelph.

Summer, W. (1971) *Odour Pollution of Air: Causes and Control*, CRC Press, p. 310.

Swanson, G.R., Dudley, E.G. and Williamson, K.R. (1980) In *Handbook or Organic Waste Conversion* (ed. M.W.M. Bewick), Van Nostrand Reinhold, New York, p. 281.

Thostrup, P. (1985) In *Composting of Agricultural and Other Wastes* (ed. J.K.R. Gasser), Elsevier Applied Science, London, p. 167.

Thostrup, P., Jespersen, L.M. and Jensen, P.E.B. (1987) In *Compost Production, Quality and Use* (eds. M. de Bertoldi, M.P. Ferranti, P. L'Hermite and F. Zucconi), Elsevier Applied Science, London, p. 147.

University of California (1953) *Reclamation of Municipal Refuse by Composting*. Technical Bulletin no. 9, Berkeley, CA.

Van der Hoek, K.W. and Ooesthoek, J. (1985) In *Composting of Agricultural and Other Wastes* (ed. J.K.R. Gasser), Elsevier Applied Science, London, p. 271.

Voorburg, J.H. (1988) *Pig production and environment.* Institute of Agricultureal Engineering, Wageningen, The Netherlands, Xnr 11890, 14.
Waksman, S.A. (1938) *Humus*, Williams and Wilkins Co., Baltimore.
Walker, J.W.M. (1991) *Biocycle*, September, 50–5.
Wiley, J.S. (1957) In *Proceedings of the 12th Industrial Wastes Conference*, Purdue University, USA, Engineering Bulletin, Vol. 17, 596.
Wiley, J.S. and Spillane, J.T. (1962) *Compost Sci.*, **2**, 18.
Williams, T.O. and Miller, F.C. (1992a) *Biocycle*, October, 72–7.
Williams, T.O. and Miller, F.C. (1992b) *Biocycle*, November, 75–8.
Witter, E. and Lopez-Real, J.M. (1987) *Biol. Agric. Hort.*, **5**, 1.
Witter, E. and Lopez-Real, J.M. (1988) *Biol. Wastes*, **23**, 279–4.
Zhan, W., Fernandes, L., Patni, N. and Jui, P. (1992) *Proceedings of the 2nd Annual Meeting of the Composting Council of Canada*, Ottawa, pp. 233–54.
Zucconi, F. and De Bertoldi, M. (1987) In *Compost: Production, Quality and Use* (eds M. De Bertoldi, M.P. Ferranti, P. L'Hermite and F. Zucconi), Elsevier Science, London, p. 30.

Part 2
Bioconversion Applications

5 Bioprocessing of agro-residues to value added products

V.S. BISARIA

5.1 Introduction

Vast quantities of agricultural and agro-industrial residues that are generated as a result of diverse agricultural and industrial practices represent one of the most important energy-rich resources. Accumulation of this biomass in large quantities every year results not only in deterioration of the environment but in a loss of potentially valuable material which can be processed to yield a number of value added products such as food, fuel, feed and a variety of chemicals. A few of these potential residues are listed in Table 5.1. While cellulose, hemicellulose and lignin are the major constituents of these lignocellulosic residues, minor quantities of protein, pectin, soluble sugars, vitamins and minerals are also present. The composition of these residues, which depends on the age of the plant, is also different in different parts of the same crop, such as stalks, stems, straws, hulls and cobs. A correct analysis of the carbohydrates and other chemical components of lignocellulosic residues is of utmost importance for the evaluation of effectiveness of pretreatment method, product yield and process design. The limitations involved in comparison of data from different laboratories have been discussed by Saddler and Makie (1990). Using recent methods for their analysis, the carbohydrate components are first hydrolysed to monomeric sugars which are then estimated by gas chromotography (GC) or high-pressure liquid chromatography (HPLC). Based on the monomeric sugars, the relative amounts of the corresponding polymers from which they are derived, i.e., cellulose and hemicellulose, are calculated stoichiometrically (Karr *et al.*, 1991; Puls, 1993; Hayn *et al.*, 1993). This procedure also allows the determination of relative amounts of soluble sugars in the lignocellulosic biomass. The composition of a few lignocellulosic residues is summarized in Table 5.2.

The complete utilization of lignocellulosic residues continues to remain an intractable problem. Effective chemical procedures exist which can fractionate the lignocellulose into its major constituent components and further hydrolyse them into its essential building blocks, i.e. glucose from cellulose, pentoses and hexoses from hemicellulose, and phenylpropanoid units from lignin. However, the chemical processes, besides being energy

Table 5.1 A few potential biodegradable agricultural and agro-industrial residues

	Source	Residues
1.	Agricultural crops (rice, wheat corn, maize, tapioca, coconut, bamboo, etc.)	Rice straw, wheat straw, maize stalks, castor stems, tapioca stem, banana stem, coconut stem, cotton sticks, corn cobs, bamboo dust, barley straw, guar straw
2.	Agro-industrial wastes and by-products:	
	Rice milling industry	Rice husk, rice bran
	Sugar industry	Bagasse, pith
	Cotton processing industry	Cotton linters, cotton seed hulls, cotton gin wastes
	Jute industry	Jute sticks, jute mill wastes
	Sawmill industry	Sawdust, wood chips, bark, shavings
	Coconut industry	Coconut husk, shell and pith

Source: Bisaria (1991).

Table 5.2 Typical composition for a few biomass materials

Biomass	Cellulose (%)	Hemicellulose (%)	Lignin (%)
Wheat straw	30.0	50.0	15.0
Rice straw	35.0	25.0	17.0
Corn cobs	45.0	35.0	15.0
Corn stalks	35.0	25.0	35.0
Guar straw	28.0	23.0	16.0
Maize straw	32.0	28.0	13.0
Cane bagasse	33.0	22.0	14.0
Oat straw	49.3	25.0	18.0
Spruce	42.0	26.5	28.5
Aspen	50.89	28,7	15.5
Red maple	39.0	33.0	23.0
Hardwood	45.8	30.7	20.3
Softwood	43.8	24.5	29.5

Sources: Ladisch *et al.* (1983), Hawley *et al.* (1986), Bisaria *et al.* (1987a), Parisi (1989).

intensive, require expensive corrosion-resistant equipment, extensive washing and disposal of chemical wastes. Severe conditions of chemical treatments may also result in partial degradation of desired end-products to toxic byproducts, e.g. furfural in acid hydrolysis. Enzymatic treatment of the residues, on the other hand, results in specific release of end-products and can be achieved at moderate conditions of pH and temperature, obviating the need for expensive equipment. However, it is not realistic at present to consider the enzymatic processes as the competing ones, because of their low catalytic efficiencies, especially towards crystalline cellulosic substrates. A future possibility may lie in the development of a hybrid process which uses a chemical treatment to accomplish the fractionation of the residues and an enzymatic treatment for the conversion of a specific component (such as cellulose) to a desirable product, e.g. glucose.

In a few cases, it may be more important and economical to recover a useful product, which may be a minor constituent of the whole crop, before processing the lignocellulose for bioconversion. This seems to be applicable to a number of 'botanochemical' crops which can be processed to give useful products such as soluble polyphenols, whole plant oils, isoprene polymers, paper-making fibres, animal feed and fermentation substrates. A few botano-chemical crops and their components which can be extracted for various applications have been reported (Buchanen *et al.*, 1980; Dale, 1987).

Because of the drawbacks associated with acid hydrolysis of lignocellulosics, extensive research efforts have been directed towards their enzymatic hydrolysis. The enzymes catalysing the breakdown of cellulose (i.e. cellulases), hemicellulose (i.e. hemicellulases such as xylanases, mannanases, etc.) and lignin (i.e. lignin peroxidase, manganese peroxidase and other oxidases) possess high specificity and act at moderate temperatures of 30–50 °C. A number of fungi and bacteria produce these enzymes in extracellular fluid which can be employed in various bioconversion processes to get useful products.

Before using any bioprocessing step for utilization of cellulose, it is important to realize that cellulose constitutes generally less than half of the total lignocellulose and, therefore, any bioconversion process based on cellulose utilization alone is not likely to be attractive and economical. This is because of (1) the lost revenue from hemicellulose or lignin (or other constituents), and (2) the added waste disposal costs. Hemicellulose and lignin and their degradation products find a number of uses which are listed in Table 5.3 along with those of cellulose. Cellulose utilization should accordingly be linked to the effective use of the lignocellulose constituent of highest possible product value, so that glucose production does not carry the entire processing and raw material costs. In the utilization of lignocellulosic wastes, therefore, an integrated approach may aim at removal of hemicellulose (which is easily recoverable by mild acid hydrolysis), and conserving the value of lignin and protein which have higher base value than sugars. It was shown by Dale (1987), for example, that the various protein products available from integrated conversion of alfalfa were approximately twice as valuable as ethanol derived from hydrolysis and fermentation of alfalfa cellulose.

The range of products that can be derived from the agricultural residues are many, depending on the type of bioprocessing steps employed for their conversion. Figure 5.1 outlines the preferred major processing routes for

Table 5.3 Applications of products derived from cellulose, hemicellulose and lignin

	Products	Applications
I	**From cellulose**	
1.	Glucose	Ethanol Fructose (sweetener) Single cell protein Protein concentrate Yeast glycan Nucleic acids Autolysates Chemicals 2,3-Butanediol Acetone/butanol
2.	Cellobiose/cellodextrins	Research chemicals
3.	Ethanol	Chemicals Solvents Beverages Biological energy (ATP) Petrochemical synthesis (via ethylene) Gasoline diluent and octane enhancer Food and feed
II.	**From hemicellulose**	
1.	Xylose	Ethanol Xylitol(sweetener) SCP
2.	Xylobiose/xylodextrins	Research chemicals Reduced-calorie bulking agents Water retention in food products
3.	Furfural and 2,3-butanediol	Chemicals and solvents (such as methyl ethyl ketone)
III.	**From lignin**	
1.	Polymeric lignins	Boiler fuel Adhesive Dispersant Complexing agent Stabilizer Antioxidant Precipitant Deflocculant Coagulant Pesticide carrier Emulsion stabilizer Soil conditioner Rubber reinforcement Phenol–formaldehyde resin and extender Urea–formaldehyde resin and extender Polyurethane foam

Table 5.3 *Continued*

	Products	Applications
2.	Chemical intermediates	Phenols Cresols Catechols Vanillin Vanillic acid Syringaldehyde Syringic acid Dimethyl sulfide Dimethyl sulfoxide Methyl mercaptan Ferulic acid Lignosulfonates (additives in stabilizers, grinding aids, binders) Butanol–lignin slurry (furnace and diesel fuel)
IV.	**From lignocellulose** (Untreated or pretreated)	Edible mushrooms Animal feed Methane (biogas) Fertilizer

Sources: Bisaria and Ghose (1981), Hawley *et al.* (1986), Linder and Wegener (1990), Viikari *et al.* (1993).

conversion of these residues into food (mushrooms), fuel (methane), feed and chemicals. Consideration of each one of them is clearly out of the scope of the present article. The discussion is, therefore, confined to the production of glucose, ethanol and a few other chemicals of industrial importance, which can be derived from cellulosic residues. Also included in the discussion are the properties, production and applications of cellulases and xylanases which modify and hydrolyse lignocellulosic residues. It is worth noting that, amongst the various microbial or enzymatic processes utilizing these residues, the ones involving mushrooms are the only processes which are commercially viable at present and utilize either unmodified (as in the case of *Pleurotus* or *Volvariella* cultivation) or partially modified (as in the case of *Agaricus* cultivation) residues to convert them into food (Zadrazil and Grabbe, 1983; Bisaria *et al.*, 1987b).

5.2 Characteristics of lignocellulosic materials and their pretreatment

5.2.1 Lignocellulosic materials

As mentioned above, cellulose is one of the three major components of lignocellulosic residues along with hemicellulose and lignin; other compounds such as plant oils, proteins and ash make up the remaining fraction

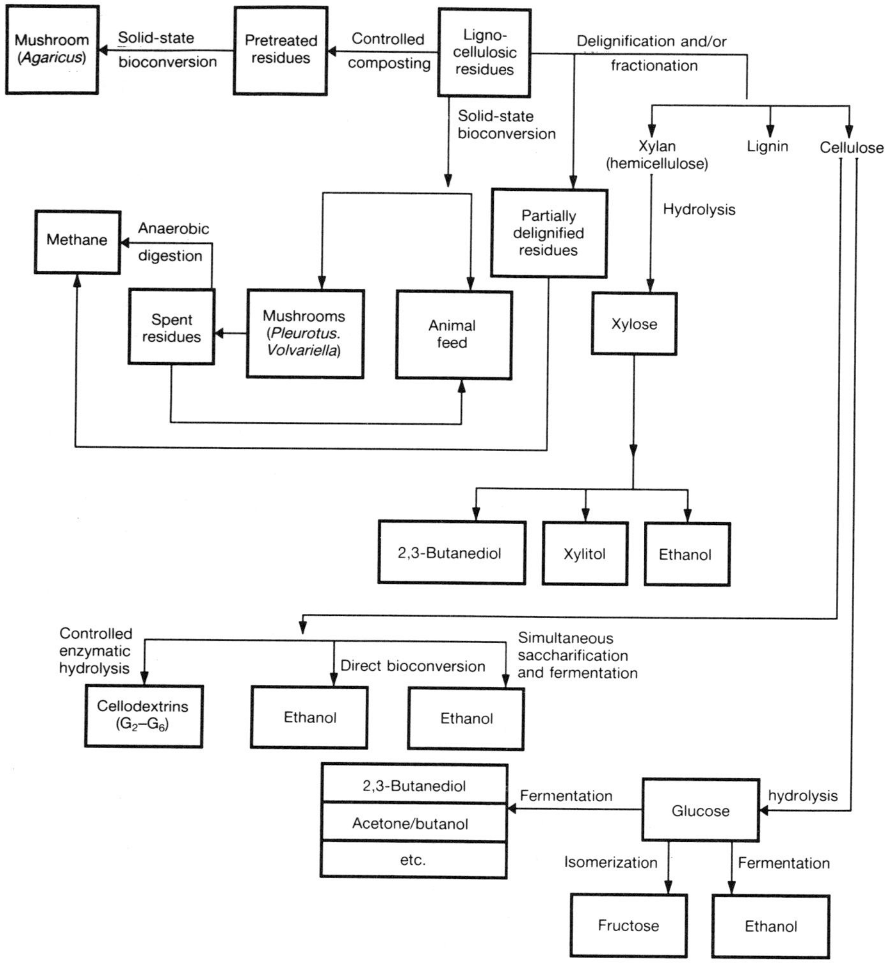

Figure 5.1 The preferred major processing routes for bioconversion of agro-residues into food, feed, fuels and chemicals. Source: Bisaria (1991).

of the biomass (Wyman and Goodman, 1993; Wyman, 1994). Cellulose is a linear polymer of D-glucose residues (building blocks), held together by β1–4-glucosidic linkages. The number of D-glucose units in a cellulose molecule may be as large as 15 000. The average degree of polymerization (DP) may vary from about 50 to a few thousand depending on its source and prior treatment.

The β1–4 linkage causes the polyglucose molecules to adopt a fully extended linear configuration which is attained by flipping of each glucose unit 180° relative to the preceding one. The glucosidic linkage acts as a functional group and this along with the hydroxyl groups mainly determines the chemical properties of cellulose. Also, the intramolecular hydrogen bonds stabilize each cellulose chain. The linear cellulose molecules pack together to form rod-like structures known as elementary fibrils or protofibrils or micelles, which are stabilized by intermolecular hydrogen bonds between adjacent cellulose chains. Micelles, the smallest structural unit, are packed into microfibrils which are each about 10 nm in diameter and up to several micrometers in length. About 10 microfibrils in turn are packed together to give rise to macrofibrils (approximately 50 nm in diameter) which can be seen by an optical microscope.

Lignocellulosic residues also contain hemicellulose and lignin. It has been proposed (Fengel and Wegner, 1984) that the microfibrils, which are about 12 nm in cross-section and about 30 nm in length, are surrounded by hemicelluloses and lignin in wood. The elementary fibrils are also surrounded by monolayers of hemicelluloses. The precise manner in which the native cellulose molecules are aggregated to form the elementary fibril structure is not yet well understood. In the fringe-micellar model, the extended cellulose crystallites have been segmented into crystalline and paracrystalline (or amorphous) regions, whereas in the folding-chain model, each cellulose chain is not extended through a series of several crystalline and paracrystalline regions but is predominantly folded within a single cellulose crystallite (Chang, 1971; Rowland and Roberts, 1972). Pretreatment of cellulosic materials, aimed at degrading the protective layers of hemicelluloses and lignin around elementary fibrils and microfibrils, is essential to rapid enzymatic hydrolysis.

The molecular structural features of cellulose, the elementary fibril and the microfibril are important for enzymatic degradation of cellulose. The proportion of crystalline and amorphous regions ranges from 50% to 90% in different materials; about 70% crystallinity exists in native cellulose. Generally, the amorphous regions in the fringe micelles are attacked first in a partially crystalline cellulose fibril, leading to an enrichment of crystalline regions which are, in turn, gradually solubilized after loosening of their peripheral parts. The ability of the cellulolytic microorganisms or that of enyzmes to hydrolyse cellulose varies greatly with the nature of the cellulosic materials.

The second major constituent of lignocellulosic biomass is hemicelluloses, the amorphous polysaccharides made up of xylose, galactose, mannose, arabinose, glucose and their uronic acids (Dekker, 1985; Visser *et al.*, 1992). These occur as heteroglycans containing different types of sugar residues, often as short monosaccharide or oligosaccharide appendages linked to the main backbone chain. The proportions of these pentoses and hexoses vary widely in the hemicelluloses of different biomass origin. Coniferous woods, for example, have a higher proportion of glucomannans, and deciduous woods a higher proportion of xylans and arabinoxylans. Agricultural residues, though more complex in hemicellulose composition, in general, have an important bearing on further processing of their enzymatic hydrolysis product. Mannose, for example, is readily fermentable by ordinary yeasts whereas xylose and arabinose can be utilized only by a few yeasts. Further, hemicelluloses appear to be associated with both cellulose and lignin in the microfibrillar structure (Fengel and Wegner, 1984).

Lignin, the third major constituent of lignocellulosic residues, is an amorphous three-dimensional polymer composed of three phenylpropanoid units – coniferyl alcohol, sinapyl alcohol and *p*-coumaryl alcohol (Crawford, 1981). Hemicellulose has been presumed to act as a molecular bonding agent between cellulose and lignin (Torrie, 1991). The monomers in the lignin molecule are not linked in the same way. Some monomers may be involved in several linkages, others in only one or two, and the intermonomeric bonds may link the monomers at different positions through C–C, C–O ether (or sometimes ester) bonds. Overall, the lignins are heterogeneous and much-branched polymers of indefinite structure and size; their molecular weights range from a few thousand to over a million. The proportion of the three monomeric residues varies with the species. The lignin in conifers is richer in coniferyl alcohol-derived residues whereas that from deciduous species has more sinapyl alcohol-derived residues. As a consequence, the lignin in conifers is more condensed and less tractable to degradation than the lignin in hardwoods. Agricultural residues have relatively less lignin but are relatively high in *p*-coumaryl derivatives.

5.2.2 *Physical and chemical constraints in enzymatic hydrolysis of cellulose*

The susceptibility of cellulosic substrates to enzymatic hydrolysis is determined largely by their accessibility to cellulolytic enzymes (Tanaka *et al.*, 1988a). Direct physical contact (adsorption) of cellulase enzymes onto the cellulosic substrate is a prerequisite to enzymatic degradation (Ghose and Bisaria, 1979; Castanon and Wilke, 1980; Kyriacou *et al.*, 1989), and hence any structural feature that limits the accessibility of cellulosic substrates to enzymes will retard its susceptibility to hydrolysis. These features are listed in Table 5.4 and have been elaborately discussed by Cowling (1975) and Fan *et al.* (1980).

5.2.3 *Pretreatment of lignocellulosic residues*

Since glucose is an easily fermentable sugar, it remains one of the most sought-after products of enzymatic degradation of cellulosic materials. However, a number of factors in terms of the characteristics of both the substrate and the enzyme limit the hydrolysis of cellulose to glucose to an

Table 5.4 Structural features of cellulosic residues that determine their susceptibility to enzymatic degradation

1. Moisture content of the fiber (water-swollen fibers significantly increase surface area by opening up the capillary structure).
2. Capillary structure of cellulosic fibers (this determines the surface area of cellulosic fibers as defined by size, shape and surface properties of the microcrystalline capillaries within the fiber).
3. Crystallinity of cellulose.
4. Unit cell dimensions of cellulose.
5. Conformation and steric rigidity of anhydroglucose units within the crystalline regions.
6. Degree of polymerization of cellulose.
7. The manner of association of cellulose with hemicellulose and lignin (hemicellulose surrounds the cellulose fibers and intrudes into cellulose through pores. Lignin surrounds and strengthens the cellulose–hemicellulose framework. Hemicellulose probably acts as a molecular bonding agent between cellulose and lignin.)
8. The nature, concentration and distribution of substituent groups (methyl, ethyl, carboxymethyl, hydroxymethyl) on cellulose molecule. (The substitution affects the solubility and crystallinity of the cellulose and also, at a high concentration, activity of the enzyme.)

Sources: Cowling (1975), Fan *et al.* (1980), Torrie (1991).

appreciable extent. The physical and chemical constraints imposed by cellulosic substrates have been mentioned in section 5.2.2. In order to overcome these constraints and to enhance the susceptibility of the residues to enzyme action, pretreatment is an essential prerequisite. Numerous pretreatment methods capable of significantly enhancing the rate and extent of hydrolysis of pure celluloses as well as celluloses found in biomass materials have been described in the literature (Hsu, 1996). These methods along with the underlying phenomenon used for increasing the susceptibility and representative examples are summarized in Table 5.5. Most pretreatment processes have substantial energy requirements depending on the severity of the process. Severe mechanical and thermochemical processes, in general, have energy inputs which, when combined with the energy requirements of the product separation, can make cellulose bioconversion processes very uneconomical (Kosaric *et al.*, 1983; Klyosov, 1986). The mechanical pretreatments give fine substrates which result in higher slurry concentrations of higher bulk density, thereby reducing the reactor volume. However, lignin remains associated to act as a significant inhibitor to enzyme accessibility. Chemical pretreatments have been used extensively to remove lignin and to modify the structure of lignocellulose. Conventional chemical pulping processes, however, such as alkali (Kraft) or acid (Sulfite) processes, which remove lignin and preserve the quality of cellulose for paper manufacture, are not used for bioconversion because of their very high cost. The most common chemical pretreatment has been caustic swelling. Pretreatment with caustic soda results in increased surface area owing to swelling and disruption of lignin

Table 5.5 Pretreatment methods for enhancing the susceptibility of cellulose to cellulase action

Pretreatment method	Underlying phenomenon	Representative examples	References
1. Mechanical	Utilize shearing and impacting forces to yield a fine substrate of low crystallinity index and high specific surface	Ball milling, two-roll milling, hammer milling, colloid milling, vibro energy milling, extrusion	Ghose (1969), Mandels *et al.* (1974), Tassionari and Macy (1977), Millet *et al.* (1979), Fan *et al.* (1981), Ladisch and Svarczkopf (1991)
2. Physical	Increase in pore size and partial hydrolysis of hemicelluloses (steaming); extensive depolymerization (irradiation); depolymerization, oxidation or dehydration (pyrrolysis)	Steaming, wetting, pulping; freezing/thawing; irradiation; pyrolysis	Neese *et al.* (1977), Macdonald and Mathews (1979), Han and Ciegler (1982)
3. Chemical	Removal of lignin and/or hemicellulose, modification of the structure of lignocellulose, reduction in crysallinity, increase in surface area	Swelling agents (NaOH, NH_3); dilute acids (HCl, H_2SO_4, H_3PO_4); oxidizing agents (peracetic acid, O_3, H_2O); gases (SO_2, NO_2, ClO_2); organosolv (methanol, ethanol, butanol, glycerol, dioxane, phenol, ethylene glycol in the presence of a catalyst such as Lewis acids); cellulose-dissolving solvents; concentrated mineral acids (H_2SO_4, HCl), ammonia-based solvents (NH_3, hydrazine), aprotic solvents (DMSO, sulfur oxides) and metal complexes (cadoxen, cuozam)	Neese *et al.* (1977), Grethlein (1978), Binder *et al.* (1980), Ben-Ghedalia and Miron (1981), Gould (1984), Hamilton *et al.* (1984), Neely (1984), Chou (1986), Grohmann *et al.* (1986), Lynd and Grethelin (1987), Stockberger (1993), Saddler *et al.* (1993), Wyman (1994), von Sivers and Zacchi (1995)

4. Biological	Preferential removal of lignin modification of lignocellulose structure	White-rot fungi; soft-rot fungi; bacteria; cellulaseless mutants of white rot-fungi	Hatakka (1983), Srivastava (1989), Ghosh and Singh (1993), Messner and Srebotnik (1994)
5. Enzymatic	Selective removal of lignin, hemicelluloses	Lignin peroxidase enzymes; xylanases	Ghose and Bisaria (1979), Kaushik and Bisaria (1989), Bajpai and Bajpai (1996, 1997)
6. Combined pretreatments	Depends on the combination of pretreatments used, e.g. steam explosion loosens the cell wall structure, increases surface area, reduces DP of cellulose and separates the individual components	Steam explosion/RASH; ammonia freeze–explosion; high-temperature milling; alkali plus ball milling; SO_2 plus steaming; NO_2 plus irradiation	Dale and Moreira (1982), Saddler *et al.* (1982), Dekker and Wallis (1983), Brownell and Saddler (1986, 1987), Schultz *et al.* (1989), Ropars *et al.* (1992)

structure. Acid pretreatment uses dilute acids to remove hemicelluloses by hydrolysis. Oxidizing agents such as ozone and hydrogen peroxide attack lignin in preference to cellulose. They also disrupt the association between the carbohydrate polymers and lignin, producing a residue with enhanced susceptibility to cellulases.

Organosolv pretreatments involve the use of an aqueous solvent (ethanol, butanol, etc.) in the presence of a catalyst. The catalyst and water hydrolyse many of the carbohydrate bonds in the hemicellulose as well as the lignin–carbohydrate and lignin–lignin bonds. The solvent dissolves lignin while cellulose remains as a solid. Organic acids that are liberated from the substrate during the pretreatment process accelerate delignification and result in concurrent solubilization of the hemicelluloses. The organosolv pretreatment processes are attractive because the solvents can be recovered and recycled. Also, this process separates the lignocellulose into a solid cellulose, a solid lignin which has undergone few condensation reactions, and a liquid containing hemicellulose hydrolysate (Sarkarsen, 1980). A lot of effort has been directed for the development of acid-catalysed organosolv (ACOS) process which effects complete hydrolysis of cellulose into fermentable sugars. The process has been claimed to convert cellulosic substrates to glucose at high rates and excellent overall yields (Paszner and Cho, 1988). Chemical pretreatments, however, have serious disadvantages in terms of the need of specialized corrosion-resistant equipment, extensive washing, and disposal of chemical wastes.

Cellulose solvents pertaining to the following four major categories have also been employed for pretreatment of lignocellulosics:

1. strong mineral acids;
2. quaternary ammonium bases;
3. nonaqueous organic solvents such as dimethyl sulfoxide (DMSO) and dimethyl formamide (DMF);
4. metal complexes.

They act by reducing crystallinity and disrupting lignin–carbohydrate association, resulting in dissolution of cellulose and hemicellulose. Some of these solvents transform native cellulose (Cellulose I) to the Cellulose II form which has been found to be highly susceptible to enzyme action (Hamilton *et al.*, 1984).

Biological pretreatments are primarily concerned with the use of fungi, especially white-rot fungi, and bacteria that can degrade lignin preferentially. The complete selective removal of lignin has not so far been possible, as the microorganisms, which are unable to utilize lignin as a growth substrate, degrade cellulose and/or hemicellulose for their growth. Cellulase-less mutants of *Sporotrichum pulverulentum* have been isolated which partially delignify wood without affecting its cellulose (Eriksson *et al.*, 1980). However, these mutants cause a significant loss of hemicellulose or externally added glucose while degrading lignin. Biological pretreatments suffer from the serious drawback of lengthy (2–5 weeks) pretreatment time (Hatakka, 1983; Srivastava, 1989). Pretreatment processes for enzymatic solubilization of lignin (Leisola and Fiechter, 1985) and hemicellulose (Ghose and Bisaria, 1979) have also been reported. However, they seem to be too expensive owing to the low activity and low stability of the lignin peroxidases (Kaushik and Bisaria, 1989).

Out of the combined pretreatment processes, the steam-explosion process seems to be the most promising. Steam pretreatment at high pressure (4 MPa) and temperature (185–240 °C) results in the partial decomposition of some of the hemicellulose to uronic acids and other acids (mainly acetic acid) which catalyse the deploymerization of the hemicellulose and lignin. After steaming, the residue can either be explosively discharged by a sudden drop in pressure or emptied out after a gradual 'bleed-down' (Wood and Saddler, 1988). This pretreatment, which is primarily a chemical autohydrolysis reaction, is attractive because it removes hemicellulose and to some extent lignin (Brownell and Saddler, 1987). While the recovered lignin and hemicelluloses can be suitably converted into useful products, cellulose can be enzymatically hydrolysed with very high yields. The effectiveness of this pretreatment has been demonstrated by, amongst others, Grous *et al.* (1986) who showed that poplar chips treated by steam explosion could be 90% hydrolysed in 24 h, as compared to 15% hydrolysis of untreated chips. The major advantages of steam explosion are:

1. increased enzymatic susceptibility of the lignocellulosic residue;
2. lower energy requirements compared to mechanical pretreatment methods;
3. no recycling or environmental costs that are associated with chemical pretreatments;
4. reduced consumption of water and ability to use larger wood chips as compared to dilute-acid prehydrolysis pretreatment (Saddler *et al.*, 1993).

It has been shown that rapid decompression does not greatly improve the accessibility of cellulose (Brownell and Saddler, 1986, 1987). Based on this, a process termed rapid steam hydrolysis (RASH) has been developed which also continuously extracts the gaseous and soluble products (Schultz *et al.*, 1989). Steam treatment under alkaline conditions enhances accessibility of the cellulose (Dekker and Wallis, 1983). Treatment of both hardwood and softwood with SO_2 prior to explosion also resulted in a residue with increased susceptibility to cellulases (Clark and Mackie, 1986). The agricultural residues and a few hardwoods can also be made highly susceptible by subjecting them to the ammonia freeze–explosion (AFEX) process which treats them with liquid ammonia under pressure followed by sudden release (Dale and Moreira, 1982; Holtzapple *et al.*, 1992).

5.3 Properties, production and applications of cellulolytic enzymes

5.3.1 Properties of cellulases

Cellulases are synthesized by a large number of microorganisms which include fungi, actinomycetes, gliding bacteria and true bacteria. While many bacteria utilize cellulose by cell-bound enzymes, fungi secrete many of these enzymes in the growth medium. The extracellular cellulases from the fungi *Trichoderma reesei* (Mandels, 1975; Nevalainen *et al.*, 1980; Ryu and Mandels, 1980), *T. koningii* (Wood, 1985; Wood *et al.*, 1988), *T. viride* (Gritzali and Brown, 1979), *Penicillium pinophilum* (Wood *et al.*, 1988), *Penicillium* spp. (Castellanos *et al.*, 1995), *Sclerotium rolfsii* (Lachke and Deshpande, 1988), *Fusarium solani* (Wood and McCrae, 1977), *Phanerochaete chrysosporium* (Eriksson and Wood, 1985), *Talaromyces emersonii* (Moloney *et al.*, 1985) and *Neocallimastix frontalis* (anaerobe from bovine rumen) (Pearce and Bauchop, 1985; Wood *et al.*, 1988) are amongst the best known examples of cellulases that can degrade the native crystalline cellulose. Most other extracellular cellulases from fungi have little action on crystalline cellulose but can hydrolyse amorphous celluloses and soluble derivatives of cellulose such as carboxymethyl cellulose (CMC). The fungal cellulases do not seem to form physical complexes with each other, but they act in strong synergy where cooperative action is greater than the sum of the action of the individual enzyme components (Wood, 1985).

All cellulases cleave the same chemical bond, viz. the β1–4-glucosidic bond, but there is variation in the microenvironment of these bonds in the cellulosic substrates, as the same cellulose molecule contains both crystalline and amorphous structure. Moreover, as mentioned above, the presence of lignin and hemicellulose in cellulosic materials further complicates the cellulose matrix. Moreover, as cellobiose is the repeating unit in the cellulose crystallite, the consecutive glucosidic linkages between glucose residues are quite different stereochemically (Wood *et al.*, 1988). This complexity of structure is probably one of the reasons why the cellulolytic microorganisms produce a family of different cellulases with different substrate specificities. It is also worth mentioning that the specificities of many of the cellulases may be much broader than the present nomenclature and traditional classification would suggest.

Fractionation of cellulase proteins present in the culture filtrate of many of the fungi mentioned above have revealed that there are at least three major types of enzymes involved in hydrolysis of native cellulose into glucose. These are:

1. Endo-β1–4-glucanases (endo-β1–4-D-glucan 4-glucanohydrolase, EC 3.2.1.4).
2. Exo-β1–4-glucanase, present as cellobiohydrolase (exo-β1–4-D-glucan 4-cellobiohydrolase, EC 3.2.1.91). Some cellulase systems also contain exo-β1–4-D-glucan 4-glucohydrolase (EC 3.2.1.74) as a minor constituent.
3. β-Glucosidase (EC 3.2.1.21).

Despite this commonality, different cellulolytic microorganisms have evolved to produce an array of cellulases which, through different modes of action and substrate specificity, allowed them to utilize a complex heterogenous substrate, cellulose. Various explanations have been provided to explain the occurrence of multiple forms of cellulases, such as differential glycosylation, partial proteolysis and nonspecific release of membrane-bound enzymes on autolysis (Bisaria and Mishra, 1989). However, since gene cloning has shown the presence of a limited number of cellulase genes, the observed cellular multiplicity is most likely to be due to post-biosynthetic modifications (Goyal *et al.*, 1991).

Valuable information has been obtained on structure–function relationships of a few cellulases by using molecular approaches of gene cloning and improved methodologies of enzyme purification and characterization (Henrissat, 1991; Gilkes *et al.*, 1991; Reinikainen *et al.*, 1992; Davies *et al.*, 1993; Henrissat and Bairoch, 1993; Poole *et al.*, 1993; Divne *et al.*, 1994). At least four enzymes have been crystallized and the details of the tertiary structure of cellobiohydrolase (type II) are revealed (Rouvinen *et al.*, 1990; Claeyssens *et al.*, 1993). *Trichoderma reesei*, the most well-studied fungus for cellulase, systems has been shown to possess two genetically distinct endoglucanases (EG I and EG III), two cellobiohydrolases (CBH I and CBH II) and one β-glucosidase. Besides, one low molecular weight endoglucanase (25 kDa) is also present whose main function is defibrillation of cellulose microbibrils (Sprey and Bochem, 1991). Molecular architecture of these enzymes revealed that a particular region (AB or BA), approximately 60 amino acids long, is very well conserved. It is located either at the N-terminus (as in EG III and CBH II) or at the C-terminus (as in EG I and CBH I).

Trichoderma reesei, as well as a majority of other fungal cellulase proteins, have been found to consist of two distinct domains – a catalytic core domain and a cellulose binding domain (CBD), separated from each other by a protease-sensitive hinge region (Claeyssens *et al.*, 1993). Whereas the core domain can catalyse the breakdown of β1–4-glucosidic bonds of soluble cellulosic substrates, the whole enzyme molecule is required to catalyse the hydrolysis of insoluble cellulosic substrates, because of the involvement of CBD in binding to insoluble cellulose prior to catalysis by the core domain. Based on the amino-acid sequence of catalytic domains of various cellulases, they have been classified into ten families. Aspartic and glutamic acid residues have been implicated to be involved in catalysis based on their high degree of conservation in core domain (Tomme *et al.*, 1995). Similarly, nine distinct families of CBDs have been elucidated based on amino-acid sequences. Aromatic amino-acid residues appear to be responsible for binding of cellulases to insoluble cellulose (Tomme *et al.*, 1995; Duff and Murray, 1996).

On the basis of the mode of action of the three main cellulase components on various cellulosic substrates, the following generalizations about their specificities can be made (Wood, 1988; Bisaria and Mishra, 1989; Warren, 1993; Duff and Murray, 1996).

Cellobiohydrolases degrade cellulose by removing cellobiose from the nonreducing end of the chain and have very limited action on substituted celluloses such as CMC. They readily hydrolyse 'swollen' partially degraded amorphous cellulose and cellodextrins. They are, with a few exceptions, considered to be inactive against crystalline cellulose but exhibit highly cooperative synergistic action in the presence of endoglucanases. The endoglucanases hydrolyse CMC or swollen cellulose in a random fashion, as a result of which there is a rapid decrease in chain length together with a slow increase in reducing groups. They also act on cellodextrins and convert them to cellobiose and glucose. β-Glucosidases hydrolyse the degradation products of cellulose (i.e. cellobiose and cellodextrins) to glucose. In spite of these well-known specificities of the cellulase components, there is still no clear picture of events that take place during hydrolysis of crystalline cellulose. This is primarily due to the complex properties of the enzymes and also the inherent difficulties in analysing the various insoluble products of cellulose degradation. Accordingly, one finds in the literature that the action of 'purified' cellobiohydrolase on cotton and Avicel™ cellulose from a variety of fungal sources has been reported to be both extensive (Chanzy *et al.*, 1983; Nummi *et al.*, 1983) and negligible (Wood and McCrae, 1979). A detailed discussion on the properties and mode of action of cellulase components on solubilization of cotton and other crystalline cellulose can be found in the literature (Bisaria and Ghose, 1981; Mandels, 1981; Wood, 1985; Enari and Niku-Paavola, 1987; Coughlan and Ljungdahl, 1988; Wood, 1988; Wood *et al.*, 1988; Kubicek, 1992; Claeyssens *et al.*, 1993; Hayn *et al.*, 1993; Wong, 1995; Duff and Murray, 1996).

It is also important to realize that additional enzymes besides cellulases may also be required for degradation of cellulose. Thus, whereas the purified and reconstituted mixture of cellobiohydrolases, endoglucanases and β-glucosidases from the culture filtrates of *Pencillium pinophilum*, *Fusarium solani*, *Trichoderma koningii* and *Talaromyces emersonii* are capable of synergistically degrading cellulose to glucose, those from *Trichoderma reesei*, *Phanerochaete chrysosporium*, *Polyporus adustus* and *Myrothecium verrucaria* require certain oxidative enzymes also for hydrolysis of cellulose (Eriksson, 1981). In *T. reesei*, lactonase (EC 3.1.1.17) has been shown to act synergistically in solubilizing cellulose (Bruchmann *et al.*, 1987). Cellobiose oxidase (EC 1.1.99.18) and cellobiose (quinone oxidoreductase; EC 1.1.5.1) have also been reported to be involved in cellulose degradation in a number of cellulolytic organisms (Dekker, 1980; Eriksson, 1981; Coudray *et al.*, 1982; Morpath and Jones, 1986).

5.3.2 Production of cellulases

In the enzymatic conversion of lignocellulosic materials to liquid fuels, the cost of cellulase enzyme has been shown to have a major impact on overall economic viability of the process. It has been computed that nearly 27–40% of the ethanol cost is attributed to cellulase production (Vallander and Eriksson, 1990; Foody and Foody, 1991; Ladisch and Svarczkopf, 1991; Lynd *et al.*, 1991; Coughlan, 1992a). Various approaches have been suggested as a possible way of lowering the enzyme-associated costs, including recycling of the enzymes (Deshpande and Eriksson, 1984; Puls *et al.*, 1985) in the production process of useful metabolites from cellulose. Although complete cellulase is secreted in extracellular fluid by a number of fungi mentioned in section 5.3.1, *Trichoderma reesei* has been subjected to extensive studies for improving the yield and productivity of cellulases synthesized by it. This has been achieved by using various approaches, such as:

1. Genetic improvement through mutagenesis by overcoming the control mechanisms of cellulase biosynthesis. This has resulted in several hyperproducing mutants of *T. reesei* (Montenecourt and Eveleigh, 1979; Durand *et al.*, 1988; Duff and Murray, 1996).
2. Optimization of fermentation conditions (Sternberg and Dorval, 1979; Pourquie and Warzywoda, 1993).
3. Affecting the physiology by noninducing sugars (Nanda *et al.*, 1986).
4. Use of different reactor configurations (Allen and Andreotti, 1982; Mohagheghi *et al.*, 1990).
5. Use of mixed fungal cultures (Duff *et al.*, 1987; Panda *et al.*, 1987).
6. Addition of surfactants and charged colloid materials (Duff *et al.*, 1985; Esterbauer *et al.*, 1991).

Cellulase synthesis in *Trichoderma* is governed by both induction as well as catabolite repression. While various soluble inducers such as cellobiose, sophorose, lactose and cellobionolactone induce cellulase synthesis, the insoluble substrate cellulose has been the best inducer in the sense that a complete array of cellulase enzymes, capable of degrading crystalline cellulose most effectively, are synthesized only in its presence (Sternberg and Mandels, 1979; Bisaria and Mishra, 1989). The inductive formation of cellulases and β-glucosidase is also subject to catabolite repression by glucose and other readily metabolizable substrates. Therefore, most studies on its production have been carried out on insoluble carbon sources such as Solka Floc^{TM}, steam-exploded biomass and even untreated agricultural residues (Doppelbauer *et al.*, 1987).

By overcoming the regulatory mechanisms that control the synthesis of cellulases, largely through mutagenesis, several hyperproducing mutants of *T. reesei* have been obtained. The genealogy of such mutants developed independently in many laboratories is given by Durand *et al.* (1988) and Kadam (1996). The increased cellulase activity in some hyperproducing mutants such as MCG 80 has been attributed to enhancement of secretion activities of the mutants. In almost all cases except for a few Cetus mutants (Shoemaker *et al.*, 1981), Rutgers' mutants, Rut C-30 and Rut-P37 (synonym RL-P37) (Eveleigh, 1981; Sheir-Neiss and Montenecourt, 1984) and Espoo mutants, VTT-D-79125 (Bailey and Nevelainen, 1981), the specific cellulase activity (enzyme units per milligram of extracellular protein) has remained the same at about 0.8 IU mg^{-1} protein. This suggests that the improvements in the efficiency of fungal cellulases are possible by specifically affecting the regulation of the secretory pathway. It may be noted that the specific activity of the best cellulase producer, VTT-D-99125, in terms of filter paper units, is approximately 3.6 IU mg^{-1} protein while the commercial amylase preparations have specific activities of about 100 IU mg^{-1} protein (Bailey and Nevelainen, 1981). In these mutants (RL-P37 and VTT-D-79125), the increased specific activity seems not to be due to the synthesis of cellulase enzymes demonstrating improved catalytic activity, but rather to the increased synthesis of cellulase proteins at the expense of other noncellulase proteins (Bailey and Nevelainen, 1981; Montenecourt, 1983).

Trichoderma reesei mutants capable of producing cellulase on soluble inducers have also been developed with a view to solve the mixing and mass transfer problems associated with solid cellulosic substrates. Thus, the mutants MCG 80, Rut-C 30, C-5 and CL-847 all have been found to produce complete cellulase when grown on lactose as the sole carbon source (Allen and Andreotti, 1982; Pourquie *et al.*, 1988; Esterbauer *et al.*, 1991; Chaudhuri and Sahai, 1993). It is also worth noting that, although xylose did not support good cellulase production [0.4 filter paper units (FPU) ml^{-1}], crude water extracts of steam-exploded straw-yielded good cellulase activity of 10.1 FPU ml^{-1} and productivity of 72 FPU l^{-1} h^{-1}, presumably due to inducing ability of oligosaccharides of hemicellulose-derived sugars (Pourquie *et al.*, 1988). The cellulase and hemicellulases produced on xylose-containing media were found to be more effective in hydrolysing pretreated hardwoods (Pourquie and Warzywoda, 1993).

In spite of the low specific activity of most cellulases, several studies have been carried out in batch, continuous, fed-batch and immobilized cell reactors under varying substrate and environmental conditions to maximize the enzyme titer (i.e. activity in IU ml^{-1}) and productivity (IU l^{-1} h^{-1}) in *T. reesei* strains. A few of these studies on production characteristics of cellulase fermentations are summarized in Table 5.6. As seen from this table, there is wide variation in the activity and productivity values, which may be due to a number of factors such as the strain, growth conditions, concentration and type of inducers, the method of fermentation, and also the assay procedure used for activity determination (Ghose, 1987; Esterbauer *et al.*, 1991). However, in a study on the effects of strain and fermentation conditions on production of cellulase by three strains, viz. QM6a, QM9414 and Rut-C30, of *T. reesei* in continuous and fed-batch cultures, Allen and Roche (1989) reported that, when glucose repression was removed by carbon limitation, the parent strain QM6a could produce as much as and even higher activity than that of mutants on cellulose hydrolysate. The hydrolysate contained the strong inducer sophorose besides glucose, xylose, cellobiose, gentibiose and laminaribiose. Keeping in view the similarity of the cellulase systems of these strains, these results are significant for further genetic improvement studies, as they suggest that the mutations may not affect the cellulase genes directly but probably affect the mechanisms controlling the synthesis and secretion of cellulase proteins. A detailed discussion on the factors which govern the biosynthesis and secretion of cellulase enzymes in various mutants and recombinant cells is given in a review by the author (Bisaria and Mishra, 1989).

In recent years, there have been numerous reports on cloning and expression of cellulase genes from *T. reesei* and other cellulolytic organisms in *Escherichia coli*, *Saccharomyces cerevisiae* and other hosts. The aim of these studies is to understand the gene structure and expression which would be helpful in obtaining the enzyme components in pure form and in quantity for effecting solubilization of native cellulose in ligno-cellulosic residues. Protein engineering of cellulases may enable alteration of catalytic sites to produce hypercellulolytic mutants with greatly enhanced specific activities (Knowles *et al.*, 1987; Claeyssens *et al.*, 1993; Wong, 1995).

Cellulolytic fungi other than *Trichoderma* which produce complete cellulase enzyme systems (section 5.3.1) have also been subjected to mutagenic treatment for isolation of hyperproducing mutants. One such potential mutant seems to be *Penicillium pinophilum*. NTG III/6 which produced 9.8 FPU ml^{-1} with a productivity of 137 FPU l^{-1} h^{-1} on 3% Avicel in 12 l batch fermentation (Brown *et al.*, 1987). *Phanerochaete chrysosporium*, a fungus studied most extensively for its lignin degrading

Table 5.6 Cellulase production by various strains of *Trichoderma reesei* in reactors of different configurations[a]

Process	Strain	Particulars of inducer(s) and control variables[b]	Filter paper activity (IU ml^{-1})	Productivity (IU l^{-1} h^{-1})	References
1. Batch	D 1-6	6% cellulose, 150 l fermenter	11.0	57	Ghosh *et al.* (1982b)
	MCG-77	6% cellulose, bi-level pH control/biotin supplement	11.5	88	Gallo (1982)
	Rut C-30	As above	13.3	95	Gallo (1982)
	MCG-80	3% cellulose	17.2	142	Gallo (1982)
	Rut C-30	5% Solka Floc	10.1	108	Sheir-Neiss and Montenecourt (1984)
	CL-847	5% CC41 cellulose, 2 l fermenter	17.5	125	Pourquie *et al.* (1988)
	CL-847	5% Avicel + 2% wheat bran	20.0	119	Pourquie *et al.* (1988)
	C-5	4% lactose	2.8	31.5	Chaudhuri and Sahai (1993)
	QM9414	1% cellulose + 1% xylose	2.0	–	Schaffner and Toledo (1991)
2. Semi-continuous	E-12	(a) 6.3% Avicel, 7 l fermenter	8.6	47.8	Ghose, T.K. (personal communication, 1981)
	E-12	(b) 5.2% Solka Floc BW 200 (initial conc.), 5 l fermenter, pH cycling between 3.2 and 5.4. Semicontinuous feeding of cellulose	17.5	97.3	Ghose, T.K. (personal communication, 1981)
3. Continuous	MCG-77	Lactose, 2-stage culture, 15 l fermenters, $D = 0.026$–0.028 h^{-1} in second stage, run for 1350 h	2.6	90	Ryu *et al.* (1979)
	MCG-80	5% lactose, $D = 0.028$ h^{-1}, pH 3.5, 5 l fermenter	6.0	160	Allen and Andreotti (1982)
	C-5	4% lactose, 1-stage, $D = 0.029$ h^{-1}	–	61.2	Chaudhuri and Sahai (1993)

4. Repeated fed batch	MCG-80	2% lactose (initial conc.), pH 3.5, 5 l fermenter	10.0	100	Allen and Andreotti (1982)
5. Fed batch	QM 6a	Cellulose hydrolysate of BW 200, 5 l fermenter	11.8	55	Allen and Roche (1989)
	MCG-80	Cellulose hydrolysate of BW 200, 5 l fermenter	9.7	–	Allen and Roche (1989)
	Rut C-30	Solka Floc BW 200, 5 l fermenter, addition rate 2.5 g cellulose h^{-1}, total final addition of 100 g of cellulose per liter	22.4	133	Hendy *et al.* (1984)
	Rut C-30	Solka Floc, 8%, 10.5 l fermenter	31.0	160	McLean & Podruzny (1985)
	Rut C-30	Industrial hardwood pulp, 25.2%, 11 l fermenter	57.0	201	Watson *et al.* (1984)
	NREL-1	Solka Floc 2% and lactose feed at 5% after 48 h, 5 l fermenter	25.0	149	Kadam (1996)
	CC 4/30	Lactose, 2%	10.8	83	Turker and Mavituna (1987)
	CL-847	Lactose, 8.2%, 3000 l fermenter	20.0	144	Pourquie *et al.* (1988)
	SVG-17	Wheat straw, 6%, 15 000 l fermenter	6.5	61	Pokorny *et al.* (1990)
6. Continuous (immobilized cells)	Rut C-30	Lactose, celite II support, loading 8%	1.9	134	Montenecourt (1983)

[a]A few more production results can be found in Kosaric *et al.* (1983), Esterbauer *et al.* (1991), Kadam (1996) and Duff and Murray (1996).
[b]D = dilution rate (h^{-1}).

ability, also possesses a complete cellulase system. Using fed-batch cultivation, Szabo *et al.* (1996) successfully optimized conditions where supply of soluble sugars (glucose and cellobiose) was controlled to avoid catabolite repression phenomenon for production of cellulase by this fungus. However, since extensive studies on this and other cellulolytic strains have not been conducted in large-scale bioreactors under controlled environmental conditions, the full biosynthetic potential of all such seemingly better (than *Trichoderma*) strains remains to be seen.

A number of bacteria belonging to the genera *Clostridium*, *Cellulomonas*, *Bacillus*, *Thermomonospora*, *Ruminococcus*, *Bacteriodes*, *Erwinia*, *Acetovibrio*, *Microbispora* and *Streptomyces* have also been studied for their cellulase enzyme sytem (Hiltner and Dehority, 1983; Giuliano and Khan, 1984; MacKenzie *et al.*, 1984; Lee and Pack, 1987; Chippaux, 1988; Yablonsky *et al.*, 1988; Demain and Wu, 1989; Gilbert and Hazelwood, 1993) and extensive studies have been conducted at the genetic level in a few organisms to understand their regulatory mechanisms (Lee and Pack, 1987; Beguin, 1990; Warren, 1993; Beguin and Aubert, 1994). In several cases, it has been found that, although the purified cellulases possessed specific activities equal to or higher than those of fungal origin, they produced smaller amounts of cellulase enzymes than fungi (Gilkes *et al.*, 1991). Furthermore, relatively few of the bacteria produce an extracellular cellulase that can completely hydrolyse crystalline cellulose. An extensive survey of properties and genetics of cellulases of bacterial origin has been made (Robson and Chambliss, 1989; Beguin and Aubert, 1994). Thus, in the context of glucose production through enzymatic hydrolysis of cellulose, the cellulases of fungal origin and especially those of *Trichoderma* spp. have been extensively employed.

5.3.3 Properties of hemicellulases

A large number of hemicellulases are required for complete degradation of hemicellulose component of lignocellulosic residues owing to their very complex structure. Hemicellulases belong to three general categories:

1. The endo-acting hemicellulases cleave internal glycosidic bonds of the polysaccharide with very little activity on short oligomers.
2. The exo-acting hemicellulases attack progressively from either the reducing or the nonreducing termini.
3. The 'ancillary' or debranching enzymes remove side-group substituents such as arabinose, glucuronosyl, acetyl, coumaryl and feruloyl groups from the main backbone. These enzymes overcome the steric hinderances of side-chain substituents by facilitating the access of main endo- and exo-acting enzymes for hydrolysis of main backbone of hemicelluloses.

Because of the complex structure of hemicelluloses, several enzymes are required to completely degrade them (Brigham *et al.*, 1996). One of the main hemicellulases deploymerizing the hemicellulose backbone is endo-β1–4-xylanase and has been studied most extensively (Visser *et al.*, 1992). Xylans which possess β1–4-linked xylopyranose backbone are present in all terrestrial plants and comprise up to 30% of the cell wall material of annual plants, 15–30% of hardwoods and 7–10% of softwoods. Endoxylanases show highest activity against polymeric xylan which decreases with decrease in chain length. Endoxylanases also exhibit two types of activities – the debranching and nondebranching types. The debranching endo-xylanases release the substituent side chains, especially arabinose and simultaneously cleave the main chain. Such action possibly involves two catalytic subunits owing to the dissimilarity between the arabinofuranosyl and xylopyranosyl linkages (Coughlan, 1992b). The nondebranching endoxylanases cleave the main chain only and leave the branches intact. The main products formed on xylan hydrolysis are xylobiose, xylotriose and substituted oligomers containing 2–4 xylosyl residues. The chain length and the type of products depend on the mode of action of individual xylanases. Thus, while one xylanase isoenzyme of *Trichoderma reesei* (pI > 9; M_r 22 kDa) produces xylose and xylobiose, other isoenzyme (pI 9, M_r 20 kDa) releases xylobiose to xylohexaose (Lappalainen, 1986; Tenkanen *et al.*, 1992). Details about mode of action of other xylanases can be found in a recent review by Viikari *et al.* (1993).

Some xylanases, like cellulases, have also been reported to possess transferase activity. Xylanases from a few sources such as *T. harzianum* exhibited synergistic behavior while the two xylanases from *T. reesei* showed no such action in hydrolysing xylan (Wong *et al.*, 1986; Tenkanen *et al.*, 1992).

5.3.4 Production of xylanases

A number of microorganisms produce xylanases which are required in enzymatic hydrolysis of xylan component of agro-residues to xylose. Because of the requirement of cellulase-free xylanases in prebleaching of pulps in paper industry, much efforts have been devoted in this direction by using various approaches such as (1) controlling cultivation conditions, (2) manipulation of genetic system, or (3) developing efficient downstream processing techniques for separation of xylanase from cellulase-containing culture broth. Fungi, such as *Trichoderma reesei*, *Aspergillus awamori*, *Aureobasidium pullulans* and *Schizophyllum commune*, are amongst those which produce xylanases in high quantities but with cellulases (Bailey *et al.*, 1993; Haltrich *et al.*, 1993; Smith and Wood, 1991; Leathers, 1989). On the other hand, some organisms, such as *Thermomyces lanuginosus* and *Melanocarpus albomyces*, are potential producers of xylanases which

Table 5.7 Production of xylanases by various microorganisms in submerged fermentation

Organism	Carbon source	Xylanase activity ($IU\ ml^{-1}$)	Reference
Aureobasidium pullulans	Oat spelt xylan	725	Leathers (1989)
Bacillus circulans	Xylan	400	Ratto *et al.* (1992)
Trichoderma harzianum	Larchwood xylan	450	Saddler *et al.* (1982)
T. harzianum	Oat spelt xylan	166	Senior *et al.* (1989)
T. viride	Sulfite pulp	188	Gomes *et al.* (1992)
Thermomyces lanuginosus	Xylan	524	Hoq *et al.* (1994)
	Corn cobs	425	
	Wheat bran	173	
Melanocarpus albomyces	Xylan	100	Saraswat and Bisaria (1997)
	Wheat straw	172	
	Corn cobs	74	

produce negligible or no cellulase (Maheshwari and Kamalam, 1985; Hoq *et al.*, 1994; Saraswat and Bisaria, 1997).

Disaccharides and trisaccharides are the best natural soluble inducers of hemicellulolytic enzymes, but the best overall inducers are synthetic analogs called gratuitous inducers (Coughlan and Hazlewood, 1993). Positional dimers of xylobiose such as β1–3-xylobiose and β1–2-xylobiose are reported to be better inducers than xylobiose of xylanases in *Cryptococcus albidus* (Biely and Petrakova, 1984). In most cases, however, polymeric substrates such as xylans and even lignocellulosic materials such as wheat straw have been found to be the best xylanase inducers (Saraswat and Bisaria, 1997). In a few cases like *Streptomyces lividans*, xylose has been found to induce more xylanases than xylan (Kluepfel *et al.*, 1990). Table 5.7 gives the xylanase titers obtained in culture broths of a few important xylanase producers on various carbon sources. Since the researchers have followed different procedures for assay of xylanase activity, the values given in this table are to be compared with caution.

5.3.5 *Application of cellulases and xylanases*

Although a lot of research efforts have been directed towards developing a process for efficient enzymatic hydrolysis of cellulosic residues to useful products, primarily glucose and ethanol, this goal has not yet been realized, but continues to remain the subject of intense academic, research and development efforts. The enzymes, cellulases and xylanases, find a number of applications in various industries which take advantage of their limited or extensive action on cellulosic and hemicellulosic components of the residues. A few of the current and potential applications of cellulases and xylanases are given in Tables 5.8 and 5.9, respectively. It may be mentioned that a few industrial applications may require the use of other enzymes besides cellulases and xylanases to derive maximum advantage. Currently, these enzymes find largest use in removal of crude fibers of cell walls for various end-uses in different industries. Cellulases, in many cases, are used together with cell-separating enzymes (CSE) which possess

Table 5.8 Some current and potential applications of cellulase enzymes

	Industry	Applications
1.	Food	To increase soaking efficiency and homogeneous water absorption of rice Removal of external soybean coat to produce unicellular soybean. This presents coloration of fermented soybean foods such as soysauce and miso Isolation of proteins from soybean and coconut Isolation of starch from corn and sweet potato Gelatinization of seaweeds (such as *Laminaria*, *Phaeophyceae*) to increase digestibility Extraction of agar-agar from seaweeds (such as *Rhodophycaea*, *Gelidium*) Extraction of various components of green tea such as tannin, caffeine, saponin and aroma To digest ball-milled lignocellulose to obtain lignin with very large surface area, which can be used as a food additive
2.	Brewing	Purification of water-soluble α-glucan from milled barley
3.	Wine and fruit juice	To break down cellulose and other insoluble plant materials in fruit pulps
4.	Paper and pulp	Increasing the tensile strength of high α-sulfite pulp Removal of ink during secondary fiber processing For enhanced water removal on paper machine
5.	Feed	As a supplement in feed for cattle and poultry
6.	Alcohol/solvent	To hydrolyse cellulose into glucose which can be fermented to ethanol, 2,3-butanediol, butanol, etc.
7.	Textile	In reduction of pilling, softening and aging used in textile processing To impart stone-washed effects on jeans
8.	Laundry	To remove protruding fibrils from clothes
9.	Other uses:	
	(a) Research and development	Elucidating the structure of complex polysaccharides To produce protoplasts of higher plants
	(b) Medical	As digestive aids in antiflatulence tablets
	(c) Pollution abatement	Clarification of sludge to improve the efficiency of digestion in a septic tank

Sources: Toyama (1969), Ghose and Pathak (1973), Mandels (1985), Jurasek *et al.* (1987), Stark and Yin (1987), Cavaco-Paulo and Almeida (1994), Tolan (1996).

Table 5.9 A few current and potential applications of hemicellulases, primarily xylanases

Industry	Application
1. Paper and pulp	To enhance the extractability of lignins by bleaching chemicals For enzymatic modification of fibers which results in: (a) Decreased energy consumption in production of thermomechanical pulp (b) Improved drainage of recycled pulp (c) Increased beatability (d) Production of dissolving pulps For enhanced water removal on paper machines
2. Food	Processing of cereals Production of food thickeners Providing different textures to bakery products To enhance recovery of starch from wheat flours
3. Feed	To enhance digestibility of cattle and poultry feed
4. Alcohol/chemicals	To hydrolyse xylan into xylose which can be fermented to ethanol (fuel) or converted to xylitol (sweetener)

Sources: McCleary *et al.* (1986), Buchert *et al.* (1992), Viikari *et al.* (1993).

protopectinolytic activity. This enzyme removes protopectin present as an intercellular substance in vegetable tissues and exposes the cell walls which can be degraded by cellulases. One of the potential applications of xylanase enzymes lies in the pretreatment of kraft pulp in the paper industry. Traditionally the residual lignin (which imparts a brown color) is removed in a multistage bleaching procedure using chlorination and extraction stages, resulting in severe environmental pollution owing to the formation of numerous chlorinated phenols and dioxines. The use of a xylanase preparation in the pretreatment step is reported to substantially reduce the requirement of chlorine and to elevate the brightness of paper without adversely affecting the strength and yield (Viikari *et al.*, 1993). Various potential applications of lignin peroxidase and related enzymes have also been reported in biopulping, biobleaching, treatment of kraft-bleaching effluents, removal of environmental pollutants, improving the feed value of lignocellulosic residues, modification of lignochemicals for use as adhesives and resins, and in specific biotransformations (Crawford, 1981; Kirk and Farrell, 1987; Kaushik and Bisaria, 1989).

5.4 Bulk chemicals from cellulose and hemicellulose

5.4.1 Glucose and xylose

A comparison of the salient features of the acidic and enzymatic processes (Table 5.10) for hydrolysis of cellulose would reveal that the enzymatic route is preferred, especially due to its specificity, mild reaction conditions and absence of side products. Although none is economical at present, significant breakthroughs, e.g. in terms of improved enzyme properties, are still possible in the enzymatic route. A closer examination of the properties of *Trichoderma* cellulases, which have been used in almost all large-scale hydrolysis investigations, would reveal its drawbacks and consequently the need for having or designing better cellulose hydrolysing agents.

The enzyme properties that affect the hydrolysis of crystalline cellulose are (1) catalytic efficiency, (2) thermal stability, (3) adsorption, (4) end-product inhibition and (5) shear inactivation. Table 5.11 shows the

Table 5.10 Salient features of acid and enzymatic catalysis of cellulosic materials for glucose production

Feature	Acid catalysis	Enzymatic catalysis
1. Specificity	Nonspecific catalysts hydrolyse cellulose; may also degrade lignin	Highly specific
2. Rate of hydrolysis	High	Low
3. Reaction conditions	Harsh conditions (high temperature and pressure; corrosion-resistant equipment)	Mild conditions (atmospheric pressure, temperature 40–60 °C and pH 4.8–7.0 depending on the enzyme source)
4. Effect on DP	Considerable depolymerization without appreciable weight loss	Slight decrease in DP with appreciable weight loss
5. Catalyst requirement to achieve same degree of hydrolysis	Very high compared to enzyme (approx. 10^5 times)	Very much less
6. Cost of catalysts	Not generally high (require catalyst recovery for costly chemicals)	High (approx. 40% of the total cost) (requires recovery and/or recycling)
7. Side reactions	Generally results in unwanted side products (e.g. furfural from xylose and 5-hydroxymethyl furfural from glucose)	No side reaction, products of hydrolysis can be directly utilized for subsequent fermentations
8. Yield of glucose	Overall yield is low due to degradation	Overall yield is high primarily due to effective pretreatment step

Sources: Gilbert and Tsao (1983), Kosaric *et al.* (1983), Duff and Murray (1996), Schell and Duff (1996).

quantitative values of the properties of a few endoglucanases from *Trichoderma* and a few other organisms. The catalytic efficiency of endoglucanases on crystalline cellulose is very low, being of the order of 0.051 s^{-1} for the most active component (Table 5.11). Even against filter paper (a pretreated substrate), which is used for measuring its activity, its catalytic constant k_{cat} is only approximately 1.0–1.2 s^{-1} considering its specific activity of 1.2 IU mg^{-1} protein and molecular weight of 50 000 Da. Glucoamylase, on the other hand, has a k_{cat} of 58 s^{-1} with specific activity about 100 times higher (≈100 IU mg^{-1} protein). However, the same value is shown by endoglucanase against soluble cellulose CMC. This indicates that catalytic activity of cellulase is low for crystalline cellulose, owing to the inherent structural features of cellulose outlined in section 5.2.2. Table 5.11 also reveals that *T. reesei* cellulases suffer from the following disadvantages under actual operating conditions of cellulose hydrolysis:

1. low thermostability which requires replenishment of the lost activity at prolonged usage;
2. a high degree of product inhibition which results in cessation of hydrolysis at increased concentration of glucose and cellobiose;
3. inactivation by shear;
4. low adsorption of some components on crystalline cellulose which results in its washing away in, for example, a flow-through rector for continuous hydrolysis.

The implications of these properties of cellulase on hydrolysis of crystalline cellulose under process conditions have been elaborately discussed by Klyosov (1989). It seems that none of the cellulases discovered so far meet the requirement of these properties for 'biotechnological' applications, although a few of them do have some desirable properties, e.g. thermal stability of cellulases from *Clostridium thermocellum*, *Rhodothermus marinus* and *Myceliophthora thermophila*. There is, therefore, a need to develop cellulases from either *Trichoderma* or other organisms which possess these properties. The desirable attributes of the cellulases and those of the organisms producing them are summarized in Table 5.12.

In spite of the shortcomings of the cellulases available so far, several reports have appeared on their application on the hydrolysis of crystalline and/or pretreated cellulose in reactors of different configurations, with a view to maximizing the yield of sugars, especially glucose. The important parameters in the hydrolysis such as the yield and concentration of sugars, the duration of hydrolysis and the enzyme loading are all interrelated, and obviously depend on both the characteristics of cellulosic residue and the enzyme. Yields are higher in dilute systems owing to decreased inhibition by sugars. High enzyme loadings can increase the yield and sugar concentration to some extent but will require higher quantities of the costly

Table 5.11 Properties of endoglucanase from *Trichoderma* and a few other courses

Enzyme source	Adsorption constant on crystalline cellulose, K_A (l g^{-1})	Thermal stability at 65 °C[a], $t_{1/2}$ (min)	Inhibition constant against cellobiose K_I (g l^{-1})	Catalytic efficiency against crystalline cellulose at 40 °C, k_{cat} (s^{-1})
1. *Trichoderma reesei*	0.02–2.8[b]	14–40[b]	51–83	0.027 and 0.051;[b] 1.6(AC); 58(CMC)
2. *Trichoderma viride* (Meiji Seika)	0.6 and 8.5	20	31	–
3. *Myceliophthora thermophila*	2.7	10 200	35	0.013; 1.4(AC)
4. *Aspergillus wentii*	0.02 and 0.75	150	116	–
5. *Aspergillus* sp. (NOVO)	0.045	65	86	–
6. *Clostridium thermocellum*	0.01–5.5	40–36 000	87–114	–
7. *Rhodothermus marinus*	3.5 h(100°C) 16.0 h (80°C)	–	–	–

[a]Thermal stability of enzyme was determined in the absence of substrate.
[b]Where more than one value is given, they represent the values exhibited by different multiple forms of the enzyme from the same source or by the enzyme from different strains.
AC = amorphous cellulose; CMC = carboxymethyl cellulose.
Source: Klyosov (1989), Hreggvidsson *et al.* (1996).

Table 5.12 Desirable attributes of cellulases and of the organisms producing them for efficient cellulose hydrolysis

	Attributes	Comments
1.	Ability to produce complete cellulase system in quantity (hypercellulase producers)	Hyperproducing strains of *Trichoderma reesei* such as Rut C-30,RL-P37 and E-12 are available
2.	Cellulase synthesis on soluble sugars (constitutive mutants)	Constitutive strains of *T. reesei* (e.g. C-5) are available. Still considerable scope for improvement
3.	Synthesis not subject to catabolite (glucose) repression (catabolite repression-resistant mutants)	Catabolite repression-resistant mutants such as *T. reesei* D 1-6 and Rut C-30 are available
4.	High specific activity	Enzymes of high specific activity (e.g. those from RL-P37 and VTT-D-79125) can significantly improve the economics of cellulose hydrolysis
5.	High catalytic efficiency against crystalline cellulose	All cellulases available so far seem to have very low catalytic efficiency. Little information available on this important hydrolysis parameter
6.	Thermostable at hydrolysis temperature	A few cellulases (e.g. from *Clostridium thermocellum* and *Rhodothermus marinus*) are very stable
7.	Decreased susceptibility of enzyme to inhibition by cellobiose and glucose (end-product inhibition resistant mutants)	Cellulase (especially the cellobiohydrolase component) is severely inhibited by cellobiose (Holtzapple *et al.*, 1990). Inhibition constants need to be 2–3 orders of magnitude higher (Klyosov, 1989)
8.	Stability against shear forces	Shear-resistant cellulases will be useful where agitation must be provided to suspend the solid cellulose in hydrolysis reactor (Reese and Ryu, 1980; Mukataki *et al.*, 1988)

Note: At present, none of the four properties mentioned against points 5–8 seem to be possessed by any cellulase from a single source.

enzyme (Sattler *et al.*, 1989). It has been found that enzyme loadings of more than, say, 20 FPU g^{-1} cellulose do not significantly enhance the rate and extent of cellulose hydrolysis (Mandels *et al.*, 1981). It may be noted that, although cellulose adsorption is a prerequisite to cellulose hydrolysis, the rate of hydrolysis is not proportional to the concentration of the adsorbed enzyme (Converse, 1993). In practice, enzyme loadings may vary from 7 to 33 FPU g^{-1} substrate depending on the type of substrate being hydrolysed (Wright, 1988). A study on process economics of separate hydrolysis and fermentation for ethanol production showed that higher savings are attained with enzyme loading of approximately 12 FPU g^{-1} cellulose (Wright *et al.*, 1986). Addition of supplemental β-glucosidase, preferably in immobilized form, to a cellulase preparation (which is deficient in β-glucosidase) can also signficantly enhance the hydrolysis of cellulose (Sternberg *et al.*, 1977; Fadda *et al.*, 1989); the optimum ratio of β-glucosidase to filter paper cellulase is approximately 1.25–1.5:1 (Srivastava, 1985). The addition of enzymes such as xylanase in hydrolysing lignocellulosic residues at increased rates can be beneficial (Ghose and Bisaria, 1979). Cooperative action of esterases and xylanases on enhanced degradation of acetyl xylan has also been reported (Biely *et al.*, 1986). Hydrolysis of cellulose has shown to be improved on addition of surfactants; the rate enhancement, however, depended on the type of surfactant as well as cellulosic substrate (Helle *et al.*, 1993; Duff *et al.*, 1995).

The economics of the hydrolysis process can also be enhanced if the enzymes could be stabilized, recovered and reused. Thermal stability of the cellulase and β-glucosidase has been reported to increase by treatment of cellulase at pH 10 and by chemical modification of β-glucosidase with the crosslinking reagent glutaraldehyde (Woodward *et al.*, 1981). Cellulase can be removed by employing an ultrafiltration membrane system and then reused but the system seems to be impractical owing to high costs (Ghose and Kostick, 1970; Sundstrom *et al.*, 1981; Tan *et al.*, 1987; Tanaka *et al.*, 1988b). The most promising system for recovery of the enzyme seems to be through adsorption of cellulase on cellulose owing to its high affinity for the substrate (Wilke *et al.*, 1976; Bisaria and Ghose, 1978; Klyosov, 1986; Singh *et al.*, 1991). A novel approach has been developed by Taniguchi *et al.* (1989) which makes use of a soluble immobilized preparation of cellulase for cellulose hydrolysis. The immobilized preparation, which was made by using enteric coating polymer Eudragit L, can exist in both soluble and insoluble forms, depending on the pH of the medium, and hence can be recovered easily by simply changing the pH and reused to hydrolyse fresh cellulosic feed.

Several types of reactor configurations have been used for enzymatic hydrolysis of cellulose. These include batch and continuous reactors, the fluidized bed reactor using immobilized cellulase, and the ultrafiltration membrane reactor (Ladisch *et al.*, 1983; Gusakov *et al.*, 1987; Ishihara *et al.*, 1991). The use of the attrition bioreactor in which simultaneous wet milling and enzymatic hydrolysis take place in the same vessel has been found to enhance the hydrolysis rate compared to that achieved in the normal stirred batch reactor (Deeble and Lee, 1986). A novel approach to enzymatic hydrolysis has been reported by Tjerneld *et al.* (1986) who combined an aqueous two-phase system and ultrafiltration for enzyme recycling during a semicontinuous cellulose hydrolysis process. These authors achieved a sugar concentration of 7.4% using 8.0% cellulose slurry with no intermittent additions of cellulase for a period of 50 days. This seems to be the longest time reported for any cellulose hydrolysis process using only the initially added enzymes, indicating the high stability of the cellulases in aqueous two-phase systems. For the product concentration, the highest reported value seems to be that of Chen and Gong (1982) who obtained a solution containing 19% glucose and 1% cellobiose from hydrolysis of 20% cellulose within 72 h. It must be emphasized, however, that almost all studies which have reported a conversion of, say, more than 80% have employed pretreated cellulose of one type or the other. Although several reactor configurations have been proposed, none of them could be discarded as being inappropriate, and the ultimate reactor design will be governed by the characteristics of both the enzyme and the substrate.

(a) Xylose. As mentioned earlier, the utilization of hemicellulosic component of agro-residues is necessary for their economic bioconversion to utility products such as ethanol. Hemicellulose (mainly xylan) hydrolysate contains xylose along with minor quantities of monosaccharides such as glucose, mannose, galactose and arabinose, as well as oligosaccharides. The composition of a typical acid hydrolysate of hardwood hemicellulose shows the xylose content as 39 g l^{-1} out of total reducing sugars of 59 g l^{-1} (Beck, 1993). A secondary dilute acid hydrolysis step after primary pretreatment may be used to convert the oligosaccharides to monomeric sugars (McMillan, 1996).

For complete enzymatic hydrolysis of hemicellulose, several enzymes are required depending on the chemical composition of the substrate. For complex substrates, a complete xylanolytic enzyme system ensures maximal xylose production. This includes xylanases, β-xylosidases and side-chain cleaving enzymes. For example, for hydrolysis of steamed birchwood xylan, synergism amongst xylanases, β-xylosidase and acetyl esterase has been demonstrated by Poutanen and Puls (1988). Xylose is the only sugar which can be converted to xylitol, an important sweetener (Hallborn *et al.*, 1991).

5.4.2 Ethanol

(a) Conversion of glucose to ethanol. There are essentially three major routes for the production of ethanol from cellulosic residues using biochemical means. In the conventional two-step process, also known as separate hydrolysis and fermentation (SHF), cellulose is enzymatically hydrolysed to glucose which can subsequently be fermented to ethanol using either the yeasts (*Saccharomyces*) or bacteria (*Zymomonas*). The

major advantage of this process lies in the fact that both the steps can be carried out at their respective optimum conditions. The major disadvantage of the process is that the sugars released by enzymatic action severely inhibit cellulase activity during hydrolysis which means that lower cellulose concentration and higher enzyme loadings must be used to obtain reasonable ethanol yields. The technology for fermentation of glucose into ethanol by the yeasts is well developed and can be used for cellulose hydrolysates as long as inhibitory compounds are not present. It must be mentioned that although *Z. mobilis* has not replaced *S. cerevisiae* for industrial ethanol production, it is likely to do so in the near future in view of a number of unique features possessed by it, such as higher specific growth rate, higher sugar uptake rate, higher ethanol production rate, tolerance to low pH and to inhibitors present in lignocellulose hydrolysates, higher ethanol tolerance, lower heat generation, lower by-product formulation, higher osmotic tolerance and higher yield (Rogers *et al.*, 1980; Doelle and Doelle, 1989; Zhang *et al.*, 1995). Further, owing to its higher optimum temperature (37 °C), it is more compatible with cellulose hydrolysis for the simultaneous saccharification fermentation process which is described below.

The single-step processes for conversion of cellulose to ethanol can be of two types depending on the use of enzymes or the bacteria. In the process employing cellulase enzymes, the saccharification of cellulose to glucose and subsequent fermentation of glucose to ethanol by yeast or bacteria take place in the same vessel and the process is known as the simultaneous saccharification and fermentation (SSF) process. The direct microbial conversion (DMC) processes employing bacteria make use of either a single cellulolytic ethanologen (monoculture fermentation) or two organisms, a cellulolytic organism and an ethanologen (co-culture fermentation). It must be mentioned that pretreated cellulose has normally been used in all these processes owing to the resistance of native cellulose to enzymatic or microbial attack (section 5.2).

Several reports have appeared on the SSF process, which utilizes the cellulase enzyme (primarily from *Trichoderma* spp.) and yeast or bacterial species for bioconversion of cellulose to ethanol. In view of the properties of cellulases (section 5.3.1), the following advantages of the SSF process are clear (Takagi *et al.*, 1978; Ghose *et al.*, 1984; Szczodrak, 1989; Philippidis, 1996):

1. An enhanced rate of cellulose hydrolysis owing to the removal of sugars which inhibit cellulase activity.
2. Lower enzyme loading.
3. Higher product yield.
4. Limitation of the initial cellulose concentration can also limit the concentration of ethanol to a level which is not inhibitory to the yeast or the cellulase.

5. A decrease requirement for aseptic conditions.
6. A considerable reduction in process time.
7. The use of a single reactor.

However, in spite of these advantages, the SSF process must fulfil the following requirements to produce ethanol at competitive prices (Ghose and Tyagi, 1979; Bailey *et al.*, 1982):

1. Compatible conditions of saccharification and fermentation.
2. A high ethanol yield and rate of production.
3. A high ethanol tolerance of the culture.
4. A minimal inhibition of the enzyme by ethanol.
5. An optimal level of ß-glucosidase in cellulase to minimize accumulation of cellobiose.
6. The resistance of ethanologen to the lytic effects of cellulase.
7. The resistance of cellulase to proteolytic enzymes of the ethanologen.

A number of studies on SSF have been reported in which the yeasts such as *Candida brassicae*, *Saccharomyces cerevisiae* and *S. carlsbergensis* have been used to produce ethanol from cellulose hydrolysis mixture (Takagi *et al.*, 1978; Ghosh *et al., 1982a; Philippidis et al.*, 1993). The bacterium *Zymomonas mobilis* and a few other yeasts have also been studied in the SSF process for ethanol production (Viikari *et al.*, 1981; Saddler *et al.*, 1983; Ghose *et al.*, 1984; Spangler and Emert, 1986). Various approaches have been used to increase the volumetric productivity of ethanol. These approaches include techniques of keeping cells in the bioreactor by using immobilization and recycle systems (Guidoboni, 1984; Kolot, 1984; Roukas, 1994). Other approaches employed were directed at keeping the ethanol concentration low in the reactor by removing it through solvent extraction or an enriched vapor stream (Ghose *et al.*, 1984; Mairoella *et al.*, 1984; Crabbe *et al.*, 1986; L'Italien *et al.*, 1989).

As mentioned above, one of the most important requirements for a successful SSF process is the compatibility of the saccharification and fermentation system with respect to temperature, pH and substrate concentration. Since the optimum temperature of *T. reesei* cellulase activity is 50 °C while that for fermentation using *S. cerevisiae* is approximately 30 °C, a compromise between these two temperatures must be made. Since the cellulase activity is still quite high at 40–45 °C, the use of thermotolerant yeasts or bacteria would have obvious advantages in an SSF process. A few of the potential advantages and disadvantages associated with the use of thermotolerant/thermophile ethanologens have been elaborated by Slapack *et al.* (1987). Isolation and use of a few thermotolerant yeasts which have been used along with cellulase in an SSF process at 40–45 °C have been reported (Ghose *et al.*, 1984; Szczodrak and Targonski, 1988; Spindler *et al.*, 1989). Szczodrak and Targonski (1988) screened 58 yeast strains for their ability to produce ethanol and reported that their best isolate, *Fabospora fragilis* CCY 51-1, was able to accumulate 56 g l^{-1} ethanol from 140 g l^{-1} glucose at 43 °C in less than 48 h. Using a number of thermotolerant yeasts at 37 °C in SSF process, Spindler *et al.* (1988) found that *S. cerevisiae* and a mixed culture of *S. cerevisiae* and *Brettanomyces clausenii* gave 70% ethanol yield in 4 days with 7.5% Sigmacell 50 cellulose at a low enzyme loading of a 7.0 FPU g^{-1} cellulose. Using a cellobiose fermenting yeast, *Candida lusitaniae* Y-5394, and *Saccharomyces uvarum* in a mixed-culture SSF process, the same group (Spindler *et al.*, 1989) reported better performance of the mixed culture than that of either single constituent culture, as it combined the cellobiose fermenting capability with the high ethanol tolerance and rapid glucose fermentation of conventional *Saccharomyces* species. However, the fermentation took a long time: 8 days to accumulate 40 g l^{-1} ethanol in the mixed-culture SSF process.

If the cellulase preparation is deficient in β-glucosidase and a non-cellobiose fermenting yeast is used, it will result in accumulation of cellobiose which is a strong inhibitor of cellulase enzyme (Sternberg *et al.*, 1977). This can be overcome by supplementation of cellulase with higher β-glucosidase activity enzyme from another source such as *Aspergillus niger* (Szczodrak, 1988) or by using cellulase from β-glucosidase hyperproducing mutant of *T. reesei* (Szczodrak, 1989). When a cellulase preparation from a β-glucosidase hyperproducing mutant of *T. reesei* was employed with *Kluyveromyces fragilis* in an SSF process with 10% pretreated wheat straw, the ethanol concentration increased from 2.5% (w/v) to 3.4% (w/v) in 24 h instead of 48 h (when the parent *T. reesei* cellulase was used) (Szczodrak, 1989). The use of whole unfiltered cellulase broth has also been reported to enhance ethanol production in SSF system presumably due to the presence of attached β-glucosidase to the cell wall of *T. reesei* mycelia (Bisaria *et al.*, 1986; Schell *et al.*, 1990). The highest productivity of ethanol in an SSF process using a thermotolerant yeast seems to have been reported by Ghose *et al.* (1984). Using cellulase from *T. reesei*, β-glucosidase from *Aspergillus wentii* and *Candida acidothermophilum* at 40 °C, and operating the SSF process under vacuum recycling with intermittent substrate feeding, cellulose utilization of 98% and ethanol productivity of 4.5 g l^{-1} h^{-1} was achieved (Ghose *et al.*, 1984). The SSF process has been tested at the 1250 l scale using Solka Floc and pulp mill waste (Philippides *et al.*, 1993). For a more elaborate coverage of various factors that affect the interpretation and performance of the SSF process, such as (1) analytical methods used for measurement of cell, product and enzyme activity, and (2) effect of pH, temperature, inoculum size, substrate concentration, media composition, enzyme loading and β-glucosidase supplementation, the reader is referred to a recent review by Grohmann (1993).

DMC of cellulose to ethanol can be affected by using a cellulolytic ethanologen such as *Clostridium thermocellum* (Ng *et al.*, 1977; Brener and Johnson, 1984; Hogsett *et al.*, 1992). This bacterium has a low tolerance to fermentation products; the inhibitory effects of ethanol and organic acids on cell viability, growth and cellulase activity have been reported (Kundu *et al.*, 1983). However, since this bacterium is the only one amongst many thermophiles which produces a true cellulase system and which has many advantages in fermentation processes, a large body of scientific literature has been devoted to various aspects of this thermophile such as characterization of the cellulase system, genetics and molecular cloning of its various cellulase genes (Demain and Wu, 1989; Beguin, 1990; Beguin and Aubert, 1994).

Although *C. thermocellum* breaks down cellulose and hemicellulose to glucose, cellobiose and xylose, it is unable to utilize xylose. Furthermore, cellobiose and glucose also inhibit its cellulase activity, resulting in low ethanol yields. Thus, by co-culture of these highly cellulolytic bacteria with ethanol-producing bacteria capable of fermenting xylose, cellobiose and glucose, the advantages of the SSF process can be realized and ethanol can be produced in higher yields. Still, this single-step mixed culture process employing cellulolytic and ethanologenic bacteria for conversion of cellulose to ethanol suffers from two major drawbacks: (1) ethanol toxicity owing to low ethanol tolerance, and (2) the rate-limiting step of cellulose hydrolysis. Some of the co-cultures employed for direct conversion of cellulose to ethanol include *C. thermocellum* with *C. thermohydrosulfuricum* (Ng and Zeikus, 1982), *C. thermocellum* with *C. thermosaccharolyticum* (Wang *et al.*, 1983), *C. thermocellum* and *Thermoanaerobacter ethanolicus* (Ljungdahl and Wiegel, 1981) and *C. thermocellum* with *Z. mobilis* or *Z. anaerobia* (Saddler and Chan, 1982). Attempts to increase the ethanol tolerance of these bacteria have resulted in the isolation of a few mutants (Tailliez *et al.*, 1989; Klapatch *et al.*, 1994). Most of them are, however, still unable to produce more than approximately 3% (w/v) in co-culture processes using cellulosic substrates. The physiological aspects of thermophilic bacteria and ethanol tolerance have been elaborately discussed by Stewart and his co-workers (Slapack *et al.*, 1987).

Notwithstanding the disadvantages associated with thermophilic bacteria (such as ethanol tolerance and stability of high ethanol-yielding strains), a study on economic evaluation of thermophilic bacteria for ethanol production shows that they might compete with the available yeast technology if more efficient technologies to recover ethanol from dilute solutions are applied (Lynd, 1989).

(b) Conversion of xylose to ethanol. Whereas glucose present in the cellulose component of agro-residues can be efficiently fermented to ethanol by several microorganisms, the pentose sugars in the hemicellulose

component, particularly xylose, remains a bottleneck in the biomass-to-ethanol conversion process. Since the feedstock can represent up to 40% of the process costs, the efficient conversion of xylose is required for a process to be economical (Schell *et al.*, 1991; Picataggio and Zhang, 1996). Thus, as is the case with the yeast *Saccharomyces cerevisiae*, high ethanol yield and concentration are required in xylose fermentation also for a process to be economical. Many microorganisms including *S. cerevisiae* and *Z. mobilis* are unable to utilize xylose. However, since *S. cerevisiae* can ferment xylulose, an isomer of xylose, several attempts were made to isomerize xylose to xylulose for its conversion to ethanol (Lastick *et al.*, 1990) but without much success.

Subsequently, considerable efforts that were made to isolate xylose-fermenting organisms led to the identification of several bacteria and yeasts; notable amongst them are the three yeasts, *Pichia stipitis*, *Pachysolen tannophilus* and *Candida shehatae* (Hahn-Haggerdal *et al.*, 1993). These yeasts successfully ferment pure xylose but not the aqueous hemicellulose streams generated by pretreatment processes, probably due to the presence of inhibitors such as acetic acid, uronic acids, furfural, hydroxy methyl furfural and lignin degradation products (McMillan, 1994). These yeasts, unlike bacteria, use a two-step reaction pathway for conversion of xylose to xylulose: (1) reduction of xylose to xylitol by NAD(P)H-dependent xylose reductase, and (2) oxidation of xylitol to xylulose by NAD-dependent xylitol reductase. Xylulose is subsequently phosphorylated to xylulose-5-phosphate by xylulokinase. Xylulose-5-phosphate is then metabolized by use of pentose phosphate and Embden–Meyerhoff–Parnas (EMP) or Entner–Doudoroff (ED) pathway to pyruvate via glyceraldehyde-3-phosphate. Pyruvate decarboxylase and alcohol dehydrogenase bring about the formation of ethanol from pyruvate. This scheme produces 5 moles of ethanol from 3 moles of xylose, neglecting NAD(P)H balance. The theoretical yield of ethanol is thus 1.67 moles of ethanol per mole of xylose (for glucose, it is 2.0 moles of ethanol per mole of glucose), although on a mass basis it is the same as for glucose, i.e. 0.51 g ethanol g^{-1} xylose. The reduced molar yield indicates poorer energetics for xylose fermentation to ethanol compared to glucose.

These xylose-fermenting yeasts have been reported to produce ethanol at 78–94% of theoretical yield at concentrations of 3–5% (w/v). The volumetric productivities are also low (0.3–0.9 g l^{-1} h^{-1}), especially in the absence of oxygen (0.1–0.2 g l^{-1} h^{-1}). Lower yields are obtained primarily because of growth and xylitol accumulation, which presumably results from inhibition of xylitol dehydrogenase by excess NADH in the absence of sufficient respiration (Skoog and Hahn-Haggerdal, 1990). The major limitations in the use of xylose-fermenting yeasts for ethanol production are due to following factors (Picataggio and Zhang, 1996):

1. the low level of oxygen requirement (2 mmoles l^{-1} h^{-1}) to maintain cell viability (and NADH balance);
2. the low ethanol tolerance;
3. the low volumetric productivity;
4. the low yields;
5. the poor performance on lignocellulose hydrolysates owing to inhibition of cell growth by various acids and aromatic compounds;
6. the reassimilation of ethanol;
7. the difficulty in controlling microaerophilic conditions on an industrial scale.

A discussion of other factors influencing the performance of xylose fermentation, such as media composition, pH and temperature, as well as methods available for detoxification of lignocellulose hydrolysates, can be found in a recent review by McMillan (1996).

Recombinant DNA technology has been applied to develop *S. cerevisiae*, *E. coli* and *Z. mobilis* strains for xylose utilization. The genes encoding the enzymes necessary for xylose assimilation, xylose reductase and xylitol dehydrogenase from *P. stipitis* have been cloned in *S. cerevisiae* (Kotter *et al.*, 1990; Takuma *et al.*, 1991; Hallborn *et al.*, 1991). The ethanol yields were, however, low primarily due to accumulation of xylitol.

The xylose-fermenting *E. coli* has been developed by introducing pyruvate decarboxylase and alcohol dehydrogenase genes of *Z. mobilis* which produced ethanol with 96% of theoretical yield at volumetric productivity of 0.7 g ethanol l^{-1} h^{-1} in a complex medium (Ingram *et al.*, 1987; Ohta *et al.*, 1990). These strains are attractive as they can utilize a wide range of sugars present in lignocellulose hydrolysates such as glucose, xylose, cellobiose, galactose and mannose. The main drawback of the recombinant *E. coli*, however, was its low ethanol tolerance and inhibition by products of dilute acid hydrolysates.

Considerable research efforts have gone in to developing xylose-fermenting *Z. mobilis* strains (Skotnicki *et al.*, 1980; Reynen *et al.*, 1990; Feldmann *et al.*, 1992). The group of Picataggio has recently reported on the development of a very potent strain of *Z. mobilis*, capable of growth on xylose for efficient ethanol production (Zhang *et al.*, 1995); the strain was developed through coordinate expression of the *E. coli* xylose isomerase, xylulokinase, transketolase and transaldolase genes. The recombinant strain produced ethanol with 86% of theoretical yield on xylose as the sole source of carbon and with 95% of theoretical yield on a mixture of glucose and xylose. The performance of this promising recombinant strain, however, remains to be ascertained on lignocellulose hydrolysates. Reviews by Schneider (1989) and McMillan (1994) may be referred for more details on hemicellulose hydrolysate conversion to ethanol.

5.4.3 Acetone–butanol

This fermentative pathway is much more complicated than that of ethanol and is a branched one. Depending on the energy status of the cell and the fermentation conditions such as pH and substrate level, the cells produce either the solvents or the acids. It has been shown (Monot *et al.*, 1982; Votruba *et al.*, 1986) that at low pH (*c.* 4.5) and high glucose concentration, the main products are acetone and butanol, whereas only acids are produced at high pH and low glucose concentration by the strict anaerobe *Clostridium acetobutylicum*. The concentration of resulting solvents (acetone, butanol and ethanol) normally does not exceed 18 g l^{-1} owing to the inhibitory effect of butanol, although improvement of strains through recombinant DNA technology has resulted in accumulation of solvents up to 23 g l^{-1} (Parisi, 1989). Approaches to increase the solvent concentration and productivity have included the use of continuous and immobilized reactors, nutrient cycling, liquid–liquid extraction, adsorption, hollow-fiber fermenter-extractor and spin-filter perfusion bioreactor (Dadgar and Foutch, 1986; Soni *et al.*, 1986; Park *et al.*, 1989; Shukla *et al.*, 1989; Mulchandani and Volesky, 1994). The use of an immobilized cell trickle-bed reactor resulted in butanol and acetone concentrations of 8.82 and 5.22 g l^{-1} with butanol and total sovents yields of 19.4 and 34.1%, respectively, from 60 g l^{-1} initial glucose concentration. The solvent productivity, which seems to be the highest reported so far, was 4.2 g l^{-1} h^{-1}, ten times higher than that obtained in batch fermentation using free cells and 2.76 times higher than that of an immobolized continuous stirred tank reactor (CSTR) (Park *et al.*, 1989). Mulchandani and Volesky (1994) used a spin-filter perfusion bioreactor for production of acetone-butanol by *C. acetobutylicum*. This reactor employed an *in-situ* spinning filter to achieve high cell density and continuous removal of inhibitory products. The use of a spin-filter bioreactor eliminated: (1) the diffusion limitations imposed by an immobilized cell bioreactor, and (2) the problem of high shear encountered in a cell recycle bioreactor. Using the reactor in a continuous mode at a dilution rate of 0.089 h^{-1}, they obtained approximately 8 g l^{-1} butanol, 4.2 g l^{-1} acetone and 0.4 g l^{-1} ethanol with a productivity of 1.14 g l^{-1} h^{-1} at 49 g l^{-1} glucose feed concentration (Mulchandani and Volesky, 1994).

The potential of *C. acetobutylicum* and *Klebsiella pneumoniae* for converting biomass substrates into acetone, butanol, ethanol and 2,3-butanediol has been investigated by Yu and Saddler (1987). Using a combined hydrolysis and fermentation approach, the production of butanol and acetone was 3.4 and 1.0 g l^{-1}, respectively, from 50 g l^{-1} xylan, and 6.5 and 3.1 g l^{-1} respectively from 50 g l^{-1} Solka Floc™ cellulose (Yu and Saddler, 1987). The low butanol yields from xylan were attributed to lower efficiency of xylose utilization. A large-scale process for conversion of lignocellulose into acetone–butanol which involved pretreatment of the corn cobs, cellulase enzyme production, enzymatic hydrolysis of pretreated material and acetone–butanol fermentation of the hydrolysate has been reported by Marchal *et al.* (1992). The 48 000 l scale fermentation run resulted in 7.8 g l^{-1} acetone, 12.4 g l^{-1} butanol and 0.3 g l^{-1} ethanol with a solvent productivity of 0.45 g l^{-1} h^{-1} and yield of 0.31 g g^{-1} based on the initial content of monomeric sugars. The physiological consequences of end-product inhibition by butanol and various possible approaches to overcome this inhibition have been discussed by Lindley and Durand (1989).

5.4.4 2,3-Butanediol

2,3-Butanediol is also an industrially important chemical that can be produced from a variety of carbohydrates such as glucose, xylose and disaccharides present in the hydrolysates of agro-residues. The bacterial species that have been used include *Klebsiella oxytoca*, *K. pneumoniae* and *Bacillus polymyxa*. In addition to 2,3-butanediol, most species also produce ethanol, acetate, acetoin, lactate, etc. The most important variable affecting the kinetics of its production appears to be the oxygen transfer rate (Mas *et al.*, 1988; Nigam, 1989). A higher oxygen availability favors the formation of biomass at the expense of butanediol. On the other hand, decreased oxygen availability increases butanediol yield but results in a decreased overall conversion rate owing to lower cell concentration. Butanediol also strongly inhibits growth at higher concentrations but does not particularly affect the specific rate of butanediol production (Sablayrolles and Goma, 1984). The use of immobilized cells of *K. oxytoca* in a continuous reactor was found to result in low product concentration and productivity, primarily due to limited availability of oxygen to the cells (Nigam, 1989). The best results were obtained in a fed-batch culture system by maintaining the glucose concentration below the inhibition level of 50 g l^{-1} and by regulating the oxygen supply rate at 11.2 mmol l^{-1} h^{-1}. Under these conditions, the product concentration of approximately 80 g l^{-1} and productivity of 2.08 g l^{-1} h^{-1} have been achieved (Nigam, 1989). When the water-soluble hemicellulose fraction of steam-treated aspen wood was used as a substrate, both growth and butanediol production was severely retarded in *K. pneumoniae* owing to the presence of several inhibitors such as furfural (Nishikawa *et al.*, 1988). The toxic effects of possible chemicals present in pretreated agro-residues should, therefore, be considered before using them for butanediol production. The possibility of utilizing lignocellulosic residues for butanediol production has also been studied by Yu and Saddler (1987). Using xylan and Solka Floc™ at a 100 g l^{-1} level, butanediol concentrations of 18.5 g l^{-1} and 22.6 g l^{-1} have been reported in a combined hydrolysis and fermentation approach (Yu and Saddler, 1987). The effect of various parameters on production of 2,3-butanediol has recently been reviewed (Garg and Jain, 1995).

5.5 Future prospects

Agro-residues are a potential resource for bioconversion to glucose, ethanol and a number of other chemicals. However, owing to the crystalline nature of cellulose in the residues, they have to be made susceptible to enzymatic action by subjecting them to some form of pretreatment. At present, organosolv, steam explosion and dilute acid prehydrolysis pretreatment processes appear to be promising for increasing the susceptibility. Despite the potential of these methods, there is rather a limited understanding as to how these pretreatments result in enhanced enzymatic hydrolysis. A more in-depth understanding of the phenomena is needed to devise technically feasible and cost-effective pretreatment methods.

A number of microbial mutants exist which secrete a fairly large amount of cellulase, but none of them possesses the properties that are optimal under actual process conditions of cellulose hydrolysis. Further, the specific activity of cellulases secreted by even hyperproducing mutants is low. Isolation of cellulases with high specific activity from new sources or developing them through modern techniques of recombinant DNA and protein engineering that can function efficiently in hydrolysing crystalline cellulose would be a worthwhile approach in this direction.

For efficient bioconversion of cellulose to ethanol, the simultaneous saccharification and fermentation process needs to be improved by isolation of thermotolerant yeasts or bacteria (along with ethanol tolerance), so that compatibility problems could be minimized. The potential of thermotolerant bacteria in lignocellulose-based bioconversion processes, in view of their ability to utilize both the cellulose and hemicellulose fractions, is large. Development of technologies for efficient conversion of both glucose and xylose present in lignocellulose hydrolysate will have a large impact on the economics of ethanol production process. Although significant progress has been made in developing xylose-utilizing yeasts, they need to be more tolerant to ethanol and produce ethanol with a higher yield. More fundamental studies concerning their physiology are required so that their ethanol tolerance can be enhanced and stable high-ethanol yielding mutants can be obtained. Successful incorporation of xylose-utilizing capability in *Z. mobilis* promises to convert both glucose and xylose to ethanol efficiently. Its performance on lignocellulose hydrolysate, however, remains to be ascertained. Genetic modifications of both glucose- and xylose-utilizing microorganisms that redirect carbon flow to ethanol will have a substantial impact on economic feasibility of ethanol fermentation. Simultaneous developments in process-oriented studies which include the development of different types of reactors and the removal of ethanol from dilute fermentation broths will also be highly valuable in developing a commercial biomass-to-ethanol process. Advances in these and related areas of both scientific and engineering aspects of lignocellulose bioconversion are expected to result in realization of the fuller potential of the microbes and the enzymes in various biotechnological applications.

Acknowledgement

The author thanks Dr Vibhor Saraswat, a scientist at the Department, for his help in the preparation of the manuscript.

References

Allen, A.L. and Andreotti, R.E. (1982) In *Fourth Symposium on Biotechnology in Energy Production and Conservation, Biotechnology and Bioengineering Symposium No. 12* (ed. C.D. Scott), John Wiley, New York, p. 451.

Allen, A.L. and Roche, C.D. (1989) *Biotechnol. Bioeng.*, **33**, 650.

Bailey, M.J. and Nevelainen, K.M.H. (1981) *Enzyme Microbiol. Technol.*, **3**, 153.

Bailey, M.J., Buchert, J. and Viikari, K. (1993) *Appl. Microbiol. Biotechnol.*, **40**, 224.

Bailey, R.B., Benitez, T. and Woodward, A. (1982) *Appl. Environ. Microbiol.*, **44**, 631.

Bajpai, P. and Bajpai, P.K. (1996) *Adv. Biochem. Eng. Biotechnol.*, **56**, 1.

Bajpai, P. and Bajpai, P.K. (1997) *Adv. Biochem. Eng. Biotechnol.* (in press).

Beck, M.J. (1993) In *Bioconversion of Forest and Agricultural Plant Residues* (ed. J.N. Saddler), CAB International, Oxford, p. 211.

Beguin, P. (1990) *Ann. Rev. Microbiol.*, **44**, 219.

Beguin, P. and Aubert, J.-P. (1994) *FEMS Microbiol. Rev.*, **13**, 25.

Ben-Ghedalia, D. and Miron, J. (1981) *Biotechnol. Bioeng.*, **23**, 823.

Biely, P. and Petrakova, E. (1984) *FEBS Lett.*, **178**, 323.

Biely, P., MacKenzie, C.R., Puls, J. and Schneider, H. (1986) *Bio/Technology*, **4**, 731.

Binder, A., Haltmeir, T. and Fiechter, A. (1980) In *Proceedings of the Symposium on Bioconversion and Biochemical Engineering*, Vol. 1 (ed. T.K. Ghose), Indian Institute of Technology, New Delhi, p. 315.

Bisaria, V.S. (1991) In *Bioconversion of Waste Materials to Industrial Products* (ed. A.M. Martin), 1st edn, Elsevier Applied Science, London, p. 187.

Bisaria, V.S. and Ghose, T.K. (1978) In *Proceedings of the Symposium on Bioconversion of Cellulosic Substances into Energy, Chemicals and Microbial Protein* (ed. T.K. Ghose), Indian Institute of Technology, New Delhi, p. 155.

Bisaria, V.S. and Ghose, T.K. (1981) *Enzyme Microbiol. Technol.*, **3**, 90.

Bisaria, V.S. and Mishra, S. (1989) *Crit. Rev. Biotechnol.*, **9**, 61.

Bisaria, V.S., Nanda, M. and Ghose, T.K. (1986) *J. Gen. Microbiol.*, **132**, 973.

Bisaria, R., Madan, M. and Bisaria, V.S. (1987a) *Biol. Wastes*, **19**, 239.

Bisaria, R., Gujral, G.S. and Bisaria, V.S. (1987b) *Crit. Rev. Biotechnol.*, **7**, 17.

Brener, D. and Johnson, B.F. (1984) *Appl. Environ. Microbiol.*, **47**, 1126.

Brigham, J.S., Adney, W.S. and Himmel, M.E. (1996) In *Handbook on Bioethanol* (ed. C.E. Wyman), Taylor and Francis, Washington, DC, p. 119.

Brown, J.A., Falconer, D.J. and Wood, T.M. (1987) *Enzyme Microbiol. Technol.*, **9**, 169.

Brownell, H.H. and Saddler, J.N. (1986) *Biotechnol. Bioeng.*, **28**, 792.
Brownell, H.H. and Saddler, J.N. (1987) *Biotechnol. Bioeng.*, **29**, 228.
Bruchmann, E.E., Schach, H. and Graf, H. (1987) *Biotechnol. Appl. Biochem.*, **9**, 146.
Buchanen, R.A., Otey, F.H. and Hamerstrand, G.E. (1980) *Ind. Eng. Chem. Prod. Res. Div.*, **19**, 489.
Buchert, J., Ranua, M., Kantelinen, A. and Viikari, L. (1992) *Appl. Microbiol. Biotechnol.*, **37**, 825.
Castanon, M. and Wilke, C.R. (1980) *Biotechnol. Bioeng.*, **22**, 1037.
Castellanos, O.F., Sinitsyn, A.P. and Vlasenko, E.Y. (1995) *Biores. Technol.*, **52**, 119.
Cavaco-Paulo, A. and Almeida, L. (1994) *Biocatalysis*, **10**, 353.
Chang, M. (1971) *J. Polym. Sci.*, Part C, no. 36, 343.
Chanzy, H., Henrissat, B., Voung, R. and Schulein, M. (1983) *FEBS Lett.*, **163**, 113.
Chaudhuri, B.K. and Sahai, V. (1993) *Enzyme Microbiol. Technol.*, **15**, 513.
Chen, L.-F. and Gong, C.-S. (1982) In *Fourth Symposium on Biotechnology in Energy Production and Conservation, Biotechnology Bioengineering Symposium No. 12* (ed. C.D. Scott), John Wiley, New York, p. 57.
Chippaux, M. (1988) In *Proceedings of a Symposium and Genetics of Cellulose Degradation* (eds J.-P. Aubert, P. Beguin and J. Millet), Academic Press, New York, p. 219.
Chou, Y.C. (1986) In *Eighth Symposium on Biotechnology for Fuels and Chemicals, Biotechnology Bioengineering Symposium No. 17* (ed. C.D. Scott), John Wiley, New York, p. 19.
Claeyssens, M., Henrissat, B. and Tomme, P. (1993) In *Bioconversion of Forest and Plant Agricultural Residues* (ed. J.N. Saddler), CAB International, Oxford, p. 117.
Clark, T.A. and Mackie, K.L. (1986) In *Biotechnology in the Pulp and Paper Industry, 3rd International Conference*, Stockholm, 16–19 June, p. 101.
Converse, A.O. (1993) In *Bioconversion of Forest and Agricultural Plant Residues* (ed. J.N. Saddler), CAB International, Oxford, p. 93.
Coudray, M.R., Canavascini, G. and Meier, H. (1982) *Biochem. J.*, **203**, 227.
Coughlan, M.P. (1992a) *Biores. Technol.*, **39**, 107.
Coughlan, M.P. (1992b) In *Xylan and Xylanases* (eds J. Visser, G. Beldman, M.A. Kusters-van Someren and A.G.J. Vorgen), Elsevier, New York, p. 111.
Coughlan, M.P. and Hazlewood, G.P. (1993) *Hemicellulose and Hemicellulases*, Portland, London.
Coughlan, M.P. and Ljungdahl, L.G. (1988) In *Proceedings of a Symposium on Biochemistry and Genetics of Cellulose Degradation* (ed. J.-P. Aubert, P. Beguin and J. Millet), Academic Press, London, p. 11.
Cowling, E.B. (1975) In *Biotechnology and Bioengineering Symposium No. 5, Cellulose as a Chemical and Energy Resource* (ed. C.R. Wilke), John Wiley, New York, p. 163.
Crabbe, P.G., Tse, C.W. and Munro, P.A. (1986) *Biotechnol. Bioeng.*, **28**, 939.
Crawford, R.L. (1981) *Lignin Biodegradation and Transformation*, Wiley Interscience, New York.
Dadger, A.M. and Foutch, G.L. (1986) In *Seventh Symposium on Biotechnology for Fuels and Chemicals, Biotechnology Bioengineering Symposium No. 15* (ed. C.D. Scott), John Wiley, New York, p. 611.
Dale, B.E. (1987) *Trends Biotechnol.*, **5**, 287.
Dale, B.E. and Moreira, M.J. (1982) In *Fourth Symposium on Biotechnolgy in Energy Production and Conservation, Biotechnology Bioengineering Symposium No. 12* (ed. C.D. Scott), John Wiley, New York, p. 31.
Davies, G.J., Dodson, G.G., Hubbard, R.E. *et al.* (1993) *Nature*, **365**, 362.
Deeble, M.F. and Lee, J.M. (1986) In *Seventh Symposium on Biotechnology for Fuels and Chemicals, Biotechnology Bioengineering Symposium No. 15* (ed. C.D. Scott), John Wiley, New York, p. 277.
Dekker, R.F.H. (1980) *J. Gen. Microbiol.*, **120**, 309.
Dekker, R.F.H. (1985) In *Biosynthesis and Biodegradation of Wood Components* (ed. T. Higuchi), Academic Press, New Delhi, p. 505.
Dekker, R.F.H. and Wallis, A.F.A. (1983) *Biotechnol. Bioeng.*, **25**, 3027.
Demain, A.L. and Wu, J.H.D. (1989) In *Bioprocess Engineering; The First Generation* (ed. T.K. Ghose), Ellis Horwood, Chichester, p. 68.
Deshpande, M.V. and Eriksson, K.E. (1984) *Enzyme Microbiol. Technol.*, **6**, 338.

Divne, C., Stahlberg, J., Reinikainen, T. *et al.* (1994) *Science*, **265**, 524.
Doelle, H.W. and Doelle, M.B. (1989) *Austral. J. Biotechnol.*, **3**, 218.
Doppelbauer, R., Esterbauer, H., Steiner, W., Lafferty, R.M. and Steinmuller, H. (1987) *Appl. Microbiol. Biotechnol.*, **26**, 485.
Duff, S.J.B. and Murray, W.D. (1996) *Biores. Technol.*, **55**, 1.
Duff, S.J.B., Cooper, D.G. and Fuller, O.M. (1985) *Appl. Environ. Microbiol.*, **49**, 934.
Duff, S.J.B., Cooper, D.G. and Fuller, O.M. (1987) *Enzyme Microbiol. Technol.*, **9**, 47.
Duff, S.J.B., Moritz, J.W. and Cosavant, T.E. (1995) *Biotechnol. Bioeng.*, **45**, 239.
Durand, H., Baron, M., Calmels, T. and Tiraby, G. (1988) In *Proceedings of a FEMS Symposium on Biochemistry and Genetics of Cellulose Degradation* (eds J.-P. Aubert, P. Beguin and J. Millet), Academic Press, London, p. 135.
Enari, T.N. and Niku-Paavola, M.L. (1987) *Crit. Rev. Biotechnol.*, **5**, 67.
Eriksson, K.-E. (1981) In *Trends in the Biology of Fermentations for Fuels and Chemicals* (ed. A. Hollaender), Plenum Press, New York, p. 19.
Erikkson, K.-E. and Wood, T.M. (1985) In *Biosynthesis and Biodegradation of Wood Components* (ed. T. Higuchi), Academic Press, New York, p. 469.
Eriksson, K.-E., Gruenewald, A. and Vallender, L. (1980) *Biotechnol. Bioeng.*, **22**, 363.
Esterbauer, H., Steiner, W., Labudova, I., Hermann, A. and Hayn, M. (1991) *Biores. Technol.*, **36**, 51.
Eveleigh, D.E. (1981) In *Alcohol Fuels Process R/O Newsletter (Summer)*, Solar Energy Research Institute, Colorado, p. 50.
Fadda, M.B., Dessi, M.R., Rinaldi, A. and Satta, G. (1989) *Biotechnol. Bioeng.*, **33**, 777.
Fan, L.T., Lee, Y.-H. and Beardmore, D.H. (1980) In *Advances in Biochemical Engineering*, Vol. 14 (ed. A. Fiechter), Springer-Verlag, Berlin, p. 101.
Fan, L.T., Lee, Y.-H. and Beardmore, D.H. (1981) In *Proceedings of a Symposium Bioconversion and Biochemical Engineering*, Vol. 1 (ed. T.K. Ghose), Indian Institute of Technology, New Delhi, p. 233.
Feldman, S.D., Sahm, H. and Sprenger, G.A. (1992) *Appl. Microbiol. Biotechnol.*, **38**, 354.
Fengel, D. and Wegner, G. (1984) *Wood: Chemistry, Ultrastructure and Reactions*, Walter de Gruyter, New York.
Foody, B.E. and Foody, K.J. (1991) In *Energy from Biomass and Wastes* (ed. D.L. Klass), Institute of Gas Technology, Chicago, p. 1225.
Gallo, B.J. (1982) Paper presented at National Meeting of American Institute of Chemical Engineers, Orlando, 28 February–3 March.
Garg, S.K. and Jain, A. (1995) *Biores. Technol.*, **51**, 103.
Ghose, T.K. (1969) *Biotechnol. Bioeng.*, **11**, 239.
Ghose, T.K. (1987) *Pure Appl. Chem.*, **59**, 257.
Ghose, T.K. and Bisaria, V.S. (1979) *Biotechnol. Bioeng.*, **21**, 131.
Ghose, T.K. and Kostick, J.A. (1970) *Biotechnol. Bioeng.*, **12**, 921.
Ghose, T.K. and Pathak, A.N. (1973) *Process Biochem.*, **8**(5), 20.
Ghose, T.K. and Tyagi, R.D. (1979) *Biotechnol. Bioeng.*, **21**, 1387.
Ghose, T.K., Roychoudhury, P.K. and Ghosh, P. (1984) *Biotechnol. Bioeng.*, **26**, 377.
Ghosh, P. and Singh, A. (1993) *Adv. Appl. Microbiol.*, **39**, 295.
Ghosh, P., Pamment, N.B. and Martin, W.R.B. (1982a) *Enzyme Microbiol. Technol.*, **4**, 425.
Ghosh, V.K., Ghose, T.K. and Gopalkrishnan, K.S. (1982b) *Biotechnol. Bioeng.*, **24**, 241.
Gilbert, H.J. and Hazelwood, G.P. (1993) *J. Gen. Microbiol.*, **139**, 187.
Gilbert, I.G. and Tsao, G.T. (1983) In *Annual Reports on Fermentation Processes*, Vol. 6 (ed. G.T. Tsao), Academic Press, New York, p. 323.
Gilkes, N.R., Henrissat, B., Kilburn, D.G., Miller, R.C. Jr and Warren, R.A.J. (1991) *Microbiol. Rev.*, **55**, 303.
Giuliano, C. and Khan, A.W. (1984) *Appl. Environ. Microbiol.*, **48**, 446.
Gomes, I., Gomes, J., Steiner, W. and Esterbauer, H. (1992) *Appl. Microbiol. Biotechnol.*, **36**, 701.
Gould, J.M. (1984) *Biotechnol. Bioeng.*, **26**, 46.
Goyal, A., Ghosh, B. and Eveleigh, D. (1991) *Biores. Technol.*, **36**, 37.
Grethlein, H.E. (1978) *Biotechnol. Bioeng.*, **20**, 503.
Gritzali, M. and Brown, R.D. (1979) In *Hydrolysis of Cellulose: Mechanisms of Enzymatic*

and Acid Catalysis, Advances in Chemistry Series 181 (eds R.D. Brown, Jr and L. Jurasek), American Chemical Society, Washington, DC, p. 237.
Grohmann, K. (1993) In *Bioconversion of Forest and Agricultural Plant Residues* (ed. J.N. Saddler), CAB International, Oxford, p. 183.
Grohmann, K., Torget, R. and Himmell, M. (1986) In *Eighth Symposium on Biotechnology for Fuels and Chemicals, Biotechnology Bioengineering Symposium No. 17* (ed. C.D. Scott), John Wiley, New York, p. 135.
Grous, W.R., Converse, A.O. and Grethlein, H.E. (1986) *Enzyme Microbiol. Technol.*, **8**, 274.
Guidoboni, G.E. (1984) *Enzyme Microbiol. Technol.*, **6**, 194.
Gusakov, A.V., Sinitsyn, A.P. and Klyosov, A.A. (1987) *Biotechnol. Bioeng.*, **29**, 898.
Hahn-Haggerdal, B., Hallborn, J., Jeppsson, H. *et al.* (1993) In *Bioconversion of Forest and Agricultural Plant Residues* (ed. J.N. Saddler), CAB International, Oxford, p. 231.
Hallborn, J., Walfridsson, J., Airaksinen, U. *et al.* (1991) *Biotechnology*, **9**, 1090.
Haltrich, D., Preiss, M. and Steiner, W. (1993) *Enzyme Microbiol. Technol.*, **15**, 854.
Hamilton, T.J., Dale, B.E., Ladisch, M.R. and Tsao, G.T. (1984) *Biotechnol. Bioeng.*, **26**, 781.
Han, Y.W. and Ciegler, A. (1982) In *Fourth Symposium on Biotechnology in Energy Production and Conservation, Biotechnology and Bioengineering Symposium No. 12* (ed. C.D. Scott), John Wiley, New York, p. 73.
Hatakka, A.I. (1983) *Eur. J. Appl. Microbiol. Biotechnol.*, **18**, 350.
Hawley, M.C., Downey, K.W., Selke, S.M. and Lamport, D.T.A. (1986) In *Cellulose: Structure, Modification and Hydrolysis* (eds R.A. Young and R.M. Rowell), John Wiley, New York, p. 297.
Hayn, M., Steiner, W., Klinger, R. *et al.* (1993) In *Bioconversion of Forest and Agricultural Plant Residues* (ed. J.N. Saddler), CAB International, Oxford, p. 33.
Helle, S.S., Duff, S.J.B. and Cooper, D.G. (1993) *Biotechnol. Bioeng.*, **42**, 611.
Hendy, N., Wilke, C.R. and Blanch, H. (1984) *Enzyme Microbiol. Technol.*, **6**, 73.
Henrissat, B. (1991) *Biochem. J.*, **280**, 309.
Henrissat, B. and Bairoch, A. (1993) *Biochem. J.*, **293**, 781.
Hiltner, P. and Dehority, B.A. (1983) *Appl. Environ. Microbiol.*, **46**, 642.
Hogsett, D.A., Ahn, H.-J., Bernardez, T.D., South, C.R. and Lynd, L.R. (1992) *Appl. Biochem. Biotechnol.*, **34/35**, 527.
Holtzapple, M., Cognate, M., Shu, Y. and Hendrickson, C. (1990) *Biotechnol. Bioeng.*, **36**, 275.
Holtzapple, M.T. Lundeen, J.E., Sturgis, R., Lewis, J.E. and Dale, B.E. (1992) *Appl. Biochem. Biotechnol.*, **34/35**, 5.
Hoq, M.M., Hampel, C. and Deckwer, W.-D. (1994) *J. Biotechnol.*, **37**, 49.
Hreggvidsson, G.O., Kaiste, E., Holst, O. *et al.* (1996) *Appl. Environ. Microbiol.*, **62**, 3047.
Hsu, T.-A. (1996) In *Handbook on Bioethanol: Production and Utilization* (ed. C.E. Wyman), Taylor and Francis, Washington, DC, p. 179.
Ingram, L.O., Conway, T., Clark, D.P., Sewell, G.W. and Preston, J.F. (1987) *Appl. Environ. Microbiol.*, **53**, 2420.
Ishihara, M., Uemura, S., Hayashi, N. and Shimizu, K. (1991) *Biotechnol. Bioeng.*, **37**, 948.
Jurasek, L., Paice, M.G., Yaguchi, M. and O'Leary, S. (1987) In *Biomass Conversion Technology: Principles and Practice* (ed. M. Moo-Young), Pergamon Press, New York, p. 131.
Kadam, K.L. (1996) In *Handbook on Bioethanol: Production and Utilization* (ed. C.E. Wyman), Taylor and Francis, Washington, DC, p. 213.
Kaushik, V. and Bisaria, V.S. (1989) *J. Sci. Ind. Res.*, **48**, 276.
Karr, W.E., Cool, L.G., Merriman, M.M. and Brink, D.I. (1991), *J. Wood Chem. Technol.*, **11**, 447.
Kirk, T.K. and Farrell, R.L. (1987) *Ann. Rev. Microbiol.*, **41**, 465.
Klapatch, T.R., Hogsett, D.A., Baskaran, S., Pal, S. and Lynd, L.R. (1994) *Appl. Biochem. Biotechnol.*, **45/46**, 209.
Kluepfel, D., Vats-Mehta, S., Aumont, F., Shareck, F. and Morosoli, R. (1990) *Biochem. J.*, **267**, 45.
Klyosov, A.A. (1986) *Appl. Biochem. Biotechnol.*, **12**, 249.

Klyosov, A.A. (1989) In *Bioprocess Engineering: The First Generation* (ed. T.K. Ghose), Ellis Horwood, Chichester, p. 59.
Knowles, J., Lehtovarra, P. and Terri, T. (1987) *Trends Biotechnol.*, **5**, 255.
Kolot, F.B. (1984) *Process Biochem.*, **19**, 7.
Kosaric, N., Wieczroek, A., Cosentino, G.P., Magee, R.J. and Prenosil, J.E. (1983) In *Biotechnology*, Vol. 3 (ed. H. Dellweg), Verlag-Chemie, Weinheim, p. 257.
Kotter, P., Amore, R., Hollenberg, C.P. and Ciriacy, M. (1990) *Curr. Genet.*, **18**, 493.
Kubicek, C.P. (1992) *Advances in Biochemical Engineering Biotechnology*, Vol. 45 (ed. A. Fiechter), Springer Verlag, Berlin, p. 1.
Kundu, S., Ghose, T.K. and Mukhopadhyay, S.N. (1983) *Biotechnol. Bioeng.*, **25**, 1109.
Kyriacou, A., Neufeld, R. and MacKenzie, C.R. (1989) *Biotechnol. Bioeng.*, **33**, 631.
Lachke, A. and Deshpande, M.V. (1988) *FEMS Microbiol. Rev.*, **54**, 177.
Ladisch, M.R. and Svarczkopf, J.A. (1991) *Biores. Technol.*, **36**, 83.
Ladisch, M.R., Lin, K.W., Voloch, M. and Tsao, G.T. (1983) *Enzyme Microbiol. Technol.*, **5**, 82.
Lappalainen, A. (1986) *Biotechnol. Appl. Biochem.*, **8**, 437.
Lastick, S.M., Mohagheghi, A., Tucker, M.P. and Grohmann, K. (1990) *Appl. Biochem. Biotechnol.*, **25/25**, 431.
Lee, D.S. and Pack, M.V. (1987) *Enzyme Microbiol. Technol.*, **9**, 504.
Leisola, M.S.A. and Fiechter, A. (1985) *Adv. Biotechnol. Processes*, **5**, 59.
Leathers, T.-D. (1989) *J. Ind. Microbiol.*, **4**, 341.
Lindner, A. and Wegener, G. (1990) *J. Wood Chem. Technol.*, **10**, 331.
Lindley, N.D. and Durand, G. (1989) In *Bioprocess Engineering: The First Generation* (ed. T.K. Ghose), Ellis Horwood, Chichester, p. 324.
L'Italien, Y., Thibault, J. and Le Duy, A. (1989) *Biotechnol. Bioeng.*, **33**, 471.
Ljungdahl, L.G. and Wiegel, J.K.W. (1981) Anaerobic thermophilic culture system, US Patent 4292406.
Lynd, L.R. (1989) In *Advances in Biochemical Engineering/Biotechnology*, Vol. 38 (ed. A. Fiechter), Springer-Verlag, Berlin, p. 1.
Lynd, L.R. and Grethlein, H.E. (1987) *Biotechnol. Bioeng.*, **29**, 92.
Lynd, L.R., Cushman, J.H., Nicols, R.J. and Wyman, C.E. (1991) *Science*, **251**, 1318.
Macdonald, D.G. and Mathews, J.F. (1979) *Biotechnol. Bioeng.*, **21**, 1091
MacKenzie, C.R., Bilous, D. and Johnson, K.G. (1984) *Can. J. Microbiol.*, **30**, 1171.
Maheshwari, R. and Kamalam, P.T. (1985) *J. Gen. Microbiol.*, **131**, 3017.
Mairoella, B.L., Blanch, H.W. and Wilke, C.R. (1984) *Biotechnol. Bioeng.*, **26**, 1003.
Mandels, M. (1975) In *Cellulose as a Chemical and Energy Resource, Biotechnology Bioengineering Symposium No. 5* (ed. C.R. Wilke), John Wiley, New York, p. 81.
Mandels, M. (1981) In *Annual Reports in Fermentation Processes*, Vol. 5, (ed. G.T. Tsao), Academic Press, New York, p. 35.
Mandels, M. (1985) *Biochem. Soc. Trans.*, **13**, 414.
Mandels, M., Hontz, L. and Nystrom, J. (1974) *Biotechnol. Bioeng.*, **16**, 1471.
Mandels, M., Medeiros, J.E., Andreotti, R.E. and Bissett, F.H. (1981) *Biotechnol. Bioeng.*, **23**, 2009.
Marchal, R., Ropars, M., Pourquie, J. and Vandescasteele, J.P. (1992) *Biores. Technol.*, **42**, 205.
Mas, C., Jansen, N.B. and Tsao, G.T. (1988) *Biotechnol. Bioeng.*, **31**, 366.
McCleary, B.V., Gibson, T.S., Allen, H. and Gams, T.C. (1986) *Starke*, **38**, 433.
McLean, D. and Podruzny, M.F. (1985), *Biotechnol. Lett.*, **7**, 683.
McMillan, J.D. (1994) In *Bioconversion for Fuels* (eds M.E. Himmel, J.O., Baker and R.P. Overend), ACS Symposium Series, 566, American Chemical Society, Washington, DC, p. 411.
McMillan, J.D. (1996) In *Handbook on Bioethanol – Production and Utilization* (ed. C.E. Wyman), Taylor and Francis, Washington, DC, p. 287.
Messner, K. and Srebotnik, E. (1994) *FEMS Microbiol. Rev.*, **1**, 351.
Millet, M.A. Effland, M.J. and Caulfield, D.F. (1979) In *Hydrolysis of Cellulose: Mechanisms of Enzymatic and Acid Catalysis*, Advances in Chemistry Series No. 181 (eds R.D. Brown, Jr and L. Jurasek), American Chemical Society, Washington, DC, p. 71.
Mohagheghi, A., Grohmann, K. and Wyman, C.E. (1990) *Biotechnol. Bioeng.*, **35**, 211.

Moloney, A.P., McCrae, S.I., Wood, T.M. and Coughlan, M.P. (1985) *Biochem. J.*, **225**, 365.

Monot, F., Martin, J.R., Petitdemange, H. and Gay, R. (1982) *Appl. Environ. Microbiol.*, **44**, 1318.

Montenecourt, B.S. (1983) *Trends Biotechnol.*, **1**, 156.

Montenecourt, B.S. and Eveleigh, D.E. (1979) In *Hydrolysis of Cellulose: Mechanisms of Enzymatic and Acid Catalysis,* Advances in Chemistry Series No. 181 (eds R.D. Brown, Jr and L. Jurasek), American Chemical Society, Washington, DC, p. 289.

Morpath, F.F. and Jones, G.D. (1986) *Biochem. J.*, **236**, 221.

Mukataki, S., Kobayashi, N., Sato, S. and Takahashi, J. (1988) *Biotechnol. Bioeng.*, **32**, 760.

Mulchandani, A. and Volesky, B. (1994) *J. Biotechnol.*, **34**, 51.

Nanda, M., Bisaria, V.S. and Ghose, T.K. (1986) *J. Gen. Microbiol.*, **132**, 2761.

Neely, W.C. (1984) *Biotechnol. Bioeng.*, **26**, 59.

Neese, N. Wallick, J. and Harper, J.M. (1977) *Biotechnol. Bioeng.*, **19**, 323.

Nevalainen, K.M.H., Palva, E.T. and Bailey, M.J.B. (1980) *Enzyme Microbiol. Technol.*, **3**, 59.

Ng, T.K. and Zeikus, J.G. (1982) *J. Bacteriol.*, **150**, 1391.

Ng, T.K., Weimer, P.J. and Zeikus, J.G. (1977) *Arch. Microbiol.*, **114**, 1.

Nigam, M. (1989) Studies on the kinetics of 2,3-butanediol formation by *Klebsiella oxytoca*. Ph.D. thesis, Indian Institute of Technology, New Delhi.

Nishikawa, N.K., Sutcliffe, R. and Saddler, J.N. (1988) *Biotechnol. Bioeng.*, **31**, 624.

Nummi, M., Niku-Paavola, M.-L., Lappalainen, A., Enari, T.-M. and Rounio, V. (1983) *Biochem. J.*, **215**, 677.

Ohta, K., Alterthum, F. and Ingram, L.O. (1990) *Appl. Environ. Microbiol.*, **56**, 463.

Panda, T., Bisaria, V.S. and Ghose, T.K. (1987) *Biotechnol. Bioeng.*, **30**, 868.

Parisi, F. (1989) In *Advances in Biochemical Engineering/Biotechnology*, Vol. 38, (ed. A. Fiechter), Springer-Verlag, Berlin, p. 53.

Park, C.-H., Okos, M.R. and Wankat, P.C. (1989) *Biotechnol. Bioeng.*, **34**, 18.

Paszner, L. and Cho, H.J. (1988) *Energy Exploit. Explor.*, **6**, 39.

Pearce, P.D. and Bauchop, T.D. (1985) *Appl. Environ. Microbiol.*, **49**, 1265.

Philippidis, G.P. (1996) In *Handbook on Bioethanol – Production and Utilization* (ed. C.E. Wyman), Taylor and Francis, Washington, DC, p. 253.

Philippidis, G.P., Smith, T.K. and Wyman, C.E. (1993) *Biotechnol. Bioeng.*, **41**, 846.

Picataggio, S.K. and Zhang, M. (1996) In *Handbook on Bioethanol – Production and Utilization* (ed. C.E. Wyman), Taylor and Francis, Washington, DC, p. 163.

Pokorny, M., Zupancic, S., Steiner, T. and Kreiner, W. (1990) In *Trichoderma reesei Cellulases: Biochemistry, Genetics, Physiology and Applications* (eds C.P. Kubicek, D.E. Eveleigh, H. Esterbauer, W. Steiner and E.M. Kubicek-Pranz), The Royal Society of Chemistry, Cambridge, p. 168.

Poole, D.M., Hazlewood, G.P., Huskisson, N.S., Virden, R. and Gilbert, H.J. (1993) *FEMS Microbiol. Lett.*, **106**, 77.

Pourquie, J. and Warzywoda, M. (1993) In *Bioconversion of Forest and Agricultural Plant Residues* (ed. J.N. Saddler), CAB International, Oxford, p. 107.

Pourquie, J., Warzywoda, M., Chevron, F., Thery, M., Lonchamp, D. and Vandercasteele, J.P. (1988). In *Proceedings of the FEMS Symposium Biochemistry and Genetics of Cellulose Degradation* (eds J.-P. Aubert, P. Beguin and J. Millet), Academic Press, New York, USA, p. 71.

Poutanen, K. and Puls, J. (1988) *Appl. Microbiol. Biotechnol.*, **28**, 425.

Puls, J. (1993) In *Bioconversion of Forest and Agricultural Plant Residues* (ed. J.N. Saddler), CAB International, Oxford, p. 13.

Puls, J., Poutanen, K., Kroner, H.U. and Viikari, L. (1985) *Appl. Microbiol. Biotechnol.*, **22**, 416.

Ratto, M., Poutanen, K. and Viikari, L. (1992) *Appl. Microbiol. Biotechnol.*, **37**, 470.

Reese, E.T. and Ryu, D.Y. (1980) *Enzyme Microbiol. Technol.*, **2**, 239.

Reinikainen, T., Ruohonen, L., Nevanen, T. *et al.* (1992) *Proteins*, **14**, 475.

Reynen, M., Reipen, I., Sahm, H. and Sprenger, G.A. (1990) *Molec. Gen. Genet.*, **223**, 335.

Robson, L.M. and Chambliss, G.H. (1989) *Enzyme Microbiol. Technol.*, **11**, 626.

Rogers, P.L., Lee, K.J. and Tribe, D.E. (1980) *Process Biochem.*, **15**, 7.

Ropars, M., Marchal, R., Pourquie, J. and Vendecasteele, J.P. (1992) *Biores. Technol.*, **42**, 197.
Roukas, T. (1994) *Biotechnol. Bioeng.*, **43**, 189.
Rouvinen, J., Bergfors, T., Teeri, T., Knowles, J.K.C. and Jones, T.A. (1990) *Science*, **249**, 380.
Rowland, S.P. and Roberts, E.J. (1972) *J. Polym. Sci. Part A-1*, no. 10, 2447.
Ryu, D.D.Y. and Mandels, M. (1980) *Enzyme Microbiol. Technol.*, **2**, 91.
Ryu, D.D.Y., Andreotti, R., Mandels, M., Gallo, B. and Reese, E.T. (1979) *Biotechnol. Bioeng.*, **21**, 1887.
Sablayrolles, J.M. and Goma, G. (1984) *Biotechnol. Bioeng.*, **26**, 148.
Saddler, J.N. and Chan, M.K.-H. (1982) *Eur. J. Appl. Microbiol. Biotechnol.*, **16**, 99.
Saddler, J.N. and Makie, K. (1990) *Biomass*, **22**, 293.
Saddler, J.N., Brownell, H.H., Clermont, L.P. and Levitin, N. (1982) *Biotechnol. Bioeng.*, **24**, 1389.
Saddler, J.N., Mes-Hartee, M., Yu, E.K.C. and Brownell, H.H. (1983) In *Fifth Symposium on Biotechnology for Fuels and Chemicals, Biotechnology Bioengineering Symposium No. 13* (ed. C.D. Scott), John Wiley, New York, p. 225.
Saddler, J.N., Ramos, L.P. and Brenil, C. (1993) In *Bioconversion of Forest and Agricultural Plant Residues* (ed. J.N. Saddler), CAB International, Oxford, p. 73.
Sahm, H. (1989) In *Bioprocess Engineering: The First Generation* (ed. T.K. Ghose), Ellis Horwood, Chichester, p. 303.
Saraswat, V. and Bisaria, V.S. (1997) *J. Ferment. Bioeng.*, **83**, 316.
Sarkarsen, K.V. (1980) In *Progress in Biomass Conversion*, Vol. 2 (eds K.V. Sarkarsen and D.A. Tillman), Academic Press, New York, p. 127.
Sattler, W., Esterbauer, H., Glatter, O. and Steiner, W. (1989) *Biotechnol. Bioeng.*, **33**, 1221.
Schaffner, D.W. and Toledo, R.T. (1991) *Biotechnol. Bioeng.*, **37**, 12.
Schell, D.J. and Duff, B. (1996) In *Handbook on Bioethanol – Production and Utilization* (ed. C.E. Wyman), Taylor and Francis, Washington, DC, p. 381.
Schneider, H. (1989) *Crit. Rev. Biotechnol.*, **9**, 1.
Schultz, T.P., Rughani, J. and McGinnis, J.D. (1989) *Appl. Biochem. Biotechnol.*, **20/21**, 9.
Schell, D.J., Hinman, N.D., Wyman, E.C. and Werdene, P.J. (1990) *Appl. Biochem. Biotechnol.*, **24/25**, 287.
Schell, D., Torget, R., Power, A. *et al.*, (1991) *Appl. Biochem. Biotechnol.*, **28/29**, 87.
Senior, D.J., Mayers, P.R. and Saddler, J.N. (1989) *ACS Symposium Series*, Vol. 399, 641.
Sheir-Neiss, G. and Montenecourt, B.S. (1984) *Appl. Microbiol. Biotechnol.*, **20**, 46.
Shoemaker, S.P., Raymond, J.C. and Burner, R. (1981) In *Trends in the Biology of Fermentations for Fuels and Chemicals* (ed. A. Hollaender), Plenum Press, New York, p. 89.
Shukla, R., Kang, W. and Sirkar, K.K. (1989) *Biotechnol. Bioeng.*, **34**, 1158.
Singh, A., Kumar, P.K.R. and Schugerl, K. (1991) *J. Biotechnol.*, **18**, 205.
Skoog, K. and Hahn-Haggerdal, B. (1990) *Appl. Environ. Microbiol.*, **56**, 3389.
Skotnicki, M.L., Tribe, D.E. and Rogers, P.L. (1980) *Appl. Environ. Microbiol.*, **40**, 7.
Slapack, G.E., Russell, I. and Stewart, G.G. (1987) *Thermophilic Microbes in Ethanol Production*, CRC Press, Boca Raton, FL.
Smith, D.C. and Wood, T.M. (1991) *Biotechnol. Bioeng.*, **38**, 883.
Soni, B.K., Das, K., Soucaille, P. and Goma, G. (1986) In *Eighth Symposium Biotechnology for Fuels and Chemicals, Biotechnology Bioengineering Symposium No. 17* (ed. C.D.Scott), John Wiley, New York, p. 591.
Spangler, D.J. and Emert, (1986) *Biotechnol. Bioeng.*, **28**, 115.
Spindler, D.D., Wyman, C.E., Mohagheghi, A. and Grohmann, K. (1988) *Appl. Biochem. Biotechnol.*, **17**, 279.
Spindler, D.D., Wyman, C.E. and Grohmann, K. (1989) *Biotechnol. Bioeng.*, **34**, 189.
Sprey, B. and Bochem, H.P. (1991) *FEMS Microbiol. Lett.*, **78**(2–3), 183.
Srivastava, S.K. (1985) Kinetics of β-glucosidase production and its role in the enzymatic hydrolysis of cellulose. Ph.D. thesis, Indian Institute of Technology, New Delhi.
Srivastava, A.K. (1989) Bioconversion of cellulosic residues into microbiol biomass protein by lignocellulolytic fungus *Coriolus hirsutus*. Ph.D. thesis, Indian Institute of Technology, New Delhi.

Stark, J.R and Yin, X.S. (1987) *Enzyme Microbiol. Technol.*, **9**, 156.

Sternberg, D. and Dorval, S. (1979) *Biotechnol. Bioeng.*, **21**, 181.

Sternberg, D. and Mandels, G.R. (1979) *J. Bacteriol.*, **139**, 761.

Sternberg, D., Vijay Kumar, P. and Reese, E.T. (1977) *Can. J. Microbiol.*, **23**, 139.

Stockberger, P. (1993) *Tappi J.*, **76**, 71.

Sundstrom, D.W., Klei, H.E., Coughlin, R.W., Biederman, G.J. and Brouwer, C.A. (1981) *Biotechnol. Bioeng.*, **23**, 473.

Szabo, I.J., Johansson, G. and Pettersson, G. (1996) *J. Biotechnol.*, **48**, 221.

Szczodrak, J. (1988) *Biotechnol. Bieng.*, **32**, 771.

Szczodrak, J. (1989) *Biotechnol. Bioeng.*, **33**, 1112.

Szczodrak, J. and Targonski, Z. (1988) *Biotechnol. Bioeng.*, **31**, 300.

Tailliez, P., Girard, H., Millet, J. and Beguin, P. (1989) *Appl. Environ. Microbiol.*, **55**, 207.

Takagi, M., Abe, S., Suzuki, S., Emert, G.H. and Yata, N. (1978) In *Proceedings of the Symposium on Bioconversion of Cellulosic Substances into Energy, Chemicals and Microbial Protein* (ed. T.K. Ghose), Indian Institute of Technology, New Delhi, p. 551.

Takuma, S., Nakashima, N., Tantirungkij, S. *et al.* (1991) *Appl. Biochem. Biotechnol.*, **28/29**, 327.

Tan, L.U.L., Yu, E.K.C., Mayers, P. and Saddler, J.N. (1987) *Appl. Microbiol. Biotechnol.*, **26**, 21.

Tanaka, M., Ikesaka, M., Matsuno, R. and Converse, A.O. (1988a) *Biotechnol. Bioeng.*, **32**, 698.

Tanaka, M., Fukuii, M. and Matsuno, R. (1988b) *Biotechnol. Bioeng.*, **32**, 897.

Taniguchi, M., Kobayashi, M. and Fujii, M. (1989) *Biotechnol. Bioeng.*, **34**, 1092.

Tassinari, T. and Macy, C. (1977) *Biotechnol. Bioeng.*, **19**, 1321.

Tenkanen, M., Puls, J. and Poutanen, K. (1992) *Enzyme Microbiol. Technol.*, **14**, 566.

Tjerneld, F., Persson, I., Albertsson, P.-A. and Hahn-Haegerdal, B. (1986) In *Seventh Symposium on Biotechnology for Fuels and Chemicals, Biotechnology Bioengineering Symposium No. 15* (C.D. Scott), John Wiley, New York, p. 419.

Tolan, J.S. (1996) In *Industrial Enzymology*, 2nd edn (ed. T. Godfrey and S. West), Macmillan Press, London, p. 327.

Tomme, P., Warren, R.A.J. and Gilkes, N.R. (1995) *Adv. Microbiol. Physiol.*, **37**, 1.

Toyama, N. (1969) In *Cellulases and their Applications*, Advances in Chemistry Series, Vol. 95 (ed. R.F. Gould), American Chemical Society, Washington, DC, p. 359.

Torrie, J. (1991) Extracellular β-D-mannanase activity from *Trichoderma harzianum* E 58, Ph.D. thesis, University of Ottawa, Canada.

Turker, M. and Mavituna, F. (1987) *Enzyme Microbiol. Technol*, **9**, 739.

Vallender, L. and Eriksson, K.E. (1990) *Advances in Biochemical Engineering Biotechnology*, Vol. 42 (ed. A. Fiechter), Springer Verlag, Berlin, p. 63.

Viikari, L., Nybergh, P. and Linko, M. (1981) In *Advances in Biotechnology*, Vol. 2 (eds. M. Moo-Young and C.W. Robinson), Pergamon Press, New York, p. 137.

Viikari, L., Sundquist, J. and Kettunen, J. (1991) *Paper Timber*, **73**, 384.

Viikari, L., Tenkanen, M., Buchert, J. *et al.* (1993) In *Bioconversion of Forest and Agricultural Plant Residues* (ed. J.N. Saddler), CAB International, Oxford, p. 131.

Visser, J., Beldman, G., Kusters-van Someren, M.A. and Voragen, A.G.J. (1992) In *Xylans and Xylanases*, Elsevier, New York.

von Sivers, M. and Zacchi, G. (1995) *Biores. Technol.*, **51**, 43.

Votruba, J., Volseky, B. and Yerushalmi, L. (1986) *Biotechnol. Bioeng.*, **28**, 247.

Wang, D.I.C., Avgerinos, G.C., Biocic, I., Wang, S.-D and Fang, H.-Y (1983) *Phil. Trans. Roy. Soc. Lond. Ser. B.*, **300**, 323.

Warren, R.A.J. (1993) *Curr. Opin. Biotechnol.*, **4**, 469.

Watson, T.G., Nelligan, I. and Lessing, L. (1984) *Biotechnol. Lett.*, **6**, 667.

Wilke, C.R., Yang, R.D. and Von Stockar, U. (1976) In *Enzymatic Conversion of Cellulosic Materials: Technology and Applications, Biotechnology Bioengineering Symposium No.6* (eds E.L. Gaden, Jr, M.H. Mandels, E.T. Reese and L.A. Spano), John Wiley, New York, p. 155.

Wong, D.W.S. (1995) *Food Enzymes: Structure and Mechanism*, Chapman & Hall, New York.

Wong, K.K.Y., Tan, L.U.L. and Saddler, J.N. (1986), *Enzyme Microbiol. Technol.*, **8**, 617.

Wood, T.M. (1985) *Biochem. Soc. Trans.*, **13**, 407.

Wood, T.M. (1988) In *Biodeterioration 7* (eds D.R. Houghton, R.-N. Smith and H.O.W. Eggins), Elsevier Applied Science, London, p. 333.

Wood, T.M. and McCrae, S.I. (1977) *Carbohydr. Res.*, **57**, 117.

Wood, T.M. and McCrae, S.I. (1979) In *Hydrolysis of Cellulose: Mechanism of Enzymatic and Acid Catalysis,* Advances in Chemistry Series no. 181 (eds R.D. Brown, Jr and L. Jurasek), American Chemical Society, Washington, DC, p. 181.

Wood, T.M. and Saddler, J.N. (1988) *Meth. Enzymol.*, **160** (Part A), 3.

Wood, T.M., McCrae, S.I., Wilson, C.A., Bhat, K.M. and Gow, L.A. (1988) In *Proceedings of the Symposium on Biochemistry and Genetics of Cellulose Degradation* (ed. J.-P. Aubert, P. Beguin and J. Millet), Academic Press, London, p. 31.

Woodward, J., Whaley, K.S., Zachry, G.S. and Wohlpart, D.L. (1981) In *Third Symposium on Biotechnology in Energy Production and Conservation, Biotechnology Bioengineering Symposium No. 11* (ed. C.D. Scott), John Wiley, New York, p. 619.

Wright, J.D. (1988) *Chem. Eng. Prog.*, **8**, 62.

Wright, J.D., Power, A.J. and Douglas, L.J. (1986) In *Eighth Symposium on Biotechnology for Fuels and Chemicals, Biotechnology Bioengineering Symposium No. 17* (ed. C.D. Scott), John Wiley, New York, p. 285.

Wyman, C.D. (1994) *Biores, Technol.*, **50**, 3.

Wyman, C.E. and Goodman, B.J. (1993) *Appl. Biochem. Biotechnol.*, **39/40**, 41.

Yablonsky, M.D., Bartley, T., Elliston, K.O. *et al.* (1988) In *Proceedings of the Symposium Biochemistry and Genetics of Cellulose Degradation* (eds J.-P. Aubert, P. Beguin and J. Millet), Academic Press, New York, p. 249.

Yu, E.K.C. and Saddler, J.N (1987) In *Biomass Conversion Technology: Principles and Practices* (ed. M. Moo-Young), Pergamon Press, New York, p. 103.

Zadril, F. and Grabbe, K. (1983) In *Biotechnology*, Vol. 3 (ed. H. Dellweg), Verlag Chemie, Weinheim, p. 145.

Zhang, M., Eddy, C., Deanda, K., Finkelstein, M. and Picataggio, S. (1995) *Science*, **267**, 240.

6 Use of photosynthetic bacteria for the production of SCP and chemicals from organic wastes

KEN SASAKI, TOHRU TANAKA AND SHIRO NAGAI

6.1 Introduction

Photosynthetic bacteria are anoxygenic phototrophs distributed widely in nature in such habitats as fields, lakes, ponds, rivers and oceans. Bacteria have the capability to utilize sulfate, thiosulfate and hydrogen as electron donors as well as various organic compounds, and are able to fix carbon dioxide and molecular nitrogen under anaerobic–light conditions (Pfenning, 1978; Hoshino and Kitamura, 1984; Truper and Pfenning, 1989; Sasaki *et al.*, 1995a). Hence, the bacteria play an important role in the circulation of sulfur and nitrogen in the acquatic environment (Kobayashi, 1984).

Photosynthetic bacteria are inherently different from algae and plants in regard to their photosynthetic organs because these bacteria harvest energy (ATP, adenosine triphosphate) only via photosystem I (Morita and Shimada, 1984) compared with the two electron-harvesting systems of algae and plants, i.e. photosystems I and II. Recently, Cyanobacteria were classified into the photosynthetic bacterial group as oxygenic photosynthetic bacteria (Truper and Pfenning, 1989). These organisms have the two electron-harvesting systems (photosystems I and II) and are able to decompose water to obtain electrons for the electron transport system. At the same time, oxygen is produced. This group is sometimes classified into the algae group. Therefore, to prevent confusion, only conventional anoxygenic phototrophs will be described as photosynthetic bacteria in this chapter.

Photosynthetic bacteria essentially grow under anaerobic–light conditions: however, some of them can grow heterotrophically under aerobic–dark conditions in a manner similar to that of other heterotrophic microorganisms. Further, some photosynthetic bacteria can reduce nitrate and sulfate to generate ATP (Truper and Pfenning, 1989).

6.1.1 General characteristics of photosynthetic bacteria

The photosynthetic bacteria are classified into green sulfur, purple sulfur and purple nonsulfur bacteria, and cyanobacteria (Table 6.1). Among these photosynthetic bacteria, several bacteria in *Rhodospirillaceae*, such as *Rhodobacter (Rb.)*, *Rhodospirillum (Rsp.)*, *Rhodocyclus (Rc.)* and *Rhodopseudomonas (Rp.)*, are attractive for their application to single cell protein (SCP) and chemical productions because they can grow aerobically and/or photosynthetically utilizing organic substrates. *Rhodospirillum* and *Rhodobacter* are being studied for biofuel (hydrogen) production from carbonaceous wastes (Takahashi, 1984; Miyake and Asada, 1996). In addition, as the bacterial cells contain a large amount of protein with valuable vitamins and photopigments, the cells are promising as a potent SCP source for animal feed. Some photosynthetic bacteria, e.g. *Rhodocyclus* spp. capable of hydrolysing starch and protein (Noparatnaraporn, 1987), are also useful to treat agro-industrial wastes as well as being a source of SCP.

As *Chromatiaceae*, *Ectothiorhodospiriraceae*, *Chlorobiaceae* and *Chloroflexaceae* grow only photosynthetically or photoautotrophically under anaerobic–light conditions, they do not seem to be a potent SCP source, owing to their slow growth under light illumination. Cyanobacteria are conventionally called blue–green algae, e.g. *Spirulina*, *Anabaena* and *Synechococcus*. *Spirulina* cells have already been commercialized as healthy human food and as feedstock for fish (Kato, 1991). The applications of these algae have been described elsewhere (Yata *et al.*, 1996).

In this chapter, *Rhodospirillaceae* capable of growing under aerobic–dark and/or anaerobic–light conditions with relatively rapid growth rates are considered for their use as bacteria for the production of SCP and chemicals from organic wastes, including agro-industrial waste.

6.1.2 Application of photosynthetic bacteria for SCP and chemical production from organic wastes

A large quantity of organic waste is discharged throughout the world, e.g. in Japan about 250 000 000 tons year^{-1} of organic waste including agricultural waste and 77 208 000 tons year^{-1} of animal waste were discharged in 1991 (Nakayama, 1993). Environmental pollution caused by piling up these wastes causes problems in local country districts.

For the treatment or recycling of such agro-industrial waste, methane fermentation (biogas production) is widely applied for chemical oxygen demand (COD) reduction together with energy recovery from the waste. Compost production is also applied using animal and agricultural solid waste (Nakasaki, 1995; Yata *et al.*, 1996).

The utilization of the photosynthetic bacteria from agro-industrial waste to produce SCP has been studied (Shipman *et al.*, 1975; Kobayashi and Kurata, 1978; Sasaki *et al.*, 1995). The advantages are that these bacteria can achieve a high growth rate under aerobic and anaerobic light

Table 6.1 Characteristics and classification of anoxygenic photosynthetic bacteria

Bacteria	Color	Bacterio-chlorophyll	Main carbon source	Main electron donor	Growth condition	Typical genera
Rhodospirillaceae	Red Violet Brown	Bchl a or b	Organic matter CO_2	Organic matter, H_2 (H_2S)[a]	Anaerobic–light Aerobic–dark Aerobic–light	*Rhodobacter* *Rhodospirillum* *Rhodopseudomonas* *Rhodocyclus*
Chromatiaceae	Red Pink Violet	Bchl a or b	Organic matter CO_2	H_2S, $S_2O_3^{2-}$ (S, H_2)[a]	Anaerobic–light	*Chromatium* *Thiocyctis* *Thiocapsa*
Ectothiorhodo-spiracaea	Red Pink Green	Bchl a or b	Organic matter CO_2	Organic matter H_2S, H_2	Anaerobic–light	*Ectothiorhodo-spira*
Chlorobiaceae	Green Brown	Bchl a or c, d, e	CO_2	H_2S, H_2 $S_dO_3^{2-}$	Anaerobic–light	*Chlorobium* *Pelodictyon* *Chlorochromatium*
Chloroflexaceae	Green Orange	Bchl a or c, d, e	Organic matter CO_2	Organic matter H_2S, $S_2O_3^{2-}$	Anaerobic–light	*Chloroflexus* *Chloronema* *Oscillochloris*

Modified from Kitamura (1988).
[a]Only a few bacteria can utilize this.

conditions. For example, *Rhodospirillaceae* can grow at a similar rate to that of SCP yeasts, however, the protein content of the bacterial cells (60–70%) is relatively high compared with that of yeasts (40–60%). In addition, these bacteria contain various vitamins, especially vitamin B_{12}. For example, using photosynthetic bacteria in animal feedstock, the addition of the cells to hen feed (0.01–0.04%) increased the egg-laying rate by 10% and improved the quality of the egg yolk (Kobayashi and Kurata, 1978). In some fish culture in Japan, photosynthetic bacterial cells have been widely used as a source of vitamins and trace minerals for cultivating healthy fish. Photosynthetic bacterial cells have been effective for the treatment of fish with viral infections (Tamai *et al.*, 1996). Thus, it is expected that photosynthetic bacteria cells will be promising for use in feedstuffs as a supplementary protein source containing valuable materials.

6.2 SCP production from waste

6.2.1 Pineapple waste

Pineapple canning industries, which are widely distributed in the tropical countries such as Thailand, produce a large quantity of solid and liquid waste. These wastes are high organic loading wastes (COD = *c.* 10 000 mg l^{-1}) which contain mainly carbohydrates and little protein. The liquid waste is usually treated by aerated lagoon or oxidation pond processes, whereas the semisolid wastes are normally discarded without proper treatment. Therefore, the semisolid wastes, consisting of the peel and core, are the major cause of serious pollution.

Photosynthetic bacteria as a source of SCP were used to treat pineapple waste water and the nutritive quality of the cell mass production. In this section, *Rhodobacter sphaeroides* strain P_{47} was used for cell mass production from pineapple peel waste.

(a) Chemical composition. The chemical composition of pineapple peel waste is shown in Table 6.2. The composition varies considerably depending on the season, area and canning process of the industry. The waste contains mainly sucrose, glucose and fructose, while dextrin, raffinose and galactose are minor constituents. Although the waste contains little nitrogen and soluble protein, the necessary elements for growth of photosynthetic bacteria, i.e. Fe, Ca, Mg, Mn and Co, are present in adequate concentrations.

The growth and sugar utilization of *Rb. sphaeroides* P_{47} dark-cultivated aerobically on pineapple waste medium are shown in Figure 6.1 using a 2.5 l fermenter (1.5 l working volume). The details of this experiment have been described in Sasaki *et al.* (1991). The growth pattern was diauxic.

Table 6.2 Chemical composition (g l^{-1})[a] of pineapple peel waste

	Filtrate sterilization[a]	After
COD	100.8	103.7
Reducing sugar	39.2	41.2
Total sugar	100.0	100.9
Dextran	1.5	1.5
Raffinose	2.6	1.5
Sucrose	40.1	40.1
Glucose	23.6	23.6
Galactose	1.7	2.1
Fructose	14.0	15.6
Soluble protein	0.9	n.a.
Kjeldahl nitrogen	0.2	n.a.
Trace elements (mg l^{-1})		
Fe	5.43	n.a.
Ca	3.31	n.a.
Mn	13.97	n.a.
Mg	62.50	n.a.
Co	0.07	n.a.
Cu	2.02	n.a.
Cd	0.03	n.a.
Na	8.61	n.a.
SO_4^{2-}	169.7	n.a.
PO_4^{3-}	223.8	n.a.
pH	4.0	4.0

[a]The variation of sugar content and fraction depended on the harvesting season, the area and the canning process used.
[b]Autoclaved at 15 lb $inch^{-2}$ for 15 min.
n.a. = not analysed.
Reproduced with permission from Noparatnaraporn, N. *et al.*, *J. Ferment. Technol.*, **64**, 137–143; published by the Society for Fermentation and Bioengineering, Japan, 1986.

After a 40 h culture, when glucose and fructose had been completely utilized, the remaining sucrose (45 g l^{-1}) was then utilized rapidly and the second phase could be observed up to 56 h of cultivation. The phenomena observed for sugar utilization would suggest that sucrase synthesis might be repressed while the glucose remained in the culture medium. The culture yielded 26.5 g dry cells l^{-1} after 60 h of cultivation with a maximum specific growth rate of 0.31 l h^{-1} and a growth yield of 0.45 g cell g^{-1} sugar. COD removal was 85.3%.

Based on basic experiments, such as that shown in Figure 6.1, the following estimations can be considered. Daily, 150 tons of waste from a medium-sized pineapple cannery produces 75 tons of juice (*c.* 100 g l^{-1} of total sugar; pH 4.0). Supplements of nitrogen, phosphorus and essential vitamins (nicotinic acid, thiamine–HCl and biotin) together with pH adjustment to 6.8–7.0 are necessary before inoculation. After 56 h of cultivation, a cell mass of 2.5 tons of dry cells day^{-1} should be recovered

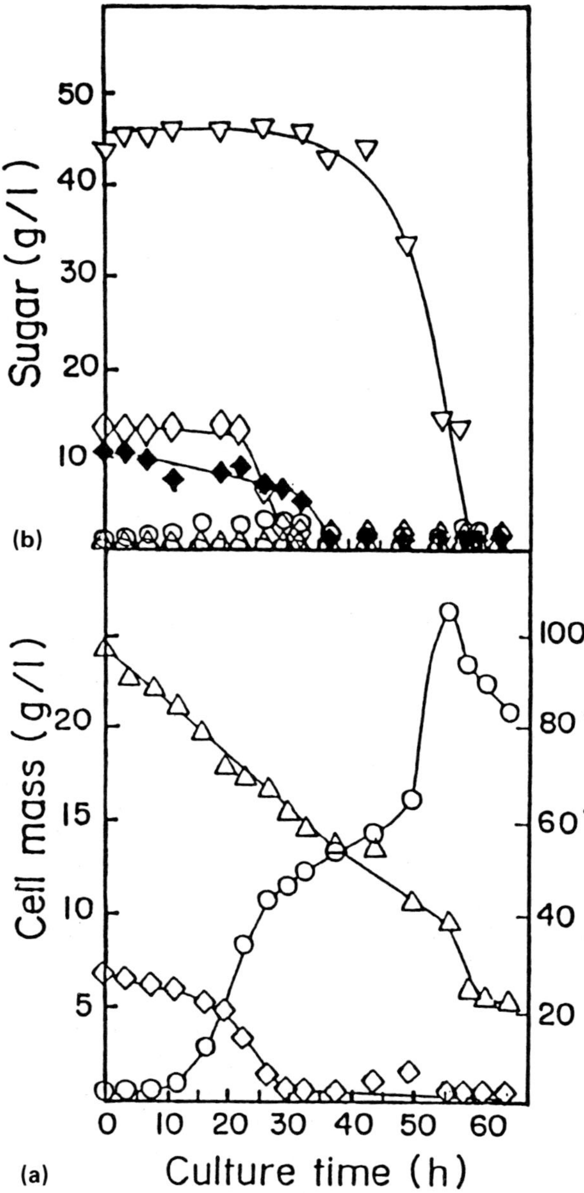

Figure 6.1 Growth, sugar consumption and uptake of sugar components of *Rb. sphaeroides* P_{47} on pineapple peel waste medium (initial total sugar *c.* 100 g l^{-1}). (a) growth and sugar consumption; (b) uptake of sugar components. (a): ○ = cell mass; △ = total sugar; ◊ = reducing sugar. (b): ▽ = sucrose; □ = galactose; ◊ = glucose; △ = raffinose; ♦ = fructose; ○ = dextran.

and the remaining effluent after the cell recovery containing approximately 23 g l^{-1} of total sugar (COD 14.8 g l^{-1}) should be further treated by ordinary treatment processes (Noparatnaraporn *et al.*, 1986a; Sasaki *et al.*, 1991). It can be concluded that *Rb. sphaeroides* P_{47} might be applicable for cell mass production from pineapple waste.

6.2.2 *Soybean waste*

In soybean curd and *miso* (fermented soybean in Japan) manufactures, a high COD waste water (COD = *c.* 20 000 mg l^{-1}), consisting of mainly carbohydrates and protein, is discharged from a steaming process of soaking soybean. Usually, the cooked soybean runoff has to be diluted 10–20 times with the process water before entering the activated aerobic sludge treatment. This causes an increase in the total volume of waste water which requires a large-scale plant with a high running cost.

An attempt was made to treat the cooked soybean runoff directly using a photosynthetic bacterium, *Rhodocyclus gelatinosus*, which is capable of utilizing various carbohydrates and protein. This would not only treat the COD but also harvest the bacterial cells for SCP (Sasaki *et al.*, 1981).

Chemical composition. The chemical composition of cooked soybean runoff is shown in Table 6.3. This shows that the runoff contains a large amount of carbohydrates, mainly consisting of stachyose, sucrose and raffinose, while galactose, fructose and glucose are minor components. Other than carbohydrate, the waste contains 10.4 g l^{-1} of soluble protein. *Rc. gelatinosus* was applied to purify such waste (Sasaki *et al.*, 1981).

Growth of *Rc. gelatinosus* on cooked soybean runoff medium (Figure 6.2) shows that COD was reduced from 20.5 to 3.5 g l^{-1} in which *c.* 73% of the total sugar and 59% of the soluble protein were consumed for a 40 h culture in a 3 l fermenter (working volume 1.2 l) under aerobic–dark conditions. The optimum temperature for the growth and COD removal was found to be 35 °C from among the culture temperatures tested (30–45 °C).

During cultivation (Figure 6.2), sucrose, glucose, galactose, fructose and raffinose were almost completely consumed, although *c.* 50% of stachyose still remained. This suggested that the organism could utilize raffinose and stachyose for which α-galactosidase might be induced to hydrolyse them. The cell mass obtained contained 62–63% of protein and 33 μg g^{-1} dry cells of vitamin B_{12}, suggesting that the cell might be applicable as an SCP source.

Some advantages are found using *Rc. gelatinosus* for soybean processing manufactures (COD 20 000 mg l^{-1}) in place of activated sludge treatment. These are summarized in Figure 6.3. From the figure, 22 m^3 of cooked soybean runoff is discharged per day. For the activated sludge treatment,

Table 6.3 Chemical composition of cooked soybean runoff

	Cooked drain	Supernatant[f]
pH	6.0	6.0
COD (mg l^{-1})	2.2×10^4	1.8×10^4
Suspended solid (mg l^{-1})	2.24×10^3	0
Reducing sugar (g glc l^{-1})	4.24	3.93
Total sugar[a] (g glc l^{-1})	24.4	24.0
Stachyose[b] (g l^{-1})	n.a.	7.27
Raffinose[b] (g l^{-1})	n.a.	3.42
Sucrose[b] (g l^{-1})	n.a.	5.99
Galactose[b] (g l^{-1})	n.a.	1.46
Fructose[b] (g l^{-1})	n.a.	1.37
Glucose[b] (g l^{-1})	n.a.	0.79
Mannose[c] (g l^{-1})	n.a.	Trace
Arabinose[c]	n.a.	Trace
Xylose[c]	n.a.	Trace
Rhamnose[c]	n.a.	n.d.
Uronic acid[c]	n.a.	n.d.
Cellobiose[c]	n.a.	n.d.
Total nitrogen[d] (g l^{-1})	1.85	1.56
Soluble protein[e] (g l^{-1})	10.4	10.4

[a]Phenol-sulfuric acid method.
[b]Quantitatively analysed by high-pressure liquid chromatography.
[c]Qualitatively analysed by thin-layer chromatrography.
[d]Kieldahl method.
[e]Folin–Lowry method.
[f]The cooked drain was centrifued at 8000 *g* for 30 min.
n.a. = not analysed; n.d. = not detected.
Reproduced with permission from Sasaki, K. *et al.*, *J. Ferment. Technol.*, **59**, 471–7; published by the Society for Fermentation and Bioengineering, Japan, 1981.

the cooked runoff has to be diluted *c.* 20 times and, consequently, about 440 m^3 of waste water has to be treated. However, in the case of *Rc. gelatinosus*, after a 40 h culture of intact cooked soybean runoff (22 m^3) under aerobic conditions, the initial COD might be reduced from 20 000 mg l^{-1} to 4000 mg l^{-1}. Then, after diluting it four times (COD = *c.* 1000 mg l^{-1}), this (88 m^3) enters activated sludge treatment. Although economical assessment is required for the selection process mentioned above, a key point will be the chemical value of the cell mass produced with the appropriate amount of the cooked soybean runoff discharged.

6.2.3 *Cassava solid waste*

Cassava solid waste is normally used as a carbohydrate source of animal feed owing to its low cost and abundance. Since *Rc. gelatinosus* can utilize starch directly, its growth on cassava starch was tested.

Chemical composition. The chemical composition of cassava waste extract (Table 6.4) indicates that the filtrate appears to be a good

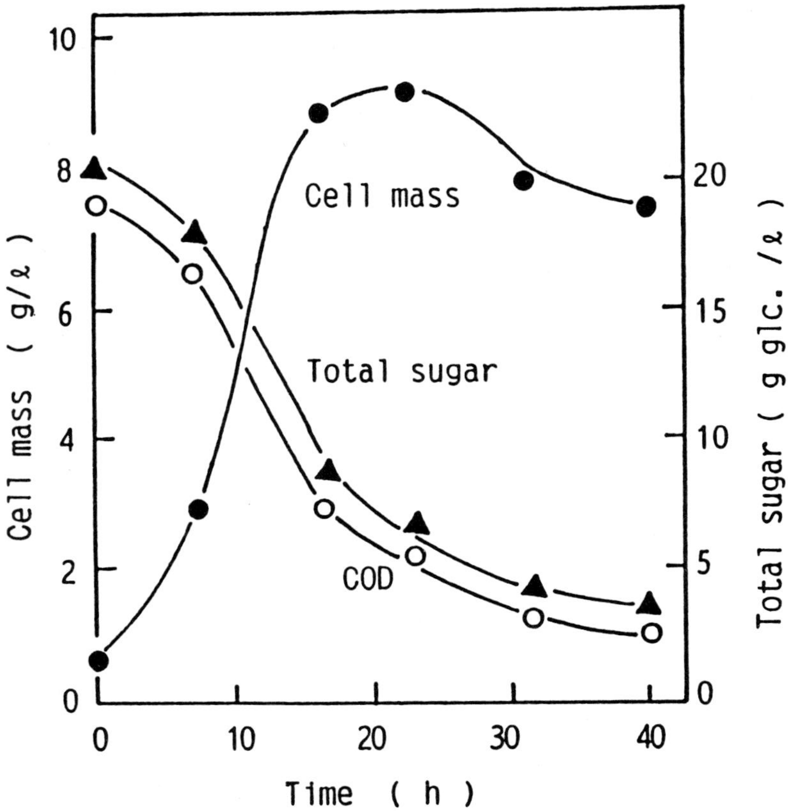

Figure 6.2 Typical growth, total sugar and COD reduction of *Rc. gelatinosus* on soybean waste under aerobic–dark conditions (35 °C).

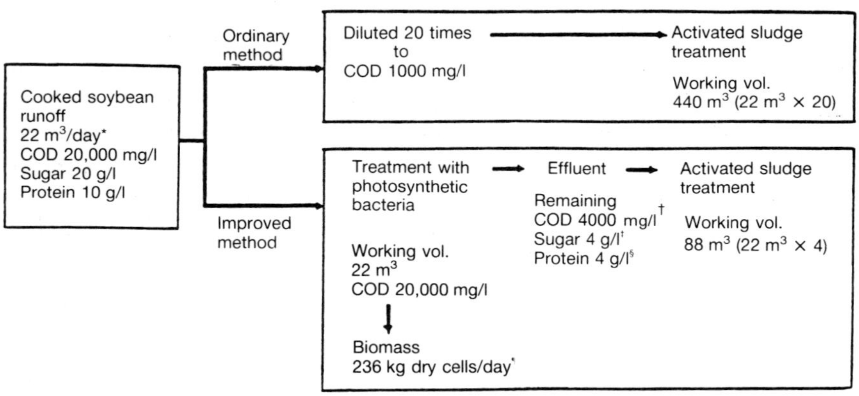

Figure 6.3 Diagrammatic scheme for applying *Rc. gelatinosus* to the treatment of cooked soybean runoff. * Data obtained from a *miso* factory in the Hiroshima area (Japan). †,‡ Calculated from COD removal of 81% at 35 °C culture. § Calculated from soluble protein removal of 59% at 35 °C. ¶ Calculated from $Y_{X/S}$ of 0.67 g cell g^{-1} glucose at 35 °C culture (total sugar was reduced from 20 to 4 g l^{-1}).

Table 6.4 Chemical composition (g l^{-1}) of the extract solution[a] of sun-dried cassava waste

	Filtrate	
	Before sterilization	After sterilization
Total sugar	11.3	12.0
Reducing sugar	0.2	0.2
Soluble protein	0.1	0.1
Sugar concentration		
Glucose	0.11	0.33
Maltose	0	0
Maltotriose	0.22	0.19
pH	6.0	5.5

[a]25 g l^{-1} dried waste were autoclaved (15 lb, 15 min) and filtrated.

carbohydrate source for *Rc. gelatinosus* and that it is also a source of nitrogen as well as other essential nutrients. Details of the experiments performed are described elsewhere (Sasaki *et al.*, 1991).

Growth and sugar utilization of *Rc. gelatinosus* on cassava waste medium is shown in Figure 6.4. Sugar analyses revealed that the abundant reducing sugar at 20 h was mainly maltose and glucose, and that both were rapidly utilized after 25 h. Also the total liquefying activity for starch tended to increase when the reducing sugar in the culture was reduced. This characteristic suggests that α-amylase might be inhibited during the culture of *Rc. gelatinosus* by the presence of a large amount of reducing sugar. The vitamin B_{12} and carotenoid content of the cells was 23 μg g^{-1} cell and 0.09 mg g^{-1} cell, respectively, suggesting the usefulness of the *Rc gelatinosus* cell as a source of SCP.

6.2.4 Mandarin orange peel

Orange juice manufacture (juice yield = *c*. 50%) produces a large quantity of orange waste (*c*. 400 000 tons $year^{-1}$ in Japan). Although part of the peel waste is converted to peel powder for use as an animal feed additive, most of this semisolid waste has to be treated by physical (heat), or chemical or biological means causing pollution.

An acidogenic fermentation of orange waste has been successful in eliminating methanogenesis by making the retention time of the culture broth shorter than the useful time for methane fermentation (Nishio *et al.*, 1982). As a result, volatile fatty acids, such as acetic and propionic acid, etc., have been accumulated as end-products, which are suitable for use by photosynthetic bacteria. Hence, photosynthetic bacteria can make use of the culture medium obtained after acidogenic fermentation of orange waste.

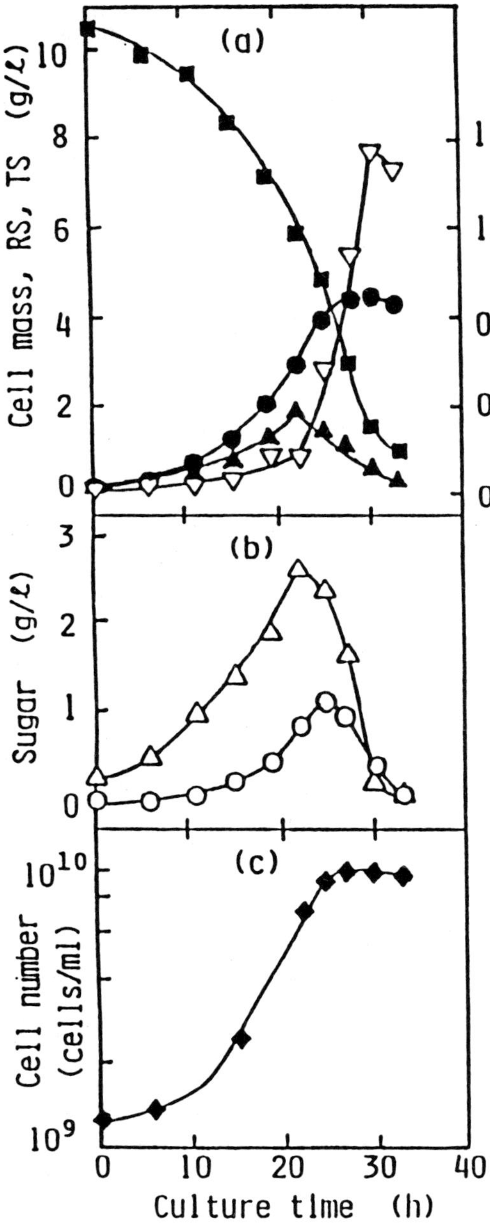

Figure 6.4 Growth and sugar consumption of *Rc. gelatinosus* on cassava waste medium under aerobic–dark conditions at 40 °C. ● = cell mass; ■ = total sugar; ▲ = reducing sugar; ▽ = liquefying activity; ○ = glucose; △ = maltose; ⧫ = cell number.

Chemical composition. The chemical composition of mandarin orange peel is shown in Table 6.5. This waste mainly consists of complex sugars including pectin, hemicellulose and cellulose, with remarkably many soluble sugars (sucrose = *c.* 0.2 g g^{-1} dry peel). The level of nitrogenous compounds such as crude protein is quite low compared with the total sugar (C/N ratio = 36:8). For the cultivation of photosynthetic bacteria, the nitrogen source, e.g. $(NH_4)_2SO_4$, should be supplemented even if acidogenesis of the waste has been carried out. The conditions and procedures of the acidogenesis of orange peel and cultivation of photosynthetic bacteria, *Rb. sphaeroides S* have been described in detail elsewhere (Sasaki *et al.*, 1991).

An acidogenic fermentation from mandarin orange peel medium (Figure 6.5) indicates that during a 5-day digestion, the total sugar was consumed almost completely by 3 days of culture, and mainly acetic acid (4.5 g l^{-1}) was accumulated with 2 g l^{-1} of propionic acid and 1 g l^{-1} of butyric acid. The supernatant from the acidogenic fermentation was used for further cultivation of *Rb. sphaeroides* S.

As shown in Figure 6.6, the cultivations of *Rb. sphaeroides* S under aerobic–dark, microaerobic–light and anaerobic–light conditions in a 10 l fermenter (working volume 6 l) are summarized. In aerobic–dark and microaerobic–light cultures, growth and COD decreases were remarkably different compared with anaerobic–light culture. In the fermenters, volatile fatty acids were also quickly consumed. In microaerobic–light culture, the maximum cell mass attained was relatively high (3.7 g l^{-1}) compared with 2.0 g l^{-1} observed in aerobic–dark culture, although the growth rate in microaerobic–light culture was relatively slow.

Table 6.5 Chemical composition of mandarin orange peel (dry basis)

	Orange peel (g g^{-1} peel)	Orange peel acid hydrolysis (g g^{-1} peel)
Total sugar	0.37	0.57
Reducing sugar	0.19	0.39
Sucrose	0.18[b]	0.18[b]
Glucose	0.04	0.13
Total uronate	0.11	0.21
Crude protein	0.07	–
Crude fat	0.02	–
Crude fiber	0.12	–
Ash	0.02	–

[a]Treated at 1 kg cm^{-2} for 15 min.
[b]Total sugar minus reducing sugar.

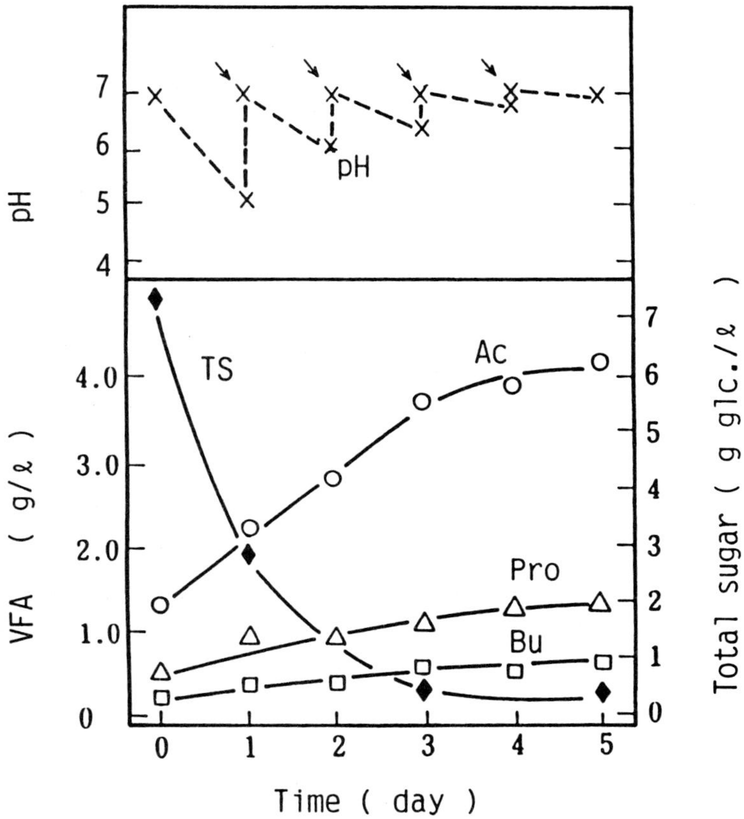

Figure 6.5 Production of volatile fatty acids (VFA) during an anaerobic fermentation of mandarin orange peel. Arrows = pH adjustment; ♦ = TS, total sugar; ○ = Ac, acetic acid; △ = Pro, propionic acid; □ = Bu, butyric acid.

COD removal was *c*. 85% in both aerobic–dark and microaerobic–light culture, indicating that the growth yield from COD consumed in the latter condition was 0.83 g cell g^{-1} COD, which was 1.6 times higher than that in aerobic–dark conditions. This suggests that in microaerobic–light culture, photosynthetic carbon fixation and aerobic assimilation of organic matter occurred simultaneously. In anaerobic–light culture, remarkably slow growth and COD reduction were observed compared with both aerobic cultures (Figure 6.6) owing to the limitation of light illumination (3 klux).

These facts are significant for the use of photosynthetic bacteria for SCP production by simultaneous utilization of organic waste and light energy utilization.

Thus, from a practical point of view, a microaerobic–light culture of *Rb. sphaeroides* S may be more promising for SCP production from volatile

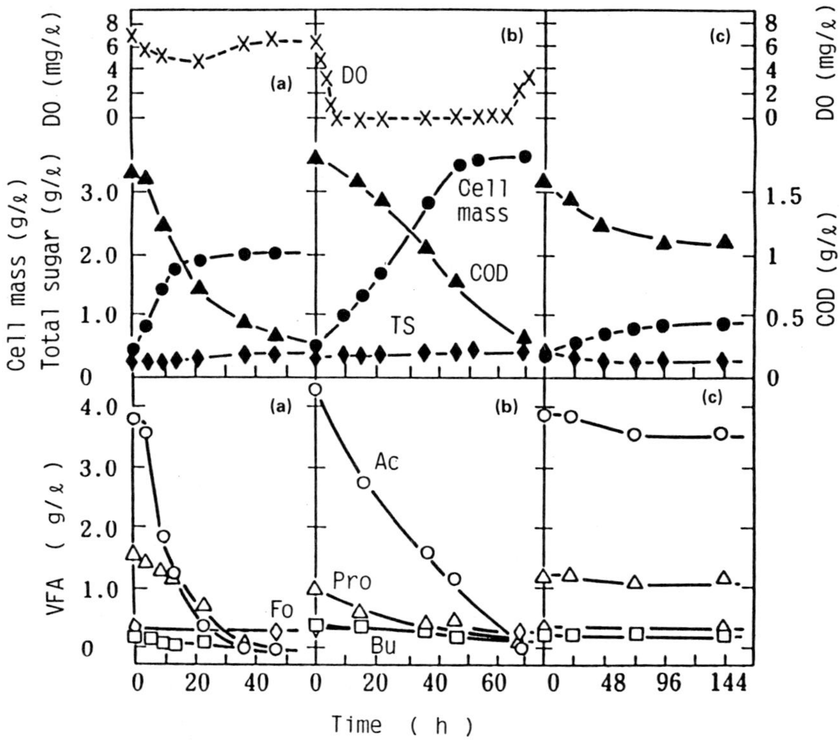

Figure 6.6 Growth and volatile fatty acids consumption by *Rb. sphaeroides* S on the medium obtained from anaerobic fermentation of mandarin orange peel. (a) aerobic–dark, (b) microaerobic–light (3 klux) and (c) anaerobic–light (3 klux). ● = cell mass; × = dissolved oxygen concentration (DO); ♦ = total sugar (TS); ◊ = Fo, formic acid; ○ = Ac, acetic acid; △ = Pro, propionic acid; □ = Bu, butyric acid; ▲ = COD.

fatty acids than an aerobic culture, since the cell yield per COD consumed was 1.6-fold higher than the aerobic assimilation. Besides, as the photosynthetic and aerobic cells contain appreciable amounts of chlorophyll, carotenoids and vitamin B_{12} (see section 6.2.6), *Rb. sphaeroides* S may become a potent source of SCP.

6.2.5 *Swine and cow dung waste*

The potential annual production of livestock discharges in Japan is *c.* 5 500 000 tons year^{-1} (Nakayama, 1993). Part of the waste is used as a fertilizer after composting treatment and another part is treated by methanogenic fermentation. However, the effluent from anaerobic digesters still contains abundant organic matter, e.g. biological oxygen demand (BOD) = 3000–4000 mg l^{-1}, mainly consisting of volatile fatty acids (VFA). To treat the discharged effluent, a method of aerobic degradation of the activated sludge is commonly applied. Sometimes, dilution of the discharged effluent with water is required before treating by aerobic degradation.

The application of photosynthetic bacteria to the secondary treatment of the effluent after anaerobic digestion of livestock discharges is described below.

Chemical composition. The chemical composition of cow and swine discharges and the effluent from anaerobic digestion of the discharges is summarized in Table 6.6. For swine discharges, a large amount of organic matter in terms of COD or BOD is present with a high nitrogen source. After anaerobic digestion of the discharges, the COD decreased one-tenth to 4000 mg l^{-1}, which mainly consists of VFA. Since VFA are useful substrates for photosynthetic bacteria, such as *Rb. sphaeroides* or *Rb. capsulatus*, the application of these strains for the treatment of VFA and SCP production are described.

The experimental conditions for the anaerobic digestion of swine (Sasaki *et al.*, 1987a) and cow dung waste (Vrati, 1984) have been described, and details of the cultivation of photosynthetic bacteria on the anaerobic digestion liquor of such wastes after anaerobic digestion are given in Sasaki *et al.* (1987a, 1991).

An aerobic cultivation of *Rb. sphaeroides* S cells on the medium prepared from post-anaerobic digestion liquor of swine discharges (Figure 6.7) indicated that, for a 20 h culture, the cell mass attained up to 1.8 g dry cells l^{-1} and the COD was reduced from 4.4 g l^{-1} to 0.9 g l^{-1} (80% removal), which corresponds to the utilization of VFA. It is interesting to

Table 6.6 Chemical composition (g l^{-1}) of cow and swine discharges and the effluent after anaerobic digestion of the discharges

	Cow[a]		Swine[b,c]	
	Discharge	Effluent after anaerobic digestion	Discharge	Effluent after anaerobic digestion
pH	–	–	7.0–7.3	7.5–8.0
BOD	–	–	62.7	–
COD	–	–	35.0	3.9
Total nitrogen	3.75	2.50	4.7	3.0
Protein	3.00	2.00	0.8	0.8
NH_4^+-N	–	–	2.8	2.5
Acetic acid	0.50	0.39	–	0.48
Propionic acid	0.50	0.32	–	1.17
Butyric acid	0.10	0.20	–	0.28

Sources: [a]Vrati (1984); [b]Sasaki *et al.* (1987a); [c]Wagai (1987).

note that *Rb. sphaeroides* S can grow faster on such an unsterilized medium than the microorganisms originally existing from methane fermentation. The predominance of *Rb. sphaeroides* S cells after 20 h of culture (Figure 6.7) was *c.* 90% under microscopic observation. This suggests that this organism may be suitable for this process. Growth yield from COD removal was 0.52 g g^{-1} COD. Continuous cultivations of *Rb. sphaeroides* S under aerobic conditions using the same digestion liquor medium indicated that, when the dilution rate of continuous culture was less than 0.125 h^{-1}, almost one month of stable operation was achieved with a COD reduction of 70–80%, and the predominance of *Rb. sphaeroides* was 80–90%. If 20 ton day^{-1} of swine discharges was produced, 42 kg dry cells day^{-1} could be produced from this process.

On the other hand, *Rb. capsulatus* cells were cultivated on a medium prepared from cow dung discharges (Vrati, 1984). After 6 days cultivation under anaerobic–light conditions, 4.56 g wet cells l^{-1} (*c.* 0.91 g dry wt l^{-1}) was obtained. This result enabled us to estimate that *c.* 64 kg of *Rb. capsulatus* cells may be harvested from the effluent from anaerobic digestion of 10 tons of cow dung.

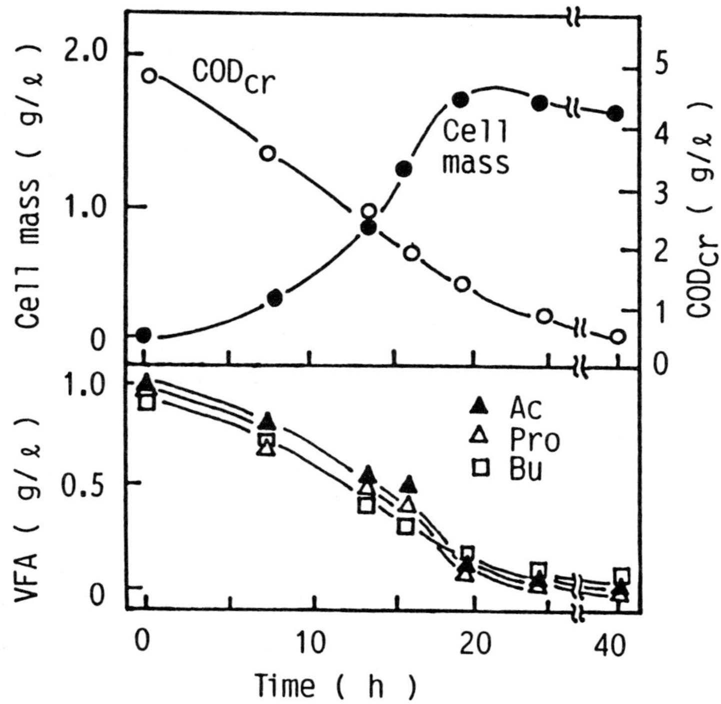

Figure 6.7 Time course of cell growth, COD and volatile fatty acids (VFA) disappearance by aerobic culture of *Rb. sphaeroides* S in the medium prepared from the effluent of anaerobic digestion of swine discharges. Ac = acetic acid; Pro = propionic acid; Bu = butyric acid.

Thus, SCP production using photosynthetic bacteria from swine waste and cow dung seems to be possible, although these discharges must be treated by anaerobic digestion before applying this process, followed by the separation of the solid to obtain the medium available.

6.2.6 *Cell yields and composition of PSB*

In this section, the growth characteristics of photosynthetic bacteria (PSB) for various waste materials together with their nutritional quality for useful animal feeds are considered.

The growth characteristics of PSB in terms of growth rate, cell mass yields from carbon substrates, maximum cell mass attained and COD removal from several agro-industrial wastes are summarized in Table 6.7. From this table, the maximum cell mass obtained was 26.5 g l^{-1} after a 56 h aerobic culture in pineapple waste. All the cell mass yields ($Y_{X/S}$), which ranged from 0.40 to 0.67 g g l^{-1}, were acceptable for cell mass production. The COD removal, which ranged from 79% to 95.6%, also indicated advantages for the use of PSB in subsequent treatment. From Table 6.7, it appears that PSB, particularly *Rb. sphaeroides* P_{47}, may be useful microbial resources for SCP production on a commercial scale, since the cell mass attained in commercial productions should reach at least 10–20 g dry cells l^{-1}.

The nutritional quality of the PSB cells from various wastes is shown in Tables 6.8 and 6.9. The proximate analyses indicated that the cell mass contained crude protein ranging from 56% to 67%, which is comparable to

Table 6.7 Growth yields and growth charactersitics of photosynthetic bacteria from various wastes

	Wastes (strain, culture, conditions)				
	Pineapple (P_{47}/AD)	Soybean (R.g./AD)	Mandarin orange peel[a] (S/MAL)	Cassava (PSBVI/ MAD + MAL)	Swine (S/AD)
X_{max} (g l^{-1})[b]	26.5	9.3	3.7	4.7	1.8
T_m (h)[c]	56	22	50	264	19
$Y_{X/S}$ (g g^{-1})[d]	0.45 (g g^{-1} TS)	0.67 (g g^{-1} TS)	2.48 (g g^{-1} COD)	0.40 (g g^{-1} TS)	0.53 (g g^{-1} COD)
COD removal (%)	85	81	84	95.6	79

P_{47} = *Rb. sphaeroides* P_{47}; R.g. = *Rc. gelatinosus*; S = *Rb. sphaeroides* S; AD = aerobic–dark; MAL = micro-aerobic–light; MAD = micro-aerobic–dark.
[a]Post-anaerobic digestion liquor (see section 6.2.4).
[b]Maximum cell mass attained.
[c]Time required until maximum cell mass attained.
[d]Growth yield from total sugar (TS) or COD.

Table 6.8 Cell composition (%) of photosynthetic bacteria compared with other SCP sources

	Rc. gelatinosus[a]	*Rb. sphaeroides* P_{47}[b]	*Rb. capsulatus*[c]	Esso-Nestle (yeast)[d]	*Chlorella vulgaris*[c]	Soybean[d]	FAO/WHO scoring pattern[e]
Crude protein	56	66.6	66.0	54.0	55.5		
Crude fat	2.45	1.88	7.0	10.0	8.07		
Carbohydrate	26.4	24.9	23.0	26.0	21.0		
Crude fiber	n.a.	2.95	n.a.	n.a.	12.1		
Ash	3.21	3.62	4.0	7.0	3.28		
Lysine	3.04	2.57	2.86	3.76	2.71	2.58	5.5
Histidine	1.15	0.96	1.25	0.90	1.06	n.a.	n.a.
Threonine	2.04	2.87	2.70	2.63	2.28	1.62	4.0
Valine	3.44	2.68	3.51	3.20	3.02	1.86	5.0
Methionine	1.92	1.47	1.58	0.51	0.27	0.43	1.9
Isoleucine	2.73	1.78	2.64	2.63	2.44	1.80	4.0
Leucine	5.84	3.90	4.50	3.54	4.46	2.70	7.0
Phenylalanine	3.08	2.36	2.60	2.20	2.65	1.98	3.0
Arginine	3.42	3.55	3.34		3.24		
Aspartic acid	4.92	5.18	4.56		4.74		
Serine	2.10	2.33	1.68		2.12		
Glutamic acid	5.84	6.22	5.34		4.62		
Proline	2.40	2.02	2.80		2.12		
Glycine	2.58	3.18	2.41		2.28		
Alanine	4.45	5.06	4.65		2.98		
Tyrosine	1.50	1.70	1.71		0.96		

[a]Grown on cassava starch medium (aerobic–dark culture, see section 6.2.3).
[b]Grown on pineapple waste (aerobic–dark culture, see section 6.2.1).
n.a. = not analysed.
Sources: [c]Kobayashi and Kurata (1978); [d]Kanamori (1984); [e]Noparatnaraporn (1987).

Table 6.9 Vitamin content (μg g^{-1} dry cells) of photosynthetic bacterial cells

	Rc. gelatinosus	*Rb. sphaeroides* P_{47}	*Rb. capsulatus*[a]	Esso-Nestle (yeast)[b]
Vitamin				
B_1	n.a.	n.a.	12	11–13
B_2	33.2	13.0	50	110–130
B_6	n.a.	n.a.	5	4.8–7.6
B_{12}	33	78	21	Trace
E	51	210	n.a.	n.a.
Carotenoid	90	800	n.a.	n.a.
Nicotinic acid	136	58	125	165–200
Folic acid	7.2	1.0	60	1.8–2.4
Pantothenic acid	n.a.	n.a.	30	14–23
Biotin	8.3	6.3	65	110–130

n.a. = not analysed.
Sources: [a]Kobayashi and Kurata (1978); [b]Kanamori (1984).

that of yeast and algae, and that the amino acid patterns were also comparable to those of yeast and algae. In particular, the content of lysine, methionine, leucine and phenylalanine (four essential amino acids) was appreciable in comparison to other SCP cells. In particular, the methionine content of PSB cells was much higher than that of other SCP and plant proteins, e.g. soybean (0.43%). Since methionine is one of the limiting essential amino acids in animal feedstuff, the PSB cells may be a useful complement for feeding animals. Therefore, PSB cells should be recommended for use as a supplement to the basic diet rather than as the sole protein source.

In addition, PSB cells contain considerable amounts of essential vitamins for feeding animals (see Table 6.9). *Rc. gelatinosus* cells from cassava starch waste were rich in vitamin B_2 (33.2 mg kg^{-1} cell) and niacin (135.8 mg kg^{-1} cell), while those of *Rb. sphaeroides* P_{47} from pineapple waste were rich in vitamin B_{12} (78 mg kg^{-1} cell) and vitamin E (210 mg kg^{-1} cell). It is clear that all of the four vitamins essential for animal feedstuff were contained in PSB cells in appreciable quantities.

The PSB cells also contain carotenoid pigment at values of 0.09–0.68 mg g^{-1} cell. As carotenoids have been reported to be useful as color-intensifying substances for egg-yolk, chicken flesh and aquarium-fish skin and, in aquaculture, to increase the viability and decrease the mortality of teleost eggs, PSB cells can be useful for animal feeds (Noparatnaraporn *et al.*, 1987b).

Therefore, based on the nutritional quality fo the PSB cells, these bacteria, cultivated on various agro-industrial wastes, would be proposed to be a good potential source of multipurpose supplement for animal feedstuffs.

Recently, the production of biocompatible plastics, such as poly-β-hydroxybutylate (PHB) by PSB has been reported. Barandl *et al.* (1991) found that PSB can produce various kinds of bioplastic in the cells. Suzuki *et al.* (1995) reported that PHB was produced intracellularly from the cells of *Rb. sphaeroides* RV by acetic acid as a carbon source. Acetic acid is a volatile fatty acid and a suitable substrate for the culture of PSB. This seems to be advantageous for recycling organic waste by bioconversion to the useful materials.

6.3 Vitamin production

6.3.1 Vitamin B_{12}

Vitamin B_{12} is characterized as a carrier for methyl transfer and as a coenzyme in the enzyme system, for example, glutamate mutase, methyl malonyl-CoA mutase, etc. (Fukui, 1980). Vitamin B_{12} is not only widely used as a treatment for pernicious anemia and nervous system disorders, but also as a supplement for animal feed to enhance its nutritional value. The production of vitamin B_{12} is carried out by the fermentation of sugars, e.g. by propionic acid bacteria.

As shown in Figure 6.8, vitamin B_{12} synthesis is closely correlated with the biosynthesis of other tetrapyrrols (porphyrin, heme and chlorophyll). For photosynthetic bacteria, such as *Rhodobacter* spp., tetrapyrrols are mainly synthesized from glycine and succinate via 5-aminolevulinic acid (ALA) and Urogen III as common intermediates (Shemin pathway) (Lascelles, 1978; Hoshino and Kitamura, 1984). However, in the algae, higher plants and methanogenic bacteria, tetrapyrrols are synthesized from glutamate via ALA (C-5 pathway) (Jaenchen *et al.*, 1891; Klein and Porra, 1982; Sasaki *et al.*, 1991), In *Rb. sphaeroides*, the Shemin pathway of total tetrapyrrols synthesis seems mainly to operate at *c.* 90% compared with that of the C-5 pathway (Klein and Porra, 1982). In the Shemin pathway, ALA synthetase seems to be a key enzyme and plays an important role in tetrapyrrol synthesis. This enzyme is regulated by a heme compound (see Figure 6.8) by feedback inhibition (Lascelles, 1978).

Dissolved oxygen concentration and light illumination during cultivation of photosynthetic bacteria affected the formation of ALA synthetase and, consequently, a remarkable difference in tetrapyrool accumulation was observed (Noparatnaraporn *et al.*, 1986a). For example, under low oxygen tension and low light illumination, ALA synthetase activity was enhanced with an increase in bacteriochlorophylls in *Rb. sphaeroides* (Sasaki and Nagai, 1979).

In corrinoid formation in *Rb. sphaeroides*, cobalamin is the main product (Sasaki *et al.*, 1978) and incomplete corrinoids such as cobirinic acid and cobinamide (FB) are undetectable intracellularly.

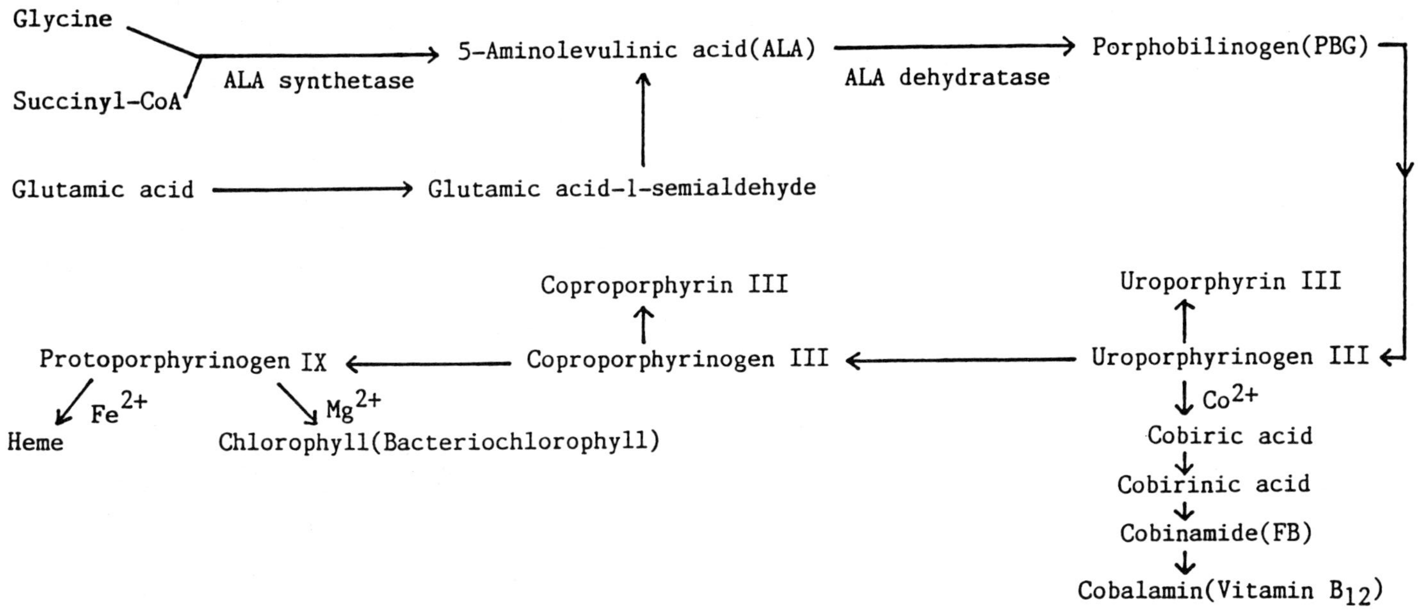

Figure 6.8 Vitamin B, biosynthesis pathway in photosynthetic bacteria.

Table 6.10 Vitamin B_{12} production by photosynthetic bacteria from agro-industrial wastes

Source of waste	Organism	Vitamin B_{12} produced		
		Culture conditions	μg g^{-1} dry cell	μg l^{-1} broth
Pineapple[b]	*Rb. sphaeroides* P_{47}	AD	75	1578
Soybean[c]	*Rc. gelatinosus*	AD	33	307
	Rc. gelatinosis	AD	23	104
Cassava starch[d]	Mixed culture (*Rc. gelatinosus*: *Rb. sphaeroides* P_{47} 1:1)	AD	44	274
Mandarin orange peel[a,e]	*Rb. sphaeroides* S	AD (DO > 4 mg l^{-1})	37	74
		MAL (DO = 0, 3 klux)	79	260
		ANL (3 klux)	87	67

[a]Supernatant after anaerobic digestion of orange waste.
AD = aerobic–dark (DO > 4 mg l^{-1}); MAL = micro-aerobic–light; ANL = anaerobic–light.
Sources: [b]Noparatnaraporn *et al.* (1986a); [c]Sasaki *et al.* (1981); [d]Noparatnaraporn *et al.* (1987a); [e]Sasaki *et al.* (1991).

In the application of photosynthetic bacteria for the production of SCP from agro-industrial wastes, it is important whether or not vitamin B_{12} can be produced intracellularly, since vitamin B_{12} containing SCP is more valuable for animal feeds than vitamin B_{12} without SCP.

From Table 6.10, the vitamin B_{12} production from agro-industrial wastes by photosynthetic bacterial cells is 23–87 $\mu g\ g^{-1}$ dry cells irrespective of the culture conditions. These values are appreciable compared with commercial SCP, e.g. ESSO-Nestle yeast (0.11–0.17 $\mu g\ g\ l^{-1}$ cell).

The intracellular content of vitamin B_{12} produced from orange peel waste (Table 6.10) increased under both microaerobic–light or anaerobic–light conditions. However, in anaerobic–light culture, the growth rate and cell mass attained were relatively lower than that in aerobic–dark cultivation. To overcome this drawback, it is possible to enhance the low vitamin B_{12} content of the aerobically grown cells by maintaining the dissolved oxygen concentration at an extremely low level (Figure 6.9) (Noparatnaraporn *et al.*, 1986b).

In the aerobic–dark culture of *Rc. gelationsus* on cassava starch medium, the first stage of the cultivation was conducted under sufficiently aerobic conditions (Dissolved oxygen (DO) >4 mg l^{-1}) to support a high growth rate with a high substrate uptake rate, then after the cell mass attained its maximum level, the aeration and agitation were drastically reduced (see the second stage in Figure 6.9) to establish a microaerobic condition where a redox potential in the broth was *c.* −200 mV. As shown in the figure, the

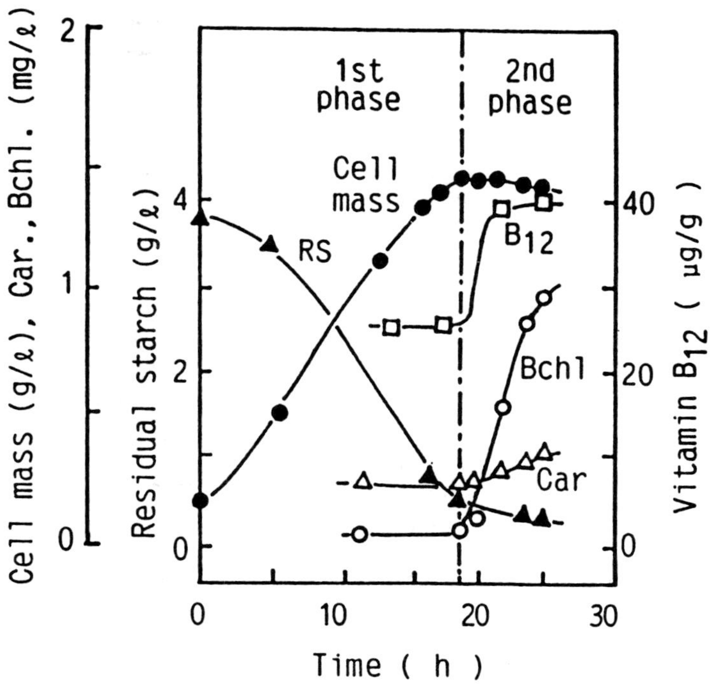

Figure 6.9 Enhancement of photopigments and vitamin B_{12} formations by *Rc. gelatinosus* growing on cassava starch medium after the culture condition was suddenly changed from aerobic to microaerobic conditions (Noparatnaraporn *et al.*, 1986b). pH = 7, temperature = 40 °C. *1st phase*: aerobic–dark, dissolved oxygen (DO) > 4 mg l^{-1} oxidation–reduction potential (ORP) > + 110 mV, aeration 1 volume/volume/minutes (vvm), agitation 500 rpm. *2nd phase*: microaerobic–dark, DO = 0 (ORP = −200 ± 10 mV), aeration 0.1 vvm, agitation 200 rpm. B_{12} = vitamin B_{12}; Car = carotenoid; Bchl = bacteriochlorophyll; RS = residual starch.

vitamin B_{12} content increased from 22 to 39 $\mu g\ g^{-1}$ cell after 6 h incubation without light illumination.

The bacteriochlorophyll and carotenoid content increased together with vitamin B_{12} under the microaerobic conditions (Noparatnaraporn *et al.*, 1986b), but the protein content of the cells was unchanged (62–63%) throughout aerobic and microaerobic conditions.

This simple technique may be applicable for the production of enriched SCP from the photosynthetic bacteria.

6.3.2 Ubiquinone

Ubiquinone consists of the structure 2,4-dimethoxy-5-methylbenzoquinone with a polyisoprenoid at the 6th position of the benzene ring. The natural ubiquinones are Q-6, Q-7, Q-8, Q-9 and Q-10, where the number refers to the carbon number of the isoprene unit of the side chain. These compounds play an important role in electron transfer in living systems. Q-11 and Q-12 are observed in the higher plants (Morimoto, 1971), and dimethylmenaquinone and menaquinone are also observed in bacteria (Hiraishi, 1988).

Among the ubiquinones in nature, Q-10 is widely used as a drug, being effective for anemia, hypertension, periodontal disease, and other organ or tissue disorders. Photosynthetic bacteria are potent producers of ubiquinone, which plays an essential role in the electron transport for both respiration and bacterial photosynthesis (Morita and Shimada, 1984). Hiraishi (1989) reported that ubiquinone profiles were applicable for the classification of photosynthetic bacteria and other bacteria.

Industrial production of ubiquinone Q-10 using photosynthetic bacteria has been carried out in Japan. Yamada *et al.* (1991) established a large-scale aerobic cultivation of photosynthetic bacteria using a fuzzy control system to enhance the ubiquinone accumulation in the cells.

Ubiquinone synthesis of photosynthetic bacteria is closely related to carotenoid synthesis via the common pathway of terpenoid synthesis (Figure 6.10). The benzoquinone moiety of ubiquinone is synthesized from phosphoenolpyruvate via shikimate and *p*-hydroxybenzoate. The intracellular accumulation of ubiquinone was frequently determined by the culture conditions, such as oxygen tension, light illumination as well as medium composition, as in the bacteriochlorophyll synthesis of photosynthetic bacteria.

In Table 6.11, the ubiquinone, carotenoid and bacteriochlorophyll formed by *Rb. sphaeroides* P_{47} grown under anaerobic–light conditions indicate that the intracellular ubiquinone (Q-10) content was *c.* 3 mg g^{-1} cell which was twice that of aerobically grown cells. The different carbon source did not affect the ubiquinone content (Table 6.10) in this strain. This would suggest that ubiquinone production might be possible for use in the culture medium of acidogenic fermentation which contains volatile fatty acids (sections 6.2.4 and 6.5). The carotinoid content in terms of spheroiden and spheroidenone in *Rb. sphaeroides* P_{47} (Table 6.11) was a rather high amount.

It may be possible to use the photosynthetic bacteria for the production of ubiquinone and carotenoids simultaneously from agro-industrial waste not only for SCP but also for other useful materials.

6.4 5-Aminolevulinic acid production

5-Aminolevulinic acid (ALA, $NH_2CH_2COCH_2CH_2COOH$) is widely distributed among plants and animals. ALA is believed to be an intermediate in the biosynthesis of tetrapyrrols, such as chlorophyll, heme

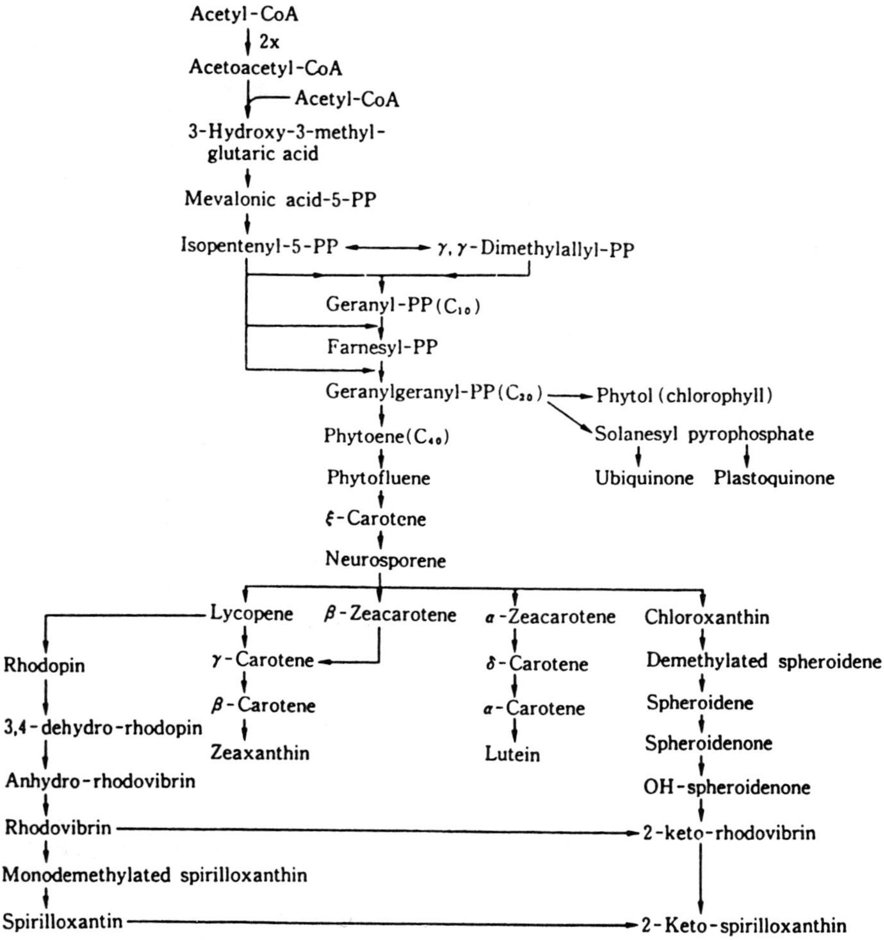

Figure 6.10 Carotenoid and ubiquinone biosynthesis.

Table 6.11 Ubiquinone and photopigment formation of Rb. sphaeroides P_{47} grown under anaerobic–light conditions (5 klux, 35 °C)

Carbon substrate	Uniquinone (Q-10)		Carotenoid (mg g^{-1} cell)		Bacterio-chlorophyll
	(mg g^{-1} cell)	(mg l^{-1} broth)	Spheroiden	Spheroidenone	(mg g^{-1} cell)
Glutamate	2.7	2.2	3.8	1.0	33
Acetate	2.8	2.7	3.0	2.8	32
Propionate	3.0	1.7	3.0	4.7	30

and vitamin B_{12}, and to play an important physiological role in such tetrapyrrol biosynthesis, as described in Figure 6.8. However, except for these functions, its physiological activity has not been studied in detail so far. Recently, ALA has been reported as a new herbicide which damages weeds but does not harm crops, humans or animals (Rebeiz *et al.*, 1984). As discussed later in section 6.4.4, the application of ALA has been studied widely in the medicine, e.g. in the diagnosis and treatment of cancer. In addition, some research has been carried out in agricultural applications, e.g. as a herbicide, insecticide and growth promoting factor, etc. (Tanaka *et al.*, 1997).

However, chemical synthesis of ALA is relatively difficult because many steps are required for synthesis and separation. This results in high costs for industrial-scale production. Therefore, the biological process of ALA production becomes attractive.

With respect to ALA formation by microorganisms, it has been found that a bacterium, *Methanobacterium thermoautotrophicum* (Jaenchen *et al.*, 1981), and algae, such as *Agnemellum quadruplicum*, *Anacystis marina* and *Chlorella vulgaris* (Beale, 1970), can produce ALA. Recently, *Chlostridium thermoautotrophicum* (Koesnanadar *et al.*, 1989). *Methanosarcina barkeri* (Lin *et al.*, 1989) and *Chrorella* sp. (Sasaki *et al.*, 1995b) have been reported as ALA producers.

Rhodobacter sphaeroides, having a tetrapyrrol synthesis pathway (Figure 6.8), seems to be a potent source of ALA not only because this organism has a high ability to produce ALA but also because it is able to utilize a variety of organic substrates. In fact, abundant ALA excretion from the growing cells of *Rb. sphaeroides* was observed (Sasaki *et al.*, 1987b).

In this section, the utilization of *Rb. sphaeroides* on a medium prepared from swine waste and sewage sludge is described as an example of the recycling of organic wastes for the production of ALA. In addition, ALA production by aerobic fermentation which does not require light energy for production is described.

6.4.1 ALA production from swine waste

If ALA can be produced by *Rb. sphaeroides* from waste such as the anaerobic digestion liquor from swine waste containing VFA, this will be an advantage because the liquor can be used directly as a fertilizer having herbicidal properties. When ALA dehydratase activity can be inhibited by levulinic acid (LA), an ALA dehydratase inhibitor, ALA may be excreted into the culture broth. In fact, ALA is accumulated extracellularly when ALA dehydratase was inhibited by LA in *Rb. sphaeroides* (Figure 6.11), provided that glycine and succinate are supplied in the medium (Sasaki *et al.*, 1987b, 1990). In particular, glycine addition is more effective for ALA extracellular formation (Sasaki *et al.*, 1990).

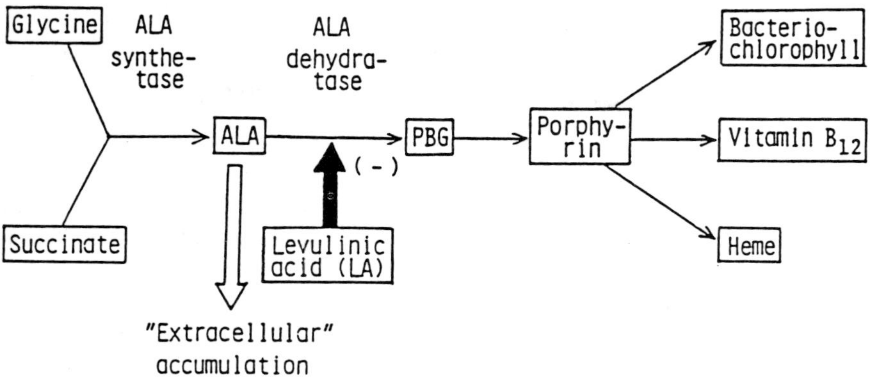

Figure 6.11 5-Aminolevulinic acid (ALA) extracellular accumulation by levulinic acid addition in *Rb. sphaeroides*. PBG: porphobilinogen.

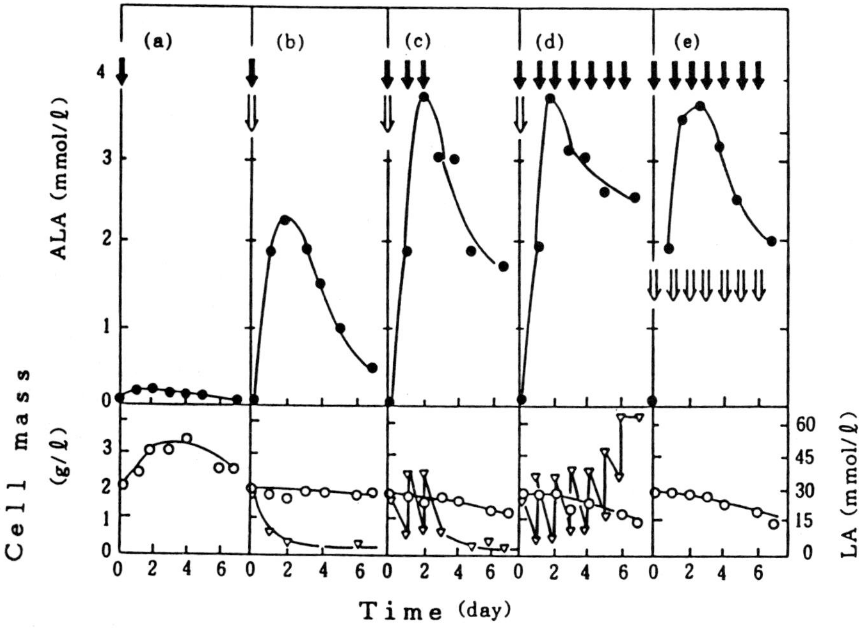

Figure 6.12 ALA production of *Rb. sphaeroides* (IFO12203) from the supernatant of post-anaerobic digestion of swine discharges (10 000*g*, 20 min) without sterilization (Sasaki *et al.*, 1990). Culture conditions: anaerobic–light illumination (5 klux). Initial pH = 6.5 and culture temperature = 30 °C. ➡ = LA (30 mM) addition, ×3 indicates 3 times; ⇨ = glycine (60 mM) addition, ×1 indicates once. ● = ALA; ○ = cell mass; ▽ = residual LA.

ALA production by *Rb. sphaeroides* from the post-anaerobic digestion liquor is shown in Figure 6.12. Details of the post-anaerobic digestion liquor obtained by centrifuging the digestion broth (without sterilization) are given in section 6.2.5. The addition of LA alone (a) did not produce

ALA, but the addition of LA plus glycine (b) resulted in a high level of ALA excretion from the cells. Repeated addition of LA (c) accelerated ALA accumulation up to 3.9 mM. It was suggested that the repeated addition of LA might maintain the ALA dehydratase activity at a low level. However, when LA was added seven times (d), ALA accumulation was no higher compared with (c). The repeated addition of LA plus glycine seven times (e) did not improve ALA accumulation compared with (c). The accumulated ALA was reutilized when LA had been consumed.

The time course of ALA, cell mass, LA and glycine during culture (corresponding to Figure 6.12c) is shown in Figure 6.13. ALA accumulation (*c.* 4 mM) was twice that produced by *Chlorella vulgalis* (Beale, 1970).

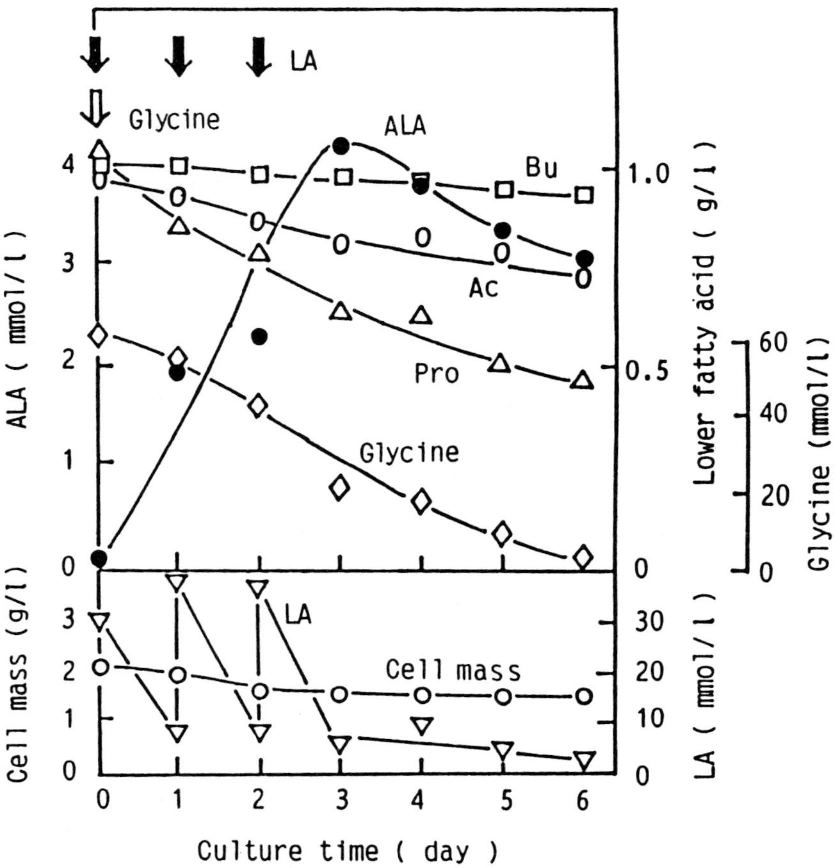

Figure 6.13 Profiles of ALA, LA, cell mass, glycine and volatile fatty acids (substrates) during culture of *Rb. sphaeroides* IFO12203 on the medium prepared from post-anaerobic digestion liquor of swine discharges (corresponding to Figure 6.12c). Ac= acetic acid; Pro = propionic acid; Bu = butyric acid. ➡ = LA (30 mM) addition, ×3 indicates 3 times; ⇨ = glycine (60 mM) addition, ×1 indicates once. ● = ALA; ○ = cell mass; ▽ = residual LA.

Propionic acid in the medium was mainly utilized together with acetic or butyric acid. It has been suggested that propionic acid might play an important role in ALA formation as a source of succinyl-CoA supply via the methylmalonyl-CoA pathway (Kitamura, 1988; Sasaki *et al.*, 1978).

This culture broth (containing 4 mM ALA) has been tested as a herbicide (Rebeiz *et al.*, 1984). The culture broth showed effective herbicidal activity within 3 days after spraying it directly on the leaves and stems of *Trifolium repense* (a clover), a common plant in fields. In addition, a similar herbicidal effect could be observed using this culture broth for worm wood, day flower and creeping woodsorrel, which are common weeds in fields, but it was less effective for monocot weeds, such as crabgrass and goosegrass (Sasaki *et al.*, 1991, 1995a).

6.4.2 ALA production from sewage sludge

Sewage sludge is produced daily in vast amounts by municipal waste water plants. This waste is mainly discharged after heat treatment such as drying or incineration. Of the 1 410 000 tons (dry wt) of excess sludge produced in 1990, about 61% was discharged after heat treatment (Watanabe *et al.*, 1993). Anaerobic digestion of sewage sludge is widely carried out to reduce the amount of waste. Up to 3–5 g l^{-1} of VFA are contained in the anaerobic digestion liquor of sewagae sludge. These VFA may be utilized for ALA production like post-anaerobic digestion liquor of swine waste described above. ALA has been produced using VFA from the anaerobic digestion liquor from sewage sludge (Tanaka *et al.*, 1994b). The VFA medium used here was the supernatant from the digestion broth (10 000 g, 20 min) without sterilization and any supplements.

As shown in Figure 6.14, when LA was added three times, ALA production reached *c.* 9 mM. Cell growth (initially *c.* 2.0 g l^{-1} was slightly retarded by the addition of LA but gradually increased. Among VFA, propionic acid was preferentially utilized as also observed in swine waste (Figure 6.12). Propionic acid appears to be an important substrate for ALA production in *Rb. sphaeroides*.

Rb. sphaeroides cannot grow in unsupplemented VFA medium from post-anaerobic digestion of swine waste or mandarin orange peel (Sasaki *et al.*, 1991) owing to contamination in the non-sterile system and the low concentration of growth factors, such as organic sources, in the crude medium. However, the VFA medium prepared from sewage sludge was relatively rich in amino acids and *Rb. sphaeroides* was able to grow directly in this VFA medium without any supplements. As shown in Figure 6.15a, cell growth was observed in VFA medium utilizing VFA. By adding LA and glycine, *c.* 2.2 mM of ALA was produced. However, ALA accumulation reached 9.3 mM if propionic and acetic acid were also added (Figure 6.15b). This was advantageous since cells could be obtained directly from

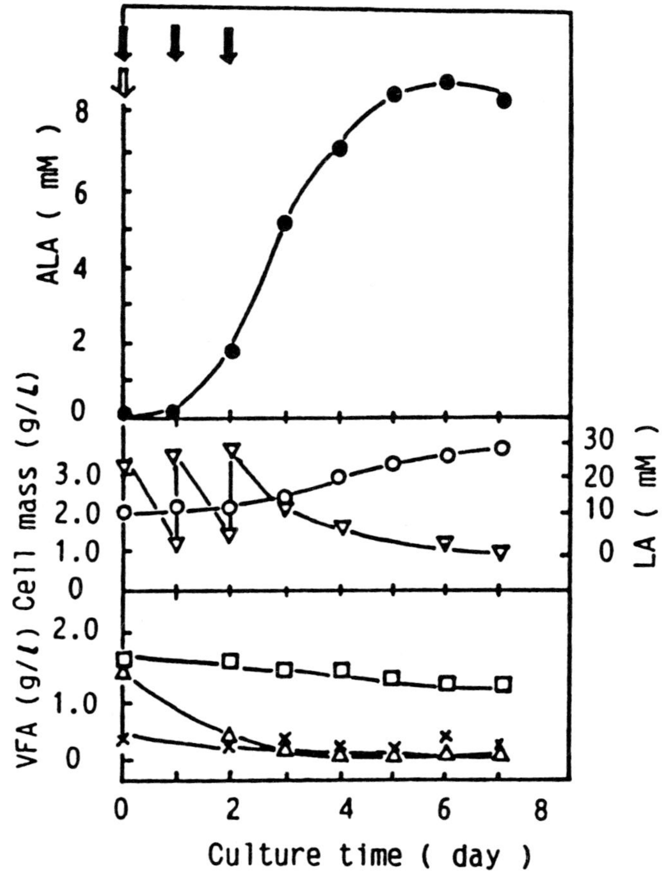

Figure 6.14 Profiles of ALA, cell mass, LA and VFA after dense inoculation of *Rb. sphaeroides* (5 klux, 30 °C, pH 7.0 ± 0.1) on VFA medium prepared from post-anaerobic digestion liquor of sewage sludge. Sewage sludge was anaerobically digested at 35 °C for 5 days. Supernatant by centrifugation (10 000 *g*, 20 min) was used as VFA medium without sterilization. ⬇ = LA (30 mM), ×3 indicates 3 times repeatedly; ⇩ = glycine (60 mM), ×1 indicates once. ● = ALA; ○ = cell mass; ▽ = LA; □ = acetic acid; △ = propionic acid; × = *n*-butyric acid.

the VFA medium without preculture of the glutamate–malate medium (Sasaki *et al.*, 1990). Thus, ALA production appears to be possible using VFA from post-anaerobic digestion liquor from sewage sludge.

6.4.3 *ALA production by aerobic fermentation*

As mentioned above, ALA can be produced in the VFA medium from post-anaerobic digestion liquor. However, ALA production was light dependent, i.e. it required 5–10 klux of light energy continuously. The

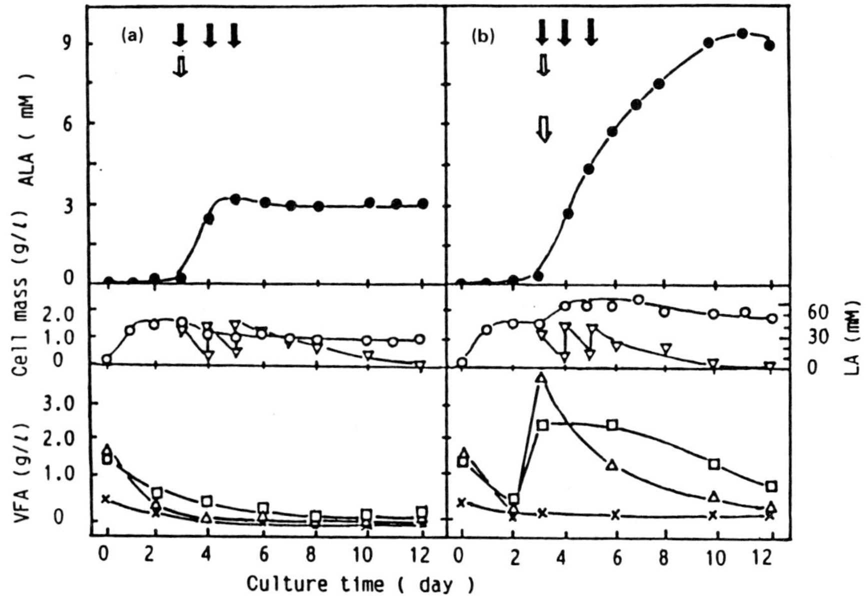

Figure 6.15 ALA production (●) without a dense inoculum (initial cell concentration 0.1 g l^{-1}) and profiles of cell mass (○), LA (▽), and acetic (□), propionic (△) and *n*-butyric (×) acids during culture of *Rb. sphaeroides* (5 klux, 30 °C, pH 7.0 ± 0.1) on VFA medium prepared from post-anaerobic digestion liquor of sewage sludge as described in Figure 6.14. (a) LA (⬇, 30 mM, ×3 indicates 3 times) and glycine (⇩, 60 mM, ×1 indicates once) were added after 3 days culture. (b) LA (30 mM; ⬇), glycine (60 mM, ⇩) and VFA (5.19 g C_2H_8COONa and 2.73 g CH_3COONa g l^{-1}) were added to the broths from day 3.

supply of light energy leads to high costs for production, therefore, an aerobic cultivation process for ALA production must be considered.

Unfortunately, the wild type of *Rb. sphaeroides* could not produce ALA extracellularly under aerobic–dark conditions even if the cells grew well (Sasaki *et al.*, 1987b), therefore, a mutant strain was selected. After several mutation processes from wild type of *Rb. sphaeroides* IFO12203 (Tanaka *et al.*, 1994a), *Rb. sphaeroides* CR-386 strain was selected as a potent ALA producer under aerobic–dark conditions.

ALA production under aerobic–dark conditions using the CR-386 strain (2.5 l fermenter, working volume 1.0 l) is shown in Figure 6.16. About 4.2 mM of ALA was produced at 300 rpm of agitation after 120 h of LA and glycine additions, but accumulation was retarded at high agitation of 500 rpm in the fermenter. It was suggested that ALA synthetase activity might be inhibited or repressed by high oxygen tension in the cells as observed in wild-type *Rb. sphaeroides* by Lascelles (1979). However, it is worthwhile noting that such large amounts of ALA can be produced under aerobic–dark cultivations. Recently, *Chlorella* sp. was reported to be able

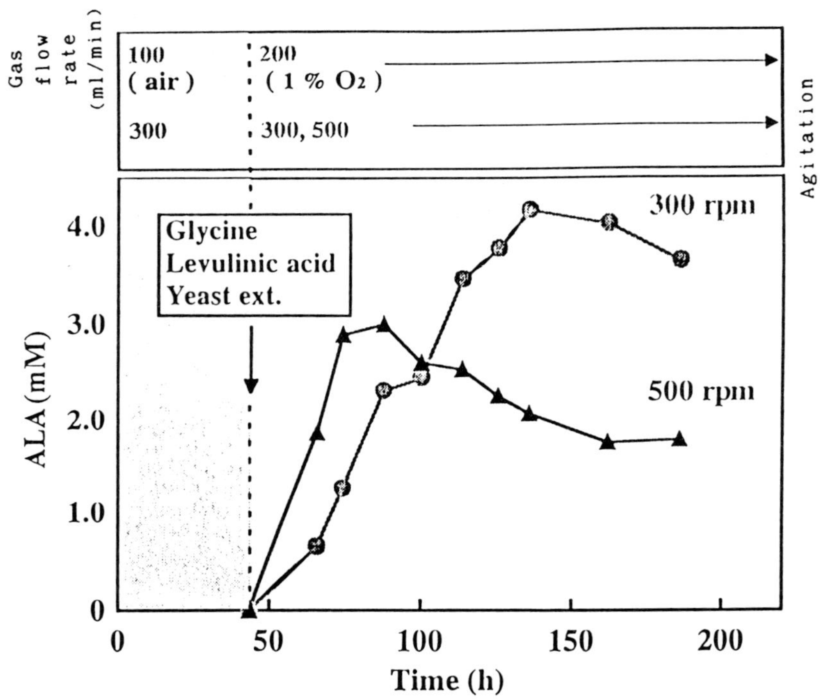

Figure 6.16 ALA extracellular production by a mutant from *Rb. sphaeroides* CR-386 under aerobic–dark conditions. CR-386 was cultured aerobically with air (100 ml min^{-1}) for 2 days on glucose–yeast extract medium in a 2.5 l fermenter (working volume 1 l). After glycine, levulinic acid and yeast extract addition, 1% oxygen gas (200 ml min^{-1}) was supplied into the fermenter to establish micro-aerobic conditions. Agitation was 300 and 500 rpm.

to produce about 2 g l^{-1} of ALA extracellularly under aerobic heterotrophic growth conditions (without light illumination) (Sasaki *et al.*, 1995b) but ALA production by CR-386 was about twice that of *Chlorella* sp.

The ALA production shown in Figure 6.16 was performed in a glucose medium containing yeast extract and glutamate. This strain can utilize VFA, therefore, it seems to be possible to produce ALA aerobically (i.e. without illumination) using post-anaerobic digestion liquor from organic wastes such as sewage sludge.

6.4.4 Applications of ALA

There have been many studies using ALA as a precursor of tetrapyrrols, however, applicable data for ALA as the physiological active substance have not been fully reported so far. This is because ALA is still considered to be an expensive and chemically unstable reagent for biochemical use, although ALA is actually stable enough under acidic conditions.

Recently, the physiological activity of ALA and the regulatory mechanism of biosynthesis of ALA have been studied and some interesting applications reported. The major applications are described in this section.

(a) Herbicide and herbicide accelerator. Regarding the use of ALA as a herbicide, studies by Rebeiz *et al.* (1984) are representative of work in this field. Figure 6.17, shows the herbicide mechanism proposed by Rebeiz *et al.* In their theory, plants which are treated with a high concentration of ALA, accumulate excessive amounts of protoporphyrin IX (PPIX) in the stage preceding chlorophyll biosynthesis. When such plants are exposed to light, the excess PPIX produces active oxygen, which damages the plant. This mechanism was called photodynamic herbicide by Rebeiz *et al.* (1984). ALA is attractive for such use because it is minimally toxic and completely biodegradable. An ALA solution of 3–5 mM, when sprayed on

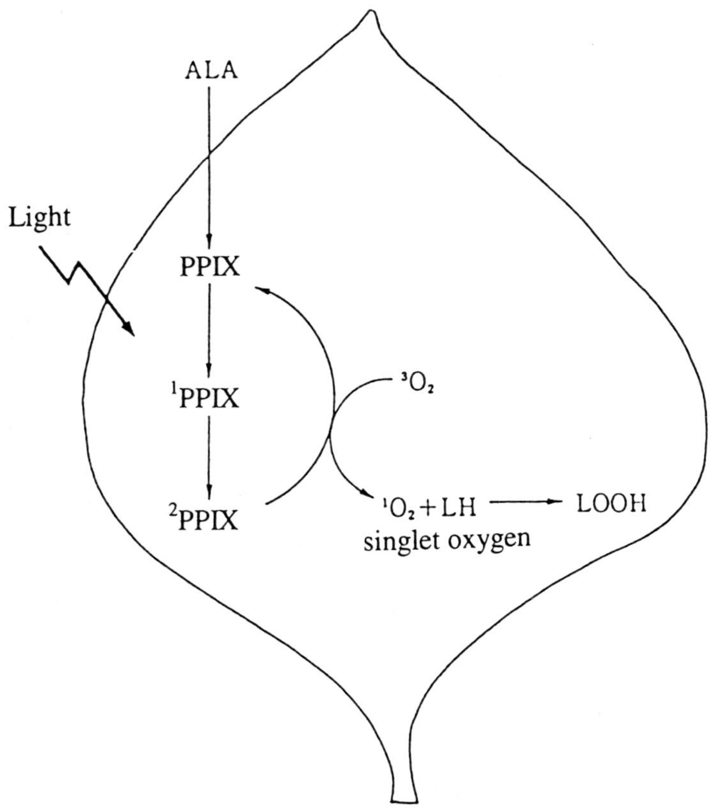

Figure 6.17 Herbicide mechanisms of ALA to weeds (based on information from Rebeiz *et al.*, 1984). ALA = 5-aminolevulinic acid; PPIX = protoporphyrin IX; LH = unsaturated fatty acid; LOOH = peroxide lipid.

the leaves and stems of plants, actually has been shown to have relatively strong herbidical effects for cucumber and several dicots plants (Rebeiz *et al.*, 1984). ALA is also able to serve as a substitute chemical for highly toxic herbicides such as paraquat owing to its similar function as a herbicide and its safety characteristics. However, as large quantities of ALA are required, its cost has prevented its practical use (Rebeiz, 1984). Reibeiz *et al.* (1984) have suggested that combination with a chemical compound, called a modulator, such as 2,2′-dipyridyl would effectively reduce the amount of ALA needed for herbicidal activity.

ALA is also applicable as a herbicide accelerator. Diphenyl ether-type herbicides such as NIP (2,4-dichlorophenyl-4-nitrophenyl ether) have function similar to the ALA photodynamic herbicide effect. The inhibition of protoporphyrinogen IX oxidase (Protox) activity is an important function of the diphenyl ether-type herbicide, the effects of which are caused by light irradiation (Lyndon and Duke, 1988; Matringe *et al.*, 1989). This diphenyl ether-type herbicide inhibits Protox and then accumulates protoporphyrinogen IX in the organs. Accumulated protoporphyrinogen IX leaks to outside of the cells and is oxidized to PPIX by autooxydation (Figure 6.18). PPIX causes the formation of singlet oxygen (active oxygen) by irradiation with light and then kills the plants (Figure 6.17). Thus, it is easy to understand the advantage of combining the diphenyl ether-type herbicide and ALA, as ALA will accelerate the herbicide effect.

As shown in Table 6.12, ALA was observed to accelerate the herbicidal effect of NIP (Tanaka, 1993). The detailed mechanism of how ALA accelerates the herbicide effect is still unclear; however, it is attractive because it is a relatively safe herbicide complex consisting of NIP (relatively toxic herbicide) and ALA, which has a higher herbicidal activity compared with ALA alone.

(b) Insecticide. Rebeiz *et al.* (1988) showed the insecticidal effect of ALA for a harmful insect such as *Trichopusia ni*. They suggested the same mechanism as the herbicide (photodynamic effect) for the insecticidal effect.

(c) Diagnosis of heavy-metal poisoning. It has been known for a long time that ALA accumulates in the urine of patients with heavy-metal poisoning and ALA concentration is frequently measured to detect and screen for heavy-metal poisoning (Fujita and Harada, 1994). Measurement of ALA in the urine for workers treating heavy metals in the factories and mines is compulsory. Thus, accurate ALA detection is important for the early diagnosis of heavy-metal poisoning.

ALA dehydratase (ALAD), which condenses two molecules of ALA to porphobilinogen (PBG), is inhibited by heavy metals, therefore, accumulation of ALA occurs (Fujita and Harada, 1994). Many patients with

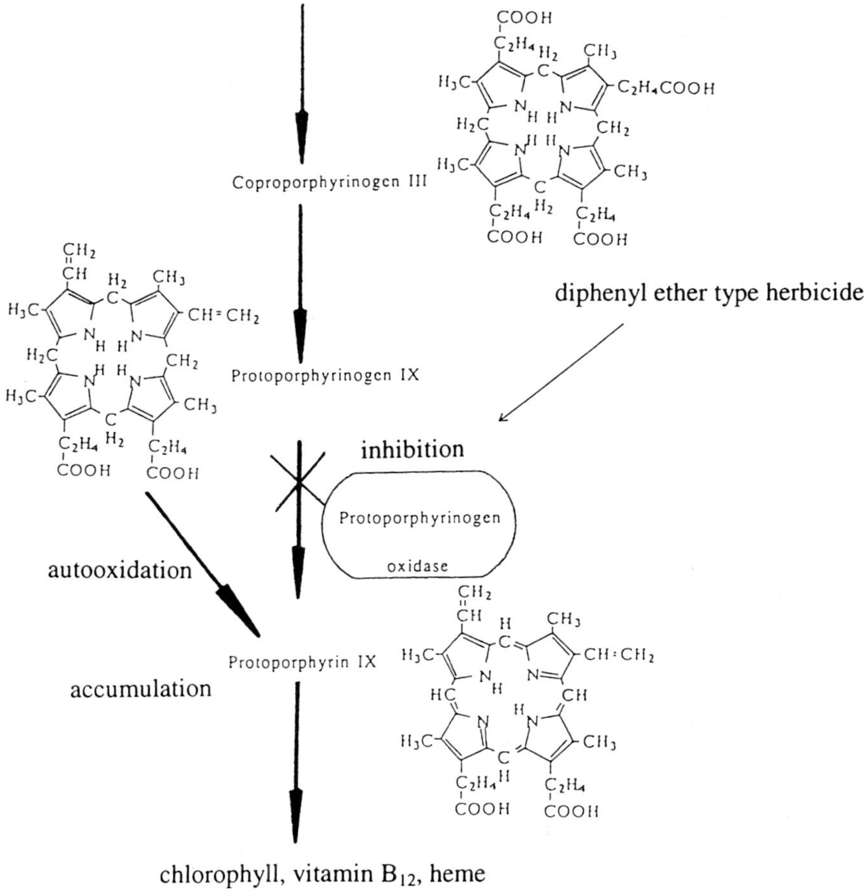

Figure 6.18 Herbicide mechanisms of diphenyl-ether type herbicide.

Table 6.12 Herbicide accelerator effects of ALA on NIP[a] observed in cucumber plants (% death) (Tanaka, 1993)

ALA	NIP concentration ($g\ l^{-1}$)					
($mg\ l^{-1}$)	0	0.1	0.3	1	3	10
1000	<5	10	50	90	90	100
300	0	<5	20	80	90	100
100	0	<5	10	70	90	100
30	0	<5	5	10	90	90
0	0	<5	<5	<5	50	90

[a]2,4-Dichlorophenyl-4-nitrophenyl ether.

heavy-metal poisoning have an abnormal heme metabolism and also accelerated ALA biosynthesis. Recently, it was found that some porphyria patients had defective ALAD (Akagi *et al.*, 1992). Porphyria and heavy-metal poisoning can sometimes cause a serious mental disorder. This fact enables us to consider the toxicity of ALA for human and animals. Therefore, many studies have been performed to test the neurotoxic effects of ALA (Brennan, 1979; Percy *et al.*, 1981; Gorchein, 1989). The results of these studies are summarized as follows:

1. Mental derangement is not observed on venous administration of ALA in animal experiments.
2. Mental derangement is observed on direct injection of ALA into the brain in animal experiments. However, ALA cannot pass the brain stem.
3. The concentration of ALA which causes mental derangement by direct injection in animal experiments is ten times greater than the concentration of ALA in patients who have a mental derangement.

These results suggest that it is difficult to regard ALA as the cause of mental derangement. These results also show the low toxicity of ALA adversely for human and animals with low concentration of ALA administration, e.g. ALA seems to be a safe medicine for both human and animals.

(d) Medical treatments for cancer. Kennedy (1991) proposed a medical application for ALA in photodynamic therapy (PDT) for cancer patients. He applied a cream containing ALA to skin cancer lesions and irradiated the lesions using a laser for a few hours. The cancer cells were killed by the singlet oxygen produced by PPIX as shown in Figure 6.19. The mode of action of ALA is similar to that for the herbicide and insecticide (see Figure 6.17).

Wide supplementary studies for cancer treatment using ALA were

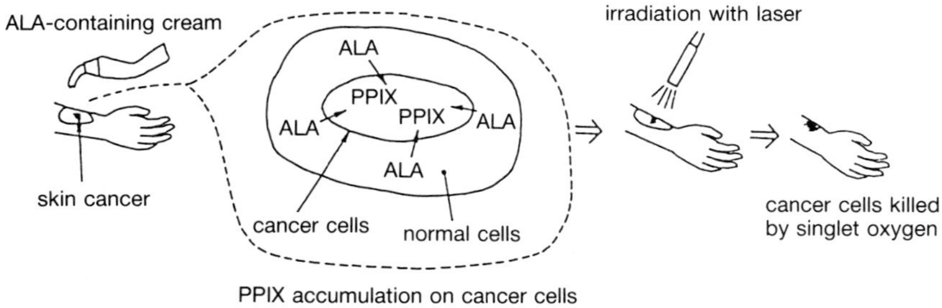

Figure 6.19 Photodynamic therapy for skin cancer (based on information from Kennedy, 1991).

performed. Excellent results have been shown for skin cancer, with over 86% of such cancers (Bowen's disease) responding to treatment. PDT using ALA is advantageous for its short relapse time after treatment. Its practical use will be considered in near future.

Grant *et al.* (1993) found that PPIX is accumulated in cancer of the larynx by oral administration. The accumulation of ALA in the cancer cells reached a maximum level after 3–4 h and returned to a normal level after 12 h (Figure 6.20). Laser light was irradiated 3–4 h after ALA administration, and cancer cells were killed without damage to normal cells. Thus, ALA is applicable for the treatment of several cancers and these applications are summarized in Table 6.13.

It has been known for a long time that porphyrins are accumulated in cancer cells and hematoporphyrin derivatives are used frequently for PDT therapy (Hillegersberg *et al.*, 1994). However, there is a difference in the level of accumulation for hematoporphyrin and PPIX (from ALA administration) in the cancer cells. The time course of the accumulation of these porphyrins is shown in Figure 6.20. PPIX accumulates and metabolizes rapidly in the cancer cells compared with normal cells. This tendency is reversed in the case of hematoporphyrin derivatives administration. This suggests that light injury as a side effect of hematoporphyrin therapy which is a serious problem in PDT will be improved by ALA therapy.

(e) Inhibition effects of peptidase. Decomposition of physiologically active peptides or proteins by peptidase enzyme is a serious problem in the medical field. Ruize *et al.* (1990) found that ALA had an inhibitory effect on peptidase. They suggested the application of ALA in a drug derivative system such as medicines for colds.

(f) Growth-promoting factor for plants. In plants, ALA plays an important biological role as an intermediate in the biosynthesis of chlorophyll. ALA biosynthesis is a rate-limiting step in chlorophyll biosynthesis. The concentration of ALA in plant cells is strictly controlled at a low level.

Recently, we found that a low concentration of ALA increased chlorophyll content and accelerated the growth of plant tissue and rice seedlings (Tanaka *et al.*, 1992). Among various investigations, foliage, soil and root treatment with ALA was considered to be effective at every treatment for increasing the yield of crops. A yield-increasing effect is observed with a variety of plants such as leafy vegetables, potatoes, etc. The effect was particularly strong for underground crops, such as potato and garlic, compared with other crops. These experimental results are shown in Figures 6.21 (potato), 6.22 (garlic) and 6.23 (rice). Other results of yield-increasing tests for various plants are summarized in Table 6.14. The concentration of ALA used was in the range of 10–100 mg l^{-1} (ppm)

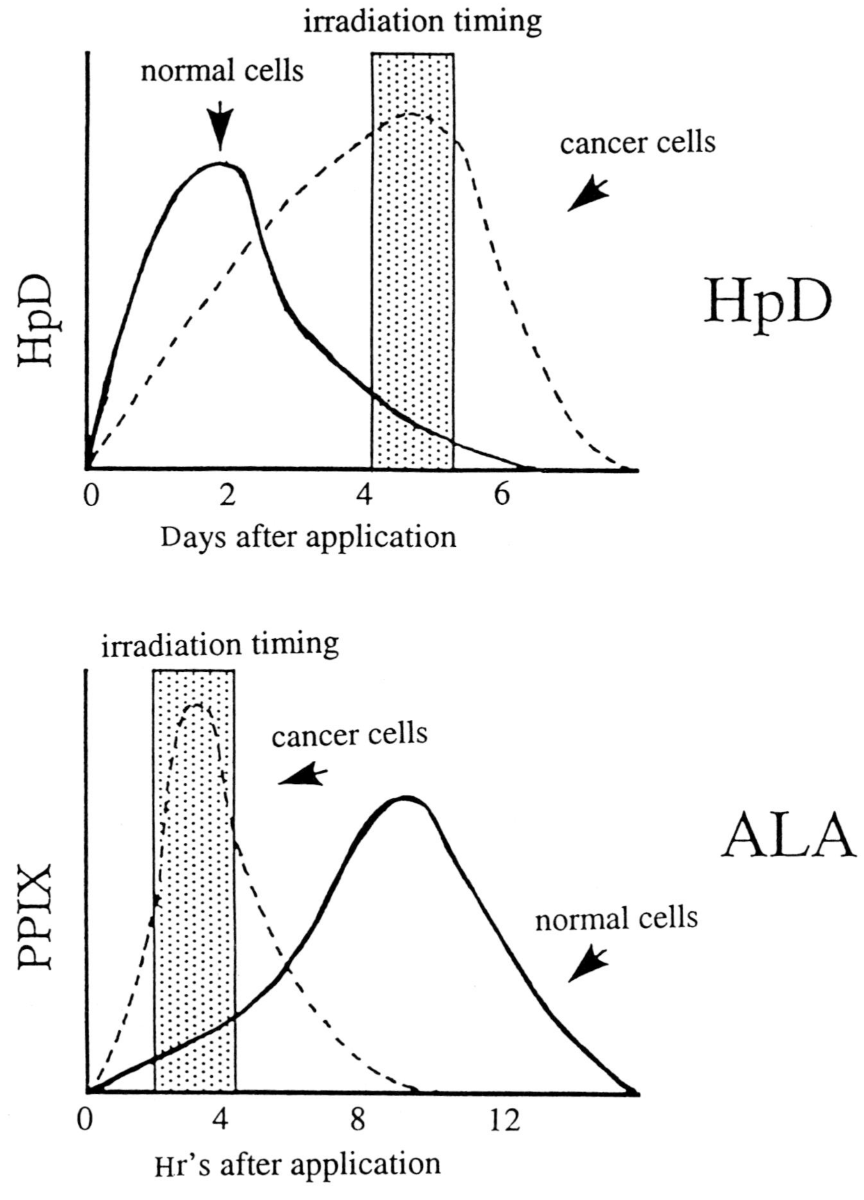

Figure 6.20 Difference of accumulation time for HpD and PPIX from ALA. HpD = hematoporphyrin dimer. In cancer cells, HpD is accumulated after 4–5 days of HpD administration; but PPIX is accumulated after 2–4 days of ALA administration. The accumulation patterns of HpD and PPIX are quite different. Razer irradiation timing should be considered.

Table 6.13 Applications of photodynamic therapy for several cancers

Cancer	Experimental level
Skin cancer	Practical
Oral cancer	Clinical
Esophageal cancer	Clinical
Colon cancer	Clinical
Duodenal cancer	Clinical
Pancreas cancer	Animal
Bladder cancer	Animal

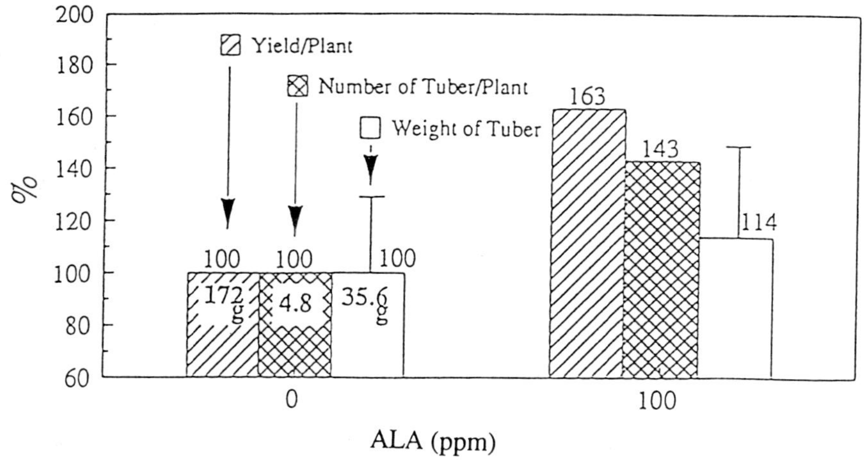

Figure 6.21 The effect of ALA on yield of potato. Potatoes were planted on March 26 and the plants were sprayed with ALA solution at the tuber formation stage on July 2. The yield was determined on July 23. Data was averaged for 40 plants.

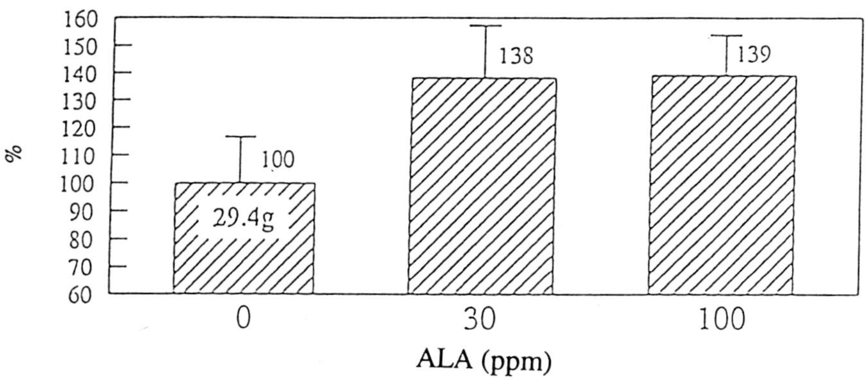

Figure 6.22 The effect of ALA on yield of garlic. Garlic bulbs were planted in an experimental field on October 5 and the plants were sprayed with ALA solution at the 5–7 leaf stage on May 13. The bulb yield of garlic was determined on June 5. Data was averaged for 200 plants.

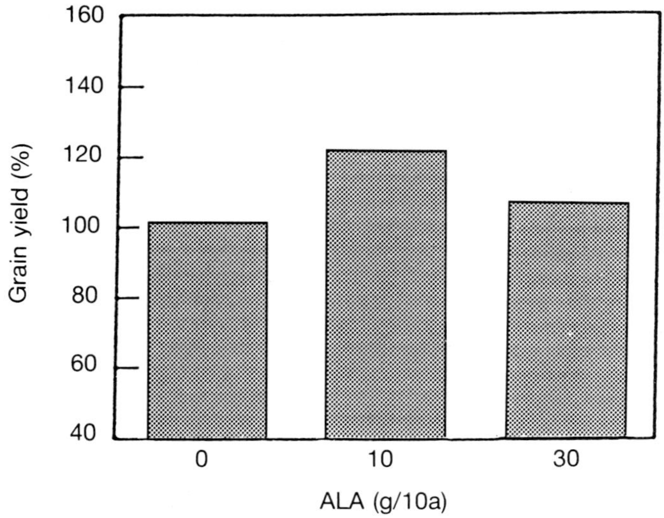

Figure 6.23 The effect of ALA on yield of paddy rice. Eight paddy rice seedlings were planted in pot (0.05 m^2). The plants were soil treated with ALA solution on September 4. Grain weight was measured on October 4. Data was averaged for 136 plants.

Table 6.14 Yield enhancement for several crops and vegetables by ALA application

Crop	Yield (%)[a]	ALA conc.[b]
Spinach (*Spinacia oleracea* L.)	140	100 ppm
Rape (*Brassica campestris* L. subsp. *napus*)	140	100 ppm
Garlic (*Allium satiyum* L.)	139	100 ppm
Potato (*Solanum tuberosum* L.)	163	100 ppm
Radish (*Raphanus sativas* var. *radicula* DC.)	145	100 ppm
Barley (*Hordium vulgar* L.)	141	30 ppm
Wheat (*Triticum aestivum* L.)	108	30 ppm
Rice (*Oriza sativa* L.)	121	100 g/10a[c]

Foliar application for all crops but rice where soil application was done.
[a]When the yield of the control experiment (without ALA application) was defined as 100%.
[b]ALA concentration of ALA application solution.
[c]100 g of ALA was applied per 10a field.

when applied to the stems and leaves, and 0.01–10 mg l^{-1} when applied to the root (Tanaka *et al.*, 1992).

ALA is considered to be important in photosynthesis energy harvesting system of plants. Table 6.15 shows the results of ALA on the photosynthetic activities of radish. Treatment with ALA enhanced the rate of CO_2 gas uptake and photosynthetic activities under light conditions. Conversely, the rate of CO_2 gas production corresponding to dark respiration was suppressed by ALA under dark conditions. For the growth promotion and yield increasing effect, it is suggested that ALA enhances photosynthetic activity and suppress respiration, e.g. net CO_2 fixation is enhanced

Table 6.15 Effect of ALA on photosynthetic and respiratory activity in radish

5-ALA conc. (ppm)	Photosynthetic activity (%) 0 days	2 days	5 days	Respiratory activity (%) 0 days	2 days	5 days	Yield (DW,g) 26 days
0 (control)	100	112	114	100	128	200	2.97 (100)
30	100	126	115	100	96.3	139	3.83 (129)
100	100	122	120	100	114	153	3.53 (119)

CO_2 concentration was measured by infra-red absorption (SPB type-03 and IRA type-102 Simazu Co. Ltd). Data was the average of 3 pots.
Reproduced with permission from Tanaka, T. *et al. Proceedings of the 19th Annual Meeting of the Plant Growth Regulator Society of America*, San Francisco, p. 242, 1992.

(Tanaka *et al.*, 1992). This enhancement of photosynthesis by ALA was also observed in algae (*Spirulina*) (Sasaki *et al.*, 1995c).

In connection with this, ALA accelerates nitrate reductase activity which is a rate-limiting step of nitrogen accumulation in plants (Mishra and Srivastava, 1993), and the rate of nitrate per all nitrogen in spinach was decreased by ALA supply (Yoshida *et al.*, 1994). This result is applicable for the improvement of vegetable quality.

It is projected that a serious food shortage may arise in the 21th century due to the continuous increase in the world's population. In addition, greenhouse effects caused by the discharge of CO_2 into the atmosphere are causing serious concern. It seems to be significant that ALA has the effect of increasing yield for crops and vegetables. The enhancement of CO_2 fixation by ALA should be appled for CO_2 reduction in the atmosphere using ALA's functions for plants and algae.

(g) Color-intensifying effects for apples. ALA was found to have a color-intensifying effect on anthocyanin-containing apples at the stage where coloring starts (Yokota *et al.*, 1994). Apple color is one of the most important aspects of apple marketing. Color treatment involves labor-intensive work, such as placing a bag on each apple and removing some of the leaves. Therefore, creating such effects with ALA is practically attractive. Some kinds of color-intensifying reagents are already known; however, almost all such reagents lead to the problem of softening of the fruit by promoting ethylene production. ALA does not relate to ethylene synthesis and has no effects on ripeness and hardness. The mode of action is still unclear, But, light irradiation enhanced the ALA effects. Some ALA derivatives are expected to be involved in such color intensifying.

(h) Increasing salt tolerance. Salt accumulation on farmland is increasing as a result of irrigation. About 6 000 000 ha of farmland is lost and

Table 6.16 Increasing salt tolerance of cotton by ALA

Salt	ALA	Root (wt %)					Aerial part (wt %)				
(%)	(ppm)	Na	K	Mg	Ca	Si	Na	K	Mg	Ca	Si
0.0	0	0.49	4.68	0.42	0.39	–	0.07	2.49	0.54	2.03	–
	3	0.52	2.30	0.39	0.37	0.30	0.09	1.96	0.49	2.15	0.02
	10	0.51	2.70	0.42	0.41	–	0.08	1.91	0.46	2.16	–
	30	0.52	2.70	0.42	0.37	0.27	0.10	2.06	0.48	2.14	0.03
	100	0.47	1.80	0.40	0.64	0.23	0.09	2.09	0.50	2.19	0.03
1.0	0	1.03	2.47	0.34	0.38	–	1.5	2.10	0.50	2.16	–
	3	0.84	2.30	0.36	0.38	0.34	1.27	2.27	0.49	2.20	0.07
	10	0.68	1.77	0.35	0.47	–	0.91	2.04	0.50	2.16	–
	30	0.58	1.90	0.31	0.53	0.23	1.70	2.10	0.476	2.26	0.02
	100	0.71	2.20	0.38	0.46	0.16	1.31	2.00	0.50	2.25	0.03
1.5	0	1.01	2.05	0.33	0.50	–	3.26	1.34	0.43	1.68	–
	3	0.56	2.70	0.35	0.51	0.23	2.26	1.60	0.47	2.10	0.08
	10	0.61	1.95	0.29	0.45	–	1.34	2.07	0.46	2.10	–
	30	0.52	1.20	0.33	0.62	0.20	1.66	1.52	0.49	2.11	0.07
	100	0.47	1.90	0.33	0.45	0.20	0.72	1.92	0.47	2.07	0.06

becomes desert every year (Kyuma, 1992). In deserts, most plants cannot grow because of salt stress even if water is supplied when farmland in the desert is used for a long time. Many studies have been carried out for breeding plants with a high salt tolerance but, salt-tolerant crops have not yet been found (Tarzynski, 1993; Takabe, 1996).

ALA was found to increase salt tolerance in cotton (Kuramoti *et al.*, 1996). Cotton seedlings treated with ALA were able to grow in soil containing 1.5% (w/w) NaCl. The Na ion concentration in roots treated with ALA was suppressed at a low concentration. However, the concentration of Ca and Si in the roots was not affected by ALA application (Table 6.16). These results suggest that ALA may raise the osmotic pressure and suppress the absorption of Na into the roots.

This observation is quite significant for expanding farmlands close to desert areas. Further studies are required to clarify the mode of action of this phenomenon. In connection with this, increasing cold tolerance and an increase of fuructan content by ALA application are observed (Hotta *et al.*, 1995; Yoshida *et al.*, 1995).

Thus, the application of ALA in agricultural field is quite promising.

6.5 Problems and future prospects

6.5.1 Problems

The application of photosynthetic bacteria for SCP and chemical production from various organic wastes seems to be promising. However, there are some problems in using these bacteria from the practical point of view.

1. The waste needs to be discharged in sufficient quantities and concentrations for the processing to be economically viable.
2. Seed cultivation systems should be considered to maintain the dominancy of photosynthetic bacteria in the reactor operations. This system is costly from the practical point of view. It is recommended that the waste water discharged from heat-treatment processes, e.g. the food industry, is used for this purpose as the supplier of seed cells.
3. Bacterial cell harvest is generally costly compared with the SCP yeast cells. However flock-forming photosynthetic bacteria have been isolated recently (Watanabe *et al.*, 1996). In addition, the conditions and mechanisms for making the flock of bacterial cells by adding Mg and Ca ions were elucidated. Therefore, the cell harvesting problem may be solved in the near future.
4. In some cases, research and development for photobioreactor construction are required, in which the efficiency of the light illumination to the cells, the maintenance of homogenous mixing and the optimum temperature have to be considered.

Economic assessments should be made when plant-scale production is planned with these problems kept in mind.

6.5.2 Future prospects

As future prospects for developing the utilization of the photosynthetic bacteria for the conversion of wastes to useful materials, the following points should be considered.

(a) Screening for superior strains. Photosynthetic bacteria are widely distributed in natural environments. Since the bacteria isolated so far are limited in number, a new strain may indicate a new ability to produce physiologically active substances with high growth rates.

(b) Improving strains. The photosynthetic bacteria can be improved using mutation and genetic engineering techniques. For example, when the gene for β-carotene synthesis was inserted into a photosynthetic bacterium, this resulted in a high β-carotene accumulation, which is valuable as a physiologically active pigment (not spheroiden and spheroidenone) (Sunarjo *et al.*, 1994). Mutation technology has been widely applied to photosynthetic bacteria in which ubiquinone (Q-10) accumulation has been increased by more than 10 times. The aerobic production of ALA has also become possible by the use of mutation techniques as shown in section

6.4.3. This technology may be applied for the production of several physiological active substances.

(c) Improving culture system. In SCP and chemical production, a high cell density after the cultivation is required from the practical point of view. In aerobic culture, this can be achieved, e.g. by means of a fed-batch culture. However, achieving a dense culture is very difficult in anaerobic–light culture owing to the limitation of light energy supply. Recently, a special photobioreactor system using a optic fiber to provide light energy to the inside of the reactor has been reported for developing a new application for photosynthetic organism (Mori *et al.*, 1987; Matsumoto and Fukui, 1996). By using this system, the fixation of CO_2 to materials on metabolic pathway may become possible at a relatively high rate compared with algae and other plants. The mass production of useful chemicals origining from photosynthetic bacteria may be realized in the near future.

References

Akagi, R., Meguro, K. and Doss, M. (1992) *Porphyrins*, **1**, 276.
Barandl, H., Gross, R.A., Lenz, R.W. and Fuller, R.(1990) In *Advances in Biochemical Engineering*, **41**, p. 77.
Beale, S.I. (1970) *Plant Physiol.*, **45**, 504.
Brennan, M.J. (1979) *South African J. Sci.*, **75**, 335.
Fujita, H. and Harada, K. (1994) In *Toxicology Today* (ed. Y. Sato), Kinhoudo, Kyoto, p. 129.
Fukui, S. (1980) In *Vitaminology II* (ed. The Vitamin Society in Japan), Tokyo Kagaku Dojin, Tokyo, p. 477.
Grant, W.E., Hopper, C., MacRobert, A.J., Speight, P.M. and Brown, S.G. (1993) *Lancet*, **342**, 147.
Gorchein, A. (1989) Biochem. Soc. Trans., **17**, 577.
Hillegersberg, R., Kort, W.J. and Wilson, P. (1994) *Drugs*, **48**, 510.
Hiraishi, A. (1988) *J. Appl. Bacteriol.*, **64**, 103.
Hiraishi, A. (1989) In *Proceedings of the 5th International Symposium on Microbial Ecology*, Tokyo, p. 663.
Hoshino, Y. and Kitamura, H. (1984) In *Kougousei-Saikin [The Photosynthetic Bacteria]* (eds H. Kitamura, S. Morita and J. Yamashita, Gakkai Shuppan Center, Tokyo, p. 3.
Hotta, Y., Tanaka, T., Yoshida, R., Takeuchi, Y. and Konnai, M. (1995) In *Proceedings of the 2nd Asian Crop Science Conference*, Fukui, Japan, p. 524.
Jaenchen, R., Harald, H.G. and Thauer, R.K. (1981) *FEMS Microbiol. Lett.*, **12**, 167.
Kanamori, M. (1984) In *Kougousei-Saikin [The Photosynthetic Bacteria]* (eds H. Kitamura, S. Morita and J. Yamashita), Gakkai Shuppan Center, Tokyo.
Kato, T. (1991) *Fuudokemikaru (Foodchemical)*, 1991–8, p. 30.
Kennedy, J. (1991) International Patent, WO91/01727.
Kitamura, H. (1988) *Biseibutu (Japan)*, **2**, 37.
Klein, O. and Porra, R.J. (1982) *Hoppe-Seyler's Z. Physiol. Chem.*, **363**, 551.
Kobayashi, M. (1984) In *Kougousei-Saikin [The Photosynthetic Bacteria]* (eds H. Kitamura, S. Morita and J. Yamashita), Gakkai Shuppan Center, Tokyo, p. 339.
Kobayashi, M. and Kurata, S. (1978) *Proc. Biochem.*, **13**, 27.
Koesnandar, A.S., Nishio, N. and Nagai, S. (1989) *Biotechnol. Lett.*, **11**, 567.
Kuramoti, H., Watanabe, K., Tanaka, T. *et al.* (1996) In *Proceedings of the 31th Plant Growth Regulator Society of Japan*, Kyoto, Japan, p. 88.

Kyuma, K. (1992) *Energy Nat. Resources*, **13**, 26.
Lascelles, J. (1978) In *The Photosynthetic Bacteria* (eds R.K. Clayton and W.R. Sistrom), Plenum Press, New York, London, p. 795.
Lin, D., Nishio, N. and Nagai, S. (1989) *J. Ferment. Bioeng.*, **68**, 88.
Lyndon, J. and Duke, S.O. (1988) *Pest. Biochem. Physiol.*, **31**, 74.
Matringe, M., Camadro, J.M. Labbe, P. and Scalla, R. (1989) *Biochem. J.*, **260**, 231.
Matsumoto, M. and Fukui, Y. (1996) Japan Patent Koukai, 8-38157.
Mishra, S.N. and Srivastava, H.S. (1993) Experimenta, **39**, 1118.
Miyake, J. and Asada, Y. (1996) *J. Hydrogen Energy Syst. Soc. Japan*, **21**, 23.
Mori, K., Ohya, H., Matsumoto, K. and Furuune, H. (1987) *Adv. Space Res.*, **7**, 447.
Morimoto, H. (1971); *Tanpakushitu Kakusan Kouso*, **16**, 183.
Morita, S. and Shimada, K. (1984) In *Kougousei-Saikin [The Photosynthetic Bacteria]* (eds H. Kitamura, S. Morita and J. Yamashita), Gakkai Shuppan Center, Tokyo, p. 293.
Nakasaki, K. (1995) *Bio Industry (Japan)*, **12**, 26.
Nakayama, T. (1993) In *Haikibutu-Syori Saisigenka Handobook*, Kensetusangyou-Tyosakai, Tokyo, p. 69.
Nishio, N., Kitaura, S. and Nagai, S. (1982) *J. Ferment. Technol.*, **60**, 423.
Noparatnaraporn, N. (1987) Ph.D. thesis, Faculty of Engineering, Hiroshima University, Hiroshima, p. 92.
Noparatnaraporn, N., Wongkornchawalit, W., Kanchote, D. and Nagai, S. (1986a) *J. Ferment. Technol.*, **64**, 137.
Noparatnaraporn, N., Sasaki, K., Nishizawa, Y. and Nagai, S. (1986b) *Biotechnol. Lett.*, **8**, 491.
Noparatnaraporn, N., Trakulnaleumai, S., Silveira, R.G., Nishizawa, Y. and Nagai, S. (1987a) *J. Ferment. Technol.*, **66**, 11.
Noparatnaraporn, N., Trakulnaleumsai, S. and Duangsawat, S. (1987b) *J. Sci. Soc. Thailand*, **13**, 15.
Percy, V.A., Lamm, M.C.L. and Taljaard, J.J.F. (1981) *J. Neurochem.*, **36**, 69.
Pfenning, N. (1978) In The Photosynthetic Bacteria (eds R.K. Clayton and W.R. Sistrom), Plenum Press, New York, London, p. 3.
Rebeiz, C.A. (1984) *Chem. Week*, **24**, 34.
Rebeiz, C.A. Montazer, Z.A. Hopen, H.J. and Wu, S.M. (1984) *Enzyme Microbiol. Technol.*, **6**, 390.
Rebeiz, C.A. Jubik, J.A. and Rebeiz, C.C. (1988) *Pest. Biochem. Physiol.*, **30**, 11.
Ruize, J. Peater, E.G. and Hshiel, H.L. Japan Patent Koukai, 2-115130.
Sasaki, K. and Nagai, S. (1979) *Euro. J. Appl. Microbiol. Biotechnol*, **7**, 201.
Sasaki, K., Hayashi, M. and Nagai, S. (1978) *J. Ferment. Technol.*, **56**, 200.
Sasaki, K., Noparatnaraporn, N., Hayashi, M., Nishizawa, Y. and Nagai, S. (1981) *J. Ferment. Technol.*, **59**, 471.
Sasaki, K., Kidani, T., Emoto, Y. and Hamaoka, T. (1987a) *J. Soc. Agric. Struct. (Japan).*, **18**, 38.
Sasaki, K., Ikeda, S., Nishizawa, Y. and Hayashi, M. (1987b) *J. Ferment. Technol.*, **65**, 511.
Sasaki, K., Tanaka, T., Nishizawa, Y. and Hayashi, M. (1990) *Appl. Microbiol. Biotechnol.*, **32**, 727.
Sasaki, K., Noparatnaraporn, N. and Nagai, S. (1991) In *Bioconversion of Waste Materials to Industrial Products*, 1st edn. (ed. A.M. Martin), Elsevier Science Publishers, New York, London, p. 225.
Sasaki, K., Tanaka, T. and Hotta, Y. (1995a) *Mizushyori-Gijutu*, **36**, 27.
Sasaki, K., Watanabe, K., Tanaka, T. and Hotta, Y. (1995b) *World J. Microbiol. Biotechnol.*, **11**, 361.
Sasaki, K., Marquez, F.J. Nishio, N. and Nagai, S. (1995c) *J. Ferment. Bioeng.*, **79**, 453.
Satoh, K. and Shimizu, S. (1988) *Vitamins (Japan)*, **62**, 69.
Shipman, R.H., Kao, I.C. and Fan, L.T. (1975) *Biotechnol. Bioeng.*, **17**, 1561.
Sunarjo, J., Burgess, J.G., Yamada, A. *et al.* (1994) In *Proceedings of the Annual Meeting of the Society for Fermentation and Bioengineering, Japan*, p. 153.
Suzuki, T., Tsygankov, A.A., Miyake, J., Tokiwa, Y. and Asada, Y. (1995) *Biotechnol. Lett.*, **17**, 395.
Takabe, T. (1996) *Biosci. Ind.*, **54**, 273.

Takahashi, H. (1984) In *Kougousei-Saikin [The Photosynthetic Bacteria]* (eds H. Kitamura, S. Morita and J. Yamashita), Gakkai Shuppan Center, Tokyo, p. 196.

Tamai, T., Nakajima, H. and Oda, K. (1996) Japan Patent Koukai, 8-325158.

Tanaka, T. (1993) Japan Patent Koukai, 5-117110.

Tanaka, T., Takahashi, K., Hotta, Y., Takeuchi, Y. and Konnai, M. (1992) In *Proceedings of the 19th Annual Meeting of Plant Growth Regulator Society of America*, San Francisco, p. 237.

Tanaka, T., Watanabe, K., Hotta, Y. *et al.* (1994a) *Seibutu-Kougakukaisi*, **72**, 461.

Tanaka, T., Sasaki, K., Noparatnaraporn, N. and Nishio, N. (1994b) *World J. Microbiol. Biotechnol.*, **10**, 677.

Tarzynsky, M.C. (1993) *Science*, **259**, 508.

Truper, H.G. and Pfenning, N. (1989) In *Bergey's Mannual of Systematic Bacteriology*, Vol. 3, Williams and Wilkins, Baltimore, p. 1635.

Vrati, S. (1984) *Appl. Microbiol. Biotechnol.*, **19**, 199.

Wagai, B. (1987) *Treatment and Utilization of Animal Discharges*, Yokendo, Tokyo, p. 1.

Watanabe, H., Okuzawa, K. and Ochi, S. (1993) In *Annual Report of Sewage Research and Investigation in 1993*, Public Works Research Institute, Ministry of Construction, Tokyo, p. 87.

Watanabe, M., Sasaki, K., Noparatnaraporn, N., Nakasimada, Y. and Nishio, N. (1996) In *Proceedings of the Annual Meeting of the Society for Fermentation and Bioengineering, Japan*, p. 243.

Yamada, Y., Haneda, K., Murayama, S. and Shiomi, S. (1991) *J. Chem. Eng. Japan*, **24**, 94.

Yata, E., Kawaguchi, H. and Sasaki, K. (1996) In *Bioconversion of Waste Materials – Recycle Use of Organic Waste Materials*, Chijin Shyokan, Tokyo, p. 15.

Yokota, K., Tanaka, T. and Hotta, Y. (1994) United States Patent, 5, 318, 788.

Yoshida, R., Tanaka, T. and Hotta, Y. (1994) In *Proceedings of the 24th International Horticultural Congress*, Kyoto, p. 19.

Yoshida, R., Tanaka, T. and Hotta, Y. (1995) In *Proceedings of the 15th International Congress on Plant Growth Substances*, Minneapolis, P. 185.

7 Utilization of starch industry wastes

SUDIP K. RAKSHIT

7.1 Introduction

Starch and flour, presumably from barley or emmer (a primitive variety of wheat), have been utilized from antiquity for the production of beer and bread (Demain and Solomon, 1981). Reliefs obtained from the fifth dynasty Egyptian tombs from about 2400BC representing scenes of baking and brewing are now kept in the museum of antiquities in Lieden. From these first applications, the use of starch has grown both in quantity and quality in a large number of applications.

Starch is the second largest compound produced by plants in nature after cellulose (Knight, 1969). It has a number of distinct advantages over cellulose which include: possible consumption by humans, greater susceptibility to partial and total hydrolysis leading to a number of useful products, and easy modification to compounds of value in a variety of industries.

The world's **total production of starch** is estimated to have been 25.1 million tons in 1990 (Sanders, 1996). About 60% of this goes into the production of starch sugars. Technological advances in the use of immobilized enzyme **isomerases** have led to its conversion to high fructose syrups used in large quantities in food and beverage industries.

The extraction of starches is accompanied by a large number of liquid and solid wastes. For example, in the production of tapioca starch, 5–10 m^3 of water is used per ton of input roots, i.e. 30–50 m^3 per ton of starch produced (Zia, 1992). This varies with the process followed and is accompanied by large quantities of solid wastes from the peels. A number of methods have been suggested for treatment of the waste waters and the use of these solid residues.

In this chapter the major sources of starch and their extraction procedures for use in food, paper and fermentation industries will be reviewed along with technological alternatives to utilization of starch industry by-products.

7.2 Nature of cereal and tuber starches

Both starch and cellulose are polymers of glucose. However, the major difference in the properties of the two carbohydrates is due to the nature of the glycosidic bonds between the monomers. Cellulose has the β 1–4 linkage, which is very difficult to digest or hydrolyse. Starch on the other hand is made up of two types of polymers made up of a skeleton of α 1–4 bonds between glucose moieties. The **amylose** fraction with molecular weights varying from several thousand to half a million are straight-chain polymers, which are water insoluble, and typically constitutes 20% starch. The larger bulk of starch is **amylopectin** and is chiefly distinguished from amylose by a considerable amount of 1–6 branching from the linear molecule. Amylopectin molecules are much larger than amylose with molecular weights varying from one to two million. It is soluble in water and forms gels by absorbing water. Another important difference between amylose and amylopectin is the difference in their **iodine-binding capacity**, which is 19–20% in the case of amylose and 1–2% in the case of amylopectin. While amylose produces a blue color, amylopectin produces a purple color.

Various types of starch occur in higher plants as cereals (wheat, corn, rice, barley, sorghum, rye, millet, oats), tubers and roots (potato, tapioca or cassava, arrow root, yam), fruits (banana, plantain), stem, pith and seeds. The starch available from various sources vary in the size and shape of the granule, composition of amylose and amylopectin, and other functional properties like pasting temperature, swelling power, etc. (Table 7.1).

These differences are reflected in their application in industry. In our laboratory, the direct bioconversion of cassava and corn starch to **lactic acid** was compared using *Lactobacillus amylovorous* for the fermentation (Wang and Rakshit, 1996). It was found that protein composition of the corn played a major role in the bioconversion and that the cassava medium had to be suitably modified for equivalent production. The difference in properties of starch obtained from different sources result in differences in extraction procedures and applications.

7.3 Starch-based industrial products

A major portion of starch is consumed without modification in the food industries in the form of baked products, etc. The products of starch obtained by modification are mainly used in the following industries:

- hydrolytic products and sweeteners;
- food applications;
- paper and textile applications;
- fermentation products, etc.

Table 7.1 Variation in properties of starch types

Starch properties	Potato	Maize	Sweet potato	Tapioca
Composition of raw materials (%)				
Starch	17	60	23	26
Moisture	78	16	68	66
Protein as $N \times 6.25$	2	9	1.5	1
Lipids	0.1	4	0.3	0.3
Fiber	1	2	–	1
Starch on dry substance	77	71	72	77
Starch granule properties				
Type	Tuber	Cereal	Root	Root
Size, diameter (μ)	5–100	2–30	5.25	4–35
Size, diameter (average μ)	28	10	15	16
Size, diameter; weight average (μ)	40	15	–	25
Shape	Oval, spherical	Round, polygonal	Polygonal	Oval, truncated
Gelatinization of native starches				
Pasting temperature (°C)	60–65	75–80	65–70	60–65
Peak viscosity range (BU)	1000–5000	300–1000	–	500–1900
Peak viscosity average (5%) (BU)	3000	600	–	1000
Properties of pastes of native starch				
Paste viscosity	Very high	Medium–low	High	High
Paste texture	Long	Short	Long	Long
Paste clarity	Clear	Opaque	Translucent	Very clear
Rate of retrogradation	Medium	High	Medium	Low
Chemical composition of granule				
Lipid (%)	0.1	0.8	–	0.1
Protein (%)	0.1	0.35	–	0.10
Phosophorus (%)	0.08	0.02	–	0.01
Amylose (%)	21	28	–	17
DP[a] (amylose)	4000	1000	–	4000
Amylopectin (%)	79	72	–	83
DP[a] ($\times 10^6$) (amylopectin)	2	2	–	2

[a]Degree of polymerization = average number of glucose residues per molecule.

7.3.1 Hydrolytic products and sweeteners

Starch on complete hydrolysis using acid and/or enzyme can be split quantitatively to glucose. Glucose syrups are hydrolysates of starch in which the reactions are stopped at intermediate levels leaving a mixture of glucose, maltose and a mixture of higher sugars. These are commercially available at various **dextrose equivalent (DE)** levels, defined as the total reducing sugars expressed as dextrose (anhydrous glucose) as a percentage of the total initial dry substance.

Beginning from low DE values of 25–37 through values of DE from 58 and higher to completely hydrolysed syrups of dextrose are now produced by the combination of amylases like α-amylase, β-amylase, glucoamylase and the debranching enzymes isoamylase or pullulanase (Rakshit, 1996; Schenck and Hebeda, 1992).

A major market for the dextrose obtained by complete hydrolysis is for the conversion into **high fructose syrups** (HFS). Because of their high sweetening property this product has replaced sucrose in a large number of food and beverage industries. A historical perspective of the **isomerase** enzymes for this important bioconversion is given by Casey (1977).

In 1972, batch production was replaced by a continuous process utilizing **isomerase** enzyme immobilized on an insoluble carrier. This was the first large-scale application of the immobilized enzyme system in the world and resulted in a significant reduction in production costs. The design and operation of this system is described extensively by Oestergaard and Knudson (1976) and Venkatasubramanian (1978). The acceptance of these sweeteners and its economic impact on the starch industry has been substantial.

According to Coker and Venkatasubramanian (1985), the price differential between **high fructose corn syrup** (HFCS) and sugar (sucrose) offered strong incentive for the substitution of HFCS for sugar in soft drinks, canned fruits, ice cream and bakery products. A 28.5% decline in sugar usage in 1981 from its value of consumption of 5.12 billion pounds in 1978 in the beverage industry is a good indicator of the changing scenario. It was predicted that future growth in HFCS consumption would occur only in direct proportion to population growth and hence sales was expected to go up at a decreasing rate.

Some of the hydrolysis products of starch are further modified to obtain useful products in the food, industrial and cosmetic industries (Figure 7.1). **Maltitol** obtained from the hydrogenation of high maltose syrup can be used as bulk sweetener without any cooling sensation or aftertaste in the mouth. **Lycasin** and **mannitol** are obtained by hydrogenation of glucose and fructose, respectively. Pfizer has produced a water-soluble compound, **polydextrose**™ as a partial or complete fat replacer. It has been marketed as a bulking agent, like many other polysaccharides, but is reported to have some unfavorable effects, e.g. it acts as a laxative and produces gas in the larger intestines (Mercier and Colonna, 1988).

The hydrolysis products constitute one of the largest users of starch. Schenck and Hebeda (1992) give a very good review on the large number of hydrolytic products, the equipment required for the production, design, quality control and waste treatment systems in operational plants.

7.3.2 Food applications

The use of **modified starch** as a food additive is not based on its nutritional value but on its functional properties, which confer to the food, for

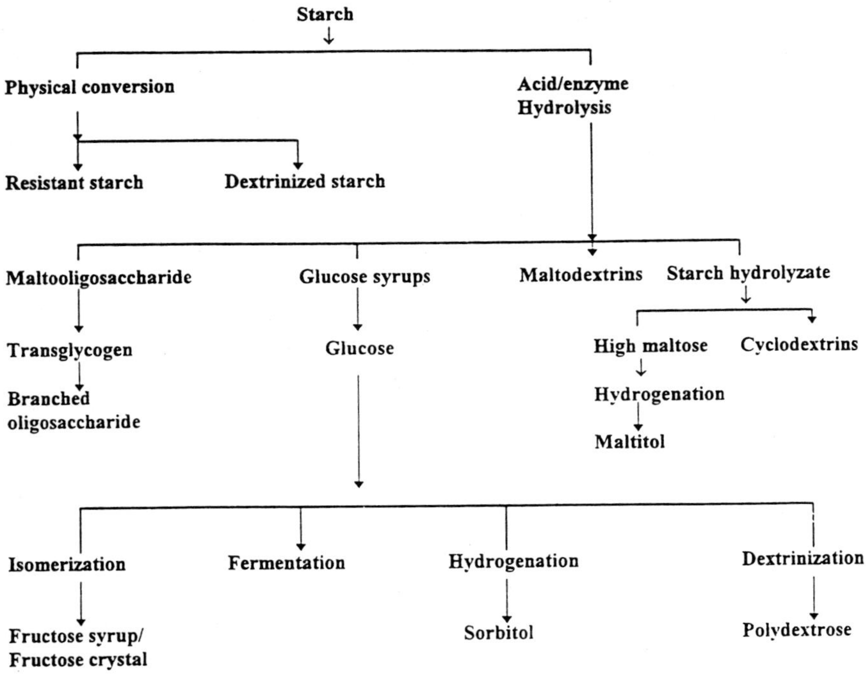

Figure 7.1 Products of physical conversion and acid/enzyme hydrolysis of starch.

example, better mouth feel, thickening or gelling characteristics and stability. Hence modification is done to obtain the final product with the required characteristic.

Potato and cassava products yield pastes with clarity and a bland flavor whereas the small amounts of proteins and lipids present in corn and wheat impart off-flavors to the food products.

Properties like pasting temperature, solid–viscosity relationships, gelatinization and cooking characteristics, resistance of starch pastes to mechanical breakdown, mechanical shear, retrogradation tendencies, ionic characteristics, etc. have been modified. Starch are also obtained from physical modifications like the production of pregelatinized starches obtained by simultaneous cooking and drying (by drum or spray driers or extruders) are applied in instant puddings, pie fillings, dry gravy, etc. **Resistant starch** also obtained by physical modification can play a role of dietary fiber with an additional health benefit of being of low calorific value. The slow dispersion in cold water and rapid dispersion of dextrinised starch provides commercial use in infant food.

Chemical modification is also used to produce a large number of products. Oxidation leads to low viscosity starches and soft gels with no retrogradation that can be used in salad dressings, mayonnaise or as starch batters in fish, meat and breaded products. Crosslinked starch with improved properties like increased resistance to swelling and gelatinization, high viscosity, etc. are used in bakery custard, cream pie filling and toppings, soup powders, salad dressings, etc. Stabilized starches obtained by etherification or esterification to products like starch acetate, mono starch phosphate, etc., are other forms of modified starch used in the food industry.

A combination of treatments for the production of **modified starch** has also been used. Attempts have been made to use forms of starch as fat substituents. These include malto-dextrins, acid-degraded starch and some pregelatinized starches. A potato-based starch mimetic, Paselli, has been marketed by AVEBE, The Netherlands. This has good gel characteristics that can be used as an emulsifier in the food industry (Mercier and Colonna, 1988).

The use of modified starch in the food industry and a guide to appropriate starch selection has been suggested by Langan (1986). Unmodified starch has no functional role in the food industry as the granules swell with relative ease and rupture with minimum force to produce weak-bodied, cohesive pastes or undesirable gels. Marketing requirements, product form, aesthetics, organoleptic considerations, shelf stability and the product utilization method (bake, fry, microwave, etc.) has to be taken into account in the production of such products. A good review of the application of modified starch in the food industry has been made by Light (1990).

While in the USA the Food and Drug Administration (FDA) regulation allowed use of modified starch in nonstandardized and some standardized foods without limit, such is not the case in all countries (Wurzburg and Vogel, 1983). Maher and Cremer (1986) indicate that ruling on the safe use of the modified starch is set in the FDA Code of Federal Regulations, Title 21 (Paragraph 178.3520 Industrial-starch modified and paragraph 172.892). These regulations require food containing modified starch to have labels bearing the name of the additive as 'industrial starch-modified'. It is expected that, in the future, the use of chemical conversions for food purposes will be more strictly regulated and the possible use of bioconversion routes considered.

7.3.3 Paper industry applications

The paper industry is the single largest consumer of modified starch. The bond strength of starch to fibers of pulp used in paper making is greater than the fiber–fiber bonds. There are four stages in paper manufacture in which starch is applied:

1. wet-end addition;
2. sizing selection;
3. calendar stack;
4. binder for pigmented coating.

In the wet-end process, cellulosic fibers from various sources are distributed in a large volume of water. The area where the wet-formed sheet may be smoothed and finished by addition of surface coating is known as the **sizing** section. In the calendar section, the sheets pass through a series of pressure rolls which gives the sheets a high degree of compaction and gloss. In case colored paper is to be produced, pigments along with binders like starch are added finally.

Sizing of paper gives mechanical strength to the paper and retards the rate of penetration of water into the finished sheet. All stages of paper production make use of physically or chemically modified starches. **Cationic starches** have a number of advantages over unmodified starches. Cationic starches are retained more fully and consistently. This results in a reduced loss to the waste water stream and hence less biological oxygen demand (BOD) in the effluent (Halabisky, 1977). They also are required in very low quantities and are absorbed on and retained on a variety of fillers. Wet-end addition also results in reduced linting and dusting, better dye utilization, more efficient rosin sizing and improved printability as additional benefits (Greif and Gaspar, 1980). Even in the sizing process, cationic starches offer some unique advantages owing to their electrochemical attraction to cellulosic fibers. Less starch is required for a given strength (Heutin and Thomin, 1975) and, because of the tenacity of the ionic binding, these modified starches are not removed during repulping of broke. As a result the BOD and chemical oxygen demand (COD) of mill effluents are lower than with other sizing agents.

With 1986 as the reference year it is expected that paper consumption by the end of the century will increase by 140% on a world-wide basis and by about 180% in the Far East with the development of a number of economies in the region (Spalding, 1996). Starch binders have a number of good chemical properties for application in the paper industry. Besides, they are cheaper than other chemical binders and are obtained from a renewable source.

7.3.4 Fermentative products from starch

Carbohydrates, including monosaccharides, cellulose and starch, are renewable sources that can not only serve as food energy, but also serve as raw materials for fermentation to many valuable chemicals and pharmaceuticals. Alternatives, like the use of hydrocarbons for fermentation, were tried in the early 1970s. However, the monomer of starch and cellulose, glucose, has always been the primary carbon for bioconversion.

Beet and cane molasses have been the least expensive substrates, while lactose from whey can be used to a limited extent. Starch substrates ranging from grains of corn and wheat, extracted starches from grain and tuber/root sources, partially hydrolysed starches and dextrose, have been the principal energy sources in the fermentations. Starch carbohydrate derived from corn continues to enjoy expanding use in the USA, where a combination of 41 kilotonne (kt) of starch and 54 kt of cornflour-derived carbohydrates are used as fermentation substrates (Dasinger *et al.*, 1985). The reasons for this have been cited to be owing to the increasing demand for molasses in livestock feeds, the dramatic swings in sugar prices, the development of enzymatic and starch saccharification technology, and the purity of the starch in terms of consistent quality as compared to molasses. Bioconversion of cellulose to other fermentation products is limited owing to the difficulty of enzymatic hydrolysis and the relatively high cost of enzymes (Mandels and Andreotti, 1978; Rakshit and Sahai, 1989).

Very few microorganisms can convert starch directly to other useful products, although many organisms can produce enzymes for the hydrolysis to glucose and then consumes it for their growth. There are some reports of the simultaneous saccharification and fermentation of starch by **lactic acid** bacteria producing their own amylolytic enzyme (Zhang and Cheriyan, 1991; Wang and Rakshit, 1996). As the cost of commercial starch enzymes are low, it is not a bottleneck, as in the case of cellulose hydrolysis. Most starch-based fermentations, therefore, either have a prehydrolysis stage or utilize a combination of organisms (one for providing the hydrolytic enzyme system and the other for bringing about the conversion) as will be discussed later in this chapter. The major fermentation product obtained from starch aside from ethanol is citric acid, which was estimated to require 181 kt of carbohydrate sugars in the USA alone (Dasinger *et al.*, 1985).

Other commercial products using this substrate include **organic acids**, such as acetic, itaconic, gluconic, 2-ketogluconic and ascorbic acids, enzymes and the penicillin antibiotics. Production of **ethanol** by fermentation of corn and other carbohydrates has been well reviewed by Maiorella (1985). Potatoes have been used directly for the production of ethanol in pre-World War II Germany (Jacobs, 1950), while nearly a tenth of Brazilian alcohol was produced from cassava (Yand and Trinidade, 1979). Use of these raw material will depend on the cost of the tuber. In case anhydrous alcohol is required for use as in gasohol, the cost of the separation becomes even more critical. The availability of better strains like *Zymomonas mobilis* with higher productivities (Rogers *et al.*, 1980; Siva Kesava *et al.*, 1995) have led to renewed interest. However, as discussed later, the use of starch industry effluents for ethanol fermentation is limited owing to the low sugar level in that stream.

An excellent review of the methods of starch modification and use in a number of industries is given by Wurzburg (1986).

7.4 Extraction procedures and starch industry waste streams

7.4.1 General extraction procedure

The extraction of starch from potato, cassava, sweet potato and other such tubers/roots follow procedures with slight variances depending on the composition and special characteristics. The general extraction procedure of starch from cassava along with the waste streams is given in Figure 7.2.

On supply of the tubers of starch from surrounding areas, care has to be taken to insure that the roots that are harvested first are consumed first. In

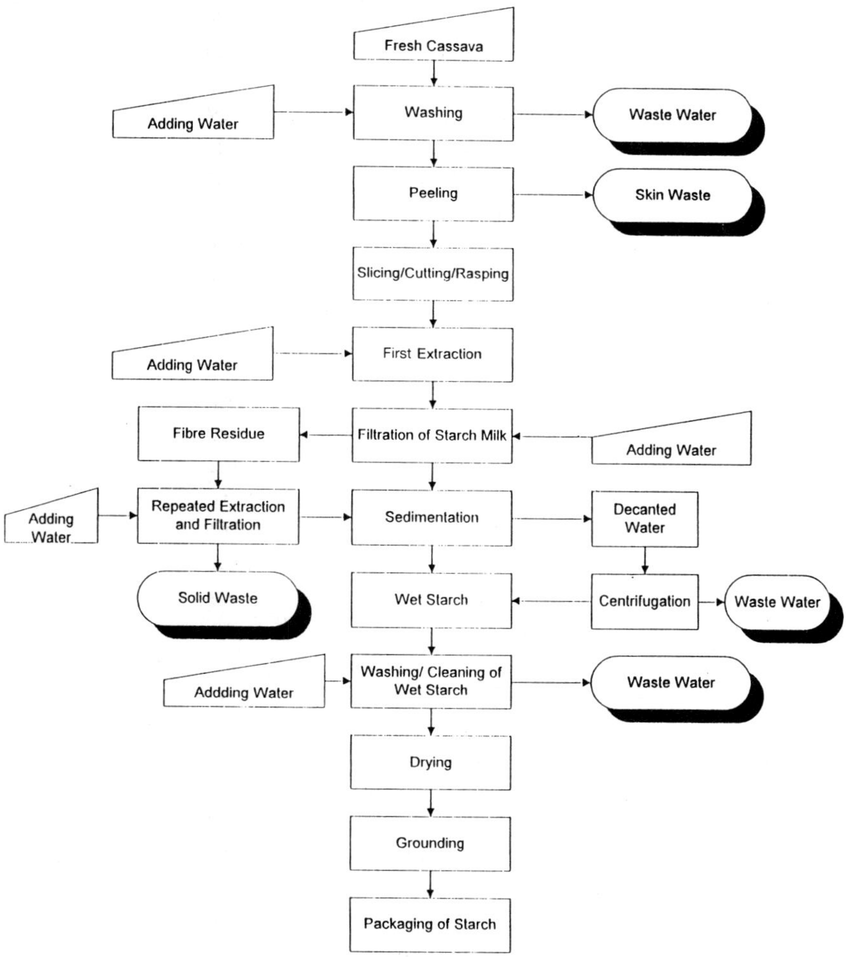

Figure 7.2 Processing flow chart for cassava starch extraction.

the case of cassava there is rapid depletion of starch content on harvest and this has to be avoided (Balagopal *et al.*, 1988). After washing thoroughly with water to remove the dust and mud, the outer skin or corky portion is removed from the cassava tubers. The inner part of the peel has starch that is recoverable in modern processes. The washed peeled roots are then chopped into small slices of 30–50 mm in thickness and sent to a rasping unit. This is an impacting machine which disintegrates the flesh and ruptures the cell walls. Efficient disintegration is necessary to obtain high yields. The **cyanide** in cassava is released at this stage and dissolved in the wash water. Stainless steel materials are used at this stage to avoid the reaction of iron with acids leading to the formation of ferrocyanide which imparts a blue color.

The disintegrated pulp is then passed to the screening section. In this section the fibers are retained on the screens while the starch passes through forming the crude starch milk. Counter-current washing is done to wash the fiber and extract all starch present. The crude starch milk is passed through degritting units to continuous centrifuges where the starch is separated from fine fibers and solubles. These bowl-type nozzle centrifuges discharge the starch from the periphery of the bowl while the fine fibers and solubles are released along the axis of rotation. The starch obtained may be reslurried and centrifuged for further separation. Fresh water introduced near the nozzles replace the impure water which is moved to the fiber washer and root washing sections. Sulfur dioxide is used during the counter-current washing in each centrifugation operation to control microbial action, for color removal and to prevent darkening. The starch-containing stream is then passed to the dewatering and drying section. Vacuum filters or basket-type centrifuges are used for dewatering and drum, belt, tunnel or flash dryers for the removal of moisture.

The removal of fiber and washing out of extractable starch are of importance in extraction for all starch sources. Bound starch in the pulp of potato is not capable of being extracted. It is held in the unopened cells of the potato tubers. However, the free unbound starch can be separated by extra washing.

While cyanide removal is a factor to be taken into account in the case of cassava, the protein content of the wash streams in potato processing has to be taken into account. The juice streams can be coagulated for the separation of proteins. The starch content of extracted sweet potato has been found to be variable and this may be due to some differences in tuber composition (Tian *et al.*, 1991). In the case of cassava and other such tubers the amount of starch extracted depends on the variety being used, the type of growing area, the time of harvest, the age of the roots, the soil fertility and the rainfall during the period (Corbishey and Miller, 1984). Whistler *et al.* (1984) give a good description of the different starch sources and their extraction procedures.

7.4.2 *By-product and effluent streams*

The waste streams in a potato processing plant include dirt, raw pieces, raw pulp, cooked pulp, dissolved solids, etc. (Talburt and Smith, 1975). The dirt or silt adhering to the surface obtained in the initial washing contribute to suspended and fixed solids, and normally are treated separately from the other process water. These sand- and clay-type materials can be removed in shallow ponds or clarifiers, while the water containing the soluble solids can be treated with the other water streams. Raw pieces which are not suitable for starch processing range in size from whole potatoes to small fragments. These materials are normally firm enough for easy separation and are used directly as cattle feed. Finely divided raw potato is defined as raw pulp, while cooked pulp is the agglomerates of cells that are released during the peeling or processing steps.

The main secondary products of potato starch factories are pulp and **fruit waters**. Another secondary product leaving such factories is sludge. Pulp is the major by-product obtained in potato factories. A total of 100 kg of potatoes yield 40–55 kg of potato pulp (Lisinska and Leszczynski, 1989). This is the residue obtained after fruit water separation (in a decanter) followed by starch separation (by extractors). It is a white–grey, thin, semiliquid substance. Fresh pulp contains 95% water. Raw fibers (insoluble nonstarch substances) comprise 50% dry matter of the potato pulp. Lisinska and Leszczynski (1989) have reported that the potato pulp may contain as much as 20–40% starch depending on the rasping method used. The pulp can be used as low-grade cattle fodder and can be enriched with other waste products following fermentation in silos to obtain silage.

Fruit water is also a serious problem in the starch industry. It contains 3–4% dry matter which is low for economical recovery processes, but high enough to cause pollution problems. Use of this water, in its diluted form, for irrigation causes long-term problems as it clogs the soil with fibrous substances and starch granules reducing permeability, and also leads to excessive nitrogen and potassium concentration in the soil. Besides this, the vast expanse of farm land required in the vicinity of the starch factory can make it uneconomical. Separation and utilization of the constituents is more reasonable if it can be done at affordable costs.

Water consumption and waste water generation are major problems in the starch extraction industry. While modern extraction methods attempt to reduce water requirement by using counter-current flows and more efficient equipment, other steps that can be practiced include reuse of waters containing minimum quantities of dissolved or suspended solids. For example, the overflow from cooking water, blanching water, cooling water and surge tanks can be used in the lye or steam treatment of waters during peel removal. Several other streams may be appropriately used in this manner. In severe cases of water shortage, conveying potatoes by water systems has been replaced by other means of conveying. Improved cleaning facilities for equipment and floors such as shut-off nozzles for hoses and high-pressure, low-volume spray units also have been used (Pailthorp *et al.*, 1975).

In the cassava industry, the combined waste water obtained from a plant amounts to 30–50 m^3 ton^{-1} of starch product (Polprasart, 1989). The pH of the solution ranges from 3.8 to 5.2 resulting from the addition of sulfuric acid in the extraction process and also from the release of prussic acid by the tapioca root. Tapioca waste waters are highly organic but have relatively large quantities of nitrogen and phosphorus concentrations. The ratio of the BOD to soluble COD in the settled separator waste is 0.6–0.8, indicating that the waste is biologically degradable.

The different waste streams from the starch industry, including solid and liquid streams, have necessarily to be suitably treated or utilized for the separation and conversion to useful by-products to prevent serious environmental hazards.

7.5 Utilization and treatment of starch industry wastes

Food processing waste streams could be used as substrates for fermentation processes for the production of useful chemicals, food ingredients or pharmaceutical products. In countries where there is surplus of wheat, corn, milk, etc., there is a great deal of interest in the application of emerging technologies for nonfood purposes (Rogers and Fleet, 1989).

Starch wastes, whether in solid or liquid forms, are a far more attractive proposition for microbial processing than cellulosic wastes. This is because of the easy hydrolysis of starch to glucose, owing both to the large number of microorganisms that produce amylolytic enzymes and to the cheap commercial sources of enzymes.

The utilization of starch industry wastes could be divided into the separation of useful materials, such as proteins present in potato wastes, bioconversion to useful chemicals or biomass for food, feed or other applications and treating the liquid streams for the production of methane, ethanol and other fuels, etc. The following sections review the methods used for these purposes.

7.5.1 Production of single cell proteins

The term single cell protein (SCP), coined at the Massachusetts Institute of Technology, refers to the microbial biomass used as food and feed additives. It does not necessarily mean extraction of the proteins of cells and very often implies the whole cells themselves. Traditional foods, such as cheese and vinegar, and substances like koji, yeast and blue–green algae like *Spirullina* has been consumed in small quantities. The use of SCP for food has been considered owing to the insufficient world supply of protein, to the higher growth rate of the microorganisms as compared to other sources of protein of animal or plant origin, and to the high protein content. In their comparison of protein content, Cruegar and Cruegar (1982) have indicated that while soya or milk proteins have raw protein content of 45% and 34%, respectively, the protein content for alkane grown yeasts, methanol bacterium, alga and fungus is 60%, 80%, 72.6% and 31.7%, respectively. The major benefit of converting some industrial wastes to SCP that can be used as feed should also be taken into account. The safety aspects that one has to consider for human consumption include the high nucleic acid content that may be hazardous to health, carcinogenic substances absorbed from the microbial growth substrate or produced by the microorganisms (like aflatoxins) and the slow digestion of microbial cells that may cause indigestion and allergic reactions.

The substrates that have been used for the production of SCP include alkanes, methane, methanol, cellulose and other waste streams like those obtained from food processing industries. The production of SCP from starch and starch wastes is especially important because of the easier hydrolysis to glucose. Yeasts used in the breweries, alcohol and bakery industry very often have starch as the raw material. The main advantage of using yeasts is that they have amylolytic enzymes. A number of studies on the screening of yeasts capable of enhanced production of extracellular starch-hydrolysing enzymes and hence suitable for SCP production have been reported (Hattori, 1961; Spencer-Martins and Van Uden, 1977; Lemmel *et al.*, 1979; Oteng-Gyang *et al.*, 1980; Correira and Van Uden, 1981). However, the processes that have been applied in industry for the production of SCP as a by-product from starch wastes have often used a combination of organisms, one that produces large quantities of hydrolytic enzymes and another that can utilize the sugars produced for growth.

The **Symba process** is one such bioconversion procedure for the production of SCP. The process was originally developed by the Swedish Sugar Company and Chemap Co. in Switzerland (Skogman, 1976). In processes using the preferred yeasts *Candida* and *Saccharomyces*, productivity was low as they produced little amylolytic enzymes. In the Symba process the symbiotic relationship of *Endomycopsis fibulger* and *Candida utilis* was exploited for the production of SCP. The enzymes required for the hydrolysis of starch are produced by the *Endomycopsis* species. These sugars are then quickly consumed by the faster growing *Candida utilis* species. The extremely high amylase productivity of the *Endomycopsis* strain and the growth of the *Candida* species limits the amount of the former to less than 4%. This is favorable to the quality of the final product considering the composition of the two yeasts. This procedure also results in the consumption of other components of the system, such as organic acids, proteins, amino acids, etc. and reduces BOD by 90%. The original Symba process for the utilization of starch industry wastes was able to process 20 m^3 h^{-1} of 3% dry matter composition, reduce BOD at the rate of 4–8 tons day^{-1} and produce 250–300 kg h^{-1} of Symba yeast. Low nucleic acid content and good amino acid levels make them good food quality yeasts. Wastes from products like potato french fries, chips, cubes, instant rice, wheat, maize, cassava, sweet potato starch industries and bakery products can be processed by this process. Skogman (1976) in his presentation of this process indicated that the economic feasibility of the process required that the minimum waste to the process should be 2000–4000 tons $year^{-1}$, the dry matter content of waste should be greater than 2% and that the production season should be long.

A number of reports involving enzymes for the hydrolysis of the starch and then exploiting the hydrolysate for SCP production have also been reported (Moreton, 1978; Lee and Simard, 1984). The slow consumption of starch by pure cultures of *Saccharomyces* and *Candida* species and the low cost of the commercial amylolytic enzymes were the rationale for these experiments. Attempts to overcome the low productivity of this process were made using debranching enzymes in addition to amylase and glucoamylase (Errate and Stewart, 1981), using a Crabtree negative yeast (De Deken, 1966), addition of cellulases and pectinases (Davids *et al.*, 1987) and the use of fed batch and semicontinuous cultures (Kombila *et al* , 1988). Recently, Ejiofor (1996) suggested the possibility of using hydrolysed waste cassava starch for the production of baking quality yeast as a low-cost replacement for using glucose as substrate.

Bacterial biomass as a form of SCP has been reviewed by Litchfield (1985). The faster growth rates of these organisms have a major advantage for these organisms. Photosynthetic and nonphotosynthetic bacteria can be used. Kobayashi and Kurata (1978) have reported the use of *Rhodopseudomanas capsulata* for the bioconversion of starch and food industry wastes where no pretreatment was required. In a patent, Busta *et al.* (1977) described the use of *Bacillus* and *Lactobacillus* species which required pretreatment in the form of sequential bacterial degradation.

The use of bacteria for SCP production has been evaluated in a number of animal studies for protein efficiency ratio (PER) and biological value (BV) (Felix D'Mello, 1978; Abbey *et al.*, 1980). The major disadvantage of bacterial species is the high content of nucleic acids which may be as high as 16%. Human consumption of more than 2 g equivalent of nucleic acid per day could lead to stone formation and gout. Calloway (1974) has reported the use of *A. eutropis* resulted in no gastrointestinal disturbances, while a number of products like Probion, developed by Hoechst-Uhde, have been produced by extraction of nucleic acids (Daly and Ruiz, 1974; Mogren, 1979).

The use of fungi as flavor enhancers in the cheese industry has been long known. The major advantages of using fungi as a protein source has been its capacity to break down a wide range of complex substrates, possibility of fermentation at low pH which helps resist infection, easy filtration and handling. The disadvantages are its slower rate of growth compared to bacteria and yeast, lower protein content, production of mycotoxins, etc. A majority of investigations of using fungi for SCP production have been for substrates from agricultural and food industry wastes, such as starch, cellulose and whey (Rolz and Humphrey, 1982). Proximate analysis of fungal biomass has to be done with care to obtain the true protein content owing to the presence of nitrogen as *n*-acetylglucosamine in chitin present in the cell wall. Using cassava extract as substrate Gregory *et al.* (1977) obtained a true protein content of 37% using *Rhizopus chinesis*, while Khor *et al.* (1976) obtained 31.5% and 26% protein using *Asperigillus fumigatus* and *Sporotrichum thermophile*. The use of cassava starch wastes have been reported by Balagopal and Maini (1977) and Nga *et al.* (1977) while **protein enrichment** of cassava meal low in protein has been described by Muindi and Hanssen (1981) using fungal strains.

The use of solid-state fermentation (SSF) or koji culture applied to solid wastes has the advantage of requiring little or no liquid waste and low investment. The disadvantages other than being labor intensive are the difficulty in characterization of substrate and product and the process kinetics.

Using *Aspergillus hennebergii* spores and solid wastes from cassava, banana and potato, Raimbault (1981) and Senez *et al.* (1980a) and Senez (1983) were able to obtain products containing 20% protein and 20–25% residual sugars. The results indicate that the possibility of improving protein content of cassava meals low in protein is possible.

Upgrading of agricultural and agro-industrial by-products by biotechnological routes has been reviewed by Macris (1989). Chopped cassava leaves were fermented to a product containing 26% protein and used in animal rations. Cassava solid by-products were used at the University of Geulph, Canada, to obtain a fermented end-product of 27% and 37% true and crude protein content protein supplement that could be used as a protein supplement (Food and Agricultural Organization, 1982). Production of fungal feeds using 200 and 3000 l fermenters and rasped cassava were reported by Santos *et al.* (1983). Upgrading cassava starch sources for protein owing to their low concentration has been the focus of considerable research (Noparatnaraporn *et al.*, 1983; Sales *et al.*, 1983). The protein-enriched fermentation feeds (PEFF) process using methods suitable for operation at village and farm level has been shown to produce 150–200 kg of product with substantial protein content in 30 h using starchy substrates such as cassava, potato and banana (Senez *et al.*, 1980b).

7.5.2 Protein extraction from potato processing

Kentie (1946) during the World War II had analysed potato protein and found it to be a remarkable source of vegetable protein. This was confirmed later by amino-acid analysis by De Noord (1974). However, isolation of this protein on a commercial scale had to take in account economic feasibility owing to the very low concentrations of the solutions (2–3% dependent on variety) and the consequent large volumes that had to be processed per unit weight of protein produced, the problems of evaporation of the protein-containing streams due to severe fouling of equipment, the inherent difficult of separation of the coagulated protein due to the low sedimentation rates and poor filterability, and the market-dependent price of the protein (Oosten, 1976).

Early attempts at the turn of this century (Sjollema, 1907) and later in Germany for a technical-scale process were not successful. However, with the pressures to ensure environmental safety and the availability of improved technology, a number of processes are now available. Improved extraction procedures leads to lower water requirement and a subsequent higher protein content in the fruit water. Oosten (1976) refers to the Koninklijke Scolten Honig (KSH) process which involves a coagulation of the protein by heating the fruit water by steam injection. The end-product obtained contains 8% humidity and 75–80% protein on dry substance. This process has the disadvantage of high energy requirements and is complicated by the tendency of the solution to foam, its small fiber content, its mud content and various bacteria and other contaminants. Attempts at using a ultrafiltration unit yield all proteins and require 40% lower energy than the earlier process. While suggesting the possible use of **reverse osmosis**, this report also suggests the possible isolation of specific amino acids like asparagine in high purity for higher returns.

Pepper and Orchard (1981) describe a reverse osmosis process used in a number of starch factories for separating proteins. The liquid coming out of the rasping section is sent to the reverse osmosis unit for concentration and separation. The resultant permeate, containing about 400 mg COD l^{-1}, can be used in potato hydrotransportation and washing operations. The concentrated juice after protein coagulation is evaporated to a dry matter content of 60%.

Peters (1972) described a patented process for protein recovery from the fruit water stream which was reported to be in commercial application in three Dutch, one Rumanian and two Japanese starch plants. The composition of the product obtained was reported to be 70–80% crude protein, 6–12% carbohydrates, 1.5% ash, 0.5–1.0% lipids and 10–12% moisture (deKoe, 1968). The large quantities of starch industry effluents that contain this by-product and the possible use of the recovered protein for human consumption have been investigated extensively (Knorr and Sinskey, 1986).

7.5.3 Energy recovery from liquid streams

Regulations controlling surface waste waters and ground water pollution are established by country, state and federal agencies. In developing countries, waste treatment could offer solutions to sanitation, besides disease and infection control by improving water quality. In more general terms, it helps the reuse of a scarce resource, it may have substantial positive environmental benefits and it could also provide some returns by the production of methane, etc.

The unit sequence in a full-scale waste disposal plant consists of pretreatment (screening) followed by primary (sedimentation and settling tanks) and secondary treatment (trickling filters, activated sludge, anaerobic contact and pond system) followed by advanced treatment processes (reverse osmosis, ion exchange, etc.) where required. While screening and the primary treatment process removes larger particles, the secondary processes require an environment suitable for biological organisms. Aerobic processes include activated sludge processes, biological filters, activated biofilters, stabilization ponds, rotating bio-discs, etc. Activated sludge treatment of potato wastes has been accompanied with some operation problems, primarily settling in secondary clarifiers. The high carbohydrate wastes have been subject to filamentous and bacterial growths if the aeration is not maintained in the proper range. The filamentous nature of sludge leads to **bulking** and leads to constraints in waste treatment especially in carbohydrate streams (Shin *et al.*, 1992). This results in a reduced ability to settle, frequently resulting in solids being lost in the effluent with subsequent low BOD removal of 60–80% compared to 95% and above in many cases (Pailthorp *et al.*, 1975). These authors made a summary of pilot-scale potato waste secondary treatment designs, both aerobic and anaerobic, available at the time and concluded that there was little likelihood that these procedures would be applied owing to the associated costs and the lack of benefits. Since then attempts have been made using anaerobic treatment and a combination of processes.

The advantages of the anaerobic systems, which involve organisms that consume organic matter in the absence of oxygen, are the low costs of designs, low operation costs, production of fuel methane by-product and low sludge quantities. However, the disadvantages of the process are the susceptibility to process failure, relatively low treatment efficiencies (50–80%), odor problems and slow peaking efficiencies.

The interest in anaerobic treatment of potato wastes started in the 1960s with research efforts at the Idaho Potato Processors Association. Subsequently, this process was studied at Sweden by A.B. Sorigona, in the Netherlands by the Agricultural University of Wageningen, at the Universities of Idaho, South Dakota and New Brunswick and the ADI Federation in Canada (Viraraghaven *et al.*, 1987). The anaerobic systems used in the attempted processes include the **upflow anaerobic sludge**

blanket process (UASB) (Lettinga *et al.*, 1979; Sax *et al.*, 1981), the **anaerobic contact process** (Parker *et al.*, 1980), **anaerobic filters** (Lin and Brown, 1980) and the **unified anaerobic fermenter–filter system** (Landine *et al.*, 1982).

Some of the systems have been used in pilot- or full-scale processes. Jackson *et al.* (1979) reported on an **anaerobic lagoon** that has been working at an Idaho potato products factory from which no sludge had been removed in more than 15 years of operation. Feed COD and BOD were reduced by 83% and 75%, respectively. The Frites Specialist plant in the Netherlands using an UASB system could treat 25 m^3 h^{-1} of waste water, removing 75–85% COD, while producing 30–35 m^3 of biogas per hour containing 80% methane. Several systems (Pette and Verspville, 1982; Landine *et al.*, 1983, 1984; Christensen *et al.*, 1984) in operation at the time of review have been reported by Viraraghaven *et al.* (1987). While most of the processes discussed above relate to the developed potato industry, reports on work done on the large cassava industry in Thailand indicate the importance attached to treatment processes here (Sundhagul and Athasampunna, 1983; Bunchueydee, 1984; Yuthavong and Gibbons, 1994). Morgan (1980) working with effluents from a wheat starch factory describes the changes that had to be made in the design concept and concluded that, although an anaerobic system seemed better, one had to be prepared for the unexpected that could arise.

Among anaerobic processes, the UASB process seems to be the most widely used (Viraraghavan *et al.*, 1987). The amount of methane produced in most of the processes used is very near the theoretical value of 0.35 m^3 kg COD removed which was corrected for the COD removed in the accumulated sludge. The biogas composition depends on a number of factors and varies from 60% and 80%. This by-product of the industry is most often used in a boiler to produce steam.

Taking into account the problems of either a aerobic or an anaerobic process, it has been suggested (Viraraghavan *et al.*, 1987) that a two-stage anaerobic–aerobic process would be the best combination and also be economical. A payback on investment owing to the large amounts of biogas produced will be possible. Pilot-scale studies taking into account the constraints of the type of waste to be treated, the geographical and climatic conditions, production levels, etc. are recommended.

The production of a liquid fuel such as ethanol by fermentation could also be a possible method of utilizing the starch industry waste. Microbial production of ethanol has been practiced from ancient times. It became a focus of a considerable research owing to the possibility of gasohol production and a number of other applications (Maiorella, 1985). However, in all such circumstances the substrate used was usually carbohydrates from surplus carbohydrate-rich crops. In the consideration of starch industry wastes for the production of ethanol, Kirsop and Hilton

(1981) warn that energy costs of distillation need to be taken into account. The problem here being that, as the concentration of ethanol in the fermentation broth decreases below 5%, the cost of separation by distillation increases rapidly. Above 10% ethanol concentration the energy cost remains almost constant up to the **azeotropic mixture**. Ethanol concentrations above these levels required for blending with gasoline require further separation (Rakshit *et al.*, 1992). Concentrations above 10% carbohydrate are required to achieve a high enough ethanol level in the fermentation broth. Hence effluents from starch industries that have low glucose concentrations are not good substrates for an economic ethanol production process.

7.5.4 Miscellaneous

Solid residues containing carbohydrates from a number of industries are used as feeds. Most of the products discussed above, such as SCP and protein recovered from the fruit water, are also used as feed. Careful analysis has to be done before they are considered for human consumption. Cassava residues, broken rice, barley, maize and cassava meal are commonly used as feed in Thailand. Potato pulp can be used as low-grade cattle fodder and can be enriched with other waste products and minerals and fermented in silos to obtain silage (Lisinska and Leszczynski, 1989). However, long-distance shipment of the potato pulp is not profitable owing to the high water content and low nutritive value. It is recommended that the weight and volume of the pulp are reduced by dewatering and drying. Waste biological solids grown as part of the secondary treatment process also have the potential to be used as cattle feed. These solids have a protein content of 35%. Attempts can be made to recover the starch and cellulose present in cassava industry solid wastes. Although solid starch wastes can theoretically be converted to SCP or biogas (with or without the liquid waste), little effort has been directed to find such a process because, unlike liquid wastes, the starch wastes do not pollute larger streams and it is not economical to covert them to other products when they can be used directly as feed. It has also been suggested that dry potato pulp could be used as a source of pectin (Tegge, 1984). However, this has not been practiced as yet. Starch wastes can be converted enzymatically to glucose and then fermented to lactic acid to be used for the production of **biodegradable plastics** which in turn could replace the large quantities of nondegradable plastics generated each year (Coleman, 1990). Use of the fruit water for irrigation purposes presents long-term problems because it affects water flow as indicated earlier. It also suffers from the problems of requiring large amounts of agricultural land and being seasonal in its availability.

7.6 Conclusion

The methods discussed in this chapter take into account most of the waste streams that are produced in a starch industry and the possible use of these materials for the production of by-products. However, additional study of the specific source, type and variety of the raw material, starch extraction procedure, etc. need to be taken into account before deciding on a choice of waste utilization systems with economic returns.

There is a considerable focus in the world to apply cleaner technology methods with zero effluents for sustainable development. The introduction of the ISO 14000 certification takes into account not only the quality of the raw material and finished product, but also the policy decisions and technological methods necessary for long-term environmental, social and economic gains with continuous reassessment. To achieve these goals, newer technology to obtain starch roots with good extractable properties, a reduction in waste streams, enzymes with higher activity and tailor-made engineered organisms may be required.

References

Abbey, B.W., Boorman, K.N. and Lewis, D. (1980) *J. Sci. Food Agric.*, **31**, 421–31.
Balagopal, C. and Maini, S.B. (1977) *J. Root Crops*, **3**, 33.
Balagopal, C., Padmaja, G., Nanada, S.K. and Moorthy, S.N. (1988) *Cassava in Food, Feed and Industry*, CRC Press, Boca Raton, FL.
Bunchueydee, P. (1984) *Industrial biogas: a feasibility study of waste utilization from agro-industry in Thailand.* Report submitted to the National Energy Administration, Ministry of Science and Technology, Thailand.
Busta, F.F., Schmidt, B.E. and McKay, L.L. (1977) US Patent, 4 018 650.
Calloway, D.H. (1974) The place of SCP in a mans diet. In *Single Cell Protein* (ed. P. Davis), Academic Press, New York, pp. 129–46.
Casey, S.P. (1977) *Dye Starch*, **29**, 196–204.
Christensen, D.R., Gerick, J.A. and Eblen, J.E. (1984) *J. Water Pollution Control Fed.*, **56**, 1984.
Coker, L.E. and Venkatasubramanian, K. (1985) Starch conversion processes, in *Comprehensive Biotechnology*, Vol. 3 (ed. M. Moo Young). Pergamon Press, pp. 777–89.
Coleman, R. (1990) *Agric. Eng.*, **71**, 20–2.
Corbishey, D.A. and Miller, W. (1984) Tapioca, arrowroot and sago starches: production, in *Starch: Chemistry and Technology* (eds R.L. Whistler, J.M. Bemiller and E.F. Paschall), Academic Press, New York, pp. 469–478.
Correira, S.A. and Van Uden, N. (1981) *Eur. J. Appl. Microbiol. Biotechnol.*, **13**, 24.
Crueger, W. and Crueger, A. (1982) *Biotechnology, A Textbook of Industrial Microbiology* (English edition) (ed. T.D. Brock), Sinauer Associates, Massachussetts/Science Tech, Inc., Madison.
Daly, W.H. and Ruiz, L.P. (1974) *Biotech. Bioeng.*, **16**, 285–7.
Dasinger, B.L., Fenton, D.M., Nelson, R.P., Roberts, F.F. and Truesdell, S.J. (1985) Enzymatic and microbial processes in the conversion of carbohydrates derived from starch, in *Starch Conversion Technology* (eds G.M. Van Beyum and J.A. Roels), Marcel Dekker, New York, pp. 237–62.
Davids, S.J., Lee, B.H. and Simard, R.E. (1987) *Microbiol. Aliments Nutr.*, **5**, 197.
De Deken, R.H. (1966) *J. Gen. Microbiol.*, **44**, 149.

deKoe, W.J. (1968) *Waste Water Treat. J.*, **12**, 55–7.
Demain, A.L. and Solomon, N.A. (1981) *Sci. Am.*, **245**, 67–75.
De Noord, K.G. (1974) *Chem. Weekblad*, **77**, 10.
Ejiofor, A.O. (1996) *Enzyme Microbiol. Technol.*, **18**, 519–25.
Errat, J.A. and Stewart, G.G. (1981) *Dev. Ind. Microbiol.*, **22**, 557.
Felix D'Mello, J.P. (1978) *J. Sci. Food Agric.*, **29**, 453–60.
Food and Agricultural Organization (1982) Agricultural residues – compendium of technologies, Agricultural Series Bulletin, Vol. 33, Rev. 1, Food and Agricultural Organization, Rome.
Gregory, K.F., Reade, A.E. Santos-Nunez, J. *et al.* (1977) *Ann. Feed Sci. Technol.*, **2**, 7–19.
Greif, D.S. and Gaspar, L.A. (1980) Cationic starch as a wet end additive, in *Dry Strength Additives* (ed. W.F. Reynolds), Tappi Press, Atlanta, Chapter 4.
Halabisky, D.D. (1977) *Tappi*, **60**, 125.
Hattori, Y. (1961) *Agric. Biol. Chem.*, **25**, 737.
Heutin, G. and Thomin, W.H. (1975) *Wochenbl. Papierfabr.*, **103**, 329.
Jackson, M.L., Lewis, J.C. and Wilson, D.E. (1979) Paper presented at the *Pacific North-West Pollution Control Association Meeting*, Spokane, Washington.
Jacobs, P.B. (1950) *Industrial Alcohol*, Miscellaneous Publication No. 695, US Department of Agriculture, Washington, DC.
Kentie, A. (1946) *Voeding*, **6**, 214.
Khor, G.L., Alexander, J.C., Santos-Nunez, J., Reade, A.E. and Gregory, K.F. (1976) *J. Inst. Can. Sci. Technol. Aliment.*, **9**, 139–43.
Kirsop, B.H. and Hilton, M.G. (1981) Research into utilization of food industry wastes, in *Food Industry Wastes: Disposal and Recovery* (eds A. Herzka and R.G. Booth), Applied Science Publishers, London, pp. 210–18.
Knight, J.W. (1969) *The Starch Industry*, Pergamon Press, Oxford.
Knorr, D. Sinskey, A.J. (1986) Biotechnology in food production and processing, in *Biotechnology: The Renewable Frontier* (ed. D.E. Koshland Jr), The American Association for the Advancement of Science, pp. 253.
Kobayashi, M. and Kurata, S.-I. (1978) *Process Biochem.*, **13** (9), 27–30.
Kombila, E.M., Lee, B.H. and Simard, R.E. (1988). *Microbiol. Aliments Nutri.*, **6**, 239.
Landine, R.C., Brown, G.J., Cocci, A.A. and Viraraghavan, T. (1982) *Water Pollution Res. J. Can.*, **17**, 63.
Landine, R.C., Brown, G.J., Cooci, A.A. and Viraraghavan, T. (1983) *Agric. Wastes*, **7**, 111.
Landine, R.C., DeGarie, C.J., Cocci, A.A. *et al.* (1984) Presented at *Proceedings of the 38th Industrial Waste Conference* (ed. J.M. Bell), Butterworths, London, pp. 789.
Langen, R.e. (1986) Food industry, in *Modified Starches: Properties and Uses* (ed. O.B. Wurzburg), CRC Press, Boca Raton, FL, pp. 200–11.
Lee, B.H. and Simard, R.E. (1984) *Dev. Ind. Microbiol.*, **25**, 459.
Lemmel, S.A., Heimsch, R.C. and Edwards, L.L. (1979) *Appl. Environ. Microbiol.*, **37**, 227.
Lettinga, G., Van Velson, A.F.M., de Zeeuw, W. and Hobma, S.W. (1979) Paper in *Proceedings of the 1979 National Conference on Environmental Engineering*, American Society of Civil Engineers, New York, pp. 35.
Light, J.M. (1990) *Cereal Food World*, **35** (11), 1081–92.
Lin, K.C. and Brown, G.J. (1980) *Can. J. Civil Eng.*, **7**, 373.
Lisinska, G. and Leszczynski, W. (1989) Potato starch processing in *Potato Science and Technology*, Elsevier Applied Science, London, pp. 281–348.
Litchfield, J.H. (1985) Bacterial biomass, in *Comprehensive Biotechnology*, Vol. 3 (ed. M. Moo-Young), Pergamon Press, New York, pp. 463–82.
Macris, B.J. (1989) Upgrading of agricultural and agoindustrial by-products to food and feed by biotechnology, in *International Biosystems*, Vol. II (ed. D. Wise), CRC Press, Boca Raton, FL.
Maher, S.L. and Cremer, C.W. (1986) Use of modified starch: paper industry, in *Modified Starches: Properties and Uses* (ed. O.B. Wurzburg), CRC Press, Boca Raton, FL, pp. 213–27.
Maiorella, B.L. (1985) Ethanol, in *Comprehensive Biotechnology*, Vol. 3 (ed. M. Moo-Young), Pergamon Press, New York, pp. 861–911.
Mandels, M. and Andreotti, R.E. (1978) *Process Biochem.*, **13**, 6–13.

Mercier, C. and Colonna, P. (1988) Starch and enzymes: Innovation in the products, processes and uses. Paper presented at the 8th IBS, Paris, France, pp. 1042–54.
Mogren, H. (1979) *Process Biochem.*, **14**, 2–4, 7.
Moreton, R.S. (1978) *J. Appl. Bacteriol.*, **44**, 373.
Morgan, H. (1980) The development of an anaerobic process for the treatment of a wheat starch factory effluent, in *Food Industry Wastes: Disposal and Recovery* (eds A. Herzka and R.G. Booth), Applied Science Publishers, London, New Jersey, pp. 92–108.
Muindi, P.J. and Hanssen, J.F. (1981) *J. Sci. Food Agric.*, **32**, 632.
Noparatnaraporn, N., Nishizawa, Y., Hayshi, M. and Nagai, S. (1983) *J. Ferment Technol.*, **61**, 515.
Nga, B.H., Eswaran, S., Tan, E.L. and Lee, T.K. (1977) *Malys. Appl. Biol.*, **6**, 123.
Oosten, B.J. (1976) Protein from potato starch mill effluent, in *Food from Waste* (eds G.G. Birch; K.J. Parker and J.T. Worgan), Applied Science Publishers, London, pp. 196–204.
Oestergaard, J. and Knudson, S.L. (1976) *Die Starch*, **28**, 350.
Oteng-Gyand, K., Moulin, G. and Galzy, P. (1980) *Acta Mikrobiol. Hung.*, **27**, 155.
Pailthorp, R.E., Filbert, J.W. and Ritcher, G.A. (1975) Waste disposal, in *Potato Processing* (eds W.F. Talburt and O. Smith), The AVI Publishing Co. Inc., Westport, CT.
Parker, J.G., Lyons, B.J. and Parker, C.D. (1980) Paper presented at *Third International Congress on Industrial Wastewater and Wastes*, Stockholm, Sweden.
Pepper, D. and Orchard, A.C.J. (1981) *Starch/Starke*, **33**, 271–274.
Peters, H. (1972) *Pure Appl. Chem.*, **29**, 129–41.
Pette, K.C. and Verspville, A.I. (1982) Application of UASB concept for waste water treatment, in *Anaerobic Digestion* (ed. D.E. Hughes), Elsevier, Amsterdam.
Polprasart, C. (1989) *Organic Waste Recycling*, John Wiley and Sons, Chichester, pp. 15–62.
Raimbault, M. (1981) Fermentation in milieu solids. *ORSTOM*, pp. 291.
Rakshit, S.K. (1996) Starch enzymes, in *Advanced Post-academic Course on Tapioca Starch Technology*, AIT, Bangkok, 19–23 February.
Rakshit, S.K. and Sahai, V. (1989) *J. Gen. Appl. Microbial*, **35**, 441–50.
Rakshit, S.K., Ghosh, P. and Bisaria, V.S. (1992) *Bioprocess Eng.*, **8**, 279–82.
Rogers, P.L., Phil, D., Lee, K.J., Tribe, D.E. and Tribe, M.E. (1980) *Process Biochem.*, **15**, 7–11.
Rogers, P.L. and Fleet, G.H. (1989) *Biotechnology and the Food Industry*, Gordon and Breach, London, p. 14.
Rolz, C. and Humphrey, A.E. (1982) *Adv. Biochem. Eng.*, **21**, 1–54.
Sales, A.M., Demenezes, T.J.B., Lajolo, F.M. and Iaderaosa, M. (1983) *Rev. Bras. Mendiaco*, **2**, 77.
Sanders, J.P.M. (1996) Starch manufacturing in the world, in *Advanced Post-academic Course on Tapioca Starch Technology*, AIT, Bangkok, 22–26 January.
Santos, G., Gomez, G. and Alexander, J.C. (1983) *Anim. Feed Sci. Technol.*, **8** (4), 313.
Sax, R.I., Holz, M. and Roberts, J.E. (1981) Paper presented at *Symposium on Energy from Biomass and Wastes V*, Lake Buena, Florida.
Schenck, F.W. and Hebeda, R.E. (eds) (1992) *Starch Hydrolysis Products, Worldwide Technology, Production and Applications*, VCH Publishers, New York.
Senez, J.C. (1983) In *Production and Feeding of Single Cell Protein*, (eds M.P. Ferranti and A. Fietcher), Applied Science, New York, pp. 101.
Senez, J.C., Raimbault, M. and Deschamps, F. (1980a) *FAO World Animal Rev.*, **35**, 36.
Senez, J.C., Raimboult, M. and Deschamps, F. (1980b) *VI International Fermentation Symposium*, London, Ontario.
Shin, B.S., Carter, A. and Malcolm, A. (1992) Waste treatment, in *Starch Hydrolysis Products, Worldwide Technology, Production and Applications* (eds F.W. Schenck and R.E. Hebeda), VCH Publishers, New York, pp. 573–604.
Siva Kesava, S., Rakshit, S.K. and Panda, T. (1995) *Process Biochem.*, **30**, 41–7.
Sjollema (1907) *Chem. Weekblad*, **4**, 637. Cited in Ooosten (1976).
Skogman, H. (1976) Production of Symba-yeast from potato wastes, in *Food from Waste* (eds G.G. Birch, K.J. Parker and J.T. Worgan), Applied Science Publishers, London, pp. 167–79.
Spalding, S.J. (1996) Marketing starch and starch products in Asia, in *Advanced Post-academic Course on tapioca starch Technology II*, AIT, Bangkok, 19–23 February.

Spencer-Martins, I. and Van Uden, N. (1977) *Eur. J. Appl. Microbiol. Biotechnol.*, **4**, 29.
Sundhagul, M. and Atthasampunna, P. (1983) Bioconversion of carbohydrate residues in Thailand, in *The Use of Organic Residues in Rural Communities* (eds C.A. Shacklady), The United Nations University, pp. 74–79.
Talburt, W.F. and Smith, O. (1975) *Potato Processing*, AVI, Westport, CT.
Tegge, G. (1984) *Starke und Starkederivate*, Behr's, Hamburg.
Tian, S.J., Rickard, J.E. and Blanshard, J.M.V. (1991) *J. Sci. Food Agric.*, **57**, 459–91.
Venkatasubramanian, K. (1978) in *Enzyme, The Interphase between Technology and Economics* (eds J.P. Dansky and B. Wolnak), Marcel Dekker, New York, p. 35.
Viraraghavan, T., Landine, R.C., Cocci, A.A. and Brown, G.J. (1987) Anaerobic treatment of potato processing wastewaters, in *Global Bioconversions*, Vol. 3 (ed. D. Wise), CRC Press, Boca Raton, FL.
Wang Xiadong and Rakshit, S.K. (1996) Paper presented at the *10th IBS*, Sydney, Australia, August.
Whistler, R.L., Bemiller, J.N. and Paschall, E.F. (eds) (1984) *Starch Chemistry and Technology*, Academic Press, New York.
Wurzburg, O.B. (ed.) (1986) *Modified Starches: Properties and Uses*, CRC Press, Boca Raton, FL.
Wurzburg, O.B. and Vogel, W.F. (1983) Modified food starch – safety and regulatory aspects, in *Gums and Stabilizers for the Food Industry II: Application of Hydrocolloids* (eds G.O. Phillips, D.J. Wedlock and P.A. Williams), Pergamon Press, Oxford, pp. 405–10.
Yand, V. and Trindade, S. (1979) *Chem. Eng. Prog.*, **75**, 11.
Yuthavong, Y. and Gibbons, G. (1994) Paper presented at *ASEAN Subcommittee Meeting on Biotechnology, on Problems of Developing Countries Addressable by Biotechnology: Environment, Health and Related issues*, Bangkok.
Zhang, D.X. and Cheriyan, M. (1991) *Biotech. Lett.*, **13** (10), 733–8.
Zia, M. (1992) Starch wastewater treatment alternatives. Masters thesis no. Ev-92-10, Asian Institute of Technology, Bangkok.

8 Bioconversion of food processing wastes

G.TH. KROYER

8.1 Introduction

Today's society, in which there is a great demand for appropriate nutritional standards, is characterized by rising costs and often decreasing availability of raw materials together with much concern about environmental pollution. Consequently, there is considerable emphasis on the recovery, recycling and upgrading of wastes. This is particularly valid for the food and food processing industry in which wastes, effluents and by-products can be recovered and can often be upgraded to higher-value and useful products.

The food industry produces large volumes of wastes, both solids and liquids, resulting from the production, preparation and consumption of food. These wastes pose increasing disposal and potentially severe pollution problems and represent a loss of valuable biomass and nutrients. In the past they have often been dumped or used without treatment for animal feed or as fertilizers. In the last few years, however, owing to the increasing necessity to take into consideration aspects aimed at preventing pollution of the environment as well as for economic motives, and the need to conserve energy and raw materials, new methods and policies for waste handling and treatment have been introduced in the recovery, bioconversion and utilization of valuable constituents from food processing wastes. Beside their pollution and hazardous aspects, in many cases, food processing wastes might have a potential for recycling raw materials or for conversion into useful products of higher value as a by-product, or even as raw material for other industries, or for the use as food or feed after biological treatment. Particularly, the bioconversion of food processing wastes is receiving increased attention regarding the fact that these wastes represent a possible and utilizable resource for conversion to useful products.

This chapter investigates the biological and biotechnical approaches to the feasibility of bioprocessing food processing waste materials for the recovery of bioresources and for the production of useful products of higher added value, with emphasis on the recycling objective of the bioconversion.

8.2 Characteristics of food processing wastes

Wastes emerging from food processing operations are extremely varied, ranging from solids to highly contaminated waste waters, and their composition depends on both the nature of the product and the production technique employed. Each sector of industry will have a typical type of waste stream associated with it, however, there are certain characteristics of the particular industries which may be useful in the selection of suitable methods for waste treatment. For example, wastes from meat processing plants will contain a high fat and protein content, whilst waste from the canning industry will contain high concentrations of sugars and starches (Isaac and Anderson, 1974). Thus, they are rich in easily degradable organic matter but generally they are free from metals, pesticides and other toxic substances, and low in inorganic solids (James and Addyman, 1974). Also the waste may not only differ from site to site but it also may vary from one time of the year to another. Futhermore, the volume and concentration of the waste material will not be constant. This may present some problems for managing a working process consistently owing to fluctuations in the nature, composition and quantity of the raw materials (Kirsop and Hilton, 1981).

In general, wastes from the food processing industry have the following characteristics (Litchfield, 1987):

- large amounts of organic materials, such as proteins, carbohydrates and lipids;
- varying amounts of suspended solids depending on the source;
- high biochemical oxygen demand (BOD) or chemical oxygen demand (COD).

Vast quantities of water are used as a whole in the food processing industry, of which a small proportion is for consumptive use. Furthermore, the rapid growth in the production of factory-processed food (canned, bottled, frozen, deyhydrated food) and the many other developments of modern food technology have led to a fast growth in liquid food waste production, which is characterized by dilute streams containing proteins, sugars, starches and fats (Isaac and Anderson, 1974). However, for the utilization of food processing wastes, solids as well as liquid wastes are applicable to bioconversion to useful products.

The appraisal of food processing wastes as animal feeds or human foodstuff requires information about its nutritional value and toxicological safety. Suggestions are made regarding protein and non-protein content, oil and fatty acid composition, crude fibre and vitamin content, microbial content and toxic compounds (Tacon, 1980).

In general, the major types of food processing industries that can be regarded as having wastes that can be converted into useful products by biotechnological processes are:

- the meat and fish processing industry;
- the fruit and vegetable industry;
- the dairy industry;
- the fermentation industry.

8.3 Biotechnological processes in food processing waste treatment

The food industry can be considered in many aspects as the largest, oldest and most potential form of biotechnology. The application of biotechnology to the modification of food components, in new methods for the production of food and food components, as well as in modern methods of food analysis is becoming increasingly important in the food industry. Therefore, it is not surprising that biotechnology has also found its application in the treatment of wastes arising from the food processing industry. In this way, food wastes could be utilized by modifying them to produce an upgraded material of higher value than the waste itself or to produce new products, either by microbe- or enzyme-aided processes. In particular, the production of completely different products from waste, usually by applying biotechnological methods, has gained great attention.

Reviews have been made on the recovery of useful substances from waste water and waste solids from food manufacturing processes and utilization thereof in the food industry by Sakai (1994), on the bioconversion of organic solid substrates into useful products by microbial biomass by El-Nawawy (1992), and concerning the treatment of food processing waste water by Walsh *et al.* (1993). Further reviews on the enzymic hydrolysis and bioconversion to value added products of pectin-rich residues generated in processing of citrus fruits, apples, and sugar beets (Grohmann and Bothast, 1994), on the reuse of agro-alimentary wastes generated in different sectors of the food industry (Maes, 1993), and on liquid, solid or gaseous wastes from the food processing industry with emphasis on modern waste-processing technologies, resource recovery and waste upgrading (Niranjan and Shilton, 1994) are documented in literature.

The simplest biological treatment of waste is the use of aerobic filters, trickle beds or activated sludge plants. Thereby only detoxification of the waste is achieved and usually no by-products are obtained, at least, not on an economic scale (Hacking, 1988).

Under anaerobic conditions, food processing wastes containing carbohydrates, lipids and proteins can be digested to yield biogas (Litchfield, 1987). Anaerobic digestion combines detoxification of waste materials by upgrading it to its use as animal feed or fertilizer, and the production of biogas. Biogas varies in composition according to the composition of the waste treated. The content of methane generally lies between 60% and 80%, with carbon dioxide and hydrogen sulphide as major residues (Hacking, 1988). Biogas is used mainly as an energy source. The anaerobic treatment processes can be applied to all segments of the food processing industries with particular reference to concentrated waste streams. Most of the full-scale installations are within the sugar processing, meat rendering, brewing and distilling, bottling and soft drinks, and wheat-starch and potato processing industries (Oleszkiewicz and Olthof, 1982). Also ethanol or organic acids can be produced from carbohydrate-containing waste materials by anaerobic microbiological processes (Litchfield, 1987).

Much research work has been focused on upgrading waste material to more valuable products using appropriate biotechnological processes. Of these the production of biomass, namely, single cell protein (SCP), and the production of ethanol seems to be most interesting and promising (Hacking, 1988).

Consideration of the substantial need for energy in the world as well as the increasing costs thereof makes the production of energy from waste in the form of ethanol very attractive. Biotechnological ethanol production on an industrial scale is already well known and large quantities are being produced from carbohydrate sources in many countries.

Single cell protein, or biomass, produced from food waste, can be used either as animal feed or human food. This imposes particular regulations regarding its consumption as a diet component. Furthermore, the product must not be contaminated with other microorganisms, otherwise sterilization of the waste becomes inevitable. Rigorous tests are usually necessary to confirm the purity of the biomass. These steps are expensive and may be limiting factors in practice for its general application as a widespread method for utilizing food industry wastes for the production of biomass.

Production of solvents, chemical feedstocks or liquid fuels by microbial digestion requires either aerobic or anaerobic digestion to be incomplete. Such incomplete digestions for the production of ethanol, acetone, butanol or butanediol are limited to the fermentation of specific substrates such as carbohydrates, require selective conditions, and recovery is costly (Pirt, 1978). Overviews on the microbial production, microorganisms, biochemistry and processes of 2,3-butanediol fermentation (Maddox, 1988), as well as on the microbial production of butanol–acetone (Bahl and Gottschalk, 1988) are cited in literature.

8.4 Production of biomass from food processing wastes

Owing to a renewed awareness of the increasing world population, the increasing costs of energy and raw materials as well as the increased concern about environmental pollution, intensive research is being directed towards the use of fermentation processes to convert waste materials into useful and profitable products. In particular, the direct conversion of waste materials into microbial biomass suitable for animal feed and human consumption is one of the main targets of studies well documented in literature, and various waste materials from different types of agricultural and food processing industries have been evaluated for use as substrates in biomass production (Mateles and Tannenbaum, 1968; Tannenbaum and Wang, 1975; Birch *et al.*, 1976; Gray and Berry, 1980; Herzka and Booth, 1981; Litchfield, 1983, 1987; Solomons, 1983; Goldberg, 1985; Grant, 1986; El-Nawawy, 1992).

In general, the composition of food wastes is such that they are suitable as growth mediums for microorganisms. This can be exploited either by conventional waste treatment using a heterogeneous population of bacteria, fungi and protozoa to convert the organic matter, or by growing food yeasts or other edible microorganisms on the waste to produce almost pure cultures of microbial mass (James and Addyman, 1974).

Biomass, or SCP, is the name given to the total dry matter of bacteria, yeasts, moulds and higher fungi when they are used for animal feed or human food (Kirk and Othmer, 1980). The use in feed is related primarily to considerations of nutritional value and balance of the ingredients, whilst the level of utilization in food is certainly limited by the concentration of nucleic acid in the final product (Cooney *et al.*, 1980).

For liquid wastes of an appropriate chemical composition, large quantities are produced in the food processing industry to be used as a substrate for microbial fermentations. It has been estimated, generally, that for most of the food industries, for every tonne of food produced, an equal amount of carbohydrate waste and effluent is produced (Imrie and Righelato, 1976). In the past this waste often caused pollution by being disposed into rivers or the sea, or increased the costs of the process by requiring treatment in an effluent plant. For example, effluents which contain simple sugars (beet processing wastes) are easier to handle than those containing high molecular weight polysaccharides such as starch processing wastes, where it is necessary to break them down to assimilable monosaccharides and disaccharides before they can be metabolized by the microorganisms (Anderson *et al.*, 1975).

A wide variety of microorganisms has been studied as possible organisms for biomass production, including several bacteria and algae, or filamentous fungi and yeasts (Gray and Berry, 1980). In general, bacteria are not as useful as yeasts and moulds, particularly where separation of the cell product from the residual waste is necessary, and cell recovery by centrifugation turns out to be uneconomic (Litchfield, 1987).

For the establishment of the microbial cultures, different designs of fermenters are used (e.g. stirred tank reactors, tubular or tower fermenters, multistage reactors), resulting in different microbial growth kinetics and rates in the fermenter, and characterized by different temperature effects, oxygen requirements and substrate conversion efficiencies (Gray and Berry, 1980).

Regarding biomass production, it is important that the yields of cell mass and protein per unit of growth-limiting substrate consumed are maximized. It is necessary for growth to be carbon-limited, to prevent accumulation of carbon, which would result in conditions of nitrogen limitation (Sucuru *et al.*, 1975). Furthermore, the initiation of secondary metabolism which may lead to toxin production is largely avoided under conditions of carbon-limited growth (Litchfield, 1977).

To obtain an optimal yield of biomass and protein, it is essential that all major nutrients which are required for growth are present together in an adequate proportion. However, generally, the effects of the various nutritional and physiological parameters for the whole process on the behaviour of the microorganisms are very complex, and, for each combination of organism and substrate, the different parameter combination should be tested at an experimental and a production-scale level in order to determine the optimum conditions. Values of substrate conversion efficiencies and yields of biomass of different organisms for their growth on different food processing wastes obtained from several authors are documented in the literature (Gray and Berry, 1980).

Above all, microbial biomass should be acceptable as food for animal or human consumption for which it has been produced. Animals normally accept microbial biomass in their diet; however, further treatment of the biomass is mostly necessary if it is to be used for human consumption (Gromov, 1967). For the production of animal-feed-grade SCP from waste, a clean but non-sterile mode of operation is sufficient. For food-grade products, however, it may be necessary to apply sterile conditions regarding the substrate and fermenter (Litchfield, 1987).

The nutritional quality of the microbial biomass depends not only on the high level of protein but, furthermore, has to have a good amino-acid profile, i.e. a high content of nutritionally valuable, essential amino acids. The levels of essential amino acids in protein from a range of filamentous fungi have been given in detail (Anderson *et al.*, 1975). According to these authors, fungal protein compares well with more conventional foods, although most fungal protein is deficient in the sulphur amino acids.

A high content of nucleic acids, which are present in all microbial biomass, can present a nutritional problem, since this can cause gout-like conditions in humans (Kihlberg, 1972). Values of RNA between 3.2% w/w and 4.7% w/w up to 10% w/w have been reported in filamentous fungi (Worgan, 1976). Techniques have been investigated to reduce the level of RNA in biomass after it has been harvested or by controlling the fermentation process (Sinskey and Tannenbaum, 1975).

For safety reasons, harmful substances should not be present in the substrate, which can become incorporated into the final product (e.g. heavy metals, herbicides, toxins, etc.). Furthermore, care must be taken to prevent contamination by toxic materials (mycotoxins) produced by the organism during the fermentation. Therefore, procedures for carrying out preclinical tests on novel foods for human consumption have been recommended by the Protein Advisory Group of the United Nations, which includes aspects of safety, nutritional value, sanitation, acceptability and technological properties (Tannenbaum adn Wang, 1975).

The possibility of using activated sludge for the production of biomass as animal feed has been described (Vriens *et al.*, 1989). In the activated sludge process, waste water containing biogradable organic compounds is brought into contact with a fluidized mixed culture of microorganisms in an aerobic environment. An assimilative process in an aeration tank then removes from the water a portion of organic carbon, chemically bound nitrogen, phosphorus and other micronutrients. Biomass is formed and other organic and inorganic compounds are absored simultaneously by various mechanisms. The biomass is then separated from the liquid-phase supernatant in a settling tank.

The activated sludge process for the treatment of waste water is in fact analogous to the production of SCP, with the exception that a mixed culture of microorganisms is used instead of pure cultures. Activated sludge grown on food processing waste water may be used successfully as a feed material because normally significant concentrations of toxic heavy metals, as in sewage sludge, would not be expected to be present, and contamination of faecal wastes or pathogenic bacteria could be prevented. Effluents from many food processing industries contain mainly simple sugars readily assimilable by microorganisms, and activated sludge treatment is considered to require less capital investment and the operating costs are also lower than the conventional SCP production. Extensive studies on different animals have proved that activated sludge from the brewing, dairy, citrus fruit canning, and slaughterhouse industries is suitable for this type of treatment. Some problems may exist concerning toxicological aspects and safety tests, but these could probably be solved on the basis of present technology. The feasibility of this process seems to be a matter of economics. The use of activated sludge from food processing waste water treatment might therefore be expected to contribute indirectly to the nutrition of the population of the world in the near future (Vriens *et al.*, 1989).

8.5 Meat and fish processing wastes

Owing to the increasing demand for protein, there is a great need to develop processes to upgrade high-protein meat and fish processing wastes which currently may be used as animal feed or are lost from the food chain altogether. Several studies have reported on the utilization of meat and fish processing wastes, which generally produce a high level of pollution. Different microorganisms, including yeasts, filamentous fungi, bacteria and algae, can be used to produce SCP (Litchfield, 1977; Hang, 1979). These processes are very energy and capital intensive, and it may be necessary to supplement the feedstock nutritionally (Litchfield, 1977). The production of biogas, and the recovery of protein and fat from meat, fish and poultry wastes using biological and biotechnological treatment methods have been reviewed by McComis and Litchfield (1989). However, for most of these processes, further scaled-up studies would be required (Litchfield, 1987).

For meat processing wastes, the utilization of collagen for the production of SCP was studied on a laboratory scale (Bough *et al.*, 1972). Biomass containing 70% gross protein with amino-acid levels which meet most animal requirements was recovered from a thermophilic aerobic process for the treatment of pig slaughterhouse and meat processing effluents (Couillard and Zhu, 1993). Studies on an alternative to traditional processing by biological conversion of poultry processing waste to valuable SCP by both direct and indirect routes are documented by Najafpour *et al.* (1994).

Meat-packing plants may also consider anaerobic digestion processes for the production of biogas, namely methane from their waste materials (Hansen, 1983). Slaughterhouse blood, which is a protein-rich residue of the meat production process, can be utilized for the recovery of applicable protein preparates. About 90% of the thick blood protein can be recovered, and the iron-containing haem fraction could be used as a dietetic ingredient for specialist food or clinical products, and as a feed additive (Kobald and Holley, 1989). For the recovery of applicable protein preparations, decolorization of the thick blood could be performed by enzymatic hydrolysis and separation of the coloured haem from the colourless globin protein fraction (Drepper and Drepper, 1981). Further investigations were carried out for the utilization of cattle and food waste leached with seawater for the improvement of γ-linoleic acid production by *Spirulina platensis* (Wang, 1995).

Experimental studies were performed on the production of SCP from different fish processing wastes (Reva-Moissev and Carroad, 1981; Cosio *et al.*, 1982; Welsh and Zall, 1984). By cultivation of *Endomycopsis fibuliger* on a medium based on waste effluents from the industrial processing of mussels, SCP was produced at levels of 6 g l^{-1} (Murado *et al.*, 1986). *Lactobacillus plantarum* was used successfully for lactic acid fermentation of hydrolysed cod gurry with the addition of the optimum amount of molasses (Giurca and Levin, 1992).

8.6 Fruit and vegetable processing wastes

Large amounts of fruit and vegetable processing wastes are produced from packing plants, canneries, etc., which may be disposed in various ways, including their immediate use for landfill or as animal feed, or which, alternatively, may be processed biotechnologically in order to produce a more valuable product. In recent times, industry has continued to make progress in solving the waste problem through recovery of by-products and waste materials, such as peel, pulp or molasses, using fermentation or other related biological processes.

The protein content of fruit and vegetable processing wastes with an adequate level of fermentable carbohydrates can be increased to 20–30% by using a solid substrate fermentation process. The production of SCP from these wastes is thought to be economically attractive if a local market exists for the product and fuel costs can be minimized (Davy, 1981).

Solid-state fermentation, in which a moist water-insoluble solid substrate is utilized by microorganisms in the absence of free water, has been applied for the enrichment of the protein content of orange peel, banana and carrot wastes (Davy, 1981), coffee pulp (Penaloza *et al.*, 1985), and citrus peel (Rodriguez *et al.*, 1985).

Developments in the citrus fruit processing industry have brought about handling problems for wastes remaining after the extraction of juice or after making other products from the fruits which are rich in carbohydrates, and account for about 45–60% of the weight of the processed fruits. By culturing microorganisms on citrus fruit peel, valuable by-products can be produced (Wicker *et al.*, 1989). Studies on the utilization of orange waste as a substrate for microbial protein production by the fungus *Sporotrichum pulverulentum* were performed, and biomass containing about 32% protein was obtained (Karapinar and Okuyan, 1982). SCP production from crude orange waste was successfully performed using *Sporotrichum thermophile* cultivated under optimal conditions. The oil-free orange waste yielded maximal SCP outputs (49%) after 20 days of incubation. A high percentage (41%) of crude proteins was achieved containing all the essential amino acids (El-Refai *et al.*, 1990). Using solid-state fermentation, SCP production was investigated using orange peel as a substrate. Among several fungal cultures tested, a selected strain of *Fusarium culmorum* on a mixed substrate containing lyophilized orange peel powder and wheat straw, supplemented with NH_4^+ salts and corn steep liquor, and aerated with sterile humidified air gave the best protein yields of up to about 16 g kg^{-1} (Rossi *et al.*, 1988). Further investigations of the utilization of orange peel by the fungal species *Memnoniella echinata* and *Fusarium roseum* emphasized the technical feasibility of microbial protein production from untreated citrus waste, possibly achieving crude protein yields greater than 16% and protein enrichment of final biomass ranging from 40% to

46% (Clement *et al.*, 1985). Citrus fruit pulp from the fruit canning industry was used as a substrate for single-cell production which showed that the fungi *Trichoderma koningii* yielded the maximum total protein productivity (Ghai *et al.*, 1979).

In the cane and sugar beet processing industries, molasses left after the sugar extraction can be utilized for growing yeasts. The molasses supply carbohydrates, mineral elements and amino nitrogen for the fermentation process, but it may be necessary to add nitrogen and phosphorus. An aerobic fermentation process has been applied for the biosynthesis of yeast protein using *Saccharomyces cerevisiae*. For the production of 100 kg yeast dry matter containing 50% protein, 400 kg molasses, 25 kg aqueous ammonia, 15 kg ammonium sulphate, 7 kg mono-ammonium phosphate and 4810 m^3 air was required (James and Addyman, 1974). Starch-containing wastes, such as effluents from the processing of potatoes, corn and rice, have been treated by combined enzymatical hydrolysis and yeast fermentation with *Candida utilis* (Tveit, 1967; Jarl, 1969). Studies with different species of yeasts showed that *Schwanniomyces castellii* can utilize potato processing waste water for single cell production, and that it is suitable for direct and efficient bioconversion of the starch substrate into cell mass with a favourable amino-acid profile for salmonid fish feed (Park *et al.*, 1991). The bioconversion of agricultural wastes, such as potato peel, banana peel and green bean pod, by *Aspergillus foetidus* yielded 25–40% SCP, which could be further increased by the addition of corn steep liquor and wort (El-Shimi *et al.*, 1987). *Cephalosporium eichorniae* (*Acremonium alabamense*) could be successfully used for the production of microbial biomass protein from potato processing wastes for use as animal feed. The substrate was supplemented with $NH_4H_2PO_4$, NH_4OH and Fe^{3+}, yielding about 0.3 g crude protein per gram of carbohydrate supplied (Stevens and Gregory, 1987).

Cassava processing industry waste water was evaluated for the production of SCP and the monoculturing of *Candida tropicalis* gave the highest biomass yield together with a 50% decrease in COD (Jamuna and Ramakrishna, 1989).

The utilization of grape marc as a substrate for single cell production after pretreatment with Na_2CO_3 and NaOH at 120 °C by fermentation with *Aspergillus* sp., *Geotrichum candidum* and *Trichoderma viride* was investigated for feeding purposes. It was suggested that for profitable utilization the waste should be submitted to consecutive alkaline and acidic treatments to obtain valuable products (Vaccarino *et al.*, 1992).

Further studies were carried out for the utilization of agricultural wastes, such as millet husk, sorghum husk and banana stem waste, for essential amino-acid synthesis (lysine, histidine) by *Penicillium expansum* (Dahot and Abro, 1994). Onion juice, obtained as a by-product or waste of the dehydrated onion industry, was used, after hydrolysis, as a carbon and nitrogen source for the cultivation of *Saccharomyces cerevisiae* strains yielding up to 46% protein (Abou-Zeid *et al.*, 1979).

Palm oil processing effluents were utilized as substrates for microbial protein production by the fungus *Aspergillus oryzae* yielding a biomass of about 500 g kg^{-1} organic matter, with a BOD reduction of 85% and a COD reduction of 75–80% (Barker and Worgan, 1981). As a means of controlling environmental pollution, wastes from palm oil production were used as the substrate with the addition of starch for butanol fermentation using *Clostridium acetobutylicum* (Kwon *et al.*, 1989).

Olive oil extraction effluents are major pollutants of the environment. Their utilization in producing useful by-products by fermentation could limit environmental pollution, since the effluents contain various amounts of carbohydrates depending on the variety of olives, growth conditions and extraction procedures. Some yeast strains of *Torulopsis* sp. *MK-1*, *Saccharomyces norbiensis MC-1*, *Saccharomyces oleaceus MC-2* and *Saccharomyces oleaginosus* could be grown well on the effluents and fermented the sugar into alcohol, but relatively low final alcohol concentrations in the fermentation broth could be achieved (Bambalov *et al.*, 1989). *Candida krusei*, *Saccharomyces chevalieri* and *Saccharomyces rouxii* were investigated for the production of SCP from olive mill waste water, which were selected because they tolerate polyphenolic substances present in the substrate. They can be used to reduce BOD and may become a source of SCP for the supplementation of animal feed (Ghasallah, 1993).

Wastes from potato and wheat-starch processing factories can be fermented to ethanol. Potato pulp is, however, difficult to ferment to ethanol owing to its content of gelling materials, such as pectins and hemicelluloses, and hydrolysis processes have been developed to obtain hydrolysates high in dry matter content with respect to the subsequent fermentation (Marihart, 1982).

Batch fermentation with molasses produces 2,3-butanediol by using *Klebsiella* cultures with a diol yield of 96–100% of the theoretical value (Afschar *et al.*, 1990a). Several variants of a fermentation system for the continuous production of butanol and acetone from high-test or invert molasses were developed involving economic batch fermentation, and using a continuous inoculation culture and a modern two-stage continuous culture with cell recycling in the second stage (Afschar *et al.*, 1990b). The optimum fermentation conditions for acetone–butanol production by *Clostridium acetobutylicum* from molasses with the addition of $CaCO_3$ and $(NH_4)_2SO_4$ have been investigated (Donmez *et al.*, 1990).

Anaerobic digestion can be used to convert solid wastes from fruit and vegetable processing into methane fuel (Lane, 1979). In the citrus processing industry, large amounts of solid wastes are generated, the disposal of which has caused an enormous pollution problem. However, the presence of peel oil makes citrus peel toxic to anaerobic digestion, which makes reduction of the oil content prior to the fermentation process necessary (Lane, 1980). A practical means for reducing the peel oil content for the subsequent utilization of the peels as feedstock for anaerobic digestion was established, where yields of 0.5 m^3 biogas kg^{-1} (50–55% methane) total solids and conversion rates of solids to gas up to 100% could be achieved (Lane, 1984). Wastes from bean bleaching, and pear and potato peeling were used for the production of methane by an anaerobic contact process, and more recently by an upflow sludge-bed reactor process, with high efficiences in converting organic solids to methane (Van den Berg and Lentz, 1977; Van den Berg *et al.*, 1983). An upflow sludge blanket process was also used for treating beet-sugar refining wastes (Pette *et al.*, 1981).

The possibility of the production of antibiotics from plant food processing wastes has been described. Egyptian blackstrap molasses obtained from different sugar cane manufacturers, and freeze-dried olive water obtained from the pressing and separation of olive oil was used for production of oxytetracycline by *Streptomyces rimosus* (Abou-Zeid *et al.*, 1980).

An interesting approach for the utilization of food processing wastes is shown in studies performed with solid or liquid wastes, such as from molasses from alcohol fermentation, from the corn industry, from waste water from coconut processing, but also from the malting and brewing industry, and from residues from the bakery industry, employing semisolid or submerged fermentation processes to produce *Bacillus thuringiensis* endotoxins in order to produce bacterial insecticides (Moraes *et al.*, 1989). The feasibility of using agricultural residues and waste water from the food and beverage industry as substrates for the production of microbial insecticides by fermentation with *Bacillus thuringiensis* in submerged and semisolid fermentations has been studied since 1970 in Brazil, and corn steep liquor and sugar cane molasses have been detected as the principal components of the culture media (Moraes *et al.*, 1994).

Production of fat from sweet potato processing wastes from starch manufacture and from molasses was performed by fermentation processes with microorganisms from *Lipomyces* sp. The sugar sources were treated with diluted sulphuric acid before fermentation, and ammonium sulphate or peptone was used as a nitrogen source, yielding about 20% fat based on the consumption of the sources (Watanabe, 1974). In regard to the microbiological conversion of agro-industrial waste materials to high-valued oils and fats, a number of yeasts and moulds can produce a range of oils, but only a few can be used as high-value-added products. Among them, the production of a cocoabutter-like oil from *Cryptococcus curvatus* and a number of fungal oils containing polyunsaturated fatty acids were evaluated (Ratledge, 1994).

An alternative treatment of vegetable oil refinery waste water might be the microbiological conversion of the unsaturated fatty acids in the waste vegetable oil into hydroxy- and keto-fatty acids. Biotransformation of oleic acid and linoleic acid by *Rhodococcus rhodochrous* was performed for the production of the corresponding hydroxy- and keto-stearic acids (Litchfield and Pierce, 1986).

Almost any substrate which can be used to produce SCP could be used to produce single cell oil (SCO). Suitable substrates are molasses (sugar beet and cane), whey, starch, and wastes such as date extracts and blanching liquors from vegetable processing units. After extraction of the oil from the microorganism, a protein-rich residue remains which is then saleable as SCP. Thus, a microorganism grown for oil is producing two high-value products (Ratledge, 1978). Factors affecting the production of SCO from different microorganisms (temperature, incubation period, C/N ratio, spore count) were studied using molasses and carrot wastes (Hassanien *et al.*, 1986).

Wastes from different categories of fruit and vegetables, from wine and fruit juice manufacture as well as from other food industry wastes (alcohol industry, dairy industry) could be used for microbial fermentation by *Aspergillus* sp., *Saccharomyces* sp., *Acetobacter* sp. and *Lactobacillus* sp. for the production of organic acids, such as citric acid, pyruvic acid, lactic acid, acetic acid, etc. Technological approaches for their production and use in the food industry as well as in other industries (pharmaceuticals, cosmetics) have been reviewed by Galkina *et al.* (1990). Citric acid was produced from beet molasses by immobilized *Aspergillus niger* cells entrapped in calcium alginate gel beads in repeated batch fermentations (Roukas, 1991). In submerged fermentation *Aspergillus niger* was used for the production of citric acid from cane molasses without pretreatment. The average yield of citric acid was 13.5% and the conversion rate of total fermentable sugar to total acid was 91.3%. The purity of citric acid in the fermentation broth reached 98.7% (Zhu *et al.*, 1991). *Aspergillus niger* was successfully used for citric acid production from mandarin orange peels with citric acid production much higher in semisolid culture than in submerged culture. The addition of NH_4NO_3, $MgSO_4$, MeOH or EtOH promoted citric acid production with almost all of the citrus peel consumed during the fermentation (Kang *et al.*, 1989). In solid-state fermentation by *Aspergillus niger*, pineapple waste was also used as the raw material for the production of citric acid. The citric acid production was 132 g kg^{-1} of the dry waste at 6 days fermentation, at 29 °C, in a medium containing 4% methanol (de Lima *et al.*, 1995). By applying a similar solid-state fermentation process, kiwi fruit peel was used as the substrate for the production of citric acid by *Aspergillus niger*, and about 100 g citric acid was produced per kilogram of peel fermented in the presence of 2% methanol at 30 °C in 4 days, yielding more than 60% based on the amount of fermentable sugar consumed (Hang *et al.*, 1987).

Sugar cane molasses, corn steep liquor and corn cooking water were used as the substrate for lactic acid production by *Lactobacillus delbrueckii*, and the highest lactic acid conversion was 98% using 95 g $^{-1}$ of H_2SO_4-treated molasses (Martinez-Gonzales *et al.*, 1988). In regard to the utilization of beet molasses for the production of lactic acid by *Lactobacillus delbrueckii*, optimization studies were carried out which resulted in the highest yields and conversion rates at a pH value of about 6 and total soluble solids of about 18% (El-Sherbiny *et al.*, 1986).

Glutamic acid can be produced with high yield (88 g l^{-1}) from palm waste hydrolysates by *Brevibacterium lactofermentum* compared with that produced from pure glucose as a C source, which may be due to the fact that the microorganism can convert other sugars present in the hydrolysate. Glutamic acid was recovered from the fermentation broth by ion-exchange resin column chromatography (Das *et al.*, 1995).

Hydrolysed hemicellulosic fractions of agro-industrial wastes were used for xylitol production of *Candida guilliermondii* under semiaerobic conditions. The highest substrate uptake rate was attained in sugar-cane bagasses, while the highest xylitol production was achieved in sugar-cane bagasse hydrolysates (Roberto *et al.*, 1995).

8.7 Dairy industry wastes

Wastes from the dairy industry are of four major types (Jelen, 1983):

- milk from flushing and spills,
- dairy products from machinery malfunctions and retail returns;
- whey from cheese/casein production;
- ultrafiltration permeate from the production of cheese and whey protein.

Research on the utilization of whey has attracted great attention. Whey is the residue which drains from the curd in cheese production and has a high organic strength and a high COD, and therefore often causes disposal problems (Moulin and Galzy, 1984). It is composed of approximately 95% water, 4–5% lactose, 0.7% protein, 0.5–0.6% salts, 0.3% fat, with slight variations depending on the source. About 50–60% is used, either directly or converted into spray-dried whey powder, as a food ingredient or as an additive to feedstuff. Lactose, the main constituent of whey, still causes some problems because it is only metabolized poorly, if at all, by most industrial microorganisms, and it is only poorly soluble in water at ambient temperature (Hacking, 1988). However, 80% of lactose-containing whey residue is already used in technical and pharmaceutical products (Kobald and Holley, 1989).

Anaerobic digestion of raw cheese whey has been reported in the literature, and may help to solve the problem by reducing COD with the production of biogas, namely methane (Van den Berg, 1984; Clanton *et al.*, 1985; Yan *et al.*, 1988). Recent studies indicate that anaerobic digestion of cheese whey using an upflow anaerobic blanket reactor (UASB) would reduce pollution strength and produce energy. In general, over 98% removal of oxygen demand was achieved, the highest methane production rate of 9.57 l CH_4 l^{-1} feed day^{-1} being obtained at a loading rate of 5.96 g COD l^{-1} and an influent concentration of 28.8 kg COD l^{-1} (Yan *et al.*, 1988).

Whey can also be used as raw material for different fermentation processes. Yeasts from the *Kluyveromyces* strain (*Kluyveromyces fragilis*) are used for ethanol fermentation (Hacking, 1988; Bernstein *et al.*, 1977; Maiorella and Castillo, 1984). Thereby, the BOD reduction achieved was about 90%, yielding up to 9% ethanol. It was further concluded that, presently, the production of fuel ethanol from cheese whey or whey ultrafiltration permeate does not seem attractive in comparison with petroleum prices, but premises may alter in times of petroleum shortage or restriction, or in regard to environmental pollution aspects (Litchfield, 1987).

Whey is also proposed for SCP production. In general, bacteria, yeasts and fungi can be used for converting cheese whey to SCP (Butany and Ingledew, 1973; Bernstein and Plantz, 1977; Kim and Lebeault, 1981; Moulin *et al.*, 1983). However, at present, processes are still in marginal operation; only processes based on the yeast *Kluyveromyces fragilis* are practised on commercial scale. Whey containing different lactose concentrations was used as a substrate for SCP production by *Kluyveromyces fragilis* observing the highest protein content at an early stage of growth (Zayed, 1991). *Fragilis* yeasts have been granted food additive approval by the Food and Drug Administration in the USA (Litchfield, 1987).

In preparation the whey is heated to a sterilizing temperature by steam injection, and supplementation of the whey is necessary for maximum yields (James and Addyman, 1974). When the yeast is separated, most of the milk protein is carried with it. New technology had devised methods of ultrafiltration and gel filtration which can extract the milk protein in its undenaturated form from the whey streams to leave essentially a lactose solution (O'Sullivan, 1972). Although yeast grown conventionally on whey is generally produced in batch culture, experiments have been made with a continuous propagator (Amundson, 1967).

Studies have been carried out for the utilization of cheese whey both alone or mixed with bakery bread waste by *Saccharomyces cerevisiae* and *Kluyveromyces fragilis* for the production of protein and ethanol. The best ethanol yields could be obtained by fermenting sweet whey containing 18% solids at pH 6.0 and acid whey containing 21% solids at pH 4.5 with *Kluyveromyces fragilis* alone, whilst the best protein yields could be obtained by fermenting sweet whey alone or in mixture with bakery waste at both pH 4.5 and 6.0 (Collins and Rajagopal, 1983).

Another waste material from the dairy industry which received great attention for its utilization in recent times is the ultrafiltration permeate generating from the production of cheese and whey protein by ultrafiltration, which, however, creates an even worse problem than the whey (Jelen, 1983). The main problem of both the whey and the ultrafiltration permeate is the lactose content. The enzymatic hydrolysis of lactose by β-galactosidase (free or immobilized) seems to be the most promising method available for managing this problem, whereas the most economically feasible process is considered to be the application of immobilized enzymes (Coton, 1979, 1981). One of the most interesting uses for lactose is its hydrolysis into sweetening sugars, which has already become industrially applicable (Coton, 1981; Jelen, 1983). Hydrolysis can be accomplished in the deproteinated ultrafiltration permeate as well as in the whole whey, only additional demineralization is required for conversion into sweetening syrups. Hydrolysed lactose syrups will have enhanced value based on their enhanced sweetness, and represent potential sweetening agents for use in the food industry, such as in ice creams, confectionary and fruit products, for yeast production, and as an additive in beer production (Jelen, 1983).

Bioconversion of whey as well as olive oil mill waste water was performed for the production of organic acids. Whey showed high conversion yields (80%) with butyric acid as the main fatty acid production while, in comparison, olive oil mill waste water showed lower conversion yields with acetic acid as the main constitutent (Beccari *et al.*, 1992). Acetic acid was obtained directly from the lactose content of whey by fermentation with selected *Propionibacterium* sp. which converted about 90% of the lactose into acetic acid and about 10% into propionic acid; it is possible to obtain about 60 kg acetic acid from 100 kg lactose in whey (Sobczak and Konieczna, 1987). By sterilizing high concentrations of cheese whey using ultra high temperature, a medium was obtained which was conducive to microbial growth and propionic acid production. While *Propionibacterium freudenreichii shermanii*, grown with pH control in 12% whey solids and 1% yeast extract sterilized by ultra high temperature, produced about 1.9% propionic acid, the yield could be significantly increased to about 6.5% by using a mixed culture of *Propionibacterium shermanii* and *Lactobacillus casei* and raising the concentration of whey solids to about 18% (Bodie *et al.*, 1987).

Various lactic acid bacteria (*Lactobacillus casei*, *Lactobacillus helveticus*, *Lactobacillus bulgaricus*) can be utilized for the production of lactic acid from whey ultrafiltration permeate by the application of different fermentation systems using batch fermenters, membrane recycle reactors or immobilized cells (Tuli *et al.*, 1985; Mehaia and Cheryan, 1986, Roy *et al.*, 1986). Continuous production of lactic acid from whey permeate, a waste product of the dairy industry, by *Lactobacillus helveticus* was investigated in two chemostats in series by studying the effect of the dilution rate on the production of lactic acid, showing that the total residence time could be reduced by nearly 50% compared with a single chemostat for the same nearly complete level of substrate conversion (Aeschlimann *et al.*, 1990). Likewise, the continuous production of L-lactic acid from whey permeate as substrate together with supplements, which enables exponential growth of the microorganism, was performed using the bacterium *Lactobacilus casei casei*. Continuous fermentation in a stirred-tank reactor and in a fluidized-bed reactor with immobilized biomass on porous sintered glass beads was used (Krischke *et al.*, 1991).

Furthermore, whey permeate can also be used as a substrate for single-cell production; however, the cell product must be washed extensively because of the relative high ash content of the permeate or the whey must be demineralized (Lane, 1977; Shay and Wegner, 1986). The pretreatment of ultrafiltration whey permeate by an anaerobic process with 80% BOD reduction and recovery of 0.3 m^3 methane kg^{-1} COD has been reported by Lie *et al.*, (1982).

In general, by the utilization of whey using biotechnical processes the following products could be obtained: ethanol, butanol, methane, SCP, enzymes (β-galactosidase), antibiotics, vitamins (L-ascorbic acid, B_{12}, B_6, B_2), organic acids (lactic acid, citric acid) and food gums (Kobald and Holley, 1989).

8.8 Wastes from the fermentation industry

Recoverable by-products from the malting, brewing and distilling industries include spent grains, surplus yeasts, carbon dioxide and fermentation residues from the distillery, which are mostly utilized directly for animal or human consumption, or recycled (Isaac and McFiggans, 1981). Development work on upgrading these by-products has been carried out for the utilization of brewery wastes for the production of ethanol (Litchfield, 1987) and for the utilization of distilling wastes for the production of biogas (Szendray, 1983; Henry, 1985).

Processing water from the brewery industry was submitted to an activated sludge treatment for the production of animal feed. The product was shown to have feed values comparable to soybean meal, and the high vitamin content led to defining the product as a vitamin B_{12} supplement (Vriens *et al.*, 1989).

Liquid brewery waste was used as a substrate for riboflavin production by *Candida guilliermondii* in a medium supplemented with biotin. A riboflavin yield of 220 $\mu g\ ml^{-1}$ was achieved from the substrate containing 22 mg total carbohydrates ml^{-1}, of which about 60% was reducing sugar. Furthermore, methionin as an N source stimulated riboflavin production (Contasti and Bahar, 1988).

The economic and environmental aspects as well as the management and utilization possibilities of brewery residues and wastes have been reviewed by Keller-Reinspach (1990) and Huige (1995).

8.9 Conclusion and future outlook

Many of the problems found in treating food processing wastes arise from the fact that they are usually dilute, and processed seasonally and in plants which are often in scattered locations and which entail high transport costs. Furthermore, food-processing waste treatment processes are characterized by high investment costs for research work, and for the development and establishment of adequate process units with the appropriate technical equipment, which must be evaluated in relation to the economic value and benefit of the recovered and upgraded products. It should also be kept in mind that waste for disposal or dumping also needs prior treatment, owing to environmental constraints and to environmental legislation requirements which are becoming increasingly important. Wherever it is economically and technologically possible, emphasis should be laid on the utilization of waste materials for recycling, recovering and upgrading purposes, even if improved technologies in waste disposal are available.

Bioconversion of food processing wastes is receiving increased attention with the realization that the wastes contain valuable components which can be utilized for conversion into useful and high-value products. Current research for methods of food-processing waste treatment is focusing on marketable products. A number of appropriate biotechnological processes are available to produce food or feedstuffs, supplements or additives, chemicals or products which are commercially exploited for their energy value. In a food process, the recovery of waste components and bioconversion into higher added value products can contribute significantly to improvements in overall profitability, and enables the food processor to reduce the unit costs of the production process. In its simplest form, product recovery can be regaining materials for use as a low-grade product for food or feed purposes, but it is more effective if the waste components can be upgraded mostly using biotechnological processes.

It is reasonable to expect that SCP will ultimately contribute a major part of the world's future food needs. A worldwide protein deficit makes the production of SCP from food wastes particularly attractive since a much needed food supplement would be produced from sources which otherwise would be disposed of and wasted. The crux of the recent advances being made in the biotechnological processes for bioconversion and upgrading waste materials to useful products is the scaling up of new biological techniques to commercial applications at an industrial level in a cost-effective and convenient manner.

With regard to future works and trends in the bioconversion of food processing wastes to upgraded and useful products, emphasis should be laid on the following aspects:

- Much greater efforts are required to recover the large and increasing amounts of energy present in food processing wastes.
- Improved methods should be developed to solve the problems associated with the worldwide supply of protein by the production of SCP, with special emphasis on the development of effective low-technology and small-scale processes for the utilization of food processing wastes, particularly in developing countries.
- Further evaluation of the microbiological aspects of food industry waste treatment processes should be carried out with emphasis on genetically engineered microorganisms. Genetic engineering can be used to improve the economics of fermentation processes by increasing productivity, final product concentrations and production rates, but also for correction of the deficiences and problems in the utilization of SCP, e.g. by applying mutants with a lower nucleic acid content.
- Further research work is needed on scaling laboratory-tested processes to commercial application for upgrading food processing waste materials to useful products based on improved understanding of the mechanisms of the fermentation process. In order to achieve commercial success, food processing waste treatment must perform the required technical role reliably, must be cost effective and must be acceptable to the regulatory authorities. The scale-up from laboratory investigations to a commercially viable level seems to be the most difficult step in the development of new biotechnologically based processes, but other problems are expensive and time-consuming safety testing procedures of these products made using new technologies.
- Applied research work should be aimed at the development of tailored waste treatment equipment suitable for the management of specific food process systems, and, should determine which processes offer the most benefit and advantages to the different types of food processing industries.
- Strict legislative control regarding food processing waste treatment seems to be necessary concerning both environmental aspects and evaluation of the novel products obtained.
- Increased efforts should be focused on marketing and propagating the

upgraded waste-derived products. Marketing of some products and processes may be hindered by legislative regulations. Further research and development in these areas would provide evidence needed for the demonstration of the safety of these products and processes to regulation authorities.

- New policies should be initiated in favour of waste treatment aimed at the bioconversion of waste materials to useful products rather than at disposal, discarding or simple downstream processes.
- Data on the chemical and physical properties of the waste-derived products and processing parameters should be collected to help to develop potential markets.
- Models should be developed to predict profitable and economic processes for bioconversion of food processing waste materials.
- Additional research is needed for optimization of the processes which would require minimal simple steps from the waste material to the useful end-product.
- Lastly, increased efforts should be made to motivate consumers, producers and authorities to conserve energy, water and raw materials, and to minimize the generation of waste materials by reutilization and upgrading techniques, and, consequently, to prevent pollution of the environment.

References

Abou-Zeid, A.Z.A., Abd El-Fattah, A.F. and Farid, M.A. (1979) Utilization of waste products of dehydrated onion industry for production of fodder yeast. *Agric. Biol. Chem.*, **43** (9), 1977–80.

Abou-Zeid, A.Z.A., El-Diwany, A.I., Shaker, H.E.D. and Salem, H.M. (1980) Utilization of food industry by-products in the production of tetracycline by Streptomyces rimosus. *Microbiol. Agric. Wastes*, **2** (4), 293–301.

Aeschlimann, A., Di Stasi, L. and Von Stockar, U. (1990) Continuous production of lactic acid from whey permeate by *Lactobacillus helveticus* in two chemostats in series. *Enzyme Microbiol. Technol.*, **12** (12), 926–32.

Afschar, A.S., Bellgardt, K.H., Rosell, C.E.V., Czok, A. and Schaller, K. (1990a) Fermentative production of 2,3-butanediol from molasses. *Dechema Biotechnol. Conf. 1990*, **4**, 733–6.

Afschar, A.S., Vaz Rosell, C.E. and Schaller, K. (1990b) Bacterial conversion of molasses to acetone and butanol. *Appl. Microbiol. Biotechnol.*, **34** (2), 168–71.

Anderson, C., Longton, J., Maddix, C., Scammell, G.W. and Solomons, G.L. (1975) The growth of microfungi on carbohydrates, in *Single Cell Protein*, 2nd edn (eds S.R. Tannenbaum and D.I.C. Wang), MIT Press, Cambridge, MA, pp. 314–29.

Amundson, C.H. (1967) Increasing protein content of whey. *Am. Dairy Food Manufact. Rev.*, **29** (7), 22, 23, 96, 99.

Bahl, H. and Gottschalk, G. (1988) Microbial production of butanol/acetone, in *Biotechnology 6b* (eds H.-J. Rehm and G. Reed), VCH, Weinheim, pp. 1–30.

Bambalov, G., Israilides, C. and Tanchev, S. (1989) Alcohol fermentation in olive oil extraction effluents. *Biol. Wastes*, **27**, 71–5.

Barker, T. and Worgan, J.T. (1981) The utilization of palm oil processing effluents as substrates for microbial protein production by the fungus Aspergillus oryzae. *Eur. J. Appl. Microbial. Bio-technol.*, **11** (4), 234–40.

Beccari, M., Campanella, L., Majone, M. and Rolle, E. (1992) Bioconversion of highly polluted wastes to organic acids. *Environ. Prot. Eng.*, **16** (2), 5–14.

Bernstein, S. and Plantz, P.E. (1977) Ferments whey into yeast. *Food Eng.*, **48** (11), 74–5.

Bernstein, S., Tzeng, C.H. and Sisson, D. (1977) The commercial fermentation of cheese whey for the production of protein and/or alcohol. *Biotechnol. Bioeng. Symp.*, **1977** (7), 1–9.

Birch, G.G., Parker, K.J. and Worgan, J.T. (1976) *Food from Waste*, Applied Science Publishers, London.

Bodie, E.A., Anderson, T.M., Goodman, N. and Schwartz, R.D. (1987) Propionic acid fermentation of ultra-high-temperature sterilized whey using mono- and mixed-cultures. *Appl. Microbiol. Biotechnol.*, **25** (5), 434–7.

Bough, W.A., Brown, W.L., Porsche, J.D. and Doty, D.M. (1972) Utilization of collagenous by-products from the meat-packing industry: production of single-cell protein by the continuous cultivation of Bacillus megaterum. *Appl. Microbiol.*, **24**, 226–35.

Butany, G. and Ingledew, W.M. (1973) Whey utilization by bacteria. *Can. Inst. Food Sci. Technol. J.*, **6**, 291–3.

Clanton, C.J., Goodrich, P.R., Morris, H.A. and Backus, B.D. (1985) Anaerobic digestion of cheese whey, *Proceedings of the 5th International Symposium on Agricultural Wastes*, pp. 475–82.

Clementi, F., Moresi, M. and Rossi, J. (1985) Effect of medium composition on microbiol utilization of citrus waste by mixed fungal culture. *Appl. Microbiol. Biotechnol.*, **22**, 26–31.

Collins, J.A. and Rajagopal, S.N. (1983) Microbial utilization of bread waste and cheese whey. *Cult. Dairy Prod. J.*, **18** (2), 22–4, 35.

Contasti, V. and Bahar, S. (1988) Riboflavin production by Candida guilliermondii from liquid brewery waste. *Acta Cient. Venez.*, **39** (1), 69–74.

Cooney, C.L., Rha, Ch. and Tannenbaum, R. (1980) Single-cell protein: engineering, economics, and utilization in foods. *Adv. Food Res.*, **26**, 1–52.

Cosio, I.G., Fisher, R.A. and Carroad, P.A. (1982) Bioconversion of shellfish chitin waste: waste pretreatment, enzyme production, process design, and economic analysis. *J. Food Sci.*, **47**, 901–5.

Coton, G. (1979) The utilization of permeates from the ultrafiltration of whey and skim milk, presented at *IDF Technical Session*, Geneva.

Coton, G. (1981) Utilization of lactose and modified lactose products in the food industry, presented at *Seminar on Dairy Ingredients in the Food Industry*, Luxembourg.

Couillard, D. and Zhu, S. (1993) Thermophilic aerobic processes for the treatment of slaughterhouse effluents with protein recovery. *Environ. Pollut.*, **79** (2), 121–6.

Dahot, M.U. and Abro, A.Q. (1994) Biosynthesis of lysine and histidine by Penicillium expansum using agricultural waste as a carbon source. *Sci. Int. (Lahore)*, **6** (1), 63–6.

Das, K. Anis, M., Azemi, B.M.N.M. and Ismail, N. (1995) Fermentation and recovery of glutamic acid from palm waste hydrolyzate by ion-exchange resin column. *Biotechnol. Bioeng.*, **48** (5), 551–5.

Davy, C.A.E. (1981) Recovery of fruit and vegetable waste, in *Food Industry Wastes: Disposal and Recovery* (eds A. Herzka and R.G. Booth), Applied Science Publishers, London, pp. 219–30.

de Lima, V.L.A.G., Stamford, T.L.M. and Salgueiro, A.A. (1995) Citric acid production from pineapple waste by solid-state fermentation using Aspergillus niger. *Arq. Biol. Technol.*, **38** (3), 773–83.

Donmez, S., Ozcelik, F. and Pamir, H.H. (1990) Optimum conditions for acetone–butanol production from molasses. *Doga Turk Tarim Ormancilik Derg*, **14** (2), 71–81.

Drepper, G. and Drepper, K. (1981) Neue Eiweißprodukte aus Schlachttierblut für Lebensmittel. *Fleischwirtschaft,* **61** (9), 1393–5.

El-Nawawy, A.S. (1992) Microbial biomass production from organic solid substrate, in *Solid Substrate Cultivation* (eds H.W. Doelle, D.A. Mitchell and C.E. Rolz), pp. 247–68, 403–54.

El-Refai, A.H., Ghanem, K.M. and El-Sabaeny, A.H. (1990) Single cell protein production from orange waste by Sporotrichum thermophile cultivated under optimal conditions. *Microbios*, **63** (254), 7–16.

El-Sherbiny, G.A., Rizk, S.S. and Yusef, G.S. (1986) Utilization of beet molasses in the production of lactic acid. *Egypt. J. Food Sci.*, **14** (1), 91–100.

El-Shimi, N.M., Mohsin, S.M. and El-Magied, M.M.A. (1987) Bioconversion of some agricultural wastes by Aspergillus foetidus. *Egypt J. Food Sci.*, **15** (1), 121–31.

Galkina, G.V., Ilranionova, V.I., Glazunova *et al.* (1990) Alimentary organic acids, acidulants, and different types of vinegar from unusual starting materials. *Pishch. Prom-st.*, **8**, 27–8.

Ghai, S.K., Kahlon, S.S. and Sood, S.M. (1979) SCP from canning industry wastes: Citrus-fruit pulp as substrate. *J. Res. (Punjab Agric. Univ.)*, **16** (1), 64–8.

Gharsallah, N. (1993) Production of single cell protein from olive mill wastewater by yeasts. *Environ. Technol.*, **14** (4), 391–5.

Giurca, R. and Levin, R.E. (1992) Optimization of the lactic acid fermentation of hydrolyzed cod gurry with molasses. *J. Food Biochem.*, **16** (2), 83–97.

Goldberg, I. (1985) *Single Cell Protein*, Springer Verlag, New York.

Grant, R.A. (1986) Biomass from wastes, in *Biotechnology 8* (ed. W. Schoenborn), VCH, Weinheim, pp. 335–47.

Gray, P. and Berry, D.R. (1980) The production of feedstuff biomass from liquid organic wastes by fermentation, in *Handbook of Organic Waste Conversion* (ed. M.W.M. Bewick), Van Nostrand-Reinhold Company, New York, pp. 330–82.

Grohmann, K. and Bothast, R.J. (1994) Pectin-rich residues generated by processing of citrus fruits, apples, and sugar beets. Enzymatic hydrolysis and biological conversion to value-added products. *ACS Symp. Ser.*, **566**, 372–90.

Gromov, B.V. (1967) Main trends in experimental work with algal cultures in the USSR, in *Man, Algae and the Environment* (ed. D.F. Jackson), Syracuse University Press, Syracuse, NY, pp. 294–78.

Hacking, A.J. (1988) Economic and commercial factors influencing the role of biotechnology in the food industry, in *Food Biotechnology 2* (eds R.D. King and P.S.J. Cheetham), Elsevier Applied Science, London, pp. 25–58.

Hang, Y.D. (1979) Production of single-cell protein from food processing wastes, in *Food Processing Waste Management* (eds J.H. Green and A. Kramer), AVI, Westport, CT, p. 442.

Hang, Y.D., Luh, B.S. and Woodams, E.E. (1987) Microbial production of citric acid by solid state fermentation of kiwifruit peel. *J. Food Sci.*, **52** (1), 226–7.

Hansen, C. (1983) Methods for animal waste recovery and energy conservation. *Food Technol.*, **37** (2), 77–80.

Hassanien, F.R., Ragab, M. and El-Makhzangy, A. (1986) Studies on the possibility of producing fats from food wastes by using micro-organisms. I: Factors affecting fat production from different fungi. *Fette-Seifen-Anstrichmittel*, **88** (1), 33–8.

Henry, M. (1985) Industrial performance of a fixed-film anaerobic digestion process for methane production and stabilization of sugar distillery and piggery wastes. *Energy from Biomass and Wastes IX*, Lake Buena Vista, Florida, Institute of Gas Technology, Chicago.

Herzka, A. and Booth, R.G. (1981) *Food Industry Wastes: Disposal and Recovery*, Applied Science Publishers, London.

Huige, N.J. (1995) Brewery byproducts and effluents. *Food Sci. Technol. (NY)*, **64**, 501–50.

Imrie, F.K.E. and Righelato, R.C. (1976) Production of microbial protein from carbohydrate wastes in developing countries, in *Food from Waste* (eds G.G. Birch, K.J. Parker and J.T. Worgan), Applied Science Publishers, London, pp. 79–97.

Isaac, P.C.G. and Anderson, G.K. (1974) Waste disposal problems in the food and drink industry: a review. *Proceedings of Symposium: Treatment of Wastes from the Food and Drink Industry* (eds The Institute of Water Pollution Control and The University of Newcastle upon Tyne), paper no. 1, pp. 7–11.

Isaac, P.C.G. and McFiggans, A. (1981) Nature and disposal of effluents from malting, brewing and distilling, in *Food Industry Wastes: Disposal and Recovery* (eds A. Herzka and R.G. Booth), Applied Science Publishers, London, pp. 144–60.

James, A. and Addyman, C.L. (1974) By-product recovery from food wastes by microbial protein production. *Proceedings of Symposium: Treatment of Wastes from the Food and Drink Industry* (eds The Institute of Water Pollution Control & The University of Newcastle upon Tyne), paper no. 18, pp. 149–54.

Jamuna, R. and Ramakrishna, S.V. (1989) SCP production and removal of organic load from cassava starch industry waste by yeasts. *J. Ferment. Bioeng.*, **67** (2), 126–31.

Jarl, K. (1969) Production of microbial food from low cost waste starch effluents through the Symba Yeast process. *Food Technol.*, **23**, 1009.

Jelen, P. (1983) Reprocessing of whey and other dairy wastes for use as food ingredients *Food Technol.*, **37**, 81–84.

Kang, S.K., Park, H.H., Lee, J.H. *et al.* (1989) Citric acid fermentation from mandarin orange peel by Aspergillus niger. *Sanop Misaengmul Hakhoechi*, **75** (5), 510–18.

Karapinar, M. and Okuyan, M. (1982) The utilization of citrus waste as substrate for microbial protein production by the fungus Sporotrichum pulverulentum. *J. Chem. Technol. Biotechnol.*, **32**, 1055–8.

Keller-Reinspach, H.W. (1990) Brewery residues and wastes and their recycling and disposal. *Brauwelt*, **130** (42), 1827–34.

Kihlberg, R. (1972) The microbe as a source of protein. *Ann. Rev. Microbiol.*, **26**, 427–66.

Kim, J.H. and Lebeault, J.M. (1981) Protein production from whey using Penicillium cyclopium; growth parameters and cellular composition. *Eur. J. Appl. Microbiol. Biotechnol.*, **13**, 151–4.

Kirk, R.E. and Othmer, D.F. (1980) Foods, non conventional, in *Encyclopedia of Chemical Technology*, 3rd edn, Vol. II, J. Wiley & Sons, New York.

Kirsop, B.H. and Hilton, M.G. (1981) Research into utilization of food industry wastes, in *Food Industry Wastes: Disposal and Recovery* (eds A. Herzka and R.G. Booth), Applied Science Publishers, London, pp. 210–18.

Kobald, M. and Holley, W. (1989) Biotechnical utilization of food manufacturing residues. *5th International Congress on Engineering and Food*, Cologne, Germany.

Krischke, W., Schroeder, M. and Troesch, W. (1991) Continuous production of L-lactic acid from whey permeate by immobilized Lactobacillus casei subsp. casei. *Appl. Microbiol. Biotechnol.*, **34** (5), 573–8.

Kwon, G.S., Kim, B.H. and Ong, A.S.H. (1989) Studies on the utilization of palm oil wastes as the substrates for butanol fermentation. *Elaeis*, **1** (2), 91–102.

Lane, A.G. (1977) Production of food waste from whey ultrafiltrate by dialysis culture. *J. Appl. Chem. Biotechnol.*, **27**, 165–9.

Lane, A.G. (1979) Methane from anaerobic digestion of fruit and vegetable processing waste solids. *Food Technol. Aust.*, **31**, 201–7.

Lane, A.G. (1980) Production of aromatic acids during anaerobic digestion of citrus peel. *J. Chem. Technol. Biotechnol.*, **30**, 345–50.

Lane, A.G. (1984) Anaerobic digestion of orange peel. *Food Technol. Aust.*, **36** (3), 125–7.

Li, A., Sutton, P.H. and Corrudo, J.J. (1982) Energy recovery from pretreatment of industrial wastes in the anaerobic fluidized bed process. *Proceedings of the 1st International Conference on Fixed-film Biological Processes (G.B.)*, Vol. 3, p. 1521.

Litchfield, J.H. (1977) Single-cell protein processes. *Adv. Appl. Microbiol.*, **22**, 267–305.

Litchfield, J.H. (1983) Single-cell proteins. *Science*, **219**, 740–6.

Litchfield, J.H. (1987) Microbiological and enzymatic treatments for utilizing agricultural and food processing wastes. *Food Biotechnol.*, **1** (1), 29–57.

Litchfield, J.H. and Pierce, G.E. (1986) Microbiological synthesis of hydroxy-fatty acids and keto-fatty acids, US Patent, 4,582,804.

Maddox, I.S. (1988) Microbial production of 2,3-butanediol, in *Biotechnology 6b* (eds H.-J. Rehm and G. Reed), VCH, Weinheim, pp. 31–50.

Maes, M. (1993) Agro-alimentary waste: a prodigious pantry ripe for direct reuse. *Eau Ind. Nuissances*, **166**, 78–81.

Maiorella, B.L. and Castillo, F.J. (1984) Ethanol, biomass and enzyme production for whey waste abatement. *Process Biochem.*, **19** (4), 157–61.

Marihart, J. (1982) Production of ethanol from starch industry by-products, especially potato-pulp. *Starch*, **34** (9), 290–3.

Martinez-Gonzales, Y., Quiroz-Camacho, M.H., Ledezma-Perez, A.S. and Jaramillo-Coronado, J.C. (1988) Lactic acid production by Lactobazillus delbrueckii from pretreated sugarcane molasses. *Rev. Latinoam, Microbiol.*, **30** (23), 209–14.

Mateles, R.J. and Tannenbaum, S.R. (1968) *Single Cell Protein*. MIT Press, Cambridge, MA.

McComis, W.T. and Litchfield, J.H. (1989) Meat, fish, and poultry processing wastes. *J. WPCF*, **61** (6), 855–8.
Mehaia, M.A. and Cheryan, M. (1986) Lactic acid from whey permeate in a membrane recycle bioreactor. *Enzyme Microbiol. Technol.*, **8**, 289–92.
Moraes, I.O., Capalbo, D.M.F. and Moraes, R.O. (1989) Solid and liquid waste utilization in fermentation processes to get bacterial insecticide. *5th International Congress on Engineering and Food, Cologne.*
Moraes, I.O., Capalbo, D.M.F. and Moraes, R.O. (1994) Byproducts from food industries: utilization for bioinsecticide production. *Developments in Food Engineering, Proceedings of the 6th International Congress on the Engineering of Food 1993*, pp. 1020–2.
Moulin, G. and Galzy, P. (1984) Whey, a potential substrate for biotechnology. *Biotechnol Genet. Eng. Rev.*, **1**, 347–74.
Moulin, G., Malige, B. and Galzy, P. (1983) Balanced flora of an industrial fermentor: production of yeast from whey. *J. Dairy Sci.*, **66**, 21–8.
Murado, M.A., Gonzales, M.P., Montemayor, M.I., Reiriz, M.J.F. and Franco, J.M. (1986) Unicellular protein in effluents from processed mussels. *Invest. Pesq.*, **50** (4), 571–605.
Najafpour, G.D., Klasson, K.T., Ackerson, M.D., Clausen, E.C. and Gaddy, J.L. (1994) Biological conversion of poultry processing waste to single cell protein. *Bioresource Technol.*, **48** (1), 65–70.
Niranjan, k. and Shilton, N.C. (1994) Food processing wastes – their characteristics and an assessment of processing options. *AIChE Symp. Ser.*, **300**, 1–7.
Oleskiewicz, J.A. and Olthof, M. (1982) Anaerobic treatment of food industry wastewaters. *Food Technol.*, **36** (6), 78–82.
O'Sullivan, A.C. (1972) Nomenclature proposed for single-cell protein derived from fermentation of whey and whey streams. *Dairy Ind.*, October.
Park, E.Y., Crawford, D.L., Korus, R.A. and Heimsch, R.C. (1991) Yeast single-cell protein production using potato processing wastewater. *J. Microbiol. Biotechnol.*, **1** (3), 212–19.
Penaloza, W.M., Molina, R., Brenes, R.G. and Bressani, R. (1985) Solid state fermentation: an alternative to improve the nutritive value of coffee pulp. *Appl. Environ, Microbiol.*, **49**, 388–93.
Pette, K.C., de Vletter, R., Wind, E. and van Gils, W. (1981) Full-scale anaerobic treatment of beet-sugar waste water, in *Proceedings of the 35th Industrial Waste Conference*, Purdue University, Ann Arbor Science Publishers, Ann Arbor, MI, pp. 635–42.
Pirt, S.J. (1978) Aerobic and anaerobic microbial digestion in waste reclamation. *J. Appl. Chem. Biotechnol.*, **28**, 232–6.
Ratledge, C. (1978) Grow fats from wastes. *Symposium on Resources Conservation by Novel Processes*, London, May.
Ratledge, C. (1994) Microbial conversions of agro-waste materials to high-valued oils and fats. *Inst. Chem. Eng. Symp. Ser.*, **137**, 25–33.
Reva-Moissev, S. and Carroad, P.A. (1981) Conversion of the enzymatic hydrolysate of shell fish waste chitin to single cell protein. *Biotechnol. Bioeng.*, **23**, 1067–78.
Roberto, I.C. Felipe, M.G.A., de Mancilha, I.M. *et al.* (1995) Xylitol production by Candida guilliermondii as an approach for the utilization of agroindustrial residues. *Bioresource Technol.*, **51** (2, 3), 255–7.
Rodriguez, J.A., Echevarria, J., Rodriguez, F.J. *et al.* (1985) Solid state fermentation of dried citrus peel by Aspergillus niger. *Biotechnol. Lett.*, **7**, 577–80.
Rossi, J., Clementi, F., Gobetti, M. and Ribaldi, M. (1988) Fungal protein production from citrus waste solid substrates. *Biotechnololgy in the Food Industry, Proceedings of an International Symposium 1987* (eds J. Hollo, D. Torley and D.A. Kiado), Budapest, pp. 545–55.
Roukas, T. (1991) Production of citric acid from beet molasses by immobilized cells of Aspergillus niger. *J. Food Sci.*, **56** (3), 878–80.
Roy, D., Goulet, J. and LeDuy, A. (1986) Batch fermentation of whey ultrafiltration by Lactobacillus helveticus for lactic acid production. *Appl. Microbiol. Biotechnol.*, **24**, 206–13.
Sakai, S. (1994) Trends in the development of environmentally friendly food. *Gekkan Fudo Kemikaru*, **10** (5), 60–6.

Shay, L.K. and Wegner, G.H. (1986) Nonpolluting conversion of whey permeate to food yeast protein. *J. Dairy Sci.*, **69**, 676–83.

Sinskey, A.J. and Tannenbaum, S.R. (1975) Removal of nucleic acid in SCP, in *Single-cell Protein*, 2nd edn (eds S.R. Tannenbaum and D.I.C. Wang), MIT Press, Cambridge, MA, pp. 158–79.

Sobczak, E. and Konieczna, E. (1987) Possibilities of using propionic bacteria in the production of acetic acid from whey. *Acta Aliment. Pol.*, **13** (4), 351–8.

Solomons, G.L. (1983) Single cell protein. *CRC Crit. Rev. Biotechnol.*, **1**, 21–58.

Stevens, C.A. and Gregory, K.F. (1987) Production of microbial biomass protein from potato processing wastes by Cephalosporium eichorniae. *Appl. Environ. Microbiol.*, **53** (2), 284–91.

Sucuru, G.A., Engelbrecht, R.S. and Chian, E.S.K. (1975) Thermophilic microbiological treatment of high strength waste waters with simultaneous recovery of single cell protein. *Biotechnol. Bioeng.*, **17**, 1639–62.

Szendray, L.M. (1983) The Bacardi Corporation digestion process for stabilizing rum distillery wastes and producing methane. *Energy from Biomass and Wastes, VI Conference*, Lake Buena Vista, Florida, Institute of Gas Technology, Chicago, IL.

Tacon, A.G. (1980) Nutritional evaluation of animal and food processing wastes. *Effluent Treatment in Biochemical Industry, Conference Paper, 3rd Process Biochem. International Conference 1979*, Paper no. 13.

Tannenbaum, S.R. and Wang, D.I.C. (1975) *Single Cell Protein II*. MIT Press, Cambridge, MA.

Tuli, A., Sethi, R.P., Khanna, P.K. and Marwaha, S.S. (1985) Lactic acid production from whey permeate by immobilized Lactobacillus casei. *Enzyme Microbiol. Technol.*, **7**, 164–8.

Tveit, M. (1967) Industrial aspects of microbiol food production with special reference to the potential of the new Symba Yeast process, in *Biology and the Manufacturing Industry*. Academic Press, London, p. 3.

Vaccarino, C., Lo Curto, R.B., Tripodo, M.M., Patane, R. and Ragno, A. (1992) Grape marc as a source of feedstuff after chemical treatments and fermentation with fungi *Bioresour. Technol.*, **40** (1), 35–41.

Van den Berg, L. (1984) Developments in methanogenesis from industrial waste water. *Can. J. Microbiol.*, **30**, 975.

Van den Berg, L. and Lentz, C.P. (1977) Methane production during treatment of food plant wastes by anaerobic digestion. *Food, Fertilizers and Agricultural Residues, Proceedings of the 9th Cornell Agricultural Waste Management Conference* (ed. R.C. Loehr), Ann Arbor Science Publishers, Ann Arbor, MI, pp. 381–93.

Van den Berg, L., Kennedy, K.J. and Hamoda, M.F. (1983) Effect of type of waste on performance of anaerobic fixed film and upflow sludge bed reactors. *Proceedings of the 38th Industrial Waste Conference*, Purdue University, Ann Harbor Science Publishers, Ann Harbor, MI, pp. 686–92.

Vriens, L., Nihoul, R. and Verachtert, H, (1989) Activated sludge as animal feed: a review. *Biol. Wastes*, **27**, 161–207.

Walsh, J.I., Ross, C.C. and Valentine, G.E. (1993) Food processing waste. *Water Environ. Res.*, **65** (4), 402–7.

Wang, H.-E. (1995) Improvement of γ-linolenic acid production by Spirulina based on mixed food and cattle wastes leached with seawater. *Huanjing Boahu (Taipeh)*, **18** (1), 77–88.

Watanabe, D. (1974) Production of fat by microorganisms (Lipomyces species). II. Production of fat from diluted sulfuric acid-treated liquor of rice hulls, from waste liquor of starch manufactured from sweet potato, and from molasses and beet molasses. *Hakko Kyokaishi*, **32** (2), 62–70.

Welsh, F.W. and Zall, R.R. (1984) Single cell protein from waste fishery refrigerator brines. *Process Biochem.*, **19** (3), 122–3.

Wicker, L., Hart, H.E. and Parish, M.E. (1989) Recovery and utilization of byproducts from citrus processing wastes. *ACS Symp. Serv.*, **405**, 368–80.

Worgan, J.T. (1976) Wastes from crop plants as raw materials for conversion by fungi to food or livestock feed, in *Food from Waste* (eds C.G. Birch, K.J. Parker and J.T. Worgan), Applied Science Publishers, London, pp. 23–41.

Yan, J.Q., Liao, P.H. and Lo, K.V. (1988) Methane production from cheese whey. *Biomass*, **17**, 185–202.

Zayed, G. (1991) Utilization of whey from reconstituted skim milk for single-cell protein production. *Cult. Dairy Prod. J.*, **26** (1), 22, 24, 26, 28.

Zhu, H., Hao, Y., Yao, H. *et al.* (1991) Submerged fermentation production of citric acid from cane molasses. *Gongye Weishengwu*, **21** (1), 1–4.

9 Bioconversion of cheese whey to organic acids

R.D. TYAGI AND D. KLUEPFEL

9.1 Introduction

Whey is a by-product of cheese production, when fat and casein are removed, which is generated at a rate of about 9 kg for every 1 kg of cheese or 6 kg for every 1 kg of cottage cheese (91% of the milk processed). One hundred kilograms of whey are equivalent to the sewage produced by 45 people (Jelen, 1979). At this rate, a medium sized cheese plant discharging 50–150 $\times$ 10^3 kg of whey day^{-1} constitutes a polluting strength equal to the sewage from 22 500–67 500 people on a daily basis (Modler, 1982). In many countries, including Canada, most of the whey is discarded as waste creating severe pollution problems owing to its biological oxygen demand (BOD) (35 000–40 000 ppm) (Kosikowski, 1976; Scott, 1981). This BOD is due mainly to the lactose which is preent at concentrations of between 4.5% and 5% (Kosikowski, 1976; Garoutte *et al.*, 1983).

Whey is a very rich source of minerals, calcium, phosphorus, potassium, sodium, copper and iron. These minerals have very good digestive qualities. It is also a very good source of vitamins of the B-complex group, riboflavin and pantothenic acid (Nickerson, 1974), but their availability varies considerably. The typical composition of cheese whey is shown in Table 9.1.

Food and dairy wastes, in fact, should be viewed as a rich, inexpensive potential source of raw material from which valuable products can be produced. An additional advantage of these wastes is that they are available all year round at a fairly constant rate.

9.2 Production of whey

The annual production of whey in the USA and Canada is 16 $\times$ 10^9 kg and 1.5 $\times$ 10^9 kg, respectively (Singh and Ghaly, 1984). The world production of cheese has been increasing steadily and in 1981 reached 11.6 $\times$ 10^6 tons. The total amount of whey generated in this activity was nearly 10.4 $\times$ 10^7 tons. In Québec, the production of cheese has increased from 32 535 tons in 1965 to 76 181 tons in 1984 (Tyagi, 1986). In 1985, the

Table 9.1 Composition of cheese whey and whey permeate

	Whey	Whey permeate
pH	4.7–6.1	6.1
Dry matter (%)	6.58–6.88	5.7
Lactose (%)	4.5–6.0	5.0
Protein (%)	0.4–0.6	0.01
Fat (%)	0.05–0.25	<0.01
Lactate (%)	0.15–0.61	0.15
Ash (%)	0.05–0.8	–
Minerals (mg 100 g^{-1})		
Calcium	45–103	140
Phosphorus	40–80	75
Magnesium	8–10	–
Potassium	140–145	–
Iron	600–800	–
Sodium	40–60	50
Vitamins (μ 100 g^{-1})		
Vitamin C	850–950	–
Thiamine	30–40	–
Pentothenate	340–370	~360
Riboflavin	140–150	–
Vitamin B6	40–50	–
Vitamin B12	240–250	–
Biotin	40–50	–
Folic acid	4	–

cheddar and speciality cheese production in Canada was 178 711 tons and that in the province of Québec 60 675 tons. This generated 5 746 422 tons and 1 697 755 tons of sweet whey in Canada and Québec, respectively.

9.3 Pollution control

In the past, the application of pollution control technology to the process effluents from the cheese making industry has pursued three alternatives: direct discharge to the municipal system; pretreatment at the cheese plant followed by discharge to the municipal system; and full treatment at the cheese plant.

The high organic content in whey coupled with a relatively balanced nutrient composition indicates that some method of biological treatment (activated sludge with its many process variations: extended aeration, contact stabilization, the oxidation ditch and, more recently, the pure oxygen injection systems and deep shaft) is appropriate, but experience has shown that this is not entirely without problems. Even though whey is nutrient-balanced, the lactose fraction causes the bacterial population to swing towards filamentous organisms which are unable to use the proteins present as a nitrogen source. Protein molecules, because of their large size, are unable to pass through the bacterial cell wall. The extracellular enzymes (proteases) break down the large protein molecules but the overall reaction rates are low (Riddle and Chandler, 1974). Moreover, the high BOD of wastes makes conventional aerobic treatment very difficult and expensive. Anaerobic treatment of waste to generate methane is possible, but in many cases gives a poor return on investment.

A further consideration is the effect which whey discharge may have on an existing biological treatment plant. In many cases, a cheese plant will contribute an organic load many times greater than that from the community. Also, because of the operational nature of a cheese plant, the loading rate to the treatment plant will not be constant unless load equalization is practiced. Experience has shown that, for successful treatment of cheese plant waste, the influent 5-day BOD (BOD_5) concentration to the biological system should not exceed 1000 mg l^{-1}. At concentrations higher than this, severe bulking problems have been experienced. The impact on a community's waste treatment facility can be reduced by a pretreatment at the cheese plant (screening followed by flow equalization prior to continuous discharge to the municipal system is one possible solution).

It should also be noted that most municipalities have a sewer discharge by-law and impose a sewer surcharge to industry if certain maximum values are exceeded. While the economics of waste treatment are site specific, it is readily apparent that opportunities for significant cost saving through pretreatment can be realized.

The problem of whey treatment demands simple economical solutions, particularly in the case of small cheese plants in which oxidative disposal methods may be too costly to install (Groves, 1972; Ripley, 1979). The large amount of whey among fewer factories can have both advantages and disadvantages. If satisfactory methods of whey disposal are not available, additional supplies of whey can mean greater economic returns (economy of scale). The larger volumes can make it economically feasible to adopt processing techniques which are not practical for small quantities of whey. In other words, if disposal is a problem, additional volume can increase returns.

9.4 Current disposal methods of whey

The most common methods of disposal and recovery are:

1. drying to produce whey powder;
2. returning whey to farmers for animal feedings;
3. selling it to processors;

4. paying processors to collect it;
5. dumping it as a waste or sewage.

The most widely used method is evaporation and drying of whey to produce whey powder which is used in bakery and other food preparations. The increase in cheese manufacture has occurred mainly in large plants that are equipped with whey-drying equipment. Although installations for drying and processing might be economical for large plants, their costs were reported to be prohibitive for medium and small plants (Muller, 1979). Also whey utilization requires adequate cooling and storage facilities. Thus only large facilities with an output exceeding 150×10^3 kg whey day^{-1} are able to realize a return on investment from processed whey products (Modler, 1982). Almost 50% of Canada's cheese-producing plants fall into the small-scale category (Table 9.2). The main source of waste whey in North America is also due to smaller plants (Jelen, 1979).

In 1978, about 35.6% of whey was disposed into the environment, whereas in 1976 the volume of whey discharged in the environment was only 28% of the total production. This amounted to 507 000 tons of whey in Canada and 161 000 tons of whey in Québec. In 1976, only 1% of the total volume was used in liquid form for animal feeding. Muller (1979) reported that, of the 1.2×10^9 kg of liquid whey produced in 1979 in Canada, 49% was used for whey powder, 8% was fed to pigs, 17% was dumped into sewers, and 26% was disposed of on land. In 1984 about 7.36×10^6 l of cheese whey was produced in Québec; about 8% was discharged into the environment, about 16% was consumed by animals and about 76% was dried.

Cheese whey has been considered as a feed to animals. Provided there are enough pigs, the cheese whey can be substituted in the pigs' diet for up to 25–30% of the total. It has been estimated that pigs can consume up to 1000 l of whey per animal annually, but, owing to transport costs, liquid whey can only be used economically for pig feeding within a radius of 30–40 km. In many cases, however, the pig population is too small in relation to the number of cheese plants (Muller, 1979).

A recent approach for whey utilization is ultrafiltration to separate the

Table 9.2 Distribution of cheese production plants

Capacity (kg d^{-1})	Number of plants
<11 000	90
11 001–24 000	28
24 001–45 000	38
45 001–91 000	14
91 001–227 000	7
>227 000	5

proteins from the so-called permeate containing mostly lactose. The protein fraction is used for food and to replace egg white while the permeate is normally disposed of without treatment or recovery (see section 9.6). A few small cheese plants are selling their whey to industries within a radium of 80 km, for recovery of the proteins by ultrafiltration.

9.5 Global utilization of whey

On a global basis, a divergence of whey disposition practices exist. In Denmark no whey is wasted and 90% of the whey is fed to animals. In Sweden it is illegal to discharge whey into the sewerage system, whereas in Israel, 84% of whey produced is wasted and not used for animal feed.

By-product recovery for proteins and lactose specifically is accomplished using concentration technologies ranging from spray drying to ultrafiltration coupled with gel filtration. Most countries practice some degree of whey conversion into useful products. The Netherlands leads the way with 93% of their whey being processed. In Canada, 48% is processed. In New Zealand, casein whey was used mainly for spray irrigation and pig feeding, whereas cheese whey was used for lactose manufacture and pig feeding (Galpin, 1981). Forty-five of the plants dispose of effluents by spray irrigation, 38 by discharge into natural waterways and municipal sewers, and 1 by means of a biological treatment unit operated by the dairy. Food uses of whey are increasing and have encompassed pharmaceutical products, dairy products, bakery products, confections and coatings, frozen and canned foods as well as powder foods. For example, production of whey powder in Canada has increased form 41×10^6 kg in 1977 to 57×10^6 kg in 1981 (Dairy Market Review, 1982). The choice of recovery technology is a function of the end-product use criteria.

9.6 Management strategies

A correct strategy for industrial pollution consists of minimizing feedstock leakage, optimizing product conversion efficiency, maximizing recovery and recycling of process losses, and optimizing by-product generation. The objective is to reduce the amount of waste material which results from the particular industry and thus to exert the minimum economic burden to that company by way of waste treatment requirements. There are many examples in industry where by-product generation and recovery have led to economic spinoffs which more than paid for the cost of pollution control (Campbell and Glenn, 1982).

The cheese producer does have a number of options when addressing the whey problem and some of them are discussed in section 9.4. Recovery of sugars, proteins and minerals in the waste effluents by crystallization, evaporation and spray drying are practiced in many food and dairy plants. However, the increased energy cost, especially that of natural gas, has made many of the above operations economically unattractive.

One management approach applies the principles of pollution control and resource recovery and involves application of ultrafiltration or hyperfiltration and reverse osmosis for concentrating the whey solids (Kosaric and Asher, 1985). Ultrafiltration has provided the cheese maker with a means of channeling the component of whey with the greatest value into the market place. Whey protein concentrate (WPC) is an item of worldwide commerce, with a significant dollar value. That is the bright side. The dark side is that for each kilogram of WPC, 2.3 kg of permeate solids are obtained, with a lactose content of approximately 85% of the whole whey.

A number of options have been proposed to convert whey permeate (WP) to value added products and other more profitable alternatives are constantly being evaluated. These alternatives include alcohols, organic acids, solvents and proteins, all of which have been discussed in a number of review articles (Kosaric and Wieczorek, 1984; Kennedy, 1985; Kosaric and Asher, 1985; Goursaud, 1986; Moebus and Teuber, 1986; Tyagi, 1986).

The absolute dependence of Canadian fermentation industries on sugar-cane molasses associated with problems like import, fluctuating prices, variations in chemical and microbiological qualities, clarification burden, storage, pumping and high residual BOD after fermentation has resulted in the search for an alternative raw material. Among a number of options proposed to convert the whey permeate to a value added product, replacement of molasses for the production of a number of chemicals (alcohols, organic acids, solvents, proteins, vitamins, enzymes) seems to be most attractive. Utilization of whey or WP as a raw material for fermentation industries is not only a waste treatment/recovery alternative, but also reduces those problems encountered in the use of molasses. Thus lactic acid produced from whey/permeate may serve as a raw material for the fermentation industry while reducing the BOD of whey by 90–95%.

The following section is devoted to a review of the organic acid production from cheese whey and WP.

9.7 Lactic acid

The art of lactic acid fermentation stems from antiquity and the biotechnology now is well advanced. Lactic acid, a natural acid, has long been of use in the pharmaceutical, chemical and food industries, primarily as an acidulant and as a preservative. About half the total world production of lactic acid is for non-food uses. The fermented product containing lactic acid has potential commercial use as a nitrogen-enriched feed supplement for ruminant animals (Gerhardt and Reddy, 1978; Marriott, 1985) or as a substitute for molasses in bakers' yeast production. Homofermentative lactic acid bacteria produce no carbon dioxide from sugars, and thus the substrate carbon is more productively utilized (Tuli *et al.*, 1985). A yield of 0–9 g lactic acid g^{-1} lactose or better should be obtained, compared to typically half that with ethanol (0.45–0.5 g g^{-1}). The lactic acid fermentation of non-concentrated whey and whey permeate has been studied in batch, continuous and immobilized cell processes using different bacterial strains.

9.7.1 *Microorganisms involved in lactic acid fermentation*

Microorganisms producing lactic acid are found in four genera (Kandler, 1982): *Streptococcus*, *Pediococcus*, *Lactobacillus* and *Leuconostoc*. The biochemical characteristics of these organisms have been discussed by Buchta (1983). The most important producers of lactic acid belong to the family of Lactobacillaceae.

Lactobacilus helveticus is especially advantageous for the following reasons:

1. It provides an alternative solution to the phage contamination in dairy industries. This problem is generally encountered using *L. bulgaricus*.
2. It produces almost twice the amount of lactic acid in milk as compared to other common lactic acid bacteria, including *Streptococcus thermophilus* and *L. bulgaricus* (Roy *et al.*, 1986; Audet *et al.*, 1992).
3. It provides a racemic lactic acid (D,L) as compared to only dextro-rotatory lactic acid (D) produced by *L. lactis*, *L. delbrueckii* and *L. bulgaricus*.
4. It can ferment lactose and all its breakdown products (glucose and galactose) (Bergey's Manual, 1974).
5. It is more advantageous for the preparation of culture starters from milk-base media.
6. It produces only lactic acid (Roy *et al.*, 1987).

L. helveticus cells do not have a preference for utilizing glucose or galactose in free cell system; however, for an unknown reason, galactose is metabolized less when the same microorganism is entrapped in calcium alginate gel, especially at low dilution rates. During batch fermentation, *L. helveticus* gave inferior productivity compared to *L. bulgaricus* (Roy *et al.*., 1986) but a higher conversion of substrate was observed using *L. helveticus* (Roy *et al.*, 1987).

Streptococcus thermophilus has two drawbacks:

1. Only a few strains of *S. thermophilus* are able to ferment galactose, one of the two sugars of lactose molecule (Somkuti and Steinberg, 1979);
2. *S. thermophilus* requires some growth factors (peptide) for lactic acid production in a milk-based medium owing to its limited amount of active cell-bound proteinase (aminopeptidases) (Roy *et al.*, 1986).

Other homofermentative lactic acid bacteria (*L. delbrueckii*, *L. lactis*, *L. casei* and *S. lactis*) produces other organic compounds (acetic acid, formic acid, ethanol) depending on the fermentation conditions.

Most of the strains used for lactic acid fermentation, produce D,L-lactic acid. It is desired to produce only the L(+) form, since only L-lactate dehydrogenase, the enzyme that metabolizes L-lactic acid, is found in mammalian systems. Elevated levels of D(−)-lactic acid can not be tolerated and will lead to hyperacidosis (Buchta, 1983). Lactic acid strains (*L. delbrueckii* NRRL B-445, *L. casei rhamnosus* ATCC 7469, *Streptococcus lactis* 5085, *Streptococcus cremoris* ATCC 19257 and *Streptococcus cremoris* 2487) were compared in a batch using WP supplemented with yeast extract (Mulligan *et al.*, 1991). *S. cremoris* was found to be superior in terms of lactic acid productivity. *S. cremoris* 2487 and ATCC 19257 were found to give similar results and 92% of lactic acid formed was of the L form. However, *S. cremoris* 2487 was found to be more suitable for industrial use.

By using fermentation temperatures of 43–45 °C, pH 5.5 and lactic-acid-tolerant organisms such as *L. bulgaricus*, unsterilized and unpasteurized whey can be used (Keller and Gerhardt, 1975; Reddy *et al.*, 1976). The unsterilized system has been used for as long as 42 days by these authors without encountering puterifactive spoilage.

9.7.2 *Batch process*

Studies on lactic acid fermentation in a batch process have been conducted by a number of workers (Campbell, 1953; Hanon and Tsao, 1972; Marshall and Earle, 1975; Reddy *et al.*, 1976; Tewari *et al.*, 1985; Roy *et al.*, 1986).

Lactobacillus bulgaricus, *L. casei*, *L. fermentii* and *L. plantarum* were screened on paneer whey (a type of cheese whey) for their ability to produce lactic acid. The acid production was efficient with a 10% inoculum of *L. bulgaricus*, 4.5% whey lactose, pH 5.5 and a fermentation period of 24 h at 30 °C (Tewari *et al.*, 1985). Lactic acid production was induced by glutamic acid. Lactose in unpasteurized whey was fermented to lactic acid and by *L. bulgaricus* at a temperature of 43 °C and pH 5.5 controlled with NH_4OH (Reddy *et al.*, 1976). The fermented product when concentrated by evaporation yielded 70% of total solids and adjusted to pH 6.8 with the addition of ammonia. The concentrated product contained about 5.5%

crude protein (approximately 6–8% of the crude protein was derived from bacterial cells, 17% from whey proteins, and 75–77% from ammonium lactate). The efficiency of conversion of lactose to lactic acid exceeded 95%. Whey containing lactose at concentrations up to 7% could be fermented efficiently, but higher concentrations of lactose were fermented incompletely (Reddy *et al.*, 1976).

WP concentrated up to four times were fermented using batch cultures and a maximum lactic acid concentration of 95 g l^{-1} was attained, but residual sugar indicated a possible limitation in growth factors. On concentrated permeates *L. helveticus* proved to be insensitive to high-level lactic acid concentration (up to 85–95 g l^{-1}) in batch fermentation (Aeschlimann and Stockar, 1989).

When using WP or whey ultrafiltrate (WU) as substrate for lactic acid fermentation, the medium must be supplemented with some easily digestible N compounds. The lactic acid bacteria showed low lactic acid production on unsupplemented whey or WU (Aeschlimann and Stockar, 1990; Mistry *et al.*, 1987).

The addition of yeast extract (YE) at a concentration of 1.5% (w/v) in the whey filtrate, as a source of nitrogen/growth factors, gave the highest maximum productivity of lactic acid of 2.7 g l^{-1} h^{-1} (Roy *et al.*, 1986). The supplementation of YE up to 20 g l^{-1} increased the productivity of lactic acid (Arasaratnam *et al.*, 1996). The increase in the lactic acid productivity was attributed to substances such as amino acids, peptides, vitamins and several organic acids including pyruvic and glyceric acid in YE. Supplementing the whey with YE up to 20 g l^{-1} and maintaining the total sugar concentration 30 g l^{-1} by adding glucose led to a further increase in the lactic acid production (Arasaratnam *et al.*, 1996).

It has also been reported that supplementation of 30 g l^{-1} of YE to enzyme-thinned corn starch was the best source of nitrogen and growth factors with respect to rate of lactic acid production, but not the lactic acid concentration by *L. amylovorus* (Cheng *et al.*, 1991). YE concentration above 30 g l^{-1} did not improve the lactic acid productivity. The reduced requirement of YE (20 g l^{-1}) observed by Arasaratnam *et al.* (1996) was attributed to the higher content of assimilable protein in whey than that in the enzyme-thinned corn starch observed by Cheng *et al.* (1991).

Many published reports, as discussed above, have shown that lactic acid production increases effectively with the concentration of added YE (Roy *et al.*, 1987; Aeschlimann and Stockar, 1990). However, addition of YE during large-scale fermentation of whey is unrealistic owing to the extra cost introduced for the fermentation process, in combination with the low value of lactic acid. Even at as low a concentration as 2.0 g l^{-1} YE, the cost could be as much as 32% (Mulligan *et al.*, 1991).

This prompted many researchers to look for an alternative of YE, such as protein hydrolysate, corn steep liquor (CSL) or malt sprout extract

(Boyaval, 1987; Roy *et al.*, 1986; Cox and MacBean, 1977). Few studies on minimization of YE requirement in whey permeate fermentation to lactic acid have also been reported (Krischke *et al.*, 1991; Lacroix *et al.*, 1990). Hydrolysed whey (lactose 4% w/v, fresh whey and suspension prepared from whey powder) was fermented for the production of lactic acid in a batch as well as in continuous process using *L. helveticus*, *L. bulgaricus* and two strains of *S. thermophilus* in pure as well as in mixed cultures (Lund *et al.*, 1992). Fresh whey gave the highest yield of lactate in batch culture. The yield of lactate (in grams of lactate produced per gram of carbohydrates consumed) varied between 0.54 and 0.81 $g\ g^{-1}$. Yield was highest in mixed culture using fresh whey (Lund *et al.*, 1992).

Few workers have considered other cheaper supplements, such as CSL, in order to find one capable of supporting both bacterial growth and acid production in WU (Roy *et al.*, 1986; Tuli *et al.*, 1985; Vahvaselka and Linko, 1987). The addition of CSL or tryptone gave inferior results. The lactic acid production, lactose utilization and maximum productivity of lactic acid increased linearly with pH (in the range of 4.7–6.30), and the specific lactic acid activity of the CSL-grown cells was observed to be higher than that of cells grown on skimmed milk. However, cell yield and cell growth were lower in CSL medium (Roy *et al.*, 1986. The addition of peptide and mustard oilseed cake to whey improved acid production; $FeCl_3$ (5 ppm) also increased the acid fermentation (Tewari *et al.*, 1985). The fermentation time was greatly reduced on the addition of 0.2% YE or 0.1% CSL as a source of growth factors (Reddy *et al.*, 1976).

Supplementation of whey with glucose, vitamin B complex and different nitrogen sources [YE, peptone, soya flour and $(NH4)_2SO_4$] to improve lactic acid production using *L. delbrueckii* was studied by Arasaratnam *et al.* (1996). Supplementing the nitrogen sources (other than YE) and vitamin B complex did not show any improvement in the lactic acid production. However, by increasing the elemental nitrogen ratio of ammonium sulfate : YE to 3 : 1, a similar substrate utilization efficiency and lactic acid production was observed as using 20 $g\ l^{-1}$ YE (Arasaratnam *et al.*, 1996).

The lactic acid fermentation of WU supplemented with molasses and YE using *L. helveticus* was compared by Chiarini *et al.* (1992). The following factors were studied: the molasses concentration (0.5, 1, 1.5% w/v), the influence of inoculum volume, the effect of YE with and without supplementation, and the effect of preculture media (inoculum grown in peptonized milk and molasses-supplemented WU). High lactose consumption (94.09%) together with good lactic acid production and yield were obtained in WU supplemented with 1% (w/v) beet molasses, with a 10% (v/v) inoculum using peptonized milk as the medium in which to grow the inoculum.

L. casei did not grow in pure whey permeate (Krischke *et al.*, 1991).

Therefore, in batch experiments, several components of the medium were tested with regard to their growth-promoting efficiency. Manganese was found to be an essential growth factor. Complex medium components such as YE, peptone and CSL showed growth-stimulating characteristics. Whey retentate, the remaining product of the ultrafiltration of whey, was showed to be an effective source of nitrogen. Fermentations supplemented with hydrolysed retentate or CSL resulted in better growh compared to experiments with hydrolysed supplements. The best growth was observed, with an even shorter fermentation time, with a combination of YE and hydrolysed retentate (Krischke *et al.*, 1991).

Inoculum prepared in skim-milk medium or in lactose synthetic medium resulted in the best performance in the fermentation of WU (in terms of growth, lactic acid production, lactic acid yield and maximum productivity of lactic acid), compared to that prepared in glucose synthetic medium (Roy *et al.*, 1986). Reddy *et al.* (1976) have shown that a 2.5% inoculum produces a fermentation which is just as rapid and contamination-free as that obtained with a 10% inoculum. The effect of preculture and culture media formulation on the lactic acid production by *L. helveticus* was studied (Amrane and Prigent, 1993). Maximum lactic acid productivity was obtained on hydrolysed whey.

Using YE as a supplement, different inocula of 1% and 10% did not significantly influence the lactic production (Chiarini *et al.*, 1992). Preculture (inoculum) media using WU supplemented with molasses gave inferior production of lactic acid and lactose consumption compared to the inoculum prepared in peptonized milk. It was also shown that the preculture (inoculum) media had a more marked effect than the inoculum size. The results obtained using the inoculum grown on peptonized milk (10%) and using WU supplemented with molasses were comparable to the results using WU supplemented with YE in terms of lactose utilization, maximum lactic acid concentration and yield of lactic acid. It was also shown that inoculum grown on peptonized milk exhibited a 2.5 times higher specific lactic acid productivity ($0.02\ g\ l^{-1} \times 10^9$ inoculated cells) in WU supplemented with molasses than the inoculum grown in molasses-supplemented WU (Chiarini *et al.*, 1992).

The batch method for the production of lactic acid by fermentation of cheese whey/permeate suffers from one or more of the following disadvantages:

1. There is a long lag period to active lactic acid fermentation.
2. There are unusually long fermentation times that required greater fermenter capacity and increased operational costs.
3. The manual procedures used for the addition of ammonium or calcium ion to neutralize the lactic acid produced during the fermentation are tedious and contribute to the increased product variability.

9.7.3 Continuous process

The continuous bioconversion process has the advantage of high productivities achievable with the selected microbial strains. For example, at a dilution rate (D = working volume/feed inflow rate) of 0.2 h^{-1} a single 6500 l continuous bioreactor could replace two 30 000 l bioreactors used in batch-operated plants. The continuous process in a free cell system has been studied by several workers (Hanson and Tsao, 1972; Keller and Gerhardt, 1975; Cox and MacBean, 1977; Stieber and Gerhardt, 1979a; Major and Bull, 1985), and as a dialysis continuous process (Friedman and Gaden, 1970; Stieber and Gerhardt, 1979b, 1981a,b). *Lactobacillus helveticus* was used for continuous fermentation of cheese whey, yeast extract and permeate medium. A maximum productivity of 9.7 g l^{-1} h^{-1} at a dilution rate of 0.352 h^{-1} was obtained with a 37.4 g l^{-1} feed lactose. Under high dilution rates, the cells were elongated to several times their normal size, resulting in a wall growth (Roy *et al.*, 1987).

The effect of dilution rate on the growth of *L. helveticus* and lactate production in a two-stage chemostat system using WP supplemented with 1% YE was investigated by Aeschlimann *et al.* (1990). A 98% conversion of lactose with lactic acid concentration of 43.7 g l^{-1} (with a lactic acid productivity of 8.27 g l^{-1} h^{-1}) at a dilution rate of 0.06 h^{-1} was observed. The fraction of L-lactate varied from 55% to 70%. Young cells produced more L-lactate than older ones. The results of Aeschlimann *et al.* (1990) compared well with Steffen *et al.* (1973) who reported, for two *L. helveticus* strain, 68% and 69% L-lactate and a decreasing L-lactate ratio with increasing fermentation time. It was shown that, to convert 49.3 g l^{-1} of lactose, the total hydraulic retention time (HRT = $1/D$) can be approximately halved in a two-stage system (8.8 h) compared with that in a single-stage system (16.3 h) (Aeschlimann *et al.*, 1990).

Lund *et al.* (1992) also studied the continous process for lactic acid production using non-sterile and unsupplemented whey at HRT varying from 2.5 to 10 h. Lactate productivity increased with dilutation rate and the highest productivity was obtained using *L. lactis*. A maximum concentration of 25 g l^{-1} lactic acid was obtained at an HRT of 10 h.

For a product inhibition reaction such as lactate production, the use of a plug flow reactor (Levenspeil, 1962), which can be approximated by several continuously stirred tank reactors (CSTRs) in series, would be advantageous. In a two-stage system (Aeschlimann *et al.*, 1990), long retention times or an elevated concentration of YE were required to achieve over 90% lactose utilization. The use of a multistage system has been suggested due to:

1. its higher productivity compared to a single- or a two-stage system;
2. the fact that is allows microorganisms in different physiological state to be maintained in each reactor;

3. the ease of pH control;
4. the fact that stability can be maintained without the cell washout.

In a search of a reduction of fermentation time, increased utilization of lactose and high yield, concentration and productivity of lactic acid, a three-stage system was investigated by Mulligan *et al.* (1991) at varying HRTs (5–10 h). The product yield was constant irrespective of HRT, and lactose utilization decreased whereas productivity increased with increased HRT. Lactose utilization and lactate production in a particular stage depends on HRT. At lower HRT, lactose utilization and lactic acid production shifts to the second and third stage of the system. A significant reduction in lactose utilization and lactic acid productivity was observed when 10–25% of the exit stream from the third stage was recycled to the first stage. Inhibition by the lactate in the recycled stream was given as one of the reasons for decreased productivity. Attempts were also made to replace YE with a Universal Foods brand (Amberex 1003) and no significant difference in terms of lactose utilization (88.4%), product yield (0.86 g g^{-1}) and lactic acid productivity (4.45 g l^{-1} h^{-1}) at an HRT of 7.5 h was observed (Mulligan *et al.*, 1991). By using *S. cremoris* 2487, the inoculation could be provided directly in to the production reactor, thus obviating the need for a seed fermenter and minimizing the problems of contamination during transfer of the medium between reactors. In order to avoid the propagation of bacteriophages, which could destroy the fermentation process, the alternate use of *S. cremoris* 2487 and ATCC 19257 strains during different runs was suggested. These strains behave identically except for their resistance towards bacteriophages (Mulligan *et al.*, 1991).

Roy *et al.* (1987) obtained a 9.7 g l^{-1} h^{-1} productivity in a single-stage continuous system using *L. helveticus* milano with WP supplemented with 15 g l^{-1} YE. Aeschlimann *et al.* (1990) obtained a productivity of 4 g l^{-1} h^{-1} with a two-stage system and 10–25 g l^{-1} YE supplementation with *L. helveticus*. The productivities for the two-stage reactor system using *L. bulgaricus* were 1.77 g l^{-1} h^{-1} (without supplementation) (Keller and Gerhardt, 1975) and 3.45 g l^{-1} h^{-1} (2 g l^{-1} YE) (Steiber and Gerhardt, 1979a,b). The latter value was comparable with the two-stage *S. cremoris* system (5 g l^{-1} YE) of Mulligan *et al.* (1991). For a single-stage reactor, productivities of 4 and 6 g l^{-1} h^{-1} for YE concentrations of 2 and 5 g l^{-1} were observed (Mulligan *et al.*, 1991). A decrease in growth and lactic acid productivity was observed with a decrease in YE concentration in batch and continuous fermentation processes (Aeschlimann and Stockar, 1990; Silva and Yang, 1995).

The production of lactic acid with *L. helveticus* in a cell-recycling system was investigated using WP supplemented with 1% (w/v) YE or WP supplemented with 0.4% YE and 0.65% skim milk powder (Aeschlimann and Stockar, 1991). A biomass concentration five times higher (21.3 ± 2.7 g l^{-1}) than without a cell-recycling system was achieved. Increasing the dilution rate from 0.18 to 1.05 h^{-1} approximately doubled the productivity from 7.8 to 15.8 g l^{-1} h^{-1}. The maximum productivity of 15.8 g l^{-1} h^{-1} was twice the productivity (8.3 g l^{-1} h^{-1}) at an optimal dilution rate 0.4 h^{-1} without cell recycling (Aeschlimann *et al.*, 1990). Lactose conversion was decreased from 93% to 34% when the dilution rate increased from 0.18 to 1.05 h^{-1}. The lactic acid yield was less affected by this change (99%–88%). The specific lactose consumption rate and specific lactic acid production rate were mainly affected in the lower range of dilution rate. The lactic acid yield (86%) was not affected by the cell-recycling procedure and compared well with that obtained without cell recycling (87%). The specific lactic acid productivity (0.62 h^{-1}) was about one-third that obtained without a cell-recycling system and was attributed to the accumulation of dead cells and mechanical stress in the cell-recycling reactor. The specific lactic acid productivity (0.62 h^{-1}) compared well with that reported for *L. delbrueckii* (0.5 h^{-1}); however, it was only half that reported for *L. bulgaricus* (1.3 h^{-1}) (Aeschlimann and Stockar, 1991). The productivity of lactic acid was decreased by 25% when YE (1%) was replaced with a mixture of YE (0.4%) and skim milk powder (0.6%).

The L(+)-lactic acid fraction increased with increasing dilution rate and was approximately 66% (Aeschlimann and Stockar, 1991) and these results were consistent with those reported by Steffen *et al.* (1973). Maximum productivity was obtained at 42 °C (productivity = 11.3 g l^{-1} h^{-1}; lactic acid concentration = 28.1 g l^{-1}; $D = 0.4$ h^{-1}). At 45 °C, the performance of the system decreased significantly (Aeschlimann and Stockar, 1991). By contrast Boyaval *et al.* (1988) found a better performance at 45 °C.

The productivity of the fermentation of whey permeate to lactic acid could be vastly improved using a membrane recycle bioreactor. Table 9.3 gives a comparison of several types of bioreactors used for the production of lactic acid. The batch productivity values do not take into account the time needed between batches for cleaning, sterilization, etc. It is important that comparison be made considering the feed, microorganism, product concentration and substrate utilization. In this regard the superiority of the membrane recycle fermenter is remarkable. This is a direct result of much higher cell concentrations used in these bioreactors. According to Mehaia and Cheryan (1986), the membrane recycle configuration operated in the CSTR mode has several advantages over immobilized cell and hollow-fiber systems that are operated in the plugflow mode. The control of the improvement, especially the pH, is much easier and substrate-cell contact is much better due to the completely mixed conditions. With such vast improvements in productivity, the additional cost of the membrane module will quickly repay itself.

Table 9.3 Comparison of lactic acid productivity in different bioreactor systems

Bioreactor	Organism	PD	*D*	LA	Reference
Batch	*L. bulgaricus*	3–5	–	44	Mehaia and Cheryan (1986)
Continuous	*L. bulgaricus*	11	0.334	33	Cox and MacBean (1977)
		1.77	0.032	55.3	Keller and Gerhardt (1975)
		5.0	0.17	29.4	Stieber and Gerhardt (1979b)
	L. helveticus	9.7	0.352	27.5	Roy *et al.* (1987)
Dialysis	*L. bulgaricus*	10.7	0.136	78.6	Stieber and Gerhardt (1979a)
Immobilized	*L. casei*	0.5	0.015	33	Tuli *et al.* (1985)
	L. bulgaricus	1.2–1.6	0.083	14.4–19.2	Stenroos *et al.* (1982)
	L. helveticus (milano)	2.6	0.144	18.0	Roy *et al.* (1987)
	L. bulgaricus	17.0	0.55	30.9	Mazali (1992)
		17.0	1.6	10.5	
	L. helveticus (ATCC-10797)				
Hollow fiber	*L. bulgaricus*	2.0	0.166	12	Mehaia and Cheryan (1986)
Membrane	*L. bulgaricus*	25.0	0.89	28	Mehaia and Cheryan (1986)
		85	11.97	43	
Batch	*L. delbrueckii*	0.27–0.6	–	12–24.5	Arasaratnam *et al.* (1996)
Continuous	*L. lactis*, *L. helveticus*,				
	L. bulgaricus	2.4–4.1	0.1–0.4	24–10.2	Lund *et al.* (1992)
Batch	*L. helveticus*	0.3–3.25	–	2.7–30.4	Chiarini *et al.* (1992)
Continuous	*L. helveticus*	10.9–5.8	0.26–0.01	41.9–15.6	Aeschlimann and Stockar (1991)
		22.0	0.88	25	Boyaval *et al.* (1987)
	L. delbrueckii	40.0	0.95	42	Ohleyer *et al.* (1985)
		10.1	0.3	33.6	Major and Bull (1989)
CSTR	*L. casei*	5.5	0.22	25	Krischke *et al.* (1991)
CFBR	*L. casei*	10	0.4	25	Krischke *et al.* (1991)
Fibrous bed	*L. helveticus*	5–6	0.167	30	Silva and Yang (1995)
Continuous	*L. helveticus*	28.5	1.21	23.5	Nortan *et al.* (1994b,c)

PD = productivity ($g\ l^{-1}\ h^{-1}$); *D* = dilution rate (h^{-1}); LA = lactic acid ($g\ l^{-1}$).

9.7.4 Product inhibition in lactic acid fermentation

In lactic acid fermentation, there is an inhibitory effect caused by the product, lactic acid (Friedman and Gaden, 1970; Keller and Gerhardt, 1975; Reddy *et al.*, 1976; Hongo *et al.*, 1986), on the lactic acid productivity. The specific growth rate, μ, varies as a function of product concentration. Different mathematical models have been proposed to describe product inhibition for growth, lactic acid production and substrate consumption (Luedeking and Piret, 1959; Hanson and Tsao, 1972; Keller and Gerhardt, 1975; Tsao and Hanson, 1975; Vick Roy *et al.*, 1982, 1983; Nicolas and Bull, 1985).

Keller and Gerhardt (1975) assumed that the specific growth rate of *L. bulgaricus*, as a function of product concentration, exhibited the following linear relationship:

$$\mu = \mu_{max} (1 - P/P_m) \quad (9.1)$$

where μ is the specific growth rate at a given product concentration, μ_{max} is the maximum specific growth rate (when $P = 0$), P is the product concentration, and P_m is the product concentration at which μ first equals zero.

In the pH range 4.8–5.6, μ is approximately inversely proportional to product concentration. The inhibition of lactic acid production was reported to be non-competitive in nature (Silva and Yang, 1995). Leudeking and Piret's (1956) data on lactic acid inhibition at different values of pH used by Keller and Gerhardt (1975) (equation 9.1) were reassessed according to the following product-inhibition equations proposed by different researchers:

$$\mu = \mu_{max} K_1/(K_1 + P) \quad (9.2)$$

$$\mu = \mu_{max} e^{-K_2 P} \quad (9.3)$$

$$\mu = \mu_{max} [1 - (P/P_m)^n] \quad (9.4)$$

where K_1, K_2 are constants of inhibition.

The data were found to give a best fit using equation 9.4 as shown in Figure 9.1. The values of the constants thus obtained are expressed in Table 9.4. The values of n (which represents the nature of inhibition) at different values of pH are close to unity (except at pH 4.8) which shows a linear relationship as represented by equation 9.1. The values of n at pH 4.8 is 0.73 (<1) which represents a parabolic inhibition curve.

Lactic acid, calcium lactate, sodium lactate and ammonium lactate were added to the medium at different concentrations at the time of inoculation to find their inhibitory effects on the growth of *L. delbrueckii* (Hongo *et al.*, 1986). At a concentration of 0.5% lactic acid, cell growth was 20% compared to that in the absence of lactic acid. At a concentration of 1.0%,

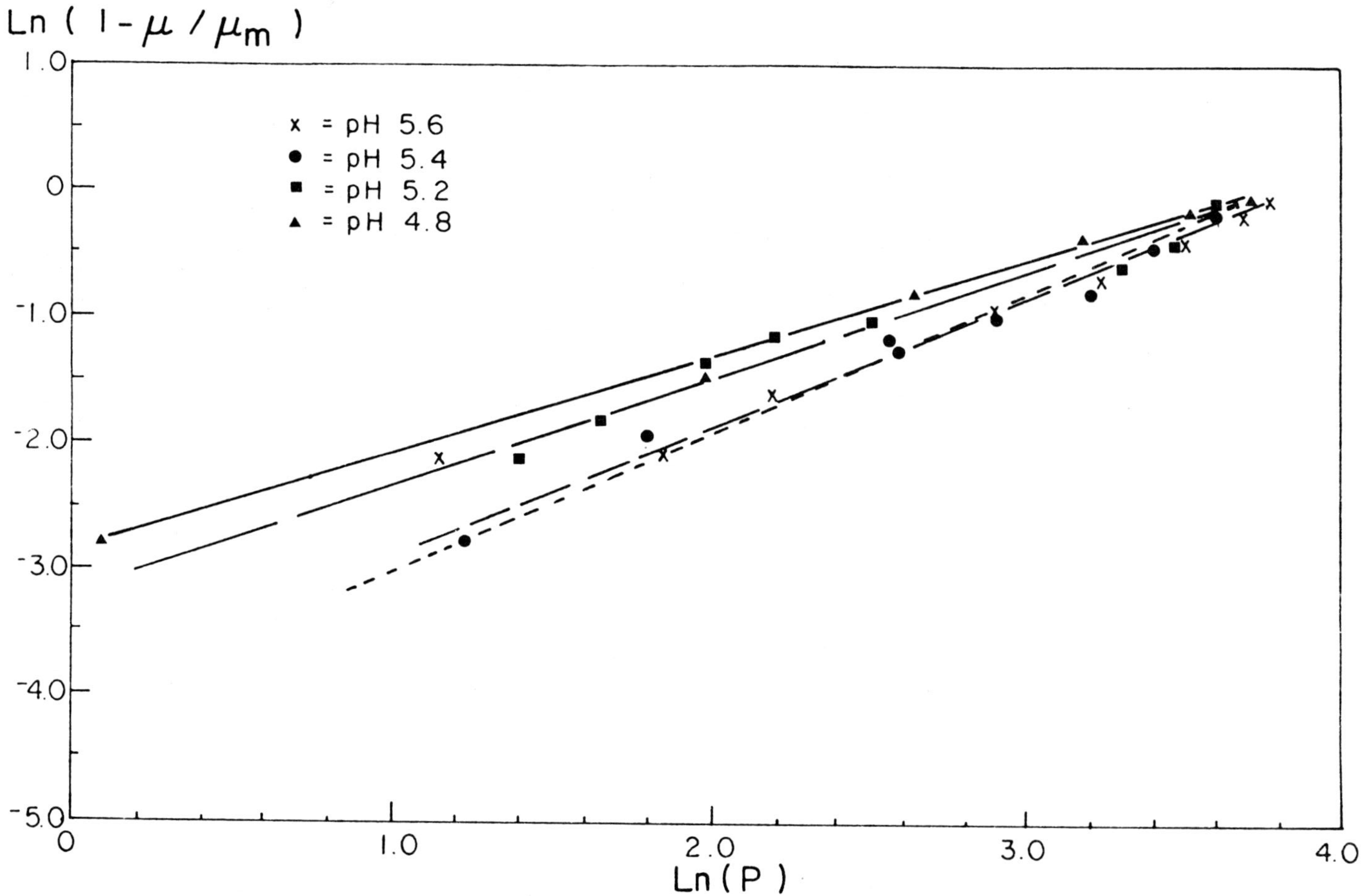

Figure 9.1 Linear plot of the effect of product (lactate) concentration on the specific growth rate in batch fermentation.

Table 9.4 Values of constants

pH	P_m (g l^{-1})	n	r^2
5.6	47.2	0.997	0.991
5.4	46.7	0.97	0.9
5.2	43.8	0.99	0.88
4.8	41.8	0.995	0.995

cell growth ceased entirely. With the addition of 1.0% of ammonium lactate, sodium lactate or calcium lactate, reductions of 50%, 67% and 72% cell growth were observed, respectively. Thus lactic acid and ammonium lactate showed the most inhibitory effect on cell growth, and even neutralized lactates clearly showed inhibitory effects. However, the faster rate of lactic acid production and the best cell growth among pH was controlled with ammonium hydroxide rather than NaOH or $CaCO_3$ (Hongo *et al.*, 1986). This observation is contrary to the inhibition observed by lactic acid additions. This could be due to the fact that NH_4OH not only neutralized the pH but also served as a nitrogen source. At a given pH, greater inhibition of batch lactic acid fermentation was also observed with ammonium lactate than with calcium lactate by Keller and Gerhardt (1975).

The inhibition data of Hongo *et al.* (1986) on ammonium, calcium, sodium lactates and lactic acid give different kinetic patterns. As shown in Figure 9.2, ammonium lactate and calcium lactate followed equation 9.2, whereas sodium lactate and lactic acid followed equations 9.4 and 9.3, respectively. This behavior suggests that the different salts of lactic acid have different inhibition mechanisms. The values of constants for this analysis are given in Table 9.5.

Owing to product inhibition, many of the kinetic advantages of continuous fermentation are not fully realized in lactic acid fermentation. In continuous lactic acid fermentation, as D (dilution rate = working volume/liquid inflow rate) decreases, the product concentration increases and causes μ to decrease. Therefore, further increase in retention time is counteracted by reduction in μ. At pH 5.5, approximately the same time is required to process a given volume in a two-stage continuous fermentation as in batch fermentation (Keller and Gerhardt, 1975). However, the retention time in continuous fermentation as in batch fermentation is reduced by increasing the pH to 5.8. Except at extremely low substrate concentrations, μ is more a function of product than substrate concentration (Keller and Gerhardt, 1975).

It has been demonstrated that the concentration of undissociated lactic acid, rather than the actual pH of the medium, inhibits lactic acid bacteria (Friedman and Gaden, 1970; Keller and Gerhardt, 1975; Reddy *et al.*,

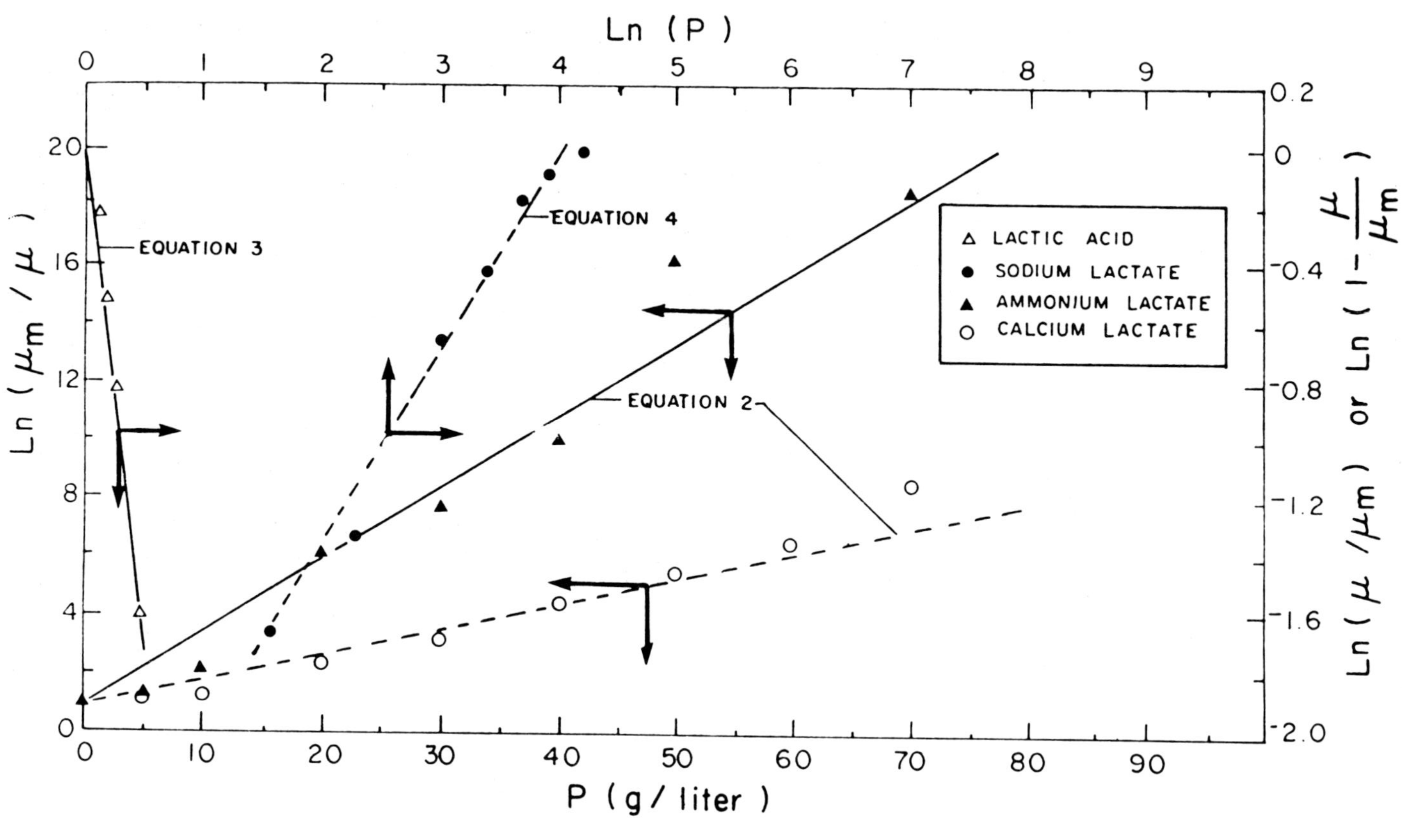

Figure 9.2 Linear plot of the effect of lactic acid and various lactates on cell growth based on data from Hongo *et al*. (1986).

Table 9.5 Inhibition constants for lactates and lactic acid

Inhibitor	Equation number	Constants (g l^{-1})
Calcium lactate	9.2	$K_1 = 10.4$
Ammonium lactate	9.2	$K_2 = 3{,}84$
Lactic acid	.93	$K_2 = 0.43$
Sodium lactate	9.4	$P_m = 57.15$ ($n = 1.51$)

1976). A control of pH 5.8 was necessary to maximize the lactic acid productivity (Mulligan *et al.*, 1991). Below pH 5.8, lactate was in the electroneutral undissociated form (lactic acid) which was more inhibitory to microorganisms (Vinegra-Gonzaliz and Gomez, 1983; Gatje and Gottschalk, 1991). *L. bulgaricus* was grown at pH 5.5 in batch cultures (Reddy *et al.*, 1976) and continuous cultures (Stieber and Gerhardt, 1979a,b). Roy *et al.* (1986) found an optimum pH 5.9 for *L. helveticus* using CSL and 5.5 using YE-supplemented WU. Vahvaselka and Linko (1987) also found an optimum pH of between 5.2 and 5.6 for milk ultrafiltrate fermentation to lactic acid in batch cultures. The effect of pH on the growth and lactic acid production of *L. helveticus* was examined in continuous culture using supplemented WU. The maximum lactate productivity of 5 g l^{-1} h^{-1} highest value of biomass (4.86 g l^{-1}) and lactic acid (31.9 g l^{-1}) occurred at pH 5.5 and a dilution rate 0.15 h^{-1} (Aeschlimann and Stockar, 1989). The optimum pH of 5.5 was obtained for *L. helveticus* (on WP supplemented with YE) with respect to the maximum specific growth rate (0.66 h^{-1}) and lactic acid productivity (7.57 g l^{-1} h^{-1}) (Norton *et al.*, 1994c). It was also shown that *L. helveticus* strain used by Norton *et al.* (1994c) exhibited high activity for pH values as low as 4.7–5.1, thus facilitating the operation under highly inhibitory conditions for potential contaminants.

Models for substrate consumption, cell growth and lactic acid production considering both substrate and product inhibition have been described (Aeschlimann and Stockar, 1989; Goncalves *et al.*, 1991). The kinetics of continuous culture could also be described adequately by the model obtained from the batch tests at low substrate concentration (Goncalves *et al.*, 1991). From the analysis of batch data, an unstructured model of the lactic acid production was also proposed, in which specific growth rate variation and lactic acid concentration with time were described by complemented logistic functions (Amrane and Prigent, 1994a). The reduction in the rate of lactic acid production at very low growth rates were also considered by these authors. Structured models were also reported to describe accurately the fermentation of lactose, glucose and galactose by *Streptococcus cremoris* (Nielsen *et al.*, 1991). An unstructured model was developed for a continuous fermenter with biomass recycling by ultrafiltration and lactic acid extraction by electrodialysis (Raucourt *et al.*, 1989a,b).

Several nitrogen supplementations and various ways of preparing seed cultures are usually compared on the basis of an overall criterion of stimulation of growth and lactic acid production, as described in section 9.7.2. The process of optimization is lengthy, cumbersome and rather uncertain. In order to minimize this problem, a mathematical model was developed to assess the nitrogen supplementation and inoculum preparation for *Lactobacillus* growing on whey and WP (Amrane and Prigent, 1994b).

The inhibitory effects of lactic acid have been alleviated to a certain extent by conducting fermentation in a continuous dialysis process (Stieber *et al.*, 1977; Stieber and Gerhardt, 1979a,b, 1981a,b), in a hollow-fiber fermenter (Vick Roy *et al.*, 1982) and in an electrodialysis system (Hongo *et al.*, 1986) (discussed in section 9.7.5).

9.7.5 *Immobilized cell process*

Increasing the cell mass concentration in the reactor is one way of increasing both the rate and extent of substrate conversion (Clausen and Gaddy, 1984). Recycling (feedback) of cells in a continuous fermentation process is used to retain cells and thereby increase the biomass concentration in the fermenter, allowing increased throughput and decreased retention time. Recycling also reduces sustrate concentration in the fermenter effluent. The cell concentration and reusability of cells in the bioreactor can also be increased by immobilizing cells to inert support. Several such studies have appeared on lactic acid production from whey and whey permeate (Stenroos *et al.*, 1982; Hamilton and Howell, 1983; Tuli *et al.*, 1985). Compared to the conventional free cell system, better results, including greatly improved reactor productivity, separation of cells from products and operation at high dilution rates without cell washout are often obtained from immobilized cell systems.

Two matrices were assessed for their ability to immobilize *L. casei* cells for lactic acid fermentation in whey permeate medium by Tuli *et al.* (1985). Agar at 2% concentration was better than polyacrylamide gel in its effectiveness for entrapping the bacterial cells to carry out batch fermentation for up to three repeat runs. Temperature and pH had no significant influence on the fermentative ability of the immobilized organism. Temperatures of 40–50 °C and pH 4.5–6.0 were optimum when fermentation was carried out under stationary conditions. In batch fermentation, aproximately 90% conversion of the substrate (lactose) was achieved in 48 h using immobilized cell gel cubes of 4 × 2 × 2 mm, containing 400 mg dry bacterial cells per flask and 4.5% (initial) whey lactose content as substrate. However, a further increase in substrate levels tested (more than 4.5%) did not improve the process efficiency. Supplementation of Mg^{2+} (1 mM) and agricultural by-products (mustard oil cake, 6%) to the whey permeate medium further improved lactate production (Tuli *et al.*, 1985).

A fixed-film unit using gelatin-coated packing cross-section was prepared by Compere and Grifith (1975). It was seeded with a kefir culture (*Lactobacilli* and lactose-fermenting yeasts) and tested using cheese whey. This packed-bed column reactor increased the lactic acid production of the whey from 1.4% to 2.1% in a single pass. Lactic acid bacteria have been immobilized in several bioreactor systems: *L. delbrueckii* and *L. bulgaricus* in a hollow-fiber fermenter (Vick Roy *et al.*, 1982; Mehaia and Cheryan, 1987) and dialysis culture (Friedman and Gaden, 1970); *L. caseii* and *L. bulgaricus* in a polyacrylamide gel lattice (Divies and Siess, 1976; Ohmiya *et al.*, 1977). Biochemical fuel cell systems based on immobilized lactic acid bacteria have also been developed (Matsunaga *et al.*, 1978; Kovach and Meyerhoff, 1982). *L. delbrueckii*, *L. bulgaricus* or *L. helveticus* was used in an immobilized form for conversion of whole cheese whey or synthetic media to lactic acid by several workers (Stenroos *et al.*, 1982; Vick Roy *et al.* 1982, 1983; Boyaval *et al.*, 1985; Champagne and Boyal, 1986; Boyaval and Goulet, 1988). Morimura *et al.* (1986) used carrageenan gel to immobilize adapted sludge for organic acid production.

Microorganisms were immobilized on porous cellulose acetate particles which have a higher mechanical strength compared with conventional carriers such as polyacrylamides, carrageenan, and polyvinyl alcohol by Nagai *et al.* (1986). *Sporolactobacillus inulins* TUA 343L was cultured in a medium containing glucose, yeast extract, polypeptone, $MgSO_4$, $MnSO_4$, $FeSO_4$, NaCl and $CaCO_3$ at 37 °C for 24 h, and to this were added sterilized porous cellulose acetate particles (particle size 5.0 mm). The culture was kept at 40 mm Hg for 2 h for immobilization. The particles were collected, washed and added to a medium containing glucose, yeast extract, $CaCO_3$, $MgSO_4.7H_2O$, $MnSO_4.4H_2O$, $FeSO_4.7H_2O$ and NaCl, which was incubated at 37 °C for 48 h for lactic acid production (Nagai *et al.*, 1986). The lactic acid concentration reached a value of 35 g l^{-1}.

Mahaia and Cheryan (1986) studied the membrane recycle bioreactor (immobilization without carrier) for the continuous production of lactic acid from cheese whey by *L. bulgaricus*. At a cell concentration of 10 g l^{-1}, optimum productivity of lactic acid was 35 g l^{-1} h^{-1}. Increasing the cell concentration to 30 g l^{-1} enabled the use of dilution rate of 1.0 h^{-1} with complete substrate utilization. At 60 g l^{-1}, productivity was over 80 g l^{-1} h^{-1} with complete substrate utilization, this is vastly superior to conventional batch and continuous fermentations.

An electrodialysis fermentation method in which lactic acid is continuously removed from the fermentation broth resulted in a continuous fermentation with a productivity threefold higher than in non-pH-controlled culture. With this method, the amount of lactic acid produced (pure) was 82.2 g l^{-1}, approximately 5.5-fold greater than that produced in non-pH-controlled fermentation. However, the fouling of anion exchange membranes by cells was observed (Hongo *et al.*, 1986). The products obtained from these processes have commercial potential, including use as a nitrogenous feed supplement for ruminants (Marriott, 1985).

A fermentation apparatus for the continuous production lactic acid from whey has been described by Prigent (1983). The apparatus contains a fermenter equipped with a mechanical stirrer, a source of NH_3 or soda to a pH regulator, as well as a means for continuous delivery of whey and supplements to the fermenter. A pump inserted between the fermenter and an ultrafiltration apparatus allows for recycling the liquor, as well processing the filtrate by electrodialysis for lactate recovery. The apparatus also allows efficient separation of lactose from lactic acid. Empirical exponential equations correlating the volume of NH_4OH solution added for pH control of the medium with the number of base additions to a batch lactic acid fermentation has been described by Borzani and Baralle (1983).

L. helveticus cells entrapped in calcium alginate were characterized by higher fermentation rates and optimum pH than free cells (Boyaval and Goulet, 1988). After a heat treatment, cell activity was higher for free cells than for immobilized cells. The volumetric productivity in a continuous packed-bed reactor was 8 g l^{-1} h^{-1} if the total volume was considered and 30.8 g l^{-1} h^{-1} if the free volume only was considered. The lactose conversion was 50%. Cell leakage was also observed by these authors. Treatment of calcium alginate beads with polyethyleneimine to increase the stability of the gel did not reduce the cell leakage and caused severe shrinkage of the beads. Incubation in glutaraldehyde has been demonstrated to kill all the beads (Boyaval *et al.*, 1985). Fewer than 1% of free cells were observed in the bioreactor. After a week's operation, a plugging in the packed bed occurred which the authors attributed to decalcification of the calcium alginate by the lactic acid produced in the bioreactor. Stenroos *et al.* (1982) proposed introducing fresh medium at the top of the bioreactor as the best feeding mode to avoid plugging. This effect was very pronounced at low dilution rates and caused higher leakage of cells. Based on volumetric productivity (free vs. immobilized cells), it was concluded that the calcium alginate bead entrapment method is not suitable for industrial purposes (Boyaval *et al.*, 1985; Boyaval and Goulet, 1988). Higher lactic acid productivity was obtained with multistage packed-bed columns (Roy *et al.*, 1987) than with a single-stage system. This is due to the fact that the maximum pH gradient is lower than in a single stage.

The immobilization of *Lactobacillus bulgaricus* on polyacrylamide and alginate beads was investigated (Lee, 1981). The most active immobilized cells were obtained by entrapment in calcium alginate beads which is in contrast to the observations of Boyaval and Goulet (1988) and Roy *et al.*

(1987). These immobilized microbial cells, when introduced into 4.5% lactose solutions and whey, showed an improved conversion to lactic acid by 28% and 18%, respectively compared to free cell cultures.

Mzali (1992) studied the production of lactic acid in immobilized system utilizing different strains of *Lactobacillus*. The productivity of lactic acid in comparison to that of other workers is shown in Table 9.3. Lactic acid productivity of 15–17 g l^{-1} h^{-1} was obtained with a lactic acid concentration of 15–30 g l^{-1}. This productivity was obtained at residence times varying between 1.8 h and 36 min.

A continuous immobilized cell process was studied by Norton *et al.* (1994a,b,c) with the objective of producing lactic acid from whey permeate with YE supplementation (0–10 g l^{-1} concentration). A second-stage free cell bioreactor was used in series to the immobilized cell stage in order to decrease the residual sugar concentration. The first stage contained *L. helveticus* cells entrapped in κ-carragenan and locust bean gum gel beads. The second stage acted as a free cell system continuously inoculated with the cells released from first stage. For the two-stage system, a total productivity of 13.5 g l^{-1} h^{-1} was reported at 1 g l^{-1} residual sugar concentration. An uncoupling effect between lactic acid and biomass production was also reported at low YE (0–1 g l^{-1}) supplementation (Norton *et al.*, 1994a).

The effects of pH on the kinetic parameters of the first-stage immobilized cell bioreactor using *L. helveticus* at various dilution rates were discussed by Norton *et al.* (1994b,c). Both free and immobilized cell continuous fermentations up to 1 g l^{-1} YE appear to have a very important effect on productivity, but further supplementation may result in a lower ratio of substrate cost to benefits. A maximum rate of lactic acid production 28.5 g l^{-1} h^{-1} was measured at pH 5.5 and dilution rate 1.2 h^{-1} for the immobilized cell bioreactor. Entrapped cells were responsible for 75–85% of the lactic acid and biomass productions within the immobilized cell fermenter at pH 4.7–6.3 and close to 90% at pH 4.3. The overall specific growth rate and the specific lactic acid production rate of entrapped cells were 12–18% those of free cells in the reactor. The low activity of the immobilized cells was attributed to the presence of high inhibition in the central part of the beads (used for immobilization of cells) as a result of substrate, product and pH gradients (Norton *et al.*, 1994c). Yabamavar and Wang (1991) also showed that the substrate transport in the immobilized cell matrix was not likely to be a limiting factor, but inhibitory product buildup and pH decrease were significant. The kinetics of immobilized cells to produce lactic acid has also been described by Audet *et al.* (1988) and Arnaud *et al.* (1992).

A CSTRs and continuous fluidized bed reactor (CFBR) were compared by Krischke *et al.* (1991) using *L. casei* and YE-supplemented WP. In the CFBR, porous sintered glass beads were used for immobilization and a maximum biomass concentration of 105 g kg^{-1} of support was observed. The specific productivity in the CFBR was lower (0.093 g g^{-1} h^{-1}) compared to CSTR (0.6 g g^{-1} h^{-1}) and was attributed to the substrate limitation caused by diffusion of the substrate. However, continuous production of lactic acid with immobilized cells in CFBR enables higher productivities compared to a conventional CSTR. In the CSTR a volumetric productivity of 5.5 g l^{-1} h^{-1} at 100% substrate conversion (D = 0.22 h^{-1}) was determined. In the CFBR a productivity of 10 g l^{-1} h^{-1} was obtained at a dilution rate of 0.4 h^{-1} (substrate conversion 93%) (Krischke *et al.*, 1991). A fast method for the determination of free and immobilized biomass in solid-rich media, based on the measurement of cellular ATP, was developed. Immobilized cells showed the best product formation in the same pH range (6–6.5) as free cells (Krischke *et al.*, 1991).

Immobilized cell systems often suffer from poor long-term stability owing to reactor-bed clogging, membrane fouling, cell degeneration or contamination. These problems have prevented their wide industrial application. A fibrous-bed, immobilized cell bioreactor was thought to circumvent these problems (Silva and Yang, 1995). In a fibrous-bed reactor during continuous lactic acid production from unsupplemented acid whey (3.7% w/v, lactose; 0.8% w/v, lactic acid) employing *L. helveticus* at pH 5.5 and 42 °C the volumetric productivity ranged from 2.6 to 7.0 g l^{-1} h^{-1} which was ten times higher than those obtained in batch fermentations with free cells (Silva and Yang, 1995). The lactate yield was >95% w/w. The reactor performance was stable for continuous, long-term operation for both sterile and non-sterile unsupplemented acid whey feeds for a 6 month period. Concentrated acid whey was also found to be an appropriate feedstock to this bioreactor. Specific productivity of free cells was 0.46 g g^{-1} h^{-1} and that of immobilized cells approximately 0.1 g g^{-1} h^{-1}, i.e. 25% of the suspended culture. The low activity was attributed to diffusion limitations, channelling, incomplete mixing (which allows pockets of poor mixing) and the probable presence of dead cells through long-term reactor use. One of the reactors operated had a cell density of 63 g l^{-1} of which 12% was in the bulk fluid, 58% weakly bound to the matrix and remaining 30% was more strongly attached to the matrix. The weak attachment of cells to the fibrous matrix allowed continuus cell regeneration in the immobilized cell reactor, thus preventing the reactor from degenerating owing to cell aging and death. The superior long-term stability, as compared to other immobilized cells systems, was thus attributed to the high viable cell density maintained in this type of reactor (Silva and Yang, 1995).

A packed-bed biofilm reactor was studied for lactic acid fermentation from various media (Denirci *et al.*, 1993) using pure and mixed culture. The support matrix was composed of blended polypropylene and various agricultural material (oat hulls, soya meals). Highest productivities of

30 g l^{-1} h^{-1} (based on liquid volume) and 15 g l^{-1} h^{-1} based on a reactor volume at 5 g l^{-1} lactic acid concentration was reported. A maximum concentration of 18 g l^{-1} lactic acid was achieved with a productivity of 4.4 g l^{-1} h^{-1}, which was comparable to that achieved in fibrous bed reactor by Silva and Yang (1995).

9.8 Acetic acid and propionic acid

The annual demand of propionic and acetic acids is approximately 1.3×10^9 kg (Tyagi, 1986). Acetic acid is used extensively in the manufacture of cellulose acetate, vinyl acetate, acetic esters, etc. Propionic acid is used in a variety of industrial processes, such as cellulose propionate, an important thermoplastic, herbicides, solvents, fruit flavor and esters which are used in the perfume industry. As a preservative, propionic acid extends the half-life of food products by inhibiting molds and some bacteria. Most of the propionic acid used by the food industry is produced by chemical synthesis. If higher yields of propionic acid could be obtained, production by fermentation might become economically competetive.

Acetic and propionic acids may be produced biologically by the fermentation of sugars using a species of *Propionibacterium* (Lee *et al.*, 1974). *Propionibacterium acidipropionici*, produces high concentrations of propionic and acetic acids at pH values of 4.1–4.9 (the natural pH of acid whey) by utilizing sugars. Approximately 2 moles of propionic acid and 1 mole of acetic acid are produced per 1.5 mole of glucose or galactose (both present in whey). The factors which limit the microbial production of propionic acid are: a very low productivity or very high fermentation time; end-product inhibition; a low concentration of propionic acid in the fermented broth; and a high separation cost due to low concentration, low volatility and the presence of acetic acid. For a conventional fermentation, at least 3–7 days in fermentation-rich medium are required. The concentration of propionic acid seldom exceeds 3% partially due to strong end-products inhibition. Hsu and Yang (1991) showed that, even if neutral pH is optimum for the growth of *P. acidi-propionic*, the propionic acid yield is low. On the other hand, in the acidic pH range, the growth rate is low, but the yield is high.

There have been many attempts to increase propionic acid yield and production rate through the development of new strains (Rehberger and Glatz, 1990; Woksow and Glatz, 1991). A propionic acid-tolerant derivative of *Propionibacterium acidipropionici* P9 was obtained by serial transfer. Growth rate, sugar utilization and acid production were monitored during batch and semicontinuous fermentations. The highest propionic acid concentration (47 g l^{-1}) was produced by the adapted strain in a semicontinuous process (Woksow and Glatz, 1991). The adapted strain produced a higher ratio of propionic acid to acetic acid, utilized sugar more efficiently and produced more propionic acid per gram of biomass.

Several studies on propionic acid fermentation have been performed using whey or whey permeate as substrate (Bodie *et al.*, 1983, 1987; Ahern *et al.*, 1985; Boyaval and Corre, 1987; Schuppert *et al.*, 1992; Lewis and Yang, 1992a,b; Colomban *et al.*, 1993; Yang *et al.*, 1994; Haddadin *et al.*, 1996). *Propionibacterium acidipropionici* is a slow-growing microorganism, requiring fermentation times of 12–14 days to achieve a 56% conversion (Prescott and Dunn, 1949; Clausen *et al.*, 1982). A batch fermentation of 8 days resulted in a conversion of only 30%. In mixed culture, the time required is about 3 days.

Whey fermentation to propionic acid in a two-step process was studied by several workers, firstly with lactic-acid-producing bacteria and, secondly, with propionic-acid-producing bacteria. Fresh pasteurized whey (6.5% solids) was inoculated with *Lactobacillus bulgaricus* and *Streptococcus thermophilus* to carry out a lactic acid fermentation at pH 4.3–6.0. To obtain propionic acid, the pH of the broth was adjusted to pH 7.0, sterilized and inoculated with *Propionibacterium shermanni*. Final yields of propionic acid were 1.6% (by weight) (Ahern *et al.*, 1985). Fermentation of supplemented whey with YE using *L. helveticus* followed by *P. acidopropionici* (free cells) for 2.5 days at 32 °C gave a broth of 5.2 g l^{-1} propionic acid and 2.4 g l^{-1} acetic acid, while immobilized cells gave a broth with 11.0 g l^{-1} propionic acid and 3.2 g l^{-1} acetic acid over 4 days (Haddadin *et al.*, 1996). Mixed-culture studies for the production of propionic acid to preserve the bakery products have also been carried by Bodie *et al.* (1983) in three stages. *P. shermanii* and *L. casei* grown together through three stages in the draw and fill mode produced about 4.5% propionic acid in 70 h and all the lactose was consumed.

Whey fermented with *P. acidipropionici* resulted in higher propionic acid yields and, consequently, greater mycostatic activity than those produced using *P. shermanii* (Anderson, 1982). A total of 25 000 gallons of sweet whey containing 7% solids and 0.5% YE was sterilized and inoculated with 2500 gallons of *P. acidipropionici* preculture and incubated for 66 h at 35 °C and pH 7.0, maintained by the periodic addition of NaOH. The fermented broth containing 0.96% propionic acid and 0.2% acetic acid (compared with 0.8 and 0.3%, respectively for *P. shermanii* fermentation) was spray-dried and packaged for use as a mycostatic agent in the manufacture of pastry, bread and other bakery products (Anderson, 1982).

There have been many attempts to improve yield and productivity of propionic acid through process development (Anderson *et al.*, 1986; Bodie *et al.*, 1987; Emde and Schink, 19990; Lewis and Yang, 1992a,b; Schuppert *et al.*, 1992). To overcome the inhibitory effects of propionic acid, a continuous process combined with cell recycling has been used (Nanba *et al.*, 1983; Colomban *et al.*, 1993). A continuous fermentation with cell recycling using ultrafiltration gave promising results with respect to productivity and product concentration (Boyaval and Corre, 1987).

Continuous fermentation of propionic acid has been studied with productivities up to 2 g l^{-1} h^{-1} (Cavin *et al.*, 1985; Boyaval and Corre, 1987; Carronodo *et al.*, 1988). A semicontinuous process has been used to improve the yield of propionic acid (Woksow and Glatz, 1991). Growth of *P. acidipropionici* was studied in YE-supplemented acid whey permeate in a three-electrode poised-potential system with cobalt sepulchrate as an artificial electron donor (Schuppert *et al.*, 1992). In this process 6.5 g l^{-1} propionic acid was accumulated with zero acetic acid concentration in the broth. Membrane bioreactors from the laboratory scale to pilot and industrial production plant using batch and continuous mode on YE-supplemented WP were studied (Colombon *et al.*, 1993). A propionic acid concentration of 30–40 g l^{-1}, a specific productivity of 0.035 h^{-1} with a productivity of 1.6 g l^{-1} h^{-1} for total acids and 1.2 g l^{-1} h^{-1} for propionic acid was achieved with no residual lactose.

The use of a fibrous-bed reactor is of great interest (Lewis and Yang, 1992a,b; Yang *et al.*, 1994). A cell concentration up to 50 g l^{-1}, 2% (w/v) propionic acid concentration from 4.2% lactose at a retention time of 34–45 h, a propionic acid yield of 46%, a ten times increase in propionic acid productivity and a constant operation for 6 months without contamination was reported. The diffusion limitation has been expected to be less severe for *P. acidopropionici* in such a reactor. The feasibility of using an extractive fermentation process for propionic acid production from lactose in a fibrous-bed reactor with increased yield and purity of product has been demonstrated (Lewis and Yang, 1992a,b).

Many new technologies have emerged and increased the volumetric productivity of propionic acid. The successful immobilization of *Propionibacterium* cells in the search for a better productivity and product concentration compared to the conventional free cell system has been reported (Iordan *et al.*, 1980; Cavin *et al.*, 1985; Champagne *et al.*, 1989; Jain *et al.*, 1991; Lewis and Yang, 1992a,b; Haddadin *et al.*, 1996).

The production of propionic acid by *P. shermanii* was studied in an immobilized batch reactor (Jain *et al.*, 1991). Cells were immobilized on an inert support. *P. shermanii* was cultivated followed by centrifugation and resuspension of cells in 1/10 volume of supernatant and immobilization. The carrier with the immobilized cells was transferred to a bioreactor of 750 ml capacity and 500 ml working volume to which whey permeate, fortified with 1% YE, was added. After the fermentation was completed, the medium was drained and the reactor was refilled with the fresh permeate and fermentation was continued.

A maximum concentration of 12.0 g l^{-1} propionic acid was obtained in an immobilized system against 2.25 g l^{-1} propionic acid in a free system in the same time. A higher concentration of acid was obtained when $CaCO_3$ was added to the fermentation medium to control pH (Table 9.6). Additions of $CaCO_3$ to the fermentation medium not only gave a higher final concentration of propionic acid (seven times that obtained in the free cell system), but also increased the productivity and the final yield of the propionic acid per unit weight of substrate consumed. In immobilized cell systems, the yield was also higher owing to the low growth of cells and, therefore, increasing amounts of energy being diverted to the maintenance of cell function, resulting in a higher yield of product. It was also noticed that, if pH was not controlled, yield did not increase even in the immobilized system (in fact it was diminished slightly). This suggests that the addition of $CaCO_3$ has a profound effect on the yield of product.

The highest yields of acetic acid in respect to lactose consumption were obtained in a free cell system or in an immobilized cell system without pH control (Table 9.6). The addition of $CaCO_3$ resulted in decreased yields of acetic acid. This suggests that the addition of $CaCO_3$ to the fermentation medium distorted the metabolism of *P. shermanii* by reducing the yields of acetic acid (0.5 times) while increasing the yield of propionic acid (2.25 times).

Propionic acid appeared in the medium as a result of fermentation which inhibited the growth and acid production activity of *P. shermanii*. A linear correlation was found between acid concentration and specific rate of acid production (v), which can be repeated by the following equation (Figure 9.3).

$$v = v_m \,(1 - P/P_m) \tag{9.5}$$

where v_m is the specific acid production at P = O. This equation is similar to equation 9.4 when μ is replaced by v and $n = 1$. This equation can be used as a design equation.

Table 9.6 Comparison of the free and immobilized cell systems

System	$CaCO_3$	pH change	PA (g l^{-1})	AA (g l^{-1})	Yield (moles mol^{-1} lactose)	
					PA	AA
Free	No	0.85	2.0	2.25	1.100	1.900
Immobilized	No	0.50	4.0	3.00	0.906	1.280
Immobilized	Yes	0.53	11.6	4.00	1.710	0.733
Immobilized	Yes	0.51	12.0	3.20	2.480	0.970

PA = propionic acid; AA = acetic acid.

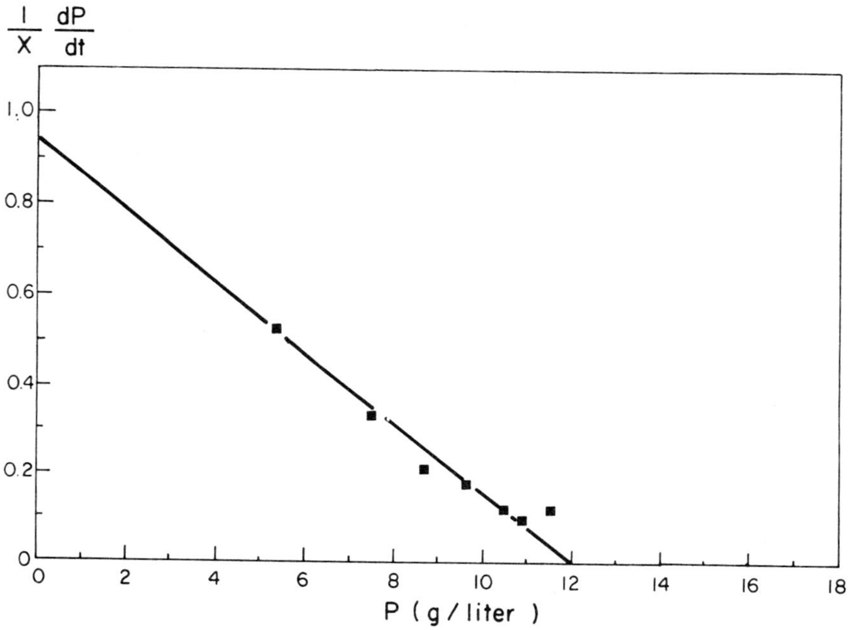

Figure 9.3 Linear plot of the effect of propionic acid on the product formation rate.

9.9 Conclusions

Cheese whey is available in large quantities and is presently not fully utilized for useful innovative products. This dairy by-product is highly nutritious and various possibilities exist for its total utilization as a fermentation substrate. The production of organic acids from whey or permeate has a good potential for several industrial uses.

Essential information on the feasibility of various microbial species for organic acid production has been reviewed. The batch and continuous organic acid fermentation processes are well documented and a state-of-art review of these methods has been provided. The rate of fermentation and microbial growth is inhibited by the product (organic acid) formed. Based on the inhibitory effects, bioreactor design equations have been presented and discussed. Various information on immobilized cell supports and immobilization methods with respect to organic acid production have been reviewed and examined with special reference to the product concentration and productivity. A decrease in pH in a packed-bed reactor (as a result of organic acid production) decreases the product fermentation rate and hence demands high residence time. Further developments are required in this area.

Acknowledgement

This work is dedicated to memory of our late colleague, Professor Denis Couillard. This research was made possible through a grant from The Natural Sciences and Engineering Research Council of Canada (grant A4984).

References

Aeschlimann, A. and Stockar, U. Von (1989) *Biotechnol. Lett.*, **11**, 195.
Aeschlimann, A. and Stockar, U. Von (1990) *Appl. Microbiol. Biotechnol.*, **32**, 398.
Aeschlimann, A. and Stockar, U. Von (1991) *Enzyme Microbiol. Technol.*, **13**, 811.
Aeschlimann, A., Stasi, L. Di and Stockar, U. Von (1990) *Enzyme Microbiol. Technol.*, **12**, 926.
Ahern, W.P., Skogerson, L.E. and Andrist, D.F. (1985) *Fermentation of Whey*, European patent application, EP 160,417, November 1985.
Amrane, A. and Prigent, Y. (1993) *Biotechnol. Lett.*, **15**, 239.
Amrane, A. and Prigent, Y. (1994a) *J. Chem. Technol. Biotechnol.*, **60**, 241.
Amrane, A. and Prigent, Y. (1994b) *Appl. Microbiol. Biotechnol.*, **40**, 644.
Anderson, T.M. (1982) Mycostatic whey, European patent application, EP 95,268 (cl. A2 3C21/02), November 1983, US application 379,841, May 1982.
Anderson, T.M., Bodie, E.A., Goodman, N. and Schwartz, (1986) *Appl. Environ. Microbiol.*, **51**, 427.
Arasaratnam, V., Senthuran, A. and Balasubramanian, K. (1996) *Enzyme Microbiol. Technol.*, **19**, 482.
Arnaud, J.P., Lacroix, C. and Choplin, L. (1992) *Biotechnol. Tech.*, **6**, 265.
Audet, P. Pacquin, C. and Lacroix, C. (1988) *Appl. Microbiol. Biotechnol.*, **29**, 11.
Audet, P., Lacroix, C. and Paquin, C. (1992) *Int. Dairy J.*, **2**, 1.
Bergey's Manual of Determinative Bacteriology 8th edn, (1974) Williams and Wilkins, Baltimore.
Bodie, E.A., Schwartz, R.D. and Anderson (1983) Fermentation of whey for bakery product preservation, European patent application EP141,642, May 1985, US application 548,170, November 1983.
Bodie, E.A., Anderson, T.M., Goodman, N. and Schwarz, R.D. (1987) *Appl. Microbiol. Biotechnol.*, **25**, 434.
Borzani, W. and Baralle, S.B. (1983) *Agric. Biol. Technol.*, **26** (4), 475.
Boyaval, P. and Corre, C. (1987) *Biotechnol. Lett.*, **11**, 801.
Boyaval, P. and Goulet, J. (1988) *Enzyme Microbiol. Technol.*, **10**, 725.
Boyaval, P., Lebrun, A and Goulet, J. (1985) *Le Lait*, **65**, 185.
Boyaval, P., Corre, C. and Terre, S. (1987) *Biotechnol. Lett.*, **9**, 207.
Boyaval, P., Corre, C. and Terre, S. (1988) *Le Lait*, **68**, 65.
Buchta, K. (1983) Lactic acid, in *Biotechnology* (ed. H.J. Rehm and G. Reed), Biomass, Microorganisms, Energy from Renewable Resources, Vol. 3, Verlag Chemie, Weinheim, p. 409.
Campbell, L.A. (1953) *Can. Dairy Ice Cream J.*, **32**, 29, 77.
Campbell, M.E. and Glenn, W.M (1982) *Profit from Pollution Prevention – A Guide to Industrial Waste Reduction of Recycling*. Pollution Probe Foundation Report, Toronto, Ontario.
Carronodo, M.J.T., Crespo. J.P.S.G. and Moura, M.J. (1988) *Appl. Biotechnol.*, **17**, 295.
Cavin, J.F., Saint, C. and Divies, C. (1985) *Biotechnol. Lett.*, **7**, 821.
Champagne, C.P. and Boyaval, P. (1986) *Techn. Lait Marketing*, **1015**, 26.
Champagne, C.P., Baillargeon-Cote, C. and Goulet, J. (1989) *J. Appl. Bacteriol.*, **66**, 175.
Cheng, P., Muller, S., Bajpai, R. and Lannotti, E.J. (1991) J. Ind. Microbiol., **7**, 27.
Chiarini, L., Mara, L. and Tabacchioni, S. (1992) *Appl. Microbiol. Biotechnol.*, **36**, 461.

Clausen, E.C. and Gaddy, J.L. (1984) *Chem. Eng. Progress*, **80**, 59.
Clausen, E.C., Shah, R.B., Najafpour, G. and Gaddy, J.L. (1982) *Biotechnology and Bioengineering Symposium Series*, no. 12 (ed. C.D. Scott), Wiley, New York, pp. 67–72.
Colomban, A., Roger, L. and Boyaval, P. (1993) *Biotechnol. Bioeng.*, **42**, 1091.
Compere, A.L. and Griffith, W.L. (1975) *Dev. Ind. Microbiol.*, **17**, 247.
Cox, G.C. and MacBean, R.D. (1977) *Aust. J. Dairy Technol.*, **32**, 19.
Coulman, R.A., Stieber, R.W. and Gerhardt, P. (1977) *Appl. Env. Microbiol.*, **34**, 725.
Dairy Market Review (1982) Marketing and Economics Branch, Agriculture Canada, Ottawa, August.
Denirci, A., Pometto, A.L., III and Johnson, K.E. (1993) *Appl. Environ. Microbiol.*, **59**, 203.
Divies, C. and Siess, M.H. (1976) *Ann. Microbiol., Inst. Pasteur*, **127B**, 525.
Emde, R. and Schink, B. (1990) *Appl. Env. Microbiol.*, **56**, 2771.
Friedman, M.R. and Gaden, E.L., Jr (1970) *Biotechnol. Bioeng.*, **12**, 961.
Galpin, D.B. (1981) *N. Zeal. J. Dairy Sci. Technol.*, **16**, 289.
Garoutte, C., Lim, J., Amundson, C.H. and Breslau, B. (1983) *Process Biochem.*, **18**, 2.
Gatje, W. and Gottschalk, G. (1991) *Appl. Microbiol. Biotechnol.*, **34**, 446.
Gerhardt, P. and Reddy, C.A. (1978) *Dev. Ind. Microbiol.*, **19**, 71.
Goncalves, L.M.D., Xavier, A.M.R.B., Almeida, J.S. and Corrondo, M.J.T. (1991) *Enzyme Microbiol. Technol.*, **13**, 314.
Goursaud, J. (1986) *Ind. Aliment. Agric.*, **103**, 349.
Groves, F. (1972) *Proceedings of the Whey Products Conference*, Agriculture Research Service, US Department of Agriculture, ERRL publication no. 3779, p. 5.
Haddadin, M.D., Al Muhirat, S.R., Batayneh, N. and Robinson, R.K. (1996) *J. Soc. Dairy Technol.*, **49**, 79.
Hamilton, K.M. and Howell, J.A. (1983) *Advances in Fermentation 83, Proceedings of a Conference on Wheatland*, Rickmansworth, Chelsea College, University of London, September, p. 171.
Hanson, T.P. and Tsao, G.T. (1972) *Biotechnol. Bioeng.*, **14**, 233.
Hongo, M., Nomura, Y. and Iwahara, M. (1986) *Appl. Environ. Microbiol.*, **52**, 314.
Hsu, S.T. and Yang, S.T. (1991) *Biotechnol. Bioeng.*, **38**, 571.
Iordan, E.P., Ikonnikov, N.P., Kovrizhnykh, V.A. and Vorob'eva, L.I. (1980) *Appl. Biochem. Microbiol.*, **40**, 465.
Jain, D.K., Tyagi, R.D., Kluepfel, D. and Agbebavi, J.T. (1991) *Process Biochem.*, **26**, 217.
Jelen, P. (1979) *J. Agric. Food Chem.*, **27**, 658.
Kandler, O. (1982) *Forum Mikrobiol.*, **5**, 16.
Keller, A.K. and Gerhardt, P. (1975) Biotechnol. Bioeng., **17**, 997.
Kennedy, P.K. (1985) *Cult. Dairy Prod. J.*, **20**, 13.
Kosaric, N. and Asher, Y.J. (1985) *Adv. Biochem. Eng. Biotechnol.*, **32**, 25.
Kosaric, N. and Wieczorek, A. (1984) *Dev. Food Sci.*, **9**, 229.
Kosikowski, F.V. (1976) *Cheese and Fermented Milk Foods*, Edward Brothers, Ann Arbor, MI.
Kovach, M.P. and Meyerhoff, M.E. (1982) *Anal. Chem.*, **54**, 217.
Krischke, W., Schroder, M. and Trosch, W. (1991) *Appl. Microbiol. Biotechnol.*, **34**, 573.
Lacroix, C., Paquin, C. and Arnaud, J.P. (1990) *Appl. Microbiol. Biotechnol.*, **32**, 403.
Lee, K.H. (1981) *Hanguk Nonghwa Hakhoe Chi (Korean)*, **24**, 149.
Lee, I.H., Fredrickson, A.G. and Tsuchiya, M.H. (1974) *Appl. Microbiol.*, **28**, 831.
Leudeking, R. and Piret, E.L. (1956) *J. Biochem. Microbiol. Technol. Eng.*, **1**, 513.
Levenspiel, O. (1962) *Chemical Reaction Engineering*, Wiley, New York.
Lewis, V.P. and Yang, S.T. (1992a) *Biotechnol. Progress*, **8**, 104.
Lewis, V.P. and Yang, S.T. (1992b) *Biotechnol. Bioeng.*, **40**, 465.
Lund, B., Norddahl, B. and Ahring, B. (1992) *Biotechnol. Lett.*, **14**, 851.
Major, N.C. and Bull, A.T. (1985) *Biotechnol. Lett.*, **7**, 401.
Major, N.C. and Bull, A.T. (1989) *Biotechnol. Bioeng.*, **34**, 592.
Marriott, T.A. (1985) *J. Soc. Dairy Technol.*, **38**, 109.
Marshall, K.R. and Earle, R.L. (1975) *N. Zeal. J. Dairy Sci. Technol.*, **10**, 123.
Matsunaga, T., Karube, I. and Suzuki, S. (1978) *Anal. Chem. Acta*, **98**, 25.
Mehaia, M.A. and Cheryan, M. (1986) *Enzyme Microbiol. Technol.*, **8**, 289.
Mehaia, M.A. and Cheryan, M. (1987) *Appl. Biochem. Biotechnol.*, **14**, 21.

Mistry, V.V., Kosikowski, F.V. and Bellamy, W.D. (1987) *J. Dairy Sci.*, **70**, 2220.
Modler, H.W. (1982) *Cult. Dairy Prod. J.*, **17**, 11.
Moebus, O. and Teuber, M. (1986) *Kiel. Milchwirtsch. Forschungsber.*, **38**, 119.
Morimura, S., Kishimoto, M. and Kida, K. (1986) Organic acid production by immobilized microbe. Japanese patent 61 56,087, March 1986.
Muller, P.G. (1979) *Economic Evaluation of Feeding Liquid Whey to Livestock*, Technical report, Food Research Institute, Research Branch, Agriculture Canada, Ottawa.
Mulligan, C.N., Safi, B.F. and Groleau, D. (1991) *Biotechnol. Bioeng.*, **38**, 1173.
Mzali, J. (1992) M.Sc. thesis. I.N.R.S.-Eau, University of Quebec.
Nagai, S., Ozaki, M., Fukunishi, K. and Yamazaki, K. (1986) Immobilized microorganisms for production of lactic acid and other substances, Japanese patent no. 61 58,588, March 1986.
Nanba, A., Nukada, R. and Nagai, S. (1983) *J. Ferment. Technol.*, **61**, 551.
Nickerson, T.K. (1974) In *Fundamentals of Dairy Chemistry*, 2nd edn (ed. B.H. Webb, A.M. Johnson and J.A. Alford), AVI, Westport, CT, p. 273.
Nicolas, C.M. and Bull, A.N. (1985) *Biotechnol. Lett.*, **7**, 401.
Nielson, J., Nikolajsen, K. and Villadsen, J. (1991) *Biotechnol. Bioeng.*, **38**, 1.
Norton, S., Lacroix, C. and Vuillemard, J.C. (1994a) *J. Dairy Sci.*, **77**, 2949.
Norton, S., Lacroix, C. and Vuillemard, J.C. (1994b) *Food Biotechnol.*, **7**, 235.
Norton, S., Lacroix, C. and Vuillemard, J.C. (1994c) *Enzyme Microbiol. Technol.*, **16**, 457.
Ohleyer, E., Blanch, H.W. and Wilke, C.R. (1985) *Appl. Biochem. Biotechnol.*, **11**, 457.
Ohmiya, K., Ohashi, H., Kobayashi, T. and Shimizu, S. (1977) *Appl. Environ. Microbiol.*, **33**, 137.
Prescott, S.C. and Dunn, C.G. (1949) *Industrial Microbiology*, 2nd edn, McGraw-Hill, New York, chapter 21.
Prigent, Y. (1983) Method and apparatus for the continuous preparation of lactic acid by fermentation of whey, French patent demande Fr 2,555,200, May 1985, application 83/18,631, November 1983.
Raucourt, A. De, Girars, D., Prigent, Y. and Boyaval, P. (1989a) *Appl. Microbiol. Biotechnol.*, **30**, 521.
Raucourt, A. De, Girars, D., Prigent, Y. and Boyaval, P. (1989b) *Appl. Microbiol. Biotechnol.*, **30**, 528.
Reddy, C.A., Henderson, H.E. and Erdman, M.D. (1976) *Appl. Environ. Microbiol.*, **32**, 776.
Rehberger, T.J. and Glatz, B.A. (1990) *Appl. Env. Microbiol.*, **56**, 864.
Riddle, M.J. and Chandler, W.D. (1974) *Proceedings of the Whey Utilization Symposium*, June, Ottawa, Ontario.
Ripley, P. (1979) *Process Biochem.*, **14** 8.
Roy, D., Goulet, J. and LeDuy, A. (1986) *Appl. Microbiol. Biotechnol.*, **24**, 206.
Roy, D., Goulet, J. and Leduy, A. (1987) *J. Dairy Sci.*, **70**, 506.
Schuppert, B., Schink, B. and Trosch, W. (1992) *Appl. Microbiol. Biotechnol.*, **37**, 549.
Scott, R. (1981) *Cheesemaking Practice*, Applied Science Publishers, London.
Silva, E. and Yang. S.T. (1995) *J. Biotechnol.*, **41**, 59.
Singh, R.K. and Ghaly, A.E. (1984) *Agricultural Waste Utilization and Management, Proceedings of the 5th International Symposium on Agricultural Wastes*, Chicago, IL, ASAE Publication 13–85, p. 546.
Somkuti, G.A. and Steinberg, D.H. (1979) *J. Food Protection*, **42**, 885.
Steffen, C., Nick, B. and Blanch, B.H. (1973) *Schweiz. Milchw. Forsch.*, **2**, 37.
Stenroos, S.L., Lindo, Y.Y. and Lenko, P. (1982) *Biotechnol. Lett.*, **4**, 159.
Stieber, R.W. and Gerhardt, P. (1979a) *Appl. Environ. Microbiol.*, **37**, 487.
Stieber, R.W. and Gerhardt, P. (1979b) *J. Dairy Sci.*, **62**, 1558.
Stieber, R.W. and Gerhardt, P. (1981a) *Biotechnol. Bioeng.*, **23**, 535.
Stieber, R.W. and Gerhardt, P. (1981b) *Biotechnol. Bioeng.*, **23**, 535.
Stieber, R.W., Coulman, G.A. and Gerhardt, P. (1977) *Appl. Env. Microbiol.*, **34**, 733.
Tewari, H.K., Seth, R.P., Sood, A. and Singh, L. (1985) *J. Res. Punjab Agricul. Univ.*, **22**, 89.
Tsao, G.T. and Hanson, T.P. (1975) *Biotechnol. Bioeng.*, **17**, 1591.
Tuli, A., Sethi, R.P., Khanna, P.K., Marwaha, S.S. and Kennedy, J.F. (1985) *Enzyme Microbiol. Technol.*, **7**, 164.

Tyagi, R.D. (1986) *Evaluation of Dairy Industry Waste (Cheese Whey) as Substrate for Bioconversion*, Scientific Report, I.N.R.S.-Eau, University of Québec.
Vahvaselka, M.I. and Linko, P. (1987) *Proceedings of the 4th European Congress on Biotechnology*, Vol. 3 (eds O.M. Neyssel, R.R. Van der Meer and K.C.A.M. Huyben), Elsevier, Amsterdam, p. 317.
Vick Roy, T.B., Blanch, H.V. and Wilke, C.R. (1982) *Biotechnol. Lett.*, **4**, 483.
Vick Roy, T.B., Mandal, D.K., Dea, D.K., Blanch and Wilke, C.R. (1983) *Biotechnol. Lett.*, **5**, 665.
Vinegra-Gonzaliz, G. and Gomez, J. (1983) In *Bioconversion Systems* (ed. D. Wise), CRC Press, Boca Raton, FL, p. 17.
Woksow, S.A. and Glatz, B.A. (1991) *Appl. Env. Microbiol.*, **57**, 2821.
Yabamavar, V.M. and Wang, D.I.C. (1991) *Biotechnol. Bioeng.*, **37**, 544.
Yang, S.T., Zhu, H., Li, Y. and Hong, G. (1994) *Biotechnol. Bioeng.*, **43**, 1124.

10 Lignocellulosic wastes: biological conversion

P.S. CHAHAL AND D.S. CHAHAL

10.1 Introduction

For the last 20 years researchers have been shifting their attentions from ways of disposing of lignocellulosic wastes to utilizing these wastes to produce useful products of higher value. Examples of these products are: single cell protein (SCP) (Chahal, 1991a) to help to meet the ever increasing demand for protein and for introducing new foods to the world; and liquid fuel (ethanol) because of the dwindling supply of fossil fuels and also to minimize the environmental effects of pollution generated by these fossil fuels (McIntyre, 1987). Complete utilization of lignocellulosic wastes is the challenge that researchers face to produce the desired products economically and to eliminate the pollution generated by them.

Lignocelluloss are usually defined as any of the several closely related substances composed of plant (wood) cell walls where cellulose is intimately associated with lignin. In addition to these compounds, lignocelluloses also contain other polysaccharides commonly known as 'hemicelluloses'. Strictly, it is difficult to define lignocelluloses as 'wastes' because every so-called waste has some uses. However, in the present context, this could be defined broadly as comprising that portion of the entire plant kingdom which is not being properly utilized for the welfare of human beings. In this chapter, therefore, the term 'lignocellulosic wastes' would include mainly the crop residues, and wood and forestry wastes. After an appropriate pretreatment, the polysaccharides are easily available for their conversion into SCP to be used as food for human consumption or as feed for animals, or into fuel ethanol and other chemicals.

It has been estimated that photosynthesis on earth results in 155 billion tons of lignocelluloses per year (Bassham, 1975). Of this, about two-thirds is on land, and one-third in the oceans, and 65.5% of the total productivity on land is in forests and woodlands. Some of these forests are, of course, the traditional source of cellulose in the form of lumber and pulp for paper. The 2.7% of the cultivated land which accounts for 5.9% of the primary productivity (mostly for food and fibre) is going to be needed entirely for agriculture (Bassham, 1975). From this, however, the lignocellulosic by-products (crop residues) would be the major source of feedstock for the production of useful products. Ishaque and Chahal (1991) have estimated that about 2.2 billion tons of cereal straw is produced annually in the world. In countries where intensive agriculture is practised, the crop residues are often burned in the fields to clear them for the next crop. Burning of crop residues in the field not only causes pollution in the atmosphere (McIntyre, 1987), but also destroys the useful microflora of the topsoil. Therefore, burning of the crop residues also reduces the fertility of the soil. The estimates of Rawat and Nautiyal (1991) indicated that about 2 billion m^3 of woody lignocellulosic materials on a sustained basis and about 300 million m^3, if just half of the wood residues generated by current felling is utilized are available annually for their bioconversion into food, feed, fuel and other useful products. Moreover, millions of tons of wood-processing residues, bark and wastes from pulp-manufacturing mills are produced every year.

Lignocelluloses seems to be the best substrate for solid-state fermentation (SSF) for the production of cellulase-system containing all the necessary enzymes (filter paper cellulase, cotton cellulase, β-glucosidase and xylanases) in the right proportion to hydrolyse cellulose into glucose, or polysaccharides (cellulose and hemicelluloses) into monomer sugars for their further bioconversion into various products. In another method of SSF, high yields of giberellic acid are achieved. The lignocelluloses can also be used as inert substrates, when impregnated with suitable nutrients for the production of spores of various microorganisms in SSF to be used as bioinsecticides and bioherbicides.

A number of processes for the bioconversion of such lignocellulosic wastes into useful products and biological alleviation of the toxic effluents from the above processes are discussed in this chapter. Although extensive literature on the degradation of lignin is available, there is no process at present by which the enormous amount of lignin could be utilized for the synthesis of new products.

10.2 Composition and structure of lignocelluloses

The degradation of wood by microorganisms is highly dependent on the chemical composition of the lignocellulosic materials. The cell type and cell-wall morphology also govern the efficiency of lignocellulosic degradation. Wood is composed mainly of cellulose (a polymer of glucose), hemicelluloses (polymers mainly of xylose and mannose) and lignin. The crop residues contain mainly hemicelluloses and less cellulose and lignin (Table 10.1). Some crop residues, such as rice straw and sugar-cane bagasse, have a high content of silica. Removal of silica was a major problem for the bioconversion of rice straw into useful products. Recently, a process has been developed to separate silica (Chahal, 1995). The hemicelluloses are mainly present as glucuronoxylan in hardwoods and as galactoglucomannans in softwoods. Although the lignin content in hardwoods and softwoods is the same, the type of lignin found in each is

Table 10.1 Chemical composition of lignocellulosic materials (% dry weight)

Lignocellulosic materials	Lignin[a]	Glucose[b]	Xylose[b]	Mannose[b]
Angiosperms	23	45	19	2
Gymnosperms	29	45	6	13
Crop residues[c]	3–13	30–45[d]	16–27[e]	

[a]Sulphuric acid lignin by the method of Effland (1977).
[b]HPLC analysis using the method of Pettersen *et al.* (1985).
[c]Data from Sloneker (1976).
[d]Reported as cellulose content.
[e]Reported as hemicelluloses content.

different. The basic building unit of lignin may be substituted in two or three positions; the addition of one methoxyl group to the phenol ring results in a guaiacyl unit and the addition of two methoxyl groups results in a syringyl unit. Hardwoods contain varying ratios of syringyl and guaiacyl types of lignin, whereas conifers have primarily guaiacyl lignin (Fenegel and Wegener, 1983).

Daniel (1994) made a detailed study of wood biodegradation using electron microscopy. Methods for labelling wood components (lignin, hemicelluloses and cellulose) *in situ* by energy-dispersive X-ray micro-analysis (EDXA) and enzyme immunogold cytochemistry have also been outlined by this author.

All the lignocelluloses are composed of plant (wood) cells with a thin primary wall that surrounds the relatively thick secondary wall (Figure 10.1). Within each secondary wall, the cellulose in each layer of the cell wall occurs as long slender bundles composed of long chains of β-D-glucopyranose residue (cellulose molecules) linked by 1–4-glucosidic bonds (Freudenberg *et al.*, 1932) called elementary fibrils with a diameter of 3.5 μm. A number of elementary fibrils, when joined laterally, form microfibrils. The primary wall is only 0.1–0.2 μm in thickness and contains a randomly and loosely organized network of cellulose microfibrils. The outer layer of the seondary wall, S_1, has a crossed fibril structure. In the S_2 layer, the main portion of the secondary wall (1–5 μm thick), the microfibrils are oriented almost parallel to the lumen axis. In the thin S_3 layer, the microfibrils are in the form of a helix. Microfibrils are covered with hemicelluloses and are also encrusted with lignin. The innermost portion, I, of the cell wall consists of the so-called warty layer, probably formed from protoplasmic debris. The central empty portion, formed after the disintegration of protoplasm at the time of ageing, is called the lumen. The primary walls of the two adjacent cells are cemented together with pectic compounds and lignin. The portion between the two adjacent cells is called the middle lamella and is filled with hemicelluloses and lignin

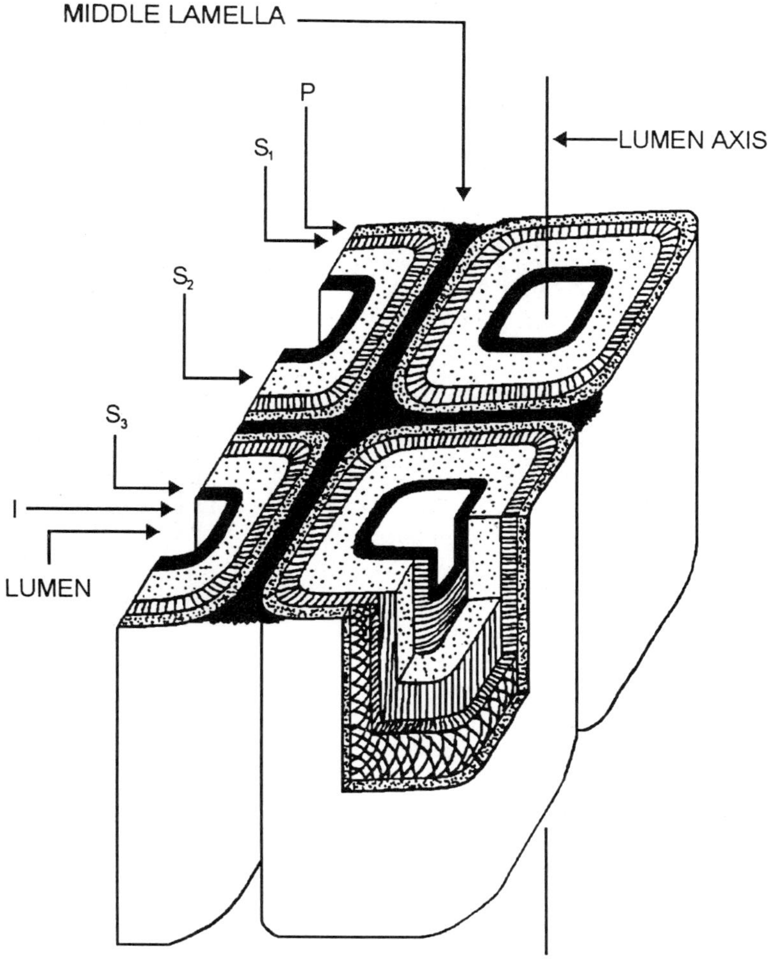

Figure 10.1 Structure of a wood cell. P = primary wall, a loosely organized network of cellulose microfibrils; S_1 = secondary wall 1, a crossed fibril structure; S_2 = secondary wall 2, microfibrils are almost parallel to the lumen axis; S_3 = secondary wall 3, microfibrils form a flat helix; I = warty layer formed from protoplasmic debris; and Lumen = the central empty portion formed after the disintegration of protoplasm at the time of ageing.

(Figure 10.1). Hemicelluloses are linked by covalent bonds with lignin (Fenegel and Wegener, 1983) (Figure 10.2). Therefore, the cellulose in nature is very well protected from biological degradation.

10.2.1 Cellulose

Cellulose was named by Payen (1838) when he recognized that cellulose and starch were isomeric products. The length of cellulose molecules in an

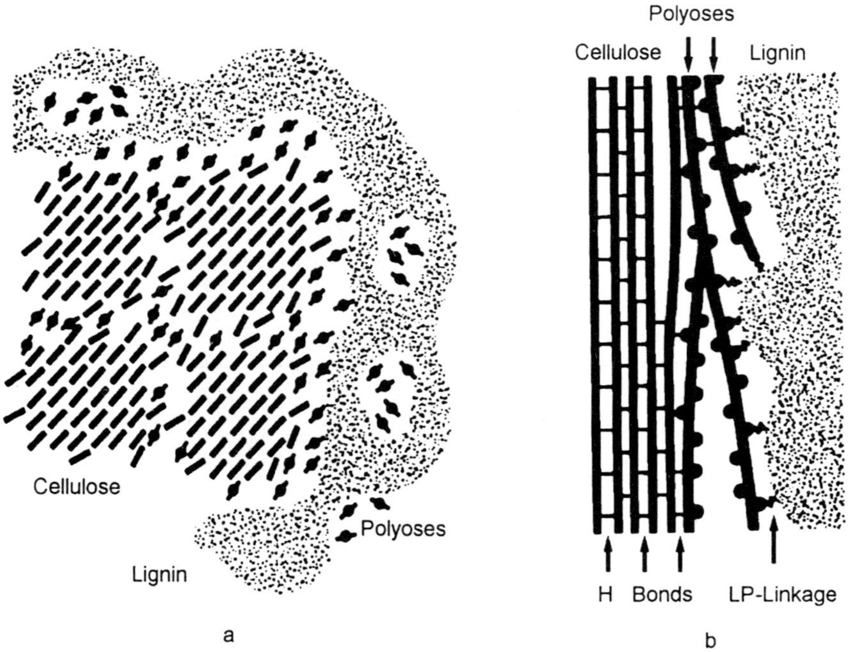

Figure 10.2 Schematic diagram showing the structural arrangements of cellulose, hemicelluloses (polyoses) and lignin in a plant (wood) cell wall. (Reproduced from Fenegel, D. and Wegener, G., *Wood Chemistry, Ultrastructure, Reactions;* by permission of the publisher Walter de Gruyter, Berlin, 1983.)

elementary fibrils varies from less than 15 β-D-glucopyranose residues in γ-cellulose to as many as 10 000–14 000 molecule^{-1} in α-cellulose. The length of the cellulose molecule is measured as the degree of polymerization (DP), that is, the number of β-D-glucopyranose residues. Within each elementary fibril the cellulose molecules are bound laterally, with adjacent molecules running in opposite directions, by hydrogen bonds. They are associated in various degrees of parallelism – regions that contain highly orientated molecules are called 'crystalline cellulose' and those with less orientated molecules are called 'amorphous cellulose' (Hess *et al.*, 1954).

According to Preston and Cronshaw (1958), the microfibril is about 0.005×0.01 μm in cross-section and consists of a crystalline core of highly ordered cellulose surrounded by a sheath, which in cotton contains mainly amorphous cellulose molecules, but in wood also contains hemicelluloses and lignin molecules (Figure 10.3). In this figure, the solid strokes represent the planes of the glucose residues in the cellulose chain molecules and the broken strokes the planes of other sugars or sugar derivatives in non-cellulosic molecular chains. The area joined into a lattice represents the solid central core. Preston and Cronshaw (1958) also

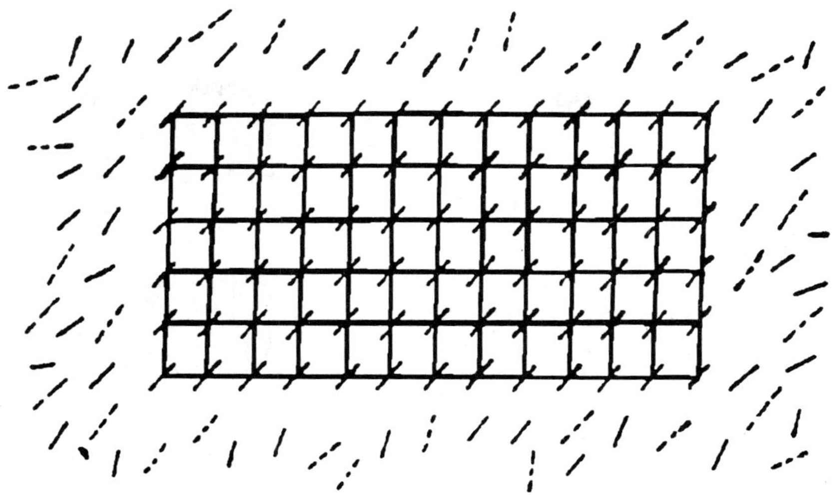

Figure 10.3 The structure of microfibrils according to Preston and Cronshaw (1958). Diagrammatic representation of a cellulose microfibril about 0.01 μm wide and about 0.005 μm thick . The solid strokes represent the planes of the glucose residues in the cellulose chain molecule and the broken strokes the planes of the other sugars or sugar derivatives in noncellulosic molecular chains. The area joined into lattic represents the solid central core. (Reproduced with permission from Preston, R.D. and Cronshaw, J. *Nature*, **181**, 248, 1958, Macmillan Magazines Ltd.)

reported that the central cyrstalline core does not, however, run uninterruptedly along the whole length of a microfibril, and that the microfibril is, therefore, hetrogeneous along its length. It was also reported that there are regions of weaknesses that are irregularly distributed along the length of the microfibrils. Manley (1964) reported that the microfibril is composed of a flat ribbon of cellulose molecules wound in the form of a tight helix (Figure 10.4), but his theory was not supported by any proof to show the existence of this structure. However, Cowling (1975) accepted Manley's concept of microfibrillar structure.

According to Rowland and Roberts (1972), the microfibrils at certain lengths contain strain-distorted tilt and twisted regions which are easily accessible for hydrolysis (Figure 10.5), and according to Gardner and Blackwell (1974), the cellulose molecules are linear and easily form intramolecular and intermolecular hydrogen and bonds. Glucan chains have a twofold screw axis of symmetry, stabilized and stiffened by intramolecular hydrogen bonds (O3-H---O5′ and O6---H-O2′) and one intermolecular hydrogen bond (O6-H---O3) (Figure 10.6). The intramolecular bonds help to maintain the rigidity of the cellulose chain, whereas the intermolecular bonds keep the cellulose chains in a tight and closely packed arrangement. The tight and closely packed arrangement

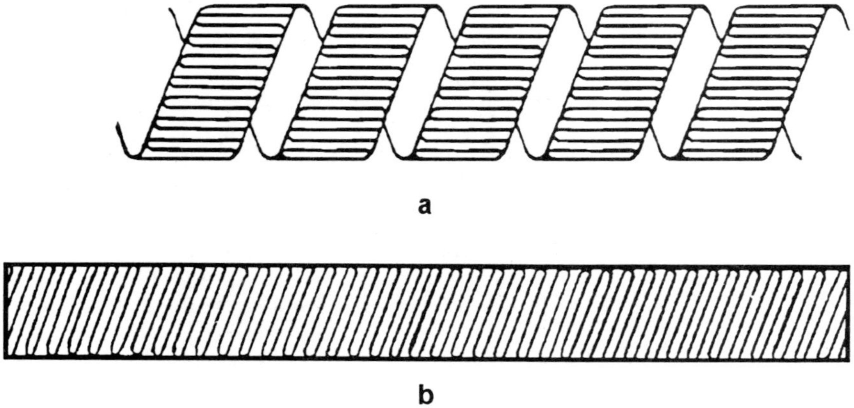

Figure 10.4 The structure of microfibrils according to Manley (1964). The microfibril is composed of a flat ribbon (b) of cellulose molecules wound in the form of a tight helix (a). (Reproduced with permission from Manley, R. St.J., *Nature*, **204**, 1155, 1964, Macmillan Magazines Ltd.)

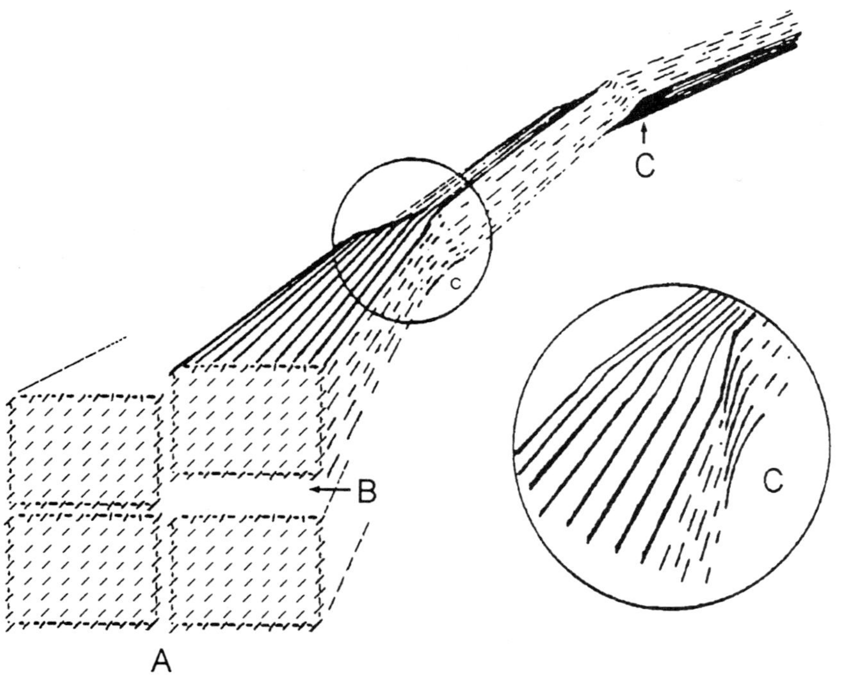

Figure 10.5 The structure of microfibrils according to Rowland and Roberts (1972). Schematic representation of the elementary fibril to illustrate the crystalline elementary fibril theory of the microstructure of cellulose and to show (A) coalesced surfaces of high order, (B) readily accessible slightly disordered surfaces, and (C) readily accessible surfaces of strain-distorted tilt and twist regions. (From Rowland, S.P. and Roberts, J.J., *Journal of Polymer Science*, part A-1, **10**, 2447, 1972, reprinted with permission of John Wiley & Sons, Inc.)

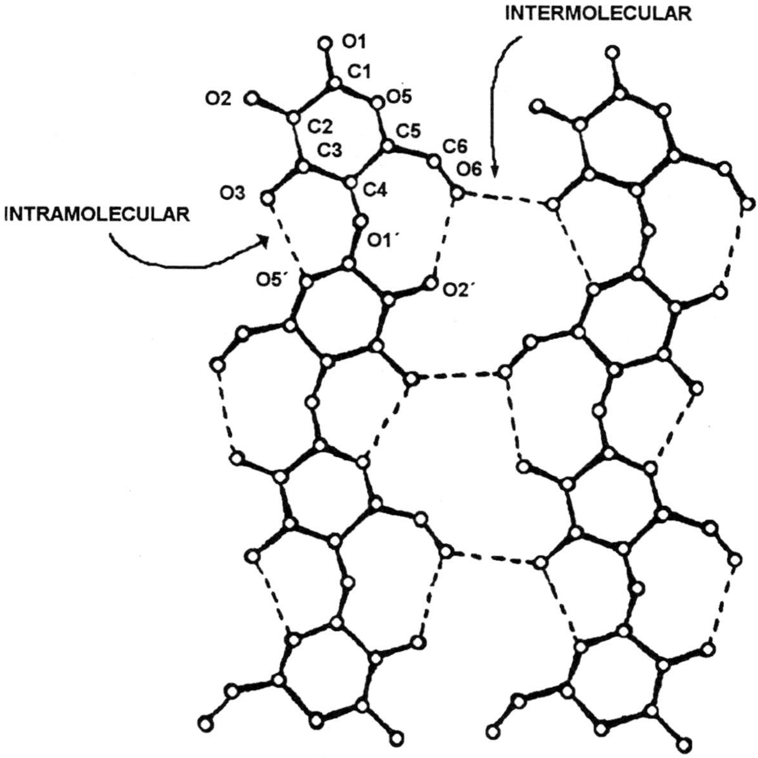

Figure 10.6 Intermolecular and intramolecular bonds in cellulose. Projection of the parallel chain model for cellulose showing the hydrogen bonding network and the numbering of the atoms. Superscript (′) refers to the atom number of the adjacent glucose molecule in the same chain. O, H and C refer to oxygen, hydrogen and carbon atoms, respectively. Each glucose residue forms one intermolecular bond (O6-H---O3) and two intramolecular bonds (O3-H---O5′ and O6---H-O2′). Redrawn from Gardner, K.H. and Blackwell, J., *Biopolymers*, **13**, 1975, 1974.)

strictly refers to the crystalline portion of the cellulose. Some researchers have reported that cellobiose, rather than glucose, is a basic structural unit of cellulose (Tönnesen and Ellefsen, 1971; Blackwell, 1982; Atalla, 1983).

Cellulose exists in several crystalline forms (cellulose I–IV) with different X-ray diffraction patterns and spectra (Blackwell, 1982; Atalla and van der Hart, 1984). Of these four forms of celluloses, the cellulose I is the native form of cellulose as it occurs in plant cell walls and the cellulose II is a regenerated form cellulose obtained by mercerization in solid state or by dissolution. The crystalline form is highly resistant to microbial and enzymatic degradation, while amorphous cellulose is hydrolysed much faster.

10.2.2 Hemicelluloses

Schulze (1891) gave the name hemicelluloses to the low-molecular-weight polysaccharides that could be extracted more readily from plants by aqueous alkali. They are also easily hydrolysed by acid. The name seemed appropriate since these polysaccharides were thought to be the intermediates in cellulose biosynthesis and were found in close association with cellulose in the cell wall. Now it is known that the hemicelluloses are not the precursors of cellulose and that they represent a distinct and separate group of plant polysaccharides which have no part in cellulose biosynthesis. Although the hemicelluloses are usually considered to be structural polysaccharides, it is convenient to include among them a few other plant polymers, such as the arabinogalactans which obviously have other functions. Hemicelluloses are built up from relatively few sugar residues, the most common of which are D-xylose, D-mannose, D-glactose, D-glucose and L-arabinose, 4-*O*-methyl-D-glucuronic acid, D-galacturonic acid and D-glucuronic acid. Other rare constituents of hemicelluloses are L-rhamnose, L-fucose and various methylated neutral sugars.

According to Timell (1967), the complete formula of hardwood xylan (*O*-acetyl-4-D-methylglucurono-xylan) is shown in Figure 10.7. The polysaccharide framework consists of approximately 200 *β*-D-xylopyranose residues, linked together in 1–4-glycosidic bonds. Some of the xylose units carry a single, terminal side chain consisting of a 4-*O*-methyl-*α*-D-glucuronic acid residue, attached directly to C-2 of the xylose. Seven out of ten xylose residues contain an *O*-acetyl group at C-2 or more frequently at C-3 (Timell, 1967). The basic framework of softwood (arabino-4-*O*-methyl-glucurono-xylan) is same as that of hardwood. However, softwood xylan also contains a *α*-L-arabino-furanose residue directly linked to C-3 of xylose. Mannan (galactoglucomannan), a major component of softwood, consists of 1–4-linked *β*-D-glucopyranose and *β*-D-mannopyranose residues distributed at random (Figure 10.8). Some of the hexose units carry a terminal residue of D-galactopyranose attached to C-6. It is probable that

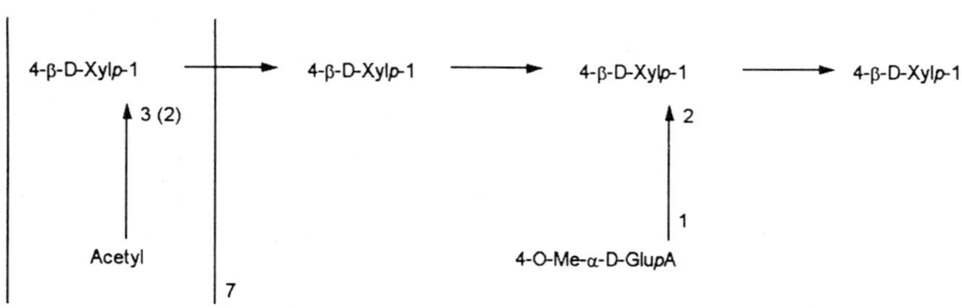

Figure 10.7 The structure and framework of xylan. (Redrawn from Timell, T.E., *Wood Science Technology*, **1**, 45, 1967.)

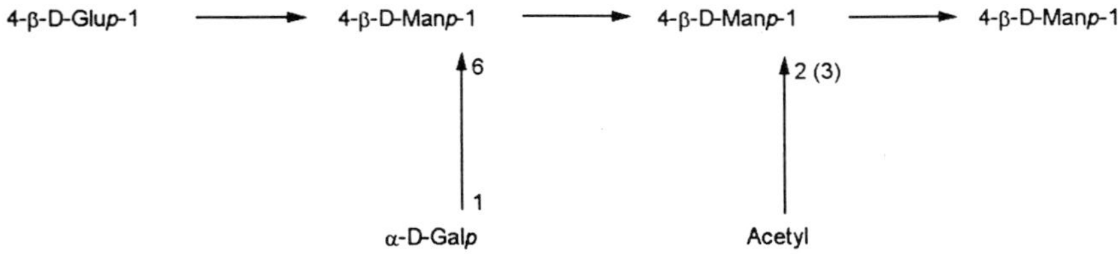

Figure 10.8 The framework of mannan. (Redrawn from Timell, T.E., *Wood Science Technology*, **1**, 45, 1967.)

all the galactoglucomannans are acetylated in their native state. The acetyl groups are attached to the mannose residues (Timell, 1967).

Kalmes (1959), Lange (1958), Meier (1958) and Sultze (1957) reported that the concentration of holocelluloses (cellulose + hemicelluloses) is approximately uniform cross the cell wall of cotton from lumen through the primary wall but decreases from the lumen toward the middle lamella in wood fibres. In both types of fibres, hemicelluloses predominate in the primary wall and diminish in concentration toward the lumen.

10.2.3 Lignin

In the late 1960s and the early 1970s, the chemical structure of lignin became clear (Freudenberg, 1965; Nimz, 1974; Adler, 1977). However, in late 1970s and early 1980s the interest in this field grew rapidly, and many reviews and books surfaced in the literature (Ander and Eriksson, 1978; Crawford and Crawford, 1980; Kirk *et al.*, 1980; Crawford, 1981; Higuchi, 1981, 1982; Zeikus, 1981; Kirk and Fenn, 1982; Janshekar and Fiechter, 1983; Palmer and Evans, 1983a,b; Kirk, 1984, Paterson *et al.*, 1984; Leisola and Fiechter, 1985; Buswell and Odier, 1987; Evans, 1987; Harvey *et al.*, 1987a,b; Kirk and Farrell, 1987; Tien, 1987; Umezawa, 1988; Eriksson *et al.*, 1990).

Lignin is concentrated mainly in the spaces between the cell walls of adjacent cells (middle lamella) and in the S_2 layer of the cell wall where it is deposited during the lignification process of the plant (wood) tissues, although some lignin is also deposited in other layers. Lignin in the cell wall not only encrusts the cellulose microfibrils in a sheath-like manner, but is also bonded physically and chemically to hemicelluloses (Higuchi, 1971; Fenegel and Wegener, 1983). Physically, lignin forms a barrier against the penetration of cellulases and hemicellulases (Kirk and Haskin, 1973).

Lignin is formed by dehydrogenative polymerization of *p*-hydroxycinnamyl alcohols. It forms an irregular noncrystalline network in the plant cell wall which is very resistant to microbial degradation. Guaiacyl lignin,

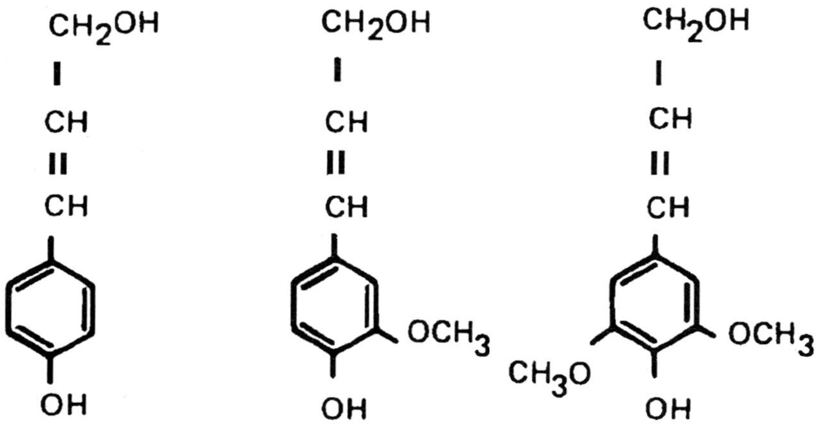

p-Coumaryl alcohol   Coniferyl alcohol   Sinapyl alcohol

Figure 10.9 The three primary monomeric precursors of lignin.

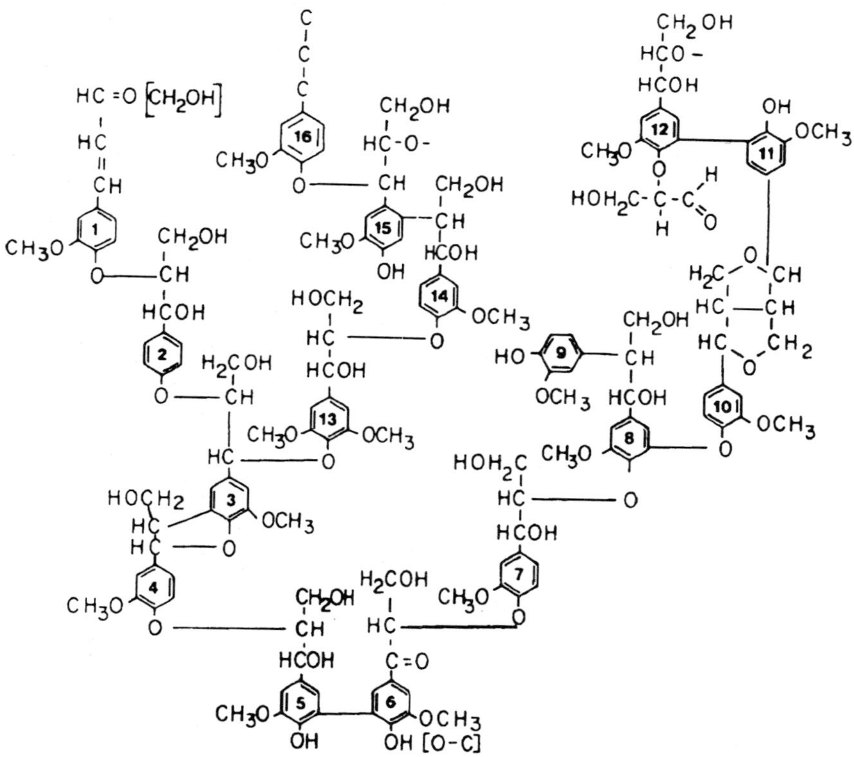

Figure 10.10 The prominent structural features of conifer lignin. (Reproduced from Adler, E., *Wood Science Technology*, **11**, 160, 1977.)

which occurs in conifers, is mainly a dehydrogenation polymer of coniferyl alcohol. Guaiacyl–syringyl lignin, which occurs in angiosperms, is composed of a mixed dehydrogenation polymer of coniferyl and synapyl alcohols. Guaiacyl–syringyl–*p*-hydroxyphenyl lignin, which is found in grasses, is composed of a mixed dehydrogenation polymer of coniferyl, sinapyl and *p*-coumaryl alcohols (Higuchi, 1980). The three primary monomeric precursors of lignin are shown in Figure 10.9. The structural features of conifer lignin (Alder, 1977) and beech lignin (Nimz, 1974) are shown in Figures 10.10 and 10.11 respectively.

The information about lignin as reported by Kalmes (1959), Lange (1958), Meier (1958), Sultze (1957) and Wardrop (1957) can be summarized as follows. The lignin is concentrated primarily in the compound middle lamella of wood cells and decreases in concentration towards the lumen. The amount of lignin in the secondary walls of coniferous woods is considerably higher than in angiospermous woods. The hemicelluloses and lignin form a matrix surrounding the cellulose. Within the microfibril, lignin and hemicelluloses may penetrate the spaces between cellulose molecules in the amorphous regions (Manley, 1964) providing rigidity to the fibrous wood structure.

As cellulose fibres are surrounded by hemicelluloses and lignin, therefore, it becomes important to expose the cellulose by various physical

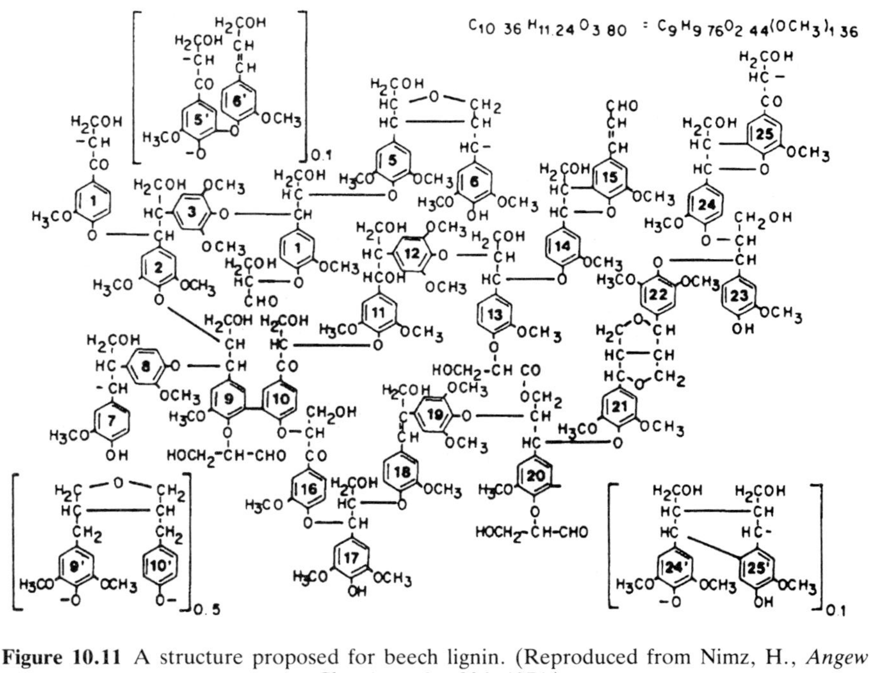

Figure 10.11 A structure proposed for beech lignin. (Reproduced from Nimz, H., *Angew Chemistry*, **86**, 336, 1974.)

and/or chemical treatments before it can be used for cellulase production or for hydrolysis.

10.2.4 Protein

Proteinacious materials are the residues of the protoplast of the cell. Although the amount is quite small (0.5%) particularly in the wood fibres (Reese, 1963), it is good for the growth of the microorganisms.

10.2.5 Extraneous materials

The small quantity of extraneous material deposited in the capillaries of the cell wall include waxes, fats, essential oils, tannins, resin and fatty acids, terpenes, alkaloids, starch, soluble saccharides and various cytoplasmic constituents (Hillis, 1962; Cowling and Merrill, 1966). The extraneous materials are in part deposited on the fibre surface and in part within the fibre wall. In cotton most of these substances are a part of primary wall. In wood they occur in lumen and within both the primary and secondary walls of the wood fibres (Kalmes, 1959; Lange, 1958; Meier, 1958; Sultze, 1957).

The presence of extraneous materials such as tannins, resins, terpenes, waxes and alkaloids may be harmful for the growth of the microorganisms used to produce useful products from the lignocellulosic wastes.

10.3 Pretreatment of lignocelluloses

Crystallinity of cellulose and encrustation of cellulose fibrils with lignin make the lignocelluloses recalcitrant. Unless lignin is deploymerized, solubilized or removed, cellulose and hemicelluloses cannot be easily hydrolysed by cellulases and hemicellulases, respectively, for their bioconversion into various products. The crystallinity of cellulose is another hindrance in the bioconversion of cellulose. Thus, pretreatment of lignocelluloses will be necessary for their efficient conversion into various products.

To discuss the various pretreatments of lignocelluloses is beyond the scope of the present chapter; however, some knowledge about them is necessary for the better understanding of the bioconversion processes. Tarkow and Feist (1969), Millett *et al.* (1976) and Chahal (1991a) provide detailed information on the pretreatment of lignocelluloses.

10.4 Biological conversions

Biological conversions are discussed under two major modes of fermentation: liquid-state fermentation (LSF) and solid-state fermentation (SSF).

10.4.1 Liquid-state fermentation

Lignocellulosic residues are not high-value feedstocks, they are classified as low-quality roughage, that is, they are high in fibre and low in protein and nutritional quality. However, these materials can be converted to high-value animal feed or food for human consumption by converting carbohydrate components of the lignocellulosic materials to protein i.e. single cell protein. One of the earliest processes for SCP production from lignocelluloses was the production of yeast from wood hydrolysate. During World Wars I and II, Germany developed a process to produce food yeast from wood sugars obtained by acid hydrolysis and sulphite waste liquor. Later, commercial plants were set up at various places in the USA and other parts of the world (Harris and Belinger, 1946; Gilbert *et al.*, 1952; Underkofler and Hickey, 1954). Similarly, in the USSR the 'hydrolysis industry' to produce sugars from wood was created in the mid-1930s and continues to the present time. The sugar obtained from wood hydrolysis is used for SCP production (Laskin, 1977).

Keeping in view the fact that lignocelluloses are constantly being replenished through photosynthesis, the attention of researchers all over the world has now been diverted to the utilization of this carbon source for SCP production. A number of processes are now available for the bioconversion of lignocelluloses into SCP. These may be grouped together into the following major processes.

(a) Indirect conversion of lignocelluloses. The lignocelluloses are pretreated with various physicochemical methods to obtain cellulose to be used in the following ways.

Chemical hydrolysis of cellulose to glucose and growth of micro-organisms on the hydrolysate. In this case wood is hydrolysed into glucose with acid by the Bergius or Scholler process, and *Candida utilis*, a yeast, is grown on the glucose thus obtained to produce SCP. The Scholler process was operated on a large scale in Germany during World Wars I and II and production reached a level of 15 000 tons year^{-1} (Laskin, 1977). Microbial protein production from wood hydrolysate could not survive because of its high cost of production. The Scholle process for hydrolysing cellulose was upgraded at the US Forest Products Laboratories and a few wood sugar plants were built in the USA (Callihan and Clemmer, 1979). Meller (1969) carried out an in-depth economic analysis on hydrolysis and growth of yeast on the resultant sugar. He determined a production cost of the order of 30 US cents kg^{-1} of yeast produced.

Han *et al.* (1976) developed a process where *Aurobasidium pullalans*, a yeast, is grown on the acid hysrolysate of ryegrass. A yield of 1.5 g cells l^{-1} is obtained from a medium containing 6 g sugar l^{-1} supplemented with

1.25 g yeast extract. The product contained 42.6% protein. Thus, about 0.64 g protein l^{-1} was obtained by adding double the quantity (1.25 g) of yeast extract, which is much more expensive than the product (SCP) itself.

Enzymatic hydrolysis of cellulose to glucose and growth of micro-organisms on the hydrolysate.

Indian Institute of Technology process. This process consists of pulverizing the cellulosic materials to a size of less than 25 μm and heating it to 200 °C in an oxidizing atmosphere followed by enzymatic saccharification at pH 4.8 and 50 °C (Das and Ghose, 1973). Residual enzyme and cellulose are recycled. About 7.3% sugar syrup is obtained by this method for SCP production with yeasts. Das and Ghose (1973) have projected production costs of 34.4, 17.1 and 14.1 US cents kg^{-1} of glucose from plant sizes of 10, 100, and 250 tons, respectively, per day. If the minimum cost of glucose production is taken as 14.1 US cents kg^{-1}, even then the cost of SCP production from such glucose syrup would still be higher than that of soybean.

University of California process. In this process (Wilke *et al.*, 1976) cellulase enzyme is produced with *Trichoderma reesei* QM 9414 on hammered mill newspaper to hydrolyse the same substrate to produce sugar syrup. About 50% hydrolysis of the substrate has been reported by that cellulase enzyme. The sugar syrup obtained is concentrated before it is fermented into SCP because the hydrolysate is too low in sugar content for economic fermentation. In this process, 59.4 tons day^{-1} of torula yeast and 81.5 tons day^{-1} of ethanol from 885 tons of pretreated newspaper were reported by Wilke *et al.* (1976).

The high cost of enzyme production, low yields (50%) of hydrolysis and low concentration of sugar syrup, which is to be concentrated before fermentation for SCP or ethanol, are the major drawbacks to this process becoming an economic enterprise.

Kyoto University process. This process, developed by Tanaka and Matsuno (1985), uses hydrolysate obtained by enzymatic hydrolysis of pretreated lignocelluloses. The hydrolysate is a mixture of glucose, xylose and disaccharides (cellobiose and xylobiose). The hydrolysate is first fermented in tank D_1 with *Saccharomyces* species to utilize glucose from the hydrolysate for production of SCP (Figure 10.12). The SCP thus produced is separated through a membrane and the residual hydrolysate containing xylose and disaccharides (cellobiose and xylobiose) is supplied back to hydrolysing tank B for further hydrolysis to monomer sugars. Part of these disaccharides is supplied to fermenter A for production of cellulases and xylanases. The immobilized β-glucosidase and β-xylosidase,

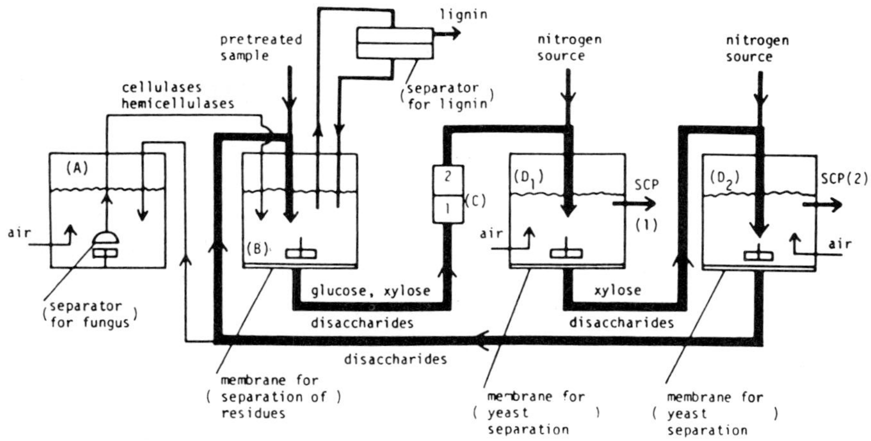

Figure 10.12 An ideal scheme for the SCP production process: A = fermenter for cellulase and hemicellulases production; B = hydrolysis tank; C = columns (immobilized enzymes: 1 = β-glucosidase; 2 = β-xylosidase); D_1 and D_2 = fermenters for SCP production. The bold lines of the liquid flow represent the larger flow rates. (Reproduced by the permission of Elsevier Science Inc. from Tanaka and Matsuno, 1985.)

column C, may be attached between the hydrolysis tank B and the fermenter D_1 to hydrolyse cellobiose and xylobiose to glucose and xylose, respectively.

The lignin after enzymatic hydrolysis must be removed by filtration as shown in Figure 10.12. This process suffers from all the drawbacks pointed out earlier. However, it would also be very difficult to synchronize the production of enzymes, the rate of hydrolysis of lignocelluloses, and the assimilation of glucose and xylose by the respective yeasts in their respective fermentation tanks.

Mixed culture. Cellulose hydrolysis by one intact organism and conversion of hydrolysis product into SCP by another intact organism is a very similar idea to that of the Symba process where *Endomycopsis fibuligera* produces amylases to hydrolyse starch and *Candida utilis* is grown for SCP production on the hydrolysis product, glucose (Wiken, 1972). However, no such process, where hydrolysis of cellulose or lignocellulose is achieved by one intact microorganism and production of SCP on their hydrolysate by another intact microorganism, is available in the literature. Nevertheless, Peitersen (1975) tried to grow a mixed culture of *Trichoderma viride* and *Saccharomyces cerevisiae* to convert cellulose into SCP without any improvement in the yield as compared to the use of the cellulolytic microorganism, *T. viride*, alone.

(b) Direct conversion of lignocelluloses. In this case lignocelluloses are also pretreated with various physicochemical methods, and hydrolysis of the substrate into utilizable compounds and their conversion into SCP is performed by the same organism.

The first attempt at conversion of lignocelluloses into animal feed with *Aspergillus fumigatus* was by Pringsheim and Lichtenstein in 1920 (Litchfield, 1968). During the last two decades a number of reports have appeared on the conversion of lignocelluloses into animal feed rich in protein (SCP) with various cellulolytic microorganisms (Chahal and Moo-Young, 1981; Rolz, 1984). The Louisiana State University process is discussed in detail here, although in this process isolated cellulose is used as the substrate instead of lignocelluloses.

Louisiana State University process. In this process a bacterium, *Cellulomonas* sp., is used (Srinivasan and Han, 1969; Dunlop and Callihan, 1973). This organism has low β-glucosidase activity, thus, there is a need for another organism, *Alcaligenes faecalis*, to remove the accumulated cellobiose which inhibits cellulase activity. Since it is endoglucanase positive and lacks exo-glucocellobiose hydrolase, it needs drastic pretreatment with high concentration of alkali (30 g NaOH 100 g^{-1} substrate at 30 psig for 4 h to remove lignin and swell the cellulose. About 68 g of cellulose is obtained by treating 100 g of biomass. On fermentation with both the bacteria, a bacterial cell mass of 18–20 g was obtained and only 40 g of cellulose was utilized. Moreover, about 32 g of hemicelluloses and lignin were obtained as the effluents from the pretreatment of 100 g of bagasse. This effluent creates a great disposal problem as it has a very high biochemical oxygen demand (BOD).

Callihan and Clemer (1979) have discussed the latest version of the above process where a yield of 0.12 g protein g^{-1} bagasse was reported. Because of the disposal problem of the effluents and the low yield of the product (SCP), the process seems to be unattractive as a commercial enterprise.

In another report, Han and Callihan (1974) obtained high yields of protein (1.76 g protein l^{-1}) by growing a mixed culture of *Cellulomonas* sp. and *A. faecalis* on a 1.5% slurry of alkali-treated rice straw. Here again the disposal problem of the effluent (hemicelluloses and lignin) has also not been discussed by them.

(c) The search of new cellulolytic fungi. All the above processes require separation of cellulose from lignocelluloses and the use of acid or enzyme for hydrolysis of cellulose into glucose. The hydrolysis of cellulose into glucose is an additional cost for SCP production from lignocelluloses. However, hemicelluloses, the second major fraction of polysaccharides of lignocelluloses, were not being utilized in these processes. Thus, they

created disposal and pollution problems. Therefore, the search continued in our laboratory to find cellulolytic fungi which could use cellulose as well as hemicelluloses for their conversion into SCP. This work, carried out by Chahal and various coworkers, was successful in finding such a fungus, which was named as *Chaetomium cellulolyticum* Chahal et Hawks (Chahal and Hawksworth, 1976). This fungus proved to be the best for bioconversion of pure cellulose, hemicelluloses, mixtures of cellulose and hemicelluloses and lignocelluloses without fractionation, both alone and also with an admixture of manures (Chahal *et al.*, 1977, 1987; Moo-Young *et al.*, 1977, 1980; Chahal and Wang, 1978; Chahal and Moo-Young, 1981; Chahal and Ishaque, 1986, 1988).

Our continuous research in this field led us to the development of a process which became popular as 'The Waterloo process for SCP production from waste biomass' (Moo-Young *et al.*, 1979). Unfortunately, it was found that the fungus, *Chaetomium cellulolyticum*, produced some toxins under certain cultural conditions (Sekita *et al.*, 1981), therefore, commercialization of the Waterloo process became doubtful. Nevertheless, Chahal at the Institut Armand-Frappier, Laval, Quebec, continued to work on this project and developed a new mutant of *C. cellulolyticum* free of the toxin problems. He also developed a process that is free of the pollution problems associated with the other processes. This is described in the following sections.

(d) Complete conversion or utilization of lignocelluloses.

Institut Armand-Frappier process. In the Institut Armand-Frappier (IAF) process *Chaetomium cellulolyticum* asporogenous mutant, nontoxin producer, and *Pleurotus sajor-caju*, an edible mushroom, are used to convert lignocelluloses into SCP. The other fungi used in this process are species of *Aspergillus* and *Penicillium*. These fungi have the ability to produce a sufficient amount of cellulases and hemicellulases to convert cellulose and hemicelluloses into SCP. Therefore, there is no need for the prehydrolysis of lignocelluloses with acid or enzymes as is required in all the other processes described earlier.

In the IAF process lignocelluloses are fractionated into cellulose, hemicelluloses and lignin. It is recommended that the cellulose fraction should be used to produce paper or other valuable products rather than for conversion into SCP. However, the cellulose fraction thus obtained is very suitable for bioconversion into SCP of high quality to be used as food for human consumption. It could also be used to produce cellulose enzyme for hydrolysis of cellulose into pure glucose syrup to be used for fermentation into ethanol fuel or other pharmaceuticals. Lignocelluloses without fractionation and the hemicullulose fractions are used to produce SCP either as food for humans or as feed for animals depending on the purification of the SCP product after harvesting. Mixtures of hemicelluloses and lignin can be fermented together, the hemicellulose fraction being converted into SCP and the lignin being exposed to the action of various enzymes produced by the fungi. Thus, lignin is degraded into compounds (oligolignols) of low molecular weight (Chahal and Hachey, 1990). The lignin obtained in this process could be used for production of adhesives and various chemicals, originally synthesized from hydrocarbons. In this process all the components are utilized for production of various products. The IAF process is also called 'an integrated process for production of food, feed and fuel from biomass' as nothing is left unutilized from the substrate; therefore, there are no pollution and disposal problems from the effluents from this integrated process.

This process is being exploited for commercialization for the production of cellulose, lignin, silica and protein-rich feed from rice straw in collaboration with the Punjab Agro Industries Corporation, Chandigarh, India, and the DC Enterprises, Inc., Laval, Quebec.

The IAF process is shown in Figure 10.13. The biomass (lignocelluloses) (1) is pretreated (2) with various methods as mentioned earlier and is fractionated into cellulose (3) and hemicelluloses (5). The pretreated biomass (2), without fractionation, is fermented with fungi (7) for production of SCP as animal feed (8). During the pretreatment some of the

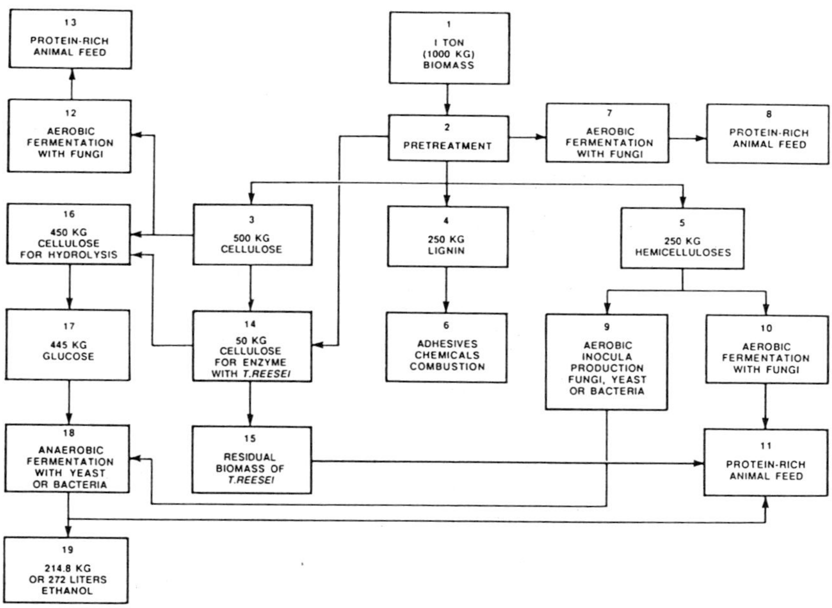

Figure 10.13 The Institut Armand-Frappier Process, an integrated process for the production of food, feed and fuel (ethanol) from biomass.

hemicelluloses and lignin are solubilized. The soluble portion is not removed as in other processes but is kept along with the cellulose in the fermentation medium. Cellulose and hemicelluloses are utilized by the organisms and lignin is left in the fermentation medium which can easily be isolated by precipitation after acidifying the spent fermentation liquor.

The fractionated cellulose (3) is used for production of paper or other cellulosic materials but it can also be fermented with fungi (12) for production of high-quality SCP to be used for human consumption or as animal feed (13). The fractionated hemicelluloses (5) are utilized for production of inocula (9) of the fungi used for fermentation at steps 7 and 12, and for fermentation of surplus hemicelluloses at step 10 for production of SCP (11). The fractionated hemicelluloses (5) are also used for production of inocula of yeasts (*Saccharomyces cerevisiae*, *Kluyveromyces* spp.) at step 9, for the fermentation of glucose (17) at step 18 into ethanol (19) or other pharmaceuticals. About a tenth of the cellulose (3) is used for enzyme production (14) with *Trichoderma reesei* QMY-1 for hydrolysis of cellulose (16) received from step 3. The inoculum of *T. reesei* QMY-1 is also produced on hemicelluloses at step 9. The residual fungal biomass obtained at steps 15 and 18 is also used as animal feed (SCP).

It is easy to obtain two fractions of lignocelluloses after pretreatment: cellulose, and a mixture of hemicelluloses and lignin. In this case the latter fraction can be fermented in the same way as the fractionated hemicelluloses are fermented for various products (SCP and inocula) as explained above. Lignin is not utilized by these fungi and becomes a part of the effluent. It can easily be isolated by acidifying the effluent.

(e) Nutritional values of single cell protein. Production of mycoprotein (SCP) with *Fusarium graminearum* on starch and sugars, a process developed by Rank Hovis McDougall of the UK, has recently been commercialized for human consumption of SCP (Newark, 1980). Similarly, Tate and Lyle, also of the UK, have developed a process to convert carob sugars, spoiled papaya, cassava, sulphite waste liquor and wastes from processing olives, palm oil, potatoes, dates and citrus into SCP (Righelato *et al.*, 1976). These two processes have been mentioned here to indicate that this type of SCP can also be produced from lignocelluloses as already explained.

The microbial food produced on pure carbohydrates (starches and sugars) contains more protein and is freer from impurities than that produced on agricultural residues or other lignocelluloses. The low protein content in biomass produced from lignocelluloses is mainly due to the presence of unutilized (residual) cellulose and lignin in the final product. The microbial biomass thus produced on such substrates would be suitable as animal feed. However, the microbial biomass produced from fractionated cellulose and from fractionated hemicelluloses (Figure 10.13) in the IAF process would be as good fungal biomass as that produced on pure carbohydrates (starches and sugars). The fungal biomass produced on lignocelluloses without fractionation contains 22–45% protein (Chahal and Mo-Young, 1981; Chahal, 1989a).

The amino-acid composition of various microorganisms used to produce food/feed from agricultural wastes and other carbohydrates is compared with that of alfalfa (a common animal fodder), soybean (a common protein source), and the FAO reference in Table 10.20. The comparison of all these organism meet the FAO reference requirement except for sulphur-containing amino acids. Anderson *et al.* (1975) have also reported similar findings by comparing the amino-acid composition of various filamentous fungi grown on soluble sugars and starch. The deficiency of sulphur amino acids can easily be met by supplementing the microbial protein with methionine. The amino-acid composition of *C. celluloyticum* aspsorgenous mutant is as good as that of the parent strain, *T. viride*, *Cellulomonas* sp. and *F. graminearum*.

Feeding trials with various cellulolytic organisms (including *C. cellulolyticum*) used to produce microbial food/feed are still in their infancy. However, most of the feeding trials have indicated that up to 20–50% of the total protein requirement can be replaced with the protein from these organisms without any pathological problems (Duthie, 1975; Han and Anderson, 1975; Peitersen, 1975; Srivastava *et al.*, 1980; Chavez *et al.*, 1988). Some feeding trials arranged by Chavez *et al.* (1988) have indicated that the protein diet for rats, chicken and piglets can be replaced with that of *C. cellulolyticum* (parent strain) supplemented with methionine.

(f) Pharmaceutical values of mushrooms. The medicinal values and health strengthening properties, including improvements in blood circulation, of various mushrooms grown in SSF and their mycelial biomass produced in LSF from crop and forestry residues have been studied in China for the last 2000 years (Yang and Yong, 1989). The pharmaceutical properties of various mushrooms and their mycelia are now gaining importance in the West also.

The mushroom biomass is composed of a linear polymer of *N*-acetyl-D-glucosamine and D-glucosamine having β1–4-glycosidic bonds. The *N*-acetylglucosamine and glucosamine have been shown chemically and enzymatically, and by infrared spectrum and X-ray differaction analyses, to be as chitin and chitosan, respectively (Novaes-Ledieu and Garcia, 1981). The chitosan has been associated with anticholestrol properties (Ibihara and Schneeman, 1989). Similarly, the mycelial biomass or mushrooms, containing *N*-acetyl-D-glucosamine and glucosamine of *Pleurotus ostreatus* (Bobek *et al.*, 1993), *Lentinus edodes* (Sannoumaru, 1996) and *Polyporus confluens* (Sugiyama *et al.*, 1992) are responsible for the reduction of cholesterol in blood serum.

Table 10.2 Essential amino-acid composition (% total true protein) of proteins in *Chaetomium cellulolyticum*, other cellulolytic organisms of SCP interest, alfalfa, soybean and FAO reference protein

Amino acid	*C. cellulolyticum*[a]	*C. cellulolyticum* asporogenous mutant[b]	*Fusarium graminearum*	*T. viride*[c]	Cellulomonas[d]	Alfalfa[e]	Soybean[f]	FAO reference
Threonine	6.14	4.78	5.1[g]	4.9	4.7	5.12	4.0	2.8
Valine	5.76	5.88	7.2[g]	4.4	6.79	6.70	5.0	4.2
Cystine	0.31	0.85	0.8[a]	1.45(?)	0.41	1.40	1.4	2.0
Methionine	2.33	2.27	2.1[a]	1.35	1.69	1.96	1.4	2.2
Isoleucine	4.70	5.07	4.3[g]	3.5	4.12	5.54	5.4	4.2
Leucine	7.54	7.93	6.5[g]	5.8	8.66	8.43	7.7	4.8
Tyrosine	3.26	3.68	4.1[g]	3.3	2.41	3.72	2.7	2.8
Phenylalanine	3.77	4.80	4.4[g]	3.7	3.69	5.75	5.1	2.8
Lysine	6.80	7.12	7.5[g]	4.4	8.00	6.70	6.5	4.2
Tryphotphan	NA	NA	NA	NA	NA	NA	NA	NA

References: [a]Moo-Young *et al.* (1977); [b]Chahal (1991b); [c]Peiterson (1975); [d]Han and Anderson (1975); [e]Livingston *et al* (1971); [f]Shacklady (1975); [g]Anderson *et al.* (1975); and Duthie (1975).
NA = not available.

The antitumor activity of various mushrooms and their mycelia have been reported in *Pleurotus sajor-caju* (Zhuang *et al.*, 1993), *Lentinus edodes* and its mycelia (Sugano *et al.*, 1982; Mori *et al.*, 1989; Mizono *et al.*, 1996), *Volvariella volvacea* (Lin and Chou, 1984; Kishida *et al.*, 1989, 1992), *Coriolus versicolor* (Dong *et al.*, 1996) and *Omphalia lapidescens* (Ohno *et al.*, 1992).

Antitumor activity has been shown to be associated with protein-containing polysaccharides. The polysaccharide has a moderately branched structure of a backbone chain of β1–3-linked-D-glucose residues, one out of five or six being substituted at O-6 with single glucosyl of β1–6-linked glucosyl groups (Kishida *et al.*, 1989). According to Zhuang *et al.* (1993) the protein-containing polysaccharides are: xyloglucan, mannoglucan, xylan, glucoxylan in *Pleurotus sajor-caju*. Glyco-chain are composed of glucose, xylose, mannose, galactose, and fucose in *Pleurotus citrinopileatus* (Zhang *et al.*, 1994). Novaes-Ledieu and Garcia (1981) have shown that neutral polysaccharides in the cell walls of *Agaricus bisporus* and *A. campestris* include alkali-soluble glucan with α1–3-linkage and β1–3- and β1–6-linked glucan. This means that the anticholesterol and anticarcinogenic properties are associated with the fibres (cell walls) of various edible mushrooms.

Antiplatelet aggregation properties of shiitake mushroom (Hokama and Hokama, 1981), *Pleurotus ostreatus* and *Lentinus edodes* (Sumi *et al.*, 1996) have been reported. Immunopharmacological effects in *Omphalia lapidescens* were reported by Ohno *et al.* (1993).

The composition of nutraceutical biomass of various edible mushrooms is as follows: crude protein 35–45%, fat about 7% (the fat has a high ratio, about 4:1, of unsaturated fatty acids to saturated fatty acids), fibre about 10–15%, and nucleic acids (less than 2%). Vetter and Rimoczi (1993) have reported that digestibility of the crude protein of *Pleurotus ostreatus* is up to 92% digestible.

10.4.2 Solid-state fermentation

Solid-state fermentation is simply defined as a process whereby an insoluble substrate is fermented with sufficient moisture but without free water. In liquid-state fermentation, on the other hand, the substrate is solubilized or suspended as fine particles in a large volume of water (Chahal, 1983, 1985a). Mudgett (1986) defined SSF as being distinguished from submerged cultures by the fact that microbial growth and product formation occur at or near the surfaces of solid materials with low moisture contents. However, this is not so because it was observed earlier (Chahal *et al.*, 1983) under an optical microscope that in SSF the organism (*Chaetomium cellulolyticun*) not only grows on the surface of the substrate (corn stover) but actually penetrates deep into the substrate and even into the cell lumen. Recently, the hyphae of *Trichoderma reesei* QMY-1 were seen to have penetrated into intercellular spaces (middle lamella) as well as intracellular spaces (cell lumens) of wheat straw particles during SSF (Chahal, 1989b). This means that the organism in SSF not only grows at or near the surface of the substrate but actually penetrates deep into the intercellular and intracellular spaces of the substrate, showing a really close contact or association with the substrate.

Although SSF has many advantages over LSF, it has its own inherent problems (Hesseltine, 1972; Chahal, 1983; Mudgett, 1986). However, because of its many advantages over LSF, it has been used for upgrading the protein values of lignocelluloses, i.e. agricultural residues (Han and Anderson, 1975; Yu *et al.*, 1976; Zadrazil, 1977; Chahal *et al.*, 1981), and for the production of aflatoxins (Hesseltine, 1972), gibberellic acid (Kumar and Lonsane, 1987), rennet (Brisk and Zuckermann, 1971; Thakur *et al.*, 1990) and spores of mycoherbicides (Silman *et al.*, 1989).

Mushroom production in SSF is one of the oldest microbiological processes known to man. It is not possible to discuss this aspect here as it is beyond the scope of this chapter. A few SSF processes where the lignocelluloses are upgraded for their protein and digestibility values, and for the production of other useful industrial compounds are discussed in the following sections.

(a) SCP production by SSF. The use of SSF for the bioconversion of lignocelluloses into SCP as animal feed, and for increasing the *in vitro* digestibility of agricultural residues for animal feed, is becoming popular, especially in developing countries owing to its low technology, low cost of dewatering of the final product and abundance of agricultural residues.

Zadrazil (1977) reported that, during SSF of wheat straw for 120 days with *Stropharia rugosannulum* or *Pleurotus cornucopiae*, *in vitro* digestibility increased by up to 60–70%. Similarly, Detroy *et al.* (1980) reported that growing *P. ostreatus* on wheat straw in SSF for 50 days increased its hydrolysis to 72% with cellulases. However, Tsang *et al.* (1987) showed that it was not practical to produce *Pleurotus* mushrooms and a highly delignified residual straw simultaneously by SSF.

There are many reports on increasing the protein values of lignocelluloses by using SSF. The protein content of newsprint was increased to 6.5% with *Sporotrichum thermophile* in 6 days (Barnes *et al.*, 1972) and sawdust to 7.7% with *Chaetomium cellulolyticum* in 9 days (Pamment *et al.*, 1978). Three such processes to enhance the protein values of lignocelluloses are described briefly as follows.

Han's process. In this process straw is chopped, mixed with three parts of 0.5 N H_sSO_4 solution, and hydrolysed in a pressure cooker under 15 lb steam for 30 min. The hydrolysed straw is neutralized with ammonia or ammonium hydroxide to raise the pH to 4.5. The treated straw contains about 20% fermentable sugars. It is inoculated with *Aurobasidium pullulans*, *Candidia utilis* or *Trichderma viride*. The fermentation chamber is revolved to provide a constant tumbling motion to the straw and to permit free exchange of air.

The straw can also be treated with NaOH (4% dry wt). It is neutralized with acid, and ammonium sulphate is added as a nitrogen source. The treated straw is inoculated with a mixture culture of *Cellulomonas* sp. and *Alcaligenes faecalis* for the production of SCP in SSF.

Protein contents of 14%, 12.4% and 10.9% dry wt in the final products were reported when acid-treated straw was fermented with *Aurobasidium pullulans*, *Candida utilis* and *Trichoderma viride*, respectively. In the case of alkali-treated straw, a protein content of only 6.8% in the final product was reported when inoculated with a mixed culture of *Cellulomonas* sp. and *Alcaligen faecalis* for 2–3 days (Han and Anderson, 1975; Han *et al.*, 1976; Yu *et al.*, 1976).

INRA-Dijon process. In this process, sugar beet pulp is fermented with a special strain of *T. viride* T.S. for the production of SCP in SSF (Durand *et al.*, 1991). A new type of pilot reactor (1 ton capacity) and a new technology of microorganism cultivation in a substrate layer 1 m deep has been developed at the INRA-Dijon research centre in France. The process developed can be easily adapted to a continuous process and can be incorporated into a sugar factory.

During 48 h of cultivation, the protein content of pulp was increased up to 20–21% on the basis of dry matter. The fermentation process was carried out in non-sterile conditions, using simple cheap mineral nutrients.

The amino-acid content of the final product increased by about 50% during the fermentation. For essential amino acids, the increase was about 60%. The nitrogen value of the enriched pulp, which takes all essential amino acids into account, was slightly lower than that of soybeans (72–76%) with limiting factors mainly in sulphur amino acids and marginally in lysine and isoleucine.

No toxic effects were observed during the 11-week feeding trial of protein-enriched pulp with lambs and rabbits. The protein-enriched pulp can replace all of the soybean supplementation in lambs and rabbits.

The estimated cost price for production of protein-enriched pulp in SSF was about 2 FFr kg^{-1} dry wt in hypothetical production at 50 t h^{-1} in the new developed reactor.

Chaetomium cellulolyticum *as SCP by SSF*. With *C. cellulolyticum*, a protein content of 9.8%, 10.7% and 19.7% in the final products was obtained from untreated, cold-ammonia pretreated and washed alkali-pretreated wheat straw, respectively, in SSF in 10–12 days (Chahal *et al.*,

Table 10.3 Comparison of SCP production on various lignocelluloses with various cellulolytic microorganisms in solid-state fermentation

Reference/organism	Substrate/treatment	Crude protein in product (% dry weight)	Incubation time (days)
Barnes *et al.* (1972)			
Sporotrichum thermophile	Newsprint (untreated)	6.5	6
Han and Anderson (1975)			
Aurobasidium pullulans	Ryegrass (0.5 N H_2SO_4, 121 °C, 30 min)	14.0	2–3
Candida utilis	Ryegrass (0.5 N H_2SO_4, 121 °C, 30 min)	12.4	2–3
Trichoderma viride	Ryegrass (0.5 N H_2SO_4, 121 °C, 30 min)	10.9	2–3
Yu *et al.* (1976)			
Cellulomonas sp. + *Alcaligenes faecalis*	Ryegrass (4% NaOH, 25 °C)	6.8	2–3
	Ryegrass (NH_3)	9.5	2–3
Pamment *et al.* (1978)			
Chaetomium cellulolyticum	Sawdust	7.7	9
	Sawdust (20% NaOH)	7.7	9
Chahal *et al.* (1981)			
Chaetomium cellulolyticum	Wheat straw (4% NaOH)	19.0	5
	Corn stover (4% NaOH)	20–24	4–5
	Corn stover (3% NH_3)	17	4
Durant *et al.* (1991)			
Trichoderma viride T.S.	Sugar beet pulp	20–21	2

1981). A high protein content was obtained when corn stower was fermented by SSF. About 17.5% (dry wt) protein was obtained in the final product when cold- or hot-ammonia pretreated corn stover was fermented with *C. cellulolyticum* in SSF for 3–4 days. The highest protein content of 20–24% dry wt was obtained when alkali-pretreated corn stover containing ammonium sulphate was fermented with this fungus in SSF for 5 days (Chahal *et al.*, 1983).

C. cellulolyticum seems to be the best organism for the production of SCP in SSF (Table 10.3). The high percentage of protein in the final product obtained in SSF with *C. cellulolyticum* is attributed to the fact that this fungus can penetrate deep into the solid substrate as well as in intercellular (middle lamella) and intracellular (cell lumen) spaces of the substrate for its bioconversion into SCP (Chahal *et al.*, 1983).

Pleurotus *as an edible mushroom by SSF*. *Pleurotus* cultivation is gaining popularity in Canada, USA, Europe and the Far East. This is rated almost as important as *Agaricus bisporus* and *Lentinus edodes* (Wood and Smith, 1987). The commercial production techniques for these mushrooms have been well documented by Tautorus (1985), and Wood and Smith (1987). The substrate is shredded, mixed with water and placed in bags or trays. There is no need to add other nutrients, because it can degrade wood or ligninocellulosic materials very easily. Coffee pulp (Guzman and Martinez, 1986), cassia (Muller, 1987) and cotton stalks (Silanikove and Levanon, 1986; Danai *et al.*, 1989; Hadar *et al.*, 1992) have been used as substrates for the production of SCP.

According to Hadar *et al.* (1992) the cotton stalks were harvested and chopped into 2–3 cm particles with a forage harvester originally designed to cut corn. The material is then taken to a 450 ton capacity concrete silo with a heavy tractor and then is covered with black plastic sheets. After 1 month of storage, the pH of the preserved cotton stalks is stabilized at 5.5 and the material is successfully utilized for commercial *Pleurotus* cultivation, up to 9 months after harvest (Danai *et al.*, 1989; Levanon *et al.*, 1988).

(b) Production of cellulase-system by SSF. The 'koji' process (SSF) is being extensively used for the production of amylases in Japan (Toyama, 1976). Solid-state fermentation, SSF, is discussed elsewhere in this book (Chapter 3). However, the application of this process was extended to the production of cellulase on wheat bran and lignocelluloses in SSF by Toyama (1976), Chahal (1985a, 1986), Deschamps *et al.* (1985) and Shamala and Sreekantiah (1986, 1987). The production of a complete cellulase-system in SSF was devised by Chahal (1985a, 1986, 1989b). Cellulase production is discussed in Chapter 5.

For complete hydrolysis of cellulose into glucose the cellulase system must contain: (1) endo-glucanases; (2) exo-glucanases; and (3) β-

glucosidase in the right proportions. It is due to their synergistic effect that cellulose is hydrolysed to glucose but the cellulase of *Trichoderma reesei* produced under most cultural conditions is deficient in β-glucosidase. Owing to this deficiency of β-glucosidase, cellobiose, an intermediate product, accumulates in the hydrolysate which inhibits further saccharification of cellulose.

During the last 12 years of research, the cellulase activity per unit volume has been increased to a great extent by producing hypercellulase mutants of *T. reesei*, e.g. MCG 77 by Gallo *et al.* (1978), Rut-C30 by Montenecourt and Eveleigh (1979), L27 by Shoemaker *et al.* (1981), C1847 by Warzywoda *et al.* (1983), D1-6 by Panda *et al.* (1983) and QMY-1 by Chahal (1985a). Similarly, the cellulase activity per unit volume has been increased by fed-batch fermentation (Hendy *et al.*, 1982; Watson and Nelligan, 1983; McLean and Podruzny, 1985) and by SSF (Chahal, 1985a). By these methods the filter paper cellulase activity has been increased up to 17–30 IU ml^{-1}.

On the other hand, to reduce the production cost of cellulases, the pure cellulose substrate has been successfully replaced with that of the crude cellulose from pretreated lignocelluloses by Chahal *et al.* (1982), Chahal (1985a) and Mes-Hartree *et al.* (1988).

This survey of the literature indicates that very little is known about the production of a complete cellulase-system containing all the enzymes as described above. To achieve this objective, SSF seems to be the most promising approach.

Comparison of cellulase production by LSF and SSF. Filter paper cellulase (FP cellulase) activity of 1.65 IU ml^{-1} and productivity of 165 IU g^{-1} substrate were obtained by growing *T. reesei* QMY-1 on 1% wheat straw in LSF. The enzyme activity increased to 6.0 IU ml^{-1} but the productivity decreased to 120 IU g^{-1} substrate when the concentration of the substrate was increased to the 5% level in LSF. On the other hand, a high enzyme activity per unit volume (8.6 IU ml^{-1}) and high activity per unit of substrate (172 IU g^{-1} substrate) were obtained under SSF conditions (Table 10.4). The FP cellulase activity obtained under SSF is higher than that produced under the best conditions of LSF at the 1% substrate level. Complete analysis of the other components of the cellulase-system produced under SSF revealed that the ratio of FP cellulase to β-glucosidase was about 1:1.2. This ratio is comparable to the 1:1 recommended by Mandels *et al.* (1981) and Chahal *et al.* (1982) to obtain high hydrolysis of cellulose. The composition of cellulase-systems produced on wheat straw and Pro-cell TM (wood) with *T. reesei* QMY-1 are presented in Table 10.5. This shows clearly that a ratio of FP cellulase to β-glucosidase close to the desired ratio, ie. 1:1, can be obtained in SSF. The xylanase activity is also very high in the cellulase-system obtained on these substrates under SSF.

Table 10.4 Cellulase production with *Trichoderma reesei* QMY-1 on wheat straw in liquid-state fermentation (LSF) and solid-state fermentation (SSF)[a]

Fermentation conditions	Fermentation time (days)	Filter paper cellulase	
		IU ml^{-1}	IU g^{-1} substrate
LSF (1%)	7	1.65	165
LSF	11	6.0	120
SSF	22	8.6[b]	172

[a]Data taken from Chahal (1985a).
[b]Five grams of original wheat straw of SSF was mixed in 100 ml of H_2O to extract the enzyme to make it comparable to 5% wheat straw in LSF.

Table 10.5 Composition of cellulase-systems of *Trichoderma reesei* QMY-1 produced on different substrates in SSF

Substrate	FP cellulase (IU ml^{-1})	β-glucosidase (IU ml^{-1})	Ratio of β-glucosidase: FP cellulase	Xylanases (IU ml^{-1})	Reference
Wheat straw	8.6	10.6	1.2	270	Chahal (1985a)
Pro-cell™	5.3	4.1	0.77	404	Chahal (1986)

[a]Five grams of the original substrate of SSF was mixed with 100 ml H_2O to extract enzymes.

Composition of various cellulase-systems produced by different cellulolytic fungi by SSF. Comparison of the composition of cellulase-systems of various cellulolytic fungi produced on SSF indicated that *T. reesei* QMY-1 proved to be the best fungus for the production of a cellulase-system containing the highest activities of all the enzymes, ie. FP cellulase, β-glucosidase and xylanases, compared with all the other cellulolytic fungi tried by other researchers (Table 10.6). The higher productivity of *T. reesei* QMY-1 could be due to the effects of species differences and the methodology used, as well as to the ability of the hyphae of this fungus to penetrate deep into the substrate as well as into intercellular spaces (middle lamella) and intracellular spaces (cell lumens) (Chahal, 1989b).

Although xylanases are normally not considered as part of cellulase-systems for hydrolysis of cellulose, these enzymes are needed when hydrolysis of cellulose and hemicelluloses, as found together in lignocelluloses, is required. For that purpose, the cellulase-system of *T. reesei* QMY-1, which is also very rich in xylanases compared to those of other fungi (Table 10.6), will be the most suitable enzyme system to hydrolyse the polysaccharides of lignocelluloses into their monomer sugars.

Hydrolytic potential of cellulase-systems. A comparison of the hydrolysis of delignified wheat straw with the cellulase-systems produced in SSF and

Table 10.6 Composition of cellulase-systems produced in SSF[a]

	Substrate	Organism	Enzyme activity (IU g^{-1} dry wt substrate)			
			FP cellulase[b]	β-glucosidase	Xylanases	References
1.	Wheat straw	*T. reesei* QMY-1	172	212	5400	Chahal (1985a)
2.	Pro-cell™ (wood)	*T. reesei* QMY-1	106	82	8080	Chahal (1986)
3.	Wheat bran	*Aspergillus ustus*	3.7	60	615	Shamala and Sreekantiah (1986)
4.	Rice straw	*Aspergillus ustus*	5.8	15.8	740	Shamala and Sreekantiah (1986)
5.	Wheat bran and rice straw (1:1)	*Aspergillus ustus*	4.6	40.7	398	Shamala and Sreekantiah (1986)
6.	Rice straw (recycled)	Various microorganisms	14	48	1431	Shamala and Sreekantiah (1986)
7.	Wheat bran (recycled)	Various microorganisms	8	81	788	Shamala and Sreekantiah (1987)
8.	Rice straw (recycled 5 times)	*Sporotrichum pulverulentum*	9	15.8	385	Shamala and Sreekantiah (1987)
9.	Wheat straw and wheat bran (80:20)	*Trichoderma harzianum*	18	–	–	Deschamps *et al.* (1985)
10.	Straw and bran (1:1)	*Aspergillus niger*	–	–	2500	Deschamps and Huet (1985)

[a]The results of cellulase production in SSF of Toyama (1976) were not included in this table as these were not in comparable units of enzymes.
[b]FP cellulase = filter paper cellulase determined according to the method of Mandels *et al.* (1976). It is due to the synergistic effect of endo- and exo-gluconase on filter paper.

LSF is given in Table 10.7. About 90% of delignified wheat straw was hydrolysed into simple sugars within 96 and 72 h of incubation with the cellulase-systems produced in SSF on wheat straw and Pro-cell TM, respectively. It is interesting to note that the quantity of cellobiose in the hydrolysate obtained with the cellulase-system produced on Pro-cell TM was higher than that obtained with the cellulase-system produced on wheat straw. This could be due to the fact that the former has a lower ratio of FP cellulase to β-glucosidase (1:0.77) than that produced on wheat straw (1:1.2). However, the Pro-cell TM cellulase-system had a faster rate of hydrolysis as it took 72 h to obtain 90% hydrolysis compared to 96 h for wheat straw cellulase-system.

The hydrolysis of delignified wheat straw with the cellulose-system produced in LSF with 5% wheat straw was slow, taking 144 h to reach 80% hydrolysis. Although the FP cellulase to β-glucosidase ratio of the cellulase-system produced in LSF was very close to that produced in SSF on Pro-cell TM, it still gave a lower rate of hydrolysis (80%) even after 144 h, double the time the Pro-cell TM cellulase-system took to reach 90%. This indicates that there may be some additional factor or enzyme produced under SSF conditions which could be responsible for the fast rate and high yields of hydrolysis.

(c) Other usages of lignocellulosic materials by the SSF technique.

Gibberellic acid production. Gibberellic acid (GA3) is a potent plant growth regulator and is extensively used in agriculture, nurseries, greenhouses, viticulture, tea gardens, etc., for the elimination of dormancy in seeds, acceleration of seed germination, induction of flowering, improvement of crop yields, overcoming of dwarfism, etc. Currently it is produced in LSF but the cost of production is very high because of extremely low yields, extensive downstream processing and consequently high capital and operating expenses. In fact, constraints on its extensive use owing to its high cost are noticeable and hence any substantial lowering in its cost of production may trigger its use on a large scale (Kumar and Lonsane, 1987).

Kumar and Lonsane (1987) have developed a SSF process for the production of GA3 by using *Gibberella fujikuroi* P-3 on wheat bran. On the basis of available carbohydrates in the medium, the conversion rates were 0.096% and 0.156% in LSF and SSF, respectively. The use of coarse wheat bran (0.3–0.4 cm) in SSF resulted in an increase of 2.5 times in the yield of GA3. A yield of 1.05–1.20 g kg^{-1} of dry moldy bran established the potential and feasibility of SSF for its production on wheat bran. Kumar and Lonsane's preliminary cost analysis shows a net saving of about 60% and 50% on fermentation medium cost and expenditure on

Table 10.7 Comparison of hydrolysis of delignified wheat straw with cellulase-systems of *Trichoderma reesei* QMY-1 produced by solid-state fermentation[a]

Substrate for the production of cellulase-system	Ratio of β-glucosidase: FP cellulase	Glucose (g l^{-1})	Cellobiose (g l^{-1})	Xylose (g l^{-1})	Arabinose (g l^{-1})	Total sugars (g l^{-1})	Hydrolysis (%)[b]
Alkali-treated wheat straw using SSF (96 h)[c]	1.2	68.18	3.19	26.71	1.67	99.75	89.77
Alkali treated Pro-cell™ using SSF (72 h)[d]	0.77	61.30	8.3	31.8	–	101.4	91.26
Alkali-treated wheat straw using LSF[e]							
24 h	0.8	13.9	6.5	10.0	0.86	31.26[f]	56.2
60 h	–	19.8	6.5	11.8	0.86	38.96[f]	70.1
144 h	–	27.3	3.4	13.2	0.86	44.76[f]	80.5

[a]Delignified wheat straw (100 g l^{-1}) was hydrolysed with 20 IU FP cellulase g^{-1} substrate for 96 h with a cellulase system produced on alkali-treated wheat straw and for 72 h with a cellulase system produced on alkali-treated Pro-cell. Wheat straw was delignified by the method described by Toyama (1972).
[b]Percentage of hydrolysis = [(Total wt of sugar produced × 0.9)/(wt of substrate)] × 100%.
[c]Data from Chahal (1985a).
[d]Data from Chahal (1986).
[e]Data from Chahal (1989b).
[f]Total sugars from 50 g l^{-1} of delignified wheat straw.

downstream processing, respectively, compared with the cost of the presently employed LSF techniques.

Citric acid production. Citric acid, a tricarboxylic acid, was originally isolated from lime juice for various uses. Wehmer (1893) first reported that citric acid can be produced by fermenting sucrose with species of the genera *Penicillium* and *Mucor*. On an industrial scale, citric acid is produced with *Aspergillus niger* under stationary or surface cultural conditions. These fermentation conditions have now been replaced by submerged fermentation (Röhr *et al.*, 1983).

The SSF for production of citric acid was introduced by Cahn as early as 1935 (Cahn, 1935). In this system the solid substrates, such as sugar-cane bagasse, potato or beet pulp, pineapple pulp, etc., were impregnated with fermentation medium and were inoculated with the fungal spores. These substrates acted as inert materials to hold the fermentation medium. Later, this method or SSF for citric acid production was patented by Hisanaga and Nishimura (1968) and Yo (1975). Lakshminarayana *et al.* (1975) reported on 80% yield of citric acid by growing *A. niger* 3/1 on sugar-cane bagasse impregnated with sucrose under SSF. The bagasse served just as an inert substrate to provide surface area for fermentation of impregnated sucrose solution in it. These yields were much higher than those obtained with surface or submerged fermentation. The production of citric acid by the koji process (SSF) on wheat bran with *A. niger* is also being practised in Japan as reported by Röhr *et al.* (1983).

Spores production. The use of spores as bioinsecticides and bioherbicides (mycoherbecides) is becoming very important in modern agriculture. The SSF is also becoming the chosen method for the production of spores of a bacterium (*Bacillus thuringiensis*) and of many fungi (*Metarhizium anisopliae*, *Hirsutella thompsonii*, *Beauvaria bassiana*, *Verticillium lencanii*, etc.) to be used as bioinsecticides (Quinlan and Lisansky, 1958). The spores of *Colletotrichum gloeosporioides* are being sold under the brand name of Collego R to be used as a bioherbicide (Silman *et al.*, 1989). Wheat bran, ground corn, cottonseed meal, barley grains and wheat grains are preferred because these solid substrates also serve as nutrients in the SSF. On the other hand, volcanic glass, diatomaceous earth, vermiculite, etc., are used as solid inert materials which are impregnated with soluble nutrients for SSF. Sugar-cane bagasse or straw could also be used as an inert substrate for SSF after impregnation with suitable media for production of spores or other products.

TeBeest (1985) and Thomas *et al* (1987) are of the opinion that production of spores in SSF is time consuming, labour intensive, prone to contamination and uneconomic. Churchill (1982) claimed that the submerged production technique (LSF) is preferable because of the availability of the technology, and the relatively easy scale-up process, but this may not be true in the case of those microorganisms which cannot produce spore in LSF. However, it is worth mentioning here that Morin *et al.* (1990) succeeded in producing spores of such microorganisms, *Phomopsis convolvulus*, in LSF in modified Richard's medium with V-8. Nevertheless, SSF is becoming a subject of intensive study these days for the production of various products, including spores, thus it seems that real competition might arise between LSF and SSF for the production of certain products in the near future.

10.5 Utilization of the lignin component of lignocelluloses

Lignins, as a group of abundant biopolymers embodying some significant diversity, occupy a pivotal position in the carbon cycle of biosphere. Therefore, lignin biodegradation has commanded attention for a considerable time. Early studies on biodegradation included only the disappearance of lignin from lignocelluloses owing to microbial activities. However, serious studies on the removal of lignin from lignocellulosic materials started during the 1970s when the research on production of biofuels started on a war footing. The discovery of ligninases (ligninolytic enzymes) in 1983 by Tien and Kirk, and Glenn *et al.* created excitement and enthusiasm among both the scientific and industrial community. Thus, most of the scientists working on biofuels jumped onto the bandwagon of biodegradation of lignin. Great expectations were laid on the capabilities of ligninolytic enzymes for the degradation and modification of lignin, particularly in the pulp and paper industries including biobleaching of pulp and detoxification of paper mill effluents.

10.5.1 Ligninases/ligninolytic enzymes

Ligninases are a family of isoenzymes, oxidases and peroxidases responsible for the oxidative depolymerization of lignin. The enzymes implicated are lignin peroxidase (LiP), manganese peroxidase (MnP) and laccase (phenol oxidase) (Tien, 1987).

In fact, there is some scepticism that the studies on lignin-model compounds have revealed very little about lignin degradation *in situ* (Lewis and Yamamoto, 1990). The LiP oxidizes nonphenolic, electron-rich aromatic rings to the corresponding radical cations at the low optimum pH 3 (Kersten *et al.*, 1985; Schoemaker *et al.*, 1985). Characteristic reactions of radical cations of lignin and lignin models include $C\alpha$–$C\beta$ cleavage and cleavage of the β–O-4 ether bond, the basis of the so-called 'depolymerizing' reactions. Other characteristic reactions include aromatic ring opening, demethoxylation, hydroxylation, decarboxylation and phenol-coupling reactions. The capability to oxidize phenolic lignin substructures efficiently is probably one of the major assets of MnP. Although LiP can oxidize phenolic compounds, this leads to a rapid inactivation of the enzyme (Harvey *et al.*, 1987b). Therefore, MnP and LiP can act synergistically. MnP, which appears somewhat earlier than LiP in ligninolytic cultures of *P. chrysosporium*, can oxidize the phenolic lignin and the phenolic reaction products, while LiP can oxidize the nonphenolic lignin substructure also. Shoemaker and Leisola (1990) have suggested that by both peroxidase-induced polymerization and peroxidase-induced depolymerization, a dynamic system is established (a so-called 'polymerization–depolymerization' equilibrium) which can be shifted towards degradation by fungal uptake of smaller fragments (Schoemaker *et al.*, 1989).

Evans *et al.* (1991) used an immunogold cytochemical labelling technique with electron microscopy of wood infected by basidiomycetes in the elucidation of the localization of enzymes that degrade lignocelluloses. Enzyme localization with an electron microscopic study of infected wood has shown that lignocellulolytic enzymes cannot penetrate into the wood structure except where the wood cell wall is already partially decayed. Most of the enzymes localized by immunogold cytochemical labelling techniques were located at the cell wall surfaces and, once the wood cell wall was partially eroded, the enzymes began to penetrate into the wall. This pattern of enzyme distribution was typical of LiP, laccase, endo-1,4-ß-glucanase and 1,4-ß-D-glucan cellobiohydrolase 1, but cellobiase (1,4-ß-glucosidase) was always located within the extracellular polysaccharide sheath surrounding the hyphae (Evans *et al.*, 1991). The above data have been supported by measurements of the pore sizes in wood which show that large molecules such as enzymes would be unable to penetrate into the cell wall (Flournoy *et al.*, 1991). Evans *et al.* (1994) have presented a hypothesis that there is regulation of the relative sequence of lignocellulolytic enzymes defining the spatial arrangement between hyphae and the wood cell with initiation of decay by low-molecular-mass mediators. Small molecules, such as hydrogen peroxide, veratryl alcohol, oxalate and manganese would be capable of diffusing into the wood cell wall structure to initiate decay, so opening up the pore size in wood to allow enzymes to penetrate to complete the degradative process (Evans *et al.*, 1994).

Koenigs (1974) reported that Fenton's reaction (Fe^{2+} + H_2O_2), which produces hydroxyl radical that is active in depolymerizing cellulose, in addition to enzymes, was thought to be a reason for the complete extraction of cellulose from the lignocellulosic matrix of wood by brown-rots. However, the Fenton's reaction mechanism has not been proven to be the primary mode of cellulose degradation in wood, mainly because the lifetime of extracellular H_2O_2 in the environment is unknown and its diffusion into the wood cell wall has not been demonstrated (Veness and

Evans, 1989). Hydrogen peroxide is necessary for the action of the ligninolytic enzymes, LiP and MnP. Its logical site of production would be in close association with them in the mucilage surrounding the hyphae to enable specific interaction. Glyoxal oxidase is thought to be a major source of extracellular H_2O_2, and is produced in the culture under the growth conditions that are identical for the production of LiP (Kersten, 1990). Veratryl alcohol is produced by many white-rot species as a secondary metabolite. It has been postulated by Harvey *et al.* (1986) that a mediator molecule such as the cation radical of veratryl alcohol, produced by interaction with LiP, could be involved as an agent in lignin depolymerization.

Oxalate has been shown to chelate cations such as Ca^{2+}, Fe^{2+} and NH_4^+ from its environment, frequently forming crystals of insoluble calcium oxalate, which has led to speculation that this is a means of environmental detoxification (Dutton *et al.*, 1993). Calcium, however, is an important constituent in plant cell walls. It can be withdrawn from calcium pectate and sequestered by oxalate. In conjunction with pectinases secreted by wood-rotting fungi, this can lead to significant changes in the cell wall structure (Volger *et al.*, 1982). The pore size within the cell wall would be enlarged by the removal of calcium ions and may permit access by enzyme molecules which were previously excluded. Another role for oxalate may be to enable the Fenton's reaction to cycle by reducing Fe^{3+} to Fe^{2+}, with the concomitant production of active hydronium ions, H_3O^+, regenerating Fe^{2+} for the reaction with H_2O_2. The exact concentrations of ions may be important in this reaction as the Fenton's reaction has been shown to be inhibited by 1 mM oxalic acid (Schmidt *et al.*, 1981).

Evans *et al.* (1994) concluded that, although many possible low-molecular-mass molecules have been suggested as candidates for a mobile factor to permeate wood cell walls and initiate decay, none has been conclusively proven as such. However, it is likely that many such agents are involved to address the degradation of the complex structure of the wood cell wall. Veratryl alcohol and oxalate are produced as a result of fungal metabolism and their secretion has enabled the fungi to colonize and degrade the wood cell wall structure more effectively than other organisms.

A small amount of polymeric [^{14}C]lignin substrates are converted to $^{14}CO_2$ by *Pseudomonas* spp. (Kaplan and Hardenstein, 1980; Odier *et al.*, 1981). Some Gram-negative bacteria, *Xanthomonas* (Odier and Monties, 1978; Kern, 1984), *Acinetobacter* (Crawford, 1975; Odier *et al.*, 1981), *Aeromonas* (Deschamps *et al.*, 1980) and *Erwinia* (Liaw *et al.*, 1988; Liaw and Srinivasan, 1989), and actinomycetes, *Actinomadura* (McCarthy and Broda, 1984), *Nocardia* (Smith and Ratledge, 1989), *Streptomyces* (Crawford, 1981; Njoku and Antai, 1987), have also been shown to degrade lignin or lignin compounds.

10.5.2 Production of ligninases

Ligninolytic enzymes or ligninases are produced by fungi, actinomycetes and bacteria, especially wood-rotting fungi and *Streptomyces*. *Phanerochaete chrysosporium* is the most studied wood-rotting fungus for production of ligninases. Carbon, nitrogen and manganese are the critical nutritional variables for production of ligninases by *P. chrysosporium* (Bonnarme *et al.*, 1991) and also for lignin degradation. Carbon limitation causes the rapid onset of lignin mineralization but it is short lived as the cells undergo autocatabolism accompanied by a rapid loss of dry weight (Jeffries *et al.*, 1981). On the other hand, nitrogen limitation also limits the ability of the organism to produce extracellular proteins (enzymes). Therefore, the supply of carbon and nitrogen is most critical in the production of ligninases in *P. chrysosporium*. Manganese is the specific effector that induces MnP and represses LiP in *P. chrysosporium* (Bonnarme *et al.*, 1991).

In the literature low yields of ligninases with *P. chrysosporium* have usually been reported so far. The reason for low yields could be attributed to insufficient nitrogen being available in the nitrogen-limited medium, the essential cultural conditions for production of ligninase in *P. chrysosporium*. Recently, a medium containing wood chips of 2.5–5 mesh size has been developed for the production of ligninases with *P. chrysosporium* and high yields of lignin peroxidase (1111 U l^{-1}) and manganese peroxidase (475 U l^{-1}) were obtained (Laplante, 1994).

Some actinomycetes grow in filamentous form very similar to that of fungi and are capable of degrading lignin and lignocellulosic materials in the soil. Ligninolytic activities have been reported in actinomycetes, especially in the genus *Streptomyces* by Pasti and Crawford (1991) by the method of decolorization of polymeric dyes. Magnuson and Crawford (1992) have confirmed the direct role of peroxidase of *Streptomyces viridosporus* T7A in lignin solubilization. Crawford *et al.* (1993) reported that a recombinant actinomycete, *Streptomyces lividans* TK 231, expressing a pIJ702-encoded extracellular lignin peroxidase, when introduced into soil in the microcosm, showed mineralization of lignin. Thus, there is a strong indication that, in addition to white-rot fungi, certain actinomycetes form another source of ligninolytic enzyme system to be explored.

It has been observed that lignin degradation by *Streptomyces badius* is greatest in the presence of high levels of organic nitrogen (Bader and Crawford, 1981). Therefore, it appears that lignin depolymerization by *S. badius* occurs during primary metabolism whereas it occurs during secondary metabolism in the case of *P. chrysosporium*.

A solid agar medium with dyes for screening the ligninolytic activity of white-rot fungi and actinomycetes has been developed (Chahal *et al.*, 1995). This method is very accurate, quick and easy for the screening of ligninolytic activity of a large number of microorganisms and their mutants or clones. By this method a new isolate of actinomycetes was found to produce equally good ligninolytic activity on the nitrogen-limited and nitrogen-complete agar medium irrespective of the type of carbon source (glucose, xylan). Chahal *et al.* (1995), while working with various mutants of *Streptomyces lividans* and a newly isolated actinomycetes, indicated that poly-blue oxidase activities of these organisms decolorized some polymeric dyes both in nitrogen-limited and nitrogen-complete media. This indicates that actinomycetes may hold some promise for producing high yields of ligninases without limiting the nitrogen or carbon in the medium.

(a) Uses of ligninolytic enzymes. Since the discovery of ligninases by Glenn *et al.*, and Tien and Kirk in 1983, they have opened up a whole array of potential uses in:

- biodelignification of lignocellulosic materials to be used as feedstock for production of fuel alcohols and other chemicals;
- delignification of wood for reducing the use of chemicals and energy for pulping;
- detoxification of pulp mills effluents;
- biobleaching of pulp (Paice *et al.*, 1988);
- improving the digestibility and nutritional values of lignocellulosic feeds (Reid, 1989; Reid and Seifert, 1982).

The application of ligninolytic enzymes for the delignification of lignocellulosic materials as a pretreatment for their conversion into biofuels and single cell proteins has been suggested by Chahal (1991a).

Although there are many uses of lignin as such, the most exciting application of ligninases and ligninolytic organisms *in situ* is for partial degradation of lignin and bioconversion of lignin into oligolignols of low molecular weight for their further chemical/biotransformation into high value chemicals (Kirk and Chang, 1981; Tien, 1987; Chahal and Hachey, 1990).

Although xylanases have been tried for the delignification and prebleaching of pulp, the use of ligninases would be more specific for delignification and that will not remove hemicelluloses, an important component of pulp. Thus, the use of ligninases would be more specific for delignification and reduction of kappa number of pulp while hemicelluloses and cellulose would be kept intact. The use of ligninases would also reduce the energy and chemical requirement for the pulping process (Tien, 1987).

The inherent ability of the ligninolytic enzyme system of white-rot fungi to cleave varieties of carbon–carbon and ether bonds in lignin suggests that such organisms may be useful for the biotransformation/biodegradation of recalcitrant environmental pollutants (Fernando and Aust, 1994). Bumpus *et al.* (1985) were among the pioneers who evaluated the ability of white-rot fungus, *Phanerochaete chrysosporium*, to degrade a variety of toxic recalcitrant chemicals: lindane (1,2,3,4,5,6-hexachlorocyclohexane), TCB (3,4,3′,4′,-tetrachlorobiphenyl), TCDD (2,3,7,8-tetrachlorodibenzo-*p*-dioxin), DDT [1,1-bis (4-chlorophenyl)-2,2,2-trichloroethane] and benzo(a)pyrene. They confirmed the involvement of the ligninolytic enzyme system of *P. chrysosporium* in the biodegradation of these xenobiotics when measured as the evolution of $^{14}CO_2$.

The large-scale application of chlorophenols in agriculture and as by-products generated from industrial plants, e.g. effluents from paper bleach plants (Huynh *et al.*, 1985), has led to the contamination of terrestrial and aquatic ecosystems. Pentachlorophenols (PCBs) have also been reported to be degraded by the ligninolytic enzyme system of *P. chrysosporium* (Mileski *et al.*, 1988). Polyaromatic hydrocarbons (PAHs) and polychlorinated hydrocarbons (mainly insecticides) can also be degraded by the ligninolytic enzyme system of *P. chrysosporium* (Haemmerli *et al.*, 1986; Hammel *et al.*, 1986).

(b) Lignin solubilization and/or degradation. Hüttermann *et al.*, (1989) reported that lignin solubilization is not necessarily correlated to the high activities of LiP or laccase. However, it was observed in their data that there was some correlation between the poly-blue oxidase and solubilization of lignin. It indicated that poly-blue oxidase activity may be responsible for solubilization of lignin.

Sarkanen (1991) declared that unprecedented reputability had been conferred on the role of lignin peroxidase in lignin degradation in 1983 because:

- detectable LiP activity is not a prerequisite for ligninolysis by the microorganism;
- LiP alone *in vitro* polymerizes rather than depolymerizes lignin preparations;
- nonoxidatively deploymerized lignin fraction can be isolated from wood that has been partially degraded by *P. chrysosporium*;
- elevated levels of LiP activity are correlated with slower rate of lignin biodegradation in *Lentinula edodus* cultures.

The quest, therefore, for the key enzyme upon which the decomposition of lignin in nature ultimately rests, promises to be as fascinating as the journey which led to the original discovery of lignin peroxidase in 1983 (Sarkanen, 1991). Nevertheless, LiP may be capable of introducing functional groups into some of the lignin monomer residues that would result in enhanced susceptibility of the macromolecular structure towards subsequent depolymerization by another enzyme (Fenn and Kirk, 1984).

10.6 Problems in bioconversion and future trends

The collection of lignocellulosic wastes and their transportation to the processing sites are two of the major problems in their bioconversion into various products. Crop residues scattered in the fields and forestry wastes left after felling may be far away from the city where bioconversion processes are to be set up. However, some wastes, such as sawdust, bark and other wood wastes, are found in large quantities near the pulp and paper mills, and lumber mills. Similarly, sugar-cane bagasse is piled up near the sugar mills. In such cases bioconversion plants can be set up close to such sites.

The other major problem is that the polysaccharides of lignocelluloses are found in recalcitrant form and are well protected with lignin. Now, a number of effective pretreatments, especially steam explosion and alkali treatment, are available to make the lignocelluloses readily available for bioconversion into various products. However, the cost of pretreatment of lignocelluloses is still a debatable question in making the bioconversion processes economic, especially for low-value products like fuel ethanol.

The lignocelluloses are composed of cellulose, hemicelluloses and lignin. On hydrolysis with chemicals or enzymes, a mixture of various sugars and lignin (in various forms) are produced. There is no bioconversion process available which can use the mixture of all these components to synthesize a product or products. To overcome such problems, the 'integrated process for production of food, feed and fuel (Ethanol) from lignocelluloses' was developed at the Institut Armand-Frappier. In this process, various microorganisms are used to ferment various fractions of lignocelluloses into food, feed and fuel (ethanol), and modified lignin is obtained as a coproduct. Thus, there is no fraction of lignocellulose left unutilized. Such integrated processes for the bioconversion of lignocelluloses into other products will be needed in the future. Currently, this process is being exploited by DC Enterprises, Inc. for its commercialization.

Lignin, in terms of its weight, is probably second only to cellulose in the lignocellulosic materials (Kirk *et al.*, 1980). It has been estimated that, for every ton of lignocelluloses used in the above IAF integrated process, approximately 250 kg of lignin will be released (Chahal, 1985b; Chahal *et al.*, 1987). It is also envisaged that enormous quantities of lignin will start accumulating as soon as the industries based on the utilization of polysaccharides from lignocelluloses for production of food, feed, fuel (ethanol) and other products are set up. Moreover, sulphite mills are already producing lignin in the form of lignosulphonate in millions of tons every year. At present only a fraction of such lignin is being used in industry. Although a voluminous literature on the degradation of lignin is available, there is no process whereby such an enormous quantity of surplus lignin could be used to synthesize new products. Therefore, there is a dire need for new processes for the chemical or biological conversion of lignin into new products before its disposal becomes a colossal problem.

In recent years tremendous progress has been made in understanding the mechanism of microbial degradation of lignocellulosic materials. However, because of the complexity of the problem, a vast amount of research remains to be done in order to fully understand all the factors involved in the biodegradation process and eventually to be able to apply this knowledge in developing commercial processes. A literature survey indicated that depolymerization/biodegradation of lignin is still not very clear. As soon as this phenomenon is understood properly, the lignin may become the most valuable feedstock for its bioconversion into high-value products and all other products that are now produced from hydrocarbons.

Solid-state fermentation of lignocelluloses for the production of cellulase and other products appears to have a great potential in the future.

References

Adler, E. (1977) *Wood Science Technology*, **11**, 160.

Ander, P. and Eriksson, K.-E. (1978) *Progress in Industrial Microbiology*, **14**, Elsevier, Amsterdam, p. 1.

Anderson, C., Longton, C., Maddix, G.W. *et al.* (1975) In *Single-cell Protein, II* (eds S.R. Tannenbaum and D.I.C. Wang), MIT Press, Cambridge, MA, p. 314.

Atalla, R.H. (1983) In *Wood and Agricultural Residues* (ed. E.J. Soltes), Academic Press, New York, p. 59.

Atalla, R.H. and van der Hart, D.L. (1984) *Science*, **223**, 283.

Bader, M.J. and Crawford, D.L. (1981) *Canadian Journal of Microbiology*, **27**, p. 859.

Barnes, T.G., Eggins, H.O.W. and Smith, E.L. (1972) *International Biodeterioration Bulletin*, **8**(3), 112.

Bassham, J.A. (1975) *Biotechnology and Bioengineering Symposium*, No. 5, John Wiley, New York, p. 9.

Blackwell, J. (1982) In *Cellulose and Other Natural Polymer Systems: Biogenesis, Structure and Degradation* (ed. R.M. Brown), Plenum, New York, p. 403.

Bobek, P., Ozdin, L. and Kuniak. I. (1993) *Nahrung*, **37**, 571.

Bonnarme, P., Perez, J. and Jeffries, J.W. (1991) *ACS Symposium Series*, **460**, 200.

Brisk, J.L. and Zuckermann, S.S. (1971) *Schweizerische Milchzeitung*, **97**, p. 544.

Bumpus, J.A., Tien, M., Wright, D. and Aust, S.D. (1985) *Science*, **228**, p. 1434.

Buswell, J.A. and Odier, E. (1987) *CRC Critical Reviews in Biotechnology*, **6**, p. 1.

Cahn, F.J. (1935) *Industrial Engineering Chemistry*, **27**, 201.

Callihan, C.D. and Clemmer, J.E. (1979) In *Microbial Biomass* (ed. A.H. Rose), Economic Microbiology, Vol. 4, Academic Press, New York, p. 271.

Chahal, D.S. (1983) ACS *Symposium Series*, **207**, 421.

Chahal, D.S. (1985a) *Applied and Environmental Microbiology*, **49**, 205.

Chahal, D.S. (1985b) *Optimization for an Improved Process for Single-cell Protein (SCP) Production from Forest Biomass*. Report submitted to Yves Leveque, Director General of Forest Industry, Government of Quebec, Ministry of Energy and Resources, Sainte-Foy, Quebec.

Chahal, D.S. (1986) *VII International Symposium on Alcohol Fuels*, Edition Technip, Paris, p. 48.

Chahal, D.S. (1989a) *Journal of Fermentation and Bioengineering*, **68**, 334.

Chahal, D.S. (1989b) *1989 International Chemical Congress of Pacific Basin Societies*, Honolulo, Hawaii, December, Abstract 431.

Chahal, D.S. (1991a) In *Food, Feed and Fuel from Biomass* (ed. D.S. Chahal), Oxford, and IBH Publishing Co., New Delhi, p. 59.
Chahal, D.S. (1991b) In *Food, Feed and Fuel from Biomass*, (ed. D.S. Chahal), Oxford, and IBH Publishing Co., New Delhi, p. 175.
Chahal, D.S. (1995) Indian patent application: Production of cellulose, hemicelluloses, lignin, silica and protein rich food from rice straw.
Chahal, D.S. and Hachey, J.M. (1990) *ACS Symposium Series*, **433**, 304.
Chahal, D.S. and Hawksworth, D.L. (1976) *Mycologia*, **68**, 600.
Chahal, D.S. and Ishaque, M. (1986) *Science et Techniques de l'Eau*, **19**, 339.
Chahal, D.S. and Ishaque, M. (1988) *Science et Technique de l'Eau*, **21**, 27.
Chahal, D.S. and Moo-Young, M. (1981) *Developments in Industrial Microbiology*, **22**, 143.
Chahal, D.S. and Wang, D.I.C. (1978) *Mycologia*, **70**, 160.
Chahal, D.S., Swan, J.E. and Moo-Young, M. (1977) *Developments in Industrial Microbiology*, **18**, p. 433.
Chahal, D.S., Vlach, D. and Moo-Young, M. (1981) In *Advances in Biotechnology, III* (ed. M. Moo-Young), Pergamon Press, Toronto, p. 327.
Chahal, D.S., McGuire, S., Pikor, H. and Noble, G. (1982) *Biomass*, **2**, p. 127.
Chahal, D.S., Moo-Young, M. and Vlach, D. (1983) *Mycologia*, **75**, p. 597.
Chahal, D.S., Kluepfel, D., Morosoli, F. *et al.* (1995) *Applied Biochemistry and Biotechnology*, **51/52**, 137.
Chahal, D.S., Ishaque, M., Brouillard, D. *et al.* (1987) *Journal of Industrial Microbiology*, **1**, p. 355.
Chavez, E.R., Touchburn, S.P. and Moo-Young, M. (1988) *Animal Feed Science Technology*, **22**, p. 23.
Churchill, B.W. (1982) In *Biological Control of Weeds with Plant Pathogens* (eds R. Charudattan and H.L. Walker), John Wiley, New York, p. 139.
Cowling, E.B. (1975) *Biotechnology and Bioengineering Symposium*, **5**, p. 163.
Cowling, E.B. and Merill, W. (1966) *Canadian Journal of Botany*, **44**, p. 1539.
Crawford, D.L. and Crawford, R.L. (1980) *Enzyme and Microbial Technology*, **2**, p. 11.
Crawford, D.L., Doyle, J.D., Wang, Z. *et al.* (1993) *Applied and Environmental Microbiology*, **59**, 508.
Crawford, R.L. (1975) *Canadian Journal of Microbiology*, **21**, p. 1654.
Crawford, R.L. (1981) *Lignin Biodegradation and Transformation*, Wiley, New York.
Crawford, R.L. and Crawford, D.L. (1984) *Enzyme and Microbial Technology*, **6**, p. 434.
Danai, O., Levanon, D. and Silanikov, N. (1989) *Mushroom Science*, **12**, p. 81.
Daniel, G. (1994) *FEMS Microbiology Reviews*, **13**, p. 199.
Das, K. and Ghose, T.K. (1973) *Journal of Applied Chemistry and Biotechnology*, **23**, p. 829.
Deschamps, F. and Huet, M.C. (1985) *Applied Microbiology and Biotechnology*, **22**, p. 177.
Deschamps, F., Mahoudeau, G. and Leeault, J.M. (1980) *European Journal of Applied Microbiology and Biotechnology*, **9**, p. 45.
Deschamps, F., Giuliano, C., Asther, M., Huet, M.C. and Roussos, S. (1985) *Biotechnology and Bioengineering*, **27**, p. 1385.
Detroy, R.W., Lindenfelser, L.A., St Julian, G. Jr and Orton, W.L. (1980) *Biotechnology and Bioengineering Symposium*, **10**, 135.
Dong, Y., Kwan, C.Y., Chen, Z.N. and Yang, M.M. (1996) *Research Communications in Molecular Pathology and Pharmacology*, **92**, p. 140.
Dunlop, C.E. and Callihan, C.D. (1973) *Single-cell Protein from Waste Cellulose*, Final report on grant EP00328-4 to the Federal Solid Waste Management Program, US Environmental Protection Agency.
Durand, A., Grajek, W. and Gervais, P. (1991) In *Food, Feed and Fuel from Biomass* (ed. D.S. Chahal), Oxford, and IBH Publishing Co., New Delhi, p. 123.
Duthie, I.F. (1975) In *Single-cell Protein II* (eds S.R. Tannenbaum and D.I.C. Wang), MIT Press, Cambridge, MA, p. 505.
Dutton, M.V., Evans, C.S., Atkey, P.T. and Wood, D.A. (1993) *Applied Microbiology and Biotechnology*, **39**, p. 5.
Effland, M.J. (1977) *Technical Association of the Pulp and Paper Industry,* **60** (10), 143.
Eriksson, K.E., Blanchette, R.A. and Ander, P. (1990) *Microbial and Enzymatic Degradation of Wood and Wood Components*, Springer-Verlag, New York.

Evans, C.S. (1987) *Process Biochemistry*, **22**(4), 102.

Evans, C.S., Gallagher, I.M., Atkey, P.T. and Wood, D.A. (1991) *Biodegradation*, **2**, 93.

Evans, C.S., Dutton, M.V., Guillen, F. and Veness, R.G. (1994) *FEMS Microbiology Reviews*, **13**, p. 235.

Fenegel, D. and Wegener, G. (1983) *Wood Chemistry, Ultrastructure, Reactions*, Walter de Gruyter, Berlin.

Fenn, P. and Kirk, T.K. (1984) *Journal of Wood Chemistry Technology*, **4**, p. 131.

Fernando, T. and Aust, S.D. (1994) In *Biological Degradation and Bioremediation of Toxic Chemicals* (ed. G.R. Chaudhry), Dioscorides Press, Portland, OR, p. 386.

Flournoy, D.S., Kirk, T.K. and Highley, T.L. (1991) *Holzforschung*, **45**, p. 383.

Freudenberg, K. (1965) *Science*, **48**, p. 595.

Freudenberg, K., Friedrich, K., Baumann, I. and Soff, K. (1932) *Annalen der Chemie (Justus Liebigs)*, **494**, 41.

Gallo, B.J., Andreotti, R., Roche, C., Ryu, C. and Mandels, M. (1978) *Biotechnology and Bioengineering*, **8**, p. 89.

Gardner, K.H. and Blackwell, J. (1974) *Biopolymers*, **13**, p. 1975.

Gilbert, N., Hobbs, I.A. and Levine, J.D. (1952) Industrial Engineering and Chemistry, **44**, p. 1712.

Glenn, J.K., Morgan, M.A., Mayfield, M.B., Kuwahara, M. and Gold, M.H. (1983) *ACS Symposium Series*, **389**, 127.

Guzman, G. and Martinez, D. (1986) *Mushroom Newsletter for the Tropics*, **6**, 7.

Hadar, Y., Kerem, Z., Gorodecki, B. and Ardon, O. (1992) *Biodegradation*, **3**, p. 189.

Haemmerli, S.D., Leisola, M.S.A.M., Sanglard, D. and Fiechter, A. (1986) *Journal of Biological Chemistry*, **261**, p. 6900.

Hammel, K.E., Kalyanaraman, B. and Kirk, T.K. (1986) *Proceedings of the National Academy of Sciences of USA*, **83**, p. 3708.

Han, Y.W. and Anderson, A.W. (1975) *Applied Microbiology*, **30**, p. 930.

Han, Y.W. and Callihan, C.D. (1974) *Applied Microbiology*, **27**, p. 159.

Han, Y.W., Cheeke, P.R., Anderson, A.W. and Lekprayoon, C. (1976) *Applied Environmental Microbiology*, **32**, p. 799.

Harris, E.E. and Belinger, E. (1946) *Industrial Engineering and Chemistry*, **38**, p. 890.

Harvey, P.J., Schoemaker, H.E. and Palmer, J.M. (1985) *Annual Proceedings of the Phytochemistry Society of Europe*, **26**, p. 249.

Harvey, P.J., Schoemaker, H.E. and Plamer, J.M. (1986) *FEBS Letters*, **195**, p. 242.

Harvey, P.J., Schoemaker, H.E. and Palmer, J.M. (1987a) *Plant Cell Environment*, **10**, p. 709.

Harvey, P.J., Schoemaker, H.E. and Palmer, J.M. (1987b) *Proceedings of an International Seminar on Lignin, Enzymic and Microbial Degradation*, Paris, April INRA (Les Colloques de l'INRA, no. 40), p. 145.

Hendy, N., Wilke, C. and Blanch, H. (1982) *Biotechnology Letters*, **4**, 785.

Hess, K., Mahi, H., and Guther, E. (1954) *Kolloid-Zeitschrift*, **155**, 1.

Hesseltine, C.W. (1972) *Biochemical and Bioengineering*, **14**, p. 517.

Higuchi, T. (1971) *Advances in Enzymology*, **34**, p. 207.

Higuchi, T. (1980) In *Lignin Biodegradation: Microbiology, Chemistry and Potential Application* (eds T.K. Kirk, T. Higushi and Hou-Min Chang), CRC Press, Boca Raton, FL, p. 1.

Higuchi, T. (1981) *Wood Research (Kyoto)*, **67**, p. 47.

Higuchi, T. (1982) *Experientia*, **38**, p. 159.

Hillis, W.E. (1962) *Wood Extractives and their Significance to the Pulp and Paper Industries*, Academic Press, New York.

Hisanaga, S. and Nishimura, Y. (1968) Japanese patent 6820708, 5 September.

Hokama, Y. and Hokama, J.L. (1981) *Research Communications in Chemical Pathology and Pharmacology*, **31**, p. 177.

Hüttermann, A., Milstein, D., Nicklas, B. *et al.* (1989) *ACS Symposium Series*, **397**, p. 361.

Huynh, V.B., Chang, H.M., Joyce, T.W. and Kirk, T.K. (1985) *Technical Association of the Pulp and Paper Industry*, **68**, 98.

Ibihara, K. and Schneeman, B.O. (1989) *Journal of Nutrition*, **119** (B), 1100.

Ishaque, M. and Chahal, D.S. (1991) In *Food, Feed and Fuel from Biomass* (ed. D.S. Chahal), Oxford, and IBH Publishing Company, New Delhi, p. 19.

Janshekar, H. and Fiechter, A. (1983) *Advances in Biochemical Engineering/Biotechnology*, **27**, p. 117.
Jeffries, T.W., Choi, S., and Kirk, T.F. (1981) *Applied and Environmental Microbiology*, **42**, 290.
Kalmes, O. (1959) The distribution of constitutents across the wall of unbleached spruce sulfite fibers. Doctoral thesis, Institute of Paper Chemistry, Appleton, WI.
Kaplan, D.L. and Hardenstein, R. (1980) *Soil Biology and Biochemistry*, **12**, p. 65.
Kern, H.W. (1984) *Archives of Microbiology*, **138**, p. 18.
Kersten, P.J. (1990) *Proceedings of the National Academy of Sciences of the USA*, **87**, 2936.
Kersten, P.J., Tien, M., Kalyanaraman, B. and Kirk, T.K. (1985) *Journal of Biological Chemistry*, **260**, p. 2609.
Kirk, T.K. (1984) In *Microbial Degradation of Organic Compounds* (ed. D.T. Gibson), Marcel Dekker, New York, p. 399.
Kirk, T.K. (1988a) In *Biochemistry and Genetics of Cellulose Degradation* (eds J.P. Aubert, P. Beguin and J. Millet), Academic Press, London, p. 315.
Kirk, T.K. (1988b) *ISI Atlas of Science: Biochemistry 0894-3753*, p. 71.
Kirk, T.K. and Chang, H.-M. (1981) *Enzyme Microbial Technology*, **3**, 189.
Kirk, T.K. and Farrell, R.L. (1987) *Annual Review of Microbiology*, **41**, p. 465.
Kirk, T.K. and Fenn, P. (1982) In *Decomposer Basidiomycetes* (eds M.J. Swift, J. Frankland, and J.N. Hedger), *British Mycological Society Symposium 4*, Cambridge University Press, Cambridge, p. 67.
Kirk, T.K. and Haskin, I.M. (1973) *American Society of Chemistry, Engineering Symposium Series*, Vol. 69, p. 124.
Kirk, T.K., Higuchi, T. and Chang, H.-M. (1980) In *Lignin Biodegradation: Microbiology, Chemistry and Potential Applications*, Vol. 241 (eds T.K. Kirk, T. Higuchi and Hou-Min Chang), CRC Press, Boca Raton, FL, p. 255.
Kishida, E., Sone, Y. and Misaki, A. (1989) *Carbohydrate Research*, **193**, p. 227.
Kishida, E., Kinoshita, C., Sone, Y. and Misaki, A. (1992) *Bioscience, Biotechnology and Biochemistry*, **56**, p. 1308.
Koenigs, J.W. (1974) *Archives of Microbiology*, **99**, p. 129.
Kumar, P.K.R. and Lonsane, B.K. (1987) *Process Biochemistry*, **22**, p. 139.
Lakshminarayana, K., Chaudhary, K., Ethiraj, S. and Tauro, P. (1975) *Biotechnology and Bioengineering*, **17**, 291.
Lange, P.W. (1958) *Paper and Pulp Magazine (Canada)*, **59**, p. 210.
Laplante, S. (1994) Optimisation de la production de ligninases avec le champignon basidiomycete phanerochaete chrysosporium ATCC-24725, M.Sc. thesis, Institut Armand-Frappier, University of Quebec, Laval, Quebec.
Laskin, A.I. (1977) *Annual Reports on Fermentation Processes*, **1**, 151.
Leisola, M.S.A. and Fiechter, A. (1985) *Advances in Biotechnology Processes*, **5**, 59.
Levanon, D., Danai, O. and Masaphy, S. (1988) *Biological Wastes*, **26**, p. 341.
Lewis, N.G. and Yamamoto, E. (1990) *Annual Review of Plant Physiology*, **41**, p. 455.
Liaw, J.H. and Srinivasan, V.R. (1989) *Applied Environmental Microbiology*, **55**, p. 2220.
Liaw, J.H., Jones, K.L. and Srinivasan, V.R. (1988) *Abstracts of Annual Meeting of the American Society of Microbiology*, Miami Beach.
Lin, J. and Chou, Y. (1984) *Journal of Biochemistry*, **96**, p. 35.
Litchfield, J. (1968) In *Single-cell Protein* (eds R.I. Mateles and S.R. Tannenbaum), MIT Press, Cambridge, MA, p. 304.
Livingston, A.L., Allis, M.E. and Kohler, G.O. (1971) *Journal of Agricultural Food Chemistry*, **19**, 947.
Magnuson, T.S. and Crawford, D.L. (1992) *Applied and Environmental Microbiology*, **58**, p. 1070.
Mandels, M., Andreoti, R. and Roche, C. (1976) *Biotechnology and Bioengineering Symposium*, No. 6, p. 21.
Mandels, M., and Medeiros, J.E., Andreoti, R. and Bissett, F.H. (1981) *Biotechnology and Bioengineering*, **23**, 2009.
Manley, R. St.J. (1964) *Nature*, **204**, 1155.
McCarthy, A.J. and Broda, P. (1984) *Journal of General Microbiology*, **130**, p. 2905.
McIntyre, T.C. (1987) In *Biomass Conversion Technology: Principles and Practice* (ed. M. Moo-Young), Pergamon Press, Toronto, p. 45.

McLean, D. and Podruzny, M.F. (1985) *Biotechnology Letters*, **7**, p. 683.
Meier, H. (1958) *Svensk Papperstidning*, **61**, 633.
Meller, F.H. (1969) *Conversion of Organic Solid Wastes into Yeast*, Bureau of Solid Waste Management, Washington, DC.
Mes-Hartree, M., Hogan, C.M. and Saddler, J.N. (1988) *Biotechnology and Bioengineering*, **31**, p. 725.
Mileski, G.J., Bumpus, J.A., Jurek, M.A. and Aust, S.D. (1988) *Applied and Environmental Microbiology*, **54**, p. 2885.
Millett, M.A., Baker, A.J. and Sattar, L.D. (1976) *Biotechnology and Bioengineering Symposium*, No. 6, p. 125.
Mizono, M., Minato, K. and Tsuchida, H. (1996) *Biochemistry and Molecular Biology International*, **39**, 679.
Montenecourt, B.S. and Eveleigh, D.E. (1979) In *Fuel from Biomass Symposium* (eds W.W. Schuster and H.R. Bungay), Department of Energy Publication, Rensselaer Polytechnic Institute, Troy, New York, p. 613.
Moo-Young, M., Chahal, D.S., Swan, J.E. and Robinson, C.W. (1977) *Biotechnology and Bioengineering*, **19**, p. 527.
Moo-Young, M., Dougulis, A.J., Chahal, D.S. and Macdonal, D.G. (1979) *Process Biochemistry*, **14**(10), p. 38.
Moo-Young, M., Chahal, D.S., Vlach, D. and Stickeny, B. (1980) *Biotechnology and Bioengineering*, **23**, p. 2407.
Mori, K., Toyomasu, T., Nanba, H. and Kuroda, H. (1989) *Mushroom Science*, **12** (Part I), 653.
Morin, L., Watson, A.K. and Reeleder, R.D. (1990) *Canadian Journal of Microbiology*, **36**, p. 86.
Mudgett, R.E. (1986) In *Manual of Industrial Microbiology and Biotechnology* (eds A.L. Demain and N.A. Soloman), American Society for Microbiology, Washington, DC, p. 66.
Muller, J. (1987) *Mushroom Journal of Tropics*, **7**, p. 89.
Newark, P. (1980) *Nature*, **287**, p. 6.
Nimz, H. (1974) *Angew Chemistry*, **86**, 336.
Njoku, C.C. and Antai, S.P. (1987) *Letters in Applied Microbiology*, **4**, 133.
Novaes-Ledieu, M. and Garcia, M.C. (1981) *Canadian Journal of Microbiology*, **27**, p. 779.
Odier, E. and Monties, B. (1978) *Annales de l'Institut Pasteur, Microbiology*, **129A**, 361.
Odier, E., Janin, G. and Monties, B. (1981) *Applied and Environmental Microbiology*, **41**, p. 337.
Ohno, N., Saito, K., Nemoto, J. *et al.* (1993) *Biological and Pharmaceutical Bulletin*, **16**, p. 414.
Ohno, N., Miura, T., Saito, K. *et al.* (1992) *Chemical and Pharmaceutical Bulletin*, **40**, p. 2215.
Paice, M.G., Bernier, R. and Jurasek, L., (1988) *Biotechnology and Bioengineering*, **32**, p. 235.
Palmer, J.M. and Evans, C.S. (1983a) *Philosophical Transactions of the Royal Society of London, Series B*, **300**, 293.
Palmer, J.M. and Evans, C.S. (1983b) *Proceedings of International Symposium on Wood Pulp Chemistry*, Vol. 3, Tsukuba, p. 19.
Pamment, N., Robinson, C.W., Hilton, J. and Moo-Young, M. (1978) *Biotechnology and Bioengineering*, **20**, p. 1735.
Panda, J., Bisaria, V.S. and Ghose, T.k. (1983) *Biotechnology Letters*, **5**, 767.
Pasti, M.B. and Crawford, D.L. (1991) *Canadian Journal of Microbiology*, **37**, p. 902.
Paterson, A., McCarthy, A.J. and Broda, P. (1984) In *Microbiological Methods for Environmental Biotechnology* (eds J.M. Grainger and J.M. Lynch), Academic Press, London, P. 33.
Payen, M. (1838) *Academy of Science*, **7**, p. 1052.
Peitersen, N. (1975) *Biotechnology and Bioengineeering*, **17**, p. 1291.
Pettersen, R.C., Schwandt, V.H. and Effland, M.J. (1985) *Journal of Chromatography Science*, **22**, p. 478.
Preston, R.D. and Cronshaw, J. (1958) *Nature*, **181**, p. 248.
Quinlan, R.j. and Linsansky, S.G. (1958) In *Biotechnology*, Vol. 3 (ed. H. Dellweg), Verlag Chemie, Deerfield Beach, FL, p. 233.

Rawat, J.K. and Nautiyal, J.L. (1991) In *Food, Feed and Fuel from Biomass* (ed. D.S. Chahal), Oxford, and IBH Publishing Company, New Delhi, p. 27.

Reese, E.T. (1963) *Advances in Enzymatic Hydrolysis of Cellulose and Related Materials*, MacMillan Company, New York.

Reid, I.D. (1989) *Enzyme and Microbial Technology*, **11**, p. 786.

Reid, I.D. and Seifert, K.A. (1982) *Canadian Journal of Botany*, **60**, p. 252.

Righelato, R.C., Imerie, F.K.E. and Vlitos, A.J. (1976) *Resources, Recovery and Conservation*, **1**, 257.

Röhr, M., Kubicek, C.P. and Kominek., J. (1983) In *Biotechnology*, Vol. 3. (ed. H. Dellweg), Verlag Chemie, Deerfield Beach, FL, p. 419.

Rolz, C. (1984) In *Fermentation Processes*, Vol. 7 (ed. G.T. Tsao), Academic Press, New York, p. 213.

Rowland, S.P. and Roberts, J.J. (1972) *Journal of Polymer Science, Part A*-1, **10**, p. 2447.

Sannoumaru, Y. (1996) *Journal of Nutritional Science and Vitaminology*, **42**, p. 97.

Sarkanen, S. (1991) *ACS Symposium Series*, **460**, 247.

Schmidt, C.J., Whitten, B.K. and Nicholas, O.O. (1981) *Proceedings of American Wood Preservation Association*, **77**, p. 157.

Shoemaker, H.E. and Leisola, M.S.A. (1990) *Journal of Biotechnology*, **13**, 101.

Schoemaker, H.E., Harvey, P.J., Bowen, R.M. and Palmer, J.M. (1985) *FEBS Letters*, **183**, p. 7.

Schoemaker, H.E., Meijer, E.M., Leisola, M.S.A. *et al. ACS Symposium Series*, **399**, 454.

Schulze, E. (1891) *Chemische Berichte*, **24**, 2277.

Sekita, S., Yoshihira, K. and Natori, S. (1981) *Canadian Journal of Microbiology*, **27**, p. 766.

Shacklady, C.A. (1975) In *Single Cell Protein, II* (eds S.R. Tannenbaum and D.I.C. Wang), MIT Press, Cambridge, MA, p. 489.

Shamala, T.R. and Sreekantiah, K.R. (1986) *Enzyme and Microbial Technology*, **8**, 178.

Shamala, T.R. and Sreekantiah, K.R. (1987) *Enzyme and Microbial Technology*, **9**, p. 97.

Shoemaker, S.P., Raymond, J.C. and Bruner, R. (1981) In *Trends in the Biology of Fermentation for Fuels and Chemicals* (eds A. Hollander *et al.*), Plenum, New York, p. 89.

Silanikove, N. and Levanon, D. (1986) *Biomass*, **9**, p. 101.

Silman, R.W., Nelson, T.C. and Bothast, R.J. (1989) *1989 International Chemical Congress of Pacific Basin Societies*, Honolulu, Hawaii, December, Abstract 404.

Sloneker, J.H. (1976) *Biotechnology and Bioengineering Symposium*, No. 6, p. 235.

Smith, M.R. and Ratledge, C. (1989) *Applied Microbiology and Biotechnology*, **30**, p. 395.

Srinivasan, V.R. and Han, Y.W. (1969) *Advances in Chemistry Series*, **95**, 447.

Srivastava, V.K., Mowat, D.N., Moo-Young, M., Daugulis, A.J. and Chahal, D.S. (1980) *VIth International Fermentation Symposium*, London, Ontario, July.

Sugano, N., Hibino, Y., Choji, Y. and Maeda, H. (1982) *Cancer Letters*, **17**, p. 109.

Sugiyama, K., Kawagishi, H., Tanaka, A. *et al.* (1992) *Journal of Nutritional Science and Vitaminology*, **38**, p. 335.

Sultze, R.F., Jr (1957) *Technical Association of the Pulp and Paper Industry*, **40**, 985.

Sumi, H., Yatagai, C. and Matsubara, K. (1996) *Nippon Shokuhin Kagaku Kogaku Kaishi [Journal of the Japanese Society of Food Science and Technology]*, **43**, 318.

Tanaka, M. and Matsuno, R. (1985) *Enzyme and Microbial Technology*, **7**, p. 197.

Tarkow, H. and Feist, W.D. (1969) *Advances in Chemistry Series*, **95**, 197.

Tautorus, T.E. (1985) *Advances in Biotechnology Processes*, **5**, 227.

TeBeest, D.O. (1985) *Journal of Agricultural Entomology*, **2**, 123.

Thakur, M.S., Karanth, N.G. and Nanad, K. (1990) *Applied Microbiology and Biotechnology*, **32**, 409.

Thomas, K.C., Kachatourians, G.G. and Ingledew, W.M. (1987) *Canadian Journal of Microbiology*, **33**, p. 12.

Tien, M. (1987) *CRC Critical Review in Microbiology*, **15**, p. 141.

Tien, M. and Kirk, T.K. (1983) *Science*, **221**, p. 661.

Timell, T.E. (1967) *Wood Science Technology*, **1**, p. 45.

Tönnesen, B.A. and Ellefsen, Ö. (1971) In *Cellulose and Cellulose Derivatives* (eds N.M. Bikales and L. Segal), Vol. 5, Part IV, Wiley, New York, p. 265.

Toyama, N. (1972) In *Fermentation Technology Today* (ed. G. Terui), Society of Fermentation Technology, Japan, Osaka, p. 743.

Toyama, N. (1976) *Biotechnology and Bioengineering Symposium*, No. 6, p. 207.
Tsang, L.J., Reid, I.D. and Coxworth, E.C. (1987) *Applied and Environmental Microbiology*, **53**, p. 1304.
Umezawa, T. (1988) *Wood Research*, **75**, p. 21.
Underkofler, L.A. and Hickey, R.J. (1954) *Industrial Fermentation*, Vol. 1, Chemical Publishing Co., New York.
Veness, R.G. and Evans, C.S. (1989) *Journal of General Microbiology*, **135**, p. 2799.
Vetter, J. and Rimoczi, I. (1993) *Zeitschrift fur Lebensmittel-Untersuchung und-Forschung*, **197**(5), 427.
Volger, C., Hesse, C. and Vogt, A. (1982) *European Journal of Forest Pathology (Berlin)*, **12**, 59.
Wardrop, A.B. (1957) *Technical Association of the Pulp and Paper Industry*, **40**, p. 225.
Warzywoda, M., Ferré, V. and Pourqui, J. (1983) *Biotechnology and Bioengineering*, **25**, p. 3005.
Watson, T.G. and Nelligan, I. (1983) *Biotechnology Letters*, **5**, p. 25.
Wehmer, C. (1893) French patent 288554, 11 March 1893. Cited in Prescott, S.C. and Dunn, C.G. (1959) *Industrial Microbiology*, McGraw-Hill, New York, p. 533.
Wiken, J.O. (1972) In *Fermentation Technology Today* (ed. G. Terui), Society of Fermentation Technology, Japan, Osaka, p. 569.
Wilke, C.R., Cysewski, G.R., Yang, R.D. and von Stockar, U. (1976) *Biotechnology and Bioengineering*, **8**, 1315.
Wood, D.A. and Smith, J.F. (1987) In *Essays in Agriculture and Food Microbiology* (eds J.R. Norris and G.L. Pettipher), Wiley, New York, p. 309.
Yang, Q.Y. and Yong, S.C. (1989) *Mushroom Science*, **12** (Part I), 631.
Yo, K. (1975) Japanese Patent 75154487, 12 December.
Yu, P.L., Han, Y.W. and Anderson, A.W. (1976) *Proceedings of the Western Section of American Society of Animal Science*, **27**, 189.
Zadrazil, F. (1977) *European Journal of Applied Microbiology*, **4**, p. 273.
Zeikus, J.G. (1981) *Advances in Microbial Ecology*, **5**, 211.
Zhang, J., Wang, G., Li, H. *et al.* (1994) *Bioscience, Biotechnology and Biochemistry*, **58**, p. 1195.
Zhuang, C., Mizuno, T., Shimada, A. *et al.* (1993) *Bioscience, Biotechnology and Biochemistry*, **57**, p. 901.

11 Bioconversion of waste water from the pulp and paper industry

K. EL HAJI, V. SACHDEVA AND R.D. TYAGI

11.1 Introduction

The pulp and paper industry is one of the most important sectors of the Canadian economy. Although beneficial, this industry is associated with numerous environmental problems. Every year this industry uses approximately 80 millions tons of chemical products and an enormous quantity of fresh water (100–170 m^3 ton^{-1} of produced pulp). This industrial sector is one of the biggest water polluters and hence is potentially harmful to aquatic ecosystems. Increasing demands for improvement in pulp quality and environmental safety standards forces this industry to make changes continuously. At present, the pulp and paper industry more than ever needs new technologies in order to minimize the production of hazardous substances.

Wood contains minor parts of fatty and resin acids and other organic compounds, which protect it from microorganisms and insects. When wood is processed, these substances introduce a certain toxicity into the waste water. During conventional bleaching, complex reactions occur involving the chlorination, oxidation and demethylation of residual lignin. The major products of these reactions are adsorbable organic halides (AOX). Extracts of bleached, Kraft-mill effluent (BKME) have been shown to contain mutagenic activity as well as to induce biochemical responses in fish, such as increased activity of the mixed-function oxygenase (MFO) enzyme system (Rao *et al.*, 1995). Also, bleaching effluents contain toxic, chlorinated, phenolic compounds and recalcitrant, chlorinated, lignin fragments of higher molecular weight (Heizle *et al.*, 1992). It is shown that the chromophoric and aromatic lignin derivatives of the waste waters from the bleaching stage are toxic and are highly resistant to biodegradation by conventional treatment methods (Bergbauer *et al.*, 1992).

Although efforts are continuing (Table 11.1), for these environmental reasons, the waste water from the pulp and paper industry must be controlled (in terms of quantity and quality), treated and new environmentally efficient pulping, bleaching processes must be developed. The pulp and paper industry must finance several expensive operations to respond to all the changes required (Table 11.1).

Table 11.1 The main legislation concerning waste water controls applied to the pulp and paper industry since the 1960s

Time period	Legislation/environmental remediation
1960s	Removal of settleable solids
1970s	Primary and secondary treatment
1980s	Tremendous growth of primary and secondary treatment
1990s	Toxicity caused by absorbable organic halides (AOX)
2000	Zero effluent device

In this chapter, an attempt is made to provide a perspective on the use of biotechnology in the pulp and paper industry, particularly in waste water treatment and/or the production of products from such waste waters. Before this, however, a brief overview of the industry is necessary to understand the complex and diverse nature of industrial processes involved.

In this work, our principal objectives are to provide an overview of the principal practices in a pulp and paper plant in order to determine the variety as well as the complexity of these effluents, characterizing the effluents which may lead to identifying those suitable for the bioconversion processes. The problem areas mentioned above can be tackled by either treating the wastewater originating from the pulp and paper industry effectively or by reducing the quantity of waste water produced by using more efficient processes involving pulping, bleaching and paper manufacture. These aspects have been presented and discussed.

Biotechnology may be applied in the pulp production, bleaching, bioconversion and effluent treatment before the effluents are released into the environment and is discussed in this paper.

11.2 Source of effluent from the pulp and paper industry

The effluents coming from a pulp and paper industry originate from different steps of pulp and paper manufacture. The volume and the chemical composition of these effluents vary from one plant to another depending on a number of parameters. Among these are the type of pulping process, the type of product coming from the process, or the type of wood used and its age, etc. To distinguish between these physical and chemical variations, we will limit ourselves to citing the major pulping processes. Similarly, before tackling waste water treatment problems, a good understanding of the principal steps in pulp and paper production is necessary to identify the sources of pollutants.

The principal target of this industry is the conversion of raw material (wood) into different varieties of paper and substantially reducing the discharge of pollutants into the environment without compromising their economical competitiveness. The transformation of wood into pulp and paper involves the following principal steps (Cossette, 1991):

1. wood treatment;
2. pulping of treated wood chips by different processes (chemical and/or mechanical);
3. bleaching pulp;
4. dissolving, drying and pressing the bleached pulp in different forms.

11.2.1 Pulping process

The wood transformation into pulp consists essentially of separating the wood into individual components (cellulose, hemicellulose, lignin) which will be suitable for paper and/or related products. Once separated, the cellulosic fibers are then dissolved to produce a pulp that is bleached in certain cases before it is transformed into paper. Several methods can be used in this process. Pulping processes are divided in two main categories, mechanical and chemical processes.

(a) Mechanical process. In the mechanical process, the cellulosic fibers are isolated from other wood components (defibration) by the application of friction energy under controlled temperature (Macleay et al., 1987). The mechanical pulp is used in paper grades, such as newsprint and magazine paper. In this process only about 8% of the wood is lost as dissolved organics and product yields (paper produced per unit wood components) range typically between 90% and 95%.

(b) Chemical process.

Alkaline process (Kraft). This technique consists of using chemical products (sodium hydroxide and sodium sulfite) which act as bond breakers between cellulosic fibers and other wood constituents under temperatures ranging from 150 to 200 °C. Kraft is by far the most widely used chemical process. The yield of this process is limited at 40% (Fiedler *et al.*, 1990) and up to 60% of the total wood mass can be transformed into soluble organics and wasted as effluent. Spent cooking liquors (the solution containing wood and the alkaline chemicals in the digester) are among the principal wastes produced by this process. They are removed from the pulp by washing and are generally treated in a chemical recovery system (Figure 11.1).

Acid process (bisulfite). In this case, the cellulosic fibers are separated by the reaction of sulfur dioxide and a metallic base under high temperature and pressure (McCubbin *et al.*, 1991). The effluent produced during this

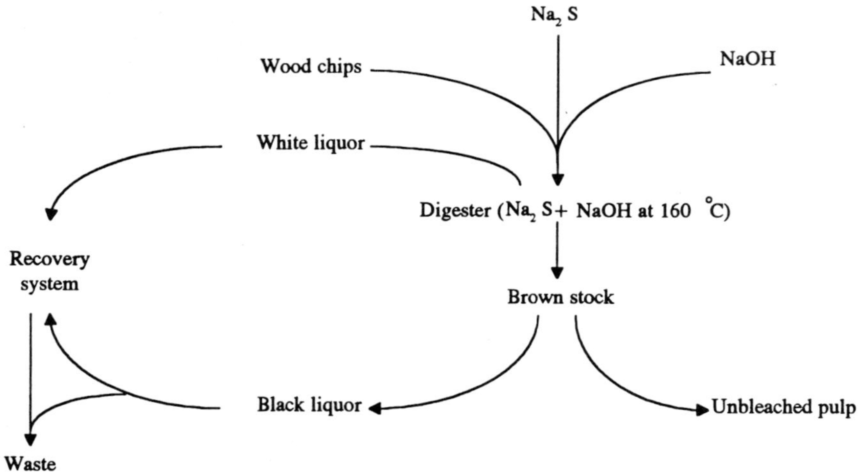

Figure 11.1 A summary of the alkaline process (Kraft).

process exhibits high toxicity towards the microorganisms, and hence possibility of any biological treatment, owing to the formation of H_2S. However, the waste water produced from this process contains some fermentable sugars (Figure 11.2).

11.2.2 Bleaching process

The two principal bleaching processes are the chemical method, which is the most used, and the biological method, which is considered as the new technology and is described later in this text (Katagari *et al.*, 1995). In the bleaching process, the lignin is first solubilized in alkali by treatment with chlorine and precipitated with sodium hydroxide. This manufacturing step requires about 100 kg chlorine ton^{-1} of pulp. This process is responsible for the production of all the organochlorines. The bleaching processes discharge large volumes of brown-colored effluents, which are associated with high biocemichal oxygen demand (BOD), total solids (TS) and total organic carbon (TOC). Residues from this proces may prove harmful to the environment.

Elimination of the bleach plant waste liquors would solve not only the problem of adsorbable organics halides (AOX) but will significantly reduce the total discharge of organic material, nutrients, non-recoverable sodium and colored substances as well (Myréen, 1994). The toxicity and mutagenicity of bleach plant effluents are well documented and are largely attributed to the chlorinated phenolics and low-molecular-weight, chlorinated, neutral compounds (Reeve and Earl, 1989; KirkPatrick, 1991).

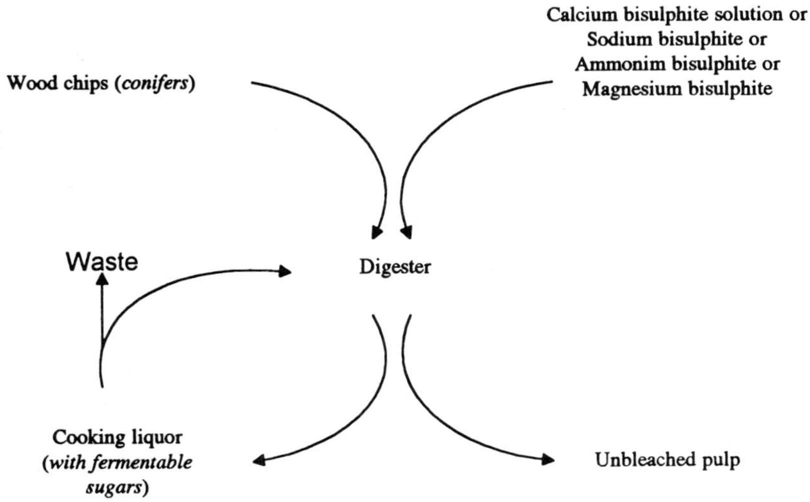

Figure 11.2 A summary of the acid process (bisulfite).

11.3 Characteristics of waste water from pulp and paper mills

The waste waters from pulp and paper industry have two principal characteristics. Firstly, the different processes (i.e. Kraft process, mechanical process, acid process) release varying volumes and compositions of wastes, and, secondly even within one process, the waste waters are produced separately at different manufacturing steps (pulping, bleaching and paper manufacturing). These wastes, all in the aqueous phase, contain different substances classified in the following three categories:

1. The biodegradable part, which is composed mainly of wood compounds (cellulose, hemicellulose, lignin and extractables).
2. Parts with difficulty in absence of biodegradability, which are represented by products from complexation of the above substances with chemicals.
3. The toxic part, which is made up of chemicals used for pulping, bleaching and paper manufacturing.

11.3.1 Biodegradable part

This part is characterized by its biological oxygen demand (BOD). The BOD charge in these pulp and paper effluents depends on the pulping as well as bleaching processes used. The bleaching step is responsible for a substantial part of the BOD of pulp and paper effluents. The BOD represents about 20% of the mass flow from standard bleaching. The physical, chemical and biological characteristics of the black liquor of the three alkaline extraction stages (dissolution of reaction products with sodium hydroxide) and combined effluent showed that the black liquor (stage 1) was the most polluted (Singh *et al.*, 1993). Black liquor is formed essentially by lignin and extract which are separated from cooked pulp by washing.

In the Kraft process, about 100 m^3 of water is used to produce one ton of pulp and its effluents contain about 35 kg of oxydizable matter. The effluents originating from the bisulfite process contain 200 kg of oxydizable matter per ton of pulp generated. The effluents from mechanical process contain only 10 kg of oxydizable matter per ton of pulp generated. Carbohydrates and phenolics (lignin) are the principal components of the water-soluble matter from the pulp and paper plant.

(a) Characterization of spent sulfite liquor. Spent sulfite liquor (SSF) is the aqueous effluent from a pulp plant operating with an acidic process. Approximately 2000 gal (9.1 m^3) of SSL are recovered per ton of pulp (Joglekar *et al.*, 1983). The liquor remains hot (i.e. at 93 °C) and, therefore, sterile. It has a specific gravity of about 1.5, a pH of 2.2, owing to the presence of residual SO_2, and contains 1.8–2.2% of fermentable sugars. Details on the composition of SSL and the fermentation condition of the sugars are given in Joglekar *et al.* (1983).

It has reported by Casey (1952) that the dry material in spent sulfite liquor contains 10–20% monosaccharides, 10–15% incompletely degraded carbohydrates while lignosulfonic acid forms about 60%. Spent sulfite liquor contains some monosaccharides as arabinose, xylose, galactose, mannose and glucose. Glucose is liberated into the liquor during a longer period of digestion at 130 °C. Arabinose is the first monosaccharide liberated at about 100 °C. Xylose and galactose are produced between 100 and 130 °C. SSL contains polysaccharides (about 3–7% of the dry matter in the liquor), while acetic and formic acids and methyl alcohol are found in different amounts (Forss, 1961).

SSL also contains a considerable amount of lignin. Several techniques are available for the isolation and purification of lignin from commercial spent pulping liquors generated in the Kraft and sulfite pulping process (Lin, 1992).

11.3.2 Wood compounds

The dissolved fraction of effluents from the pulp and paper industry are composed principally of wood compounds and their decomposed by-products. While cellulose is the component of interest in paper manufacture, some anionic substances such as hemicelluloses, lignins and extractives interfere with the production process by reducing the performance of most chemical additives, influencing the production rates and, finally, by impairing the quality of the finished paper (Oblak-Ramer *et al.*, 1993). Some of the major components of wood are discussed below.

(a) Cellulose. Cellulose is the most abundant constituent in wood. Generally, this fiber represents about 45% of the total constituents of the wood. It is largely crystalline and is a linear-chain polymer composed of D-glucose units. This polymer can be hydrolysed either by an enzymatic or an acidic catalyst to produce glucose (Linko, 1987). Thus it is considered as one of the best sources of sugars for microorganisms.

(b) Hemicellulose. Hemicelluloses are largely non-crystalline, accessible to water molecules and are easily degraded (Kennedy and Melo, 1989). Non-fibrous polysaccharides represent about 25% of the primary matter and are made up of xylose, galactose, mannose, arabinose, glucose and their uronic acids (Dekker, 1980). Hemicellulose can be readily hydrolysed by dilute acid at 140 °C to produce pentoses and hexoses. Xylose and arabinoses, however, can be utilized only by certain yeasts (Bisaria, 1991) for their bioconversion.

(c) Lignin. This compound represents 20–30% of the total constituents. It is an amorphous substance, composed of several macromolecules. Lignin is formed by dehydrogenative polymerization of *p*-hydroxycinnamyl alcohols (Chahal, 1991). Lignin is composed of three phenylpropanoid units. More than 30 million tonnes of lignin is produced annually in pulp production plants throughout the world.

(d) Extractables. These are composed of resinic acids, fatty acids, alcohols and other compounds, which are mostly water soluble. An appreciable concentration of resin acids and fatty acids is more readily found in softwood than hardwood. A high quantity of these elements is found in the Kraft process effluent, particularly in black liquor. The resin and fatty acids are toxic to fish and have an inhibitory action on several microbial activities. Thus, identification of their production source in the plant could prove helpful for the effective management of these wastes.

11.3.3 Parts with difficulty in or absence of biodegradability

A wide variety of organochlorides are formed by the reaction between chlorine and residual lignin as well as other organic matters in the bleaching process. A substantial part of these compounds is of high molecular weight and is consequently unable to cross cell membranes. The adsorbable organic halides are produced generally from the bleaching pulp process with chlorine and chlorine compounds. The AOX originating from bleaching units depend on the quantity of chlorine used as oxidizing agent. Reducing adsorbable organic chlorine is becoming the most important criterion for the efficiency of pulp-mill effluent treatment in the 1990s (Lankinen *et al.*, 1991). Dioxins (PCDD) and furans (PCDF) are also formed in the bleaching process and are generally toxic.

11.3.4 Toxic substances

These are chemicals used in pulping, bleaching, paper manufacturing and effluent treatment. Depending on the process used at different stages in the pulp and paper plants, different substances can be found in the effluent. Some of these substances are listed below.

- Pulping: for an alkaline process, the industry tends to use Na_2S, and NaOH and for an acid process. Bisulfite is most commonly used in the industry.
- Bleaching pulp: bleaching by reduction ($Na_2S_2O_4$), bleaching by oxidation (ClO_2 essentially and H_2O_2).
- Paper manufacturing: defoamers are frequently used to control foaming in pulp washing and screening areas, dyes, kaolin and $CaCO_3$, coagulants (glyoxal), bactericides and fungicides.
- Metals: the effluent from pulp and paper industry contains several metals such as aluminium, which is used to reduce toxicity and COD from effluent, zinc, cadmium, copper, nickel, vanadium and mercury. Most of these metals are found in the Kraft effluent.

11.4 Treatment technologies

Identifying and understanding the primary factors in the environmental problems generated by a pulp and paper plant is a first step in the treatment of its waste water. To reach its objective, the industry has to concentrate simultaneously on the internal and the external treatments (Figure 11.3).

11.4.1 Internal treatment

The internal treatment or treatment at source is the correctional steps taken inside the paper industry. This reduction can be achieved by changing the basic technology used to produce pulp and paper. Thus, these changes may decrease the need for effluent treatment and, at the same time, increase the pulp yield. The principal objectives of internal treatment are: reducing waste water volumes, reducing consumption of energy, reducing the generation of various chemicals and solid waste, concentrating various substances found in effluents, and inhibiting or eliminating the

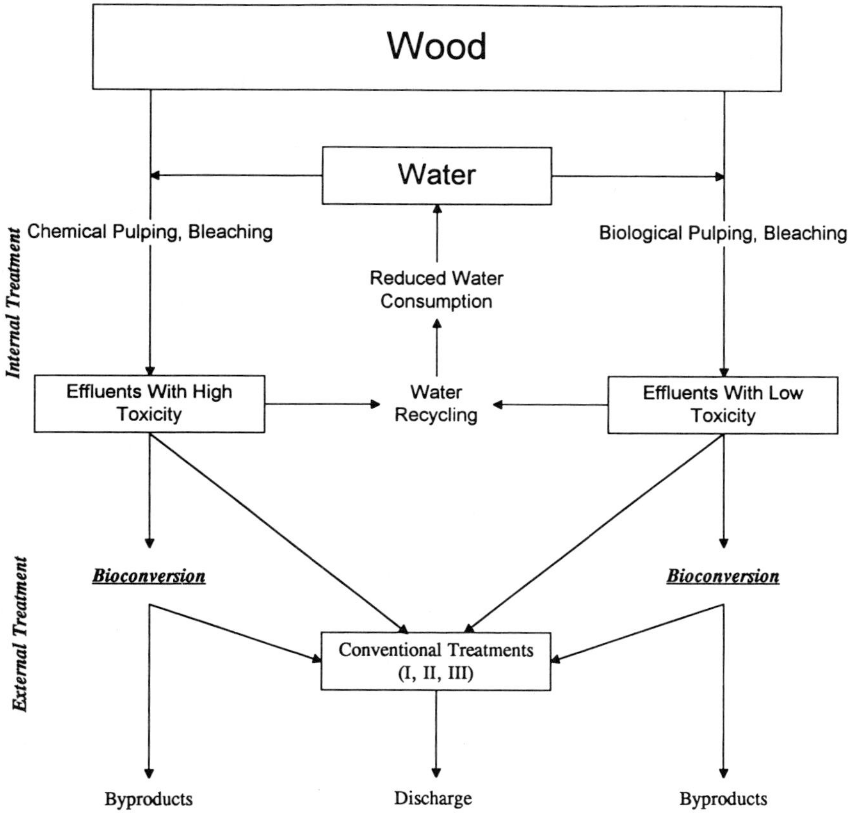

Figure 11.3 Pathways for pulp and paper waste water treatment.

formation of toxic compounds. In order to reach these objectives, the pulp and paper industry may use the following different methods.

1. Recycling white waters (is the treated and clarified mixture which composed of $CaCO_3$, NaOH, Na_2S and water from the causticizer in the Kraft process) in order to reduce the amount of effluents by recovering the cellulosic fibers and some chemical substances. A typical pulp mill of the late 1970s and early 1980s discharged 300 m^3 t^{-1} of pulp, while today's mills can reduce water flow to below 50 m^3 ton^{-1} of pulp, through the recycling of white water. Huster *et al.* (1991) studied the closure of paper-mill white water circuits by inserting an anaerobic stage with subsequent treatment. However, several problems have been encountered in white water recycling, including corrosion and increasing oxidant demands in the bleaching steps.
2. Choosing the pulping process, which has a higher yield and does not necessitate a higher chlorination for bleaching process.

3. Oxygen delignification, which reduces the use of chlorine during the bleaching pulp process. The potential of this process lies in a considerable reduction of the organochlorines concentration.

(a) Pulping process. There are several methods used now as alternatives to the conventional pulping processes. Among these methods, some are practical and economically attractive while others are still being treated on a laboratory scale. The principal techniques used in this field that have been enumerated (Gullichsen, 1991) are polysulfide pulping, anthraquinone additive, extended delignification, prenox treatment, alkaline leaching and improved washing.

(b) Bleaching process. Most organochlorides are formed in the bleach plant, mainly as a result of the molecular chlorine in the first stage of bleaching (Gullichsen, 1991). The principle target, therefore, is to reduce the amount of applied chlorine or substitute it completely by another oxidant such as chlorine dioxide. Elevating the level of chlorine dioxide substitution caused the yield of AOX to decrease linearly while the yield of total chlorophenols diminished in a non-linear manner (McFarlane *et al.*, 1991). Biological process can also be used for bleaching the pulp (KirkPatrick, 1991).

11.4.2 External treatment

External treatments are concerned with all the treatment steps realized outside pulp and paper manufacture. These steps may be physical and/or chemical and/or biological. The main aim of this approach has been to investigate technology which could directly or indirectly solve the problems generated by effluent from pulp and paper industry. External waste water treatment will always be required because the internal treatment is not sufficient to eliminate all pollutants. The treatments used most frequently for pulp and paper can be summarized as follows:

1. Primary treatment, which consists of removing the total suspended solids (primary sludge).
2. Secondary treatment, which consists of removing the organic matter and several toxic substances. Biological treatment under aerobic and/or anaerobic conditions is by far the most widely used method in secondary treatment.
3. Tertiary treatment, which involves chemical and physical treatment of the effluent from secondary treatment processes.

Biological treatment of pulp-paper effluents may require nutrient supplements such as nitrogen and phosphorus because they are deficient in these elements. Excessive concentrations of phosphorus may result in a eutrophied environment (Bothwell, 1992).

(a) Biological treatment. Two principal biological treatments are widely used in North America. The first involves activated sludge treatment (AST), which has been employed in the pulp and paper industry since the late 1970s. This treatment can be applied directly on the effluent or after its pretreatment. Among the pretreatments are precipitation, ozonation and irradiation, which are used before the application of AST (Haberl *et al.*, 1991). The results obtained are reductions by more than 95% in BOD_5, by 70–80% in COD and 70–85% in color. AST treatment can help achieve low BOD and total suspended solids (TSS) concentrations, but produces considerable quantities of sludge which should be disposed of safely. Some industries mix the sludge produced in the AST process with primary sludge followed by burning to generate steam.

The second biological treatment is the aerated stabilization basin (ASB) which operates without sludge recycling which has lower energy requirements than an activated sludge treatment (Lee *et al.*, 1993).

The use of anaerobic technology to generate methane for treating highly concentrated effluents from paper mills is well known but its use in white water circuits of paper mills is still in the planning stages (Huster *et al.*, 1991). Hâggblom and Salkinoja (1991) combined aerobic and anaerobic treatment, and obtained over 65% reduction in AOX and over 75% reduction in chlorinated phenolic compounds. However, treatment of certain paper-mill waste streams is limited by the presence of toxic and recalcitrant organic compounds, high sulfur concentrations and in some cases high waste water temperatures (Lettinga *et al.*, 1991).

During the late 1990s, the treatment processes have been treated by new technologies as alternatives to resolve several problems involving the waste water from pulp and paper industry. Among these technologies are the ultrafiltration process combined with an activated sludge treatment and the bioconversion process to obtain value added products from the waste water from the pulp and paper industry (Ek *et al.*, 1990).

(b) Ultrafiltration technology (or membrane technology). Ultrafiltration technology was combined with activated sludge as a new emergent technology in effluent treatment and has been used for two decades. In order to be feasible, this new membrane-separation technology combined with biological treatment must remove the effluent toxicity, must permit the recycling and the conservation of water, must concentrate on the targeted compounds in order to be valorized, and must be practical and economically attainable. Several important studies have been carried out in this field of treatment (Bowman *et al.*, 1990; Ek and Kolar, 1990; Ekengren *et al.*, 1991).

(c) Removal of color. The wood compounds listed in section 11.3.2, specially lignin, are responsible for the dark color of the effluent. The pulping process contributes to approximately 50–70% of the color load in the effluent. The remaining color usually results from the bleaching operation and spills (Ramanathan, 1991). To remove color from these effluents, Duran *et al.* (1993) used some white-rot fungi (*Lentinus edodes*, *Chrysonilia sitophila*, *Phanerochaete chrysosporium* and *Phaeocoriolellus trabeus*) and observed that *P. trabeus* removes the color by modifying chlorolignins without any degradation of the polymer. However, *L. edodes* exhibited 70% polymer consumption while another, *Chrysonilia sitophila*, only 30%.

11.5 Biotechnological applications in the pulp and paper industry

As already discussed, the most frequently used pulping process in the contemporary pulp and paper industry is the Kraft process. The effluents generated during this process are toxic and are inhibitory to most microorganisms of interest. At the same time, the Kraft process produces very large amounts of effluents. Hence, it would be incomplete not to discuss the role of microorganisms (bioprocesses) which lead to an effluent with good bioconversion properties (i.e. a rich substrate and an absence of inhibitory compounds) while conserving the quality of the end-product (paper).

Recently, biotechnology has emerged as potential alternative to provide the pulp and paper industry with new concepts in biopulping, biobleaching, by-product conversion and treatment of effluents. The fundamental principle in this case is the use of microorganisms, including fungi, actinomycetes, gliding and true bacteria. Waste water from pulp and paper industries can be treated at source or outside the source as indicated earlier.

11.5.1 Pulp manufacture

The disadvantages of widespread pulping process (high cost of operation and environmental problems) have forced the paper industry to look for a novel technology which can partially or fully solve these problems. A preliminary analysis suggests that biopulping is economically promising (Harpole, 1989) leading to an increased interest in this concept. Biomechanical pulping is an experimental process that uses a fungal treatment of wood chips as an alternative to chemicals prior to mechanical refining (Myers *et al.*, 1988). Treatment of wood chips with lignin-degrading fungi prior to pulping has shown to have great potential for mechanical as well as chemical pulping on a laboratory scale (Sachs *et al.*, 1989).

The utility of certain enzymes in this process is also interesting. Enzymes are produced from microbiological cultures grown under controlled conditions. These enzymes are extracellular. The main steps in this process are the isolation of the culture, cell lysis and the isolation of specific enzyme. Several enzymes are being developed that may benefit the pulp and paper manufacture and may also treat the effluent from this industry. Pulpzyme-HA is one potentially useful enzymatic product (McCubbin *et al.*, 1991). This enzyme has properties which affect the viscosity of the pulp limiting the quantity of enzymes that can be used. Xylanase is natural enzyme derived from certain species of fungi which assists in binding lignin to the cellulose. Xylanase was used in 8% of Canada's bleached Kraft pulp (830 000 m^3) in 1994 (Tolan, 1995).

Advances in enzymology can offer economical and environmentally safe ways to make paper. The introduction of enzymes into pulping processes also allows the chlorine level required for bleaching process to be decreased (Trotter, 1990; KirkPatrick, 1991) leading to decreased toxic substances in the effluents. At present, the research goal is the inhibition or elimination, or the cellulolytic fungal action produced by different approaches (biochemical, genetic, etc.). This type of research, however, has yet to be applied on an industrial scale.

11.5.2 Bleaching of pulp

Several laboratory studies have shown that enzymes are able to remove lignin from pulp, thus reducing the brown pulp colors. The use of this new biotechnology may permit reduced operational costs and eliminate undesirable waste effluents (Eriksson and Kirk, 1985). By incorporating certain enzymes into a pulp bleaching sequence it may be possible to eliminate, or at least decrease, the amount of chlorine used in pulp bleaching.

There are two approaches for increased lignin removal. The first is by the indirect action of enzymes that hydrolyse xylan (Viikari *et al.*, 1986). Biobleaching with xylanases gives a new direction to biotechnological pulp bleaching (Senior and Hamilton, 1992). The amount of chemicals required for bleaching reduced substantially (25–50%) in the biobleaching process using xylanases (Trotter, 1990; KirkPatrick, 1991).

The second approach is related to the direct action applied on lignin. The principle of this method is the application of lignolytic and hemicellulolytic enzymes to pulp to facilitate lignin extraction and consequently to decrease demand for chlorine in chemical bleaching. Biobleaching removes residual lignin from wood pulp, making it brighter and ultimately leading to savings in chlorine consumption during a conventional bleaching sequence (KirkPatrick, 1991). The degradation of lignin by white-rot fungi is an oxidation process (Reid and Seifert, 1982).

However, the degradation of the lignin by lignolytic enzyme is often accompanied by the degradation of the cellulose which is undesirable for the pulp and paper manufacture. The two most studied biobleaching microorganisms are *Phanerochaete chrysosporium* and *Coriolus versicolor*, which are principal producers of lignolytic enzymes (Paice *et al.*, 1989).

Many industries around the world have shown interest in developing enzymatic products for the pulp and paper industry. Presently, however, this technology cannot be considered as a full-scale alternative.

11.6 Evaluation of the potential for effluent use from the pulp and paper industry in bioconversion

Effluents produced during the first step (pulping) in a plant contain lower quantities of microbial growth inhibitors than those produced during bleaching process. During the bleaching process, different oxidizing (toxic) components are used (chlorine and its derivatives). The toxic products will not be present for bioconversion processes if the effluent is selected before this stage (Figures 11.4 and 11.5). Effluents from the mechanical process do not seem to be fully adapted to bioconversion given the high yield of this type of process at around 95%. This high yield reduces the concentrations of cellulose and hemicellulose and, thereby, the fermentable sugars in these effluents. As for the chemical processes, the process most feasible for the bioconversion operations is generally the one which would facilitate the hydrolysis of cellulose and hemicellulose while preserving a yield of around 55–75%. Various studies were carried out in this area, mostly focused on the effluents (spent sulfite liquor) produced from the acid process (bisulfite) (McCarthey *et al.*, 1954; Forss, 1961; Muller, 1970; Sixton and Wilkinson, 1980; Joglekar *et al.*, 1983). These effluents were

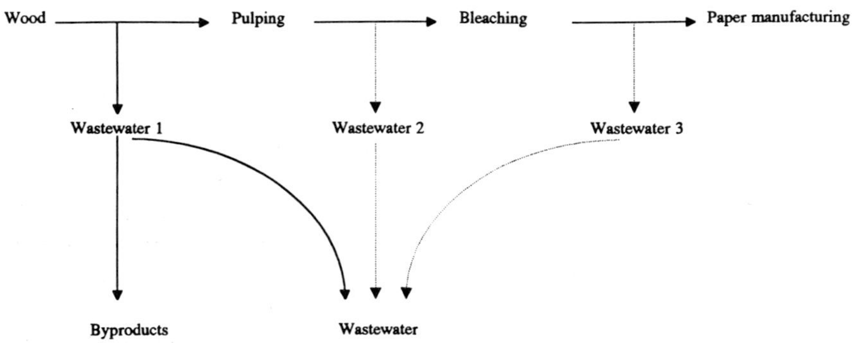

Figure 11.4 Principal steps in pulp and paper plants. Solid arrows indicate the areas most suitable for utilization of biotechnology.

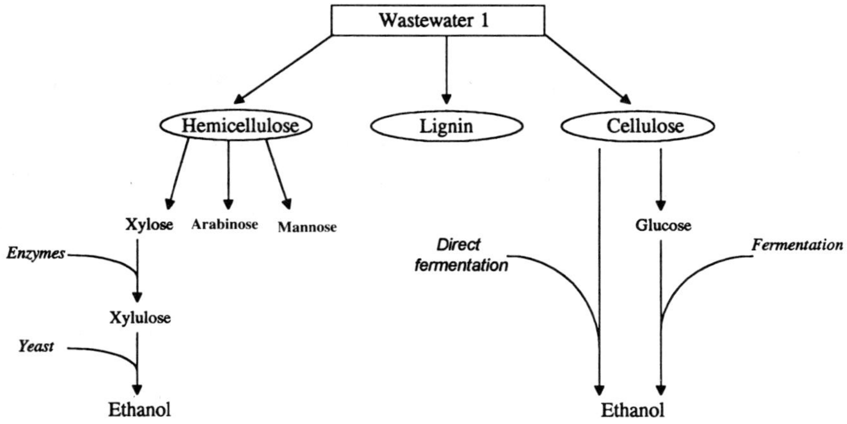

Figure 11.5 Principal methods of ethanol production from effluent produced by the pulping process.

characterized as the most favorable for bioconversion. The greatest disadvantage, which is often ignored of this process, is that a high temperature hydrolysis of cellulose (150 °C) leads to the production of furfurals, its derivatives and certain undesirable components. These by-products inhibit cellular activities during bioconversion (fermentation) processes. Hence, to maximize the exploitation of effluents generated directly from the pulping stage of an acid process, it is desirable to minimize the inhibition due to furfural by either improving the pulping process itself or to use microorganisms which can tolerate these undesirable compounds. There are number of studies reported on the potential to apply biotechnology as an alternative to treat or to convert pulp and paper effluents into useful products like xylitol, ethanol and butanol, etc. (Erikson *et al.*, 1985, Harbel *et al.*, 1991, Singh *et al.*, 1992).

11.7 Suitability of spent sulfite liquor for the bioconversion of by-products

Spent sulfite liquor creates a major pollution problem when discharged into receiving waters, since the constituents of the waste (i.e. wood sugar and lignosulfonic acid) exert a large BOD, increase turbidity and are thus toxic to aquatic life (Muller, 1970; Joglekar *et al.*, 1983) or may necessitate their treatment. SSL is the waste, from the pulp and paper industry, which is used the most for bioconversion. The principal particularity of these effluents was their high sugar concentration. The SSL is used for alcohol and yeast production. Xylose is probably the abundant renewable sugar in SSL that is not fully utilized. This sugar can be fermented to useful products, including ethanol and xylitol. Xylitol is a non-glucose sweetener and can be used in food, the leather industry and surfactants.

The SSL produced from the acidic process is the most compatible with the bionconversion process compared with effluents from the other pulping processes. The Kraft process is carried out at a temperature of 180 °C under which the glucose and other sugars liberated by hemicellulose are transformed into their acidic components (Linko, 1987). These acids have an inhibitory effect on the microorganisms responsible for most of bioconversion processes.

11.8 Effluent treatment by conversion to by-products

The bioconversion of lignocellulose, released in effluents from pulp and paper industries, to chemicals has gained considerable attention in recent years. The choice of an effluent which can be used as a rich source of microbial growth nutrients seems logical for optimum bioconversion of the waste into commercial products.

Biotechnology permits the conversion of fermentable substrates in waste to useful products, such as acetic, lactic, propionic and formic acids, acetone, butanol, ethanol, xylitol and other products (Linko, 1987). The microorganisms used in this conversion are mainly yeast (*Candida utilis* and *Candida tropicalis*) and fungi. The use of enzymes for direct effluent treatment is still in the early phases of development. Protein enrichment of effluents containing lignocellulosic materials may be economically realizable if wood-rotting fungi can be used simultaneously for the production of extracellular enzymes (Hatakka *et al.*, 1989). Past research has focused on a dilute-acid prehydrolysis process as a means of hydrolysing the hemicellulose component of effluents. Such processes provide a solid residue in effluents that is more easily hydrolysable by cellulase enzymes, as well as a hemicellulose-sugar component that can be converted by pentose-fermenting microorganisms (Elander and Hsu, 1995). The products of this hydrolysis, like xylose, arabinose and mannose, could be assimilated by certain yeasts into useful products such as ethanol, butanol, xylitol and other useful products (Linko, 1987).

The use of anaerobic technology to generate methane for treating highly concentrated effluents from paper mills is largely known but its use in white water circuits of paper mills is still in the planning stages (Huster *et al.*, 1991). Hâggblom and Salkinoja (1991) combined aerobic and anaerobic treatment, and obtained over 65% reduction in AOX and over 75% reduction in chlorinated phenolic compounds. However, treatment of certain paper-mill waste streams is limited by the presence of toxic and recalcitrant organic compounds, high sulfur concentrations and in some cases high waste water temperatures (Lettinga *et al.*, 1991).

A process known as steam classification transforms the pulp and paper materials into a fairly uniform product that appears to be highly suitable as a feedstock for conversion to fuel, fertilizer and/or fermentable sugars (Eley *et al.*, 1995).

11.8.1 Bioconversion of cellulose and lignocellulosic materials present in pulp and paper waste waters

The primary material extracted from wood for paper manufacture is cellulose which is also considered as a principal source of glucose in waste water from this industry. Thus it is more appropriate to use the waste originating from low-yield processes because the primary matter is abundant in the waste water. At the same time, it is suggested that the waste should be chosen from a process which uses the smallest amount of chemicals toxic to the microorganisms' growth.

The hydrolysis of cellulosic materials found in the waste water from the pulp and paper industry, to glucose or to other monosaccharides can be accomplished by acids, enzymes or microorganisms. The bioconversion is realized by different processes: fermentation, enzymatic activities, biodegradation coupled with ultrafiltration and other techniques. There are two types of fermentation process. One type represents the well-established anaerobic or aerobic production of simple chemicals, such as various alcohols and carboxylic acids (Raimo, 1990). The other type of fermentation process is that used for the production of more complex chemicals such as antibiotics, enzymes and hormones (Raimo, 1990). Several industrially important cytolytic fungi have been named by Adaskaveg (1993).

Several microorganisms have been reported to synthesize cellulolytic enzymes when grown in a medium containing cellulose or a cellulase inducer, such as wood pulp, or pulp and paper waste; however, they cannot convert sugars to ethanol. On the other hand, many organisms are able to convet saccharoses to ethanol but they lack the genetic information to produce the polysaccharases for the hydrolysis of cellulose. A few filamentous fungal strains have been reported recently that are able to hydrolyse and convert cellulose directly to ethanol (Singh *et al.*, 1992).

One of the principal qualities of fungi is their capacity to penetrate the components of wood by their complex system of hyphae. *Phanerochaete chrysosporium* and other white-rot fungi are also being studied as pretreatment agents for the mechanical pulping of wood with the objective of reducing energy requirements, increasing paper strength properties and decreasing pollution (Kirk, 1993). *Trichoderma reesei* has been one of the most useful microorganisms for the production of cellulases (Durand *et al.*, 1988). The filamentous fungus *Trichoderma reesei* is the predominant industrial producer of cellulolytic enzymes by secreting an enzyme system capable of degrading crystalline cellulose, which consists of several cellobiohydrolases, endoglucanases and β-glucosidases (Kubicek, 1992).

A pilot plant based on the utilization of the advanced *Trichoderma reesei* fungal enzyme systems, utilized in a fed-batch simultaneous saccharification and fermentation system, has been operated successfully in an eastern Canadian pulp and paper mill (Katzen and Monceaux, 1995).

The number of cellulases involved in the degradation of cellulose is still speculative. However, the association between cellulose and other compounds, such as hemicellulose and lignin, complicates the cellulolytic enzyme degradation. The enzymes catalysing the breakdown of cellulose (i.e. cellulases), hemicellulose (i.e. hemicellulases, such as xylanases, mannases, etc.) and lignin (i.e. lignin peroxidase, manganese peroxidase and other oxidases) possess high specificity and act at moderate temperatures of 30–50 °C. A number of fungi and bacteria produce these enzymes in their extracellular fluid which can be employed in various bioconversion processes to obtain useful products (Bisaria, 1991). Cao *et al.* (1995) enhanced the conversion of cellulosic materials (which are abundantly available in pulp and paper wastewaters) to ethanol by introducing zinc chloride as an alternative for pretreating biomass prior to the hydrolysis of cellulose. The results of this research (in 67% $ZnCl_2$ solution, a 99.5% yield of soluble sugars has been obtained at 70 °C and 0.5 M acid concentration) demonstrate that the formation of a zinc–cellulose complex during the pretreatment of cellulose improves the yield of glucose in both the enzymatic and acid hydrolysis of cellulose. It was observed in most of cases that the fungal system would not grow under anaerobic conditions, and would not produce ethanol under aerobic conditions (Singh *et al.*, 1992).

The susceptibility of cellulosic substrates to enzymatic hydrolysis is determined largely by their accessibility to cellulolytic enzymes (Tanaka *et al.*, 1988). Cellulose is degraded by a multienzyme complex involving at least three enzymes: endo-1,4-β-D-glucanase, exo-1,4-β-D-glucanase and a β-D-glucosidase also called a cellobiase. Endoclucanase cleave β1–4 chains of cellulose randomly, whereas exoglucanase releases cellobiose or glucose units from the non-reducing end of the cellulose polymer (Singh *et al.*, 1992). For the complete hydrolysis of insoluble cellulose, synergistic action among these components is required. Cellulases are produced by bacteria, actinomycetes, fungi, algae, myxobacteria, basidiomycetes and also by some higher forms like mollusks and insects.

The lignin is broken down only during secondary metabolism after growth of the organism ceases owing to nitrogen, carbon or sulfur limitation. In this phase, veratryl alcohol (IX) (3,4-dimethoxybenzyl alcohol) is synthesized via phenylalanine and excreted into the medium. The compound induces the appearance of the ligninolytic system in *Phanerochaete chrysosporium* (Tuor *et al.*, 1989).

11.8.2 Production of ethyl alcohol from cellulosic by-products

Ethanol is considered now as the best alternative to petroleum. The potential demand for this product as fuel or feedstock is higher than its production. Thus, ethanol from wood sugars contained in waste water could have a place in this market. Strict environmental norms as well as a high treatment cost have resulted in an increasing bioconversion of pulp and paper wastes into alcohol. Bioconversion of SSL into by-products, essentially ethanol, has been practiced from as early as the 1940s. Industrial ethanol production by fermentation from SSL did not develop very far because of the concurrent production of synthetic alcohol from ethylene. For example, 15.5×10^6 l of ethanol are produced in USA from sulfite liquor but at the same time 788.8×10^6 l of synthetic ethanol is produced from ethylene (Sherman and Kavasmaneck, 1980; Joglekar *et al.*, 1983).

The biological conversion of lignocellulosic materials into ethanol is a versatile process which can be used in various applications for replacing or improving petroleum products, treating wastes or reducing air pollution (Lynd, 1989).

Xylose yielded by hydrolysis of hemicellulose can be converted into ethanol and other compounds (Raimo, 1990) (Table 11.2). This conversion is accomplished via biochemical metabolic reaction sequence known as the pentose phosphate cycle. However, only a small number of yeasts can use xylose to produce alcohol in economically significant quantities (Toivola *et al.*, 1984). Xylanases enzymes have been used for selective removal of xylan from pulp (Senior *et al.*, 1988). Xylan removal was evaluated at a 20% level, a quantity which can easily be converted to ethanol by specific yeasts (Table 11.2).

Production of ethanol from pulp-mill hardwood and softwood SSLs by genetically engineered *E. coli* was also reported (Lawford and Rousseau, 1993). The best studied yeasts responsible for fermentation of xylose to produce ethanol are *Candida shehatae*, *Pachysolen tannophilus* and *Pichia stipitis*.

Ingram and Doran (1995) developed an organism through genetic engineering which utilizes pentoses and hexoses simultaneously. Recombinant strains of Gram-negative bacteria (*Escherichia coli, Klebsiella oxytoca* or *Erwinia* sp.) have been constructed in which genes encoding the ethanol pathway from *Zymomonas mobilis* (*pdc* and *adh*) were inserted in the chromosomes. Mutants resistant to comparatively high levels of acetic acid were isolated from the xylose-fermenting yeasts *Candida shahatae* and *Pichia stipitis* by adapting these cultures to increasing concentrations of acetic acid grown in shake-flask cultures (Mohandas *et al.*, 1995). The soft-rot bacteria *Erwinia carotovora* Sr38 and *Erwinia chrysanthemi* EC16 have been genetically engineered to produce ethanol and carbon dioxide

Table 11.2 Some microorganisms involved in the fermentation of xylose to ethanol

Microorganisms	Substrate	By-product	References
Candida shehatae and *Pichia stipitis*	Xylose	Ethanol	Mohandas *et al*. (1955)
Pachysolen tannophilus	Xylose	Ethanol	Lawford and Rousseau (1993)
Escherichia coli[a] and *Klebsiella oxytoca*[a]	Xylose	Ethanol	Ingram and Duran (1995)
Erwinia carotovora[a] and *Erwinia chrysanthemi*[a]	Xylose, cellobiose, and glucose	Ethanol	Beall and Ingram (1993)

[a]Genetically engineered microorganisms.

efficiently as primary fermentation products from cellobiose, glucose and xylose (Beall and Ingram, 1993) (Table 11.2). These organisms have the native ability to secrete a battery of hydrolases and lyases to aid in solubilization of lignocellulose.

Separating the end-product from the process of fermentation is one of the major problems encountered in the bioconversions, especially in the case of alcohol where the end-product (ethanol) inhibits yeast. The extractive bioconversion (extracting the end-product from a fermenting mixture of substrates) must be applied for optimizing the industrial-scale production of ethanol. Use of these techniques enables ethanol concentration by pervaporation through a water-selective membrane to >99% wt (Strathmann and Gudernatsch, 1991).

To make the fermentation economically feasible, the alcohol production facility must be integrated into the pulp and paper production. This reduces the transportation costs related to the primary matter, while the costs related to the storage of alcohol simultaneously reduce the costs attached to the treatment of the waste that would otherwise have been generated.

11.9 Major difficulties in bioconversion

The major difficulties in bioconversion are listed below.

1. The biotechnology applied in the valorization of pulp and paper waste and/or waste water loses some of its effectiveness because a substantial amount of cellulose waste is extracted from the primary matter for manufacturing paper. This shortcoming can be compensated for by using biotechnology in the processes of pulping, bleaching and waste water treatment.
2. The active chlorine used in bleaching process can react with the enzyme protein and inhibit its catalytic site.
3. If an enzyme hydrolysis process is used, the pretreatment of waste still has to be optimized. The separation of the enzymes from the products and substrates also remains a challenge in this technology.
4. If an acid hydrolysis process is used for biopulping, the problem of equipment corrosion must be resolved. Treatment of the sludge must also be realized.
5. The use of freely suspended enzymes in pulp and paper bioconversion processes has many disadvantages, such as their limited life. The enzymes can also be difficult to separate from their substrates and products. Lignin peroxidase is difficult to purify, thus the cost of process may rise.
6. The cost of isolating intracellular enzymes for a large-scale commercialization can significantly decrease their use, rendering it unattractive.

7. Biopulping is difficult to control because there are several physical problems related to bringing the enzymes into contact with the inner part of wood.
8. All the dilute-acid hydrolysis technologies are not economically feasible and can only run with government support (Blazej and Kosik, 1993).
9. Among the obstacles faced in the conversion of cellulosic materials to ethanol, the following are notable:

   - low yield of sugars;
   - the high energy consumption in pretreatment processes;
   - the difficulty of recycling the pretreatment reactants.

The separation of products from other bioreactor constituents is often a difficult and costly step in large-scale industrial bioprocesses. Particularly when products or by-products have an inhibitory effect on the production rate, their continuous selective removal generally leads to significantly increased conversion rates and improved overall process economics (Strathmann and Gudernatsch, 1991).

11.10 Conclusions

1. The pulp and paper industry is one of the most polluting industries.
2. Stringent environment regulations and deadlines have forced the pulp and paper industry to control its pollution levels.
3. Toxic compounds in the waste as well as the amount of waste generated can be minimized by using improved technology in pulp and paper production.
4. Recycling of waste water in the industry has a great potential not only in reducing the amount of waste water generated, but also in reducing the overall water consumption in this industry. The operational short-comings caused by recycling, however, are yet to be solved.
5. Biopulping and biobleaching has great potential for the future, not only in reduction of toxic pollutants and minimizing the use of chemicals, but it may also lead to effluents which are more compatible to bioconversion.
6. Bioconversion of the effluents from the industry leads to production of numerous value added products. Its potential has not been fully harnessed to bring biotechnology to a realistic level in the bioremediation of environmental impacts, and in producing fuels and valuable chemicals. Further research is needed to enhance the available data on enzymatic activities, feedstock pretreatment, bioreactors, immobilization of cells and enzymes, fuel and valuable chemicals production and hazardous waste bioremediation.

Acknowledgements

The authors wish to thank the Natural Sciences and Engineering Research Council of Canada (Grant A 9484) for supporting this research.

References

Adaskaveg, J. (1993) Taxonomy of industrially important wood-rotting fungi. *Newsletter of the Mycological Society of America*, **44**, 24.

Beall, D.S. and Ingram, L.O. (1993) Genetic engineering of soft-rot bacteria for ethanol production from lignocellulose. *J. Ind. Microbiol.*, **11**, 151–5.

Bergbauer, M., Eggert, C. and Kalnowski, G. (1992) Biotreatment of pulp mill bleachery effluents with the ceolomycetous fungus *Stagonospora gigaspora*. *Biotechnol. Lett.*, **14**, 317–22.

Bisaria, V.S. (1991) Bioprocessing of agro-residues to glucose and chemicals, in *Bioconversion of Waste Materials to Industrial Products*, 1st edn (ed. A.M. Martin), Elsevier Press, London.

Blazej, A. and Kosik, M. (1993) Environmentally acceptable conversion technology for the biochemical utilization of lignocellulosics, in *Cellulosics: Pulp, Fibre and Environmental Aspects* (eds J.F. Kennedy, G.O. Phillips and P.A. Willliams), Ellis Horwood, New York, p. 509.

Boman, B., Ek, M. and Frostell, B. (1990) Treatment of total bleachery effluents 'a comparison between biological systems in combination with ultrafiltration'. [in Swedish] Miljo 90, Report no. 18, Sweden.

Bothwell, M.L. (1992) Eutrophication of rivers by nutrients in treated Kraft pulp mill effluent. *Wat. Poll. Res. J. Canada*, **27**, 447–72.

Cao, N.J., XU, Q. and Chen, L.F. (1995) Acid hydrolysis of cellulose in zinc chloride solution. *Appl. Biochem. Biotechnol.*, **51/52**, 21–8.

Casey, J.P. (1952) *Pulp and Paper*, Vol. I, Interscience Publishers, New York, p. 128.

Chahal, D.S. (1991) Lignocellulosic wastes: biological conversion, in *Bioconversion of Waste Materials to Industrial Products*, 1st edn (ed. A.M. Martin), Elsevier Press, London.

Cossette, C. (1991) *La Préparation du Bois et les Pâtes méchaniques*, Cégep de trois rivière, Quebec, Canada, p. 241.

Dekker, R.F.H. (1980) In *Biosynthesis and Biodegradation of Wood Components* (ed. T. Higuchi), Academic Press, New York.

Duran, N., Esposito, E. and Canhos, V.P. (1993) Kraft mill effluent: biological treatment. In *Cellulosics: Pulp, Fiber and Environmental Aspects* (eds J.F. Kennedy, G.O. Phillips and P.A. Williams), Ellis Horwood, London, p. 509.

Durand, H., Baron, M., Calmels, T. and Tiraby, G. (1988) *Proceedings of the FEMS Symposium Biochemistry and Genetics of Cellulose Degradation* (eds J.-P. Aubert, P. Beguin and J. Millet), Academic Press, London, p. 135.

Ek, M. and Kolar, M.C. (1990) Reduction of AOX in bleach plant effluents by a combination of ultrafiltration and biological methods, in *Biotechnology of Pulp and Paper Manufacture* (eds T. Kent and Hou-Min Chang), Butterworth-Heinemann Press, Boston, p. 666.

Ekengren, O., Burham, J.E. and Filipsson, S. (1991) Treatment of bleach-plant effluents with membrane filtration and sorption techniques. *Wat. Sci. Technol.*, **24**(3/4), 207–18.

Elander, R.T. and Hau, T. (1995) Processing and economic impacts of biomass delignification for ethanol production. *Appl. Biochem. Biotechnol.*, **51/52**, 463–78.

Eley, M.H., Guinn, G.R. and Bagchi, J. (1995) Cellulosic materials recovered from steam classified municipal solid wastes as feedstocks for conversion to fuels and chemicals. *Appl. Biochem. Technol.*, **51/52**, 387–97.

Eriksson, K.E. and Kirk, T.K. (1985) Biopulping, biobleaching, and the treatment of Kraft bleaching effluents with White-rot fungi, in *The Principles, Applications and Regulations of Biotechnology in Industry, Agriculture and Medicine* (eds C.L. Cooney and A.E. Humphrey), Vol. 4, Pergamon Press, New York, pp. 271–94.

Fiedler, H., Hutzinger, O. and Timms, C.W. (1990) Dioxines: sources of environmental load and human exposure. *Toxicol. Environ. Chem.*, **29**, 157–234.

Forss, K. (1961) In *The Composition of a Spent Spruce Sulfite Liquor*. Åbo Akademi (University Publication), Finland.

Gullichsen, J. (1991) Process internal measures to reduce pulp mill pollution load. *Wat. Sci. Technol.*, **24**(3/4), 45–53.

Håggblom, M. and Mirja Salkinoja, S. (1991) Biodegradability of chlorinated organic compounds in pulp bleaching effluents. *Wat. Sci. Technol.*, **24**(3/4), 161–70.

Harbel, R., Urban, W. and Gehringer, P. (1991) Treatment of pulp-bleaching effluent by activated sludge, precipitation, ozonation and irradiation. *Wat. Sci. Technol.*, **24**(3/4), 229–39.

Harpole, G.B. (1989) *Proceedings of an International Mechanical Pulping Conference*, Helsinki.

Hatakka, A.I., Lundell, T.K., Mohammadi, O.K. and Terviolo-Wilo, A.L.M. (1989) Catalytic activities of lignin-degrading enzymes from white rot fungus *Plebia radiata*: lignin model compound studies in *Biotechnology in Pulp and Paper Manufacture Applications and Fundamental Investigations* (eds. K.T. Kent and H.-M. Chang), Butterworth-Heinemann, Boston.

Heizel, E. Geiger, F., Fahmy, M. and Kut, O.M. (1992) Integrated ozonation biotreatment of pulp bleaching effluents containing chlorinated phenolic compounds. *Biotechnol. Prog.*, **8**, 67–77.

Huster, R., Demel, I. and Geller, A. (1991) Closing paper mill whitewater circuits by inserting an anaerobic stage with subsequent treatment. *Wat. Sci. Technol.*, **24**(3/4), 81–90.

Ingram, L.O. and Doran, J.B. (1995) Conversion of cellulosic materials to ethanol. *Euro. J. Pharmacol., FEMS. Microbiol.-Rev.*, **16**(2/3), 235–41.

Joglekar, R., Clerman, R.J., Ouellette, R.P. and Cheremisinoff, P.N. (eds) (1983) *Biotechnology in Industry, Selected Applications and Unit Operations*. Ann Arbor Science, MI, p. 179.

Katagari, N., Tsutsumıı, Y. and Nishida, T. (1995) Correlation of brightning with cumulative enzyme activity related to lignin biodegradation during biobleaching of Kraft pulp by white rot fungi in the solid-state fermentation system. *Appl. Environ. Microbiol.*, **61**, 617–22.

Katzen, R. and Monceaux, D.A. (1995) Development of bioconversion of cellulosic wastes. *Appl. Biochem. Biotechnol.*, **51/52**, 585–92.

Kennedy, J.F. and Melo, E.H.M. (1989) Controlled bioconversion of cellulose – a biochemical industry feedstock, in *Wood Processing and Utilization* (eds J.F. Kennedy, G.O. Phillips and P.A. Williams), Ellis Horwood Press, New York.

Kirk, K.T. (1993) Lignin degradation: basic research progress, and applications in soil remediation and biopulping, in *Cellulosics: Pulp, Fiber and Environmental Aspects* (eds J.F. Kennedy, G.O. Phillips and P.A. Williams), Ellis Horwood Press, New York.

Kirk, T.K. and Chang, H.M. (1989) Overview of biotechnology in pulp and paper manufacture, in *Biotechnology in Pulp and Paper Technology* (eds T. Kirk and Hou-Min Chang), Butterworth Heinemann Press, Boston.

Kubicek, C.P. (1992) The cellulase proteins of *Trichoderma reesei*: structure, multiplicity, mode of action and regulation of formation. *Adv. Biochem. Eng. Biotechnol.*, 45, 1–27.

Lankinen, V.P., Inkeroinien, M.M., Pellinen, J. and Hattakka, A.I. (1991) The onset of lignin-modifying enzymes, decrease of AOX and color removal by white-rot fungi grown on bleach plants effluents. *Wat. Sci. Technol.*, **24**(3/4), 189–98.

Lawford, H.G. and Rousseau, J.D. (1993) Production of ethanol from pulp mill hardwood and softwood spent sulphite liquor by genetically engineered *E. coli*. *Appl. Biochem. Biotechnol.*, **39/40**, 667–85.

Lee, E.G.-H., Crowe, M.F. and Stutz, H. (1993) Anaerobic–aerobic lagoon treatment of Kraft mill effluent for enhanced removal of AOX. *Wat. Pollut. Res. J. Canada*, **28**, 549–70.

Lettinga, G., Field, J.A., Sierra-Alvares, R., Van Lier, J.B. and Rintala, J. (1991) Future perspectives for the anaerobic treatment of forest industry wastewater. *Wat. Sci. Technol.*, **24**(3/4), 91–102.

Lin, S.Y. (1992) Commercial spent pulping liquors, in *Methods in Lignin Chemistry* (eds Y.S. Lin and W.D. Carlton), Springer-Verlag, Berlin, pp. 75–80.

Linko, M. (1987) Fermentation of cellulose feed stocks, in *Wood and Cellulosic, Industrial*

Utilisation, Biotechnology, Structure and Properties (eds J.F. Kennedy, G.o. Phillips and P.A. Williams), Ellis Horwood Press, New York.

Lynd, L.R. (1989) Production of ethanol from lignocellulosic materials using thermophilic bacteria: critical evaluation of potential and review. *Adv. Biochem. Eng. Biotechnol.*, **38**, 1–52.

Macleay, D. and Associates Ltd (1987) *Aquatic Toxicity of Pulp and Paper Mill Effluent: A Review*, Report # EPS 4/PF/1, Environment Canada, Ottawa.

McCarthey, J.L. (1954) Conversion of sugar cane products into fuel and chemicals feedstocks. *Sugar J.*, August, p. 271.

McCubbin, N., Edde, H, Barnes, E. *et al.* (1991) *Best Available Technology for the Ontario Pulp and Paper Industry*, Consultant Report Ontario Ministry of Environment, Canada, p. 270.

McFarlane, P.N., Allison, R.W., Clark, T.A. and Mackie, K.L. (1991) The effects of chlorination conditions on the AOX and chlorinated phenol content of Kraft bleach plant wastewater. *Wat. Sci. Technol.*, **24**(3/4), 55–63.

Mohandas, D.V., Whlan, D.R. and Panchal, C.J. (1995) Development of xylose-fermenting yeasts for ethanol production at high acetic acid concentrations. *Appl. Biochem. Biotechnol.*, **51/52**, 307–18.

Muller, J.E. (1970) Fermentative utilization of spent sulfite liquor: a review and proposal. *Pulp Paper Mag.*, **71**(122), 72–6.

Myers, G.C., Leatham, G.F., Wegner, T.H. and Blanchette, R.A. (1988) Fungal pretreatment of aspen chits improves strength of refiner mechanical pulp. *TAPPI J.*, **71**, 105–9.

Myréen, B. (1994) Pulp and paper manufacture in transition. *Wat. Sci. Technol.*, **29**(5/6), 1–9.

Oblak-Ramer, M., Budin, D., Cerne, S. and Lipic, B. (1993) Water soluble substances from bleached paper fibers, in *Cellulosics: Pulp, Fiber and Environmental Aspects* (eds J.F. Kennedy, G.O. Phillips and P.A. Williams), Ellis Horwood, New York.

Paice, M.G., Jurasek, L., Ho, C., Bourbonnais, R. and Archibald, F. (1989) Direct biological bleaching of hardwood Kraft pulp with a fungus *Coriolus versicolor. TAPP J.*, **72**, 217–21.

Raimo, A. (1990) Conversion of cellulose-containing materials into useful products, in *Cellulose Sources and Exploitation: Industrial Utilization, Biotechnology and Physico-Chemical Properties.* (eds J.F. Kennedy, G.O. Phillips and P.A. Williams) Ellis Horwood, Chichester.

Ramananthan, M. (1991) Mills try new bleaching, washing technology to cut effluent color, in *Bleaching Technology for Chemical and Mechanical Pulps* (ed. L.K. Patrick), Miller Freeman, Inc. Press, San Francisco.

Rao, S.S., Quinn, B.A., Burnison, B.K., Hayes, M.A. and Metcalfe, C.D. (1995) Assessment of the genotoxic potential of pulp mill effluent using a bacterial, fish and mammalian assays. *Chemosphere*, **31**, 3553–66.

Reeve, D.W. and Earl, P.F. (1989) Chlorinated organic matter in bleached chemical pulp production: part I: Environmental impact and regulation of effluents. *Pulp Paper Canada*, **90**, 128–32.

Reid, I.D. and Seifert, K.A. (1982) Effect of an atmosphere of oxygen on growth, respiration, and lignin degradation by white-rot fungi, *Can. J. Bot.*, **60**, 252–60.

Sachs, I.B., Leatham, G.F., Myers, G.C. and Wegner, T.H. (1989) Biomechanical pulping of aspen chips: fungal growth pattern and effecs on cell wall, fiber, and pulp morphology, in *Biotechnology in Pulp and Paper Manufacture: Applications and Fundamental Investigations* (eds T.K. Kirk and Hou-Min Chang), Butterworth-Heinemann, Boston.

Senior, D.J. and Hamilton, J. (1992) Biobleaching with xylanase brings biotechnology to reality. Economic and environmental advantages of using xylanase in bleach plants catapult enzyme from the lab to the mill. *Pulp Pap*, **66**(9), 111–14.

Sherman, P.D., Jr and Kavasmaneck, P.R. (1980) Ethanol, in *Kirk-Othmer Encyclopedia of Chemical Technology*, Vol. 9, John Wiley and Sons, pp. 338–90.

Singh, A., Kumar, P.K. and Schugerl, K. (1992) Bioconversion of cellulosic materials to ethanol by filamentous fungi. *Adv. Biochem. Eng. Biotechnol.*, **45**, 30–55.

Singh, R.S., Marwaha, S.S., Khanna, P.K. and Kennedy, J.F. (1993) Pulp and paper mill effluent biobleaching using immobilized *Phanerochaete crysosporium* BKME 1767, in

Cellulosics: Pulp, Fiber and Environmental Aspects (eds J.F. Kennedy, G.O. Phillips and P.A. Williams), Ellis Horwood, New York.

Sixton, E.A. and Wilkinson, E.J. (1980) Spent liquor recovery at Ontario Paper. *Pulp Paper Can.*, **81**(1), 86–9.

Strathmann, H. and Gudernatsch, W. (1991) Continous removal of ethanol from bioreactor by pervaporation, in *Extractive Bioconversions* (eds B. Mattiasson and O. Holst), p. 328.

Tanaka, M., Fukuri, M. and Matsuno, R. (1988) Removal of lignin and of cellulases for continuous saccharification of lignocelluloses. *Biotech. Bioeng.*, **32**, 897.

Toivola, A., Yarrow, D., Van den Bosh, E. Van Dijken, J.P. and Scheffers, W.A. (1984) Alcoholic fermentation of D-xylose by yeasts. *Appl. Environ. Microbiol.*, **47**, 1221–3.

Tolan, J.S. (1995) Survey of xylanase enzyme usage in bleaching in Canada. 81st Annual Meeting Canadian Pulp and Paper Association (Technical Section), February 1995.

Trotter, P.C. (1990) Biotechnology in the pulp and paper industry: a review. *TAPPI J.*, **73**, 198–204.

Tuor, U.M., Haemmerli, S.D., Schoemaker, H.E. *et al.* (1989) On the metabolism of 3,4-dimethoxybenzyl alcohol and its methyl ether by *Phanerochaete chrysosporium*, in *Biotechnology in Pulp and Paper Manufacture* (eds T.K. Kirk and Hou-Min Chang), Butterworth-Heinemann, Boston.

Viikari, L., Ranua, M., Kantelinen, A., Sundquist, J. and Linko, M. (1986) *Proceedings of 3rd International Conference on Biotechnology in the Pulp and Paper Industry*, Stockholm, June, pp. 67–9.

12 Fisheries waste biomass: biconversion alternatives

A.M. MARTIN

12.1 Introduction

When compared with the rest of the food industry, it has been generally regarded that the fish processing industry has been late in introducing new technologies to its production operations, including the treatment and/or recovery of wastes. Recently, interest has been focused on the application of new technological methods to operations related to the seafood industry, with the objective of increasing its general efficiency. To this end, the effects of technology on the nutritional value of seafoods has been presented by Pigott and Tucker (1990). In this context, it has been evident that the application of biotechnology to the utilization of biomass from by-products or wastes of the seafood industry could bring about improvements in its overall economy (Martin and Patel, 1991).

Marine biomass constitutes an abundant and relatively inexpensive source of feed and food, and of potential raw materials for several industries (Bligh, 1992). Although fishing for food is the main activity relating to the exploitation of marine resources, biopolymers, enzymes and pharmaceuticals are among the other valuable products that can be extracted from marine organisms (Jefford *et al.*, 1988; Halevy, 1989). For example, enzymes extracted from fisheries biomass, specifically proteases, have several potential uses, including uses in the seafood industry itself and in other food industries (Haard and Simpson, 1994).

Increased coastal and sea pollution had led to the need for better ways to dispose of fisheries wastes among others. The exhaustion of many fishing areas and the need for increased economy in processing operations also contributes to the requirement for better and more extensive utilization of the fisheries biomass harvested. Given the potential value of many of their components, such as high-quality protein and oils, enzymes and complex polysaccharides, the recovery of much of the presently discharged fisheries wastes and by-products and their conversion to value-added products is a solution to such problems that should not be difficult to achieve. Biotechnological processes offer many possibilities for their incorporation into the processing of fisheries biomass (Martin, 1994).

Børrensen (1992) noted that the application of biotechnology to the seafood industry is mostly related to the utilization of enzymes and microorganisms. The use of microorganisms in the seafood processing industry has been developed less than has the application of enzymes to some specific fisheries processes. Indeed, the use of enzymes, and specifically the use of microbial enzymes, has a great potential in the seafood industry. In addition to the protein fraction of the fisheries biomass, the bioprocessing of fish oils and chitin may yield important new sources of raw materials. Figure 12.1 presents an overview of biotechnological processes with applications to fisheries operations.

As indicated above, fisheries, including aquaculture processes, presently constitute the main form of exploitation of marine resources; therefore, they are also the main source of marine wastes. This chapter will concentrate on those wastes and the application of biological (enzymatic and microbial) processes for their recovery, in accordance with the objectives of the book. Waste water treatments for fisheries applications can also involve biological operations, in the so-called secondary treatments. Veiga *et al.* (1994) have presented a review of these processes.

An objective that should be pursued is that, when developing new processes, attention should primarily be given to the use of fisheries biomass that is not presently used as food.

12.1.1 Antecedents of the recovery of fisheries wastes and by-products

Many of the fish species which could be caught are not consumed, owing to cultural or organoleptic factors. Those underutilized species, sometimes part of the fisheries by-catch, together with fisheries processing wastes, constitute a valuable source of raw materials, especially protein. This, if recovered, can be employed in the production of several industrial products.

In the past, physical and chemical processes have been applied with limited economic success for the recovery of some components of fisheries wastes, and the production of products with varying degrees of quality. For example, use of wastes and underutilized species for the production of fish meal is an established process. However, fish meal production requires relatively high capital investment and energy input, and generally is not economical for small operators. Other factors which hamper the recovery of wastes from fisheries are the small size and seasonal operation of a large number of seafood processing plants. Early works have reviewed fishery by-product utilization (Brody, 1965). Other general works related to fisheries wastes, their utilization and methods of disposal, include Martin (1972), Kreag and Smith (1973), Green and Mattick (1979), Martin and Patel (1991) and Martin (1994).

Some traditional processes for fish preparation have been based on the action of endogenous enzymes, such as the production of fermented fish

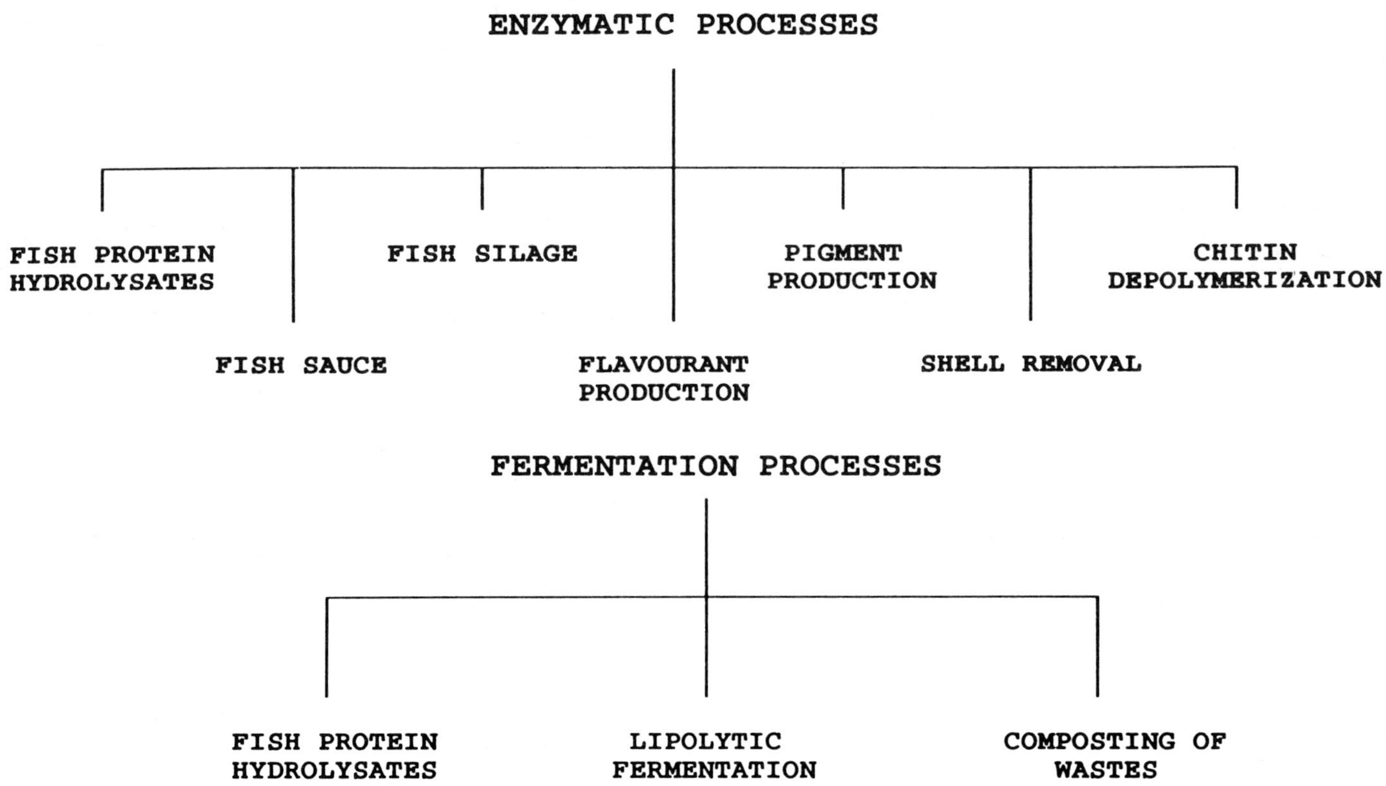

Figure 12.1 Overview of biotechnological processes with applications to fisheries operations.

sauces (Amano, 1962; van Veen, 1965; Beddows and Ardeshir, 1979a,b; Ooshiro *et al.*, 1981; Beddows, 1985; Raksakulthai *et al.*, 1986; Saisithi, 1994), and the ripening of salted herring, *Clupea harengus* (Ritskes, 1971; Ruiter, 1972; Eriksson, 1975; Stefansson and Steingrimsdottir, 1990). Rosario and Maddo (1984) studied the activity of cathepsins during the fermentation of fish sauce. Mizutani *et al.* (1992) defined the main categories of fermented fish products, studying their chemical components. Saisithi (1994) discussed the different types of traditional fermented fish products. Table 12.1 summarizes some of the main traditional food products made by bioconversion processes using fish biomass as substrate.

Fish sauce production is an option for the use of underutilized fish species. Saisithi *et al.* (1966) reported that it is part of the daily diet of hundreds of millions of people, mostly in Asia, although its demand is growing in other parts of the world. It is used mostly as a condiment, although Beauchat (1983) indicates that it is also a source of protein and calcium.

In the fish sauce process, the degradation of proteins into soluble proteins, peptides and amino acids has been primarily attributed to enzymes present in the fish biomass (McIver *et al.*, 1982). However, it has been indicated that microorganisms have an important function in flavour development in the sauce (Beauchat, 1983). The role of microorganisms in the transformations occurring were also highlighted by Amano (1962), who indicated that ammonia may arise from both the activity of endogenous enzymes in the fish and from the activity of bacterial enzymes.

The recent addition of biotechnological processes to those available for the recovery of fisheries wastes and their transformation into useful products should contribute to the advancement of the technological level of traditional fisheries processing operations to that of other food industries.

12.2 Hydrolytic processes for the recovery of fish protein

The hydrolysis of fisheries biomass for the recovery of fish protein can be catalysed by acids, alkalis or by biochemical agents. Among the last, the use of proteases and of proteolytic microorganisms present good potential for the production of an acceptable fish protein hydrolysate product.

In the past few decades, better knowledge about enzyme biochemistry has resulted in the increased utilization of enzymes in industry and other activities. Simultaneously, applications of enzymes in the seafood industry have been developed, and the use of enzymes extracted from fisheries biomass has emerged (Stefansson and Steingrimsdottir, 1990; Haard and Simpson, 1994). Most of the enzymes used in industry, specifically the food

Table 12.1 Traditional food products made by bioconversion processes using fish biomass as substrate

Product type	Product name	Characteristics	References
Sauces	Budu, garos, ketjab-ikan, mam-chau, nampla, ngapi, nuoc-mam-gau-cha, pati, pissala, shotturu	Auto-digested, liquid	Martin and Patel (1991), Mizutani *et al*. (1992), Saisithi (1994)
Pastes	Balachan, bagoong, mam-tom, ngapi, trassi, prahoc, pradec, shiokara	Auto-digested, ground or unground	Mizutani *et al*. (1992), Saisithi (1994)
Solids	Mam, narezushi, padec, pekasam, pla-ra, wadi	Auto-digested, and/or lactofermented, solid with added grains	Mizutani *et al*. (1992), Saisithi (1994)

industry, are hydrolases and an important percentage of them are proteases.

Fisheries biomass contains a number of proteolytic enzymes, mostly from the digestive organs, such as chymotrypsin, pepsin and trypsin. Fish muscle tissue also contains enzymes such as peptidases, cathepsins, transaminases, amidases, amino-acid decarboxylases and glutamic dehydrogenases (Siebert and Schmitt, 1965). Haard (1990) discussed the characteristics of enzymes from muscle food.

Exogenous enzymes also participate in the degradation of fisheries wastes, most of these being from microbial sources. In non-traditional methods for the production of fish sauce, the process could be quickened by adding enzymes. Microbial enzymes such as those from *Aspergillus oryzae* and *Monascus purpureus* have been used as accelerators in the production of fish sauce (Amano, 1962; Miyazawa *et al.*, 1979). Tatterson and Windsor (1974) discussed the stages of fermentation in the production of fish sauce.

Endogenous proteases are also responsible for the liquefaction of fisheries wastes in the process resulting in the product known as fish silage, which is generally employed as animal feed or fertilizer (Wignall and Tatterson, 1976; Gildberg *et al.*, 1984; Gildberg and Almas, 1986; Arason, 1994). Mowbray *et al.* (1988) produced solubles from dogfish (*Squalus acanthias*) by digestion and concentration of ground dogfish processing wastes under acidic and heated conditions. Espe *et al.* (1992) studied the effect of storage for up to one year on the nutritional value of ensiled capelin (*Mallotus villosus*). The authors ensiled cooked and non-cooked minced fish and conducted growth studies in rats.

Acid is generally added to accelerate the process of protein hydrolysis by creating appropriate conditions for the enzymes to work and to limit the growth of spoilage microorganisms. Oil should be removed after liquefaction to enhance the product's acceptability. The resulting product is regarded as a good animal feed with a long storage life (Tatterson and Windsor, 1974).

Several acids have been used for the fish ensiling process. Arason (1994) reviewed this technology. The reduction of pH could be also achieved by inducing a lactic acid fermentation in the ensiling process, requiring the addition of fermentable carbohydrates to it. Fagbenro and Jauncey (1993) reported on the chemical and nutritional attributes of raw, cooked and salted fish silages of tilapia (*Oreochromis niloticus*) fermented with *Lactobacillus plantarum*. Levin (1994), and Martin and Bemister (1994) studied lactic acid-aided fish silage processes. The effect of different concentrations of molasses, used as the carbohydrate source, and salt on a fermented fish ensiling process was presented by Ahmed and Mahendrakar (1995). These authors used fish viscera from tropical freshwater fish and also studied the changes in microbial population during the process

(Ahmed and Mahendrakar, 1996a). The production of fish silage is illustrated in Figure 12.2.

It is to be expected that the composition of fish silage will be similar to that of the material from which it is made, containing approximately 80% water, the rest being protein, lipids and ash from the bones. The lipid

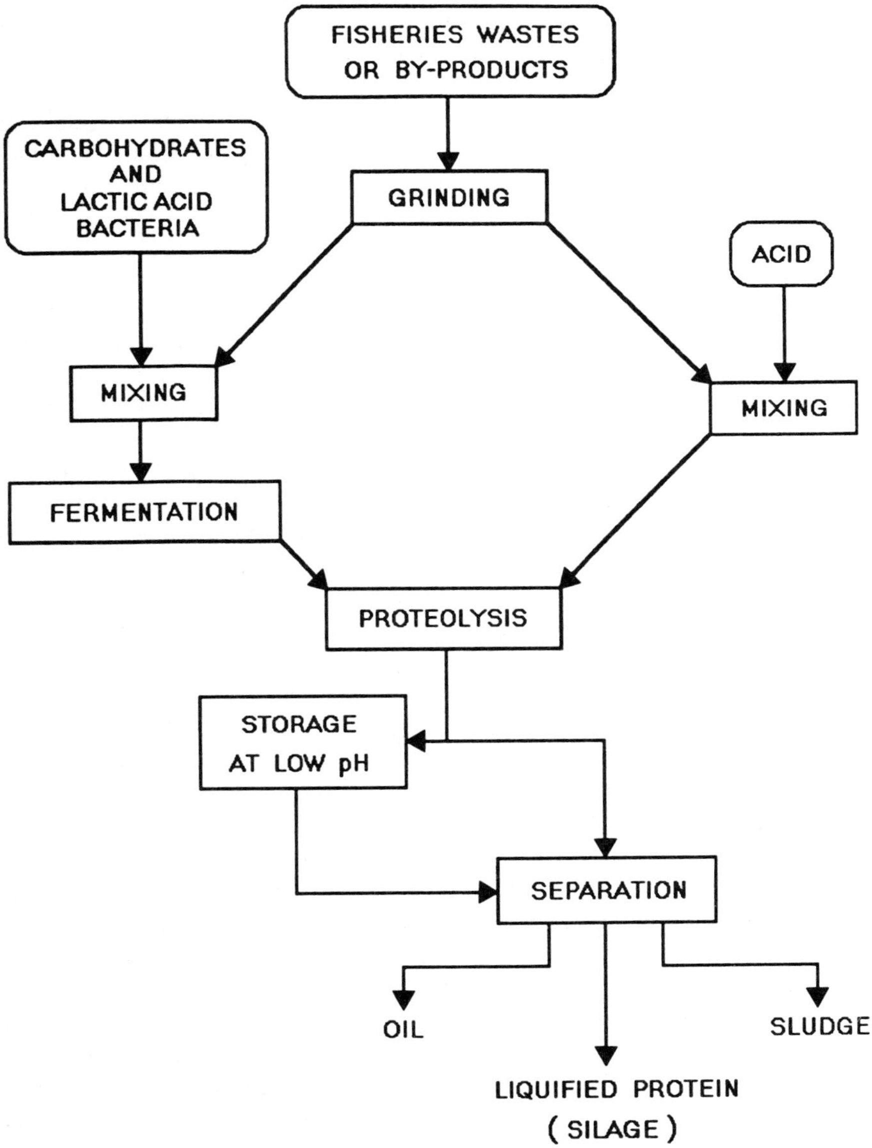

Figure 12.2 Schematic representation of the production of fish silage.

composition will depend upon the characteristics of the raw material and whether the oil had been removed during the process. Fish silage with an appropriate pH value could be kept at room temperature for at least 2 years without putrefaction (Martin and Patel, 1991).

An indication of the great number of possibilities existing for the design of silage processes is given by the studies of Samuels *et al.* (1992). The authors studied the fermentation of fish and crab processing wastes combined with low-quality roughages such as corn stover or peanut hulls. Experiments also included the addition of molasses and of wilted Johnsongrass (*Sorghum halepense* L.).

The nutritional value of fish silage in animal feeds has been studied by several authors, including Hillyer *et al.* (1976), Smith and Adamson (1976), Whittenmore and Taylor (1976), Arason (1994), Heras *et al.* (1994) and Ahmed and Mahendrakar (1996b).

12.2.1 Enzymatic methods

The use of enzymes for the processing of wastes in the food industry has been discussed by Reed (1980) and Shoemaker (1986) among other authors. Hydrolases such as amylases, pectinases, mannanases, cellulases, hemicellulases and lactases find applications in the fruit, vegetable, grain, coffee and dairy industries. In the fisheries, chitinases and proteases are the enzymes with the most potential for application.

Proteases extracted from fish biomass can be used in the same fisheries industry for the production of silage, fish sauce and fish solubles. New technologies employ fish enzymes for skin, membrane and scale removal in fish preparation, roe production, and extraction and recovery of flavourings and pigments (Wray, 1988; Voigt and Botta, 1990; Haard and Simpson, 1994).

There is a growing need for new sources of protein with appropriate functional properties to be incorporated into foods. Concentrates of fish proteins have a potential role in satisfying this need, depending on their functional and organoleptic properties. Bárzana and García-Garibay (1994) reviewed the attempts to produce fish protein concentrates through various technologies. There remains a need for the development of new technologies for the production of good-quality fish protein concentrates (Venugopal *et al.*, 1994a,b, 1995). However, in general those concentrates have not been used for human consumption, primarily due to their poor functional properties. The same problem has limited the use of fish meal. Therefore, those products have been mostly utilized for animal feed.

The production of fish protein concentrates was attempted with the main objectives of obtaining a new food source and solving some of the chronic problems of the seafood industry, such as incomplete utilization of the catch and perishability of the product, among others (Green and Mattick,

1979). Initial works were based on solvent extraction (Knobl, 1967; Finch, 1970). However, there was practically no commercial acceptance of these products because of their already mentioned deficient functional properties. More recently, work has also included the use of enzymatic hydrolysis (Quaglia and Orban, 1987; Rebeca *et al.*, 1991; Sugiyama *et al.*, 1991; Raghunath, 1993; Martin and Porter, 1995; Shahidi *et al.*, 1995; Vieira *et al.*, 1995a,b; Diniz and Martin, 1996, 1997a,b). For the food industry, it is important to produce fish protein concentrates with acceptable functional properties. The final product, after concentration and drying, should be soluble to be successfully incorporated into foods (Mohr, 1980; Mackie, 1982; Hoyle and Merritt, 1994). The use of proteases has also been reported in the protein hydrolysis of stickwater, an aqueous by-product of fish meal production (Jacobsen and Lykke-Rasmussen, 1984). In this application, energy savings resulting from the reduction in viscosity of the stickwater contribute to the overall economy of the process.

Emulsification capacity and solubility are among the functional properties of protein concentrates needed for the food processing industry. Solubilization of proteins can be accomplished by breaking them down into smaller-sized peptides by hydrolysis. The product resulting from this process is known as fish protein hydrolysates. Adler-Nissen (1986) and Venugopal (1994) discussed the production of fish protein hydrolysates by biological methods. Protease enzymes with broad specificity are preferred for commercial processes, as they are capable of splitting the proteins at random independently of the given patterns of amino acids in the proteins (Bárzana and Garcia-Garibay, 1994). The optimum pH values of proteolytic enzymes can vary from those of pepsin and some fungal enzymes (acidic) to those of bacterial proteases (alkaline or neutral). A potential market for fish protein concentrates and hydrolysates is in the production of animal feed, including pet food. An enzymatic process for converting fisheries wastes to fish-food pellets was presented by Shoemaker (1986).

Another category of products which can be obtained by enzymatic hydrolysis of fish biomass is seafood flavourings. One commercial operation has been reported in France (In, 1990). The final product is in the form of a paste or powder for incorporation into food products such as sauces, seasonings and seafood analogues. This kind of product can be used to bolster the flavour of expensive fish products such as crab, lobster and salmon, whose original flavours are reduced during storage. Pan (1990) studied the recovery of the volatile components of shrimp, responsible for its taste, after an enzymatic digestion process.

12.2.2 Methods employing microorganisms

It has been reported that microbial proteases have been found to be superior to proteolytic enzymes from other sources for the solubilization of proteins, owing to their broader specificities (Venugopal, 1994). Sources of proteolytic enzymes, including microbial proteases, and their applications in the food industry have been presented by Löffler (1986).

Venugopal *et al.* (1989) immobilized cells of *Bacillus megaterium*, *Aeromonas hydrophila* and *Pseudomonas marinoglutinosa*, and applied them to the protein hydrolysis of a low-cost fish (*Johnius dissumeri*) meat suspension in water. The cells secreted protease, which solubilized the fish meat. *B. megaterium* was found to be the most efficient hydrolysing agent, producing the solubilization of 30% of the fish protein. Figure 12.3 presents a schematic view of the technologies for the production of fish protein hydrolysates.

Biomass recovered from fisheries sources contains lipids. Also, lipids are a significant component of fishery waste waters. For many seafood processing operations, the removal and possible recovery of lipids from fish wastes cannot be justified economically. Even in fisheries processing plants having waste water treatment facilities, the disposal of fish oils is frequently a problem.

Stickwater is usually evaporated to a product known as 'fish solubles', which can in turn be processed and incorporated into animal feed. However, Green *et al.* (1976) indicated that the lipid content in fish solubles (approximately 11%) limits the use of stickwater in feeds. As is well known, lipids can develop oxidative rancidity.

Potential markets exist for fish oils of good quality. However, the technology for their recovery and processing could be expensive (Wignall and Tatterson, 1976). Biological methods could assist in economically reducing the oil content of fish biomass. Lipolytic fish fermentations have been reported successful in decreasing the lipid contents of fish and fish wastes (Burkholder *et al.*, 1968; Li *et al.*, 1970; Hottinger *et al.*, 1974a; Green *et al.*, 1976). Martin and Patel (1991) indicated that the biological removal of fish oils using microorganisms, by contributing microbial biomass protein, could improve the nutritional value of the recovered fish protein.

The studies of Burkholder *et al.* (1968) found that the yeasts *Candida lipolytica* and *Geotrichum candidum* produced a reduction of the lipid content of young menhaden (*Brevortia tyrannus*) by 30–50%. Studies on the fermentation of fish lipid *in situ* (Burkholder *et al.*, 1968), of stickwater (Green *et al.*, 1976) and of fish oils (Li *et al.*, 1970; Hottinger *et al.*, 1974a) have been conducted with those yeasts. In all cases, it was reported that the lipid content was reduced and microbial cell growth was obtained. However, no commercial applications have been developed, as yet, for this technology. More discussion on the fermentation of fish lipids is presented in section 12.6.

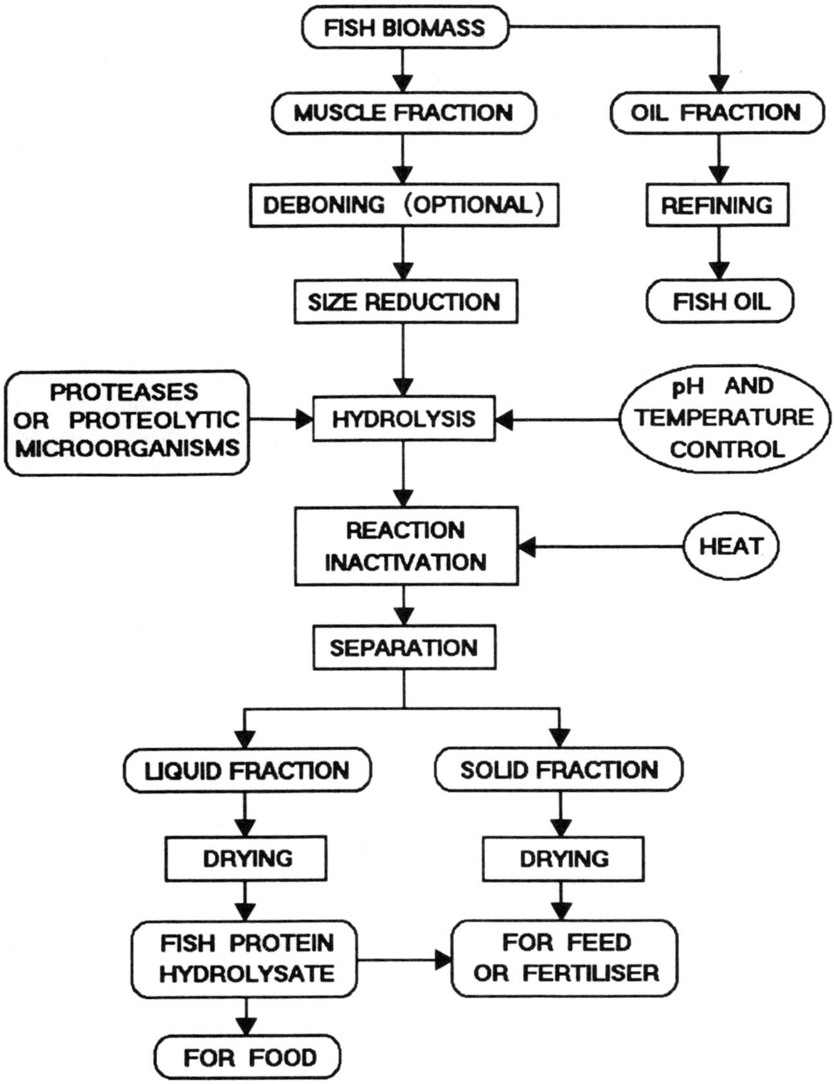

Figure 12.3 Schematic representation of the technologies for the production of fish protein hydrolysates.

12.3 Biological methods for the recovery of chitin and chitosan

The presence of relatively insoluble crustacean shells in shellfish processing wastes presents an environmental problem, which requires appropriate technological solutions. In these shells, chitin is associated with such components as lipids, pigments and protein (Simpson *et al.*, 1994). It is estimated that chitin is the most widely distributed polymer on earth, and is one of the most abundant (Austin *et al.*, 1981). Chitin is a polymer of *N*-acetylglucosamine and glucosamine residues.

Chitin and its deacetylated derivative, chitosan, are polysaccharides with interesting characteristics. Chitin is insoluble in many solvents and resists most chemical reactions; however, it is deacetylated to chitosan by hot concentrated NaOH. Chitosan is soluble in organic acid solutions and carries amine groups with positive charges. In general, it has a variety of potential applications. The main characteristics of chitin and chitosan are presented in Table 12.2.

Comprehensive studies on chitin and chitosan have been published by Skjåk-Bræk *et al.* (1989) and Brine *et al.* (1992). Hansen and Illanes (1994) and Simpson *et al.* (1994) have presented their main applications. Table 12.3 presents some of the potential uses of chitin and chitosan.

Biotechnological methods for the recovery and processing of chitin have been studied as a waste treatment alternative to the disposal of shellfish waste (Carroad and Tom, 1978; Revah-Moisseev and Carroad, 1981). Cosio *et al.* (1982) studied the conditions for crustacean chitin waste pretreatment (size reduction, deproteination and demineralization) and the production of chitinase by *Serratia marcescens*. Gagne (1993) reported on the use of bacterial protease, chymotrypsin and papain to deproteinize crustacean shells. The author found that the deproteinization achieved with chymotrypsin was similar to that produced by using NaOH, making it the most effective enzyme.

Because traditional methods of processing chitin and chitosan can produce depolymerization and deacetylation of the original compounds, it

Table 12.2 Main characteristics of chitin and chitosan

Substance	Characteristic
Chitin	Biodegradable, non-toxic natural product Insoluble in alkaline solutions and organic solvents Insoluble in dilute acid solutions
Chitosan	Biodegradable, non-toxic natural product Insoluble in alkaline solutions and organic solvents Soluble in dilute acid solutions Stable to most reagents, including aqueous alkalis Highly positively charged polyelectrolyte at acidic pH: can react with many biological compounds having negative charges such as proteins, anionic polysaccharides, nucleic acids, and others Can form complexes with metal ions: useful in waste water treatment processes

Sources: Muzzarelli (1977), Austin *et al.* (1981), Simpson *et al.* (1994)

Table 12.3 Main potential applications of chitin and chitosan

Fields	Applications	References
Agriculture	Reduces bean root-rot and radish vascular wilt (chitin)	Mitchell and Alexander (1961)
	Antifungal (chitosan)	Struszczyk *et al.* (1989)
	Protects seeds and enhances crop yield (chitosan)	Sandford (1989), Struszczyk *et al.* (1989)
Cosmetics	Abrasive for skin cleansing (chitin/chitosan)	Yanagida (1985)
	High water-holding capacity moisturizer (chitosan)	Hirano (1989)
Environmental	Purifies municipal, industrial, food processing effluents (chitosan)	Moore *et al.* (1987), Hirano (1989), No and Meyers (1989), Knorr (1991)
	Precipitates, recovers proteins (chitosan)	Landes and Bough (1976), Sandford (1989)
	Chelates, removes pollutants; recovers microorganisms (chitosan)	Knorr *et al.* (1989)
Food	Functional and direct ingredients in foods (chitin/chitosan)	Knorr (1982, 1984), Sandford (1989), Hirano *et al.* (1990)
	Food preservative (chitosan)	Hirano (1989)
	Clarifying agents (chitin/chitosan)	Ornum (1991)
Technical	In photographic films (chitin)	Ryan and Yankowski (1969)
	In microporous spray-dried particles for chromatography (chitosan)	Rodríguez-Sánchez and Rha (1981)
	In membranes for reverse osmosis, ultrafiltration (chitosan)	Yang and Zall (1984)

Table 12.3 *Continued*

Fields	Applications	References
Medicine	Facilitates wound healing, reduces blood serum cholesterol (chitosan)	Nagyvany *et al.* (1979)
	Affects blood coagulation (chitosan)	Hirano *et al.* (1985), Fradet *et al.* (1986), Muzzarelli *et al.* (1986)
	Stimulates immune system (chitin)	Nishimura *et al.* (1986)
	Controls appetite, prevents gastritis (chitin/chitosan)	Olsen *et al.* (1989)
	Inhibits thrombin hydrolytic activity (chitin)	Okei *et al.* (1986)
	In diet, decreases cholesterol levels (chitosan)	Hirano (1989)
	Tissue regeneration, vascular surgery (chitosan)	Malette *et al.* (1986)
Other biotechnological applications	Bioconversion to SCP for animal feed (chitin)	Carroad and Tom (1978), Revah-Moiseev and Carroad (1981), Cosio *et al.* (1982)
	In preparation of membranes for encapsulation of tissue and bacterial cells (chitosan)	Rodríguez-Sánchez and Rha (1981)
	Membranes used in mammalian cell culture technology (chitosan)	Kim and Rha (1989), Izume *et al.* (1989), Shioya and Rha (1989)
	In cells and cell debris recuperation; also for SCP recovery (chitosan)	Holland (1989)

has been expected that biotechnology could become the choice for their recovery, minimizing the loss of their characteristics. Simpson *et al.* (1994) studied the biologically based technologies for chitin and chitosan processing. Knorr (1986) presented their processing and biotechnological characteristics.

Biotechnological procedures can be also employed for the deacetylation of chitin to produce chitosan. The following microorganisms have been found to produce enzymes for the deacetylation of chitin: *Mucor rouxii* (Araki and Ito, 1975; Knorr and Klein, 1986); *Colletotrichum lindemuthianum* (Kauss *et al.*, 1982/83); and *Phycomyces blakesleeanus* (Knorr and Klein, 1986).

Chitosan can be hydrolysed by chitosanases, and some microorganisms are able to produce them. These include bacteria (Tominaga and Stujisake, 1975; Davis and Eveleigh, 1984), fungi (Fenton and Eveleigh, 1981), myxobacteria (Hedges and Wolfe, 1974) and actinomycetes (Price and Stork, 1975; Ohtakara *et al.*, 1984).

12.4 Biological water treatment of fisheries wastes

Seafood processing produces waste effluents generally rich in suspended solids and soluble organic matter. Therefore, their treatment should consider, as the first alternative, the recovery of some of the valuable nutrients present within them. In the case of solid wastes, two simple biological processes, composting and ensiling, are discussed elsewhere in this chapter. This section will concentrate on waste waters.

The composition of fisheries processing waste waters depends on the species processed and the kind of seafood produced. For example, it is expected that waste waters from canning will differ from those from filleting operations. Both aerobic and anaerobic processes can be used in biological fisheries waste water treatment. Veiga *et al.* (1994) discussed the characteristics of both systems when applied to the seafood industry.

Studies on the biological treatment of fisheries waste waters have involved novel processes with pure microbial cultures in addition to the more conventional mixed-culture operations. Kakuta *et al.* (1985) reported the treatment of waste waters from a dried-bonito processing factory using yeasts. One strain of *Hansenula anomala* showed the highest potential for utilizing organic substances, reducing the waste water chemical oxygen demand (COD) to 6450 ppm from 24 600 ppm, and the nitrogen content to 1050 ppm from 3610 ppm.

Some liquid wastes from fisheries operations can be recovered for the manufacturing of products. For example, Shiau and Chai (1990) recovered oyster shucking liquid wastes, including shell liquor, bled liquor and wash water, for the production of oyster soup. Other wastes can be employed as fermentation substrates, as in the use of mussel processing wastes for the production of single cell protein (Siso *et al.*, 1987). Other cases will be presented in section 12.6.1. The main product of anaerobic processes for biological waste water treatment is methane, which can be recovered for use as fuel. Hudson *et al.* (1978) reported on the anaerobic treatment of shellfish processing waste waters in packed columns, and Lema *et al.* (1987) studied the anaerobic digestion of mussel processing waste waters in mesophilic and thermophilic ranges of temperature. Other works include those of Pohland and Hudson (1976), Balslev-Olesen *et al.* (1990) and Nair (1990).

12.5 Composting of fisheries offal

Many thousands of tonnes of solid fisheries wastes are produced each year all around the world. They can be used as a valuable organic fertilizer but their foul odour discourages use of this option. Mathur (1991) discussed a novel method of fish-waste composting, incorporating peat moss into a passively aerated windrow (PAW) composting system. An important attribute of this method is the potential retention of some of the nitrogen liberated during the composting. The product obtained was of high quality with good concentrations of nutrients. This process is presented in more detail in section 4.4.1(c) of Chapter 4 of this book.

For the successful composting of fisheries wastes, a bulking agent is required. Because the composting process is aerobic, the bulking agent should allow aeration and at the same time provide an adequate carbon to nitrogen ratio for the microbial population. The appropriate design of the composting pile and its components will facilitate an efficient composting reaction without the production of foul odours (Frederick *et al.*, 1989). These authors pointed out that, in general, wood wastes such as shredded brush, shredded bark and wood chips satisfy those needs. Martin *et al.* (1993) presented a comparative study of the use of *Sphagnum* peat and sawdust in the composting of fisheries and other wastes.

Martin and Chintalapati (1989) produced liquid extracts, by chemical hydrolysis, of the fish offal–peat compost made by the process of Mathur (1991), and employed those extracts as fermentation media for the growth of the acid-tolerant fungus *Scytalidium acidophilum*. The compost hydrolysate contained higher concentrations of nitrogen than a liquid extract produced by a similar acid hydrolysis of peat, although the total carbohydrate concentration was higher in the latter. The effect of a compost of shrimp wastes and peat on the growth of barley (*Hordeum vulgare* L.) was reported by Hountin *et al.* (1995).

12.6 Other products from fisheries waste biomass

12.6.1 Fermentation substrates

The utilization of nutrient-rich fisheries wastes as substrate sources for fermentation processes could create a new incentive for their recovery.

Lipids such as fish oils can be employed as energy and carbon sources for the growth of microbial populations and the production of microbial products. Martin and Patel (1991) discussed the aspects of fish oil that could make it an attractive alternative to carbohydrates as a fermentation substrate. An example of the many existing possibilities in this regard is the production of microbial biomass protein employing fisheries wastes as substrate in submerged fermentation processes. Microbial biomass protein, also known as single cell protein (SCP), is potentially useful as a protein and vitamin supplement for animal feeds and human foods. The lipolytic yeasts *C. lipolytica* and *G. candidum*, already mentioned in this chapter, have been used for studies on SCP production (Hottinger *et al.*, 1974a,b). Using a concentration of up to 5% fish oil in the culture media, the authors reported an average crude protein content of 40% for both species of microorganisms. Hottinger *et al.* (1974a) reported that alewife (*Alosa pseudoharingus*) oil in a basal medium was as effective a nutrient supplement as yeast extract or corn steep liquor for the abovementioned yeasts. The increase in cell yield stopped at the 5% oil level. The operating conditions for this process were optimized (Hottinger *et al.*, 1974b). Those studies suggested the possibility of obtaining 800 g of dry yeast biomass with a crude protein yield of 320 g from 1 kg of fish oil using both batch fermentation and continuous fermentation processes.

The use of fish lipids for fermentation processes, owing to the specific properties of this substrate, requires further study. For example, Li *et al.* (1970) observed that growth in a fish lipid fermentation in shake flasks stopped before the expected level of consumption of the substrate was reached.

Fish oil contains a high concentration of polyunsaturated fatty acids which, in the presence of oxygen in aerobic fermentations, produce high concentrations of peroxides. These are oxidative deterioration products which could inhibit the growth of microorganisms. Higgs (1974) tested various microorganisms for their ability to metabolize fish oil. The addition of food-grade antioxidants to the fish oil prior to fermentation, at concentrations of approximately 0.5–1.0%, eliminated the growth inhibition produced by high peroxide concentrations. Zajic *et al.* (1974) discussed the additional advantages that reduced oxidative deterioration could provide during fish oil fermentation, such as better emulsification of the oil during the process, and the prevention of the formation of polymers and soaps, which decrease the fermentation yields, in the fermentation equipment.

Other seafood processing wastes, beside fish lipids, can be employed in fermentation processes. As mentioned above, Carroad and Tom (1978) proposed chitin bioconversion to yeast SCP as a waste treatment alternative to the disposal of shellfish waste. *Pichia kudriavzevii* grew well on chitin hydrolysates and yielded a microbial protein with acceptable amino-acid composition (Revah-Moiseev and Carroad, 1981). Cosio *et al.* (1982) presented process design information and economic analysis for an integrated process for the conversion of shrimp shell chitin waste to microbial biomass. The authors studied the use of *N*-acetylglucosamine, released by the chitinase reaction, as a substrate to grow yeast for use in the production of SCP. Salt brines are used as contact refrigerants aboard fishing ships. In the process, they generally become contaminated with organic matter from the fish. Those brines should be treated before being discarded and a biological process could make it possible to recycle them while yielding some useful product. To that end, a method was presented by Welsh and Zall (1984) for growing *C. utilis* yeasts on spent food processing brines to produce SCP.

The growth of microfungi and yeast in mussel processing wastes has also been studied, with the aim of developing a biological waste treatment process by means of their use as fermentation substrate (Murado *et al.*, 1994).

The use of culture media prepared from fish waste juice for the growth of yeasts was reported by Hossain *et al.* (1988). Almas (1990) discussed the production of microbial growth media from marine biomass.

12.6.2 Enzymes from fish biomass

Fisheries waste biomass is an important potential source for a variety of enzymes, some of them with unique attributes. The recovery of enzymes from fish wastes was suggested by Green and Mattick (1978, 1979). A summary of the research conducted with enzymes extracted from fish is presented in Table 12.4.

Fish are known to adapt to quite low temperatures, which could affect such enzyme properties as binding affinity, cold stability, molecular activity, specificity and thermodynamic properties in general (Simpson and Haard, 1987). The same authors have discussed the features of some enzymes from marine organisms. Regarding their proteases, it has been reported that they differ from those present in terrestrial animals and in plants. Table 12.5 summarizes some of their characteristics.

Enzymes from marine organisms may be exploited in certain food processing operations. Haard and Simpson (1994) discussed the proteases

Table 12.4 Summary of research conducted with proteolytic enzymes extracted from marine species

Source species	References
Arctic cod	Arunachalam and Haard (1985)
Arctic cod, American smelt	Haard *et al.* (1982)
Atlantic cod	Brewer *et al.* (1984), Asgeirsson *et al.* (1989)
Bonito	Kubota and Ohnuma (1970a,b)
Brook trout	Owen and Wiggs (1971)
Capelin	Hjelmeland and Raa (1982), Raksakulthai *et al.* (1986)
Capelin and herring	Kalac (1978a,b)
Dogfish	Merrett *et al.* (1969)
Fish muscle tissue	Huang and Tappel (1971), Makinonda and Ikeda (1971)
Greenland cod	Squires *et al.* (1986a), Simpson and Haard (1984)
Harp seal	Shamsuzzaman and Haard (1984)
Sardine	Noda and Murakami (1981), Noda *et al.* (1982), Van *et al.* (1983)
Smelt	Haard *et al.* (1982)
Tuna	Norris and Mathies (1953)

from aquatic organisms and their uses in the seafood industry. Other potential applications include the detergent and leather industries, and the food industry in general. For example, Brewer *et al.* (1984) reported that it is possible to prepare satisfactory cheddar cheese using proteases from marine organisms.

12.6.3 Media for the cultivation of edible mushrooms

The utilization of fisheries by-products in the growth of mushrooms has been based on the indication of a relationship between lipid metabolism and the initiation of fruiting in the cultivated mushroom *Agaricus bisporus* (Lange) Sing. (Schisler and Sinden, 1966). These authors reported that supplementation with refined and crude seed oils increased mushroom yield. It has been suggested that fish solubles, as well as other fisheries wastes, may be regarded as inexpensive nutrient supplements that would increase the yields and sizes of cultivated mushrooms (Green *et al.*, 1973). Some of the works that report on the use of fisheries wastes and by-products in the cultivation of commercial mushrooms are presented in Table 12.6.

Meyers and No (1995) studied the extraction of the carotenoid astaxanthin from Louisiana crawfish processing wastes. The authors reported that ensilage of the raw crawfish wastes before the extraction process increased the pigment concentration by 40–50%.

Table 12.5 Characteristics of some marine enzymes

Enzyme	Source	Characteristics	References
Trypsin	Pyloric caeca and intestines of Greenland cod (*Gadus ogac*)	Attributes distinct from bovine trypsin. Greater stability and activity at alkaline pH; suitable for industrial hydrolysis of fish protein	Simpson (1983)
Gastric proteases	Greenland cod (*G. ogac*)	More alkaline pH optima with protein substrates than porcine pepsin. Heat stability, substrate specificity, milk clotting ability and amino-acid composition also differ	Squires (1984), Squires *et al*. (1986a,b)
Acidic proteases	Harp seal (*Pagophilus groenlandicus*) gastric mucosa	Similar to calf chymosin: higher milk clotting to proteolytic activity ratio than porcine pepsin, clots milk up to pH 7.0, optimum pH 2.2–3.5 for haemoglobin hydrolysis. Possible rennet substitute in cheese manufacture	Shamsuzzaman (1983)

Table 12.6 Use of fisheries wastes and by-products in the cultivation of mushrooms

Substance	Comments	Reference
Menhaden fish oil	Used as a composting supplement for *Agricus bisporus*. Results similar to some plant oils	Schisler and Patton (1974)
Condensed fish solubles	Used as a composting supplement for *Agaricus bisporus*. Yields similar to other organic supplements, with average size increase	Schisler and Patton (1974)
Fish meal, fish oil and fish offal–peat compost	Used as nutrient supplements for *Pleurotus ostreatus*. Good growth resulted, except no growth with fish meal. Fish oil produced best results	Martin and Bassler (1989)

12.7 Conclusions

12.7.1 Present developments

New analytical methods are being applied to evaluate the influence of process parameters in hydrolytic reactions in the bioconversion of fisheries wastes. Ashie *et al.* (1996) studied the combined effects of, and optimum processing conditions for, controlling undesirable endogenous enzymatic activity that could result in deterioration of the texture of fisheries products. The authors applied response surface methodology and concluded that this method was effective in analysing combination treatments to control undesirable effects of enzymes in fish muscle. In addition, they mentioned that this analysis could be useful for fish products in which protease activity results in gel softening, such as surimi and minced fish products. Diniz and Martin (1996, 1997a) have also applied similar mathematical techniques for the interpretation of the results in the enzymatic hydrolysis of dogfish (*Squalus acanthias*).

The production of fish protein hydrolysates can benefit from biotechnological advances such as enzyme and microbial engineering (Bhumiratana *et al.*, 1977; Nakajima *et al.*, 1992). Also, developments in biological reactor design for these processes and in downstream operations could improve both the quality of the final product and the overall economy of the process. Some of those potential developments are presented in Table 12.7.

Product quality is paramount in the development of new products, especially foods, from fisheries biomass. Fish protein concentrates or hydrolysates from any recovered fisheries biomass, used by the food

Table 12.7 Potential developments in the production of fish protein hydrolysates by biological methods

Process component	Development
Use of proteolytic enzymes and microorganisms	Use of new species, varieties or microbial strains. Development of genetically engineered enzymes. Immobilization of enzymes or microbial cells
Reactor design	Membrane reactors allow enzyme reuse, easier separation of products, and the possibility for continous processes
Downstream operations	Use of spray drying of hydrolysate slurry. Development of inexpensive drying processes, including solar drying

industry as food additives, require satisfactory protein functional properties, as they need to compete with other sources of protein additives (Park, 1994).

The smell and taste of fish are not accepted by some populations, and work is needed to develop practical and economical ways for their removal. In the case of protein hydrolysis, bitterness tends to arise, which also occurs in hydrolysates of fish protein. Recently, important research has been taking place to reduce and inhibit bitterness in foods (Roy, 1992). It is expected that the results will contribute to the acceptance of protein concentrates from fish.

There is also a need to improve some aspects of the products derived from chitin and it is expected that biological processes, such as those previously indicated, will be successful in this because they use gentler processing conditions. It is important to mention that shellfish wastes could yield pigments and minerals in addition to chitin (Hansen and Illanes, 1994).

The use of combined chemical and enzymatic treatments to recover the process waste fisheries biomass could overcome the limitations of the individual processes (Simpson and Haard, 1985). The use of microorganisms in biological processing of wastes still requires further study, which should result in proper design for the prevention of unwanted side reactions.

In general, although the biological processing of fisheries wastes could be a low-capital operation, many of its limitations are due to the expensive downstream operations required for the final processing of the derived products. Because fish biomass has a large percentage of water, drying is one of the limiting steps, from the economical point of view, in the development of processes for the production of fish protein concentrates, hydrolysates or silage.

12.7.2 Future trends

Owing to the general development of biotechnology, it is expected to play an important role in the future development of seafood industry processing operations, including fisheries waste biomass recovery and processing. The following are areas of possible future developments in the bioconversion of fisheries wastes to industrial products:

1. the development of fish silage and fish compost processes, mostly in small fishing communities;
2. new uses and markets for fish protein concentrates and hydrolysates;
3. the use of fish oil and other nutrients from fisheries wastes as substrates for biotechnological processes;
4. the application of enzymatic and microbial techniques to the processing of chitin and chitosan from shellfish wastes;
5. the extraction of specific high value added products such as enzymes, hormones, pharmaceuticals, and other chemicals.

References

Adler-Nissen, J. (1986) *Enzymatic Hydrolysis of Food Proteins*, Elsevier Applied Science, London.

Ahmed, J. and Mahendrakar, N.S. (1995) Effect of different levels of molasses and salt on acid production and volume of fermenting mass during ensiling of tropical freshwater fish viscera. *Journal of Food Science and Technology*, **32** (2), 115–18.

Ahmed, J. and Mahendrakar, N.S. (1996a) Changes in microbial population during fermentation of tropical freshwater fish viscera. *Journal of Applied Bacteriology*, **80**, 153–6.

Ahmed, J. and Mahendrakar, N.S. (1996b) Growth and meat quality of broiler chicks fed with fermented fish viscera. *International Journal of Animal Science*, **11**, 1–5.

Almas, K.A. (1990) Utilization of marine biomass for production of microbial growth media and biochemicals, in *Advances in Fisheries Technology and Biotechnology for Increased Profitability* (eds M.N. Voigt and J.R. Botta), Technomic, Lancaster, PA, pp. 361–72.

Amano, K. (1962) The influence of fermentation on the nutritive value of fish with special references to fermented fish products of South-East Asia, in *Fish in Nutrition* (eds E. Heen and R. Kreuzer), Fishing News (Books), London, pp. 180–200.

Araki, Y. and Ito, E. (1975) A pathway of chitosan formation in *Mucor rouxii*. Enzymatic deacetylation of chitin. *European Journal of Biochemistry,* **55**, 75–8.

Arason, S. (1994) Production of fish silage, in *Fisheries Processing, Biotechnological Applications* (ed. A.M. Martin), Chapman and Hall, London, pp. 244–72.

Arunachalam, K. and Haard, N.F. (1985) Isolation and characterization of pepsin from Polar cod (*Boreogadus saida*). *Comparative Biochemistry and Physiology*, **80B**, 467–73.

Asgeirsson, B., Fox, J.W. and Bjarnason, J.B. (1989) Purification and characterization of trypsin from the poikilotherm *Gadus morhua. European Journal of Biochemistry*, **180**, 85–94.

Ashie, I.N.A., Simpson, B.K. and Ramaswamy, H.S. (1996) Control of endogenous enzyme activity in fish muscle by inhibitors and hydrostatic pressure using RSM. *Journal of Food Science*, **61**, 350–6.

Austin, P.R., Brine, C.J., Castle, J.E. and Zikakis, J.P. (1981) Chitin: new facets of research. *Science*, **212**, 749–53.

Balslev-Olesen, P., Lynggaard-Jensen, A. and Nickelsen, C. (1990) Pilot-scale experiments on anaerobic treatment of wastewater from a fish processing plant. *Water Science Technology*, **22**, 463–74.

Bárzana, E. and García-Garibay, M. (1994) Production of fish protein concentrates, in *Fisheries Processing, Biotechnological Applications* (ed. A.M. Martin), Chapman and Hall, London, pp. 206–22.

Beauchat, L.R. (1983) Indigenous fermented foods, in *Biotechnology*, Vol. 5, (ed. G. Reed), Verlag Chemie, Deerfield Beach, FL, pp. 509–11.

Beddows, C.G. (1985) Fermented fish and fish products, in *Microbiology of Fermented Foods*, Vol. 2 (ed. B.J.B. Wood), Elsevier Applied Science, London, pp. 1–39.

Beddows, C.G. and Ardeshir, A.G. (1979a) The production of soluble fish protein solution for use in fish sauce manufacture. I. The use of added enzymes. *Journal of Food Technology*, **14**, 603–12.

Beddows, C.G. and Ardeshir, A.G. (1979b) The production of soluble fish protein solution for use in fish sauce manufacture. II. The use of acids at ambient temperature. *Journal of Food Technology*, **14**, 613–23.

Bhumiratana, S., Hill, C.G. and Amundson, C.H. (1977) Enzymatic solubilization of fish protein concentrate in membrane reactors. *Journal of Food Science*, **42**(4), 1016–21.

Bligh, E.G. (ed.) (1992) *Seafood Science and Technology*, Fishing News (Books), Oxford.

Børrensen, T. (1992) Biotechnology, by-products and aquaculture, in *Seafood Science and Technology* (ed. E.G. Bligh), Fishing News (Books), Oxford, pp. 278–87.

Brewer, P., Helbig, N. and Haard, N.F. (1984) Atlantic cod pepsin – characterization and use as a rennet substitute. *Canadian Institute of Food Science and Technology Journal*, **17**, 38–42.

Brine, C.J., Sandford, P.A. and Zikakis, J.P. (eds) (1992) *Advances in Chitin and Chitosan*, Elsevier Applied Science, London.

Brody, J. (1965) *Fishery By-Products Technology*. AVI, Westport, USA.

Burkholder, L., Burkholder, P.R., Chu, A., Kostyk, N. and Roels, O.A. (1968) Fish fermentation. *Food Technology*, **22**, 1278–84.

Carroad, P.A. and Tom, R.A. (1978) Bioconversion of shellfish chitin waste: process conception and selection of microorganisms. *Journal of Food Science*, **43**, 1158–63.

Cosio, I.G., Fisher, R.A. and Carroad, P.A. (1982) Bioconversion of shellfish chitin waste: waste pretreatment, enzyme production, process design, and economic analysis. *Journal of Food Science*, **47**, 901–5.

Davis, B. and Eveleigh, D.E. (1984) Chitosanases: occurrence, production and immobilization, in *Chitin, Chitosan and Related Enzymes* (ed. J.P. Zikakis), Academic Press, New York, pp. 161–79.

Diniz, F.M. and Martin, A.M. (1996) Use of response surface methodology to describe the combined effects of pH, temperature and E/S ratio on the hydrolysis of dogfish (*Squalus acanthias*) muscle. *International Journal of Food Science and Technology*, **31** (5), 419–426.

Diniz, F.M. and Martin, A.M. (1997a) Optimization of nitrogen recovery in the enzymatic hydrolysis of dogfish (*Squalus acanthias*) protein. Composition of the hydrolysates. *International Journal of Food Science and Nutrition*, **48**, 191–200.

Diniz, F.M. and Martin, A.M. (1997b) Effects of the extent of enzymatic hydrolysis on functional and rheological properties of shark protein hydrolysate. *Food Science and Technology/Lebensmittel Wissenschaft und Technologie*, **30**, 266–72.

Eriksson, C. (1975) Method of controlling the ripening process of herring, Canadian patent 96419.

Espe, M., Haaland, H., Njaa, L.R. and Raa, J. (1992) Growth of young rats on diets based on fish silage with different degrees of hydrolysis. *Food Chemistry*, **44**, 195–200.

Fagbenro, O. and Jauncey, K. (1993) Chemical and nutritional quality of raw, cooked and slated fish silages. *Food Chemistry*, **48**, 331–5.

Fenton, D.M. and Eveleigh, D.E. (1981) Purification and mode of action of a chitosanase from *Penicillium islandicum. Journal of General Microbiology*, **126**, 151–65.

Finch, R. (1970) Fish protein for human foods. *CRC Critical Reviews in Food Technology*, **1**, 519–80.

Fradet, G., Brister, S., Mulder, D.S., Lough, J. and Averbach, B.L. (1986) Evaluation of chitosan as a new hemostatic agent: *in vitro* and *in vivo* experiment, in *Chitin in Nature and Technology* (eds R.A.A. Muzzarelli, C. Jeuniaux and G.W. Goody), Plenum Press, New York, pp. 485–91.

Frederick, L., Harris, R., Peterson, L. and Kehrmeyer, S. (1989) *The Compost Solution to*

Dockside Fish Wastes. University of Wisconsin Sea Grant Advisory Report No. WIS-SG-89-434.

Gagne, N. (1993) Optimization studies on chitin/chitosan from crustacean waste, M.Sc. thesis, McGill University, Montreal, Canada. Cited in Simpson, B.K., Gagne, N. and Simpson, M.V. (1994) Bioprocessing of chitin and chitosan, in *Fisheries Processing, Biotechnological Applications*, (ed. A.M. Martin), Chapman and Hall, London, pp. 155–73.

Gildberg, A. and Almas, K.A. (1986) Utilization of fish viscera, in *Food Engineering and Process Applications*, Vol. 2 (eds M. Le Maguer and P. Jelen), Elsevier Applied Science, London, pp. 383–93.

Gildberg, A., Espejo-Hermes, J. and Magno-Orejana, F. (1984) Acceleration of autolysis during fish sauce fermentation by adding acid and reducing the salt content. *Journal of the Science of Food and Agriculture*, **35**, 1363–9.

Green, J.H. and Mattick, J.F. (1978) Possible methods for the utilization or disposal of fishery solid wastes. *Journal of Food Quality*, **1**, 239–51.

Green, J.H. and Mattick, J.F. (1979) Fishery waste management, in *Food Processing Waste Management* (eds J.H. Green and A. Kramer), AVI, Westport, pp. 202–27.

Green, J.H., Paskell, S.L., Goldmintz, D. and Schisler, L.C. (1973) New methods under investigation for the utilization of fish solubles, a fishery byproduct, as a means of pollution abatement, in *Food Processing Waste Management; Proceedings of the 1973 Cornell Agricultural Waste Management Conference*, Syracuse, NY, pp. 51–68.

Green, J.H., Paskell, S.L. and Goldmintz, D. (1976) Lipolytic fermentations of stickwater by *Geotrichum candidum* and *Candida lipolytica*. *Applied Environmental Microbiology*, **31**, 569–75.

Haard, N.F. (1990) Enzymes from food myosystems. *Journal of Muscle Foods*, **1**, 293–338.

Haard, N.F. and Simpson, B.K. (1994) Proteases from aquatic organisms and their uses in the seafood industry, in *Fisheries Processing, Biotechnological Applications* (ed. A.M. Martin), Chapman and Hall, London, pp. 132–54.

Haard, N.F., Feltham, L.A.W., Helbig, N. and Squires, E.J. (1982) Modification of proteins with proteolytic enzymes from the marine environment, in *Advances in Chemistry Series, No. 198: Modification of Protein, Food, Nutrition and Pharmacological Aspects* (eds R.E. Feeney and J.R. Whitaker), American Chemical Society, Washington, DC, pp. 223–44.

Halevy, S. (1989) Drugs from the sea, in *Microbiology Applications in Food Biotechnology* (eds B.H. Nga and Y.K. Lee), Elsevier Applied Science, London, pp. 123–34.

Hansen, M.E. and Illanes, A. (1994) Applications of crustacean wastes in biotechnology, in *Fisheries Processing, Biotechnological Applications* (ed. A.M. Martin), Chapman and Hall, London, pp. 174–205.

Hedges, A. and Wolfe, R.S. (1974) Extracellular enzymes from Myxobacter Al-1 that exhibit both β-1,4 glucanase and chitosanase activities. *Journal of Bacteriology*, **120**, 844–53.

Heras, H., McLeod, C.A. and Ackman, R.G. (1994) Atlantic dogfish silage *vs*. herring silage in diets for Atlantic salmon (*Salmo salar*): growth and sensory evaluation of fillets. *Aquaculture*, **125**, 93–106.

Higgs, T.W. (1974) Enhanced fish oil fermentation. M.Eng. thesis, The University of Western Ontario, London, Ontario.

Hillyer, G.M., Peers, D.G., Morrison, R., Parry, D.A. and Woods, M.P. (1976) Evaluation for on-farm use of de-oiled herring silage as a protein feed for growing pigs. Paper No. 4. *Proceedings of the Torry Research Station Symposium on Fish Silage*, Aberdeen, Scotland.

Hirano, S. (1989) Production and application of chitin and chitosan in Japan, in *Chitin and Chitosan: Sources, Chemistry, Biochemistry, Physical Properties and Applications* (eds G. Skjåk-Bræk, T. Anthonsen and P. Sandford), Elsevier Applied Science, London, pp. 37–44.

Hirano, S., Tanaka, Y., Hasegawa, M., Tobetto, K. and Nishioka, A. (1985) Effect of sulfated derivatives of chitosan on some blood coagulant factors. *Carbohydrate Research*, **137**, 205–15.

Hirano, S., Hakura, C., Seino, H. *et al*. (1990) Chitosan as an ingredient for domestic animal feed. *Journal of Agriculture and Food Chemistry*, **38**, 1214–18.

Hjelmeland, K. and Raa, J. (1982) Characteristics of two trypsin type isozymes isolated from the arctic fish, capelin (*Mallotus villosus*). *Comparative Biochemistry and Physiology*, **71B**, 557–62.

Holland, C.R. (1989) Recovery of single cell protein by chitosan in a batch dissolved air flotation system, in *Chitin and Chitosan: Sources, Chemistry, Biochemistry, Physical Properties and Applications* (eds G. Skjåk-Bræk, T. Anthonsen and P. Sandford), Elsevier Applied Science, London, pp. 559–66.

Hossain, M.A., Furuichi, M. and Yone, Y. (1988) Cultivation of yeast in medium containing liquid from fish waste juice. *Nippon Suisan Gakkaishi*, **54**, 469–71.

Hottinger, H.H., Richardson, T., Amundson, C.H. and Stuiber, D.A. (1974a) Utilization of fish oil by *Candida lipolytica* and *Geotrichum candidum*. I. Basal conditions. *Journal of Milk and Food Technology*, **37**, 463–68.

Hottinger, H.H., Richardson, T., Amundson, C.H. and Stuiber, D.A. (1974b) Utilization of fish oil by *Candida lipolytica* and *Geotrichum candidum*. II. Optimization of conditions. *Journal of Milk and Food Technology*, **37**, 522–8.

Hountin, J.A., Karam, A., and Parent, L.É. (1995) Effect of peat moss-shrimp wastes compost on the growth of barley (*Hordeum vulgare* L.) on a loamy sand soil. *Communications in Soil Science and Plant Analysis*, **26**, 3275–3289.

Hoyle, N.T. and Merritt, J.H. (1994) Quality of fish protein hydrolysates from herring (*Clupea harengus*). *Journal of Food Science*, **59**, 76–9.

Huang, F.L. and Tappel, A.L. (1971) Action of cathepsin C and D in protein hydrolysis. *Biochimica et Biophysica Acta*, **236**, 739–49.

Hudson, J.W., Pohland, F.G. and Pendergrass, R.P. (1978) Anaerobic packed column treatment of shellfish processing wastes, in *Proceedings of the 33rd Purdue Industrial Waste Conference*, Ann Arbor Science, Ann Arbor, pp. 560–74.

In, T. (1990) Seafoods flavourants produced by enzymatic hydrolysis, in *Advances in Fisheries Technology and Biotechnology for Increased Profitability* (eds M.N. Voigt and J.R. Botta), Technomic, Lancaster, PA, pp. 425–36.

Izume, M., Taira, T., Kimura, T. and Miyata, T. (1989) A novel cell culture matrix composed of chitosan and collagen complex, in *Chitin and Chitosan: Sources, Chemistry, Biochemistry, Physical Properties and Applications* (eds G. Skjåk-Bræk, T. Anthonsen and P. Sandford), Elsevier Applied Science, London, pp. 653–6.

Jacobsen, F. and Lykke-Rasmussen, O. (1984) Energy savings through enzymatic treatment of stickwater in the fish meal industry. *Process Biochemistry*, **19** (5), 165–9.

Jefford, C.W., Rinehart, K.L. and Shield, L.S. (eds) (1988) *Pharmaceuticals and the Sea*, Technomic, Lancaster, PA.

Kakuta, T., Koizumi, T., Yoshizawa, K., Kodama, K. and Nojiro, K. (1985) Treatment of waste-water discharged from a dried-bonito processing factory using yeast. *Nippon Shokuhin Kogyo Gakkaishi*, **32**, 260–5.

Kalac, J. (1978a) Studies on herring (*Clupea harengus* L.) and capelin (*Mallotus villosus* L.) pyloric caeca protease II: characterization of the anion fraction of trypsin. *Biologia (Bratislava)*, **33**, 704–10.

Kalac, J. (1978b) Studies on herring (*Clupea harengus* L.) and capelin (*Mallotus villosus* L.) pyloric caeca protease III: characterization of the anion fraction of chymotrypsin. *Biologia (Bratislava)*, **33**, 939–49.

Kauss, H., Jeblick, W. and Young, D.H. (1982/83) Chitin deacetylase from the plant pathogen *Colletotrichum lindemuthianum*. *Plant Science Letters*, **28**, 231–6.

Kim, S.-K. and Rha, C. (1989) Chitosan for encapsulation of mammalian cell culture, in *Chitin and Chitosan: Sources, Chemistry, Biochemistry, Physical Properties and Applications* (eds G. Skjåk-Bræk, T. Anthonsen and P. Sandford), Elsevier Applied Science, London, pp. 617–26.

Knobl, G.M., Jr (1967) The fish protein concentrate story. *Food Technology*, **21** (8), 56–9.

Knorr, D. (1982) Functional properties of chitin and chitosan. *Journal of Food Science*, **47**, 593–5.

Knorr, D. (1984) Use of chitinous polymers in food. *Food Technology*, **38** (1), 85–97.

Knorr, D. (1986) Nutritional quality, food processing, and biotechnology aspects of chitin and chitosan: a review. *Process Biochemistry*, **21** (3), 90–2.

Knorr, D. (1991) Recovery and utilization of chitin and chitosan in food processing waste management. *Food Technology*, **44** (1), 114–22.

Knorr, D. and Klein, J. (1986) Production and conversion of chitosan with cultures of *Mucor rouxii* or *Phycomyces blakesleeanus*. *Biotechnology Letters*, **8** (10), 691–4.
Knorr, D., Beaumont, M.D. and Pandya, Y. (1989) Potential of acid soluble and water soluble chitosans in biotechnology, in *Chitin and Chitosan: Sources, Chemistry, Biochemistry, Physical Properties and Applications* (eds G. Skjåk-Bræk, T. Anthonsen and P. Sandford), Elsevier Applied Science, London, pp. 101–18.
Kreag, R. and Smith, F.J. (1973) Seafood solid waste in Oregon: disposal or recovery? *Oregon State University Extension Special Report 395*.
Kubota, M. and Ohnuma, A. (1970a) Studies on bonito pepsin – I. Purification of pepsin. *Bulletin of the Japanese Society of Scientific Fisheries*, **36**, 1147–51.
Kubota, M. and Ohnuma, A. (1970b) Studies on bonito pepsin – II. Enzymic properties of bonito pepsin. *Bulletin of the Japanese Society of Scientific Fisheries*, **36**, 1152–56.
Landes, D.R. and Bough, W.A. (1976) Effects of chitosan – a coagulating agent for food processing wastes – in the diets of rats on growth and liver and blood composition. *Bulletin of Environmental Contamination and Toxicology*, **15** (5), 555–63.
Lema, J.M., Soto, M., Veiga, M.C. and Méndez, R. (1987) Mesophilic and thermophilic anaerobic treatment of wastewaters from industrial processing of mussel, in *Biomass for Energy and Industry* (eds G. Grassi, B. Delmno, J.-F. Molle and H. Zibetta), Elsevier Applied Science, London, pp. 872–6.
Levin, R.E. (1994) Lactic acid and propionic acid fermentations of fish silage, in *Fisheries Processing, Biotechnological Applications* (ed A.M. Martin), Chapman and Hall, London, pp. 273–310.
Li, C.F., Stuiber, D., Richardson, T. and Amundson, C.H. (1970) Fermentation of fish lipids. *Food Product Development*, **4**, 28–32, 102–3.
Löffler, A. (1986) Proteolytic enzymes: sources and applications. *Food Technology*, **40** (1), 63–70.
Mackie, I.M. (1982) Fish protein hydrolysates. *Process Biochemistry*, **17** (1), 26–28, 31.
Makinonda, V. and Ikeda, S. (1971) Studies on fish muscle protease. *Bulletin of the Japanese Society of Scientific Fisheries*, **37**, 1002–8.
Malette, W.G., Quigley, H.J., Jr and Adickes, E.D. (1986) Chitosan effect in vascular surgery, tissue culture and tissue regeneration, in *Chitin in Nature and Technology* (eds R.A.A. Muzzarelli, C. Jeuniaux and G.W. Goody), Plenum Press, New York, pp. 435–41.
Martin, A.M. (ed.) (1994) *Fisheries Processing, Biotechnological Applications*, Chapman and Hall, London.
Martin, A.M. and Bassler, E.C. (1989) Production of the edible mushroom *Pleurotus ostreatus* with fisheries byproducts as nutrient supplements. *1989 Annual Meeting of the Institute of Food Technologists*, Chicago, IL.
Martin, A.M. and Bemister, P. (1994) Use of peat extract in the ensiling of fisheries wastes. *Waste Management and Research*, **12**, 467–79.
Martin, A.M. and Chintalapati, S.P. (1989) Fish offal–peat compost extracts as fermentation substrate. *Biological Wastes*, **27**, 281–8.
Martin, A.M. and Patel, T.R. (1991) Bioconversion of wastes from marine organisms, in *Bioconversion of Waste Materials to Industrial Products*, 1st edn (ed. A.M. Martin), Elsevier Applied Science, London, pp. 417–40.
Martin, A.M. and Porter, D. (1995) Studies on the hydrolysis of fish protein by enzymatic treatment, in *Food Flavours: Generation, Analysis and Process Influence* (ed. G. Charalambous), Elsevier, Amsterdam, pp. 1395–404.
Martin, A.M., Evans, J., Porter, D. and Patel, T.R. (1993) Comparative effect of peat and sawdust employed as bulking agents in composting. *Bioresource Technology*, **44** (1), 65–9.
Martin, R.E. (1972) *Mechanical Recovery and Utilization of Fish Flesh* (Oak Brook Seminar). National Fisheries Institute, Washington, DC.
Mathur, S.P. (1991) Composting processes, in *Bioconversion of Waste Materials to Industrial Products*, 1st edn (ed. A.M. Martin), Elsevier Applied Science, London, pp. 147–83.
McIver, R.C., Brooks, R.I. and Reineccius, G.A. (1982) Flavor of fermented fish sauce. *Journal of Agricultural and Food Chemistry*, **30**, 1017–20.
Merrett, T.G., Bar-Eli, E. and Vunakis, H.V. (1969) Pepsinogens A, C and D from the smooth dogfish. *Biochemistry*, **8**, 3696–702.
Meyers, S.P. and No, H.K. (1995) Utilization of crawfish pigment and other fishery

processing by-products, in *Nutrition and Utilization Technology in Aquaculture* (eds C.E. Lim and D.J. Sessa), AOCS Press, Champaign, IL, pp. 269–77.

Mitchell, R. and Alexander, M. (1961) Chitin and biological control of *Fusarium* diseases. *Plant Disease Report*, **45**, 487–90.

Miyazawa, K., VanLe, C., Ito, K. and Matsumoto, F. (1979) Studies on fish sauce. *Journal of the Faculty of Applied Biological Sciences Hiroshima University*, **18**, 55–63.

Mizutani, T., Kimizuka, A., Ruddle, K. and Ishige, N. (1992) Chemical components of fermented fish products. *Journal of Food Composition and Analysis*, **5**, 152–9.

Mohr, V. (1980) Enzyme technology in the meat and fish industries. *Process Biochemistry*, **15** (6), 18–21, 32.

Moore, K.J., Johnson, M.G. and Sistrunk, W.A. (1987) Effect of polyelectrolyte treatments on waste strength of snap and dry bean wastewaters. *Journal of Food Science*, **52** (2), 491–2.

Mowbray, J.C., Rossi, H.A. and Chai, T. (1988) Processing and quality assessment of solubles prepared from dogfish processing wastes. *Journal of Agriculture and Food Chemistry*, **36**, 1329–33.

Murado, M.A, González, M.P. and Pastrana, L. (1994) Mussel processing wastes as a fermentation substrate, in *Fisheries Processing, Biotechnological Applications* (ed. A.M. Martin), Chapman and Hall, London, pp. 311–43.

Muzzarelli, R.A.A. (1977) *Chitin*, Pergamon Press, Oxford. Cited in Simpson, B.K., Gagne, N. and Simpson, M.V. (1994) Bioprocessing of chitin and chitosan, in *Fisheries Processing, Biotechnological Applications*, (ed. A.M. Martin), Chapman and Hall, London, pp. 155–73.

Muzzarelli, R.A.A., Tanfani, F., Emanuelli, M., Chiurazzi, E. and Piani, M. (1986) Sulfated *N*-carboxymethyl chitosans as blood anticoagulants, in *Chitin in Nature and Technology* (eds R.A.A. Muzzarelli, C. Jeuniaux and G.W. Goody), Plenum Press, New York, pp. 469–76.

Nagyvany, J.J., Falk, J.D., Hill, M.L., Schmidt, M.L., Wilkins, A.K. and Bradbury, E.L. (1979) The hypolipidemic activity of chitosan and other polysaccharides in rats. *Nutrition Report International*, **20**, 677–84.

Nair, C. (1990) Pollution control through water conservation and wastewater reuse in the fish processing industry. *Water Science and Technology*, **22** (9), 113–21.

Nakajima, M., Shoji, T. and Nabetani, H. (1992) Protease hydrolysis of water soluble fish proteins using a free enzyme membrane reactor. *Process Biochemistry*, **27**, 155–60.

Nishimura, K.S., Nishimura, N., Nishi, N., Tokura, S. and Azuma, I. (1986) Immunological activity of chitin derivatives, in *Chitin in Nature and Technology* (eds R.A.A. Muzzarelli, C. Jeuniaux and G.W. Goody), Plenum Press, New York, pp. 477–83.

No, H.K. and Meyers, S.P. (1989) Crawfish chitosan as a coagulant in recovery of organic compounds from seafood processing streams. *Journal of Agriculture and Food Chemistry*, **37**, 580–3.

Noda, M. and Murakami, K. (1981) Studies on proteinases from the digestive organs of sardine. II. Purification and characterization of two acid proteinases from the stomach. *Biochimica et Biophysica Acta*, **658**, 27–34.

Noda, M., Van, T.V., Kusakabe, I. and Murakami, K. (1982) Substrate specificity and salt inhibition of five proteinases isolated from the pyloric caeca and stomach of sardine. *Agricultural and Biological Chemistry*, **46**, 1565–9.

Norris, E.R. and Mathies, J.C. (1953) Preparation, properties and crystallization of tuna pepsin. *Journal of Biological Chemistry*, **204**, 673–80.

Ohtakara, A., Ogata, H., Taketomi, Y. and Mitsutomi, M. (1984) Purification and characterization of chitosanase from *Streptomyces griseus*, in *Chitin, Chitosan and Related Enzymes* (ed. J.P. Zikakis), Academic Press, New York, pp. 147–60.

Okei, W., Nishimura, O., Somorin, N., Nishi, N. and Tokura, S. (1986) Inhibitory action of sulphated chitin derivatives on the hydrolytic activity of thrombin, in *Chitin in Nature and Technology* (eds R.A.A. Muzzarelli, C. Jeuniaux and G.W. Goody), Plenum Press, New York, pp. 453–9.

Olsen, R. Schwartzmiller, D., Weppner, W. and Winandy, R. (1989) Biomedical application of chitin and its derivates, in *Chitin and Chitosan: Sources, Chemistry, Biochemistry, Physical Properties and Applications* (eds G. Skjåk-Bræk, T. Anthonsen and P. Sandford), Elsevier Applied Science, London, pp. 813–27.

Ooshiro, Z., Ok, T., Une, H., Hayashi, S. and Itakura, T. (1981) Study on use of commercial proteolytic enzymes in production of fish sauce. *Memoirs of Faculty of Fisheries of Kagoshima University*, **30**, 383–94.

Ornum, J.V. (1991) Chitosan, in *Proceedings of the 34th Annual Conference of the Canadian Institute of Food Science and Technology*, Montreal, Canada. Cited in Simpson, B.K., Gagne, N. and Simpson, M.V. (1994) Bioprocessing of chitin and chitosan, in *Fisheries Processing, Biotechnological Applications* (ed. A.M. Martin), Chapman and Hall, London, pp. 155–73.

Owen, T.G. and Wiggs, A.J. (1971) Thermal compensation in the stomach of the brook trout (*Salvelinus fontinalis* Mitchill). *Comparative Biochemistry and Physiology*, **40B**, 465–73.

Pan, B.S. (1990) Recovery of shrimp waste for flavourant, in *Advances in Fisheries Technology and Biotechnology for Increased Profitability* (eds M.N. Voigt and J.R. Botta), Technomic, Lancaster, PA, pp. 437–52.

Park, J.W. (1994) Functional protein additives in surimi gels. *Journal of Food Science*, **59**, 525–527.

Pigott, G.M. and Tucker, B.W. (1990) *Seafood, Effects of Technology on Nutrition*, Marcel Dekker, New York.

Pohland, F.G. and Hudson, J.W. (1976) Aerobic and anaerobic microbial treatment alternatives for shellfish processing wastewaters in continuous culture. *Biotechnology and Bioengineering*, **18**, 1219–47.

Price, J.S. and Stork, R. (1975) Production, purification and characterization of an extracellular chitosanase from *Streptomyces*. *Journal of Bacteriology*, **124**, 1574–85.

Quaglia, G.B. and Orban, E. (1987) Enzymic solubilisation of proteins of sardine (*Sardina pilchardus*) by commercial proteases. *Journal of the Science of Food and Agriculture*, **38**, 263–9.

Raghunath, M.R. (1993) Enzymatic protein hydrolysate from tuna canning wastes – Standardisation of hydrolysis parameters. *Fishery Technology*, **30**, 40–5.

Raksakulthai, N., Lee, Y.Z. and Haard, N.F. (1986) Effect of enzyme supplements on the production of sauce from male capelin, *Mallotus villosus*. *Canadian Institute of Food Science and Technology Journal*, **19**, 28–33.

Rebeca, B.D., Peña-Vera, M.T. and Díaz-Casteñeda, M. (1991) Production of fish protein hydrolysates with bacterial proteases. Yield and nutritional value. *Journal of Food Science*, **56** (2), 309–14.

Reed, G. (1980) (ed.) *Enzymes in Food Processing*, Academic Press, New York, USA.

Revah-Moiseev, S. and Carroad, P.A. (1981) Conversion of the enzymatic hydrolysate of shellfish waste chitin to single cell protein. *Biotechnology and Bioengineering*, **23**, 1067–78.

Ritskes, T.M. (1971) Artificial ripening of maatjes-cured herring with the aid of proteolytic enzyme preparations. *Fisheries Bulletin*, **69** (3), 647–54.

Rodriguez-Sánchez, D. and Rha, C.K. (1981) Chitosan globules. *Journal of Food Technology*, **16**, 469–79.

Rosario, R.R. del and Maddo, S.M. (1984) Biochemistry of patis formation I. Activity of cathepsins in patis hydrolysis. *Philippine Agriculture*, **67**, 167–75.

Roy, G. (1992) Bitterness: reduction and inhibition. *Trends in Food Science and Technology*, **3**, 85–91.

Ruiter, A. (1972) Substitution of proteases in the enzymic ripening of herring. *Annals of Technology and Agriculture*, **21** (4), 597–605.

Ryan, W.H. and Yankowski, E.L. (1969) Photographic image-receiving material, German patent 1116969.

Saisithi, P. (1994) Traditional fermented fish: fish sauce production, in *Fisheries Processing, Biotechnological Applications* (ed. A.M. Martin), Chapman and Hall, London, pp. 111–31.

Saisithi, P., Kasemsarn, B.-O., Liston, J. and Dollar, A.M. (1966) Microbiology and chemistry of fermented fish. *Journal of Food Science*, **31**, 105–10.

Samuels, W.A., Fontenot, J.P., Allen, V.G. and Flick, G.J. (1992) Fermentation characteristics of ensiled seafood wastes and low-quality roughages. *Animal Feed Science and Technology*, **38**, 305–17.

Sandford, P.A. (1989) Chitosan: commercial uses and potential applications, in *Chitin and Chitosan: Sources, Chemistry, Biochemistry, Physical Properties and Applications* (eds G.

Skjåk-Bræk, T. Anthonsen and P. Sandford), Elsevier Applied Science, London, pp. 51–69.

Schisler, L.C. and Sinden, J.W. (1966) Nutrient supplementation of mushroom compost at casing. *Canadian Journal of Botany*, **44**, 1063–9.

Schisler, L.C. and Patton, T.G. (1974) The use of marine fishery products as mushroom compost additives, in *Mushroom Science IX. Part I. Proceedings of the Ninth International Scientific Congress on the Cultivation of Edible Fungi*, International Society of Mushroom Science, Tokyo, pp. 175–84.

Shahidi, F., Han, X.-Q. and Synowiecki, J. (1995) Production and characteristics of protein hydrolysates from capelin (*Mallotus villosus*). *Food Chemistry*, **53**, 285–93.

Shamsuzzaman, K. (1983) Isolation, properties and the use of a chymosin-like enzyme from harp seal (*Pagophilus groenlandicus*). Ph.D. thesis, Memorial University of Newfoundland, St John's.

Shamsuzzaman, K. and Haard, N.F. (1984) Purification and characterization of a chymosin-like protease from the gastric mucosa of the harp seal, *Pagophilus groenlandicus. Canadian Journal of Biochemistry and Cell Biology*, **22**, 699–708.

Shiau, C.Y. and Chai, T. (1990) Characterization of oyster shucking liquid wastes and their utilization as oyster soup. *Journal of Food Science*, **55**, 374–8.

Shioya, T. and Rha, C. (1989) Transmembrane permeability of chitosan/carboxymethyl-cellulose capsule, in *Chitin and Chitosan: Sources, Chemistry, Biochemistry, Physical Properties and Applications* (eds G. Skjåk-Bræk, T. Anthonsen and P. Sandford), Elsevier Applied Science, London, pp. 627–34.

Shoemaker, S. (1986) The use of enzymes for waste management in the food industry, in *Biotechnology in Food Processing* (eds S.K. Harlander and T.P. Labuza), Noyes Publications, Park Ridge, USA, pp. 259–69.

Siebert, G. and Schmitt, A. (1965) Fish tissue enzymes and their role in the deterioration changes in fish, in *Technology of Fish Utilization* (ed. R. Kreuzer), Fishing News (Books), London, pp. 47–52.

Simpson, B.K. (1983) Isolation, characterization and some application of trypsin from Greenland cod (*Gadus ogac*). Ph.D. thesis, Memorial University of Newfoundland, St John's.

Simpson, B.K. and Haard, N.F. (1984) Purification and characterization of trypsin from Greenland cod (*Gadus ogac*) I: Kinetic and thermodynamic characteristics. *Canadian Journal of Biochemistry and Cell Biology*, **62**, 894–900.

Simpson, B.K. and Haard, N.F. (1985) Extraction of carotenoprotein from shrimp processing offal with the aid of trypsin. *Journal of Applied Biochemistry*, **7**, 212–22.

Simpson, B.K. and Haard, N.F. (1987) Cold-adapted enzymes from fish, in *Food Biotechnology* (ed. D. Knorr), Marcel Dekker, New York, pp. 495–527.

Simpson, B.K., Gagne, N. and Simpson, M.V. (1994) Bioprocessing of chitin and chitosan, in *Fisheries Processing, Biotechnological Applications* (ed. A.M. Martin), Chapman and Hall, London, pp. 155–73.

Siso, M.I.G., Murado, M.A. and Franco, J.M. (1987) SCP from mussel processing wastes, II: Microfungi in pure culture and mixed with non-amylolytic yeasts, in *Proceedings of the II World Congress of Food Science and Technology*, Vol. 1, Barcelona, Spain, pp. 677–88.

Skjåk-Bræk, G., Anthonsen, T. and Sandford, P. (eds) (1989) *Chitin and Chitosan: Sources, Chemistry, Biochemistry, Physical Properties and Applications*, Elsevier Applied Science, London.

Smith, P. and Adamson, A.H. (1976) Pig feeding trial with whitefish and herring liquid protein (fish silage), Paper No. 3. *Proceedings of the Torry Research Station Symposium on Fish Silage*, Aberdeen, UK.

Squires, E.J. (1984) Isolation and characterization of gastric proteases from the Greenland cod (*Gadus ogac*). Ph.D. thesis, Memorial University of Newfoundland, St John's.

Squires, E.J., Haard, N.F. and Feltham, L.A.W. (1986a) Pepsin isozymes from Greenland cod (*Gadus ogac*). *Canadian Journal of Biochemistry and Cell Biology*, **64**, 205–14.

Squires, E.J., Haard, N.F. and Feltham, L.A.W. (1986b) Gastric proteases of the Greenland cod, *Gadus ogac*. II: Comparison of structural properties. *Canadian Journal of Biochemistry and Cell Biology*, **64**, 215–22.

Stefansson, G. and Steingrimsdottir, U. (1990) Application of enzymes for fish processing in

Iceland. Present and future aspects, in *Advances in Fisheries Technology and Biotechnology for Increased Profitability* (eds M.N. Voigt and J.R. Botta), Technomic, Lancaster, PA, pp. 237–50.
Struszczyk, H., Pospieszny, H. and Kotlinski, S. (1989) Some new applications of chitosan in agriculture, in *Chitin and Chitosan: Sources, Chemistry, Biochemistry, Physical Properties and Applications* (eds G. Skjåk-Bræk, T. Anthonsen and P. Sandford), Elsevier Applied Science, London, pp. 733–42.
Sugiyama, K., Egawa, M., Onzuka, H. and Oba, K. (1991) Characteristic of sardine muscle hydrolysates prepared by various enzymic treatments. *Bulletin of the Japanese Society of Scientific Fisheries*, **75**, 475–9.
Tatterson, I.N. and Windsor, M.L. (1974) Fish silage. *Journal of the Science of Food and Agriculture*, **25**, 369–79.
Tominaga, Y. and Stujisake, Y. (1975) Purification and some enzymatic properties of the chitosanase from Bacillus R-4 which lyses Rhizopus cell walls. *Biochimica et Biophysica Acta*, **410**, 145–55.
Van, T.V., Kusakabe, I. and Murakami, K. (1983) Purification and some properties of two aminopeptidases from sardines. *Agricultural and Biological Chemistry*, **47**, 2453–9.
van Veen, A.G. (1965) Fermented and dried seafood products in Southeast Asia, in *Fish as Food*, Vol. 3 (ed. G. Borgstrom), Academic Press, New York, p. 223.
Veiga, M.C., Mendez, R. and Lema, J.M. (1994) Waste water treatment for fisheries operations, in *Fisheries Processing, Biotechnological Applications* (ed. A.M. Martin), Chapman and Hall, London, pp. 344–69.
Venugopal, V. (1994) Production of fish protein hydrolyzates by microorganisms, in *Fisheries Processing, Biotechnological Applications* (ed. A.M. Martin), Chapman and Hall, London, pp. 223–43.
Venugopal, V., Alur, M.D. and Nerkar, D.P. (1989) Solubilization of fish protein using immobilized microbial cells. *Biotechnology and Bioengineering*, **33**, 1098–103.
Venugopal, V., Martin, A.M. and Patel, T.R. (1994a) Extractability and stability of isolated capelin (*Mallotus villosus*) muscle in water. *Food Hydrocolloids*, **8**, 135–42.
Venugopal, V., Martin, A.M., Omar, S. and Patel, T.R. (1994b) Protein concentrate from capelin (*Mallotus villosus*) by spray drying process and its properties. *Journal of Food Processing and Preservation*, **18**, 509–19.
Venogopal, V., Martin, A.M., Patel, T.R., Vasantahn, T. and Omar, S. (1995) Rheological and solubility characteristics of washed capelin (*Mallotus villosus*) muscle in water. *Journal of Food Biochemistry*, **19**, 175–90.
Vieira, G.H.F., Martin, A.M. Saker-Sampaiao, S., Sobreira-Rocha, C.A. and Goncalves, R.C.F. (1995a) Production of protein hydrolysate from lobster (*Panulirus* spp.), in *Food Flavours: Generation, Analysis and Processs Influence* (ed. G. Charalambous), Elsevier, Amsterdam, pp. 1405–15.
Vieira, G.H.F., Martin, A.M., Saker-Sampaiao, S., Omar, S. and Goncalves, R.C.F. (1995b) Studies on the enzymatic hydrolysis of Brazilian lobster (*Panulirus* spp.) processing wastes. *Journal of the Science of Food and Agriculture*, **69**, 61–5.
Voigt, M.N. and Botta, J.R. (eds) (1990) *Advances in Fisheries Technology and Biotechnology for Increased Profitability*, Technomic, Lancaster, PA.
Welsh, F.W. and Zall, R.R. (1984) Single cell protein from waste fishery refrigeration brines. *Process Biochemistry*, **19** (3), 122–3.
Whittenmore, C.T. and Taylor, A.G. (1976) Nutritive value to the growing pig of de-oiled liquefied herring offal preserved with formic acid (fish silage). *Journal of the Science of Food and Agriculture*, **27**, 239–43.
Wignall, J. and Tatterson, I. (1976) Fish silage. *Process Biochemistry*, **11**, 17–9.
Wray, T. (1988) Fish processing: new uses for enzymes. *Food Manufacture*, **63** (7), 48.
Yanagida, T. (1985) Application of chitin and chitosan in Japan. *Kokai Yokkyo Koho*, JP 61/210014 A2 (80/210014). Cited in Simpson, B.K., Gagne, N. and Simpson, M.V. (1994) Bioprocessing of chitin and chitosan, in *Fisheries Processing, Biotechnological Applications* (ed. A.M. Martin), Chapman and Hall, London, pp. 155–73.
Yang, T. and Zall, R.R. (1984) Chitosan membranes for reverse osmosis application. *Journal of Food Science*, **49**, 91–3.
Zajic, J.E., Higgs, T.W. and Kosaric, N.H. (1974) Enhanced fish oil fermentation. *GVC/AIChE Joint Meeting and Jahrestreffen 1974 der Verfahrens-Ingenieure*, Munich.

13 Production of *Bacillus thuringiensis* biopesticides using waste materials

MARIA DE LOURDES TIRADO MONTIEL, RAJESHWAR D. TYAGI AND JOSÉ R. VALÉRO

13.1 Introduction

Chemical insecticides have been used to control insect pests that damage plants in agriculture and forestry or that are vectors of human diseases, such as malaria, filaria, yellow fever and encephalitis, which represent serious health problems in many countries (Ejiofor, 1991). These products are efficient but their production costs are high and they are sources of environmental pollution (Carlton, 1990; Ejiofor, 1991). They are also harmful to non-target organisms thus creating problems for the environmental equilibrium. Their action can affect nervous systems by disruption or inhibition of certain metabolic functions (Fisher, 1993) becoming a risk to all kind of organisms. They can also be accumulated in the environment, contaminating surface and underground water, soils, agriculture products and reach the human food chain (Carlton, 1990).

Many insects have developed a remarkable ability to resist the action of chemical insecticides. About 400 insect species have shown to be resistant to some kind of chemical insecticides (Georghiou and Lagunes, 1988). This situation represents a serious problem to farmers and pest managers. Over the last couple of years, restrictions have already been imposed in many countries on the use of such hazardous chemicals and, as a result, costs for developing new products have risen. The utilization of entomopathogenic microorganisms as biological agents to control pests has represented a good option to avoid the problems caused by the use of chemical products. This group of microorganisms include a wide range of bacterium, viruses, fungi and protozoa (Aronson *et al.*, 1986). *Bacillus thuringiensis* (Bt) is the one that has been the most studied.

Products based on *Bacillus thuringiensis* are the most successful microbiological pesticides used today because of the particular characteristics of this organism: it is not harmful to predatory insects, or to other animals, including man; the activity of a strain is target specific; and it is biodegradable, and thus has little effect on environmental pollution (Rodriguez *et al.*, 1991; Valéro, 1990; Flexner *et al.*, 1986; Wilcox *et al.*, 1986). This bacterium is characterized by the production of endotoxins during the sporulation phase which are the active agent against the target insects. These endotoxins are insoluble, form crystalline inclusions and are virtually always plasmid encoded (Carlton and Gonzalez, 1985).

Ishiwata was the first to isolate *Bacillus thuringiensis* (Bt) in 1902, but it was Berliner who named the microorganism thus 10 years later. In the 1920s, Bt was used in Europe during some experimental assays as a bioinsecticide against the European corn borer. The first commercial product appeared in 1938 in France under the name of Sporeine. By the end of the 1950s other commercial formulations appeared that performed only against Lepidoptera (Rajnchapel-Messaï, 1993).

During the 1960s the discovery of Bt *kurstaki* serotype HD-1 (Dulmage, 1970a), which is active against agricultural pests, and the adoption of an international system for the standardization of Bt preparations improved the field efficiency and commercialization. Before that, all products were standardized on the basis of spore count which did not reflect the insecticidal activity of the preparations. At the end of the 1970s, strain *israelensis*, which is toxic to mosquitoes and black flies, was discovered. In the mid-1980s Bt *tenebrionis*, which is active against a few Coleoptera, was isolated (Herrnstadt *et al.*, 1986).

Bacillus thuringiensis has been successfully produced on a very large scale and formulated as an active bioinsecticide in many countries including the USA, France, Germany, the former USSR and China (Bulla *et al.*, 1980). Efforts have been made to improve its performance with the help of recombinant DNA technology and other engineering techniques. This has helped in removing many constraints which limited the extensive use of Bt products.

13.2 Characteristics of *Bacillus thuringiensis*

Bacillus thuringiensis is a rod-shaped, aerobic, Gram-positive, spore-forming, crystalliferous bacterium. The term crystalliferous is applied to those *Bacillus* species that produce a characteristic inclusion body within the sporangium in addition to the endospore (Bulla *et al.*, 1980). Its insecticidal activity is related to the production of proteinic sequences contained in this protein inclusion called parasporal crystal.

13.2.1 Taxonomy

The presence of the parasporal crystal that is formed adjacent to the spore, outside the exosporium is one criterion on which to identify *Bacillus thuringiensis* from closely related species, such as *Bacillus cereus* and *anthracis* (Andrews *et al.*, 1987; Baumann *et al.*, 1984; Claus and Berkeley, 1986). The utilization of flagellar H antigen, crystal toxin shape, size and antigens, some phenotypic properties and/or DNA plasmid profiles have been used to classify Bt. More than 40 subspecies have been identified based on classification of the H antigen (de Barjac and Bonnefoi, 1962, 1973). However, this classification is not precise, because two strains pertaining to the same subspecies can present with a different protein content in their crystals (Andrews *et al.*, 1985), having a different toxic spectrum (Norris, 1964; Krywienczyk *et al.*, 1978). There can also be metabolic differences within a serotype (Baumann *et al.*, 1984; Lynch and Baumann, 1985). Classification based on genes coding for the crystal proteins has also been used, grouping the δ-endotoxins in five different types according to their insecticidal properties and molecular relationships (Höfte and Whiteley, 1989).

13.2.2 Metabolism

Bacillus thuringiensis is a chemoheterotroph microorganism with a complex metabolism which is not yet well defined. From its central metabolism, the best known pathways are glycolysis, tricarboxylic acids cycle (TCA) and glyoxylic cycle (Rowe and Margaritis, 1987). Production of Bt biomass can be described by three phases of its cultivation: vegetative growth, transition phase and sporulation phase. During the vegetative phase, carbohydrates are used for growth (Freese and Fujita, 1976) and recent studies have shown that amino acids can also be metabolized (Sakharova *et al.*, 1984; Rowe, 1990). Sugars are mainly metabolized through the Embden–Meyerhof–Parnas pathway (Nickerson *et al.*, 1974), TCA activity is almost nil and, as a result, there is an accumulation of different kinds of metabolites like pyruvate, lactate, acetate, poly-β-hydroxybutyrate and 2,3-butanodiol (Anderson, 1990; Rowe, 1990).

During the spore and crystal formation phase, metabolism is based in the utilization of poly-β-hydroxybutyrate and amino acids, which are used as energy sources for spore and crystal maturation and cellular lysis. The metabolism of poly-β-hydroxybutyrate generates acetyl-CoA, which is metabolized by the TCA pathway and glyoxylate cycle, which continues to be active during sporulation phase (Rowe, 1990). Pyruvic and acetic acids that accumulate during vegetative growth are oxidized in the early stages of the sporulation process (Hanson *et al.*, 1964). The mechanisms by which Bt assimilates nitrogen is complex and not well understood. Nitrogen can be assimilated as ammonia by the pathways of alanine dehydrogenase and glutamate dehydrogenase (Borris and Aronson, 1969; Aronson *et al.*, 1975; Aronson, 1976) or amino acids. Because of the synthesis of both sporulation-specific enzymes and crystal proteins, Bt must utilize, during the sporulation phase, most of the nitrogen assimilated during vegetative growth. Nitrogen metabolism in this organism requires more investigation because much of the metabolic energy is spent at the expense of spores and crystals formation, both of which are abundant in nitrogen (Bulla *et al.*, 1980).

13.3 Genetic characteristics

In general, genetic information on toxic production in Bt is contained in large plasmids, which can be used as an alternative to identify specific isolates of Bt (Stahly *et al.*, 1992). Some isolates have unique plasmid size profiles that can be used to describe the strain (Gonzalez *et al.*, 1982; Kronstad *et al.*, 1983; Carlton and Gonzalez, 1985). However, the fact that Bt can lose its plasmids during extended culturing limits the use of this technique (Stahly *et al.*, 1978).

13.3.1 Localization and organization of crystal producing genes

Bt crystal protein genes are present in one or more plasmids (Gonzalez *et al.*, 1982; Faust *et al.*, 1983) which range in size from 1.5 to 180 MDa (Gonzalez and Carlton, 1980; Gonzalez *et al.*, 1981; Lereclus *et al.*, 1982) and each strain can carry from 2 to 17 plasmids. Crystal genes are transcribed by a sporulation-specific RNA polymerase (Klier *et al.*, 1983) and studies have demonstrated that the control of gene expression is exerted at the transcriptional level (Wong *et al.*, 1983).

In the classification, which is based on the structure (deduced from the DNA sequence) of genes encoding for crystal proteins as well as their host range (Höfte and Whiteley, 1989), proteins are denoted 'Cry' proteins and the genes as '*cry*' genes. There are four main classes and several subclasses; the principal gene classes are *cryI*, *cryII*, *cryIII*, *cryIV* and *cytA*.

All genes of group I encode for polypeptides with molecular weights of 130 kDa, which are lepidopteran toxic. Genes from group II (lepidopteran–dipteran toxic) and group III (coleopteran toxic) encode for polypeptides of about 70 kDa. The genes of group IV (dipteran toxic) are different from each other in that: *cryIVA* encodes for a 134 kDa protein; *cryIVB* for a 128 kDa protein; *cryIVC* for a 78 kDa protein and *cryIVD* for a polypeptide of 72 kDa. The *cyt* gene which encodes for a 27 kDa polypeptide is present in subspecies *israelensis*; these Bt proteins differ from each other in their size and cytolytic action on invertebrate and vertebrate cells. There seems to be no relation between Cyt proteins and Cry proteins (Federici, 1993). Recently, two new *cry* genes have been reported, *cryV* and *cryVI*. These genes produce proteins toxic to nematodes which are not related to toxins produced by the other genes (Feitelson *et al.*, 1992).

13.4 Toxicity (crystal–spore complex)

Bt is present in different geographical areas and can be isolated from living or dead insects, soil, grass, grain dust and water (Martin and Travers, 1989; Smith and Couche, 1991). In nature, the action of Bt endotoxins is restricted to insect larvae; larval age is another factor that has influence over toxic action: younger larvae are more susceptible than older ones (Andrews *et al.*, 1987).

13.4.1 Characteristics

There is a great correlation between the shape of the parasporal crystal and its toxicity spectrum. The lepidopteran-toxic crystals are bypiramidal or diamond shaped, the dipteran-toxic crystals are pleomorphic, and the coleopteran-toxic crystals are rectangular and flat.

The crystals from the three pathotypes share some common properties. They are all proteins and despite their antigenic diversity, they tend to have some common size ranges. For example, when crystal proteins from pathotypes I and II are separated by sodium dodecyl sulfate polyacrylamide gels, major bands appear in the molecular weight range of 120 000–140 000 and a second band in the range of 23 000–70 000. Pathotype II crystals give another band in the range of 23 000–30 000. Pathotype III crystals contain only proteins in the middle range (Stahly *et al.*, 1992).

The complexity within the crystals varies considerably because they contain more than one kind of protein (Höfte and Whiteley, 1989; Federici *et al.*, 1990). Probably the hydrogen and disulfide bonds, by another kind of complex interaction, are responsible for self-assembling and maintaining these proteins together in the inclusion body (Bulla *et al.*, 1980). Disulfide bonds are important for the stabilization of the tertiary structure of crystals and its solubility (Aronson *et al.*, 1986).

13.4.2 Synthesis

There are several reports in the literature that crystal production is linked to sporulation in Bt. First, the parasporal crystals appear in the cells in close proximity to the spores, and the time of their appearance coincides with the spore formation (Bechtel and Bulla, 1976). The crystal toxin begins to accumulate in the cells 4–6 h after the onset of sporulation (Andrews *et al.*, 1981).

The sporulation process is divided into seven stages. As development of the forespore proceeds, a well-defined program of protein synthesis is observed. Some proteins that are present in vegetative cells are turned off and new proteins, which are found only in sporulating cells, are expressed (Andrews *et al.*, 1987). The formation of crystal is initiated at stage II or III of sporulation (Bechtel and Bulla, 1976; Fitz-James and Young, 1969; Lecadet and Dedonder, 1971; Wong *et al.*, 1983). The inclusion body reaches its final size by stage V. The resulting crystal can account for 20–30% of the total protein of the sporangium (Lecadet and Dedonder, 1971).

The synthesis of crystal is controlled at the level of transcription. Synthesis of mRNA specific for crystal toxin correlates with the accumulation of crystal toxin in sporulating cells. There is evidence that the formation of mRNA specific for crystal toxin requires a unique form of RNA polymerase (Brown and Whiteley, 1988).

13.4.3 Specificity

The specificity of the action of Bt toxins has been under investigation for a long time. Some factors that could explain this specific action are:

1. the structure of the crystal protein;
2. differences in the larval midgut affecting solubilization and or processing efficiency of the protoxin contained in the crystal protein;
3. the presence of specific toxic-binding receptor sites on the insect gut epithelium.

Depending on their δ-endotoxin composition, the crystals have various forms as mentioned previously, and there is a correlation between its shape and its target insect. So, microscopic examination of the crystals produced by a strain can provide an idea of the range of susceptible target insects (Lereclus *et al.*, 1993).

Despite their homologies, there are considerable differences between the activity spectra of the eight CryI toxins which are active against several lepidopteran species. Apparently, the differences found in the activity spectra of these toxins is related to a small domain in the variable amino-terminal region of CryIA protoxin (Lereclus *et al.*, 1993).

CryII-type proteins are always found in strains that also produce CryI polypeptides and its host range specificity has been also determined. Small differences in amino-acid positions can substantially alter the specificities of CryIIA, CryIIB and CryIIC polypeptides (Lereclus *et al.*, 1993).

The crystals of Bt *israelensis* are composed of three different types of polypeptides: (1) the 130 kDa type CryIVA, CryIVB or CrIVC; (2) the CryIVD 72 kDa protein; and (3) a cytolytic factor of 28 kDa, CytA. The presence of these different proteins has complicated the identification of the protein(s) responsible for mosquitocidal activity (Lereclus *et al.*, 1993).

There are two Coleoptera-specific toxin genes which have been analysed, *cryIIA* and *cryIIIB*. The former is active against the Colorado potato beetle; there is little information about the toxicity of the latter (Lereclus *et al.*, 1993).

13.4.4 Mode of action

The number of subspecies and the number of susceptible hosts that have been studied, the production of more than one toxic material, the possible interaction between spores and crystals to impart toxicity in some larvae and evidence that the toxic materials may act at several sites in susceptible hosts all complicate the discussion about the mode of action of Bt preparations (Aronson *et al.*, 1986). Among the many potentially toxic materials produced by Bt, β-exotoxin and δ-endotoxin are the most significant in pathogenicity to insects (Aronson *et al.*, 1986). Classification of lepidopteran insects in regard to their susceptibility to δ-endotoxin, bacterial spores or mixtures of both appeared quite early in literature (Heimpel and Angus, 1959). Type I insects are those that are susceptible to crystalline δ-endotoxin, but spores do not exert any influence; type II insects are those susceptible to endotoxin but its effect is enhanced by the presence of spores; and type III insects are those that are killed by mixtures of spores and crystals.

Once the susceptible insect has ingested the Bt crystals containing either CryI or CryII proteins, the crystals are dissolved in the alkaline juice (pH range 10–12) of midgut, releasing 130–140 kDa CryI proteins or 70 kDa CryII proteins. The efficiency of this process, influencing specificity, is co-determined by the conditions present in the larval midgut and the composition of the crystals (Aronson *et al.*, 1986). Next, the crystal protein or protoxin is solubilized to produce the actual toxic fragment of 60–70 kDa, which is protease resistant and is derived from the N-terminal of the unprocessed protein. This model accounts for CryIA(c) protein, and may be generalized for other 130 kDa proteins (Aronson *et al.*, 1986; Pfannenstiel *et al.*, 1986, 1990).

Later, at least in the case of CryI class crystals, the toxic fragment binds to specific receptors present on the membranes of epithelial midgut cells, provoking K^+ pump dysfunction, pore formation, or cation and anion selective channels (Knowles and Ellar, 1986; Muthukumar and Nickerson, 1987; Valéro and Letarte, 1988; van Rie *et al.*, 1990; Schwartz *et al.*, 1993). The microvilli lose their characteristic structure within minutes. The microvillar membrane slackens and the microtubules degenerate. Subsequently, the cells and organelles such as mitochondria become vacuolated and begin to swell (Huber and Luthy, 1981). This situation continues until the cells lyse and slough from the basement membrane of the midgut epithelium. As more cells slough, the alkaline gut juices begin to leak into the hemacoel, causing a rise in hemolymph pH. This causes the paralysis and death of the insect (Heimpel, 1967).

The action of the CytA toxins is thought to form pores in the microvillar membrane, but it does not require a receptor protein for insertion. Instead, it binds to unsaturated phospholipids in the lipid bilayer of the cell plasma membranes (Gill *et al.*, 1992; Thomas and Ellar, 1983).

13.5 Effect of medium composition and operation conditions on the production of spore–crystal complex

Various reports indicate that medium composition and process operating conditions can have an influence over toxicity of Bt products. A good growth and a high spore count do not always mean that products have a good toxicity level (Rogoff *et al.*, 1969; Dulmage, 1970b; Dulmage and Rhodes, 1971). Dulmage (1970b) was one of the first to conclude the following:

1. toxic activity is not predictable by spore count, so Bt preparations cannot be standardized this way;
2. toxic activity of Bt varies depending both on the isolates and on the cultivation conditions.

Bt can grow in many standard laboratory media but none of them will give satisfactory results with every strain. Nutrional requirements of different subspecies of Bt are variable (Dulmage *et al.*, 1990), and it is not easy to define a standard culture media for Bt because of these conditions. However, it is possible to define some basic requirements for its production.

Production of the crystal protein occurs only during the sporulation phase, so maximum sporulation is an important step. Optimum oxygen supply is essential for growth and sporulation of Bt because both are affected by dissolved oxygen conditions. All the strains are able to produce amylases and proteinases, which allows them to use a wide range of raw substrates (Bernhard and Utz, 1993). The influence of these and other factors on the growth and sporulation of Bt are discussed below.

13.5.1 Temperature and pH

Growth of Bt occurs between 15 °C and 45 °C, with the optimum between 26 °C and 30 °C, the latter being the most commonly used fermentation process temperature. Higher temperatures can cause loss of plasmid (Gonzalez and Carlton, 1984). Incubation of Bt subsp. *berliner* on nutrient agar at temperatures between 12 °C and 16 °C (which is 10 °C to 12 °C below optimum) disturbs the metabolism of the bacterial cells causing a desynchronization of the crystal formation and the spore formation, which normally occur at the same time (Smirnoff, 1963). In fermentation processes, temperatures are kept at 28–30 °C. Studies with Bt subps. *israelensis* have demonstrated that the toxin yield was reduced by about 90% and 45% compared to the yield at 30 °C, when the incubation temperatures were kept at 28 °C and 33 °C, respectively (Abdel-Hameed *et al.*, 1991).

With respect to pH, optimum growth of most bacteria occurs near neutral (pH = 7). In particular, Bt is not sensitive to pH variations and growth will occur between pH 5.5 and 8.5, with the optimum between 6.5 and 7.5 (Bernhard and Utz, 1993). During growth, products of microbial metabolism often cause major shifts in pH and still further changes may occur when growth is culminated by subsequent metabolism of primary products. In fermentation media, pH variations are frequently controlled by buffers or by the automatic addition of steril alkali or acid solutions (Dulmage and Rhodes, 1971).

13.5.2 Process options for Bt production

There are two types of processes for Bt production: the submerged process and the semisolid fermentations process. As Bt is an aerobic bacteria, its production demands an intimate association of nutrients and air. The economics of large-scale production demand that the fermentation occupies as little space as possible. The semisolid fermentation is designed to meet these requirements for microorganisms in surface culture. In semisolid fermentation, nutrients are contained in a coarse porous matrix which achieves a high ratio of surface area to volume and furnishes, in a relatively small space, a large liquid-gas interface. To allow the growth of microorganisms without reducing the surface area through clumping induced by excessive moisture, humidity has to be maintained within narrow limits (Bernhard and Utz, 1993). Wheat bran, ground corn, peanut meal, oat or rice hulls, cottonseed meal, alfalfa meal are utilized as organic porous matrix to furnish nutrients to the fermentation as well as to serve as an absorption base. Inorganic material like volcanic glass, diatomaceous earth, vermiculite and pumice have also been used (Dulmage and Rhodes, 1971).

In some semisolid fermentation processes, large rectangular trays where inocula is spread on semipermeable autoclavable cellulose sheets are used. Once the endotoxin is produced, it is harvested by collecting the cellulose sheets. This technique has given promising results in terms of endotoxin yield. The second approach utilizes static growth of the bacteria in shallow trays with liquid media. To minimize anaerobiosis, some high redox potential compounds are incorporated into the media (Foda and Salama, 1986).

The advantages of the semisolid fermentation are that the equipment used is relatively simple and unsophisticated: the culture media are mostly made from cheap raw material: a wettable powder or dust formulation containing spores; and crystal endotoxin is obtained simply by drying and grinding the final bran cake. This makes the process inexpensive. However, semisolid media are difficult to sterilize and to maintain sterile during the production, and adjusting parameters such as pH is also difficult, thus limiting the control that can be maintained in the fermentation process (Dulmage and Rhodes, 1971).

Submerged cultures are, therefore, preferred for industrial mass production (Bernhard and Utz, 1993). This process is used to produce flowable suspensions of the spore crystal complex or, by a recovery procedure, it is possible to produce dry preparations (Dulmage and Rhodes, 1971).

Since δ-endotoxins are synthesized during sporulation, which occurs in the stationary growth phase, batch production is the obvious method of choice. Growth process continues until nutrients exhaust and spores begin to form. At the end, cellular lysis is completed and spore–crystal complexes are released into the culture medium. Mature spores account for at least 15% of the vegetative cell mass, whereas crystals represent 17–20% (Mettus and Macaluso, 1990; Stahly *et al.*, 1978). Industrial production of Bt is realized by this method (Rowe, 1990).

Another modality of submerged cultures is continuous culture where there is simultaneous input of nutrients and output of product. However, there are certain disadvantages of continuous culture:

1. Growth during long periods can cause morphological or biochemical alterations in the original strain, like formation of asporogenic crystalliferous variants (Sachidanandham and Jayaraman, 1993; Selinger *et al.*, 1988).
2. It is difficult to obtain the production of certain metabolites not associated with growth in just one stage (Rodriguez Monroy *et al.*, 1991). Despite these disadvantages, some authors have proposed continuous culture to increase yields (Freiman and Chupin, 1973; Kang *et al.*, 1993).

To obtain a good maturation of spores and a high toxin synthesis, a combination of continuous stirred tank reactor followed by an aerated tubular reactor with plug flow behavior has also been proposed (Dreier *et al.*, 1990; Moser, 1991). The plug flow reactor can be regarded as a batch reactor and can be employed for vegetative growth in its first stage and for spore maturation in its second stage (Dreier *et al.*, 1990).

Another culture system is a fed-batch culture. This is an open system, in which one or more nutrients are added to the bioreactor during cultivation. Product accumulates and is harvested at the end of the process (Rodriguez Monroy *et al.*, 1991). When utilized with Bt, this culture system has given good results in terms of higher spore counts of Bt subsp. *kurstaki* than those obtained in batch cultures. However, the toxicity level was the same as that found in batch cultures with the same strain (Arcas *et al.*, 1987).

A variation of fed-batch culture is intermittently fed batch culture (IFBC). This system resulted in higher spore concentrations when used with Bt subps. *kurstaki* than that obtained with continuous fed batch culture (CFBC). The increase in spore production by IFBC compared to CFBC can be attributed to higher cell growth rate during the fed-batch operation (Kang *et al.*, 1992). It has been reported that higher growth rates in batch culture favored better sporulation (Goldberg *et al.*, 1980). It is likely that cells growing fast contain enough energy reserves and other factors necessary for sporulation, while slow cells do not. Avignone-Rossa and Mignone have compared the toxicity of Bt *israelensis* (Bt) crystal-spore complex obtained by batch and fed-batch cultures. The toxicity levels obtained with batch culture measured in international toxic units (ITU) per colony forming unit (CFU), was 5050 and 5196 ITU/10^8 CFU, values which lie within the range reported by Pearson and Ward which is 3400–5800 ITU/10^8 CFU, which seems to be characteristics for batch cultures of Bti (Avignone-Rossa and Mignone, 1993; Pearson and Ward, 1988). However, when fed-batch culture was utilized, a high spore count was obtained, ranging from 4.7 to 5.3 CFU mg^{-1} biomass, but toxicity levels were lower ranging from 132–517 ITU 10^8 CFU (Avignon-Rossa and Mignone, 1993).

13.5.3 Aeration

As the production of δ-endotoxin occurs only during sporulation, good aeration is essential for the sporulation process (Holmberg *et al.*, 1980; Foda *et al.*, 1985; Bernhard and Utz, 1993). The necessity of high aeration rates has been reported for Bt (Holmberg *et al.*, 1980; Foda *et al.*, 1985; Pearson and Ward, 1988), but only the effect of dissolved O_2 on the respiration rate and growth has been studied by Moraes *et al.* (1980), who utilized Bt NCIB 9207 in their experiments.

Studies with Bt subsp. *entomocidus* cultivated under different aeration rates demonstrated that viable spore count, sporulation titers and toxicity levels were higher when a 19:1 ratio (air/liquid) was used instead of 9:1 ratio (air/liquid) (Foda *et al.*, 1985). It has been found for Bt subsp. *thuringiensis* that, under high aeration rates, small crystals with increased toxicity were obtained, whereas a reduced oxygen supply led to relatively large crystals with a low δ-endotoxin content (Scherrer *et al.*, 1973). Under O_2 limitation, δ-endotoxin concentrations and spore counts were lower than those obtained under non-limited conditions, showing the dependence of sporulation and endotoxin synthesis on good aeration (Avignone-Rossa *et al.*, 1992). These workers also showed that, even if there is an interruption of aeration in limited and non-limited oxygen fermentations, spore counts will not be affected; it seems that, once sporulation has been triggered, it will be completed. However, δ-endotoxin synthesis is affected by such an interruption and only a fraction of the expected yield can be achieved. Thus, O_2 must be continuously supplied if high δ-endotoxin concentrations are to be reached (Avignone-Rossa *et al.*, 1992).

Other works performed with Bt subsp. *israelensis* have stated that there exists an optimum range for aeration rate; however, further increase in aeration rate seemed to inhibit toxin production and working below this optimum level can disturb growth and sporulation, which led to a reduction in the toxin yield (Abdel-Hameed *et al.*, 1991). The influence of oxygen on sporulation rates and δ-endotoxin synthesis is far from being clearly understood. Further research in this area must continue.

13.5.4 Mineral elements

The requirements of minerals varies with the type of organisms as well as with the nature of the medium under investigation (Sikdar *et al.*, 1991). In the case of Bt, there is no way to generalize the requirements for all strains, for each of them, optimal conditions must be defined taking into account the kind of culture medium utilized. However, there are variations in the culture medium necessary to improve high toxicity and abundant growth. Among the elements necessary for the good development of Bt, it has been demonstrated that potassium stimulates δ-endotoxin production in subsp. *entomocidus* (Foda *et al.*, 1985), and in subsp. *kurstaki* and *aizawai* (Wakisaka *et al.*, 1982). Sporulation is stimulated by inorganic ions, particularly Ca^{2+} and Mn^{2+}. Fortification of culture media with Mg^{2+}, Cu^{2+}, Fe^{3+}, Co^{+} and Zn ions may also improve growth and sporulation, even if complex substrates are used (Bernhard and Utz, 1993).

Experiments for the production of mosquitocidal δ-endotoxin by Bt subsp. *israelensis* in a semisynthetic medium supplemented with yeast have shown that the optimum levels of K_2HPO_4 and $MgSO_4.7H_2O$ were 1 g l^{-1} and 0.3 g l^{-1}, respectively. Calcium was essential for cell growth and toxin production and, if pH is controlled near neutral, a calcium chloride concentration of 1 g l^{-1} gives a higher yield of endotoxin. Calcium carbonate reacts as calcium chloride, with the advantage that it can also act as a neutralizing agent. Molybdenum has been found to produce an inhibitory effect over δ-endotoxin production. The optimum levels of metals for cell growth and δ-endotoxin production are different. Fe, Mn, and Cu are required in concentrations of 2, 5 and 0.25 μg ml^{-1} under the culture conditions used in the experiments (Sikdar *et al.*, 1991).

The influence of mineral salts on the productivity and insecticidal activity of the spore-crystal complex of a culture of Bt IPM-1140 when cultured in a yeast-polysaccharide medium has been also studied. Addition of Fe^{3+} ions in the range of concentrations between 0.2 and 3 mg 100 ml^{-1} considerably increased the productivity of the culture in terms of spores ml^{-1} (Abrosimova *et al.*, 1986).

It is essential to note the positive influence exerted by addition of iron on the process of crystal formation with the same Bt culture IPM-1140. The ratio of the number of toxin crystals to the number of bacterial spores on the control medium, without Fe^{3+} was 1:1.3, while on a medium containing Fe^{3+} ions (20 μg ml^{-1}) it was 1:1. The presence of potassium phosphates or Mn^{2+}, Zn^{2+} and NH_4^+ ions in media, gave more synchronous spore formation, accompanied by a greater yield of endotoxin crystal than that observed with the normal culture medium (Abrosimova *et al.*, 1986).

13.5.5 Nitrogen and amino acids

(a) Nitrogen. The assimilation of nitrogen by Bt is less well known and more complex than that of carbohydrates (Rodriguez Monroy *et al.*, 1991). Nitrogen can be assimilated as ammonium or as amino acids. The presence of NH_4^+ ions can have an important influence over δ-endotoxin production by Bt subsp. *israelensis*. When this strain is cultured in a medium containing only yeast extract as the nitrogen source, production of δ-endotoxin was not as good as when it was cultured in a medium supplemented with $(NH_4)_2SO_4$. Whatever the C:N ratios used, these did not affect the metabolic pathway of the organism, and it seems that the combinations of organic and inorganic nitrogen can have a great influence over δ-endotoxin production.

This suggests that, if a production medium for Bt *israelensis* is to be optimized, it is necessary to adjust the initial NH_4^+ concentration to promote the highest endotoxin production. It would be desirable to study the role of other inorganic nitrogen sources over the production of Bt *israelensis*, so that its incorporation in the culture medium can reduce the production costs (Avignone-Rossa *et al.*, 1990).

(b) Amino acids. Most strains of Bt will not grow in glucose–mineral salt media unless they are supplemented by some source of amino acids, e.g. peptones or yeast extract (Rajalakshmi and Shethna, 1977). Since Bt produces exoproteases, amino acids may be substituted by proteins or peptides (Bernhard and Utz, 1993). Certain amino acids support growth, sporulation and crystal formation of Bt strains, while others inhibit its growth. Histidine, threonine, tyrosine, valine, isoleucine, leucine, serine and lysine did not support the growth of the organism, while arginine, aspartic acid, glycine, proline, asparagine, methionine and glutamine supported its growth with a longer lag period compared to cystine which stimulated exponential growth in 9 h after inoculation, when cultivated in a glucose–mineral salts medium also containing phosphates as buffering agents (Rajalakshmi and Shethna, 1977).

Growth was seen to be enhanced by increasing the cystine concentration in the medium. The presence of cystine in excess inhibits sporulation in Bt while vegetative growth continued. This aspect can be used to prolong the vegetative growth without affecting the spore and crystal formation in the manufacture of commercial preparations (Rajalakshmi and Shethna, 1977, 1980). Generally, valine and leucine enhanced the growth of Bt subsp. *thuringiensis*, *entomocidus* and *sotto* varieties in a citrate–salts basal medium; however, subsp. *sotto* and *entomocidus* grew less in the higher concentration of value than in the lower concentration. Isoleucine partially inhibited the growth of subsp. *thuringiensis*, although it stimulated the growth of subsp. *sotto* and *entomocidus*. The growth of Bt subsp. *sotto* was generally sparse in any of the culture media utilized. The effect of the addition of valine on the growth of Bt *thuringiensis* can differ if a different culture medium is employed (Conner and Hansen, 1967).

The growth of Bt subsp. *galleriae* is inhibited by the presence of leucine and isoleucine in a basal synthetic medium, whereas the growth of Bt subsp. *sotto* is inhibited by the presence of isoleucine in the same synthetic medium (Singer and Rogoff, 1968). For strains M1 and S128 Bt var. *israelensis*, leucine was the most effective to enhance the growth of both tested strains. On the other hand, isoleucine showed no significant effect on the growth of both these strains. Valine showed no significant effect on the growth of strain S128 (Abdel-Hameed, 1992).

When serine, threonine or glycine are added alone in the culture medium they can inhibit Bt *thuringiensis* growth. Methionine does not have this effect, and it can reverse the inhibition caused by the presence of serine but not that caused by threonine (Singer and Rogoff, 1968). For strains S128 and M1 of Bt *israelensis*, methionine significantly stimulates the growth of both strains whereas serine and glycine had no significant effect. A little enhancement effect on the growth of both strains can be observed when threonine is present in the basal medium in combination with methionine or glycine, respectively (Abdel-Hameed, 1992). It is, therefore, important to consider the effect of the presence of amino acids and the nutritional imbalance they can cause when a culture medium is adopted or optimized.

13.5.6 Carbon source

The carbon source is one of the nutrients that has been most studied, glucose being the most utilized carbon source. Media for industrial production are based on complex carbon and nitrogen sources (Priest and Sharp, 1989). Starches, molasses, dextrose, flours, and glycerol, among others have been used as the principal carbon sources. However, the best results are achieved when simple carbon sources are employed. Carbohydrate concentration influences the production of δ-endotoxin and special attention is required for commercial production.

Carbon concentration can have an influence on the size and morphology of the crystals produced by Bt. Different glucose concentrations can have an influence on the size of the crystals produced by Bt subsp. *thuringiensis* when cultured in a semisynthetic medium. Crystals were barely visible under the light microscope when they were obtained on a medium containing yeast extract as the sole carbon source, but the addition of different concentrations of glucose increased the crystal size. Optimum culture conditions for crystal formation were reached with 6–8 g l^{-1} of glucose. More than 8 g l^{-1} of glucose present in the culture medium produced amorphous crystals and some cells contained more than one granule. The presence of higher concentrations of glucose produced acid conditions in the culture medium which inhibited the growth (Scherrer *et al.*, 1973).

Protein content in the crystal can also be affected by the carbon concentrations present in the medium. A glucose-free medium yielded only 0.1 mg of protein from 5.1×10^8 parasporal bodies produced by Bt subsp. *thuringiensis* but the amount of crystal protein doubled when 1 g l^{-1} of glucose was added. The maximum crystal production was obtained with 8 g l^{-1}. Further glucose addition did not lead to an increased protein yield (Scherrer *et al.*, 1973).

Experiments made with *Pieris brassicae* larvae demonstrated that product obtained when Bt *thuringiensis* was cultured in a semisynthetic medium without glucose, a greater quantity of crystals (9000) were necessary to inhibit 100% feeding of the tested larvae. The quantity of crystals was reduced (2200) with a product obtained when the same strain was cultured in the same medium containing 8 g l^{-1} of glucose. For other media and other culture conditions, glucose requirements for optimal δ-endotoxin production would be different (Scherrer *et al.*, 1973). The inhibitory effect on growth and sporulation of high sugar concentrations have also been demonstrated for Bt subsp. *entomocidus*. A good aeration and pH control can reverse the effect (Foda *et al.*, 1985).

Operational conditions can influence the role played by carbohydrates in the growth and sporulation of Bt subsp. *kurstaki*. To enhance the cell mass and spores during CFBC, glucose-limiting conditions were assayed. Even though cell mass increased proportionally to the amount of glucose consumed, no sporulation occurred during CFBC operation. This is in contrast to the observations of the batch operation, where sporulation occurred when the culture condition became glucose limited. Therefore, it appears that glucose limitation is not the only factor that supports sporulation in fed-batch operation (Kang *et al.*, 1992). The biological activity and morphology of the crystals can differ if the carbon source is changed in a culture medium but, if the amount of a particular carbon source is varied in a culture medium, there is no effect over the crystal properties and the δ-endotoxin content in this case is proportional to the biomass of the Bt strain used (Yudina *et al.*, 1993).

A medium able to promote sporulation and production of bioinsecticide by Bt *israelensis* strain must contain high carbohydrate to protein ratios. This situation can be coupled with the utilization of a medium with a high protein content that can inhibit Bt *israelensis* sporulation (Pearson and Ward, 1988). It would be interesting to investigate whether this situation is present when other Bt strains are assayed under the same culture conditions. Owing to the fact that the growth and yield of δ-endotoxin per sporulated cell in different Bt strains is strongly influenced by the carbohydrate content and culture conditions, it is important for the commercial production of Bt to pay attention not only towards high spore yields but also to the composition of the medium.

13.6 Alternate raw materials for Bt biopesticide production

To date Bt is the most successful bioinsecticide and is estimated to account for 80–90% of all biological pest control agents sold in the world. Nevertheless, Bt insecticides have not yet made much impact and only have a 1–2% share of the total insecticide market (Berhard and Utz, 1993). Most laboratory culture media allow cell densities of only 10^8 cells ml^{-1} to 10^9 cells ml^{-1} at reasonable sporulation rates. In industrial mass production with optimized culture conditions for the specific needs of the strain to be produced, cell densities can reach more than 5×10^9 cells ml^{-1} and sporulation rates of more than 90% (Bernhard and Utz, 1993).

Production of Bt must be economical to permit its use in the control of several insect pests, but its wider use has been restricted by economic reasons. The introduction of alternate cheap raw material that can be used as culture media to promote high cell growth and good sporulation rates can help in reducing production costs.

13.6.1 Production of Bt subsp. thuringiensis *on alternate protein-rich raw materials*

The production of Bt var *thuringiensis* has been achieved using dehusked greengram powder and deffated soybean powder as protein sources with different combinations of soluble starch and/or cane sugar molasses as the major carbohydrate source in submerged fermentation (Mummigatti and Raghunathanm, 1990). Cultures that contained soybean as a major protein source could be harvested after 96 h of incubation giving a product yield of 8.7–9.1 g l^{-1} (dry weight basis), whereas culture media containing greengram powder were harvested after 120 h with a produce yield of 9.7–10.3 g l^{-1} (dry weight basis). Probably, the higher carbohydrate content of the culture medium might have prolonged the vegetative exponential phase of the organism, delaying sporulation and lysis, thus prolonging harvest time.

As discussed before, a high spore count does not assure a high toxicity level. This situation was corroborated in these experiments. The maximum viable spore count was obtained from the medium containing defatted soybean powder and soluble starch (91.3×10^6 spores mg^{-1}); this product had a potency of 35 800 IU mg^{-1}. However, the product from medium containing dehusked greengram powder and cane sugar molasses, having a lower spore count (49.5×10^6 spores mg^{-1}) recorded the highest potency of 38 300 IU mg^{-1}. One can thus infer that the quality of δ-endotoxin produced differed with the quality of the media.

13.6.2 Production of Bt subsp. entomocidus, kurstaki, aizawai, finitimus *and* galleriae *from various raw materials*

(a) Utilization of fodder yeast, vegetable extracts, ground horse bean seeds and some agricultural by-products. Owing to the proteinaceous nature of the endotoxins of Bt, raw materials with high protein contents have been used for Bt production. These products must be locally available for cheaper bioinsecticide production.

Material rich in protein content such as fodder yeast, vegetable extracts (potatoes, carrots and sweet potatoes), ground horse bean seeds (horse beans and kidney beans) and some agricultural by-products (fish meal, cotton seed meal, residues from the chicken slaughter house), wheat bran and agricultural wastes (citrus peel and date seeds) have been used to support growth, sporulation and toxin production by Bt subspecies *kurstaki*, *entomocidus*, *aizawai*, *finitimus* and *galleriae*. These strains are used to fight against some of the principal lepidopteran cotton pests *Spodoptera littoralis*, *Heliothis armigera* and *Spodoptera exigua* (Salama *et al.*, 1983c).

The higher sporulation yields from Bt subsp. *entomocidus* HD-635 grown in mono-component media of ground legume seeds (horse beans or kidney beans) or fodder yeast were obtained with 2% concentration in the media. When tested against the first instar larvae of the cotton leaf worm *Spodoptera littoralis*, the highest biological activities were obtained in media containing fodder yeast.

In order to further evaluate the efficiency of fodder yeast as a practical mono-component medium for endotoxin production, it was compared with a standard synthetic medium with respect to endotoxin yield and potency produced from Bt subsp. *galleriae* HD-129, *kurstaki* HD-251 and *entomocidus* HD-635. The fodder yeast medium at 2% final concentration in tap water produced higher yields of endotoxin for each of the subspecies tested than that of synthetic standard medium, and when tested against third instar larvae of *Spodoptera littoralis*, the mortality was very similar to that when standard medium was used (between 80% and 100%).

The supplementation of fodder yeast with ground dates seeds, minced citrus peels and wheat bran increased the endotoxin yield produced by Bt subps. *entomocidus* HD-635, *kurstaki* HD-251 and HD-1. The supplementation of fermentation media with wheat bran and ground date seeds increased the endotoxin yield 2–3 times compared to the control fodder yeast medium, without appreciably affecting the biological activity against *Spodoptera exigua* and *Heliothis armigera*. Fishmeal, cotton seed meal and residues from the chicken slaughter houses used as mono-component media at a 2% concentration level in tap water produced large yields of endotoxin complexes but their biological activity was very low compared to those obtained with fodder yeast media, indicating that these are not suitable for growth and endotoxin production for these three Bt strains.

Extracts of carrots, sweet potato roots and potato tubers were used as complete media for growth and sporulation of Bt subsp. *galleriae* HD-129, *entomocidus* HD-635, *kurstaki* HD-72 and HD-251. The strain *kurstaki* HD-251 gave the highest sporulation yields in the three alternate medium, compared to the other strains tested and gave 100% mortality against third instar larvae of *S. littoralis*.

Fodder yeast has also been tested with a new fermentation technology based on the growth and endotoxin production on semisolid media under conditions of static incubation where high redox potential compounds, namely salts of nitrate, sulfate and phosphate, were added as substitutes of aeration to the semisolid media. The highest endotoxin yields were obtained with 1% sodium nitrate for subsp. *entomocidus* and with 3% K_2HPO_4 for subsp. *aizawai*. The activity of the endotoxins produced, however, were low against *S. littoralis* at the level of 500 $\mu g\ ml^{-1}$ (40% and 48% mortality, respectively), but increased at 1000 $\mu g\ ml^{-1}$ (82%) for subsp. *entomocidus*. With *S. exigua*, the activity was also high at 1000 $\mu g\ ml^{-1}$ for both strains (Foda and Salama, 1986).

(b) Utilization of whey. Whey has a potential nutritional value that has induced extensive studies aimed at the economic recycling of this by-product for producing useful compounds and also to reduce the BOD of the liquor to disposal (Salama *et al.*, 1983a). It has also been tested for its ability to support growth, sporulation and δ-endotoxin production by Bt. Twelve cultures of Bt were tested for their ability to grow, sporulate and produce endotoxins against some of the major cotton pests namely *S. littoralis*, *S. exigua* and *H. armigera*. The strains tested were: subsp. *entomocidus*, *dendromilus*, *alesti* (two strains), *thuringiensis* (two strains), *kurstaki* (HD-1, HD-73 and HD-251 strains), *galleriae* and *finitimus* (Salama *et al.*, 1983a). Samples of sweet and salted rennet buffalo whey were used as complete media either as such or after certain simple treatment including dilution and protein precipitation. In several experiments the media were supplemented with rich protein material such as

ground seeds of kidney beans, horse beans or fodder yeast at 2% final concentration (Salama *et al.*, 1983a).

All cultures were able to grow on the sweet rennet buffalo whey, but only three subspecies, *kurstaki*, *galleriae* and *entomocidus* gave significant growth on salted whey containing 5% NaCl, whereas no growth was observed with whey medium containing 10% NaCl. Salted whey, therefore, may not be an appropriate culture medium for all Bt strains (Salama *et al.*, 1983a). Supplementation of sweet whey with urea and ammonium in enhancing sporulation did not exerted a significant effect with subsp. *entomocidus*. It was concluded that these nitrogen sources tested are not suitable for growth and sporulation. Supplementation with $CaCO_3$ did not increase the spore yield substantially with subsp. *kurstaki* HD-1 and *entomocidus* (Salama *et al.*, 1983a).

Bt subsp. *kurstaki* HD-1 and *entomocidus* have been cultured in the supernatant of sweet whey which, after autoclaving and coagulated proteins removal, was supplemented with yeast extract, $CaCO_3$, mineral salts solution, nitrogen sources and fodder yeast. In general, the spore yields obtained with subsp. *entomocidus* were higher than those obtained with subsp. *kurstaki* HD-1 on raw media. Supplementation with nitrogen sources (urea or ammonium) did not improve the spore yield obtained. On the other hand, the addition of the yeast (0.2%) or fodder yeast (1%) resulted in a remarkable increase in the yield of spores obtained in both strains tested (Salama *et al.*, 1983a).

Sweet whey diluted 1:1 with tap water and supplemented with legume seeds (kidney beans, horse beans) and fodder yeast at 2% final concentrations can be used as culture medium for Bt subsp. *galleriae* (HD-129), *kurstaki* HD-251, HD-1 and HD-73 and *entomocidus*. To determine the toxicity of the products obtained with Bt subsp. *galleriae* (HD-129), *kurstaki* HD-251 and *entomocidus*, bioassays were carried out against third instar lavae of *S. littoralis* (Salama *et al.*, 1983a).

Bt subsp. *galleriae* (HD-129) produced the higher larval mortality (100%) when cultured in sweet whey diluted and supplemented with 2% horse beans seeds. Bt subsp. *kurstaki* HD-251 produced the higher larval mortality when cultured in this medium supplemented with 2% kidney beans seeds, and Bt subsp. *entomocidus* produced the higher larval mortality when cultured in this medium supplemented with 2% fodder yeast (Salama *et al.*, 1983a). Bt subsp. *kurstaki* HD-1 and HD-73 were able to synthesize potent endotoxin against third instar larvae of *H. armigera* when using the sweet whey supplemented with fodder yeast (Salama *et al.*, 1983a).

(c) Utilization of cottonseed meal. The high protein content of cottonseed meal made it an obvious choice to be used as an alternative culture medium to enhance Bt growth and sporulation. This meal is a by-product

of oil extraction from seeds and, therefore, is a cheap alternate culture medium.

A basal medium supplemented with 2% cottonseed meal has been tested for the cultivation of two isolates of subsp. *kurstaki*, HD-1 and HD-73, and that of subsp. *entomocidus* (Salama *et al.*, 1983b). The endotoxins produced were evaluated against *S. littoralis* and *H. armigera*. Results demonstrated that subsp. *kurstaki* HD-1 had a higher activity against *H. armigera* than that against *S. littoralis*. Bt *kurstaki* HD-73 exhibited a lower but similar activity against both insect species. Subsp. *entomocidus* also showed low activity against both insect species, but this was greater against *S. littoralis* than against *H. armigera* (Salama *et al.*, 1983b).

(d) Utilization of sorter liquor. The content of starch, proteins and amino acids in sorter liquor gives it the characteristics to be used as an alternate cheap culture medium for Bt bioinsecticide production. This aqueous by-product has been tested (Salama *et al.*, 1983b) as a fermentation medium with no additives and no further dilution in water to support the growth, sporulation and endotoxin production of Bt subsp. *kurstaki* HD-1 and HD-73, and subsp. *entomocidus*. The higher spore yield was produced by subsp. *kurstaki* HD-73 (30×10^7 spores ml^{-1}) while the other subspecies gave lower yields. The toxicity level of the products was not determined.

In order to increase the performance of this by-product, sorter liquor was mixed with sweet whey in a ratio 1:1 and the solution was supplemented with mineral salts. Bt subsp. *kurstaki* HD-73 gave the highest spore yield, while *kurstaki* HD-1 and *entomocidus* also produced a higher spore yield than before the supplementation. The addition of 1% fodder yeast doubled the spore yield in the three subspecies, e.g. for *kurstaki* HD-1 (52×10^7 spore ml^{-1}), for *kurstaki* HD-73 (100×10^7 spore ml^{-1}) and, for *entomocidus* (90×10^7 spore ml^{-1}) (Salama *et al.*, 1983b).

(e) Utilization of dried beef blood. The addition of dry beef blood as a dry powder at a final concentration of 2% (w/v) to a base medium supplemented with mineral salts has been tested to produce endotoxins of Bt subsp. *kurstaki* and *entomocidus*. The spore yield obtained was 65×10^7 spores ml^{-1} with Bt subsps. *kurstaki* HD-73, 80×10^7 spore ml^{-1} for *kurstaki* HD-1 and 81×10^7 spores ml^{-1} for subsp. *entomocidus*. Bt subsp. *kurstaki* HD-73 showed a lower activity against *S. littoralis* and *S. exigua*, but exhibited 100% mortality against *H. armigera*. Thus dried beef blood proved to be efficient in supporting the biosynthesis of endotoxins with appreciable insecticidal activity against *H. armigera* by Bt *kurstaki* HD-73 and HD-1 strains (Salama *et al.*, 1983b).

(f) Utilization of chicken slaughterhouse residues. To produce endotoxins of Bt subsp. *kurstaki* HD-1 and HD-73, and *entomocidus*, residues from chicken slaughterhouses were incorporated at 2% (w/v) final concentration to a basal medium supplemented with mineral salts. Spores yields varied between 80 and 90 × 10^7 spores ml^{-1}. Bioassays were made against *S. littoralis*, *H. armigera* and *S. exigua* to determine the toxicity level of the endotoxins produced. The product obtained with subsp. *kurstaki* HD-73 was the only one which achieved good mortality of *H. armigera* (100% mortality) (Salama *et al.*, 1983b). When Bt subsp. *kurstaki* HD-251 was cultured using this raw material, it produced large crystals of endotoxin but had a poor bioinsecticide activity (Salama *et al.*, 1983c). This further demonstrates that variations exist in the performance of the different subspecies of Bt even if they are cultivated in the same medium.

(g) Utilization of leguminous seeds. Leguminous seeds are generally excellent sources of protein for many purposes. Their performance to support Bt fermentation has also been tested (Salama *et al.*, 1983b). Bt subsp. *kurstaki* HD-1 and subsp. *entomocidus* were grown in basal medium plus 2% of each of the following legume seeds: chick peas, peanuts, lima beans, horse beans, soya beans and kidney beans. Using these legume seeds, except peanuts, as a nitrogen source, subsps. *kurstaki* HD-1 products were active against *S. littoralis* larvae, killing 72–80% of the larvae at a concentration of 500 $\mu g\ ml^{-1}$ diet. This is a remarkable fact because, when this strain was grown in different culture media, it did not show toxicity against this pest (Salama *et al.*, 1983b).

Bt subsp. *entomocidus* products, irrespective of which legume was used as a supplement, consistently performed well against *S. littoralis*. Further studies on the utilization of legume seeds is necessary because they can be utilized in commercial media instead of expensive nitrogen sources (Salama *et al.*, 1983b).

13.6.3 Production of Bt subsp. israelensis *(Bti) using different raw materials*

Mosquitoes and black flies, which are common in tropical and subtropical regions, are vectors of many diseases and have been controlled since the early 1940s with chemical insecticides. However, these insects have developed resistance to chemical products making their control more difficult. Bt subsp. *israelensis* has been described as a very potent biocide against all these insects (Goldberg and Margalit, 1977) and can be used as an alternative for their control.

Production of bacteria for use as a biological control agent is far less sophisticated than the production of chemicals. The former can be undertaken in many countries in which the disease vector diptera are epidemic (Obeta and Okafor, 1984). Various raw materials used to obtain Bt *israelensis* products are described below.

(a) Utilization of dried cow blood and seeds of legumes. Seed legumes have proved to be a good and inexpensive source of nitrogen when they are used in the culture medium to support growth, sporulation and endotoxin production of different Bt subspecies (Salama *et al.*, 1983a,b).

A basal medium consisting of cow blood, $MnCl_2$, $MgSO_4$ and $CaCO_3$, supplemented with powdered seed legumes, has been tested to evaluate its capacity to sustain growth and endotoxin formation by Bt subsps. *israelensis*. The seed legumes used were groundnut cake, cow peas, soyabeans and bambara beans (Obeta and Okafor, 1984). The degree of sporulation ranged from 70% to 75% in the medium containing cow pea to 90–95% in the medium containing groundnut cake. Powders produced from the different media were effective against the larvae of *Aedes aegypti*, *C. quinquefasciatus* and *Anopheles gambiae*. The concentrations required to kill 50% of the larvae (LC_{50}) indicated that locally produced Bt subsp. *israelensis* powders compared favorably with the known standard medium. The medium containing bambara beans was found to produce the best toxicity level against *Aedes aegypti* and is recommended for further investigation and possible large-scale production (Obeta and Okafor, 1984).

(b) Utilization of organic wastes obtained after food processing. Grated coconut, soyabean, palm oil effluent, fishmeal and rice bran have been evaluated to determine their ability to support growth, sporulation and toxin production by Bt serotype H-14 (IMR-BT-8). In all media, except palm oil, there was growth and sporulation of Bt H-14. The final product obtained from fishmeal was 3.44 times more toxic than that obtained with the standard nutrient broth. Biomass production in medium utilizing fishmeal was also comparatively high. Apparently fishmeal can be a suitable fermentation medium for the mass production of IMR-BT-8 (Lee and Seleena, 1991).

Defatted mustard-seed meal (MSM) has been tested as the principal carbon and nitrogen source for growth, sporulation and toxin production of Bt subsp. *israelensis* IPS-82. Production costs can be minimized by the utilization of defatted MSM which has been found to be a potential source of carbohydrates, proteins and amino acids (mainly glutamic acid and arginine) necessary to support the growth, sporulation and endotoxin production by the Bti strain (Gangurde and Shethna, 1995).

(c) Utilization of maize and cow-pea steep liquors. Studies have demonstrated that maize and cow-pea steep liquor were able to sustain sporulation and crystal formation of Bt H-14 in a 5 l laboratory fermenter

(Ejiofor and Okafor, 1989). The whole final culture was centrifuged and the residue was resuspended in emulsified palm oil to give a spore/crystal concentration of 3.2×10^9 CFU ml^{-1}. The carrier medium was composed of gelatinized starch, powdered charcoal, molasses and emulsified palm oil which served as preservative, dispersant and adhesive. This preparation was assayed in field trials against larvae of *Aedes* and *Culex* species over a period of 2 years with high efficacy (Ejiofor and Okafor, 1991). These raw materials constitute a cheaper alternative for the production of this bioinsecticide because raw local material can be used in its production.

(d) Utilization of a by-product from a monosodium glutamate factory. A medium composed of a 4% hydrolysed liquor (HDL) by-product from a monosodium glutamate factory and 0.05% K_2HPO_4 was used for cultivating Bt subsp. *israelensis*. When comparing the yield of biomass produced with routine culture medium (nutrient broth with salts and glucose; NBSG) and HDL, the results indicated that there was not much difference. Efficacy levels against second instar larvae of *A. aegypti* were comparable to those in the NBSG medium. The economic analysis showed that the cost for cultivations of Bt *israelensis* in 10 l of alternate medium (HDL) (US$0.02) was much lower than the cost per 10 l of NBSG medium (US$7.05) (Dharmsthiti *et al.*, 1985).

The medium utilized for Bt subsp. *israelensis* has the following advantages: it is easy to prepare and requires no pretreatment, it supports high growth and toxicity of the Bt strain and its cost is extremely low since it is a by-product from a fermentation industry. The use of an inexpensive, locally available material like this for the production of Bt products could possibly lead to its efficient local production in large quantities at low cost in many countries (Dharmsthiti *et al.*, 1985).

(e) Utilization of cassava starch, cow pea, maize steep liquor and maize starch. Cassava starch, maize starch, maize steep liquor and cow pea liquor have been evaluated to determine their suitability in terms of low cost, availability and ability to support high yields of Bt *israelensis* (Ejiofor and Okafor, 1989).

The fermented cassava medium supported growth up to 5.5×10^7 CFU ml^{-1}. Sporulation was low and no crystals were formed. The fermented maize flour medium yielded 6.5×10^9 CFU ml^{-1} and supported sporulation, but crystal formation was poor. The ground maize medium also performed fairly well, probably due to a very high level of carbohydrates which could reduce the pH level (WHO, 1983; Ejiofor and Okafor, 1989). Bioassays to determine toxicity levels were not performed. These two raw materials do not seem to be able to support sporulation and crystal production of Bt subsp. *israelensis*.

On the other hand, fermented cow pea medium yielded 5×10^{10} CFU

ml^{-1} and the maize steep liquor 7.5×10^{10} CFU ml^{-1} and both supported sporulation and crystal formation of Bt H-14. Toxic potency of products obtained with these two 1:3 combination by-products were acceptable compared with the products obtained with the standard medium. A 1:3 combination of cow pea and maize gave better results in terms of toxicity (Ejiofor and Okafor, 1989).

(f) Utilization of coconut wastes. Coconut water and endosperm residues are normally waste products of the coconut oil industry. Both are rich in amino acids, sugars and salts (Child and Nathanael, 1950; Kuberski *et al.*, 1979) and may therefore be able to sustain growth, sporulation and endotoxin production of Bt. Samples of coconut water and endosperm were used as culture media for the production of Bt subsp. *israelensis*. Composition of the media was very variable: protein content varied from 1 to 4 mg ml^{-1} of medium in the endosperm samples and carbohydrate content varied from 6 to 16 mg ml^{-1} in endosperm samples to 38.4 mg ml^{-1} in coconut water (Chilcott and Pillai, 1985). The final products were tested against second instar *Aedes australis* mosquito larvae.

Results demonstrated that the endosperm media gave yields significantly less (from 0.09 to 0.17 g, dry wt) than the standard medium used (0.22 ± 0.06 g dry wt), while coconut water and endosperm extract gave yields similar to the standard. Powders produced from coconut water and from two of the endosperms samples had a larvicidal activity greater ($LC_{50} = 0.012$–0.014) than products obtained with the standard ($LC_{50} = 0.044$).

(g) Utilization of a mixture of 'spent' brewer's yeast and waste cassava starch. The wastes from breweries contain yeast that can be utilized as a nitrogen source and waste cassava starch is rich in carbohydrates. These two wastes have been tested together as alternate culture medium for the development of Bt subsp. *israelensis* (Bti) (Ejiofor, 1991).

The yeast medium and the hydrolysed waste cassava starch were mixed in the ratio 2:3 and dispensed into 1 l and 5 l culture vessel modules of a Gallenkamp modular fermenter. A 1 l fermenter was inoculated with the Bti second seed culture obtained from the inoculation of nutrient broth supplemented with yeast and mineral salts. After 48 h of agitation at 30 °C a 80 ml sample was withdrawn and used to inoculate the 5 l vessel. The culture in this vessel was aerated for 48 h at the same temperature. The potency of the preparations was determined by exposing the second and third instar larvae of *Aedes aegypti* to various concentrations of whole culture broth (Ejiofor, 1991).

The yeast medium and hydrolysed waste cassava medium supported the growth of Bti up to 4.4×10^{10} CFU ml^{-1} in a 1 l fermenter and 4.2×10^{10} CFU ml^{-1} in the 5 l vessel, since the culture in the nutrient broth used as reference gave 3.8 × 10^{10} CFU ml^{-1}. Sporulation and crystal formation ranged over 95%. Analysis of the bioassay data showed that the LC_{50} values for the flask, 1 l and 5 l fermenter were 18.5, 12 and 14.8 ppm, respectively (Ejiofor, 1991). These data indicate that the products of fermentation using this mixed medium were very potent larvicides (Dulmage, 1981; Obeta and Okafor, 1984; Ejiofor and Okafor, 1988).

The use of agro-industrial wastes from breweries in countries where these materials are available can contribute in increasing the production of this bioinsecticide and in the reduction of production costs as well.

(h) Utilization of molasses, Proflo and soya. Soya meal, Proflo (a partially defatted cooked cottonseed flour) and molasses, which are inexpensive ingredients, were assessed for growth, sporulation and δ-endotoxin production by the mosquito-toxic strains M1 and S128 of Bt H-14. Optimal conditions for production of endotoxin in a 8 l fermenter in batch culture with respect to pH, aeration, agitation and temperature were also investigated (Abdel-Hameed *et al.*, 1991). Experiments with six different combinations of molasses, soya meal and Proflo were studied. The medium containing soya (3% w/v) as the sole nitrogen source gave better toxin yield. The toxin production reached a maximum when the molasses concentration was 1% (w/v) in the presence of an optimal concentration of soya (3% w/v). Increasing the molasses concentration to 2% (w/v) did not increase the toxin yield (Abdel-Hameed *et al.*, 1991).

Operational conditions were included, with an aeration rate in the range of 0.37–0.62 (vol vol min^{-1}) to achieve high yields. It was found that further increase in aeration rate inhibited toxin production. Agitation rates higher than the optimum (200 rev min^{-1}) caused a gradual decrease in the toxin yield. Growth and sporulation were disturbed when the agitation rate was as low as 100 rev min^{-1}, causing a reduction in toxin yield (Abdel-Hameed *et al.*, 1991).

pH changes did not have an influence on toxin production when it was maintained between 6.5 and 7.5. However, when pH was maintained at a value of 8, bacterial growth and sporulation were greatly disturbed, reducing the toxin yield. On the other hand, fermentation temperature appreciably influenced toxin production. Toxin yields were about 90% and 45%, compared to the yield at 30 °C, when incubation temperatures were kept at 28 °C and 33 °C, respectively (Abdel-Hameed *et al.*, 1991).

13.7 Toxicity determinations

The only way to measure the efficacy of the action of the toxins produced by Bt over the target insects is to use *in vivo* assays. In this way, all the factors that contribute to the potency can be integrated (Burges and Thomson, 1971).

For each group of insects, a different kind of bioassay has been developed. The response to the toxic effect can be determined by the mortality of target insects (Dulmage *et al.*, 1971) or decreased feeding (Burgerjon, 1962). The choice of test insect, life stage, method of administring the toxin, parameter measured as an indicator of toxicity and environmental conditions may all affect the results of bioassays and, thus, must be carefully chosen (Dulmage, 1973).

Some *in vitro* assays have also been developed for quantifying the toxins produced by Bt. They are easier and more rapid to realize than bioassays, but results obtained did not relay exactly with the quality of the toxins, just with its quantity.

13.7.1 Bioassays

Since there is not a reliable relationship between spore count and toxicity level in Bt preparations, it is necessary to carry out a bioassay to determine its potency against the target insect (Beegle, 1990).

(a) Lepidopteran bioassays. Two potency bioassays have been used to test Bt activity against lepidopteran insects. The French bioassay based on the reference standard of Bt (E-61), with its assigned potency of 1000 IU mg^{-1} and *Anagasta* (Ephestia) *kuehniella*, the Mediterranean flour moth, as test insect, and the American bioassay based on Bt *kurstaki* serovar 3a3b strain HD-1 as the reference standard, with a potency of 16 000 IU mg^{-1} for the HD-1-S-80, and *Trichoplusia ni*, the cabbage looper, as the test insect (Dulmage *et al.*, 1971).

Basically there are four different techniques for lepidopteran bioassays: diet incorporation, surface contamination, hole-in-wax and droplet. Forced ingestion methods are not currently used in the determination of potencies (Beegle, 1990).

(b) Dipteran bioassays. Bioassays of materials against aquatic dipteran larvae have to take into account a set of problems related to keeping the test material in the feeding zone of the test species. Some species require static and others require moving water to ensure ingestion of the test material (Burges, 1982). Efficacity of Bt *israelensis* also depends upon the development stage of the target organisms, their feeding behavior, the temperature of water and other factors that made bioassays quite different from that of lepidopteran insects.

There are two standardized bioassays for Bt dipteran toxins. The first is the World Health Organization (WHO) method, designed with flexible protocols in view of the differing conditions and availability of materials throughout the world. The second is the US standard bioassay with rigid protocols for US conditions (Beegle, 1990).

(c) Coleopteran bioassays. The bioassays are developed to test the efficacy of the Bt *tenebrionis* products as this is currently the most important commercial subspecies used in the control of coleopteran pests. One of the major coleopteran pests that causes extensive damage on potatoes, tomatoes and eggplants is the Colorado potato beetle (CPB).

Several bioassay methods have been developed for CPB larvae. In order to compare different preparations using an LC_{50} value, potato leaflets are dipped in Bt *tenebrionis* suspensions and compared to a standard preparation (Keller and Langenbruch, 1993). A second method to ascertain the LD_{50} is by applying defined Bt *tenebrionis* doses to small leaf discs which are completely consumed by the larvae (Riethmüller and Langenbruch, 1989).

13.7.2 Tests in vitro

Bt researchers and producers have been working to replace insect bioassays with chemical or *in vitro* assays. Different reports have appeared for correlating the quantity of crystal protein in a preparation with its insecticidal activity (Beegle, 1990).

Immunoassay techniques have been used to determine the amount of crystal protein and its relationship to insecticidal activity as determined by insect bioassay. Winkler and collaborators use an immunoelectrophoretic method for the quantification of relative insecticidal activity in terms of ITU per milligram of Bt preparations (Winkler *et al.*, 1971).

Dubeikovskaya *et al.* (1990) described the utilization of Mancini's immunodiffusion method, Lorell's rocket electrophoresis method and enzyme immunoassay to determine the δ-endotoxin content of commercial dendrobacillin samples, with a view to standardize preparations on the basis of δ-endotoxin content. Microplate enzyme-linked immunosorbent assay (ELISA) technology has been used to determine the total CryIA toxin protein in production samples. Utilizing a synthetic peptide approach, primary antibody reagents have been developed that react equally with CryIA(a), CryIA(b) and CryIA(c) toxin proteins (Groat *et al.*, 1990). Sodium dodecyl sulfate–polyacrylamide gel eletrophoresis (SDS-PAGE) has also been developed to quantify Bt endotoxins in a variety of production samples ranging from fermentation broths to technical powders and formulated end-products.

Some workers have tested the ability of δ-endotoxins to inhibit the K^+-dependent intravescicular accumulation of amino acids to provide a rapid means to test entomocidal activity or newly synthesized toxins. A technique based in the use of erythrocytes to assess the toxicity of δ-

endotoxin of Bt *israelensis* (Bti) has also been developed (Ingle *et al.*, 1993).

Limitations of the *in vitro* assays are that they measure the quantity of endotoxin, not the quality. The potency of a Bt toxin lies in its quality and its quantity. In this regard, one of the most serious problems presented by chemical methods, is that they are not able to distinguish between active and inactive (Beegle, 1990).

Chemical methods are also unable to detect the presence of spores. Since there are some insects that only react when spores are present in the preparations (Sutter and Raun, 1966; Sommerville *et al.*, 1970; Yamvrias and Angus, 1970; McGaughey, 1978; Smirnoff and Valéro, 1970), this limits also the use of these techniques. It is clear that, at this time, there is no substitute for bioassays when assessing the quality of the toxin protein in commercial preparations (Beegle, 1990).

13.8 Applications of Bt biopesticides

The current worldwide market for Bt products is mainly for agriculture, but its use in forestry has been increasing during the past couple of years. By the end of the decade, global Bt sales were estimated to be in the range of US$50–80 million (Carlton *et al.*, 1990). It is not easy to estimate commercial production because there is insufficient information available about production in the former USSR and China, and the growing production in developing countries (van Frankenhuyzen, 1993).

13.8.1 Utilization of Bt for control of lepidopteran pests

From the 34 strains identified to date (de Barjac and Frachon, 1990), the following 22 strains are active against lepidopteran pests: *thuringiensis*, *finitimus*, *alesti*, *kurstaki*, *sotto*, *kenyae*, *galleriae*, *canadensis*, *entomocidus*, *aizawai*, *morrisoni*, *ostriniae*, *tolworthi*, *darmstadiensis*, *kyushuensis*, *thompsoni*, *tohokuensis*, *yunnanensis*, *pondicheriensis*, *colmeri*, *shandongiensis* and *wuhanensis*. Some of them also have activity against dipteran or coleopteran insect larvae.

Since 1971 the most commonly used commercial products are based on Bt subsp. *kurstaki* strain HD-1 because it is very active against more than 100 lepidopteran species. NRD12 strain has been utilized to improve the control of *Spodoptera* species that are not sensitive to HD-1 and a product based on a strain of Bt subsp. *aizawai* is used specifically to control the wax moth, *Galleria mellonella* (Navon, 1993). Some products based on Bt *kurstaki* have been used to control pests in vegetables and fruits. In Japan, only strains with low activity against the silkworm *Bombyx mori* can be utilized, owing to the importance of the silk industry (Takaki, 1985).

Applications of Bt *kurstaki* in agriculture accounted for about 60% of total sales in 1990. The largest market is for the protection of vegetable crops in North America and elsewhere against a complex of species including *Plutella xylostella*, small white butterfly, *Pieris rapae*, *P. brassicae*, *T. ni, cabbage moth, Mamestra brassicae* and *S. exigua*. It is also used in the protection of soybean, cotton, tobacco and corn. Forestry applications of Bt *kurstaki* accounted for about 20% of 1990 sales, again mostly in North America. The main target insects are spruce budworm in Canada and gypsy moth in the USA (van Frankenhuyzen, 1993).

13.8.2 Utilization of Bt for control of dipteran pests

The control of insect pest vectors with Bt *israelensis* accounted for the remaining 20% of sales in 1990, mostly in North America and West Africa (van Frankenhuyzen, 1993). The largest single market is the Onchocerciasis Control Programme in West Africa; which was initiated in 1974 by the WHO (Akpoboua *et al.*, 1989). When *Simulium damnosum*, the vector of *Onchocerca* filaria, became apparently resistant to organophosphate products, Bti was introduced.

In the Upper Rhine Valley, 90% of floodwater mosquitoes are controlled by the application of Bti (Becker and Margalit, 1993). Another place where mosquitoes including the malaria vector *Anopheles sinensis* cause serious problems is along the Yangtze River. Since the application of Bti; the mosquito density has fallen to 3.7, 2 and 1.3 mosquitos per person per hour for 1987, 1988 and 1989, respectively. Malaria cases are also reported to have diminished (Xu *et al.*, 1992). However, at the same time, there is resurgence of several vector-borne parasitic diseases in many countries.

In places where blackflies and mosquitoes do not transmit pathogens, they are still a great nuisance to habitats in the area (Molloy, 1990). In temperate zones, the ecological value is affected by the presence of these insects and, in some other places, cattle are attacked by them. With the development of effective formulations of Bti, the control of mosquitoes and blackflies is now possible (Becker and Margalit, 1993).

13.8.3 Utilization of Bt for control of coleopteran pests

Bt *tenebrionis* (Btt) is the most important commercial subspecies utilized for coleopteran pest control. Most of the known susceptible species belong to the chrysomelid family and just six species from other families have been reported as being susceptible, often at low levels, to this strain (Keller and Langenbruch, 1993).

Besides the Colorado potato beetle, which represents the most important pest of the Coleoptera group that can be controlled by Btt, other insects susceptible to the endotoxins are: *Agelastica alni*, which damages alders in Europe; *Chrysomela scripta*, the cottonwood leaf beetle; *Galerucella* (*Pyrrhalta*) *viburni*, the cranberry tree leaf beetle; *Paropsis charybdis*, the eucalyptus tortoise beetle; and *Xanthogaleruca* (*Pyrrhalta*) *luteola*, the elm leaf beetle (Keller and Langenbruch, 1993).

13.9 Conclusions

Bacillus thuringiensis has great potential as a biological control agent because of its ability to produce endotoxins that can be used to control insect pests. Many studies have been reported to understand the biochemical mechanism of toxicity and the factors that determine the extreme specificity.

The use of new, natural and genetically modified, Bt strains opens a wider spectrum for the use of Bt and, with other biocontrol agents, it can exhibit efficacy. To improve delivery and residual activity, reseachers have succeeded in expressing Bt toxin genes in plants and in plant colonizing microorganisms (Gelernter and Schwab, 1993).

For example, in Mexico, plants have been successfully transformed and have been shown to kill the larvae of tobacco hornworm, *Manduca sexta* (Salama and Morris, 1993). Other plants that have been stably transformed are potato, sugar beet, carrot, celery, lettuce, tomato, pepper, melon, cotton, sunflower, alfalfa, etc. (Ely, 1993). Among the plant colonizing microorganisms, a corn-root colonizing isolate of *P. fluorescens* has been engineered to produce the lepidopteran active CryIA(b) toxin (Obukowicz *et al.*, 1987).

With the world population increasing, the demand for vegetables, fruits and other agricultural consumable products is increasing, and the public awareness for environmental safety and health will necessarily promote the use of control measures, such as Bt products, that do not damage the environment. Sales of Bt products are expected to reach close to US$300 million by the year 2000 (McKemy, 1990).

Bt is the only bioinsecticide manufactured on an industrial scale and available on the market at prices affordable to farmers, but they are still expensive in comparison to synthetic chemical insecticides (Bernhard and Utz, 1993). The use of cheaper raw material will favor the increased production and use of Bt products. However, it is important to note that each strain performs in a different way depending on the culture medium and culture conditions.

Acknowledgements

Sincere thanks are due to the National Sciences and Engineering Research Council of Canada (Grant A4984) for providing financial assistance to carry out this work. Thanks are also due to Dr Vinod Bihari for reading the manuscript and providing useful suggestions.

References

Abdel-Hameed, A. (1992) Effect of amino acids in defined media on growth of *Bacillus thuringiensis* H-14. *Acta Pharm. Fenn.*, **101**, 221–6.

Abdel-Hameed, A., Carlberg, G., and El-Tayeb, O.M. (1991) Studies on *Bacillus thuringiensis* H-14 strains isolated in Egypt-IV. Characterization of fermentation conditions for δ-endotoxin production. *World J. Microbiol. Biotechnol.*, **7**, 231–6.

Abrosimova, L.I., Babaeva, P.V., Zubareva, G.M. and Shevtsov, V.V. (1986) Influence of mineral salts on the level of exotoxin production and productivity of a culture of *Bacillus thuringiensis*. *Mikrobiologiya*, **55**(3), 440–4.

Akpoboua, L.K.B., Guillet, P., Kurtak, D.C. and Pangalet, P. (1989) Le rôle du *Bacillus thuringiensis* H14 dans la lutte contre *Simulium damnosum* (Diptera: Simulidae), vecteur de l'onchocercose en Afrique occidentale. *Naturaliste Can.* (*Rev. Eco. Systs.*), **116**, 167–74.

Anderson, T.B. (1990) Effects of carbon:nitrogen ratio and oxygen on the growth kinetics of *Bacillus thuringiensis* and yield of bioinsecticidal crystal protein. M.E.Sc. thesis, London, Ontario, 193 p.

Andrews, R.E., Jr, Bechtel, D.B., Davidson, L.I. and Bulla, L.A., Jr (1981) Solubility of parasporal crystals of *Bacillus thuringiensis* and presence of toxic protein during sporulation, germination and outgrowth, in *Sporulation and Germination* (eds H.S. Levinson, A.L. Sonenshein and D.J. Tipper), American Society for Microbiology, Washington, DC, pp. 174–7.

Andrews, R.E., Jr, Bibilos, M.M. and Bulla, L.A. (1985) Protease activation of the entomocidal protoxin of *Bacillus thuringiensis* subsp. *kurstaki*. *Appl. Environ. Microbiol.*, **50**, 737–42.

Andrews, R.E., Jr, Faust, R.M., Wabiko, H., Raymond, K.C. and Bulla, L.A., Jr (1987) The biotechnology of *Bacillus thuringiensis*. *CRC Crit. Rev. Biotechnol.*, **6**, 163–232.

Arcas, J., Yantorno, O.M. and Ertola, R.J. (1987) Effect of high concentration of nutrients on *Bacillus thuringiensis* cultures. *Biotechnol. Lett.*, **9**, 105–10.

Aronson, A.I., Beckman, W. and Dunn, P. (1986) *Bacillus thuringiensis* and related insect pathogens. *Microbial Rev.*, March, 1–24.

Aronson, J.N. (1976) Ammonia assimilation and glutamate catabolism by *Bacillus thuringiensis*, in *Microbiology-1976* (ed. D. Schlesinger), American Society for Microbiology, Washington, DC, pp. 444–9.

Aronson, J.N., Borris, D.P., Doerner, J.F. and Akers, E. (1975) γ-aminobutyric acid pathway and modified tricarboxylic acid cycle activity during growth and sporulation of *Bacillus thuringiensis*. *Appl. Microbiol.*, **30**, 489–92.

Avignone-Rossa, C. and Mignone, C. (1993) δ-Endotoxin activity and spore production in batch and fed-batch cultures of *Bacillus thuringiensis*. *Biotechnol. Lett.*, **15**(3), 295–300.

Avignone-Rossa, C., Yantorno, O.M., Arcas, J.A. and Ertola, R.J. (1990) Organic and inorganic nitrogen source ratio effects on *Bacillus thuringiensis* var. *israelensis* delta-endotoxin production. *World J. Microbiol Technol.*, **6**, 27–31.

Avignone-Rossa, C., Arcas, J. and Mignone, C.F. (1992) *Bacillus thuringiensis* growth, sporulation and δ-endotoxin production in oxygen limited and non-limited cultures. *World J. Microbiol. Biotechnol.*, **8**, 301–4.

Baumann, L., Okamoto, K., Unterman, B.M., Lynch, M.J. and Baumann, P. (1984) Phenotypic characterization of *Bacillus thuringiensis* and *Bacillus cereus*. *J. Invertbr. Pathol.*, **55**, 329–41.

Bechtel, D.B. and Bulla, L.A. (1976) Electron microscopic study of sporulation and parasporal crystal formation in *Bacillus thuringiensis*. *J. Bacteriol.*, **127**, 1472–81.

Becker, N. and Margalit, J. (1993) Use of *Bacillus thuringiensis israelensis* against mosquitoes and blackflies, in Bacillus thuringiensis, *An Environmental Biopesticide: Theory and Practice* (eds P.F. Entwistle, J.S. Cory, M.J. Bailey, and S. Higgs), Wiley, Chichester, pp. 147–70.

Beegle, C.C. (1990) Bioassay methods for quantification of *Bacillus thuringiensis* δ-endotoxin, in *Analytical Chemistry of* Bacillus thuringiensis, American Chemical Society, pp. 14–21.

Bernhard, K. and Utz, R. (1993) Production of *Bacillus thuringiensis* insecticides for experimental and commercial uses, in Bacillus thuringiensis, *An Environmental Biopesticide: Theory and Practice* (eds P.F. Entwistle, J.S. Cory, M.J. Bailey and S.Higgs), Wiley, Chichester, pp. 255–67.

Borris, D.P. and Aronson, J.N. (1969) Relationship of L-alanine and L-glutamate dehydrogenases of *Bacillus thuringiensis. Biochim. Biophys. Acta*, **191**, 716–8.

Brown, K.L. and Whitely, H.R. (1988) Isolation of a *Bacillus thuringiensis* RNA polymerase capable of transcribing crystal protein genes. *Proc. Natl. Acad. Sci. USA*, **85**, 4166–70.

Bulla, L.A., Jr, Bechtel, D.B., Kramer, K.J. *et al.* (1980) Ultrastructure, physiology, and biochemistry of *Bacillus thuringiensis. Crit. Rev. Microbiol.*, **8**(2), 147–204.

Burgerjon, A. (1962) Relation entre l'intoxication provoquee par *Bacillus thuringiensis* Berliner et la consommation chez *Pieris brassicae. L. Ann. Epiphyt.*, **13**, 59–72.

Burges, H.D. (1982) Control of insects by bacteria. *Parasitology*, **84**, 79–117.

Burges, H.D. and Thomson, E.M. (1971) Standardization and assay of microbial insecticides, in *Microbial Control of Insects and Mites* (eds H.D. Burges and N.W. Hussey), Academic Press, New York, pp. 591–622.

Carlton, B.C. (1990) Economic considerations in marketing and application of biocontrol agents, in *New Directions in Biological Control: Alternatives for Suppressing Agricultural Pests and Diseases*, UCLA Symposia on Molecular and Cellular Biology, Alan R. Liss, Inc., New York, pp. 419–34.

Carlton, B.C. and Gonzalez, J.M., Jr (1985) Plasmids and delta-endotoxin production in different subspecies of *Bacillus thuringiensis*, in *Molecular Biology of Microbial Differentiation* (eds J.A. Hoch and P. Setlow), American Society for Microbiology, Washington, DC, pp. 246–52.

Carlton, B.C. Gawron-Burke, C. and Johnson, T.B. (1990) Exploiting the genetic diversity of *Bacillus thuringiensis* for the reaction of new bioinsecticides, in *Proceedings Vth International Colloquium on Invertebrate Pathology and Microbial Control*, Adelaide, Australia, pp. 18–22.

Chilcott, C.N. and Pillai, J.S. (1985) The use of coconut wastes for the production of *Bacillus thuringiensis* var. *israelensis. MIRCEN J.*, **1**, 327–32.

Child, R. and Nathanael, W.R.W. (1950) Changes in the sugar composition of coconut water during maturation and germination. *J. Sci. Food Agriculture*, **1**, 326–9.

Claus, D. and Berkeley, R.C.W. (1986) Genus Bacillus Cohn 1872, in *Bergey's Manual of Systematic Bacteriology*, Vol. 2 (eds P.H.A. Sneath, N.S. Mair, M.E. Sharpe and J.G. Holt), Williams and Wilkins, Baltimore, pp. 1105–207.

Conner, R.M. and Arne Hansen, P. (1967) Effects of valine, leucine, and isoleucine on the growth of *Bacillus thuringiensis* and related bacteria. *J. Invertebr. Pathol.*, **9**, 12–18.

de Barjac, H. and Bonnefoi, A. (1962) Essai de classification biochimique et serologique de 24 souches de *Bacillus* du type *thuringiensis. Entomophaga*, **7**, 5–31.

de Barjac, H. and Bonnefoi, A. (1973) Mise au point sur la classification des *Bacillus thuringiensis. Entomophaga*, **18**, 5–17.

de Barjac, H., and Frachon, E. (1990) Classification of *Bacillus thuringiensis* strains. *Entomophaga*, **35**, 233–40.

Dharmsthiti, S.C., Pantuwatana, S. and Bhumiratana, A. (1985) Production of *Bacillus thuringiensis* subsp. *israelensis* and *Bacillus sphaericus* strain 1593 on media using a byproduct from a monosodium glutamate factory. *J. Invertebr. Pathol.*, **46**, 231–238.

Dreier, P., Brauhart, I., Küng, W. and Moser, A. (1990) Continuous processing of *Bacillus thuringiensis*, in *Proceedings of the Fifth European Congress on Biotechnology* (eds C. Christianse, L. Munck and J. Villadsen), Copenhagen, July, pp. 142–5.

Dubeikovskaya, A.Z., Frank, R.I., Pakhtuev, A.I. and Shevtosov, V.V. (1990) Quantitation of protein in entomocidal *Bacillus thuringiensis* crystals by immunochemical methods. *Biotekcnologiya*, **5**, 89–92.

Dulmage, H.T. (1970a) Insecticidal activity of HD-1, a new isolate of *Bacillus thuringiensis* var *alesti. J. Invert. Pathol.*, **15**, 232–9.

Dulmage, H.T. (1970b) Production of the spore δ-endotoxin complex by variants of *Bacillus thuringiensis* in two fermentation media. *J. Invertebr. Pathol.*, **16**, 385–9.

Dulmage, H.T. (1973) Assay and standardization of microbial insecticides. *Ann. N.Y. Acad. Sci.*, **217**, 187–99.
Dulmage, H.T. (1981) Production of microbial insecticides by fermentation, in *Microbial Control of Pests and Plant Diseases* (ed. H.D. Burges), Academic Press, London, pp. 143–222.
Dulmage, H.T. and Rhodes, R.A. (1971) Production of pathogens in artificial media, in *Microbial Control of Insects and Mites* (eds H.D. Burges and N.W. Hussey), Academic Press, London, pp. 507–40.
Dulmage, H.T., Boening, O.P., Rehnborg, C.S. and Hansen, G.D. (1971) A proposed standardized bioassay for formulations of *Bacillus thuringiensis* based on the international unit. *J. Invert. Pathol.*, **18**, 240–5.
Dulmage, H.T., Correa, J.A. and Gallegos-Morales, G. (1990) Potential for improved formulations of *Bacillus thuringiensis* var *israelensis* through standardization and fermentation development, in *Bacterial Control of Mosquitoes and Blackflies: Biochemistry, Genetics and Applications of Bacillus thuringiensis israelensis and Bacillus sphaericus* (eds H. de Barjac and D. Sutherland), Rutgers University Press, New Brunswick, NJ, pp. 110–33.
Ejiofor, A.O. (1991) Production of *Bacillus thuringiensis* serotype H-14 as bioinsecticide using a mixture of 'spent' brewer's yeast and waste cassava starch as the fermentation medium. *Discovery Innovation*, **3** (2), 85–8.
Ejiofor, A.O. and Okafor (1988) The production of *Bacillus sphaericus* 2362 using fermented cowpea (*Vigna unguiculata*) medium containing mineral substitutes from Nigeria. *MIRCEN J. Appl. Microbiol. Biotechnol.*, **4**, 455–62.
Ejifor, A.O. and Okafor, N. (1989) Production of mosquito larvicidal *Bacillus thuringiensis* serotype H-14 on raw material media from Nigeria. *J. Appl. Bacteriol.*, **67**, 5–9.
Ejifor, A.O. and Okafor, N. (1991) Formulation of a flowable liquid concentrate of *Bacillus thuringiensis* serotype H-14 spores and crystals as mosquito larvicide. *J. Appl. Bacteriol.*, **71**, 202–6.
Ely, S. (1993) The engineering of plants to express *Bacillus thuringiensis* δ-endotoxins *in Bacillus thuringiensis*, in *An Environmental Biopesticide: Theory and Practice* (eds P.F. Entwistle, J.S. Cory, M.J. Bailey and S. Higgs), Wiley, Chichester, pp. 105–24.
Faust, R.M., Abe, K., Held, G.A. *et al.* (1983) Evidence for plasmid-associated crystal toxin production in *Bacillus thuringiensis* subsp. *israelensis*. *Plasmid*, **9**, 98–103.
Federici, B.A. (1993) Insecticidal bacterial proteins identify the midgut epithelium as a source of novel target sites for insect control. *Arch. Insect Biochem. Physiol.*, **22**, 357–1.
Federici, B.A., Luthy, P. and Ibarra, J.E. (1990) The parasporal body of Bti: structure, protein composition, and toxicity, in *Bacterial Control of Mosquitoes and Blackflies; Biochemistry, Genetics and Applications of* Bacillus thuringiensis *and* Bacillus sphaericus (eds H. de Barjac and D. Sutherland), Rutgers University Press, New Brunswick, pp. 16–44.
Feitelson, J.S., Payne, J. and Kim, L. (1992) *Bacillus thuringiensis*: insects and beyond. *Bio/technology*, **10**, 271–5.
Fisher, W.S. (1993) Pesticides for tomorrow. *Am. Nurseryman*, pp. 77–80.
Fitz-James, P. and Young, E. (1969) Morphology of sporulation, in *The Bacterial Spore* (eds G.W. Gould and A. Hurst), Academic Press, New York, pp. 39–72.
Flexner, J.L., Lighthard, B. and Crogt, B.A. (1986) The effects of microbial pesticides on non-target beneficial arthropods. *Agric. Ecosyst. Environ.*, **16**, 203–54.
Foda, M.S. and Salama, H.S. (1986) New fermentation technologies for the production of bacterial insecticides. *Zentralbl. Mikrobiol.*, **141**, 151–7.
Foda, M.S., Salama, H.S. and Selim, M. (1985) Factors affecting growth physiology of *Bacillus thuringiensis*. *Appl. Microbiol. Biotechnol.*, **22**, 50–2.
Freese, E. and Fujita, Y. (1976) Control of enzyme synthesis during growth and sporulation, in *Microbiology 1976* (ed. D. Schlesinger), American Society for Microbiology, Washington, DC, pp. 164–8.
Freiman, V.B. and Chupin, A.A. (1973) Aspects of continuous cultivation of spore-forming microbes from the group *Bacillus thuringiensis*. *Biotechnol. Symp. No.* 4, pp. 259–65.
Gangurde, R.P. and Shethna, Y.I. (1995) Growth, sporulation and toxin production by *Bacillus thuringiensis* subsp. *israelensis* and *B. sphaericus* in media based on mustard-seed meal. *World J. Microbiol.*, **11**, 202–5.

Gelernter, W. and Schwab, G.E. (1993) Transgenic bacteria, viruses, algae and other microorganisms as *Bacillus thuringiensis* toxin delivery system, in Bacillus thuringiensis, *An Environmental Biopesticide: Theory and Practice* (eds P.F. Entwistle, J.S. Cory, M.J. Bailey and S. Higgs), Wiley, Chichester, pp. 89–104.
Georghiou, G.P. and Lagunes, A. (1988) *The Occurrence of Resistance to Pesticides: Cases of Resistance Reported Worldwide through 1988*, Food and Agriculture Organization, Rome, p. 35.
Gill, S.S., Cowles, E.A. and Pietrantonio, P.V. (1992) The mode of action of *Bacillus thuringiensis* endotoxins. *Annu. Rev. Entomol.*, **37**, 615–36.
Goldberg, I., Sneh, B., Battat, E. and Klein, D. (1980) Optimization of a medium for a high yield preparation of *Bacillus thuringiensis* effective against the Egyptian cotton leaf worm *Spodoptera littoralis* Boisd. *Biotechnol. Lett.*, **2**, 419–26.
Goldberg, L.J. and Margalit, J. (1977) A bacterial spore demonstrating rapid larvicidal activity against *Anopheles segentii*, *Uranotaenia unguiculata*, *Culex univitattus*, *Aedes aegypti* and *Culex pipiens*. *Mosq. News*, **37**, 355–8.
Gonzales, J.M., Jr and Carlton, B.C. (1980) Patterns of plasmid DNA in crystalliferous and acrystalliferous strains of *Bacillus thuringiensis*. *Plasmid*, **3**, 92.
Gonzales, J.M., Jr and Carlton, B.C. (1984) A large transmissable plasmid is required for crystal toxin production in *Bacillus thuringiensis* variety *israelensis*. *Plasmid*, **11**, 28–38.
Gonzalez, J.M., Jr, Dulmage, H.T. and Carlton, B.C. (1981) Correlation between specific plasmids and delta-endotoxin production in *Bacillus thuringiensis*. *Plasmids*, **5**, 351.
Gonzalez, J.M., Jr, Brown, B.J. and Carlton, B.C. (1982) Transfer of *Bacillus thuringiensis* plasmids coding for δ-endotoxin among strains of *B. thuringiensis* and *B. cereus*. *Proc. Natl. Acad. Sci. USA*, **79**, 6951–55.
Groat, R.G., Mattison, J.W. and French, E.J. (1990) Quantitative immunoassay of insecticidal proteins in *Bacillus thuringiensis* products, in *Analytical Chemistry of* Bacillus thuringiensis, American Chemical Society, pp. 88–97.
Hanson, R.S., Blicharska, J. and Szulmajster, J. (1964) Relationship between the tricarboxylic acid cycle enzymes and sporulation of *B. subtilis*. *Biochem. Biophys. Res. Commun.*, **17**, 1–7.
Heimpel, A.M. (1967) A critical review of *Bacillus thuringiensis* var. *thuringiensis* Berliner and other crystalliferous bacteria. *Ann. Rev. Entomol.*, **12**, 287.
Heimpel, A.M. and Angus, T.A. (1959) The site of action of crystalliferous bacteria in lepidopera larvae. *J. Insect. Pathol.*, **1**, 152–70.
Herrnstadt, C., Soares, G.C., Wilcox, E.R. and Edwards, D.L. (1986) A new strain of *Bacillus thuringiensis* with activity against coleopteran insects. *Bio/Technology*, **4**, 305–8.
Höfte, H. and Whiteley, H.R. (1989) Insecticidal crystal proteins of *Bacillus thuringiensis*. *Microbiol. Rev.*, **53**(2), 242–55.
Holmberg, A., Sievanen, R. and Carlberg, G. (1980) Fermentation of *Bacillus thuringiensis* for exotoxin production: process analysis study. *Biotechnol. Bioeng.*, **22**, 1707–24.
Huber, H.E. and Luthy, P. (1981) *Bacillus thuringiensis* delta-endotoxin: composition and activation, in *Pathogensis of Invertebrate Microbial Diseases* (ed. E.W. Davidson), Allanheld, Osman and Co., Totowa, NJ, pp. 235–67.
Ingle, S.S., Rao, K.K. and Chhatpar, H.S. (1993) *In vitro* assay for testing the toxicity of *Bacillus thuringiensis* subsp. *israelensis* delta endotoxin using blood agarose plates. *J. Appl. Bacteriol.*, **74**, 645–8.
Kang, B., Lee, S.Y. and Chang, H.N. (1992) Enhanced spore production of *Bacillus thuringiensis* by fed-batch culture. *Biotechnol. Lett.*, **14**(8), 721–6.
Kang, B., Lee, S.Y. and Chang, H.N. (1993) Production of *Bacillus thuringiensis* spores in total cell retention culture and two-stages continuous culture using an internal ceramic filter system. *Biotechnol. Bioeng.*, **42**, 1107–12.
Keller, B. and Langenbruch, G.A. (1993) Control of Coleopteran pests by *Bacillus thuringiensis*, in Bacillus thuringiensis, *An Environmental Biopesticide: Theory and Practice* (eds P.F. Entwistle, J.S. Cory, M.J. Bailey and S. Higgs), Wiley, Chichester, pp. 171–91.
Klier, A., Parsot, C. and Rapport, G. (1983) *In vitro* transcription of the cloned chromosomal crystal gene from *Bacillus thuringiensis*. *Nucleic Acids Res.*, **112**, 3972–87.
Knowles, B.H. and Ellar, J.D. (1986) Characterization and partial purification of a plasma

membrane receptor for *Bacillus thuringiensis* var. *kurstaki* lepidopteran-specific delta-endotoxin. *J. Cell. Sci.*, **83**, 89.

Kronstad, J.W., Schnepf, H.E. and Whiteley, H.R. (1983) Diversity of location for the *Bacillus thuringiensis* crystal protein genes. *J. Bacteriol.*, **54**, 419–28.

Krywienczyk, J., Dulmage, H.T. and Fast, P.G. (1978) Occurrence of two serologically distinct groups within *Bacillus thuringiensis* 3a3b var. *kurstaki. J. Invertebr. Pathol.*, **31**, 372.

Kuberski, T., Roberts, A., Lineham, B., Bryden, R.N. and Teburae, M. (1979) Coconut water as a rehydration fluid. *N. Zeal. Med. J.*, **90**, 98–100.

Lecadet, M.M. and Dedonder, R. (1971) Biogenesis of the crystalline inclusion of *Bacillus thuringiensis* during sporulation. *Eur. J. Biochem.*, **23**, 282–94.

Lee, H.L. and Seleena, P. (1991) Fermentation of a malaysian *Bacillus thuringiensis* serotype H-14 isolate, a mosquito microbial control agent utilizing local wastes. *Southeast Asian J. Trop. Med. Public Health*, **22**(1), 108–12.

Lereculus, D., Lecadet, M.M., Ribier, J. and Dedonder, R. (1982) Molecular relationship among plasmids of *Bacillus thuringiensis*: conserved sequences through 11 crystalliferous strains. *Mol. Gen. Genet.*, **186**, 391.

Lereclus, D., Delécluse, A. and Lecadet, M. (1993) Diversity of *Bacillus thuringiensis* toxins and genes, in Bacillus thuringiensis, *An Environmental Biopesticide: Theory and Practice* (eds P.F. Entwistle, J.S. Cory, M.J. Bailey and S. Higgs) Wiley, Chichester, pp. 37–69.

Lynch, M.J. and Baumann, P. (1985) Immunological comparisons of the crystal protein from strains of *Bacillus thuringiensis. J. Invertebr. Pathol.*, **46**, 47.

Martin, P.A.W. and Travers, R.S. (1989) Worldwide abundance and distribution of *Bacillus thuringiensis* isolates *Appl. Env. Microbiol.*, **55**, 2437–42.

McGaughey, W.H. (1978) Response to *Plodia interpunctella* and *Ephestia cautella* larvae to spores and parasporal crystals of *Bacillus thuringiensis. J. Econon. Entomol.*, **41**(4), 687–8.

McKemy, C. (1990) Bt: keeps bugs from bouncing back. *The Grower*, August, pp. 22–4.

Mettus, A.M. and Macaluso, A. (1990) Expression of *Bacillus thuringiensis* δ-endotoxin genes during vegetative growth. *App. Environ. Microbiol.*, **56**, 1128–34.

Molloy, D. (1990) Progress in the biological control of blackflies with *Bacillus thuringiensis* with emphasis on temperate climates, in *Bacterial Control of Mosquitoes and Blackflies*, Rutgers University Press, New Brunswick, pp. 161–86.

Moraes, I.O., Santana, M.H.A. and Hokka, C.O. (1980) The influence of oxygen concentration on microbial insecticide production, in *Advances in Biotechnology, Proceedings of the 6th International Fermentation Symposium*, London, Ontario, Vol. 1, pp. 75–9.

Moser, A. (1991) Tubular bioreactors: case study of bioreactor performance for industrial production and scientific research. *Biotechnol. Bioeng.*, **37**, 1054–65.

Mummigatti, S.G. and Raghunathan, N. (1990) Influence of media composition on the production of δ-endotoxin by *Bacillus thuringiensis* var. *thuringiensis, J. Invertbr. Pathol.*, **55**, 147–51.

Muthukumar, G. and Nickerson, K.W. (1987) The glycoprotein of *Bacillus thuringiensis* subsp. *israelensis* indicates a lectinlike receptor in the larval mosquito gut. *Appl. Environ. Microbiol.*, **53**, 2650–5.

Navon, A. (1993) Control of Lepidopteran pests with *Bacillus thuringiensis*, in Bacillus thuringiensis, *An Environmental Biopesticide: Theory and Practice* (eds P.F. Entwistle, J.S. Cory, M.J. Bailey and S. Higgs), Wiley, Chichester, pp. 125–46.

Nickerson, K.W., St Julian, G. and Bulla, L.A., Jr (1974) Physiology of spore-forming bacteria associated with insects: radiorespirometric survey of carbohydrate metabolism in the 12 serotypes of *Bacillus thuringiensis. Appl. Microbiol.*, **28**(1), 129–32.

Norris, J.R. (1964) The classification of *Bacillus thuringiensis. J. Appl. Bacteriol.*, **23**, 439.

Obeta, J.A.N. and Okafor, N. (1984) Medium for the production of primary powder of *Bacillus thuringiensis* subsp. *israelensis. Appl. Environ. Microbiol.*, **47**(4), 863–67.

Obukowicz, M.G., Perlak, F.J., Bolten, S.L. *et al.* (1987) IS50L as a non-self transposable vector used to integrate the *Bacillus thuringiensis* delta-endotoxin gene into the chromosome of root-colonizing pseudomonas. *Gene*, **51**, 91–6.

Pearson, D. and Ward, O.P. (1988) Effect of culture conditions on growth and sporulation of *Bacillus thuringiensis* subsp. *israelensis* and development of media for production of the protein crystal endotoxin. *Biotechnol. Lett.*, **10**(7), 451–6.

Pfannenstiel, M.A., Couche, G.A., Ross, E.J. and Nickerson, K.W. (1986) Immunological relationships among proteins making up the *Bacillus thuringiensis* subsp. *israelensis* crystalline toxin. *Appl. Environ. Microbiol.*, **52**, 644–649.

Pfannenstiel, M.A., Cray, W.C., Couche, G.A. and Nickerson, K.W. (1990) Toxicity of protease-resistant domains from the delta-endotoxin of *Bacillus thuringiensis* subsp. *israelensis* in *Culex quinquefasciatus* and *Aedes aegypti* bioassays. *Appl. Environ. Microbiol.*, **56**, 162–6.

Priest, F.G. and Sharp, R.J. (1989) Fermentation of bacilli, in *Fermentation Process Development of Industrial Microorganisms* (ed. J.O. Neway), Marcel Dekker, New York, pp. 73–132.

Rajalakshmi, S. and Shethna, Y.I. (1977) The effect of aminoacids on growth, sporulation and crystal formation in *Bacillus thuringiensis* var *thuringiensis. J. Indian Inst. Sci.*, **59**, 168–76.

Rajalakshmi, S. and Shethana, Y.I. (1980) Spore and crystal formation in *Bacillus thuringiensis* var. *thuringiensis* during growth and cystine and cysteine. *J. Biosci.*, **2**(4), 321–8.

Rajnchapel-Messaï, J. (1993) *Bacillus thuringiensis*, les insectes font de la resistance. *Biofutur*, October.

Riethmüller, U. and Langenbruch, G.A. (1989) Zwei Biotestmethoden zur Prüfung von *Bacillus thuringiensis* subps. *tenebrionis* gegen Larven des Kartoffelkafers (Leptinotarsa decemlineata). *Entomophaga*, **34**, 237–45.

Rodriguez Monroy, M., de la Torre, M.M. and de Urquijo, N.E. (1991) *Bacillus thuringiensis*: caracteristicas biológicas y perspectivas de producción. *Rev. Lat-amer. Microbiol.*, **33**, 279–92.

Rogoff, M.H., Ignoffo, C.M., Singer, G., Gard, I. and Prieto, A.P. (1969) Insecticidal activity of thirty-one strains of *Bacillus thuringiensis* against five insect species. *J. Invert. Pathol.*, **14**, 122–9.

Rowe, G.E. (1990) Central metabolism of *Bacillus thuringiensis* during growth and sporulation. London, Ontario, Canada, pH.Doc. 295 p.

Rowe, G.E. and Margaritis, A. (1987) Bioprocess development in the production of bioinsecticides by *Bacillus thuringiensis. CRC Crit. Rev. Biotechnol.*, **6**(1), 87.

Sachidanandham, R. and Jayaraman, K. (1993) Formation of spontaneous asporogenic variants of *Bacillus thuringiensis* subsp. *galleriae* in continuous cultures. *Appl. Microbiol. Biotechnol.*, **40**, 504–7.

Sakharova, Z.V., Ignatenko, Yu, N., Khourychev, M.P. Likov, V.P., Rabotnova, I.L. and Shevtson, V.V. (1984) Sporulation and crystal formation in *Bacillus thuringiensis* with growth limitation via the nutrient sources. *Microbiology*, **53**(2), 221.

Salama, H.S. and Morris, O.N. (1993) The use of *Bacillus thuringiensis* in developing countries, in Bacillus thuringiensis, *An Environmental Biopesticide: Theory and Practice* (eds P.F. Entwistle, J.S. Cory, M.J. Bailey and S. Higgs), Wiley, Chichester, pp. 237–53.

Salama, H.S., Foda, M.S., Dulmage, H.T. and Sharaby, A.El. (1983b) Novel fermentation media for production of delta-endotoxin from *Bacillus thuringiensis. J. Invert. Pathol.*, **41**, 8–19.

Salama, H.S., Foda, M.S., Selim, M.H. and Sharaby, A.El. (1983c) Utilization of fodder yeast and agro-industrial by-products in production of spores and biologically active endotoxins from *Bacillus thuringiensis. Zbl. Mikrobiol.*, **138**, 553–63.

Salama, H.S., Foda, M.S., Sharaby, A.El. and Selim, M.H. (1983a) A novel approach for whey recycling in production of bacterial insecticides. *Entomophaga*, **28**(2), 151–60.

Scherrer, P., Lüthy, P. and Trumpi, B. (1973) Production of δ-endotoxin by *Bacillus thuringiensis* as a function of glucose concentrations. *Appl. Microbiol.*, **25**(4), 644–6.

Schwartz, J.L., Garneau, L., Savaria, D., Masson, L. and Brousseau, R. (1993) Lepidopteran-specific crystal toxins from *Bacillus thuringiensis* form cation- and anion-selection channels in planar lipid bilayers. *J. Membrane Biol.*, **132**, 53–62.

Selinger, L.B., Dawson, P.S.S. and Khachatourians, G.G. (1988) Behavior of *Bacillus thuringiensis* var. *kurstaki* under continuous phased cultivation in a cyclone fermentor. *Appl. Microbiol. Biotechnol.*, **28**, 247–53.

Sikdar, D.P., Majumdar, M.K. and Majumdar, S.K. (1991) Effect of minerals on the production of the delta endotoxin by *Bacillus thuringiensis* subsp. *israelensis Biotechnol. Lett.*, **13**(7), 511–4.

Singer, S. and Rogoff, M.H. (1968) Inhibition of growth of *Bacillus thuringiensis* by amino acids in defined media. *J. Invertebr. Pathol.*, **12**, 98–104.
Smirnoff, W.A. (1963) The formulation of crystals in *Bacillus thuringiensis* var. *thuringiensis* Berliner before sporulation atalow-temperature incubation. *J. Insect Pathol.*, **5**, 242–50.
Smirnoff, W.A. and Valéro, J.R. (1979) Mode d'action de *Bacillus thuringiensis* chez *Choristoneura fumiferana* (Lepidoptera-Torticidae): importance des spores. *Can. Entomol.*, **111**, 305–8.
Smith, R.A. and Couche, G.A. (1991) The phyllophane as a source of *Bacillus thuringiensis* variants. *Appl. Env. Microbiol.*, **57**, 311–5.
Sommerville, H.J., Tanada, Y. and Omi, E.M. (1970) Lethal effect of purified spore and crystalline endotoxin preparations of *Bacillus thuringiensis* on several lepidopterous insects. *J. Invertebr. Pathol.*, **16**, 241–8.
Stahly, D.P., Dingman, D.W., Bulla, L.A., Jr and Aronson, A.I. (1978) Possible origin and function of the parasporal crystals in *Bacillus thuringiensis*. *Biochem. Biophys. Res. Commun.*, **84**, 581–8.
Stahly, D.P., Andrews, R.E. and Yousten, A.A. (1992) The genus *Bacillus*-insect pathogens, in *The Prokaryotes*, 2nd edn (eds A. Balows, H.G. Trüper, M. Dworkin, W. Harder and K.H. Schleifer), Springer-Verlag, London, pp. 1697–745.
Sutter, G.R. and Raun, E.S. (1966) The effect of *Bacillus thuringiensis* components on the development of the european corn borer. *J. Invertebr. Pathol.*, **8**, 457–60.
Takaki, S. (1985) Bt preparations in Japan. *Japan Pest. Inform.*, **25**, 23–6.
Thomas, W.E. and Ellar, D.J. (1983) Mechanism of action of *Bacillus thuringiensis* var *israelensis* insecticidal δ-endotoxin. *FEBS Lett.*, **154**, 362–8.
Valéro, J.R. (1990) Microbiologie contre tordeuse, recherches à forêts Canada – region du Québec. *L'Aubelle*, 12–15.
Valéro, J.R. and Letarte, R. (1988) Diagnostique biochimique de la présence d'une intoxication par *Bacillus thuringiensis* serotype H3a3b chez deux Lepidoptères. *Can. J. Microbiol.*, **35**, 444–9.
van Frankenhuyzen, K. (1993) The challenge of Bacillus thuringiensis, in Bacillus thuringiensis, *An Environmental Biopesticide: Theory and Practice* (eds P.F. Entwistle, J.S. Cory, M.J. Bailey and S. Higgs). Wiley, Chichester, pp. 1–35.
van Rie, J., Jansens, D., Höfte, H., Deghelee, D. and van Mellaert, H. (1990) Receptors on the brush border membrane of the insect midgut as determinants of the specificity of *Bacillus thuringiensis* δ-endotoxins. *Appl. Environ. Microbiol.*, **56**, 1378–85.
Wakisaka, Y., Masaki, E. and Nishimoto, Y. (1982) Formation of crystalline δ-endotoxin or poly-β-hydroxybutyric acid granules by asporogenous mutants of *Bacillus thuringiensis*. *Appl. Environ. Microbiol.*, **43**, 1473–80.
WHO (1983) *Basic Biology of Microbial Larvicides of Vectors of Human Diseases* (ed. F. Michal), World Health Organisation, Geneva, pp. 78–83.
Wilcox, D.R., Shivakumar, A.G., Melin, B.E. *et al.* (1986) Genetic engineering of bioinsecticides, in *Protein Engineering: Applications in Science, Medicine, and Industry* (eds M. Inouye and R. Sarma), Academic Press, Orlando, FL, pp. 395–413.
Winkler, V.W., Hansen, G.D. and Yoder, J. (1971) Immunochemical analysis of parasporal crystal digests of *Bacillus thuringiensis* as an index of insecticidal activity. *J. Invertebr. Pathol.*, **18**, 378–82.
Wong, H.C., Schnepf, E. and Whitely, H.R. (1983) Transcriptional and translational start sites for the *Bacillus thuringiensis* crystal protein gene. *J. Biol. Chem.*, **258**, 1960–7.
Xu, B.Z., Becker, N., Xiao, X.Q. and Ludwig, H.W. (1992) Microbial control of mosquitoes in Hubei province, People's Republic of China. *Bull. Soc. Vector Ecol.*, **17**, 1–10.
Yamvrias, C. and Angus, T.A. (1970) The comparative pathogenicity of some *Bacillus thuringiensis* varieties of spruce budworm, *Choristoneura fumiferana. J. Invertebr. Pathol.*, **15**, 92–9.
Yudina, T.G., Salamakha, O.V., Olekhnovich, E.V., Rogatykh, N.P. and Egorov, N.S. (1993) Effect of carbon source on the biological activity and morphology of paraspore crystals from *Bacillus thuringiensis*. *Microbiology*, **61**(4), 402–7.

14 Biorecovery of metals from mining wastes

DAVID S. HOLMES

14.1 Historical perspective

It is widely accepted that the Romans made use of biological solubilization to recover copper from the Rio Tinto area of Spain nearly two thousand years ago. However, one of the first written descriptions of what was almost certainly biologically assisted metal recovery appeared in about 1510.

> This work is a wonderful one in the light of Nature, viz., that by the Magistery or the operation of the Spagyrist [alchemist] a metal which formerly existed should perish and another be produced. This fact has rendered that same Aristotle, with his ill-founded philosophy, fatuous. For truly, when the rustics in Hungary cast iron at the proper season into a certain fountain, commonly called Zifferbrunnen, it is consumed into rust and when this is liquefied with a blast-fire, it soon exists as pure Venus [copper], and nevermore returns to iron. Similarly in the mountain commonly called Kutenberg they obtain a lixivium out of marcasites, in which iron is forthwith turned into Venus of a high grade, and more malleable than the other produced by nature.
>
> [From an English translation by Arthur Edward White of *The Book Concerning the Tincture of the Philosophers*, Chapter VI, by Aureolus Phillipus Theophrastus Bombast (Paracelsus the Great) born about 1493 and died 1541.]

We would now interpret the events recorded at the Zifferbrunnen fountain as follows: copper was being solubilized from copper-bearing rock *in situ* by biological oxidation and biological production of sulfuric acid. The acid solution, rich in dissolved copper, issued as a spring at the base of the copper-bearing ores. Copper could then be recovered by cementation (precipitation) onto iron placed in the acid solution. Finally, the copper was purified from the iron by smelting. This is essentially the same process used today by mining companies to recover copper from dumps of low-grade copper ores.

By 1750, 200 tons of copper were obtained annually from the Zitterbrannen area of Hungary. Biological leaching of copper was also practiced in Germany in the 16th century and at the Rio Tinto mine in Spain by 1725 (Karavaiko and Groudev, 1985). More recently, biological leaching of copper was begun in the USSR at the end of the last century and in the USA by the 1920s. Now almost every copper mine in the world has some form of solution technology for copper recovery in which, in many cases, microorganisms play a major role.

Pioneering work by Rudolfs and Helbronner (1922), Waksman and Joffe (1922) and Carpenter and Herndon (1933) on bacteria capable of oxidizing sulfur compounds to sulfuric acid and by Colmer and Hinkle (1947) on the iron- and sulfur-oxidizing bacterium *Thiobacillus ferrooxidans* provided the foundation for subsequent research into the role of microorganisms in metal solubilization. The concept of using microorganisms for the recovery of metals dates back to the early 1950s when Bryner *et al.* (1954) described how iron pyrites and copper sulfides could be oxidized by acidophilic *Thiobacilli* from the mine drainage of Kennecott's Bingham Canyon open pit mine. These observations led to the issuance of the first patent (Zimmerly *et al.*, 1958) assigned to the Kennecott Copper Corporation for a cyclic leaching process employing acidophilic iron-oxidizing bacteria. This initial patent led to the issuance of a series of additional patents.

Despite these efforts, the belief that copper solubilization was purely a chemical process was commonly held by many mining personnel even up to the 1970s. A curious event in 1967 has been said to have contributed to the acceptance of a biological role in copper solubilization. At that time, the Anaconda Company was recovering 100 tons of copper per day from their waste dumps in Butte, Montana. Part of the copper solubilization process called for dilute sulfuric acid to be sprinkled onto the dumps. It was this acid that mining engineers believed to be responsible for the resulting copper solubilization. It was brought home sharply that microorganisms were involved in the process, when the company, during a prolonged strike, purchased sulfuric acid that was subsequently shown to be contaminated with 50 ppm mercury. Using this acid, the solubilization of copper dropped to nearly zero within 3 months. The mercury was clearly inhibiting microbial activity. The Anaconda leaching operation never reopened after this unfortunate but enlightening accident.

14.2 Significance of biomining

Worldwide reserves of high-grade ores are diminishing at an alarming rate as a result of rapid increase in the demand for metals. For example, between 1975 and 1997 the requirement for zinc is expected to rise from 6 to 12 million tons per year (tpy), copper from 8 to 24 million tpy and aluminum from 13 to 53 tpy (Ebner, 1978). It has been estimated that many high grade ore reserves will be depleted within the next 10–15 years.

There are adequate reserves in North America (including Canada) of many important strategic metals. However, these exist primarily in stockpiles of low-grade ores or as low-grade ores yet to be mined. The problem is that recovery of metals from low-grade ores using existing technology is prohibitively expensive owing to high energy and capital costs. For example, a new smelter costs about $1 billion. In addition, smelters are large energy consumers and, hence, leave relatively little room for significantly reducing variable costs of production. Production solutions to reduce the high costs of conventional metal-recovery processs are being investigated in such diverse areas as intelligent mining systems and improved techniques for mineral separations. However, these new advances can only result in an incremental improvement when the mining and metal-processing industries remain such large consumers of energy. In contrast, biomining offers a low-cost alternative for the recovery of metals from low-grade ores.

14.3 Copper dump bioleaching

The principal method used for the biological recovery of copper is dump bioleaching. Almost every large copper company has a dump formed from overburden rock stripped away to provide access to underlying high-grade ores. These dumps generally contain about 0.1–0.5% copper, too low to recover profitably by conventional procedures. Some of these dumps are huge, containing in excess of 10 million tons of waste rock. One of the largest dumps is at the Kennecott Bingham Canyon mine in Utah, USA, where as much as 200 tons of copper per day have been recovered, accounting for 25% of the total copper production of the mine. Other examples of past or present copper dump bioleaching operations are shown in Table 14.1.

To recover copper from dumps, dilute sulfuric acid is sprinkled on to the surface. The acid percolates down through the dump, lowering the pH and promoting the growth of acidophilic microorganisms. These microorganisms oxidize the copper sulfide in the ore putting it into solution in the acid. The acid runoff, often blue with dissolved copper, is collected at the bottom of the dump from where it is pumped to a recovery station. Copper is recovered from the acid runoff by solvent extraction and electrowinning (electroplating on cathodes) at the larger mining operations. Since solvent extraction is a relatively capital intensive process, smaller mining operations use a process called cementation to recover copper. This involves precipitation of the copper onto scrap iron, followed by recovery of the copper by smelting, a process that has been used for over 500 years, as indicated at the beginning of this chapter. The residual acid solution from the solvent extraction or cementation plants is recycled back to the dump to preserve water, reduce acid costs and lower environmental pollution.

Table 14.1 Example of copper dump bioleaching operations

Organism	Optimum growth physiological temperature (°C)	Special traits	References
Thiobacillus ferrooxidans	30	Oxidizes Fe^{2+}, S°, $S_2O_3^{-2}$, U(IV), metal sulfides, CuS_2	Holt (1993)
Leptospirillum ferrooxidans	30	Oxidizes Fe^{2+} but not S°	Balashova *et al.* (1974)
Thiobacillus thiooxidans	30	Oxidizes S°, $S_2O_23^{-2}$, but not Fe^{2+} metal sulfides	Holt (1993)
Sulfolobus brierleyi	70	Oxidizes Fe^{2+}, S°, MnS_2, $CuFeS_2$	Brierley and Brierley (1973)
Thiobacillus TH-1	55	Oxidizes Fe^{2+}, S°, metal sulfides, and needs organic carbon (e.g. yeast extract)	Brierley *et al.* (1978)
Sulfobacillus thermosulfidoxidans	50	Oxidizes Fe^{2+}, S°, forms endospores	Karavaiko *et al.* (1988a)
Desulfovibrio	30	Reduces SO_4^{2-}, S° and some partially oxidized S components, forms metal sulfides	Postgate (1979)
Acidophilium cryptum	30	Heterotrophic, can scavenge carbon from organic excretions of *T. ferrooxidans* in the laboratory	Harrison (1984)
Protozoa	30	Prey on bioleaching microorganisms	Ehrlich (1963)

14.3.1 Economics of dump bioleaching

Biological copper recovery worldwide is thought to exceed one billion US dollars annually in value and to account for about 25% of world copper production. In 1988, US production of copper by biological leaching exceeded 250 000 tons, valued in excess of US\$650 million.

Very few data exist on the cost of production of individual dump leaching operations. Estimates within the industry for total copper production costs can be as low as \$0.70 kg^{-1} for large operations in the US Southwest where solvent extraction is widely used. This about 50% less than conventional copper recovery techniques.

It should be noted that one of the reasons that biological recovery of copper from dumps is economically attractive is that the waste rock had to be moved in any case to provide access to the higher grade ore. Therefore, most dump-leaching operations do not bear the costs of rock removal, rock transport or comminution (rock breakage).

A breakdown of individual costs for dump leaching is largely unavailable and is considered to be proprietary information. Schlitt (1980) estimated the costs for dump leaching of copper for Kennecott's Bingham Canyon operation, where the bacterial-assisted leaching of existing waste dumps containing 0.15–0.25% Cu (predominantly as $CuFeS_2$) is credited as having produced 200 tons per day (TPD) of saleable cement copper at a 1980 operating cost of \$0.80–0.87 kg^{-1}. Of this, about 40% (34 cents kg^{-1}) is attributed to the leaching operation proper, where the major cost is for power to pump the leach solutions in three stages through the 2700 ft head between the base and the top of the dumps. The cost of copper recovery accounts for another 40% (34 cents kg^{-1}), with most being spent on detinned scrap iron for cementation. The remaining cost of \$0.12–0.19 kg^{-1} is for labor and maintenance.

There is a limited market for cement copper in the manufacture of clutch plates and in antifouling marine paints. The low purity (55–80%) cement copper is not generally considered an acceptable final product, and further upgrading by smelting and refining is required. The cost of producing cathode copper this way at Bingham Canyon was \$1.42–1.54 kg^{-1}, leaving a comfortable operating margin of between \$0.53 and \$0.99 kg^{-1} (based on copper prices of \$1.95–2.53 kg^{-1}.

Copper extracted pyrometallurgically from high-grade ore from the mine costs \$2.42–2.53 kg^{-1}. Thus, conventional operation is economically viable only because of substantial credits for gold, silver and molybdenite by-products, which reduce the net cost of copper production to about \$1.33 kg^{-1}.

These figures were calculated in 1980. Since that time, most major copper leaching operations have made the transition from cementation to solvent extraction with reported cost savings exceeding 50% in some operations. As mentioned, about 50% of the costs of dump leaching are manifested in scrap iron costs for the cementation processes. However, it should be noted that extraction is a capital-intensive process relative to cementation and, therefore, may not be suitable to scaling down to relatively small operations.

14.3.2 Microbiology

Reference will be made in this section to some of the well-established microbial reactions encountered in bioleaching operations resulting from activity of some of the better known microorganisms. In section 14.9.2 there is a discussion of recent advances in our understanding of the fauna and flora of bioleaching operations resulting from the use of novel techniques to identify and quantify microbes *in situ*.

A copper dump represents a complex and heterogeneous microbiological habitat. It contains solids ranging in size from boulders to fine sand and includes material of complex mineralogy. The interior of a dump is frequently hot (40–70°C) supporting a range of thermophilic microorganisms, and is often anaerobic or microaerophilic. The exterior of the dump is at ambient temperature and suffers changes in temperature reflecting seasonal and diurnal fluctuations.

Many different microorganisms have been isolated from copper dumps, some of which have been studied in the laboratory. These include a variety of mesophilic, aerobic iron- and sulfur-oxidizing microorganisms, thermophilic iron- and sulfur-oxidizing microorganisms, including sulfate-reducing bacteria. Some are heterotrophic bacteria which indirectly affect metal solubilization by affecting the growth and activity of metal-solubilizing bacteria and others are protozoa which prey on the different types of bacteria (Table 14.1).

The biological solubilization of copper in dumps is known to involve both a direct enzymatic oxidation of the copper and an indirect, chemical oxidation of the copper by biologically produced ferric iron. The most studied microorganism involved in these direct and indirect processes is *Thiobacillus ferrooxidans*.

T. ferrooxidans is a Gram-negative rod, a strict aerobe and an obligate chemolithoautotroph (Holt, 1993). It obtains energy by the oxidation of iron and reduced sulfur compounds. It derives carbon by carbon dioxide fixation. It has a pH optimum for growth between 2.0 and 2.5, and a temperature optimum for growth around 30°C. Being a strict autotroph, it cannot obtain energy from the oxidation of reduced carbon compounds and, in fact, some carbon compounds are inhibitory to its growth.

T. ferroxidans can directly oxidize copper enzymatically by a process still poorly understood:

$$CuFeS_2 + 4O_2 \xrightarrow{\text{Bacteria}} CuSO_4 + FeSO_4 \qquad (14.1)$$

It can also oxidize ferrous (FeII) iron and several forms of reduced sulfur found in sulfide ores such as pyrite:

$$4FeS_2 + 15O_2 + 2H_2O \xrightarrow{\text{Bacteria}} 2Fe_2(SO_4)_3 + 2H_2SO_4 \qquad (14.2)$$

T. ferrooxidans obtains energy from the process and the resulting ferric iron (Fe III) is a powerful chemical oxidant, oxidizing Cu(I) to Cu(II):

$$CuFeS_2 + 2Fe_2(SO_4)_3 + 2H_2O \xrightarrow{\text{Chemical}} CuSO_4 + 5FeSO_4 + 2H_2SO_4 \qquad (14.3)$$

Ferrous iron is regenerated in the process. This again serves as an energy source for *T. ferrooxidans* which converts it to ferric iron:

$$4FeSO_4 + O_2 + 2H_2SO_4 \xrightarrow{\text{Bacteria}} 2Fe_2(SO_4)_3 + 2H_2O \qquad (14.4)$$

thus completing a ferrous iron–ferric iron cycle. Note that the microorganisms also produce sulfuric acid (equation 14.2). This promotes the solubilization of the Cu(II) and reduces the amount of exogenous acid that must be added to the dump.

Leptospirillum ferrooxidans is another microorganism found in copper dumps. It is Gram-negative, motile and spiral shaped. It is a strict aerobe and obligate chemolithoautotroph, using iron or pyrite as an energy source. It cannot obtain energy from reduced sulfur compounds. It has a temperature and pH optimum similar to *T. ferrooxidans*. An important difference, however, is that it grows and can oxidize iron at much lower pH values than *T. ferrooxidans*, for example, between pH 1.0 and 1.5. *Thiobacillus thiooxidans* is a close relative of *T. ferrooxidans* with similar properties regarding the oxidation of sulfur compounds, but it differs in that it cannot oxidize iron.

Deep inside dumps, temperatures as high as 60–80 °C have been recorded, owing to heat liberated by biological and chemical oxidation processes. In such areas, many types of thermophiles have been noted, for example, the sulfur-, iron- and pyrite-oxidizing archaebacterium *Sulfolobus brierleyi* (Brierley and Brierley, 1973). This microorganism has a temperature optimum of around 70 °C. Other thermophiles include iron- and sulfur-oxidizing, Gram-negative *Thiobacillus* Th-1 with a temperature optimum of 50 °C, and the Gram-positive, spore forming *Sulfobacillus thermosulfidoxidans*, which can oxidize iron, sulfur and pyrite. The latter has a wide temperature optimum between 50 and 60 °C. All the thermophiles recorded so far are strict aerobes, and presumably contribute significantly to the solubilization of copper in hot regions of dump-leaching operations. There are probably thermophilic anaerobes inside dumps but none have been described, most likely due to difficulties in their isolation and culture.

An example of an anaerobe found in anoxic regions of dumps is the sulfate-reducing bacterium *Desulfovibrio* spp. These and other Gram-negative sulfate reducers, such as *Desulfotomaculum* spp. and *Desulfobacter* spp., diminish mineral recovery by forming metal sulfides that precipitate out of solution, making the metals unavailable for recovery and also coating the surfaces of minerals, thereby preventing further metal solubilization. Such activity highlights the need to build dumps in which oxygen can freely diffuse to all areas in order to optimize bioleaching.

Also found in dump-leaching operations are an enormous variety of acidophilic, heterotrophic mesophilic bacteria (e.g. *T. acidophilus*), fungi and protozoa. Such organisms are thought to affect leaching processes in at least two ways. Bacteria such as *Acidophilium cryptum* have been shown in the laboratory to promote the growth of *T. ferrooxidans* by metabolizing organic excretions that are otherwise inhibitory to the growth of *T. ferrooxidans*. A similar process is presumed to occur in dumps. Secondly, protozoa, such as amoeba, prey on the microorganisms found in dumps and presumably affect population levels of microorganisms involved in metal solubilization.

There are no reports of bacteriophages of *T. ferrooxidans*, although it is almost certain that they must exist. There is a preliminary report of bacteriophages of the acidophilic heterotroph *Acidophilium* sp. (Ward *et al.*, 1990).

Finally, there is a group of mesophilic, sulfur-oxidizing microorganisms exemplified by *Thiobacillus neapolitanus* that oxidize reduced sulfur compounds at pH values between 4 and 7. These microorganisms play an important role in the natural development of a dump in which exogenous acid is not applied by the mining company. When a freshly prepared dump is first exposed to rain and the pH is around neutral or higher, these microorganisms oxidize available reduced sulfur compounds to sulfuric acid. Ultimately, they lower the pH to a level where acidophilic microorganisms such as *T. ferrooxidans* can take over.

A picture has emerged of a complex collection of interacting microorganisms changing dynamically in time, as the dump matures from neutrality to low pH, and in space from the surface to the inside of the dump. The surface of the dump is characterized by mesophilic anaerobes while the interior contains mesophilic anaerobes and thermophilic aerobes. Microbial diversity and dynamics are widely recognized as being very important for metal bioleaching. What is not clear is that complete extent of this diversity. Nor is there a thorough understanding of all the ways in which the microorganisms interact with each other to promote or retard metal solubilization. For example, biofilms have been observed on rocks in dump-leaching operations and are presumed to play a key role in bioleaching. However, there is only a rudimentary knowledge about the microbial composition and activity of the biofilms, and, unfortunately, this important area of research is still largely neglected (Geesey and Jang, 1990).

One problem that has impeded progress in the general area of microbial ecology of dump leaching is the inadequacy of techniques for microbial strain identification and enumeration. For example, conventional techniques of strain characterization, relying on biochemical and morphological characteristics, have placed *Thiobacillus ferrooxidans*, *T. ferrooxidans* M1, *T. acidophilus*, *T. novellus*, *T. versutus* and *T. neopolitanus* in the same genus. This designation was based largely on a common ability to oxidize sulfur. However, preliminary work using total cellular DNA hybridization (Harrison, 1984) and more recently the use of cloned DNA probes (Yates *et al.*, 1986) and ribosomal RNA sequencing (Lane *et al.*, 1985) have convincingly demonstrated that these different species of *Thiobacillus* are well separated phylogenetically and should not be classified in the same genus. The recent introduction of DNA probe techniques including the polymerase chain reaction (PCR) technique and rRNA sequencing promises to revolutionize our understanding of the diversity and phylogenic relationships of bioleaching microorganisms.

A second problem concerns the strong potential that exists for the biased sampling of microbial species from bioleaching operations (Holmes, 1988). Current techniques for the collection of microorganisms from bioleaching dumps strongly favor the recovery of microbes that are more readily accessible, for example, those in solution, rather than those that are firmly bound on the ore particles. Moreover, existing methods of cultivation of microorganisms in the laboratory probably do not permit the growth (or allow only reduced growth) of some of the microorganisms from a bioleaching operation. It is quite possible, in fact almost certain, that some types of bioleaching microorganisms are being underestimated or missed altogether owing to these problems. It is hoped that DNA probe technology, coupled with PCR amplification techniques, with the ability to detect single microbial cells in environmental samples (Steffan and Atlas, 1988), will solve some of the abovementioned problems and present us with a clearer picture of the microbial diversity of a bioleaching dump. This type of information is essential for the advancement of our understanding of microbial metal solubilization in dumps and will be discussed further in section 14.9.2.

14.3.3 Problems

The construction of dumps has historically proceeded with little regard for the optimization of the biological processes. This can lead to inefficient and slow recoveries of copper. However, because of the low capital costs involved in dump construction, copper can still be economically recovered. In the future, as bioleaching becomes more prevalent, serious attention will be given to improving dump bioleaching rates and yields. For example, the following problems deserve increased attention:

- Material is dumped with no attempt to separate barren rock from low-grade copper bearing ore.
- Material is dumped in sizes ranging from large boulders to fine sand. The large material is very inefficiently bioleached owing to its small surface area to volume ratio, and the finer material can impede the flow rate of the acid solution and cause significant channeling in the dump.
- There is serious compaction of the dump by haulage trucks used in the dumping process, impeding solution flow and separation.
- Air penetrates only a few meters into a dump, so large dumps frequently have considerable areas that are anaerobic. This prevents biological oxidation processes eliminating concomitant solubilization of copper. It also promotes the growth of anaerobic sulfate-reducing microorganisms which reprecipitate dissolved metals on rock surfaces as metal sulfides.
- Although an enormous diversity of microorganisms has been identified and studied in the laboratory, a thorough understanding of their interactions in the dump has not been developed. For example, biofilms have been observed in dumps and presumably play a crucial role in metal solubilization (Karavaiko *et al.*, 1988b). However, because biofilms are difficult to study and because there has been a dearth of research funds, this area has been largely neglected.
- Existing techniques for microbial strain identification and enumeration are inadequate to describe the ecology and microbial dynamics of a dump properly.

14.3.4 Technical solutions

- Many mining companies have begun to address several of the problems outlined above. Especially important has been the implementation of finger dumps where waste material is arranged in dumps several hundred feet long and only 30–50 feet high. The enlarged surface area to volume ratio of these dumps significantly increases aeration.
- The development of DNA probes for bioleaching microorganisms has provided an important diagnostic tool for the identification and enumeration of microorganisms in dumps (Yates *et al.*, 1986). These DNA probes can distinguish between strains of microorganisms as well

as between species and genera. The probes are also quite sensitive and can detect less than 10^4 microorganisms. The sensitivity of DNA probes has been increased by the amplification by the PCR to a level where individual bacteria can be detected in environmental samples (Steffan and Atlat, 1988). In this technique, specific segments of DNA can be amplified a millionfold.

- Progress continues to be made regarding the biochemistry, physiology and genetics of microorganisms involved in copper solubilization. Information from these studies may be used to make marginal improvements in bioleaching rates and efficiency. However, it is likely that more significant returns will be obtained in the near term by engineering improvements in dump design and operation.
- The uses of genetically engineered microorganisms in dump bioleaching is not a high priority with the research community. This is mainly due to uncertainties regarding the activity and survivability of such strains in the complex and competitive microbiological milieu of the dump. Mining executives are also cautious because of regulatory issues that govern the release of genetically engineered microorganisms (GEMS) in environmental situations such as dumps.

14.4 Heap leaching

Heap leaching is similar to dump leaching except that the rock is usually ground to a much finer and more uniform size, and the heaps are built by deposition from conveyer belts so that compaction by haulage trucks is not a problem. Also, heaps are generally smaller than dumps with adequate aeration. All of this leads to much faster and more efficient leaching, and copper recovery is accomplished in days or weeks compared to years in a dump.

In heap leaching, mine development and comminution processes are attributed to the costs of copper recovery. It has been estimatd that this adds about \$0.18 lb^{-1} of copper for an ore containing 0.5% copper (Schlitt, 1980).

In heap leaching the effectiveness and economics of seeding with prepared innocula of microorganisms, including specifically developed laboratory strains could be explored. Because of the increased capital costs tied up in heap leaching, the time cost in monetary terms becomes important and, the faster the leaching proceeds, the more competitive the process becomes.

14.5 Concentrate bioleaching

Copper is conventionally recovered from high-grade ore by smelting copper concentrates. Copper concentrates are prepared by grinding the raw ore to about the size of sand. This material is mixed with organic bubbling reagents in huge tanks. Air is pumped into the tanks and bubbles formed by the organic reagents preferentially float ore particles enriched in copper. This material, termed copper concentrate, often contains as much as 20% copper. The copper concentrate is collected from the surface and shipped to a smelter. Because of the high capital and operating costs of smelting, together with environmental pollution problems associated with smelting, biological means for recovering copper from concentrates by heap bioleaching or in tanks have long been considered an attractive alternative. The technical feasibility of such a process was demonstrated by McElroy and Bruynsteyn (1978) but the industry has been slow to adopt it. The reluctance to adopt bioleaching may be because the process is only marginally more attractive economically, and biological processes generally must have cost advantages of 20% or better compared with conventional processes to realize commercial adoption within the mining industry (Holmes *et al.*, 1988a).

14.6 *In situ* bioleaching

Ore bodies that are too low-grade or too small to be mined by conventional methods have potential for *in situ* bioleaching. With virgin ore bodies the costs of shattering the rock may prove prohibitively expensive (Ismay *et al.*, 1986). However, the bioleachng of old mine workings, backfilled stopes (an excavation in the form of steps, to recover ore from vertical veins) and open pits is often an economic method for recovering additional copper when conventional minable reserves have been depleted, as was proved by the recovery of copper by *in situ* leaching in the USA at Magma Copper's Miami mine for nearly 30 years.

Like dump leaching, it is more likely that advances in *in situ* bioleaching will come, in the near term, from engineering improvements and not from advances in biology.

14.7 Uranium bioleaching

Bioleaching has been used successfully to obtain uranium from waste gold ore at Buffelsfontein in South Africa (Matthew Hall Ortech Ltd, 1979). It has been used to recover uranium from the walls and stops of several underground uranium mines in the Elliot Lake region of Canada including the Stanrock, Denison and Millikan mines. Uranium production from the Denison mines was reported to be around 33 000 kg month^{-1} in 1987 (McCready, 1988). Uranium has also been recovered by bioleaching from the Agnew Lakes Mines in Canada.

Although microorganisms such as *T. ferrooxidans* are capable of oxidizing tetravalent uranium directly into the soluble hexavalent form, it is thought that the main contribution of microorganisms to uranium bioleaching is in the production of acid ferric iron according to equation 14.2 above. The acid ferric iron can then oxidize the uranium:

$$UO_2 + Fe_2(SO_4)_3 \xrightarrow{\text{Chemical}} UO_2SO_4 + 2FeSO_4 \qquad (14.5)$$

resulting in soluble uranium sulfate and ferrous iron. These bacterial and chemical oxidation processes take place *in situ*. However, since the main role of microorganisms is to produce acid ferric iron it might be advantageous to build a bioreactor through which mine solution can be passed for the regeneration of the ferric iron. Large concentrations of microorganisms such as *T. ferrooxidans* can be obtained in a bioreactor, especially if the cells are immobilized. Also, in a bioreactor they can be properly nurtured to maximize the rate of ferrous iron bioxidation. Work in this area was pioneered by Liversey-Goldblatt who described the Bacfox process (Liversey-Goldblatt *et al.*, 1977), and has been developed further by others (Murayuma *et al.*, 1987; Nikolov and Karamanev, 1987; Grishin and Tuovinen, 1988).

14.8 Bio-oxidation of gold ore

14.8.1 Principles

All existing procedures for the recovery of gold from ore incorporate a step where the gold is complexed with cyanide. Ores that do not respond well to direct cyanidation require some method of pretreatment to render them amenable to cyanidation. Generally, these procedures involve the oxidation of sulfides to release gold from entrapment in the sulfide matrix. The pretreatment processes most frequently used have been roasting and pressure oxidation. However, microorganisms are also capable of accomplishing this step.

Recently, interest in the biological oxidation of refractory gold ores has been heightened by the successful initiation of the first commercial-scale bio-oxidation plants in Zimbabwe and South Africa, and several successful pilot-scale plants in North America. Today there are full-scale tank leaching operations at the Fairview Gold Mine (South Africa), the Sao Bento mine (Brazil), the Harbour Lights mine (Australia), the Wiluna mine (Australia) and the Ashanti mine (Ghana). There are also large-scale operations at the Mintek and Van Reefs gold mines (South Africa).

Simultaneous with these successful technical developments, the price of gold has risen and mineral companies have taken steps to increase production, improve recoveries and develop new ore bodies. Many companies are now taking a second look at deposits that were once considered uneconomical. Many of these deposits are refractory and tend to resist cyanidation. Bio-oxidation offers a new low-cost alternative for oxidizing these refractory ores.

The Giant Bay plant at Salmita in Northwest Territories in Canada has been the most successful pilot plant operating at 10 tons day^{-1} during 1987, using ore from Giant Yellowknife Company. Previously it had been possible to recover only 65–70% of the gold by physical means. In contrast, bio-oxidation was able to raise the recovery to 95%. These data are among the most reliable and available because the plant was operated with run-of-the-mine ore. Most other data available have been from bench scale, hence they have not met the rigors of actual site operation.

Capital costs are significantly lower for bio-oxidation versus the alternatives, pressure oxidation and roasting. For example, the costs of a 1000 tons day^{-1} plant at Salmita have been estimated to be (in Canadian dollars) \$3 000 000 compared to \$45 000 000 for pressure oxidation or \$35 000 000 for roasting. The estimated operating costs for bio-oxidation also showed savings (in Canadian dollars) \$42 ton^{-1} versus \$44 ton^{-1} for pressure oxidation or \$50 ton^{-1} for roasting. When these costs are compared on a discounted cash-flow basis assuming a 10 year project life, a 15% cost of capital and all tax and depreciation effects are taken into account, bio-oxidation appears to have considerable cost advantages over the conventional alternatives.

The economics of bio-oxidation become particularly advantageous for short-lived projects. Furthermore, bio-oxidation requires much less energy than conventional alternatives so that, if and when energy prices resume their upward course, the cost advantages of bio-oxidation will be further enhanced.

Existing operations for the bio-oxidation of refractory gold ore use only naturally occurring microorganisms. In this process ore or ore concentrate is conditioned (i.e. allowed to mix) in a stirring tank with water derived from a downstream concentrator (thickener). The conditioned material is passed through to a series of agitation tanks where oxidation takes place. Nutrients, usually in the form of agricultural fertilizer, are added and the tanks are sparged with air. The pH is maintained in the range of 1.0–2.0 and excess heat generated by the oxidation process is removed with cooling coils. Retention time in the bio-oxidation tanks varies from 15 to 150 h depending on whether the material used is ore or concentrate.

The oxidized slurry is sent to a thickener, where the overflow is treated for the removal of ferric iron and arsenic, and then returned to the conditioning tank. The underflow is filtered and the filtrate is recycled to the thickener. The solid residue is mixed with water to form a slurry and then advanced to a series of neutralization tanks where $Ca(OH)_2$ is added to raise the pH to the level desired for cyanidation. The slurry is then sent to the cyanidation circuit for gold recovery.

The following characteristics have been found to be limiting in reactors used for gold bio-oxidation and are most likely to be important also for copper concentrate bio-oxidation.

1. temperature;
2. oxygen and carbon dioxide transfer;
3. pulp density;
4. residence time;
5. physical attrition of microorganisms;
6. high levels of chlorides;
7. high levels of arsenic.

Whereas the control of temperature, oxygen, carbon dioxide and pulp density and their optimization is most likely to be accomplished by engineering, biology may be the most important contributor to the solution of the other aspects listed above. Examples are given below.

1. The residence time could be reduced if the microorganisms could be induced to grow faster and/or oxidize iron faster. When we know the full metabolic flow of energy and carbon in *Thiobacillus*, we can identify rate-limiting steps and, using genetic engineering, increase the rates of key reactions. For example, it was known for years that the microorganism *Haemophilus* required huge quantities of glutamate for its growth but nobody understood why. Was it due to a poor glutamate uptake system, a poor intracellular transport mechanism, or a bottleneck in the utilization of glutamate or what? When the complete sequence of all its genes was recently deduced it was suddenly discoverd that *Haemophilus* lacks a key enzyme in the tricarboxylic acid cycle, a highly surprising result, which accounted for its need for exogenous glutamate. The gene for this missing enzyme could, if desired, be introduced by genetic engineering into the chromosome of the microorganism correcting its deficiency. Something similar may be discovered when more DNA sequence information becomes available for microorganisms involved in bioleaching. For example, in the case of *T. ferrooxidans*, it is not impossible that, with the full knowledge of the genes involved in the flow of oxygen from iron to oxygen, an artificial iron-oxidizing system could be devised. One in which one or perhaps two key enzymes operated on an artificial support to oxidize iron. Such a system is purely chemical and free of the constraints of biological demands, i.e. one does not need to bother with the problem of keeping the microorganisms alive and functioning. The system can be considered a chemical catalyst and chemical engineers can bring to bear an enormous quantity of experience to manipulate and improve the catalyst.

2. The problem of attrition in tank reactors (in which the mineral particles physically break the microorganisms) might be overcome if the microorganisms could be induced to grow a tougher cell wall or perhaps by encapsulating them in a protective devise. It is difficult at the moment to see exactly how this could be achieved without compromising the rate of iron oxidation but it is more likely to be resolved with the full genetic information of relevant microorganisms than without it.
3. Salt and arsenic resistance could probable be genetically engineered into bioleaching microorganisms. There exists a considerable database of information regarding such attributes in other microorganisms that could be utilized to propose appropriate genetic changes to relevant microorganisms.

Further discussion on the feasibility of genetic engineering can be found in section 14.9.

14.8.2 Opportunities

There are several aspects of the gold recovery process that could benefit from advances in biotechnology which are listed below.

1. The principal source of microorganisms is the overflow from the thickener, together with a smaller number of microorganisms associated with the incoming ore or concentrate. The microorganisms that are returned with the thickener overflow are those that are not attached to the pulp, which is separated from the overflow at this step. Therefore, these microorganisms are constantly being selected against attachment to mineral particles. Thus, over many cycles of operation there could be a significant enrichment for microorganisms that have lost the ability to attach to ore particles. If microbial attachment to ore particles is important for the bio-oxidation of gold ore, and it is not clear yet whether it is important, then the current practice of relying on thickener overflow as a source of microorganisms may be counterproductive. A high priority should be given to determining the relative importance of attached versus unattached microorganisms for gold ore bio-oxidation and, in the event that attachment is deemed important, then alternate sources of bacterial inoculation should be sought.

2. Gold sulfide ores frequently have levels of mercury and arsenic that are toxic to microorganisms. Microorganisms that are resistant to these compounds could be developed by adaptation or by genetic manipulation (Holmes *et al.*, 1984; Rawlings *et al.*, 1984).

3. Heat is liberated in the bio-oxidation process, especially in the bio-oxidation of gold concentrates. This heat is normally carried away by cooling coils, costing energy. An alternative is to explore the use of thermophilic microorganisms, such as *Sulfolobus* instead of the mesophile

T. ferrooxidans, for high-temperature bio-oxidation. An additional advantage of using thermophiles would be the faster rates of the ferric iron oxidation at elevated temperatures.

14.9 Development of new strains of microorganisms

14.9.1 Introduction

At the present time only indigenous microorganisms, occurring naturally in dumps or mine runoff, are used in bioleaching operations. This is equivalent, in a sense, to using native strains of wild wheat for farming in which the only selection is for strains that might have a growth advantage in a manipulated situation. In the case of farming this manipulation would be the removal of competing weeds, addition of fertilizer and irrigation, etc. and, in the biomining case, would be the construction of leaching dumps using ground ore perhaps with the addition of acidified water and nutrients, etc. The strains of microorganisms that flourish best under such circumstances are not necessarily the most efficient at putting metals into solution. So, in a way, the biotechnology of dump bioleaching might be considered at only a slightly post-neolithic stage in development. A corollary of this argument is that there is no reason to think that classically genetic mutation and selection and genetic engineering could not produce strains of microorganisms that could significantly outperform their wild relatives. The problem is which microorganism(s) to focus on for genetic manipulation and how to provide a continuing selective advantage to such a microorganism if returned to the relatively 'wild' and unsterile environment of the bioleaching dump.

Improved yields or enhanced rates of metal solubilization would make existing applications of bioleaching microorganisms more attractive and could extend the use of microorganisms to totally new opportunities for metal recovery and ore beneficiation. New strains of microorganisms can be developed for commercial applications by four procedures:

- the isolation of new, naturally occurring strains of microorganisms from the environment;
- the selection and adaptation of naturally occurring strains (usually in a chemostat);
- the modification of strains by classical genetic techniques;
- genetic engineering.

14.9.2 Fauna and flora of a bioleaching operation

There are two central issues to be addressed if one wishes to develop an understanding of the ecology and biochemistry of a leaching operation:

1. what organisms are present and in what relative proportions;
2. what are the relevant interactions between these microorganisms and what governs the nature of these interactions?

It has been thought for many years that there are complex consortia of microorganisms in a bioleaching habitats (Kelly, 1988; Tuyovinen *et al.*, 1991; Goebel and Stackebrant, 1994) and it was virtually axiomatic that *T. ferrooxidans* was the most important microorganism involved in bioleaching. The reasons behind this latter assumption were:

1. *T. ferrooxidans* was found to be ubiquitous in bioleaching operations, present sometimes in as many as hundreds of thousands or millions of cells per millilitre of mine runoff.
2. It had all the necessary biochemical characteristics to account for the solubilization of metals without the need to invoke the extensive activities of other microorganisms.

However, it was pointed out several years ago (Yates and Holmes, 1986) that a knowledge of the floral and faunal composition of a leaching operation was potentially biased in favor of organisms that could be cultivated easily in the laboratory. It is only recently that molecular biology techniques have become available to investigate this possibility.

Using direct PCR amplification of DNA isolated from ores in a leaching operation (Espejo *et al.*, 1995) or rDNA from cells isolated in natural acidic communities (Ward *et al.*, 1990; Goebel and Stackebrandt, 1994), it has been possible to show that there are, as suspected, varieties of *T. ferrooxidans*, *T. thiooxidans*, *Leptospirillum* species and acididiphilic heterotrophs as well as a number of unculturable (unknown) microorganisms in such habitats. However, the variety of species of unculturable microorganisms is not as great as in many other habitats such as soils. A possible reason for the paucity of diversity is that acidic mining operations are generally very poor in organic matter and very low in pH, restricting the variety of microorganisms to those that are well adapted to such harsh conditions.

Perhaps the most surprising result of such studies is that *F. ferrooxidans* is found in very low concentrations when the ferrous ion concentration is low, a condition which is frequently encountered in dump and heap leaching operations (Espejo *et al.*, 1995, Garcia and Jerez, 1995) and in gold ore bio-oxidation tanks (Rawlings, 1995). In these cases the predominant microorganisms appear to be *L. ferrooxidans* and *T. thiooxidans*. These are preliminary results and may not provide very accurate estimates of relative cell numbers but, if confirmed, the observation that *L. ferrooxidans* and *T. thiooxidans* are much more frequent than *T. ferrooxidans* in leaching operations will have important repercussions in our thinking of the mineral leaching process.

With regard to the types of intermicrobial and microbial–substrate interactions, there remains an unresolved long-standing debate as to whether indirect or direct biochemically process are responsible for the majority of the metal solubilization. The indirect mechanism refers to the ability of microorganisms such as *T. ferrooxidans*, in the planktonic state, to produce ferric ions in solution by the oxidation of ferrous iron. The ferric ions can, in turn, oxidize reduced metals promoting their solubilization in acidic aqueous solutions. The direct mechanism refers the direct oxidation of a metal either electrolytically, or by an enzyme(s) situated on or in a microorganism requiring direct contact between the microorganism and the ore. The indirect mechanism is the oxidation of a metal by a by-product of a microorganism in the planktonic state (usually considered to be ferric ion but could theoretically be an enzyme in solution).

For a dump or heap leaching, almost certainly, both planktonic and attached bacteria contribute to leaching. However, in tank reactors with short residence times, measured in hours or a few days, the question of the relative contributions of planktonic and attached bacteria remains largely unknown, and is obviously an important parameter to consider. In natural ecological situations and in long-term dump or heap leaching operations the majority of the microorganism are found attached to the ore as a biofilm. But even here we understand little of the relative contributions of attached and unattached bacteria to the leaching process, nor do we understand very much about composition and biochemistry of the biofilm. Important work contributed from the Biofilm Engineering Center of Montana State University, Bozeman, Montana, has led to the concept that biofilms are not generally composed of relatively uniform layers of interacting microorganisms but, rather, appear as 'pods' with a rather amorphous mushroom shape of 100 μm or more in thickness.

One of the striking features of this type of organization is its complexity. An individual pod could have an exterior that has an oxygen content in equilibrium with the surrounding solution, and an interior that is depleted in oxygn or lacking oxygen altogether. The organisms that compose the pod will vary from aerobes on the outside to extreme anaerobes on the inside according to the availability of oxygen. They could also vary depending on their nutritional requirements and on the type of symbiotic and mutalistic interactions established with other members of the community. Neighboring pods could also interact with each other establishing pH, E_h and/or O_2 gradients. A consequence of this is that electrons could flow between pods down an E_h gradient via the solution or through minerals in the ore with concomitant oxidation/reduction of metals. Pods could also harbor *Thiobacillus* species and other micro-organisms that could contribute to metal solubilization. In addition, the pods could affect the activity, either positively or negatively, of micro-organisms that are situated away from the pod in solution or attached to the ore in a non-biofilm manner. It is obviously important to understand the structure, composition and activity of such biofilms, and to determine their relative contribution to the bioleaching of dumps and heaps (biofilms probably do not have the right conditions or sufficient time to develop substantially during tank bioleaching). The correc 'ecological management' of such biofilms is also important to promote bioleaching.

The concept of the biofilm also further blurs the distinction between direct and indirect mechanisms of metal solubilization; a distinction that was already beginning to be fuzzy from a practical point of view. The rate constants for indirect versus direct oxidation would be very different, but if a microorganism is attached directly to the substrate or indirectly via a biofilm, then the local concentration of biologically produced ferric ion could be very high altering the rate constant for ferric ion-induced oxidation. In fact, a complete spectrum of rate constants might be present in a leaching operation ranging from essentially zero order to pseudo second order. Not only will there be spatial variation of rate constants but they will also vary over time as the substate changes during leaching and the biofilm composition changes.

This makes life very complicated for the engineer trying to model a dump or heap leaching operation. In addition, there is concern whether kinetics derived from laboratory columns or flasks are valid in the field. Life is also more difficult for the biologist, who must now think in terms of ecological consortia, some of whose members may be very difficult to grow and/or understand because they are strict anaerobes, or because they behave differently in consortia than they do as individuals. Even in well-defined laboratory conditions, where everything, including microorganisms and substrate, is nicely in solution and using well-studied microorganisms, it is very difficult to model the activities and interactions of two or three microorganisms. So it is not surprising that it is even more difficult to model a leaching operation containing not only many different microorganisms, some of which might not be known and others poorly understood, but that it also consists of a solid and variable substrate, and with striking variations of temperature, oxygen and carbon dioxide concentration, pH and E_h.

What is also important is that many of these issues are also of concern to environmental biotechnologists who are using indigenous or laboratory strains of microorganisms to clean the environment, and also to scientists studying biofilms that corrode materials or that can be used for biotechnological applications such as in the fossil fuel industry. Hopefully, advances made in these areas regarding the nature and uses of biofilms will be applicable to the mineral industry. Meanwhile, the engineers will continue to use largely empirically derived rate constants in their models and the biologists will struggle to understand the nature of the catalysts (microorganisms), with a view to improving their performance. With all the best engineering optimization in the world the reaction rate will never exceed a certain theoretical value dictated by intrinsic properties of the catalyst. Ultimately, it will be up to the biologists to alter the intrinsic properties of the catalyst to break through this ceiling.

14.9.3 Isolation of new strains from the environment

It is only relatively recently that an effort has been made to evaluate the commercial potential of any microorganism other than those belonging to the *Thiobacilli* group. For example, there may be advantages to using *Sulfolobus* spp. for certain high temperature applications (Norris and Barr, 1988). Also, the use of *Leptospirillum* spp. for low pH (around pH 1.0) applications (Karavaiko *et al.*, 1988b) and microorganisms from deep-sea vents for manganese recovery have been suggested (Ehrlich, 1987).

It is almost certain that there exists a huge repository of microorganisms in natural habitats such as mine dumps, mines drainages, coal refuse piles, hot springs, toxic wste sites, deep-sea vents, psychrophilic habitats, etc. Some of these might make interesting new enzymes and/or exhibit properties potentially useful for metal recovery applications.

Another neglected area of research concerns the evaluation of mixed cultures of microorganisms for metal recovery. Although symbiosis and commensalism are recognized as being important for bioleaching in copper ore dumps, little effort is being made to develop mixed cultures for commercial metal recovery applications. An advantage is that mixed cultures can frequently do things that the individual strains of microorganisms cannot do alone (Tuovinen *et al.*, 1991). Unfortunately, mixed cultures can also be hard to manipulate and maintain under industrial conditions.

14.9.4 Selection and adaptation of naturally occurring strains

Starting with a naturally occurring consortia of microorganisms, one can frequently improve their performance for a given industrial application by selection and adaptation. Adaptation can occur when selective pressure is applied to the consortium in a reactor and is thought to be the result of either or both of the following:

1. the selection of a specific strain of microorganisms in the reactor whose initial concentration at the start may not have selection of a specific strain of microorganisms in the reactor whose initial concentration at the start may not have been very high;
2. the appearance of a new strain with new properties by mutation or by the interchange of genetic information between two or more of the resident strains of microorganism.

The latter can occur by transfer of genes between microorganisms by conjugation, transformation or transduction, such as the interchange of plasmids by conjugative mechanisms.

Even though an appropriate new strain may arise infrequently in a mixed culture, its appearance and propagation can be selected by culturing in a chemostat. Although chemostat selection has been widely used for microbial strain development in organic waste biodegradation, it has been neglected as a tool for the development of microbial strains for metal recovery. It was recently shown that *T. ferrooxidans* undergoes mutation at a rate that far exceeds normal familiar mutation rates increasing the probability that chemostat selection could be used successfully to select for interesting strains (Holmes *et al.*, 1988b; Holmes and Haq, 1989).

An important variation on the use of continuous culturing to select for microorganisms is the development of the 'auxostat' (Fraleigh *et al.*, 1987). The auxostat is particularly useful for selecting for fast-growing cultures and could prove of value for the development of faster growing strains of *T. ferrooxidans*.

It has been suggested that at least part of the high mutation rate in *T. ferrooxidans* could be due to the presence of different classes of repetitive DNA sequences in the chromosomes and plasmids (Yates and Holmes, 1987; Schrader and Holmes, 1988). One class of repetitive DNA sequences has been shown to be an insertion sequence (Yates *et al.*, 1988). Also, recent evidence indicates that the repetitive DNA sequences are mobile, moving within the genome. This movement can cause mutations to arise (Holmes *et al.*, 1988b; Cadiz *et al.*, 1994), some of which may create useful commercial strains (Holmes and Haq, 1989).

Much work remains to be done on evaluating the functions and the importance of these repetitive DNA sequences for the generation of commercially useful strains of *T. ferrooxidans*. There are also interesting questions to be addressed concerning the frequency and specificity of the genetic changes. For example, do the repetitive DNA sequences move at random around the genome of *T. ferrooxidans* or do they integrate into and excise only from particular regions of the genome? Can the rate of movement be influenced by external conditions including conditions known to induce physiological stress such as heat shock or exposure to toxic metals? Are there other classes of repetitive DNA elements in *T. ferrooxidans* in addition to those that have already been described?

A key consideration is that the release of natural strains of microorganisms, selected in the laboratory for their superior performance into the environment is not subject to the guidelines and regulations governing the release of genetically engineered strains.

14.9.5 Classical genetic mutation

Pioneering studies were carried out by Groudeva and Groudev (1980) using nitrosoguanidine to mutate and select mutants of *T. ferrooxidans* that exhibited faster iron-oxidation rates. Cox and Boxer (1986) also attempted to develop strategies for the mutation and selection of *T. ferrooxidans*. However, little has been done to follow up these studies.

14.9.6 Genetic engineering

The utilization of the full range of genetic engineering tools usually implies:

1. The isolation and cloning of genes from the organism in question into *E. coli* or some other suitable laboratory organism.
2. The sequencing and possible identification of the cloned gene.
3. The return of the cloned gene to the original organism to identify or validate function. This requires a technique, usually transformation, to introduce the cloned gene back into the host.
4. Frequently it also implies the *in vitro* manipulation of the cloned gene, for example, by *in vitro* mutagenesis, to change its properties, followed by the return of the modified gene to the original host.

Whereas steps (1) and (2) above have been frequently accomplished, there exists no reliable way to introduce genes back into several of the economically important bacteria used in the biorecovery of metals including *T. ferrooxidans*. There is only one published example of the *in vitro* manipulation of a gene considered important in bioleaching (rusticyanin) (Casimiro *et al.*, 1995) and no examples in which a gene has been returned to a bioleaching microorganism resulting in a strain with improved metal solubilization.

It is a relatively simple matter to isolate and clone DNA from *T. ferrooxidans* into *E. coli*. Several genes of have been cloned from *T. ferrooxidans* by selection of functions that complement mutations in *E. coli*, such as the glutamine synthetase gene (Barros *et al.*, 1995), the *nif* gene (Pretorius *et al.*, 1986), the *ntra* gene (Berger *et al.*, 1990) and an ATP synthase gene (Brown *et al.*, 1994). Other genes have been cloned from *T. ferrooxidans*, such as a group of genes conferring mercury resistance (Shiratori *et al.*, 1989), a thioredoxin-like gene (Powles *et al.*, 1996), genes encoding ribulose bisphosphate carboxylase (Kusano *et al.*, 1991), an Fe(II) oxidase (Kusano *et al.*, 1992) and the gene for rusticyanin (Bengrine *et al.*, 1995).

Of these examples only the rusticyanin gene has been manipulated by site-directed mutagensis. Casimiro *et al.* (1995) manipulated an artificially constructed gene for rusticyanin resulting in the convincing demonstration that a critically placed His85 was involved in binding the molecule of copper found associated with the protein.

The key problem at present is the lack of reliable systems to return genes to many of the microorganisms involved in bioleaching. In the early 1980s, Yankofsky *et al.* (1983) successfully transformed amino-acid auxotrophs of the facultative autotrophic *Thiobacillus thioparus* to prototrophy using sheared DNA. Plasmids have also been conjugated into the facultative thiobacilli *T. neapolitanus* (Kulpa *et al.*, 1983) and *T. novellus* (Davidson and Summers, 1983) and for heterotrophic acidophilic bacteria (Roberto *et al.*, 1989).

More recently, broad host range IncP plasmids were conjugated into *F. ferrooxidans* and *T. thiooxidans* from *E. coli* (Jin *et al.*, 1992; Peng *et al.*, 1994). Also a cassette of genes conferring mercury resistance was transformed into *T. ferrooxidans* by electroporation (Kusano *et al.*, 1992). In both the conjugation and transformation studies special strains of *T. ferrooxidans* were used pertaining to the laboratories in question and, in the case of the transformation studies, they were unsuccessful with 29 other strains tested. Therefore, it remains to be investigated whether the techniques are of general utility.

Bacteriophages for transduction have not been reported for *Thiobacilli*, although there is a report of bacteriophage-like particles in acidophilic heterotrophs (Ward *et al.*, 1989), which may prove valuable for developing transduction systems.

Advances in the genetic engineering of microorganisms involved in metal recovery, especially *T. ferrooxidans*, have been discussed by Holmes and Yates (1990) and reviewed by Rawlings and Kusano (1994). Other reviews of interest are Bell *et al.* (1987), Nicholaides (1987), and Rawlings and Silver (1995).

14.10 Conclusions

- Biohydrometallurgy has won wide acceptance for the production of copper, especially from dumps of waste ore. Although the copper industry sometimes overlooks the microbiological nature of their leaching operations, bioleaching accounts for approximately 25% of the world's copper production. It should be noted that, despite this unsung role, microorganisms were probably responsible for the financial survival of several major US copper producers during the 1982–86 period of overproduction and low prices, as several companies shut down their high-cost conventional production systems but continued to bioleach copper. Most notable of these was the Kennecott Bingham Canyon operation in Utah, which closed its mining but continued dump leaching.

- *In situ* bioleaching holds the promise of being the least costly method for copper, uranium and other metal production at new or remote deposits. Although recovery rates are considerably slower than conventional recovery, the capital investments in such operations are very low. The primary impediment to adoption of *in situ* bioleaching is not the leaching process itself but the cost of reducing the size of the ore particles to allow for adequate percolation and surface area contact. Innovative and economical comminution techniques must be developed in order to take advantage of the biological opportunities of *in situ* bioleaching.
- Bio-oxidation of refractory gold is a relatively newly developed commercial procedure. Bio-oxidation in continuously stirred tank reactors has superior economic characteristics compared to conventional roasting and pressure oxidation.
- Microorganisms can also oxidize waste gold ores from current or abandoned refractory gold mining operations and hence avoid adding the mining or comminution costs to the process. It is possible that bio-oxidation of refractory gold waste products could be an important growth area.
- It is difficult to estimate the present market for the bio-oxidation of gold ores because the data from the various operations in Zambabwe, South Africa and North America are not complete. However, the combined market is estimated to be between $10 and $50 million and is projected to rise to $2–$3 billion by 1998 if current trends in environmental standards, and labour, energy and capital costs continue to rise over the next decade. It is expected that bio-oxidation will likely do for refractory gold ore oxidation what bioleaching did for copper production, that is, cut costs in half.
- The contribution of new biological knowledge and the development of new strains of microorganisms including genetically engineered strains to future metal recovery is very dependent on the nature and the site of the proposed application.
- For dump leaching and *in situ* leaching, significant advances are likely to come in the near term from improved engineering using concepts that take into account the biology of the process.
- More attention is being paid to improving the understanding of the role of attached microorganisms and biofilms in bioleaching.
- Newly developed techniques of DNA probe analysis and PCR are being used to extend our knowledge of the types of microorganisms found in dumps and *in situ* operations.
- There may be an opportunity to develop new strains of microorganisms for the enhanced conversion of ferrous to ferric iron in bioreactors for uranium bioleaching. Such strains might also be of value for other metal recovery applications where indirect leaching with the resulting ferric iron proves to be effective.

- There is an opportunity to develop new strains of microorganisms for the bio-oxidation of refractory gold in bioreactors. Because bioreactors can be used to regulate microbiological growth conditions, laboratory strains of microorganisms have a higher chance of surviving and retaining their engineered activities than in an open environmental situation such as a dump.
- One of the most important roles of genetic engineering in the near term is as a research tool to unravel the unusual biochemistry of bioleaching microorganisms. Ultimately, such understanding will permit superior strains of microorganisms to be developed for commercial metal recovery.

14.11 Summary

An overview is presented of existing biological technologies for the recovery of copper, gold and uranium. Engineering and biological challenges and opportunities for the bio-oxidation of refractory gold ore are described. Techniques for the development of new strains of microorganisms for commercial metal recovery applications are discussed with special reference to the use of genetic manipulation for bacterial strain improvement.

Acknowledgements

This work was supported in part by a grant from Fondecyt project 1950560.

References

Balashova, V.V., Vedenina, I.Y., Markosyan, G.E. and Zavarsin, G.A. (1974). The autotrophic growth of *Leptospirillum ferrooxidans*. *Microbiologiya*, **43**, 581–5 [English trans., p. 491–4].

Barros, M.E., Rawlings, D.E. and Woods, D.R. (1985) Cloning and expression of the *Thiobacilus ferrooxidans* glutamine synthetase gene in *Escherichia coli*. *J. Bacteriol.*, **164**, 1386–9.

Bell, P.R., McBride, B.C. and Spiegelman, G.B. (1987) *Potential Genetic Engineering of* Thiobacillus ferrooxidans *to Optimize the Leaching of Mineral Values*. Canmet Report Serial No. Osq85–0023.

Bengrine, A., Guiliani, N., Appia, C. *et al.* (1995) Studies of the rusticyanin encoding gene of *T. ferrooxidans* ATCC33020, in *Biohydrometallurgical Processing* (eds C.A. Jerez, T. Vargas, H. Toledo and J.V. Wiertz), University of Chile, Santiago, pp. 75–83.

Berger, D.K., Woods, D.R. and Rawlings, D.E. (1990) Complementation of *Escherichia coli* σ54 (NtrA)-dependent formatehydrogenlyase activity by a cloned *Thiobacillus ferrooxidans ntrA* gene. *J. Bacteriol.*, **173**, 4399–4406.

Brierley, C.L. and Brierley, J.A. (1973) A chemoautotrophic and thermophilis microorganism from an acid hot spring. *Can. J. Microbiol*, **19**, 183–9.

Brierley, J.A., Norris, P.R., Kelly, D.P. and LeRoux, N. (1978) Characteristics of a

moderately thermophilic and acidophilic iron oxidizing *Thiobacillus*. *Eur. J. Appl. Microbiol. Biotechnol.*, **5**, 291–9.
Brown, L.D., Dennehy, E. and Rawlings, D.E. (1994) The F1 genes of the F1F0 ATP synthase from the acidophilic bacterium *Thiobacillus ferrooxidans* complement *Escherichia coli* F1 *unc* mutants. *FEMS Microbiol. Lett.*, **122**, 19–25.
Bryner, L.C., Beck, J.F., Davis, D.B. and Wilson, D.G. (1954) Microorganisms in leaching sulfide minerals. *Ind. Eng. Chem.*, **46**, 2587–92.
Cadiz, R., Gaete, L., Jedlicki, E., Yates, J., Holmes, D.S. and Orellana, O. (1994) Transposition of IST2 *Thiobacillus ferrooxidans*. *J. Mol. Microbiol.*, **12**, 165–70.
Carpenter, V. and Herndon, L.K. (1933) Acid mine drainage from bituminous coal mines. *West Virginia Univ. Eng. Expt. Stat. Res. Bull.*, **10**, 37.
Casimiro, D.R., Toy-Palmer, A., Blake, R.C. and Dyson, H.J. (1995) Gene synthesis, high level expression and mutagenesis of *Thiobacillus ferrooxidans* rusticyanin: His 85 is a ligand to the blue copper center. *Biochem.*, **34**, 6640–8.
Colmer, A.R. and Hinkle, M.E. (1947) The role of microorganisms in acid mine drainage: a preliminary report. *Science*, **106**, 253–6.
Cox, J.C. and Boxer, D.H. (1986) The role of rusticyanin, a blue copper protein, in the electron transport chain of *Thiobacillus ferrooxidans* grown in iron or thiosulfate. *Biotech. Appl. Biochem.*, **8**, 269–75.
Davidson, M.S. and Summers, A.O. (1983) Wide host range plasmids function in the genus *Thiobacillus*. *Appl. Environ. Microbiol.*, **46**, 565–72.
Ebner, H.G. (1978) Metal recovery and environmental protection by bacterial leaching of inorganic waste materials, in *Metallurgical Applications of Bacterial Leaching and Related Microbiological Phenomena* (eds) L.E. Murr, A.R. Torma and J.A. Brierley Academic Press, New York, pp. 195–206.
Ehrlich, H.L. (1963) Microorganisms in acid mine drainage from a copper mine. *J. Bacteriol.*, **86**, 350–2.
Ehrlich, H.L. (1987) Manganese oxide reduction as a form of anaerobic respiration. *Geomicrobiol. J.*, **5**, 423–31.
Espejo, R.T., Pizarro, J., Jedlicki, E., Orellana, O. and Romero, J. (1995) Bacterial populations in the bioleaching of copper as revealed by analysis of DNA obtained from leached ores and leaching solutions, in *Biohydrometallurgical Processing* (eds T. Vargas, C.A. Jerez, J.V. Wiertz and H. Toledo), University of Chile, Santiago, pp. 1–8.
Fraleigh, S.., Bungay, H.R. and Clesceri, L.S. (1987) Continuous culture, feedback control and auxostats. *Trends Biotechnol.*, **7**, 159–64.
Garcia, A. and Jerez, C. (1995) Changes of the solid-adhered populations of *Thiobacillus ferrooxidans* and *Thiobacillus thiooxidans* in leaching ores as determined by immunological evidence, in *Biohydrometallurgical Processing* (eds T. Vargas, C.A. Jerez, J.V. Wiertz and H. Toledo), University of Chile, Santiago, pp. 19–30.
Geesey, G. and Jang, L. (1990) Extracellular polymers for metal binding, in *Microbial Mineral Recovery* (eds H.L. Ehrlich and C.L. Brierley), McGraw-Hill, New York, pp. 223–47.
Goebel, B.M. and Stackebrant (1994) Cultural and phylogenetic analysis of mixed microbial populations found in natural and commercial bioleaching environments. *Appl. Environ. Microbiol.*, **60**, 1614–21.
Grishin, S.I. and Tuovinen, O.H. (1988) Fast kinetics of $Fe^{2}{}_{+}$ oxidation in packed bed reactors. *Appl. Environ. Microbiol.*, **54**, 3093–110.
Groudeva, V. and Groudev, S. (1980) Mutant selection of *Thiobacillus ferrooxidans* for the purpose of leaching copper sulfide minerals. *Mikrobiologiya*, **17**, 33–47.
Harrison, A.P. (1984) The acidophilic *Thiobacilli* and other acidophilic bacteria that share their habitat. *Ann. Rev. Microbiol.*, **38**, 265–92.
Holmes, D.S. (1988) Biotechnology in the mining and metal processing industries: challenges and opportunities. *Minerals Metals. Proc.*, **5**, 49–56.
Holmes, D.S. and Haq, R. (1989) Adaptation of *Thiobacillus ferrooxidans* for industrial applications, in *Biohydrometallurgy. International Symposium Proceedings* (eds J. Salley, R.G.L. McCready and P.C. Wichlacz), Jackson Hole, WY, pp. 115–28.
Holmes, D.S. and Yates, J.R. (1990) Basic principles of genetic manipulation of *Thiobacillus ferrooxidans* for biohydrometallurgical applications, in *Microbial Mineral Recovery* (eds H.L. Ehrlich and C.L. Brierley), McGraw-Hill Inc., New York, pp. 29–54.

Holmes, D.S., Yates, J.R., Lobas, J.G. and Doyle, M.V. (1984) Genetic engineering of biomining microorganisms. *Biotech.*, **84**, A64–A81.

Holmes, D.S., Debus, K. and Watson, R. (1988a) *Bioextraction, Biorecovery and Biodetoxification of Metals*, Falmouth Association, Falmouth, Maine, p. 409.

Holmes, D.S., Yates, J.R. and Schrader, J.A. (1988b) Mobile repeated DNA sequences in *Thiobacillus ferrooxidans* and their significance for biomining, in *Biohydrometallurgy, Science and Technology Letters* (eds D.P. Kelley and P.R. Norris) Kew, Surrey, pp. 153–60.

Holt, J.G. (ed.) (1993) *Bergey's Manual of Determinative Bacteriology*, 9th edn, Williams and Wilkins, Baltimore.

Ismay, A., Rosato, L. and McKinnon, D. (1986) Engineering prefeasibility for in-place bacterial leaching of copper, in *Fundamental and Applied Biohydrometallurgy* (eds R.W. Lawrence, R.M.R. Branion and H.G. Ebner) Elsevier, Amsterdam, pp. 191–213.

Karavaiko, G.I. and Grooudev, S.N. (1985) Biotechnology of metals, its history, tasks and trends of development, in *Biotechnology of Metals* (eds G.I. Karavaiko and S.M. Groudev), Centre of International Projects, GKNT, Moscow, pp. 6–23.

Karavaiko, G.I., Golovacheva, R.S., Pivovarova, T.A., Tzaplina, I.A. and Vartanjan, N.S. (1988a) Thermophilic bacteria of the genus *Sulfobacillus*, in *Biohydrometallurgy, Science and Technology Letters* (eds P.R. Norris and D.P. Kelly) Kew, Surrey, pp. 29–41.

Karavaiko, G.I., Rossi, G., Agate, A.D., Groudev, S.N. and Avakyan, Z.A. (eds) (1988b) *Biotechnology of Metals Manual.* Centre for International Projects GKNT, Moscow.

Kelly, D.P. (1988) Evolution of the understanding of the microbiology and biochemistry of the mineral leaching habitat, in *Biohydrometallurgy, Science and Technology Letters* (eds P.R. Norris and D.P. Kelly) Kew, Surrey, pp. 3–14.

Kulpa, C.F., Roskey, M.T. and Travis, M.T. (1983) Transfer of plasmid RP1 into chemolithotrophic *Thiobacillus neopolitanus. J. Bacteriol.*, **156**, 434–6.

Kusano, T., Takeshima, T., Inoue, C. and Sugawara, K. (1991) Evidence for two sets structural genes coding for ribulos bisphotate carboxylase in *Thiobacillus ferrooxidans. J. Bacteriol.*, **173**, 7313–23.

Kusano, T., Takeshima, T., Inoue, C. and Sugawara, *et al.* (1992) Molecular cloning of the gene encoding *Thiobacillus ferrooxidans* Fe(II) oxidase. *J. Biol. Chem.*, **267**, 11242–7.

Lane, D.J., Pace, B., Olsen, G.J. *et al.* (1985) Rapid determination of *I6S* ribosomal RNA sequences for phylogenetic analysis. *Proc. Natl. Acad. Sci. USA*, **82**, 6955–9.

Liversey-Goldblatt, E., Tunley, T.H. and Nagy, I.F. (1977) Pilot-plant bacterial film oxidation (Bacfox process) of recycled acidified uranium plant ferrous sulphate leach solution, in *Conference on Bacterial Leaching* (ed. W. Schwartz), Verlag Chemie, Weinheim, pp. 175–90.

Mathew Hall Ortech Ltd (1979) *Quarry Mine*, **18**, 17.

McCready, R.G.L. (1988) Progress in the bacterial leaching of metals in Canada, in *Biohydrometallurgy '87. Science and Technology Letters* (eds P.R. Norris and D.P. Kelly) Kew, Surrey, pp. 177–95.

McElroy, R.O. and Bruynsteyn, A. (1978) Continuous biological leaching of chalcopyrite concentrates: demonstration and economic analysis, in *Metallurgical Applications of Bacterial Leaching and Related Microbiological Phenomena* (eds L.E. Murr, A.E. Torma and J.A. Brierley), Academic Press, New York, pp. 441–62.

Murayuma, T., Konna, Y., Sakata, T. and Imaizumi, T. (1987) Application of immobilized *Thiobacillus ferrooxidans* for large-scale treatment of acid mine drainage. *Method. Enzymol.*, **136**, 530–40.

Nicholaides, A.A. (1987) Microbiological mineral processing: the opportunities for genetic engineering. *J. Chem. Tech. Biotechnol.*, **38**, 167–185.

Nikolov, L. and Karamanev, D. (1987) Experimental study of the inverse fluidized bed biofilm reactor. *Can. J. Chem. Eng.*, **65**, 214–17.

Norris, P.R. and Barr, D.W. (1988) Bacterial oxidation of pyrite in temperature reactors, in *Biohydrometallurgy '87. Science and Technology Letters* (eds D.P. Kelly and P.R. Norris), Kew, Surrey, pp. 532–6.

Postgate, J.R. (1979) *The Sulfate Reducing Bacteria*, Cambridge University Press, Cambridge.

Powels, R.E., Deane, G.R. and Rawlings, D.E. (1996) Molecular genetic analysis of a thioredoxin gene from *Thiobacillus ferrooxidans. Microbiol.*, **141**, 2175–85.

Pretorius I.M., Rawlings, D.E. and Woods, D.R. (1986) Identification and cloning of *Thiobacillus ferrooxidans* structural *nif* genes in *Escherichia coli. Gene*, **45**, 59–65.

Rawlings, D.E. (1995) Restriction enzyme analysis of 16S rRNA genes for the rapid identification of *Thiobacillus ferrooxidans, Thiobacillus thiooxidans* and *Leptospirillum ferrooxidans* strains in leaching environments, in *Biohydrometallurgical Processing* (eds T. Vargas, C.A. Jerez, J.V. Wiertz and H. Toledo) University of Chile, Santiago, pp. 9–18.

Rawlings, D.E. and Kusano, T. (1994) The molecular genetics of *Thiobacillus ferrooxidans. Microbiol. Rev.*, **58**, 39–55.

Rawlings, D.E. and Silver, S. (1995) Mining with microbes. *Bio/Technology*, **13**, 773–8.

Rawlings, D.E., Pretorius, I.M. and Woods, D.R. (1984) Construction of arsenic resistant *Thiobacillus ferrooxidans* recombinant plasmids and the expression of autotrophic plasmid genes in a heterotrophic cell-free system. *J. Biotech.*, **1**, 129–32.

Roberto, F.F., Glenn, A.W., Rowland, M.L. and Ward, T.E. (1989) Introduction and replication of broad-host range RP4-based plasmids in acidophilic bacteria, in *Biohydrometallurgy. International Symposium Proceedings* (eds J. Salley, R.G.L. McCready and P. Wichlacz), Jackson Hole, WY, pp. 137–48.

Rudolfs, W. and Helbronner, A. (1922) Oxidation of zinc sulfide by microorganisms. *Soil Sci.*, **14**, 459–64.

Schlitt, W.J. (1980) Current status of copper leaching and recovery in the U.S. copper industry, in *Leaching and Recovering Copper from As-mined Ores* (ed. W.J. Schlitt), SME-AIME.

Schrader, J.A. and Holmes, D.S. (1988) Phenotypic switching of *Thiobacillus ferrooxidans. J. Bacteriol.*, **170**, 3915–23.

Shiratori, T., Inoue, C., Sugawara, K., Kusano, T. and Kitagawa, Y. (1989) Cloning and expression of *Thiobacillus ferrooxidans* mercury ion resistance genes in *Escherichia coli. J. Bacteriol.*, **171**, 3458–64.

Steffan, R.J. and Atlas, R.M. (1988) DNA amplification to enhance detection of genetically engineered bacteria in environmental samples. *Appl. Environ. Microbiol.*, **54**, 2185–91.

Tuovinen, O.H., Kelly, B.C. and Groudev, S.N. (1991) Mixed cultures in biological leaching processes and mineral bioleaching, in *Mixed Cultures in Biotechnology* (eds J.G. Zeikus and E.A. Johnson), McGraw-Hill, New York, pp. 373–427.

Waksman, S.A. and Joffe, I.S. (1922) Microorganisms concerned in oxidation of sulfur in the soil. *J. Bacteriol.*, **7**, 239–56.

Ward, T.E., Rowland, M.L., Watkins, C.S., Bruan, D.F. and Roberto, F.R. (1989) Studies on a bacteriophage which infects members of the genus *Acidophilium*, in *Biohydrometallurgy. International Symposium Proceeding* (eds J. Salley and R.G.L. McCready), Jackson Hole, WY, pp. 159–70.

Ward, T.E., Weller, R. and Bateson, M.M. (1990) 16s rRNA sequences reveal numerous uncultured microorganisms in a natural community. *Nature*, **345**, 63–65.

Yanokofsky, S.A., Gurevich, R., Grimland, N. and Stark, A. (1983) Genetic transformation of obligately chemolithotrophic thiobacilli. *J. Bacteriol.*, **153**, 652–7.

Yates, J.R. and Holmes, D.S. (1986) 'The Use of Genetic Probes to Detect Microorganisms in Biomining Operations', *J. Ind. Microbiol.*, **1**, 129–135.

Yates, J.R. and Holmes, D.S. (1987) Two families of repeated DNA sequences in *Thiobacillus ferrooxidans. J. Bacteriol.*, **169**, 1861–70.

Yates, J.R., Lobos, J.H. and Holmes, D.S. (1986) The use of genetic probes to detect microorganisms in biomining operations. J. Indust. Microbiol., **1**, 129–35.

Yates, J.R., Cunningham, R.P. and Holmes, D.S. (1988) IST2: an insertion sequence from *Thiobacillus ferrooxidans. Proc. Natl. Acad. Sci. USA*, **85**, 7284–7.

Zimmerly, S.R., Wilson, D.G. and Pratter, J.F. (1958) US patent, 2 829 964.

Index

Acetic acid 328, 331, 367
Acetone 319, 326
Acetone–butanol 235–6
Acidogenic fermentation 256
Acinetobacter cells, immobilization 84
Aeration 104, 110, 111, 490–1, 494, 497, 504
  in composting 464
Aerobic waste degradation, *see* Composting
Aeromonas hydrophila 458
Agaricus bisporus 467, 469
Agriculture 461
Agroindustrial waste 248
Air and compost 167–70
Alewife 465
Alkaliphiles 11
Alosa pseudoharingus 465
American smelt 467
Amidases 454
Amino acid 452
  composition 466
  decarboxylases 454
  pattern 265, 457
5-Aminolevulinic acid (ALA) 266, 270, 272
  aerobic fermentation 276–8
  applications 278–88
  sewage sludge 275–6
  swine waste 272–5
Amylase 12, 456
  recombinant 12
  thermostable 10
Amyloglucosidase 13, 118
  thermostable 10
Amylopectin 294
Amylose 294
Anaerobic contact process 310
Anaerobic filters 310
Animal feed 462
Anoxygenic phototrophs 247
Antibiotics 327, 332
Aquaculture 450
Arctic cod 467
Aromatic amines 17
L-Ascorbic acid 332
Aspergillus awamori 67
  co-immobilized with *Saccharomyces cerevisiae* 67
Aspergillus niger, immobilization 60, 71
Aspergillus oryzae 454
Astaxanthin 467
Atlantic cod 467
Autolysis 9
Azeotropic mixture 311

Bacillus megaterium 458
Bacillus thuringiensis 480–2
Bagoong 453
Balachan 453
Batch culture 489–90, 504
Beet sugar, recovery of waste from 71
Bifunctional agents 9
Bioassays 498, 500, 502, 504–7
Bioconversion 423–4, 429, 431, 436–41, 443–4
  cellulose 123, 439
  enzyme 435–6, 438–41, 443–4
  ethanol 441
  microorganisms 426, 429, 434, 437–9, 441, 443
  SSL 437
  xylose 442
Biodegradable plastics 311
Biofilms 536
  analysis 126
Bioinsecticide 481, 494–6, 499–500, 502, 504, 508
Bioleaching
  biochemistry 523
  bioreactors 531
  concentrates 527
  gold 529
  historical perspective 517
  in situ 528
  microorganisms 520, 522
  uranium 528
Biological conversion 376, 388, 416
Biological oxygen demand 14
Biomining 518
Biopolymers 449
Bioreactors, gold biooxidation 531
Biotechnology
  advances 469
  applications 123, 449, 462
  characteristics 463
  methods 460
  processes 449–52, 463, 471
Bonito 467

Bromelain 3
Brook trout 467
Budu 453
Bulking 309
  agent, in composting 464
Butanediol 236–7
2,3-Butanediol 319, 326
Butanol 319, 326
Butteroil, glycerolysis of 24
Butyric acid 331
By-catch, fisheries 450
By-products
  fish meal 457
  fisheries 449, 450, 467
  seafood industry 449

Calcium alginate, entrapment of cells in 49
Candida lipolytica 458
Capelin 454, 467
Carbohydrates 455
Carbon, in composting 464
Carotenoid 467
Carrageenan 49, 84
  for immobilization of microalga 84
Cassava solid waste 254
Catabolite repression 6
Cathepsins 452, 454
Cationic starch 299
Cell yield 263
Cellobiohydrolase 14
Cellulase 123, 206, 210–18, 220–1, 226, 456
  system 402
  thermostable 10
Cellulose 14, 60, 376–83, 385, 387–95, 402–6, 410, 415
  bioconversion of 60
  ethanol from 60
  wastes 60
Cellulosome 14
Characteristics of waste 427
  cellulose 425, 428–9, 431, 435–6, 439–40, 444
  fermentable 426, 428, 436, 438–41, 443
  spent sulfite liquor 428, 437–8, 441
Cheese 468
  whey 342
Chitin 158, 177, 450, 459–63, 466, 470, 471
  depolymerization 451
Chitinases 456, 460, 466
Chitosan 459–63, 471
  in complexation 34
Chitosanases 463
Chloroperoxidase 18
Cholesterol 462
Chromatography 461
Chymosin 3, 4, 468
Chymotrypsin 10, 23, 454, 460
Citric acid 328, 332
Clean technology 24
Clupae harengus 452
C/N ratio 118, 492
  mature compost 186
  raw compost 163–4
CO_2 fixation 286
Colletotrichum lindemuthianum 463
Colour intensifying effect 287
Comamonas acidovorans 89
  immobilization 89
Composite granules 50
Compost
  C/N ratio of 186
  feedstocks 157–60
  fish offal–peat 464
  hydrolysate 464
  maturity 184–6
  process 103, 471
  shrimp wastes 464
  uses 104, 187–8
Composting
  additives 175–6
  aeration 167–70
  air, movement and spaces 161–2, 167
  backyard 155
  biochemistry of 176–8
  biosolids 159
  bulking agent 464
  chemistry of 176–8
  C/N ratio 163–4, 464
  compostables (feedstocks) 157–60
  confined 184
  definition of 156
  feedstocks 157–60
  fish wastes 182, 463–4
  fisheries wastes 158, 182
  heap size 171
  heat 170
  history of 154–5
  Indore method 155, 179
  in-vessel systems 184
  manure slurries 157–8
  moisture content 161, 166
  municipal wastes 182
  natural aeration static pile 180–3
  nitrogen retention 464
  odour 171–5, 464
  odour compounds 173
  open systems 179
  optimal 160–76
  organic farming and 187
  oxygen content 161, 167–70
  particle size 165–6
  passive aeration system 180–3
  passively aerated window system 464
  pathogen kill in 157, 186–7
  peat moss 464
  pH and 164–5
  phosphates and 175
  plant (facilities) closure 155

plant inhibitors and 156, 184–6
popularity of 154–5
principles of 156–7
problems and solutions 167–70
processes in 176–8
redox 172
requirements 160–76
sanitation 157, 162, 186–7
seafood wastes 158, 182
sewage sludge 159–60
slaughterhouse wastes 158
sustainable development and 154
technology of 178–84
temperature of 168, 170–1
turned windrow 179
wastes 451
water in 161
wood wastes 160
Concentrate bioleaching 527
Copper, bioleaching 519
Corynebacterium glutamicum, immobilization 43
Cosmetics 461
Crawfish 467
Cyanide 302

Deboning 459
Dehydrogenases 454
Denitrifying bacteria, co-immobilized 85
Dextrins 13
Dextrose equivalent (DE) 295
Diagnosis of heavy metal poisoning 280
Dilute acid hydrolysis 199, 228, 237
Direct microbial conversion 229, 231
Dogfish 454, 467, 469
Down-stream operations 470
Drying
solar 470
tables 461, 470
Dump leaching 519
problems 526

Edible mushrooms 393, 398, 402
Emulsification 457, 465
Encapsulation 49
tissue and bacterial cells 462
Endoglucanase 14
Endotoxin 480–2, 484–6, 488–501, 503–4, 506–7, 509
Ensiling processes 113, 138
Environmental applications 461
Enzymatic browning 133
Enzymatic hydrolysis 457, 469
Enzymatic processes 450–1, 457
Enzymes
amylases 456
bacterial 452, 457, 463
biochemistry 452
cellulases 456
chemical modification of 9
chitinases 456, 460, 466
chitosanases 463
cloned 7
crosslinking 54
deamination 12
endogenous 450, 452
engineering 469
exogenous 454
fish 456
from fish biomass 449, 452, 466
for fish sauce 450–4, 456
in fisheries processing 450
in food industry 452–4
fungal 457
genetically engineered 470
hemicellulases 456
immobilization 8, 73
inactivation 8
lactases 456
mannanases 456
from marine organisms 449, 466
microbial 449, 454
in muscle 454, 469
in organic solvents 73
overexpression 7
pectinases 456
properties 466
protein hydrolysis 454
proteolytic 454, 457–8, 470
purification of 6
recombinant 7
recovery 466
in seafood industry 449, 452
stability 7
stabilization 8
thermostable 10
from wastes 466
Erwinia herbicola, immobilized 43
Ethanol 60, 67, 228–34, 237–8, 300, 310, 319, 326, 330, 331, 332, 376, 390, 393, 395, 415
Explosives 16
Extremophiles 10

Farm manures 158, 182
Fed-batch culture 489–90, 494
Feed 449, 454, 456–8, 465
stickwater, use in 458
Fermentation 487–90, 495, 497, 499–502, 504, 506
aerobic 465
batch 215, 235, 465
co-culture 229, 232
continuous 465
equipment 466
fed batch 215, 236
fish lipid 458, 465

Fermentation *cont'd*
- fish products 452
- fish sauce 450–2,, 454
- kinetics 111
- lactic acid 132, 453–5
- lipolytic 458
- media 464
- pilot plant 124
- processes 130, 451, 465–6
- processing wastes 456
- substrate 464
- yield 466

Fermented food
- miso 103, 125
- pozol 103
- tempeh 103, 132

Fermenter
- auger tube 127
- fluidized biomass 127
- helical screw 127
- packed-column 126
- rotating drum 126
- solid substrate 129
- tray 125

Fertiliser
- from fish protein hydrolysate 459
- organic 464
- from silage 454

Ficin 3, 4

Fish
- biomass 452, 459
- compost 463–4, 471
- meal 450, 456–7
- mince 469
- muscle tissue 467
- oil 459, 469
- processing industry 449
- protein 452, 456–9, 470–1
- sauces 451–4, 456
- silage 451, 454–6, 467, 470
- solubles 469

Fish protein concentrate
- animal feed 457
- functional properties 457, 469–70
- markets 471
- technologies 456

Fish protein hydrolysate 451
- animal feed 457, 459
- biological methods 457, 470
- deboning 459
- functional properties 469–70
- industrial hydrolysis 468
- markets 471
- proteases 452, 459
- proteolytic microorganisms 452, 459, 470
- technologies 459

Fish sauce
- enzymatic process 451
- fermented 452
- production 452–4, 456
- use of enzymes 451–4, 456

Fisheries
- biomass 449–50, 452, 454, 458, 465–6, 469–70
- by-catch 450
- by-products 449–50, 467
- enzyme applications 456
- filleting 463
- industry 456
- operations 450–2, 463
- processes 450
- processing 450, 458, 463
- products 469
- waste waters 450, 463–4
- wastes 449–55, 457, 463–7, 469, 470–1

Fisheries wastes, composting 158, 182

Flavourant production 451

Flocculation, of cells 50–1

Food
- additives 470
- antioxidants 465
- applications of chitin and chitosan in 461
- bitterness 470
- from fisheries biomass 450, 469
- food industry 449, 452, 456–8, 467, 469–70
- functional properties 456–7
- gums 332
- human 465
- muscle 454
- new sources 456
- pet 457
- products 457
- solubles 454, 456, 458, 467
- source of 449
- traditional products 452

Food processing wastes
- bakery 327, 330
- banana 324, 325
- bean 325, 327
- carrot 324, 328
- cassava 325
- citrus fruit 324, 325, 326, 328
- coconut 327
- coffee berry 140
- coffee pulp 140, 324
- corn 324, 325, 327
- date 328
- grape marc 325
- kiwi peel 328
- millet husk 325
- molasses 323–9
- mussels 323
- olive oil 326, 327, 331
- onion 325
- orange peel 324, 328
- palm oil 325, 326, 329
- pear 327

pineapple 328
potato 325, 326, 327
rice 325
slaughterhouse blood 323
sorghum husk 325
sugar beet 325, 328, 329
sugar cane 141, 325, 327, 328, 329
wheat 324, 326
whey 329–32
whey ultrafiltration permeate 329, 331
wine 328
Fruit water 303
Fungi 105, 109–10, 114, 116, 12
actinomycetes 104
ascomycetes 113
basidiomycetes 113
edible 138
filamentous 113, 116
mesophilic 109
phycomycetes 113
thermophilic 109

Gadus ogac 468
α-Galactosidase, immobilized 71
β-Galactosidase 14, 332
Garos 453
Genetic engineering 118, 539
Genetic mutation 539
Geotrichum candidum 458
Gluconobacter oxydans, immobilized 71
Glucose 69, 222–4, 226–8, 237
ethanolic fermentation 69
Glucose isomerase, *see* Xylose isomerase
Glucose syrups 13
β-Glucosidase 14
Glutamic acid 329
Glutaraldehyde, in immobilization 51
Glyceraldehyde, 3-phosphate dehydrogenase 10
Gold, bioleaching 529
Green microalgae, immobilized 86
Greenland cod 467–8
Growth promoting factor 283

Haemoglobin 468
Halophiles 11
Harp seal 467–8
Heap leaching 527
Hemicellulase 456
thermostable 10
Hemicellulose 19, 61, 376–80, 384–8, 392–5, 404
conversion 61
Herbicide 279
Herring
proteolytic enzymes 467
salted 452
High fructose corn syrup 296
High fructose syrup (HFS) 13, 18, 296
Hordeum vulgare L. 464
Horseradish peroxidase 17, 18
Hydrogen, production by photosynthetic bacteria 90
Hydrolases 12, 454, 456
Hydrolysate
chitin 466
compost 464
fish protein 452, 457–9, 469–71
Hydrolysis
acidic 222
enzymatic 222–8
Hydrolytic potential 404
Hydroxy-fatty acids 328

Immobilization 29, 31, 54
advantages 29
carriers 31
cells 29
disadvantages 29
entrapment 33
enzyme 29, 470
methods 32
microbial cells 470
reactors 55
Immobilized cells 51, 78, 83, 362
animal 151
methane production 78
and waste water treatment 83
Immune system 462
In situ bioleaching 528
Indore method, composting 155, 179
Induction 6
Insecticide 280, 327
In-vessel composting 184
Invertase, encapsulation 49
Iodine binding capacity 294
Isomerase 18, 293, 296

Johnius dissumeri 458

Ketjab-ikan 453
Keto-fatty acids 328
Kinetics, first order 8, 9
Klebsiella aerogenes 5, 13
Kluyveromyces fragilis, immobilization 58
Kluyveromyces marxianus 61
IMB3, immobilization 70, 71
Koji 125
Koji process 103, 113

Laccases 121
immobilization 87
Lactase 456
immobilization 57
Lactic acid 67–9, 294, 300, 323, 328, 329, 331, 332, 347
bacteria 132–3, 455
batch process 349

Lactic acid *cont'd*
continuous process 353
fermentation 350, 453–4
inhibition 357
microorganisms 348
productivity 356
in silage processes 454
Lactobacillus plantarum 454
Lactococcus lactis 59
Lactose 58, 328, 329
Leather treatment 16
Lignin 16, 376–80, 385–8, 390, 392–5, 409–16
biodegradation 119, 121–3
peroxidase 18, 119
recombinant 17
Ligninases 119, 409, 412, 413, *see also* Ligninolytic enzyme
Ligninolytic enzyme 412–14, *see also* Ligninases
Lignocellulose 60, 376–8, 388–96, 399, 402–4, 409, 410, 415, 416
cellulose, in 379
complete conversion 393
composition 377
direct conversion 392
extraneous materials in 388
hemicelluloses, in 384
indirect conversion 389
lignin, in 385
pretreatment 388
protein, in 388
structure 377
Lignocellulosic wastes 114, 118–19, 138
as substrates 138
Lignocellulosic wastes bioconversion
Han's proces 399
Indian Institute of Technology process 390
INRA-Dijon process 400
Institut Armand-Frappier process 393
Kyoto University process 390
Louisiana State University process 392
University of California process 390
Linoleic acid 323, 328
Lipase
co-immobilization 43, 73–7
in fixed bed reactor 77
in fluidized bed reactor 77
immobilization 43, 73–7
thermostable 10
Lipids, bioconversion 76
Lipolytic fermentation 451
Lipoprotein lipase, immobilization 77
Liquid-state fermentation 103, 388, *see also* LSF
LSF 103, 105, 116, 117, 131, 142, 146, 388, 389, 396, 399, 403, 406–9
Lyases 19
Lycasin 296
Lyozyme 10

Mallotus villosus 454
Maltitol 296
Maltose syrups 13
Mam 453
Mam-chau 453
Mam-tom 453
Mandarin orange peel 256
Manganese peroxidase 18
Mannanases 456
Mannitol 296
Manure slurries 157–8, 182
Marine biomass 449, 466
Marine organisms 449, 466–7
Marine resources 449–50
Marine wastes 450
Mass transfer 9, 21
Medical treatment for cancer 282
Medicine 462
Medium composition 487
Methane 323, 326, 327, 330, 332, 464
biosynthesis 79
Methanogenic bacteria, co-immobilized 85
Methylmalonyl–CoA 275
Micelles 20
Microaerobic–dark condition 268
Microaerobic–light culture 258
Microalgae, immobilized 83
Microencapsulation 49, 50
Microorganism, identification 525
unculturable 525
Microorganisms
adaptation 537
aerobic 141
bacteria 116, 131
bioleaching 522, 533
biomass 466
consortia 536
cultures 104, 463
development of new strains 533, 537
engineering 469
enzymes 450, 454, 457–8, 463
fish protein hydrolysates 458
fish sauce 452
growth 105–6, 113, 458, 465
growth media 466
growth patterns 114
growth rate 114, 116–17
immobilization 89
industry 450
population 454, 464–5
protein 458, 465–6
proteolytic 452, 459, 470
recovery of 461
release into the environment 538
spoilage 454

techniques 471
thermophilic 113
waste treatment 470
yeasts 105, 116, 131
Milk, processing 59
Miso 103
Mixed culture 391, 392, 400
Modified starch 296, 298
Molecular biology 19
Monascus purpureus 454
Mucor rouxii 463
Municipal wastes, composting 182
Munitions 16
Mushroom 113, 123, 138, 467, 469
Mutagenesis, site-directed 20

Nampla 453
Narezushi 453
Natural aeration composting 180–3
Ngapi 453
Nitrate ester reductase 16
Nitrobacter agilis, immobilization 83
Nitrogen, *see* C/N ratio
Nitroglycerine 16
Nitrosomonas europaea, immobilization 83
Non-aqueous systems 17, 20–4
Nuoc-mam-gau-cha 453
Nutritional quality 263

Odour
from compost plants 173–5
in composting 464
foul 174
nature of 173
Oil
crude seeds 467
fish 449–50, 454, 456, 458, 465, 469, 471
Oleic acid 328
Olive oil, hydrolysis 74
Open composting methods
Indore method 155, 179
NASP 180–3
PAWS 180–3, 464
simple turned windrow 179
Oreochromis niloticus 454
Organic acid 300
Organic waste 247, 248
Organoleptic factors 450
Outer membrane 7
Oxidoreductoases 16
Oxygen in composts 167–70
Oxygen transfer 110
Oxytetracycline 327

Padec 453
Pagophilus groenlandicus 468
Papain 3
Particle size of composts 165–6
Passively aerated composting 180–3
Pati 453
Peat 464
use in composting 158–9, 182
Pectin transeliminase 19
Pectinases 456
Pekasam 453
Pentaerythriol tetranitrate 16
Pepsin 454, 457, 468
Peptidase 283, 454
Peptides 452, 457
Periplasm 7
Pharmaceutical 449, 471
Phenoloxidases, immobilized 87
Phosphates and composts 155, 175
Photosynthetic bacteria 247, 248, 250, 256
Phycomyces blakesleeanus 463
Pichia kudriavzevii 466
Pigment production 451
Pineapple waste 250
Pissala 453
Pla-ra 453
Pleurotus ostreatus 469
Pollution
coastal and sea 499
control 343
Polydextrose 296
Polyelectrolyte complex, immobilization 33
Polymer formation 22
Polyphenol oxidase 18
Polysaccharides
anionic 460
chitin 460
chitosan 460
complex 449
Pozol 103
Pradec 453
Prahoc 453
Pretreatment 376, 388, 392
acid 208, 237
biological 208–9
chemical 205, 206, 208
mechanical 205, 206, 209
organosolv 208, 237
steam explosion 209, 237
Production
citric acid 408
gibberellic acid 399, 406
spores 408
Propionibacterium 367–71
Propionibacterium shermanii, immobilization 57
Propionic acid 331, 332, 367
Proteases
acidic 468
activity 469
alkaline, thermostable 11

Proteases *cont'd*
  application in fisheries 456
  bacterial 457, 460
  co-immobilization 36
  for commercial processes 457
  endogenous 454
  for fish protein hydrolysate 452, 459
  from fisheries biomass 449, 456
  gastric 468
  for hydrolysis of stickwater 457
  immobilization 34
  in industry 454
  from marine organisms 466–7
  microbial 457–8
  sulphydryl 3
  thermostable 10
Protein
  additives 470
  amino acids 457
  conformation 8
  crude 465
  enrichment 307
  for feeds and foods 465
  fish 452, 456, 458, 470
  fish concentrates 456–7, 469–70
  fish hydrolysate 452, 457–9, 469–70
  in fish sauce 452
  in fish silage 455
  from fisheries biomass 449
  from fisheries wastes 450
  functional properties 470
  hydrolysis 58, 454, 458, 470
  hydrolysis of stickwater 457
  lipases 114
  microbial 465–6
  microbial biomass 458, 465
  milk 58
  recovery of 462
  shellfish 459
  single cell 103, 462, 464–6
  solubilization 452, 457–8
  sources 456
  splitting 457
Protein enriched fermentation feeds 307
Proteolysis 455
Protoporphyrin IX (PPIX) 279, 280, 281, 282–3, 284
Pseudomonas alcaligenes, immobilized 89
Pseudomonas marinoglutinosa 458
Psychrophiles 11
Pullulanase 5, 13
Pulp and paper 423–4, 428, 431, 433–6, 438–41, 443
  biobleaching 423–4, 426, 429–32, 434–6, 444
  wastewater 423–44
Pyruvic acid 328

Quinoline, degradation 89

Reactor
  batch 54
  continuously stirred tank 54
  design 470
  fixed-film 56
  fluidized bed 55–6
  membrane 56
  plug flow packed-bed 55
  Wortington 55, 56
Rennet 468
Resistant starch 297
Reverse osmosis 308, 461
Rhodobacter 248
Rhodobacter capsulatus 262
Rhodobacter sphaeroides 270
Rhodobacter sphaeroides P_{47} 250
Rhodobacter sphaeroides S 258, 262
Rhodocylcus gelatinosus 253, 256, 268
Rhodospirillaceae 248
Rhodospirillium ribrum, immobilized 90

Saccharomyces cerevisiae, immobilization 60, 61–2, 69, 72
Salt tolerance 287
Sardine 467
SCP 103, 376, 377, 389–95, 399–402, *see also* Single-cell protein
Scytalidium acidophilum 464
Seafood
  analogues 457
  flavourings 457
  industry 449–50, 452, 457, 463, 467, 471
  nutritional value 449
  processing 450, 458, 463, 471
  processing wastes 466
  wastes, composting 159, 182, *see also* Fisheries
Semisolid state fermentation 127
  tubular reactor in 127
Separate hydrolysis and fermentation (SHF) 228–9
Serratia marcescens 58, 460
Shell removal 451
Shellfish
  waste waters 464
  wastes 460, 466, 470–1
Shiokara 453
Shotturu 453
Shrimp
  compost 464
  shell 466
  volatile components of 457
Signal peptide 7
Silage
  composition 455
  crab processing wastes 456
  fish silage 454–6, 470
  lactic acid 454
  pH 456

raw crawfish 467
tilapia 454
use of proteases 456
Silver, recovery of 16
Simultaneous saccharification and fermentation 229–32, 237
Single cell oil 328
Single cell protein 103, 304, 319–26, 328, 330, 332, 376, 462, 464, 466
Site-directed mitogenesis 10
Sizing (in paper industry) 299
Smelt 467
Solid-state culture 110
Solid-state fermentation 377, 388, 396, 398–409
Solid-substrate fermentation 103–19, 121, 124–8, 130–2, 134–5, 141–3, 146–7
aerobic processes 112
bioreactors 112, 127
processes 105–6, 112–15, 118, 130–2, 134–5
thermal diffusivity in 109
Solubility 457
Solvent engineering 24
Sorghum halepense L. 456
Soybean runoff 253
Soybean waste 253
Sphagnum 464
Sporulation 481–5, 487, 489–92, 494–9, 501–4
Spray drying
of hydrolysate slurry 470
of particles 461
Squalus acanthias 454, 469
Starch
bioconversion of wastes 62
ethanol production 64–5
fermentation 63
one-step hydrolysis 13
production 293
saccharification 63
Static piles, compost 180–4
Stickwater 457–8
Straw 14
Structure
cellulose 379
hemicelluloses 384
lignin 385
wood cell 379
Substrate 487, 491
Subtilisin 20
Sulphydryl bridges 20
Swine waste 260
Symba process 305

Tank bioreactors 532
Tempeh 132
Temperature, composting and 168, 170–1
Tetrapyrrols 266
Thermophiles 10
Thrombin 462
Tilapia 454
Toxicity 484–7, 489–91, 496, 498–506, 509
Transaminases 454
Trassi 453
Treatment technologies 430–3
biological treatment 433
colour removal 434
ultrafiltration 433
Trypsin 454, 468
bovine 468
Tuna 467
Two-phase systems 20
Tyrosinase, immobilization 87

Ubiquinone 269
Ultrafiltration 21, 461
Unified anaerobic fermenter–filter system 310
Upflow anaerobic sludge blanket (UASB) 309
Urease, encapsulation 49

Value-added products 449, 471
Vitamin B_2 332, 333
Vitamin B_6 332
Vitamin B_{12} 266, 332

Wadi 453
Waste
agricultural and food 81
animal 157–8, 182
biological processing 470
biological treatment 466
blood 15
chitin 460, 466
in compost 161, 166
composting 464
crab processing 456
crawfish processing 467
dogfish processing 454
effluent 463
feathers 15
fermentation substrate 464
fish 449–54, 457–8, 463–7, 470–1
fish composting 464
fish processing 159, 182
food industry 456
gas, purification 89
hoof and horn 15
lactose 14
lignocellulosic 114, 118, 119, 138
lipid content 458
liquid 463–4
methane bioconversion 78
municipal 14
mussel processing 464, 466
paper mill 17

Waste *cont'd*
phenolic 23
production of fish meal 450
pulp and paper mills 159, 182
recovery 449
sawdust 14
seafood 159, 182
seafood industry 449
seafood processing 466
shellfish 466, 471
shellfish processing 459–60
shrimp 464
silk 16
slaughterhouse 159, 182
sochu 16
solid 463–4
treatment 466
turkey 15
whey 14
wood 464
Waste treatment
aerobic 463
anaerobic 463–4
pretreatment 460
secondary 450
shellfish waste 460
Waste water
biological treatment 464
bonito processing 463
chemical oxygen demand 463
fisheries 450, 458, 463
fisheries biomass 470
fisheries processing 458, 463
mussel processing 464
shellfish 470
shellfish processing 464
treatment 458
treatment with chitosan 460
Water activity 105–6, 109, 111
control 107
factor 105
Water holding capacity 161, 166
Whey 342
animal feed 345
bioconversion 57
composition 343
disposal methods 344
organic acids conversion 342–75
permeate 347, 350
protein concentrate 347
White rot fungi 17
Windrow
passively aerated (PAW) 464
systems, for composting 179–84

Xanthan 34, 37, 41, 43, 44–5
Xylanase 36, 39, 43, 219–22, 237
thermostable 10
Xylitol 329
Xylose 62, 222–4, 226–8, 232–4, 237
Xylose isomerase 13, 18

Yield 487, 491, 494–500, 504

Zymomonas mobilis 63, 65, 66

UNIVERSITY COLLEGE LONDON
LIBRARY